The Dynamics of Manifolds

Book 3 of Physics from Maximal Information Emanation, a seven-book physics series.

ISBN 979-8-9888160-3-4

The Dynamics of Manifolds

by

Stephen Winters-Hilt

ISBN 979-8-9888160-3-4

Golden Tao Publishing
Angel Fire, NM
USA

Dedication

This book is dedicated to my family that helped on this lengthy road of discovery: Cindy, Nathaniel, Zachary, Sybil, Eric, Joshua, Teresa, Steffen, Hannah, Anders, Angelo, John and Susan.

Contents

Preface to Physics Series on:

Physics from Maximal Information Emanation

"The Road goes ever on and on
Down from the door where it began.
Now far ahead the Road has gone,
And I must follow, if I can,
Pursuing it with eager feet,
Until it joins some larger way
Where many paths and errands meet.
And whither then? I cannot say"

– **J.R.R. Tolkien, The Fellowship of the Ring**

Variation, Propagation, and Emanation

This is a seven book Physics Series that starts with Classical Mechanics (Book 1 [11]), then Classical Field Theory, such as electromagnetism (Book 2 [13]), then Manifold Dynamics, such a General Relativity (Book 3 [1]). The switch to a quantum mechanics description is given in Book 4 [3], and to a quantum field theory, QED in particular, in Book 5 [9]. A 'quantum manifold theory' would be the obvious next step except it cannot be done (there is not a renormalizable Field theory for Gravitation). Instead a thermal quantum manifold theory is considered, as well as Black Hole thermodynamics in general, in Book 6 [4]. Book 7 [5] describes a new theory, Emanator Theory, that provides a deeper mathematical construct that undergirds Quantum theory, much like quantum theory can be shown to provide a deeper (complexified) mathematical construct based on the classical theory.

This is a modern exposition where subtleties of chaos theory are described in Book 1, of Lorentz Invariance in Book 2, of Covariant Derivatives (General Relativity) and Gauge Covariant Derivatives (Yang-Mills Field Theory) in Book3. Book 4 on Quantum Mechanics provides an extensive review of QM, then considers a full self-adjoint analysis on the full general relativistic solution to the spherical shell in-fall system (a result carried over from Book 3). Book 5 considers QFT basics in detail, along with alternate vacua in specific scenarios. Book 6 considers thermodynamics from the basics to the Hamiltonian thermodynamics of some Black Hole systems. Throughout, the odd recurrence of the alpha parameter is noted. In Book 7 we look to a deeper mathematical formulation from which the Quantum Path Integral formulation would result, as well as explaining the odd parameters and structures that have been discovered (such as alpha and Lorentz Invariance).

The physical description starts with the classic formulations of point particle motion. The first approach to doing this is using differential equations (Newton's 1st and 2nd Law); the second is using a variational function formulation to select the differential equation (Lagrangian variation); the third is using a variational functional formulation (Action formulation) to select the variational function formulation. Historically, it wasn't realized until much later that there are two domains for motion in many systems: non-chaotic; and chaotic.

In a description of particle motion, assuming not in a parameter domain with chaotic motion, several important limits are found to exist. Examples include: the universal constants from the aforementioned chaos phenomenon, that are still encountered in non-chaos regimes if driven "to the edge of chaos". Limits are found where scattering is defined in the asymptotic limit and perturbation theory is well-defined in the sense that it is convergent. Overall, if the evolution is described as a 'process' it is often a Martingale process, which has well-defined limits. So, we have descriptions for motion, typically reducible to an ordinary differential equation (ODE), and for which solutions (requiring limit-definitions) are typically found to exist.

The physical description then contends with field dynamics in 2D, 3D, and 4D (in Book 3 [1]). Two-dimensional ("2D") field dynamics can be described as a complex function (that maps complex numbers to complex numbers). A novelty of the 2D complex function is it also shows how to handle many types of singularities (the residue theorem), thus provides important information about fundamental structures in physics as well as fundamental mathematical techniques for solving many integrals. For the 3D field dynamics we do an analysis of the electromagnetic field in 3D. The level of coverage begins at an overview of electrostatics at the level of the graduate text Jackson [106]. Some problems from Jackson Ch's 1-3 are examined closely in developing the theory itself. For some this material (in Book 2 [13]) might provide a useful accompaniment to Jackson's text in a full course on EM (based from Jackson's text). A quick review of electrodynamics and electromagnetic wave phenomena is then given. In essence, we see many more examples of ODE problems with solutions, such as for the 3D Laplacian, usually involving separation of variables. We then review the famous transform, discovered by Lorentz in 1899 [107], that relates the EM field as seen by two observers differing by a relative velocity. With the existence of this transform, that brings in the time dimension along with the relative velocity, we effectively have a 4D theory.

From Lorentz Invariance we have, as a point transformation, rotational invariance under SO(3) or SU(2). If Lorentz Invariance is fundamental, then we should see both forms of rotation invariance, one of vector/tensor type from SO(3), and one of spinorial type from SU(2). This is the case, as gauge fields are vectorial and matter fields are spinorial. From Lorenz Invariance as a local invariance we have the Minkowski (flat) spacetime metric, which then generalizes to the Riemannian metric (in General Relativity).

As with the point particle dynamics, for the field dynamics we have three ways to formulate the behavior: (1) differential equation; (2) function variation (on Lagrangian); and (3) functional variation (on the Action). We will see similar limit phenomena as before, but also new phenomena, including (i) inevitable BH singularity formation (the Penrose singularity theorem); (ii) FRW Universe formation (from homogeneity and isotropy); (iii) the BH collapse singularity; (iv) the atomic collapse radiative 'singularity'.

Classical dynamics, thus, has two field-like formulations to describe the world: field and manifold. Such formulations can be interrelated mathematically, so what is happening is more a matter of physics emphasis and convenience. The emphasis on this difference, that appears to be no difference (mathematically), is that different physical phenomenologies are at play. Field descriptions appear to work for 'matter', where the fundamental elements are spinorial. Manifold descriptions appear to work best for geometrodynamics (GR), where the fundamental elements are vectorial (or tensorial, such as the metric). Matter fields are renormalizable, thus quantizable in the standard QFT formulation (to be described in Book 5 [9]), while gravitational manifolds are not renormalizable, and have constraints (weak energy condition and positive energy condition given the existence of spinor fields on the manifold).

The presentation in Books 1-3 [1,11,13], on 'classical' physics, is partly done to make the transition to quantum physics simple, obvious, and in some cases, trivial. Consider the functional variation (Action) formulation of the behavior (whether point-particle or field), this can be captured in integral form, as was done by D'Alembert very early [108] (then by Laplace [109]). Note the use of a large constant to effect a 'highly damped' integral for selection purposes (on variational extremum of the action). To transition to the quantum theory we also have the large constant from 1/h, and so the only difference is the introduction of a factor of 'i', to effect a 'highly oscillatory' integral for selection purposes.

After the transition to a quantum theory, for the point-particle descriptions, the classical collapse problem for atomic nuclei is eliminated. The spectral predictions have excellent agreement with theory, but there is still fine-structure in the spectra not fully explained. The theory is not relativistic and some initial corrections for this are possible (without going to a field-theory) and these indicate closer agreement and explain most of the fine-structure constant discrepancy (and reveal alpha in another place in the theory). It is shown in Book 3 [1] and Book 4 [3], that the GR singularity problem, however, remains unresolved (for the test case of spherical dust shell collapse, done in a full GR analysis, then quantized in a full self-adjoint quantization analysis [3]).

In Book 5 [9], the transition to quantum theory is continued to the field theory descriptions. A precise description/agreement of atomic nuclei is now possible with QED, and within the nuclei themselves (quark confinement) with QCD. The field theories have a small set of bothersome infinities, however, which is eventually solved by renormalization [9]. As mentioned, the quantization of manifold theories, such as GR, does not appear to be possible due to non-renormalizability. Not to be deterred, in Book 6 [4] we consider a Hamiltonian description of a GR system whose quantization would involve an energy spectrum based on that Hamiltonian, if we then use analytic continuation to take us to the thermal ensemble theory based on the partition function that results, we can consider the thermal quantum gravity (TQG) of such systems.

This last example (from Book 6), showing a consistent TQG theory if we use analyticity, is part of a long sequence of successful maneuvers involving analytic continuations in different settings. What is indicated is the presence of an actual complex structure to the stated theory. There is the trivial complex structure extension mentioned above that brought us from the standard classical physics theory to the standard path integral quantum theory. But we also see actual complex structure at the component level with time complexation (that ties to thermal version of the theory by defining the partition function), and we have complex structure as the dimension-level in the form of the successfully applied dimensional regularization procedure used in the renormalization program.

As well as covering the breadth of core physics topics at both undergraduate and graduate level (for courses taken at Caltech and Oxford), including extensive presentation of problems and their solutions, the Series also examines, in specific cases, the boundaries of the physical world "from the inside" (and then later "from the outside"). To this end exploration of spherical dust collapse to form a singularity is examined in a fully general relativistic formalism, and then carried-over to a quantum minisuperspace (quantum gravity) analysis (in Books 3 and 4 [1,3]). Also examined in-depth are the topics of black hole thermodynamics and quantum field theory with alternate vacua (part of Books 5 and 6 [4,9]). The in-depth material comprises the topics covered in my PhD dissertation [55], portions of which are published [110-113].

In recent work on machine learning, that includes statistical learning on neuromanifolds [6], we find a possible new source for a foundational element for statistical mechanics (entropy) via seeking a minimal learning process/path on a neuromanifold [6]. By the time the Series reaches thermodynamics in Book 6, therefore, the foundational thermodynamics elements have all been established from the physical descriptions discovered in Books 1-5, they just haven't been put together in a comprehensive analysis that gives us the fundamental constructs of thermodynamics and statistical mechanics. That said, it would seem that thermodynamics is, thus, entirely derivative from other, truly fundamental theories. Not so, in the joining of the parts to make thermodynamics we have something greater than the sum of the parts. In the 'system' descriptions we find that emergent phenomena exist. This, at least, is unique to thermodynamics, so it is fundamental in this "sum greater than the parts' aspect.

In Book 7 (the last) of the Series, we consider the standard physical world, described by modern physics, "from the outside." In doing this we've already eliminated part of the mystery of entropy by the geometric 'neuromanifold' description. If we can understand other oddities of the standard theory, and arrive at them naturally, then we might have an even deeper dive into modern physics, testing the limits of what is possible, and see possible future developments and unifications of the theory. This is what is described in papers [96,114-119], and organized along with current results into the final Book of the series.

Efforts in the last book of the Series involve choices and concepts identified in the prior six books of the Series, and theoretical maneuvers gleaned from the most advanced courses in physics and mathematical physics taken while at Caltech (as an undergraduate and then as a graduate) and the Oxford Mathematics Institute (as a graduate), and the University of Wisconsin at Milwaukee (as a graduate).

The broad range of topics covered in the Series is, initially, similar to the Landau & Lifshitz graduate textbook series (see [14]), with a similar exposition on classical mechanics at the start of Book 1. Even with well-established classical mechanics, however, there are significant, modern, updates, such as (modern) chaos theory. In the final two books of the Series (Books 6 and 7 [4,5]) we arrive at statistical mechanics and thermodynamics, together with modern topics such as black hole thermodynamics, thermal quantum gravity, and emanator theory.

Key constants and structures of physics, their discovery from the experimental data, and their theoretical placement in the "Grand Scheme," are emphasized throughout the Series. The constant alpha, a.k.a. the fine structure constant, appears in numerous settings so special note of the occurrence of alpha will be made in each chapter. This is the case even at the outset with Book 1, due to fundamental numerical constants appearing from chaos theory. In Book 7 we see the origin of alpha, as a maximal perturbation amount, appears naturally in a formalism for maximal information 'emanation'. But maximal perturbation in what space and in what manner? In Book 7 of the series [5] we will see a possible representation of such an information entity, and its space of existence, in terms of chiral trigintaduonions.

Thus, in the end, this is an effort to tell of a journey to a special place "where many paths and errands meet", giving rise to emanator theory and an answer to the mystery of alpha. Part of this journey is equivalent to 'finding the arkenstone' (alpha) in the most unlikely of places, the trigintaduonion emanation mathematics underpinning the emanator formalism (e.g., Smaug's Lair, described in Book 7 [5]). Why I should have wandered into such an odd place (mathematically speaking), and why I should posit a deeper form of quantum propagation using hypercomplex trigintaduonions, here called emanation, is why there is such extensive background on standard topics. This extensive background even impacts the classical mechanics description via its modern chaos theory material (due to a possible relation between C_∞ and alpha). The critical role of emergent phenomena is only understood at the end, including for manifolds in geometry and neuromanifolds in statistical mechanics, and leads to a Book 6 that goes from very basic (initial thermodynamics) to very advanced (emergent phenomena). Much is made clear with emanator theory, including how reality is both fractal and emergent. At this point in the journey, as with Tolkien, this much I can say: "The Road goes ever on and on … And whither then? I cannot say".

The seven books in the Series are as follows:

Book 1. Classical Mechanics and Chaos
Book 2. Classical Field Theory
Book 3. Classical Manifold Theory
Book 4. Quantum Mechanics and the Path Integral Foundation
Book 5. Quantum Field Theory and the Standard Model
Book 6. Thermal & Statistical Mechanics, and BH Thermo.
Book 7. Maximum Information Emanation and Emanator Theory

Overview of Book 1

Book 1 is a modern exposition of classical mechanics, including chaos theory, and including ties to later theoretical developments as well. The exposition consists, throughout, of the presentation of interesting problems with many solved, the others left for the reader. The problems are drawn from classical mechanics (CM) and mathematics courses taken at Caltech, Oxford, and the University of Wisconsin. The courses range from undergraduate level to advanced graduate level. The courses had a rich and sophisticated selection of textbook and reference material, as you might expect, and those reference texts are, similarly, drawn on here. Those classical mechanics texts, listed by author, include: Landau and Lifshitz [120]; Goldstein [121]; Fetter & Walecka [122]; Percival & Richards [123]; Arnold (ODE) [124]; Arnold (CM) [125]; Woodhouse [126]; and Bender & Orszag [127]. Notice how the first Arnold reference and the Bender and Orszag reference involve textbooks focused on ordinary differential equations (ODEs). Likewise, an analysis of the excellent, and rapid, exposition by Landau and Lifshitz, reveals that it partly progresses through the material by going through ODEs of increasing complexity (corresponding to more complicated pendulum motion, for example, such as by adding a frictional force). This strong alignment with the underlying mathematics of ODEs is continued in this exposition, so much so that an appendix is provided for a quick review of ODEs from the applied mathematics perspective.

Particle dynamics, with and without forces, are described, with all arriving at descriptions with chaotic motion, with chaos described in the latter half of Book 1 [11]. Universally it is found that systems transitioning to chaotic behavior do so with a remarkable period-doubling process and this will be described both mathematically and with computer results. In the analysis of such dynamical systems we will find that periodic physical systems can be described in terms of repeated "mappings", e.g., classic dynamic mappings [128], and when described in this way the transition to chaos is made much more mathematically evident (as will be shown). The familiar Mandelbrot set is generated by such a repeated mapping, where it's "edge of chaos" is defined by the fractal boundary of the classic Mandelbrot image.

Properties of the classic Mandelbrot set will be relevant to the physics discussed in Book 1 and Book 7, including the property that the fractal boundary has a fractal dimension of 2 (the fractal dimension of the boundary can be between 1 and 2, to get equal to 2 is special). With the Mandelbrot set we also recover the well-studied constants associated with the universal Feigenbaum constants [15]. In the Mandelbrot set we can clearly see the fundamental constant for maximum perturbation that is at maximum antiphase (negative) with magnitude C_{∞}, where the same results hold for a family of basic formulations (for a variety of Lagrangian formulations, for example).

From the Lagrangian variational formulation of 'action' for particle motion we will eventually define the path integral functional variational formulation involving that same Lagrangian to arrive at a quantum description for the non-relativistic quantum particle motion (described in detail in Book 4 [3], and relativistic in Book 5 [9]). From the quantum description we arrive at the propagator formalism for describing dynamics (this exists in the classical formulation too, but typically is not used much in that context). Complex propagators will then be found to have ties to statistical mechanics and thermodynamics properties (Book 6

[4]). The ties to statistical mechanics are further emphasized when at the "edge of chaos" but with the orbit motion still confined. This may be associated with an ergodic regime, thus an equilibrium and martingale regime, the existence of which can then be used at the start of Book 6 [4] statistical mechanics and thermodynamics derivations with the existence of equilibria established at the outset. The existence of the familiar entropy measures are already indicated in the neuromanifold description (Book 3 [1]), thus, together with equilibria, the Book 6 thermodynamics description is able to begin with a well-established foundation that is not claimed by fiat, rather claimed as a direct result of what has already been determined in the theory/experiment described in the previous books of the Series.

Overview of Books 2 & 3

When moving from a theory of point particles to a theory of fields, there's not much discussion in the core physics books on fields in a general sense, it usually just directly jumps to the main field of relevance, Electromagnetism (EM). If advanced, it may also cover General Relativity (GR), as with [129]. In what follows we will cover these topics, but we will also cover the more basic fields in 1, 2, and 3D (including fluid dynamics), as well as 4D Lorentzian Field formulations (for Special Relativity), the Gauge Field formulation (thus Yang Mills covered in a classical context), and the GR geometric and gauge formulations. This establishes the foundation for the standard forces, and upon quantization (Books 4 and 5 in the Series), lays the foundation for the standard renormalizable forces (all but gravitation).

The gravitational coupling constant 'G' is a dimensionful coupling (not like with alpha in EM), and gravitation with manifold construct can be described as a gauge field construct, although not renormalizable. Gravitation, and associated geometry/manifolds, appears to relate to its own emergent structure, as will be discussed in Book 6. From the local Lorentzian geometry and Lorentzian field descriptions we also see the first of many examples where there is system information in the complexification of some parameter, here the time component. If the Lorentzian is shifted to complex time, this shifts it to being a Euclidean field, with formally well-defined convergence properties (as occurs in statistical mechanics). Complex time also shows deep connections between classical motion and associated Brownian motion (where random walk reveals pi). Thus, it should not be surprising that an emergent manifold may have complex structure such that there is also an emergent 'thermal' manifold, possibly the neuromanifold described in Book 3 and the related partition functions examined in Book 6. Just like locally flat space-time is a natural construct in GR, so too are optimization "learning" steps on a neuromanifold such that relative entropy is selected as a preferred measure, and from it Shannon entropy and Boltzmann's statistical entropy. Thus, the manifold construct appearing at Book 3 has far reaching impact into the foundations of the thermodynamic and statistical mechanical theory described in Book 6.

Before we even get to the manifold/geometry complexities of GR, however, we have already established much with the EM field part of the theory: (i) from 'free' EM without matter we get the speed of light c, Lorentz invariance, and from that special relativity and locally flat space-time; (ii) from EM with matter we get the dimensionless coupling constant alpha.

In going over field theories to describe matter, force fields, and radiation we first describe the classical field theories (CFTs) of fluid mechanics, EM, and General Relativity, with many examples shown. This is then carried over to the quantum field theory (QFT) description in Book 5. A review of the core mathematical constructs employed in CFT and QFT is given in the Appendix. Even as the mathematical physics approach grows in sophistication, we still obtain solutions via variational extrema. Thus, determining the evolution of the system from its variational optimum now becomes the focus of the effort. System 'propagation' from one time to a later time can be described by a propagator. Although a 'propagator' formulation is possible mathematically in classical mechanics (CM) and classical field theory (CF), which are shown, this is usually not done, in favor of simpler representations for the experimental application at hand. As we move to descriptions in the quantum realm, however, the use of the propagator formalism becomes typical, and when used in the path integral formulations we arrive at a compact formulation describing both the evolution and stationary-phase solution at once.

In Book 2 the focus is on classical field theory in a fixed geometry, the main physical example is EM. In this setting alpha appears, for example, in the description of an electron-positron pair: $F = e^2/(4\pi\varepsilon a^2)$ for electron-positron distance 'a' apart, where alpha appears as the coupling constant. Later, in quantum mechanics (QM), both modern and in the early Bohr model, we have that alpha = $[e^2/(4\pi\varepsilon)]/(c\hbar)$. The appearance of alpha in these situations is occurring in bound systems. If we examine EM interactions that are unbound, on the other hand, such as with the Lorentz Force $F = q(E \times v)$, here there arises no alpha parameter, nor with the early quantum mechanical analysis of such systems such as with Compton scattering. Thus, we see an early role for alpha, but only in bound systems, thus only in systems with (convergent) perturbative expansions in system variables.

In Book 3, classical field theory with *dynamic* geometry, i.e. GR, we don't see alpha at all. Instead we see manifold constructs and the mathematics of differential geometry (and to some extent differential topology and algebraic topology). Manifold constructs are entirely encapsulated in the math background given in Book 3 and the Appendix there. An application in the area of neuromanifolds (see [6]), shows the equivalent of a geodesic path in this setting is evolution involving minimum relative entropy steps. Similar to the description of a locally flat space-time we now have a description of 'entropy' increasing/evolving according to minimum relative entropy.

General relativity (GR) stands apart from the other force fields. All the other force fields are part of an adjoint representation of the standard model vis-à-vis the stability subgroup U(1)xSU(2)$_L$xSU(3). The form of which is derivable from the chiral T one-sided products described in Book 7. The standard model is uniquely obtained in this process, and with no mention of GR. Keep in mind, however, that the adjoint representation has operation on some space (hyperspinorial in case of simple octonion right-products, for example). The 'force' due to gravity is that due to manifold curvature, where the manifold construct is possibly emergent on the space of operation. Thus, the origin of the GR force is entirely different, and it will not allow quantization like the other forces, nor will its singular solutions be resolvable via quantum physics alone, as with EM in Books 4&5, but will also need thermal physics (as will be described in Book 6).

The existence of singular GR solutions, outside of specially symmetric cases (the classic Black hole solutions), wasn't firmly established until the Penrose singularity theorem [42] (awarded Nobel prize in Physics for this in 2020). Some of this material is covered in Book 3 to show how the mathematical formalism shifts to differential topology methods to describe the singularities, with examples referencing the Hawking and Ellis classic [39] and using Penrose diagrams. This, in turn, will come in handy when describing the classic FRW cosmologies with radiation and matter dominated phases (using notes from Peebles [103], Peebles won the Nobel in Physics in 2019).

The GR development would be remiss if it didn't briefly delve into cosmological models, the classic FRW cosmologies in particular. With the GR tools developed, cosmological results are examined, starting with the entry of the cosmological constant into the formalism (a candidate for Dark energy). Various observational data on galaxy rotations and universe simulations of galaxy cluster formation both indicate the existence of Dark matter. This, then, means we have new matter, non-interacting except gravitationally, and this is actually consistent with the latest observational data on the muon g-2 value [104], where the discrepancy between theory and experiment has grown to 4.2 standard deviations, where an extension in the Standard Model appears to be in the works. This is convenient as Emanator theory (Book 7 [5]), predicts such an extension.

We can thus arrive at field equations for EM, GR, and Yang-Mills Gauge Fields (Strong and weak). We can obtain wave and vortex phenomena (as hinted in fluid dynamics). We show the classical instability for atomic matter (classical EM instability) and classical gravitational instability (leading to black hole formation with singularity). From Lagrangian formulations we can then arrive at a QFT formulation (Book 5). The QFT formulation completes the QM (Book 4) cure of "non-relativistic atomic instability" with the cure of the fully relativistic atomic description of the radiative-collapse instability. Introduction of QFT also leads to new instability or infinities, but these can be eliminated by renormalization for the EM and electroweak formulations, and the Yang-Mills strong formulation, but not the GR (gauge) formulation. The current theoretical formulation in modern physics has one glaring gap, therefore: a quantum theory of gravitation. Perhaps this is not a missing element, however, if geometry/GR is a derivative phenomenon, like the field of statistical mechanics and thermodynamics appeared as derivative phenomenon when the complexified quantum propagator gives rise to a real (quantum) partition function. The hint of a deeper emanator theory suggests emergent structures of geometry and thermodynamics are arrived at in the process of emanation, with the information emanated being that of the renormalizable quantum matter fields. In Book 7 [5] a precise mathematical meaning will be found for describing maximal information emanation.

Overview of Book 4

By 1834, with Hamilton's Principle, there was a strong foundation for what is now called classical mechanics. By 1905, with Einstein's publication on the photoelectric effect [130], the rules of classical mechanics were being superseded by the new rules of quantum mechanics. The earliest appearance of quantum mechanics, however, began with the various observations of quantization of light, starting with the strange occurrence of spectral lines for hydrogen. The hydrogen spectrum was made even stranger by a precise fit to a succinct empirical formula by Balmer in 1885 [131]. This is the beginning of an amazing period of discovery. The developments of QM from introductory to advanced roughly follows that history.

The early phase of discovery for quantum mechanics moved into the modern quantum mechanics formalism with the discovery of Heisenberg of the successful application of matrix mechanics and the resultant uncertainty principle (1925) [132]. In 1926, Schrodinger showed that the problem of finding a diagonal Hamiltonian matrix in the Heisenberg's mechanics is equivalent to finding wavefunction solutions to his wave equation [133]. An interpretation of the wavefunction was then clarified in 1927 by Born [134]. Dirac developed a manifestly relativistic formalism for the wavefunction and wave-equation for fermionic matter (1928) [135]. An axiomatic reformulation of quantum mechanics was then given by Dirac (1930) [136], laying the foundation for much of modern quantum notation and for critical issues such as self-adjointness. Dirac then described a formulation of a quantum propagation path, with quantum propagator having the familiar phase factor involving the action, in his paper "The Lagrangian in Quantum Mechanics" in 1933 [137]. In essence, Dirac had obtained a single path, in what would eventually be generalized by Feynman to all paths with the invention of the path integral formalism (1942 & 1948) [138,139]. The equivalence of a quantum mechanical formulation in terms of path integrals and the Schrodinger formalism was shown by Feynman in 1948 [139].

In a path integral description, the quantum mixture state, semiclassical physics, and classical trajectories are all given by the stationary phase dominated component. A stationary phase solution that is dominated by a single path is typical for a classical system. Thus, variational methods are fundamental to analysis of physical systems, whether it be in the form of Lagrangian and Hamiltonian analysis, or in various equivalent integral formulations.

Feynman's discovery of the path integral formalism wasn't solely based on the prior work of Dirac (1933) [137], although by appending that paper to his PhD thesis (1946) its importance was clearly emphasized. Feynman also benefited from work going as far back as Laplace [140] for selection process based on highly oscillatory integral constructions that self-select for their stationary phase component. This branch of mathematics eventually became associated with Laplace's method of steepest descents, then to the work of Stokes and Lord Kelvin, then to the work of Erdelyi (1953) [141,142].

Feynman and others then invented quantum field theory for electromagnetism (QED) during 1946-1949 (more on this later). Extension to electroweak occurred in 1959, and to QCD in 1973, and to the "Standard Model" in 1973-1975. Thus, the impact of the path integral revolution in quantum physics was felt well into the 1970's, but this was only the beginning. At their inception path integrals were examined by Norbert Wiener, with the introduction of the Wiener Integral, for solving problems in statistical mechanics in diffusion and Brownian motion. In the 1970's this led to what is now known as "the grand synthesis" which unified quantum field theory (QFT) and statistical field theory (SFT) of a fluctuating field near a second-order phase transition, and where use of renormalization group methods enabled significant advances from QFT to be carried over to SFT.

The grand synthesis is one of many instances to come where we see analytic continuation of a constant or a parameter giving rise to familiar physics in the thermodynamic and statistical mechanics domains, showing a deeper connection (still not fully understood, see Book 7). The Schrödinger equation, for example, can be seen to be a diffusion equation with an imaginary diffusion constant. Likewise, the path integral can be seen to be an analytic continuation of the method for summing up all possible random walks.

In Book 4 we also carefully examine the closest gravitational equivalent to the hydrogenic atom (dust shell collapse). What results is an incomplete formulation due to boundary conditions, where to get the time choice you must input that time choice. No specific choice of time is indicated to avoid infall-collapse. The results, however, can show stability and consistency in a "full" thermal quantum gravity description where analyticity is employed. Success in this way, and not others, suggests possible fundamental role of analyticity and thermality (Books 6&7) and also suggests that thermal quantum gravity TQG may 'exist' or be well-formulate-able, while quantum gravity QG generally might not 'exist'. These results, shown in Book 6, provide the lead-in to the Book 7 discussion on Emanator theory, where core concepts in Books 1-6 that tie to emanator theory are brought together in a new theoretical synthesis.

Overview of Book 5

In Book 5 we show QFT's in the gauge field representation, which clearly relates the choice of field theory to a choice of Lie algebra, which, in turn, can be related to a choice of group theory (such as U(1) and SU(3)). From this we can see that non-classical algebraic constructs are ubiquitous in QM and QFT, so a review of Group Theory and Lie Algebras is given in the Appendix, as well as a review of Grassman Algebras, and other special algebras needed in QM and QFT. Similarly, as regards choice of approach, we find that the Schrodinger and Heisenberg formulations often provide the only tractable way to get a solution for bound systems. In critical theoretical considerations, however, the path integral approach is best (as will be shown). In seeking a deeper theory, the more unified path integral (PI) approach provides important hints as to a deeper theory (see Book 7).

In Book 5 we get the highest precision result for the value of alpha, in its role as perturbation parameter. If a calculation of the electron magnetic moment parameter g-2 is performed, with all of the Feynman diagrams appropriate to expansions up to 5th order, we get a determination of alpha up to 14 digits, where 1/alpha=137.05999….. . This gives us one of the most precise measurements of alpha known. When a similar analysis is done for the muon g-2, given the much larger muon mass, particle production pairs of other particles have a measurable effect, and we are able to probe the lower masses of the standard model that are present. In doing this, in preliminary experiments, there is a discrepancy indicating more particles, e.g. the Standard Model will need to be extended (possibly with a type of 'sterile' neutrino). These missing particles could be the missing "Dark Matter". The prediction of such in Emanator Theory, and why there should be an imbalance between the left and right neutrinos (hint: maximum information transmission) is described in Book 7.

Part of the description of quantum field theory entails use of analyticity and other complex structures to encapsulate more of the physics in a complex extension to the space (or dimension). This often leads to formulations in terms of complex integration, with the choice of complex contour specified, such as with the Feynman propagator. One of the main renormalization methods, for example, is to use dimensional regularization, which entails analytically continuing expressions with dimensionality to dimensionality as a complex parameter. There is also the aforementioned shift to complex and to "Wick rotate" expressions with real time to expressions with pure complex time. In doing this the statistical mechanical partition function for the system is obtained, with well-defined summation. Thus, a connection between 'thermality' and complex structure, in the time dimension at least, is indicated.

The second part of Book 5 describes QFT on curved space-time (CST), where we arrive at an early analysis of Black Hole thermodynamics. Here we find that space-time curvature gives rise to thermality and particle production effects. Black Hole thermality was revealed in Hawking radiation [143], due to the causal boundary at the horizon. Such thermality is even seen in flat space-time (Book 5) if causal boundaries are induced, such as in the case of an accelerated observer [144].

QFT on CST has one further gift, critical to the statistical mechanics formalism to follow in Book 6, and that's the spin-statistics relation. This relation is usually assumed, along with other critical notions, such as entropy, and the relation between entropy and density of states. These are all shown, with the presentation path chosen in this Physics Series, to be fundamental or derivative to the formalism already established in Books 1-5 (to prepare for Book 6).

The choice of time is related to choice of vacuum, which is related to choice of field geometry or observer motion (such as constant acceleration or expansion). If you have flat spacetime QFT with a boundary, then you have thermodynamic effects (e.g., the Rindler observer). In this setting we can compare the Hawking derivation of Hawking Radiation using the Euclideanization 'trick' vs the Bogoliubov transformations of the field to the Rindler geometry from the Minkowski geometry (if chosen as the asymptotic vacuum reference). With QFT on CST we also arrive at spin-statistics as mentioned, and get the final extension of the theory by way of Grassman algebras, to arrive at thermodynamically consistent Bose and Fermi statistical descriptions on quantum matter.

Overview of Book 6

Thermodynamics is the oldest of the physics disciplines (fire), with unapologetic use of phenomenological arguments and mysterious thermodynamic potentials (entropy). Obviously, thermodynamics is still prevalent today, including in its more quantified form via statistical mechanics. How is this not a failure of the mechanistic description of the universe indicated by CM and even QM? Concepts that appeared in QM, such as probability, are now occurring again. Other new concepts appear as well, including: approximate statistical laws; equations of state; heat as a form of energy; entropy as a variable of state; existence of equilibria; ensembles/distributions; and existence of the partition function. Many of these concepts appear in the path integral descriptions with the analyticity methods/extensions mentioned previously, so there are hints of a deeper theory that arrives at much of thermodynamics/Statistical mechanics foundation from the existing quantum theory.

Book 6 has been placed after the other chapters to await identification of entropy as fundamental in that it can be identified as an intrinsic system function even before getting to thermodynamics. We also already have experience with many particle systems, via QFT (especially in CST where particle creation is almost unavoidable), without directly tackling that scenario (due to QFT effectively already being many-particle, with analytic determination of many-particle system functions, such as entropy). With entropy presented at the outset as an important system variable, the derivation of thermodynamic potentials is then a straightforward process, as will be shown. The standard SM connections to thermodynamics can then be given. Thus, in covering Thermodynamics and Statistical Mechanics we start with the foundations of the theory mostly established, such as entropy (also with equipartition equivalent to sum on paths with no weightings, etc.), with no assumptions. Everything follows directly from the theoretical discoveries outlined in the preceding books in the Series. We don't see new connections to alpha, but we do see new structures/effects, especially manifold constructs (as with GR, where we also saw no role for alpha).

The close ties between QM Complexified giving rise to a particle ensemble partition function, and QFT complexified and field ensemble partition function, is now simply a derivative aspect of the fundamental complexation posited. This complexation will be posed in Book 7 with emanation in a complexified perturbation space.

From Atomic Physics, described in Book 4, we also obtain the standard rules on electron shell completion (that is encoded in the periodic table). Similarly, we can also understand the origins of the intermolecular quantum chemistry rules. When taken to the statistical mechanics (SM) extreme we have thermodynamic equilibrium emergent from (the Law of Large Numbers (LLN) and reverse Martingale convergence. With completion of application to chemical processes we have clear phase-transition effects, as well as equilibrium and near-equilibrium effects. The familiar chemistry results, with phases of matter.

From chemical equilibrium and near-equilibrium, with 10^{23} elements that interact weakly or not at all, we have two generalizations. The first is to consider chemical near-equilibrium and directly obtain an emergent process at this level, this is the branch that gives us biology/life at its most primitive level. The second is to consider equilibrium and near-equilibrium in general when the elements interact strongly (with10^{10}elements, say), this is the branch that describes biology/life at its most advanced social level and economics. In classic shot noise, the granularity of low-current flow (due to discreteness off electron charge) leads to a noise effect. Thus, as we consider situations with fewer elements, there are more complications, not less, due to granularity noise effects, and we enter the realm of machine learning with sparse data. Noise effects can be significant in complex systems, especially in biology where it is part of what is selected (such as in hearing, for background noise cancellation).

The second part of Book 6 explores the role of thermodynamics in efforts to extend to TQFT and TQG. This is done by exploring Black Hole settings. The recognition of a role for complex structure on system variables becomes apparent in this process (on top of the generalization to non-trivial algebras as already revealed).

In Book 6, part 2, we examine the Hamiltonian thermodynamics of some black hole geometries with stabilizing boundary conditions. In this foray into directly exploring a thermal quantum gravity (TQG) solution we assume a path integral form for the GR problem and shift directly to a partition function (by 'Wick rotation' mentioned above). We see that TQG is possible, where positive heat capacity shows stability. Another encouraging result as to an eventual unifying theory comes from String theory via its explanation of BH thermodynamics and BH horizon effects with the BH fuzz solution (via use of the holographic hypothesis and the related AdS-CFT relation [145,146]).

In Book 6, part 2, we also examine the propagator to partition-function transformation upon complexation, which leads to a thermodynamic theory for some equilibrium formulation, with certain parameter settings required for stability (positive heat capacity). This is doable in a variety of settings, suggesting how such thermodynamically consistent boundary conditions may be what constrains the classical motion and BH singularity formulation by the effect of this stabilization manifesting for certain internal geometries. Successful TQG (Thermal Quantum Gravity) formulations, such as for RNadS and Lovelock spacetimes shown in Book 6, via reformulation using analyticity, and not via non-analytic approaches, suggests a possible fundamental role of analyticity once again and also suggest that TQG may 'exist' or be well-formulate-able, while QG generally might not 'exist'. These results, together with core concepts from Books 1-6 that tie to emanator theory, are brought together in a new theoretical synthesis in Book 7.

Overview of Book 7

In Books 4,5, and 6 of the Series, we explored examples of QM with imaginary time, QFT in CST, Thermal QFT, minisuperspace QG, and Thermal QG. In this effort we find the path integral, and PI propagator, to provide the most general representation. In seeking a deeper theory in Book 7 we build on the sum-on-paths with propagator formulation to arrive at a sum-on-emanations with emanator formulation.

Propagation in a complex Hilbert space, in a standard QM or QFT formulation, requires the propagator function to be a complex number (not real or quaternionic, etc., [147]). This prohibits what would otherwise be an obvious generalization to hypercomplex algebras. In order to achieve this generalization, we have to introduce a new layer to the theory, one with universal emanation involving hypercomplex algebras (trigintaduonions) that is hypothesized to project to the familiar complex Hilbert space propagation with associated fixed elements (e.g., the emanator formalism projects out the observed constants and group structure of the standard model). The 'projection' is an induced mathematical construct, like having SU(3) on products of octonions, but here it we be the standard model U(1)xSU(2)xSU(3) on products of emanator trigintaduonions. Thus, in Book 7 a unified variational formulation is posed, one that arrives at alpha as a natural structural element, among other things, uniquely specified by the condition of maximal information emanation.

In Book 7 we also make note of the implications of a fundamental mathematical operation on a space that is repeated or added. The non-GR forces are given by the form of the operation (the sequence forming an associative algebra), the GR forces are given indirectly by the form of the space, this leaves the aspect "repeated or added" to be considered with care. If a purely 'repeated' operation, or mapping, occurs we can return to the dynamical mapping discussion of Book 1, where chaos can occur and is ubiquitous. There, the primal 'phase transition', the transition to chaos, is evident. If an operation with addition is involved (in the statistical sense of multiple elements), along with repeated overall steps, we arrive at the general framework of statistical mechanics with effects from the Law of Large Numbers (LLN) and reverse Martingale convergence, among other things (Book 6). Most notable, however, is the prevalence of a new effect, that of phase transitions and the emergence of new structure (order from disorder), including the remarkable structures of chemistry and biology.

Why the recurring 'Cabbalistic formula'? was a question even in the time of Sommerfeld [148]. Now, the numerological parallel is more exact than realized at that time, so is too much a coincidence to be by chance. The non-coincidence appears to be due to the maximal nature of information transmission in a variety of circumstances (in physics, biology, and even human communication with sufficient optimization) as well as with the fractal-like repetition of key parameter sets that occurs in these different settings $\{10,22,78,137 \cong 1/alpha\}$. We see that 10 expresses the dimensionality of propagation (or nodes of connectivity), while 22 corresponds to the number of fixed parameters in the propagation (in Book 7 we explore propagation in a 10 dimensional subspace of the 32 dimensional trigintaduonion space, leaving 22 dimensions at fixed values that appear as parameters in the theory). We will see the number 78 relates to generators of the motion, and that there are 4 chiralities of motion ('doubly chiral'). We will also see that 137 is simply the number of independent tri-octonionic product terms in the general chiral trigintaduonion 'emanation'.

Synopsis – Frodo Lives

Tolkien wrote of eucatastrophes [149], perhaps he anticipated the constructive role of emergent phenomena in maximum information transmission.

Preface to Physics Series, Book #2, on:

The Dynamics of Manifolds

Geometrodynamics, or full General Relativity, is considered in this book, #3 of the Series. We start with a review of manifolds and introductory topology definitions. Derivation of Einstein's equations from the Hilbert Action is then done. The Cartan Method is used extensively to simplify analysis. The ADM and FNC formalisms are then described. The ADM formalism describes the evolution on a space-like foliation of the space-time, where that foliation is described in terms of a metric formulation in terms of "lapse" and "shift" function. The FNC formalism involves a geodesic path described in terms of an observer's proper time. A number of specialized problems are then explored with the tools developed.

The full GR solution for a dust shell will be carefully analyzed and reparametrized to have the classical formulation for which a quantization is most likely to succeed. The actual quantization analysis will then be done in Book 4 of the Series. (In Book 4 it will be shown that the shell minisuperspace quantization sees key geometric parameters as externally-set boundary conditions, so effectively "classical" or "apparatus" parameters).

Geodesics are then examined for geometries with closed time-like curves (CTCs), for which divergent blue-shifting is found (the prevention of which is conjectured in Hawking's Chronology Protection conjecture). A brief recounting of the Penrose Singularity theorem is also given. A variety of Black Hole geometries are also examined.

Cosmological GR derivations are then done and consistency with astrophysical observations places limits on both its large-scale cosmological evolution, but also in terms of the numbers of species of fundamental particles that can exist (thus constraining the form of the extended standard model that is discussed in Book 5).

Thus, this text explores the "boundaries" of the theory in the sense of: (i) actual boundaries terms on the choice of the Action; (ii) boundaries due to singularities and causal domains (the Black hole problems); (iii) boundaries on allowable geodesic topologies (e.g., CTCs forbidden); and (iv) from the cosmological theory, with constraints on particle species and existence of dark matter. Derivations are shown in detail, throughout the text, as an instructional aid.

Chapter 1. Introduction

Geometrodynamics, or full General Relativity, is considered in this book, #3 of the Series [1]. We start with a review of manifolds and introductory topology definitions, and after some short derivations (Ch. 2) we show that the Riemannian Manifolds are of interest for General Relativity. The derivation of Einstein's equations from the Hilbert Action is then done in Ch. 3. This requires a careful differential geometry analysis, so review of some of the mathematical tools used is given in Ch. 4 (the Cartan Method) and the Appendix. This is a lot of preparatory background and review of Einstein's equations of motion to have a thorough understanding of the mathematical environment. Even so, we are not done preparing. Two formalisms are used as standard problem reductions (and have been studied extensively to serve as a refences for this reason). The first formalism describes the evolution on a space-like foliation of the space-time, where that foliation is described in terms of a metric formulation in terms of "lapse" and "shift" function (e.g., the ADM [2] formalism that is described in Ch. 5). The second formalism describes tensor fields such as the metric along an observers path, typically a geodesic path. For geodesic path described in terms of an Observer's proper time, we arrive at Fermi Normal Coordinates (e.g., 'FNC' coordinates described in Ch. 6). We now have the tools to do advanced analysis of Closed Time-like Curve (CTC) producing spacetimes (Ch. 7); Black hole space-times (Ch. 8); likely Cosmological space-times (the FRW Big Bang Universes; Ch. 9); and Dust Shell Collapse (Ch. 10).

The full GR solution for Dust shell will be carefully analyzed and reparametrized to have the classical formulation for which a quantization is most likely to succeed. The actual quantization analysis will then be done in Book 4 [3] of the Series. In Book 4 it will be shown that the shell minisuperspace quantization sees key geometric parameters as externally-set boundary conditions (effectively "classical" or "apparatus" parameters). So, there is, possibly, the suggestion that geometry is "apparatus," or partly so, which goes to the heart of the matter as to whether gravity is quantum at all (e.g., observable or apparatus). In Book 6 & 7 [4,5], ways to break past the theoretical impasse are explored using thermality (Book 6) and maximal hypercomplex extensions to the formulation (Book 7). Some of the advanced features considered in Books 6 & 7 have their beginnings in the generalized manifold constructs examined at the end of this book, in Ch. 11, where a brief examination of other manifold constructs is given, including the neuromanifold that occurs in statistical learning methods [6], and the geometric algebra generalization [7].

Geodesics are examined for geometries with closed time-like curves (CTCs), for which divergent blue-shifting is found (the prevention of which is conjectured in Hawking's Chronology Protection conjecture [8]). A brief recounting of the Penrose Singularity theorem is given. This establishes that the well-defined spherically symmetric shell collapse that forms a singularity can't be dismissed as an artifact of the symmetry – BH singularity formation happens even without the symmetry. We will find that in classical GR singularity formation occurs very easily, but unlike the renormalizable EM theory (and Standard Model Forces) the same quantization approaches mentioned above (in Book 4 [3])will not rescue GR from collapse.

Ch. 9 goes into the cosmological aspects of GR and how consistency with astrophysical observations places limits on both its large-scale cosmological evolution, but also in terms of the numbers of species of fundamental particles that can exist (thus constraining the form of the extended standard model that is discussed in Book 5 [9]). This process reveals inflation, dark matter, and dark energy. In Book 5 [9] we will see that the current muon g-2 experiments reveal an incomplete set of matter particles (indicating massive sterile neutrinos as a candidate Std. Model extension, for part of the missing dark matter, at least).The FRW solutions are given for flat, anti-DeSitter (AdS) and de Sitter (dS) space-times. Ties to the AdS/CFT correspondence, and to dS/CFT with thermally stable black hole thermodynamic stability, are discussed in Book 6 [4]. Thus, this book provides the critical foundation needed in Books 6 and 7 (last) of the Series.

At the end of this book we explore unifying field formulations that are under active consideration by researchers. This includes exploring unifying field formulation in terms of generalized associative algebras (geometric algebras). We see that GA descriptions for standard gravity are possible with either a curved space-time embedding or a flat space-time (pure gauge field) embedding. The latter form brings us closer to an AdS/CFT (or barely-dS/CFT description), that can be aligned with string theory and black hole thermodynamics (Book 6 [4]). The GA formalism also provides a tractable Kerr derivation [10]. Along these lines, in Book 6 [4] is given an explicit derivation for the Reissner-Nordstrom Anti-deSitter Black Hole using Cartan Calculus (also for Lovelock space times).

At the end of this book we also explore another theoretical manifold space: 'information geometry'. An application of differential geometry in the case of a neuromanifold construct reveals a fundamental basis for the choice of Shannon Entropy (a locally flat construct that results as an optimum in a manner akin to locally flat Minkowski space-time in GR). We get entropy as a fundamental concept from an analysis of neuromanifolds, and we get the second law of entropy increase from analysis of black hole thermodynamics in Book 6 [4]. In Book 5 Part 2, on advanced quantum field theory, we also see how shifting to complex time (Euclideanization) gives rise to mathematical forms corresponding to a system partition function, where complex time also gives rise to an inverse temperature. Thus, all the main constructs of thermodynamics and statistical mechanics are already established by the time we get to a formal derivation of thermodynamic concepts and their relations in Book 6 [4]. (Book 6 will still have some surprises, however, such as emergent phenomena and phase-transitions.)

The geometrodynamics analyzed in the chapters that follow will explore the "boundaries" of the theory in the sense of: (i) actual boundaries terms on the choice of the Action; (ii) boundaries due to singularities and causal domains (the Black hole problems); (iii) boundaries on allowable geodesic topologies (e.g., CTCs forbidden), where there is clear indication of the fundamental nature of the weak energy condition (WEC). In turn, WEC becomes the positive energy condition when geometric manifolds have (consistent) spinor fields present. In geometry with matter field, in other words, local convexity requirements on kinetic energy distributions (a property from Chaos Universality taken as fundamental, see Book 1 [1]), force positive energy (see Witten [12] for original derivation along these lines), which continued in reverse (with pos. energy taken as true) then implies WEC validity. Thus CTC prevention, and the ability to place matter onto our geometry in a manner consistent with Einstein's field theory equations, both require that WEC be satisfied. Lastly, (iv) geometrodynamics, oddly enough, also explores the Standard model of Particle Physics in that the number of species of particles affects the thermodynamic equation of state used in the FRW

Universe model. Thus, in Ch. 9 on Cosmological theory, we see consistency with the minimally extended Standard Model (that has a dark matter candidate).

Chapter 2. Einstein's reification of the semi-Riemannian Manifold

In this chapter we examine Einstein's reification of the semi-Riemannian Manifold. We begin by asking why the specific mathematical structure known as a manifold, and what is that structure? So we begin with some definitions. Further math background on Differential Geometry,

Differential Topology; and Algebraic Topology, has been placed in the Appendix.

2.1 Why Manifold? Chaos Universality, Morse Theory, and Palais
Geometry is a manifold, not a field. Indirectly, it is a manifold defined by a set of fields satisfying local Lorentz Invariance, thus it will satisfy Lorentz Invariance locally as well (see Book 2 [13] for details). If classical field invariances 'imprint' on what we are calling the 'physical manifold' (that we simply call the 'geometry' of the system), what else imprints on the definition of 'physical manifold' from the classical theory? Most notable in this regard is the classical behavior known as 'chaos', which is clearly exhibited in terms of simple classical trajectories in Book 1 [11], but also is seen in fluid dynamics systems and fields in general [13,14], thus it presents a seemingly strict constraint or 'imprint' on the behavior of the manifold. Surprisingly, rather than restrict to a sub-class of Manifolds, the 'imprint' of the chaos universality constraint selects the mathematical object of the Manifold itself.

Consider the scenario where the 'system' description is 'maximal' in the sense that it includes all behavior, trajectories, up to the 'edge-of-chaos'. (Such limits are considered in situations where evaluating the maximum allowable noise perturbation/transmission the system can sustain, such as in Book 7 [5].) What this would indicate is another type of 'local behavior' other than Lorentz Invariance, here known as Chaos Universality [15]. This is because, from chaos theory, there have emerged universal constants [15,16], with system behavior describing a path to chaos via period doubling [15], and central to the derivation of the Universality property (that is observed) is the requirement of local quadratic structure [15]. A seemingly benign constraint, described as such by Feigenbaum [15], but if taken to be a strict mathematical property of the system (thus not a universal output of the system description for instances of explicit chaos, but a universal input to the system description), then from Algebraic Topology (Appendix) we arrive at the requirement that it be a manifold structure. This is because local quadratic structure, as indicated, is by Morse Theory (and the Palais Lemma) indicative of a manifold structure.

So, Geometry is a manifold not a field. Of course, all are present in a physical situation: geometry and field and matter (which can be represented as a spinor field at the elementary level). From the Witten Positive Energy [12] theorem this imposes a constraint on the manifold and field such that the WEC is satisfied (or AWEC for QFT). The WEC, in turn, must be satisfied to not have CTCs (as explored in Ch. 7). WEC is, thus, a 'given' if we are seeking the maximal system representation that be functionally described. Thus, we can posit any kind of field, locally, but we cannot do the same for the physical manifold consisting of manifold with field. This means the manifold is constrained (positive energy) in a way that the (quantizable) fields are not.

Geometry (as manifold) and Standard Model force fields are found to couple to matter very differently. Take the classic electromagnetic field coupling to matter represented by 'alpha', versus the gravitational coupling given by the Gravitational constant G. Alpha is dimensionless, while G has dimension. This alone destroy the possibility of the gravitational theory having a renormalizable quantum field theory due to the proliferation of infinities at all orders in the (dimensionful) coupling constant. Various orders of Alpha, on the other hand, are just numbers (dimensionless) and can be added, to only have one infinity (for example) to offset via renormalization.

So, 'physical' geometry is only consistently described via the mathematical object of a manifold. The manifold with matter-field is constrained as mentioned above, but what of constraints when just considering the manifold itself? The most significant of these will result from the simple observation that rigid bodies don't change length on rotation (e.g., exist). This will result in the requirement that the manifold have a Riemann metric

2.2 Why the Riemannian Metric?

Notes in this section make use of notation and examples from [17], including the use of concrete index notation (abstract index notation, by Penrose [18], is described in the Appendix).

In studying surfaces in three dimensional Euclidean space Gauss introduced a general parametric representation of surfaces ($\{u_1, u_2\}$), where intrinsic geometric features are expressed independent of surface coordinates, including distance as

$$ds = \left(\sum_{i,k}^{2} g_{ik} du_i du_k \right)^{1/2}.$$

(Eqn. 2-1)

In 1854 Riemann [19] publishes "On the hypotheses which lie at the Foundation of Geometry." Riemann points out the restriction of Gauss to the case of two surface coordinates was not necessary and only motivated by his focus on two-dimensional surfaces in a three-dimensional space. Riemann proposed the n-dimensional, but still quadratic, generalization:

$$ds = \left(\sum_{i,k}^{n} g_{ik} du_i du_k\right)^{1/2}.$$

(Eqn. 2-2)

The quadratic form is a function of the $du_i's$, e.g. $ds = F(du_i)$, but more general would be not just quadratic and also have explicit coordinate dependencies: $ds = F(u_i, du_i)$. Such general spaces are known as Finsler spaces. It was pointed out by the famous mathematician Helmholtz, however, that most Finsler geometries can be excluded since they lack symmetries. In fact, the simple existence of systems where we rotate a "rigid body" about a fixed point require transformations to exist that are length preserving. We will now show, following a discussion from [17], that requiring distance-preserving transformations to exist requires the geometry to be Riemannian.

We assume a metric function exists in which three-parameter groups of distance-preserving transformation are possible (for rigid body). Suppose near a fixed point O we have $ds = F(dx_i)$. Introduce a local coordinate system around the "origin" O. We have a three parameter family of transformations: $\tilde{x}_i = f_i(x_k, p_j)$ which depends on the three parameters p_j:

$$d\tilde{x}_i = \sum_{k=1}^{3}\left(\frac{\partial f_i}{\partial x_k}\right)_{x_k=0} dx_k = \sum_{k=1}^{3} \alpha_{ik}(p_j)dx_k\,,$$

(Eqn. 2-3)

and now $F(d\tilde{x}_i) = F(dx_i)$. So, to begin, consider:

$$F(x_i) = F\left(\sum_{K=1}^{3} \alpha_{iK} x_K\right) \rightarrow now\ \text{use}\ \varepsilon's \rightarrow F(\varepsilon_i) = F\left(\sum_{K=1}^{3} \alpha_{iK}\varepsilon_K\right)$$

(Eqn. 2-4)

Given an arbitrary plane in the ε-space, there exists still a one parameter subgroup of those transformations which carries the plane into itself. Assume $\varepsilon_3 = 0$ (no loss of generality). Since $\tilde{\varepsilon}_i = \sum_{K=1}^{3} \alpha_{iK}(P_1, P_2, P_3)\varepsilon_K$ we must have: $\alpha_{31}(P_1, P_2, P_3) = 0$ and $\alpha_{32}(P_1, P_2, P_3) = 0$, thus P_2 and P_3 can be determined as functions of the remaining free parameter $P = P_1$. We consider only $F(\varepsilon_i)$ positive definite. Now, with $\varepsilon_3 = 0$:

$$\tilde{\varepsilon}_i = \sum_{K=1}^{2} \alpha_{iK}(p)\varepsilon_K \qquad i = 1,2.$$

(Eqn. 2-5)

Let $F(\varepsilon_1, \varepsilon_2, 0) = \Phi(\varepsilon_1, \varepsilon_2, 0)$ and, so, $\Phi(\tilde{\varepsilon}_i) = \Phi(\varepsilon_i)$. We can assume without loss of generality that $p = 0$ corresponds to the identity transformation and that the matrices depend differentiably upon the parameter p:

$$\tilde{\varepsilon}_i(p) = \varepsilon_i + p\sum_{K=1}^{2} C_{iK}\varepsilon_K + O(p^2)$$

(Eqn. 2-6)

where C_{iK} is the derivative matrix of α_{iK} at $p = 0$. Now, differentiate $\Phi(\tilde{\varepsilon}_i) = \Phi(\varepsilon_i)$ with respect to p and put $p = 0$ to find the identity (also, dropping tildas):

Define the characteristic curve $\varepsilon_i(t)$ by $\frac{\partial \varepsilon_i}{\partial t} = \sum_{K=1}^{2} C_{iK}\varepsilon_K$, $i = 1,2$, then:

$$\frac{d}{dt}\Phi\varepsilon_i(t) = 0 \quad \longrightarrow \quad \sum_{i=1}^{2} \frac{\partial \Phi}{\partial \varepsilon_i}\left(\sum_{K=1}^{2} C_{iK}\varepsilon_K\right) = 0 \quad \longrightarrow \quad \Phi(\varepsilon_i) = \text{const}$$

(Eqn. 2-7)

Let $m = \min \Phi$, $M = \max \Phi$ on $\varepsilon_1^2 + \varepsilon_2^2 = 1$. Because of the homogeneity of Φ we can assert that

$$mr \le \Phi \le Mr \quad \text{for } \varepsilon_1^2 + \varepsilon_2^2 = r^2.$$

Thus, the curve $\Phi = a$ must be between the circles a/M and a/m, and is thereby bounded away from both zero and infinity.

The general solution to $\frac{\partial \varepsilon_i}{\partial t} = \sum_{K=1}^{2} C_{iK}\varepsilon_K$, $i = 1,2$ is:

$$\varepsilon_i = A_i e^{\lambda t} + B_i e^{\mu t}$$

with $\det\|C_{iK} - \lambda\delta_{iK}\| = 0$ etc. Thus, $\varepsilon_i = a_i \cos \lambda t + b_i \sin \lambda t$ where the integral curve is an ellipse. Instead of the parametrized form we have

$$\sum_{i,K=1}^{2} g_{iK}\varepsilon_i\varepsilon_K = 1.$$

(Eqn. 2-8)

Since $\varepsilon_3 = 0$ was arbitrary we have

$$\sum_{i,K=1}^{3} g_{iK}\varepsilon_i\varepsilon_K = \text{const}.$$

(Eqn. 2-9)

Thus, $ds^2 = \sum g_{iK}\, dx^i dx^K$, and the geometry is indeed Riemannian. Since ds^2 must be quadratic in the space differentials if the time coordinate differential *dt* equals zero, we expect that ds^2 will be quadratic as a space-time differential as well. The only new feature of the extended space time geometry will be the fact that ds^2 need not be positive definite.

2.3 4D Tensor analysis, transformations at a point, concrete indices

Using Adler [17] definitions and notation to establish conventions for concrete indices.

2.3.1 4D Tensors

Consider

$\overline{x}^j = f^j(x^0, x^1, x^2, x^3), \quad x^K = h^K(\overline{x}^0, \overline{x}^1, \overline{x}^2, \overline{x}^3)$

We shall deal with a Riemann space with a metric $ds^2 = \sum_{iK} g_{iK}\, dx^i dx.$

We assume $g_{iK} = g_{Ki}$ and has the signature (+1, -1, -1, -1).

Scalar Quantities: $ds^2 = \sum_{iK} g_{iK}\, dx^i dx^K$ has to keep the same numerical value under an arbitrary change of coordinates since it is a scalar. A scalar field is a point function in the space considered – invariant under transformation.

Contravariant Vectors: Defined analogous to distance differential 'vector'. So, for $\overline{x}^j$ and x^K are basis vectors in whatever coordinate system we have:

$$d\overline{x}^i = \sum_{j=0}^{3} \frac{\partial f^i}{\partial x^j} dx^j$$

(Eqn. 2-10)

Any set of four quantities $\varepsilon^i (i = 0,1,2,3)$ which transform according to $\overline{\varepsilon}^i = \sum_{j=0}^{3} \frac{\partial \overline{x}^i}{\partial x^j} \varepsilon^i$ forms a contravariant vector.

Covariant Vectors: Consider a scalar field φ. Consider Row the four quantities $A_i = \partial\varphi/\partial x^i$ transform under a change of coordinate system from (x^i) to $(\overline{x}^i)$:

$$\frac{\partial \varphi}{\partial \overline{x}^i} = \sum_{j=0}^{3} \frac{\partial x^j}{\partial \overline{x}^i} \frac{\partial \varphi}{\partial x^j}.$$

(Eqn. 2-11)

and we have:

$$\overline{A}_i = \sum_j \frac{\partial x^j}{\partial \overline{x}^i} A_j$$

(Eqn. 2-12)

where the latter form of transformation defines a covariant vector.

<u>Tensors:</u>

Once we have scalars, and covariant and contravariant vectors, we have the building blocks for more general transformational objects (rank n covariant and m contravariant), known as tensors.

$\overline{\varepsilon}^i = \frac{\partial \overline{x}^i}{\partial x^j} \varepsilon^j$ for contravariant vectors (indices above)

$\overline{\eta}_i = \frac{\partial x^j}{\partial \overline{x}^i} \eta_j$ for covariant vectors (indices below)

The inner product of a covariant vector and a contravariant vector is a *scalar* invariant:

$$\overline{P} = \sum_i \overline{A}_i \overline{\varepsilon}^i = \sum_{ijK} \frac{\partial x^K}{\partial \overline{x}^i} A_K \frac{\partial \overline{x}^i}{\partial x^j} \varepsilon^j = \sum_{jK} A_K \varepsilon^j \delta_j^K = \sum_j A_j \varepsilon^j = P.$$

(Eqn. 2-13)

Consider a multilinear form P where summation is implied on indices that match when one is covariant and one contravariant -- known as a 'contraction' (the implied summation is also known as Einstein convention):

$$P = \left(T^{i_1 i_2 \cdots i_a}_{j_1, j_2 \cdots j_b}\right)\left(\varepsilon^{j_1}_{(1)} \varepsilon^{j_2}_{(2)} \cdots \varepsilon^{j_b}_{(b)}\right)\left(\eta^{(1)}_{i_1} \eta^{(2)}_{i_2} \cdots \eta^{(a)}_{i_a}\right),$$

(Eqn. 2-14)

where $T^{i_1 i_2 \cdots i_a}_{j_1, j_2 \cdots j_b}$ is a set of n^{a+b} elements (with overall rank $(a + b)$ if a tensor). T is a tensor if, for an arbitrary change of coordinates under ε and η, T transforms such that P remains unchanged (a scalar). Thus, a tensor of rank zero is a scalar, and of rank one is either a contravariant or covariant vector. The metric g_{iK} is a second rank covariant tensor.

<u>General transformation law of a tensor:</u>

$$\left(\overline{T}^{i_1,\cdots i_a}_{j_1,\cdots j_b}\right)\left(\overline{\varepsilon}^{j_1}_{(1)}\cdots\overline{\varepsilon}^{j_b}_{(b)}\right)\left(\overline{\eta}^{(1)}_{i_1}\cdots\overline{\eta}^{(a)}_{i_a}\right)$$
$$= \left(T^{i_1,\cdots i_a}_{j_1\cdots j_b}\right)\left(\varepsilon^{j_1}_{(1)}\cdots\varepsilon^{j_b}_{(b)}\right)\left(\eta^{(1)}_{i_1}\cdots\eta^{(a)}_{i_a}\right)$$

(Eqn. 2-15)

So,

$$\overline{T}^{i_1,\cdots i_a}_{j_1,\cdots j_b}\left(\frac{\partial\overline{x}^{j_i}}{\partial x^{\beta_i}}\cdots\frac{\partial\overline{x}^{j_b}}{\partial x^{\beta_b}}\right)\left(\frac{\partial x^{\alpha_1}}{\partial\overline{x}^{i_1}}\cdots\frac{\partial x^{\alpha_a}}{\partial\overline{x}^{i_a}}\right) = \overline{T}^{\alpha_1\cdots\alpha_a}_{\beta_1\cdots\beta_b}$$

(Eqn. 2-16)

or

$$\overline{T}^{K_1,\cdots K_a}_{\ell_1,\cdots\ell_b}\left(\frac{\partial\overline{x}^{K_i}}{\partial x^{\alpha_i}}\cdots\frac{\partial\overline{x}^{K_a}}{\partial x^{\alpha_a}}\right)\left(\frac{\partial x^{\beta_1}}{\partial\overline{x}^{\ell_1}}\cdots\frac{\partial x^{\beta_b}}{\partial\overline{x}^{\ell_b}}\right) = \overline{T}^{\alpha_1\cdots\alpha_a}_{\beta_1\cdots\beta_b}$$

(Eqn. 2-17)

A typical example:

$$T^{\alpha\beta}_{\gamma} = \frac{\partial\overline{x}^{\alpha}}{\partial x^{i}}\frac{\partial\overline{x}^{\beta}}{\partial x^{j}}\frac{\partial x^{K}}{\partial\overline{x}^{\gamma}}T^{ij}_{K}.$$

(Eqn. 2-18)

2.3.2 Tensor Algebra

Let's now consider some basic properties of tensor algebra.

Equality

A and B are equal if $A^{\alpha\beta}_{\gamma} = B^{\alpha\beta}_{\gamma}$.

Additive property

Sum of two tensors with the same number of respective indices is simply

$$A^{\alpha\beta}_{\gamma} + B^{\alpha\beta}_{\gamma} = C^{\alpha\beta}_{\gamma},$$

(Eqn. 2-19)

which is a new tensor.

Multiplicative property

The product of a tensor by a scalar is again a tensor. The product of tensors is a tensor. The tensor product $T^{\alpha\beta}_{\gamma}S^{\mu v} = G^{\alpha\beta\mu v}_{\gamma}$ is often called the outer product of the two tensors T and S. The reverse can be done, where decomposition of a tensor into a sum of vector products is done (Tensor Products of Tensors of Rank 1).

Contraction of indices

$T^{i_1 i_2 \cdots i_{a-1}\sigma}_{j_1, j_2 \cdots j_{b-1}\sigma}$ is a tensor of rank $a + b - 2$ which one can denote $R^{i_1 i_2 \cdots i_{a-1}}_{j_1, j_2 \cdots j_{b-1}}$.

The Quotient Theorem
S = TA is valid if the indices count is correct. Then if any two are tensors the third must be as well.

Lowering and Raising Indices – Associated Tensors
Consider

$$T^{\alpha}_{\gamma} = g_{\gamma\beta} T^{\alpha\beta} \quad and \quad T_{\delta\gamma} = g_{\alpha\delta} T^{\alpha}_{\gamma} = g_{\alpha\delta} g_{\gamma\beta} T^{\alpha\beta}.$$

(Eqn. 2-20)

The second-rank tensor g_{iK} which we are using to lower indices can be chosen arbitrarily. However, once selected, it plays a central role in tensor calculus since it establishes a relation between contravariant and covariant tensors; it is called the fundamental tensor. In a metric space, such as the four-dimensional space of general relativity, it is quite natural to take for g_{iK} to be the metric tensor itself. Thus

$$g^{iK} g_{iK} = \delta^{i}_{j}.$$

(Eqn. 2-21)

Connection with Vector Calculus in Euclidean Space
Contravariant vector with components (μ^1, μ^2) $\quad \tilde{u} = \mu^1 \hat{e}_1 + \mu^2 \hat{e}_2$
Covariant vector with components (μ_1, μ_2) $\quad \mu_1 = u \cdot \hat{e}_1, \; \mu_2 = \hat{u} \cdot \hat{e}_2$

Connection between Bilinear Forms and Tensor Calculus
Let $\overline{X}_K = \sum_i a_{Ki} X_i$ and $\overline{Y}_K = \sum_i b_{Ki} Y_i$ and consider the bilinear forms: $F = \sum_i X_i Y_i$ and $\overline{F} = \sum_K \overline{X}_K \overline{Y}_K$:

$$\overline{F} = \sum_{ijK} a_{Ki} X_i b_{Kj} Y_j = \sum_{ij} \left(\sum_{K} a_{Ki} b_{Kj} \right) X_i Y_j$$

where

$$\overline{F} = F \text{ if } \sum_{K} a_{Ki} b_{Kj} = \delta_{ij} .$$

Using the transpose notation (if A^T is the transpose of a_{iK} then it is the matrix with a_{Ki}) we have our condition to be:

$$A^T B = I \quad \text{or} \quad AB^T = I$$

So, we have:

$$\begin{cases} \text{Original Matrix} & A \\ \text{Matrix contragradient to A} & B = (A^T)^{-1} \\ \text{Matrix contragradient to B} & C = (B^T)^{-1} = (B^{-1})^T = (A^T)^T = A \end{cases}$$

The contragradience relationship is an automorphism, i.e., it preserves the law of multiplication with the order of the factors. If, for example, $A = (B^T)^{-1}$ and $D = (E^T)^{-1}$ then $AD = [(BE)^T]^{-1}$.

2.4 4D Tensor analysis for points and vector fields on a manifold

So far transformation has been considered at a point. Now consider tensor at other points (everywhere in fact). There is now the possibility of 'moving about' and of transplanting vectors.

Vector Fields in Affine and Riemann Space

A Tensor Field is the assignment of a tensor to each point of the space.

2.4.1 Vector Transplantation and Affine Connections

Consider

$$\overline{\varepsilon}^i = \frac{\partial \overline{x}^i}{\partial x^K} \varepsilon^K$$

(Eqn. 2-22)

where vector $\overline{\varepsilon}^i$ is not a constant over the space since $\partial \overline{x}^i / \partial x^K$ is arbitrary. Covariance requirements on a vector field ε^K taken to be constant is then possible. Starting with a vector ε^i of constant components in an original coordinate system x^i, we see that $\overline{\varepsilon}^i$ in another coordinate system (generally) does not have constant components over the space. Let us now see how the components $\overline{\varepsilon}^i$ vary when we go from one point to a neighboring one in space along a curve parametrized with a parameter p; to do this we differentiate $\overline{\varepsilon}^i = \left(\partial \overline{x}^i / \partial x^K\right) \varepsilon^K$ with respect to p, remembering that the ε^K's are constant along the curve by assumption:

$$\frac{d\overline{\varepsilon}^i}{dp} = \frac{\partial^2 \overline{x}^i}{\partial x^K \partial x^\ell} \frac{dx^\ell}{dp} \varepsilon^K = \left(\frac{\partial^2 \overline{x}^i}{\partial x^K \partial x^\ell} \frac{dx^\ell}{\partial \overline{x}^m} \frac{\partial \overline{x}^m}{dp} \frac{\partial x^K}{\partial \overline{x}^j} \right) \overline{\varepsilon}^j = \overline{\Gamma}^i_{mj} \frac{\partial \overline{x}^m}{dp} \overline{\varepsilon}^j$$

(Eqn. 2-23)

where $\overline{\Gamma}^i_{mj} = \frac{\partial^2 \overline{x}^i}{\partial x^K \partial x^\ell} \frac{dx^\ell}{\partial \overline{x}^m} \frac{\partial x^K}{\partial \overline{x}^j}$. Thus,

$$d\varepsilon^i = \Gamma^i_{mj} dx^m \varepsilon^j,$$

(Eqn. 2-24)

and this defines a general law for transplantation of the vector ε^j at the point x into the quantities $\varepsilon^i + d\varepsilon^i$ at the point $x + dx$. It is a law of affine character; that is, it has invariant structure under a linear transformation of the coordinates.

So far we have achieved vector transplantation: $d\varepsilon^i = \Gamma^i_{mj} dx^m \varepsilon^j$. Now to make it coordinate-invariant. This will force certain requirements on the Γ^i_{mj}. To start, we have:

$$\varepsilon^i(x + dx) = \varepsilon^i + d\varepsilon^i = \varepsilon^i + \Gamma^i_{mj} dx^m \varepsilon^j.$$

(Eqn. 2-25)

Since we require $\varepsilon^i(x + dx)$ to be a vector this requires:

$$\overline{\varepsilon}^j(x + dx) = \varepsilon^i(x + dx)\left(\frac{\partial \overline{x}^j}{\partial x^i}\right)_{x+dx}$$

(Eqn. 2-26)

So,

$$\overline{\varepsilon}^j + \overline{\Gamma}^j_{ms} d\overline{x}^m \overline{\varepsilon}^s = \left(\varepsilon^i + \Gamma^i_{m\ell} dx^m \varepsilon^\ell\right)\left(\frac{\partial \overline{x}^j}{\partial x^i}\right)_{x+dx}.$$

(Eqn. 2-27)

Expanding using a Taylor series:

$$\left(\frac{\partial \overline{x}^j}{\partial x^i}\right)_{x+dx} = \left(\frac{\partial \overline{x}^j}{\partial x^i}\right)_x + \frac{\partial^2 \overline{x}^j}{\partial x^i \partial x^\eta} dx^\eta$$

Thus

$$\overline{\Gamma}^j_{ms} d\overline{x}^m \overline{\varepsilon}^s = \overline{\varepsilon}^j + \varepsilon^i \left(\frac{\partial \overline{x}^j}{\partial x^i}\right)_x + \varepsilon^i \left(\frac{\partial^2 \overline{x}^j}{\partial x^i \partial x^\eta} dx^\eta\right) + \Gamma^i_{m\ell} dx^m \varepsilon^\ell \left(\frac{\partial \overline{x}^j}{\partial x^i}\right)_x$$
$$+ \Gamma^i_{m\ell} dx^m \varepsilon^\ell \frac{\partial^2 \overline{x}^j}{\partial x^i \partial x^\eta} dx^\eta$$

(Eqn. 2-28)

And since

$$\left(\varepsilon^i + d\varepsilon^i\right) \frac{\partial^2 \overline{x}^j}{\partial x^i \partial x^\eta} dx^\eta \simeq \varepsilon^i \frac{\partial^2 \overline{x}^j}{\partial x^i \partial x^\eta} dx^\eta$$

We have

$$\overline{\Gamma}^j_{ms} d\overline{x}^m \overline{\varepsilon}^s = \Gamma^i_{m\ell} dx^m \varepsilon^\ell \left(\frac{\partial \overline{x}^j}{\partial x^i}\right) + \frac{\partial^2 \overline{x}^j}{\partial x^i \partial x^\eta} \varepsilon^i dx^\eta$$

$$= \left(\Gamma^i_{\alpha\beta}\left(\frac{\partial \overline{x}^j}{\partial x^i}\right) + \frac{\partial^2 \overline{x}^j}{\partial x^\beta \partial x^\alpha}\right) \varepsilon^\beta dx^\alpha$$

(Eqn. 2-29)

and using

$$\varepsilon^\beta dx^\alpha = \frac{\partial x^\alpha}{\partial \overline{x}^m} \frac{\partial x^\beta}{\partial \overline{x}^s} \overline{\varepsilon}^s d\overline{x}^m$$

this becomes:

$$\overline{\Gamma}^j_{ms} d\overline{x}^m \overline{\varepsilon}^s = \left(\frac{\partial \overline{x}^j}{\partial x^i} \frac{\partial x^\alpha}{\partial \overline{x}^m} \frac{\partial x^\beta}{\partial \overline{x}^s} \Gamma^i_{\alpha\beta} + \frac{\partial^2 \overline{x}^j}{\partial x^\alpha \partial x^\beta} \frac{\partial x^\alpha}{\partial \overline{x}^m} \frac{\partial x^\beta}{\partial \overline{x}^s}\right) d\overline{x}^m \overline{\varepsilon}^s$$

Thus,

$$\overline{\Gamma}^j_{ms} = \frac{\partial \overline{x}^j}{\partial x^i} \frac{\partial x^\alpha}{\partial \overline{x}^m} \frac{\partial x^\beta}{\partial \overline{x}^s} \Gamma^i_{\alpha\beta} + \frac{\partial^2 \overline{x}^j}{\partial x^\alpha \partial x^\beta} \frac{\partial x^\alpha}{\partial \overline{x}^m} \frac{\partial x^\beta}{\partial \overline{x}^s}$$

(Eqn. 2-30)

These coefficients are called "coefficients of affine connection". Note that the transformation law is inhomogeneous in the coefficient $\Gamma^i_{\alpha\beta}$. So not a tensor! This is fundamentally different from a tensor transformation law and the connections are not tensors. Let's consider the consequence of this non-tensorial behavior:

(1) If we restrict ourselves to linear transformations of coordinates, the term $\partial^2 \overline{x}^j / \partial x^\alpha \partial x^\beta$ vanishes and Γ^i_{mj} transforms like a tensor. In this special case constancy of components is therefore an acceptable criterion for a constant vector field.

(2) $\Gamma^i_{K\ell} - \overline{\Gamma}^i_{K\ell}$ transforms like a tensor since the inhomogeneous terms cancel.

(3) Certain axiomatic deductions can be made from the equation (these were first developed by Levi-Civita):
(a) The Γ coefficients remain unsymmetric under any change of coordinates
(b) It is impossible to find a coordinate system in which all Γ coefficients are 0 at a point.

2.4.2 Parallel Displacement – Christoffel Symbols

Now we focus our study on Riemann spaces. Consider the metric requirement that the scalar product of two vectors be invariant under transplantation:

$$\frac{d}{dS}\left(g_{iK}\varepsilon^i\eta^K\right) = 0 \rightarrow \quad \frac{\partial g_{iK}}{\partial x^\ell}\frac{dx^\ell}{dS}\,\varepsilon^i\eta^K + g_{iK}\frac{d\varepsilon^i}{dS}\,\eta^K + g_{iK}\varepsilon^i\frac{d\eta^K}{dS}$$
$$= 0$$

(Eqn. 2-31)

Using $\frac{d\varepsilon^i}{dS} = \Gamma^i_{\ell r}\frac{dx^\ell}{dS}\,\varepsilon^r$ *and* $\frac{d\eta^K}{dS} = \Gamma^K_{\ell r}\frac{dx^\ell}{dS}\,\eta^r$ this becomes:

$$\frac{\partial g_{iK}}{\partial x^\ell}\frac{dx^\ell}{dS}\,\varepsilon^i\eta^K + g_{iK}\Gamma^i_{\ell r}\frac{dx^\ell}{dS}\,\varepsilon^r\eta^K + g_{iK}\Gamma^K_{\ell r}\frac{dx^\ell}{dS}\,\eta^r\varepsilon^i = 0$$

or

$$\frac{\partial g_{iK}}{\partial x^\ell}\,\varepsilon^i\eta^K + g_{iK}\,\underbrace{\Gamma^i_{\ell n}\varepsilon^n\eta^K}_{n\to i,i\to r} + \underbrace{\Gamma^K_{\ell n}\eta^n\varepsilon^i}_{n\to K,K\to r} = 0$$

with relabeling dummy indices to get:

$$\frac{\partial g_{iK}}{\partial x^\ell} + g_{rK}\Gamma^r_{\ell i} + g_{ir}\Gamma^r_{\ell K}\,.$$

(Eqn. 2-32)

By permitting the indices $iK\ell$ two additional equations are found and solving for Γ^r_{Ki} we get:

$$\Gamma^r_{Ki} = -\frac{1}{2}g^{\ell r}\left(\frac{\partial g_{K\ell}}{\partial x^i} + \frac{\partial g_{\ell i}}{\partial x^K} - \frac{\partial g_{iK}}{\partial x^\ell}\right).$$

(Eqn. 2-33)

Define $[iK,\ell] = \frac{1}{2}\left(\frac{\partial g_{K\ell}}{\partial x^i} + \frac{\partial g_{K\ell}}{\partial x^i} - \frac{\partial g_{iK}}{\partial x^\ell}\right)$ this is a Christoffel symbol of the first kind. Also, define $\left\{ {j \atop i\;K} \right\} = g^{j\ell}\,[iK,\ell]$ which is called the Christoffel symbol of the second kind. Thus, $\Gamma^r_{iK} = -\left\{ {r \atop i\;K} \right\}$ and the law of parallel displacement in a metric space is thus:

$$d\varepsilon^i = -\left\{ {i \atop \alpha\;\beta} \right\} dx^\alpha\varepsilon^\beta.$$

(Eqn. 2-34)

To put this into context. Consider the evolution in time of a mechanical system described by generalized coordinates $x^i(t)$, $\acute{x}^i = dx^i/dt, T = \frac{1}{2}g_{ik}\dot{x}^i\dot{x}^K$, and $V(x^i)$ which gives $F_i = -\partial V/\partial x^i$. As usual in analytical dynamics, we take $Tdt^2 = ds^2$ to define a metric on the space of the generalized coordinates, which is called configuration space. Using $L = T - V$, and $\frac{d}{dt}\left(\frac{\partial L}{\partial \dot{x}^i}\right) = \frac{\partial}{\partial x^i}$ we then get:

$$g_{iK}\ddot{x}^K + \frac{\partial g_{iK}}{\partial x^\ell}\dot{x}^\ell\dot{x}^K = \frac{1}{2}\frac{\partial g_{\ell K}}{\partial x^i}\dot{x}^\ell\dot{x}^K + F_i$$

(Eqn. 2-35)

or

$$g_{iK}\ddot{x}^K + \frac{1}{2}\left[\frac{\partial g_{iK}}{\partial x^\ell} + \frac{\partial g_{i\ell}}{\partial x^K} - \frac{\partial g_{\ell K}}{\partial x^i}\right]\dot{x}^\ell\dot{x}^K = F_i$$

(Eqn. 2-36)

and using the Christoffel notation:

$$\ddot{x}^i + \left\{ {i \atop \ell\, K} \right\}\dot{x}^\ell\dot{x}^K = F^i.$$

(Eqn. 2-37)

***2.4.3 Geodesics in Affine and Riemann Space*:**

Suppose we have

$$d\varepsilon^i = \Gamma^i_{\alpha\beta}dx^\alpha\varepsilon^\beta = 0,$$

and we parametrize by q:

$$\frac{d\varepsilon^i}{dq} - \Gamma^i_{\alpha\beta}\frac{dx^\alpha}{dq}\varepsilon^\beta = 0.$$

dx^i/dq is a particular tangent vector, while a more general one is $\lambda(q)\left(\frac{dx^i}{dq}\right)$, where $\lambda(q)$ is an arbitrary function of q. In which case:

$$\frac{d}{dq}\left(\lambda(q)\frac{dx^i}{dq}\right) = \Gamma^i_{\alpha\beta}\frac{dx^\alpha}{dq}\lambda(q)\frac{dx^\beta}{dq}.$$

If we let $dp = dq/[\,\lambda(q)]$, then

$$\lambda(q)\frac{d}{dq}\left(\frac{dx^i}{dp}\right) = \Gamma^i_{\alpha\beta}\frac{dx^\alpha}{dp}\frac{dx^\beta}{dp}$$

and we get:

$$\frac{d^2x^i}{dp^2} - \Gamma^i_{\alpha\beta}\frac{dx^\alpha}{dp}\frac{dx^\beta}{dp} = 0$$

which define geodesic lines in an affine space. The defining equations for a geodesic in Riemann space become

$$\frac{d^2x^i}{dp^2} + \left\{ {i \atop \alpha\beta} \right\}\frac{dx^\alpha}{dp}\frac{dx^\beta}{dp} = 0.$$

(Eqn. 2-38)

2.4.4 Geodesics as stationary curve between two points

A geodesic can also be defined as the stationary curve between points: If $S = \int_{P_0}^{P_1} \left(g_{iK} \frac{dx^i}{dp} \frac{dx^K}{dp} \right)^{1/2} dp$:

$$\delta S = \delta \int_{P_0}^{P_1} \underbrace{\left(g_{iK} \frac{dx^i}{dp} \frac{dx^K}{dp} \right)^{1/2}}_{"L"} dp = 0$$

(Eqn. 2-39)

Choose $p = s$ arc length to get:

$$\frac{d}{ds}\left(g_{iK} \frac{dx^K}{ds} \right) - \frac{\partial g_{iK}}{\partial x^i} \frac{dx^i}{ds} \frac{dx^i}{ds} = 0$$

or

$$\frac{d^2 x^i}{ds^2} + \left\{ {i \atop \ell\, K} \right\} \frac{dx^\ell}{ds} \frac{dx^K}{ds} = 0.$$

(Eqn. 2-40)

2.5 Covariant derivatives

2.5.1 The Covariant Derivative

Let's review the (covariant) derivative operator as described by Wald [20], with additional discussion regarding the algebraic topology induced by having a consistent chain rule:

The derivative operator is defined to satisfy six conditions:

(1) Linearity
(2) Leibnitz rule
(3) Commutativity with contraction
(4) Consistency with directional derivatives on scalar fields
(5) Torsion free
(6) Chain Rule (imposed at Alg. Top. level, on tangent space mappings, see App. B.)

Any two derivative operations must agree in their action on scalar fields from (4):

$$\tilde{\nabla}_a (f w_b) - \nabla_a (f w_b) = f\left(\tilde{\nabla}_a w_b - \nabla_a w_b\right).$$

(Eqn. 2-41)

At p, $\tilde{\nabla}_a w_b$ and $\nabla_a w_b$ depend on how w_b changes as one mores away from p. However $\tilde{\nabla}_a w_b - \nabla_a w_b$ depends on w_b only at p in this regard. Suppose w_b' equals w_b at p, we can write $w_b' - w_b$ in terms of a collection of functions that vanish at p, $f_{(\alpha)}$, summed according to independent D-component scalar factors $\mu_b^{(\alpha)}$ (eventually they sum to a D-dim vector):

$$w'_b - w_b = \sum_{\alpha=1}^{n} f_{(\alpha)} \mu_b^{(\alpha)} .$$

(Eqn. 2-42)

At p we can then write:

$$\tilde{\nabla}_a(w'_b - w_b) - \nabla_a(w'_b - w_b) = \sum_{\alpha} f_{(\alpha)} \left\{ \tilde{\nabla}_a \mu_b^{(\alpha)} - \nabla_a \mu_b^{(\alpha)} \right\} = 0,$$

(Eqn. 2-43)

where equality with zero follows due to the $\mu_b^{(\alpha)}$ terms being scalars. Thus, we have:

$$\tilde{\nabla}_a w'_b - \nabla_a w'_b = \tilde{\nabla}_a w_b - \nabla_a w_b ,$$

(Eqn. 2-44)

for arbitrary w'_b, thus, $\tilde{\nabla}_a w_b - \nabla_a w_b$ depends only on the value at a given point.

Since $(\tilde{\nabla}_a - \nabla_a) w_b$ only depends on the value at a point, it defines a map on dual vectors at p to (0,2) tensors at p, and it is linear, thus it must be a tensor of type (1,2) at p, denoted C^c_{ab}:

$$\nabla_a w_b = \tilde{\nabla}_a w_b - C^c_{ab}\, w_c \quad (and\ C^c_{ab} = C^c_{ba})$$

(Eqn. 2-45)

Now,

$$(\tilde{\nabla}_a - \nabla_a)(w_b t^b) = 0,$$

(Eqn. 2-46)

since the contraction is a scalar. In turn, this then indicates that:

$$\nabla_a t^b = \tilde{\nabla}_a t^b + C^b_{ac} t^c.$$

(Eqn. 2-47)

Generalizing:

$$\nabla_a T^{b_1 \ldots b_n}{}_{c_1 \ldots c_\ell} = \tilde{\nabla}_a T^{b_1 \ldots b_n}{}_{c_1 \ldots c_\ell} + \sum C^{b_i}{}_{ad}\, T^{b_1 \ldots d \ldots b_n}{}_{c_1 \ldots c_\ell}$$
$$- \sum C^{d}{}_{ac_j}\, T^{b_1 \ldots b_n}{}_{c_1 \ldots d \ldots c_\ell} .$$

(Eqn. 2-48)

Now consider a metric with compatible derivative operator:

$$0 = \nabla_a g_{bc} = \widetilde{\nabla}_a g_{bc} - C^d_{ab} g_{dc} - C^d_{ac} g_{bd} \quad \rightarrow \quad C^c_{ab} = \frac{1}{2} g^{cd}\{\widetilde{\nabla}_a g_{bd} + \widetilde{\nabla}_b g_{ad} - \widetilde{\nabla}_d g_{ab}\}.$$

(Eqn. 2-49)

Note that g_{bc} is the metric compatible with the other derivative operator, ∇_a, so let's do a notation change and also introduce a one-parameter (line-function mapping) parameterization. So, let ${}^\lambda\nabla_a$ denote the operator associated with $g_{ab}(\lambda)$ and let ${}^o\nabla_a$ denote the operator associated with ${}^o g_{ab}$, for which the associated $C^c_{ab}(0) = 0$, but for $\lambda \neq 0$ we now have:

$$C^c_{ab}(\lambda) = \frac{1}{2} g^{cd}(\lambda)\{{}^o\nabla_a g_{bd}(\lambda) + {}^o\nabla_b g_{ad}(\lambda) - {}^o\nabla_d g_{ab}(\lambda)\}.$$

(Eqn. 2-50)

Recall that $\nabla_a\nabla_b w_c - \nabla_b\nabla_a w_c = R_{abc}{}^d w_d$, so $\left({}^\lambda\nabla_a {}^\lambda\nabla_b - {}^\lambda\nabla_b {}^\lambda\nabla_a\right) w_c = {}^\lambda R_{abc}{}^d w_d$ and using:

$${}^\lambda\nabla_a w_b = {}^0\nabla_a w_b + C^c_{ab} w_c, \quad {}^0\nabla_a w_b = {}^\lambda\nabla_a w_b + C^c_{ab}(-\lambda) w_c, \text{ and}$$
$${}^\lambda\nabla_a w_b = {}^0\nabla_a w_b - C^c_{ab}(\lambda) w_c,$$

we get:

$${}^\lambda R_{abc}{}^d w_d = {}^\lambda\nabla_a\left({}^0\nabla_b w_c - {}^\lambda C^d_{bc} w_d\right) - {}^\lambda\nabla_b\left({}^0\nabla_a w_c - {}^\lambda C^d_{ac} w_d\right)$$

$$\begin{aligned}{}^\lambda R_{abc}{}^d w_d = {} & \nabla_a\nabla_b w_c - {}^\lambda C^d_{ab}\nabla_d w_c - {}^\lambda C^d_{ac}\nabla_b w_d + \nabla_a\left({}^\lambda C^d_{bc} w_d\right) \\ & + {}^\lambda C^e_{ab}{}^\lambda C^d_{ec} w_d + {}^\lambda C^e_{ac}{}^\lambda C^d_{be} w_d - \nabla_b\nabla_a w_c + {}^\lambda C^d_{ab}\nabla_d w_c \\ & + {}^\lambda C^d_{bc}\nabla_a w_d + \nabla_b\left({}^\lambda C^d_{ac} w_d\right) - {}^\lambda C^e_{ab}{}^\lambda C^d_{ec} w_d \\ & - {}^\lambda C^e_{bc}{}^\lambda C^d_{ae} w_d\end{aligned}$$

$$\begin{aligned}{}^\lambda R_{abc}{}^d w_d = {} & {}^0 R^d_{abc} w_d - \left(\nabla_a {}^\lambda C^d_{bc} - \nabla_b {}^\lambda C^d_{ac}\right) w_d \\ & + \left({}^\lambda C^e_{ac}{}^\lambda C^d_{be} - {}^\lambda C^e_{bc}{}^\lambda C^d_{ae}\right) w_d\end{aligned}$$

Thus:

$${}^\lambda R_{abc}{}^d = {}^0 R_{abc}{}^d - 2\nabla_{[a}{}^\lambda C^d_{b]c} + 2{}^\lambda C^e_{c[a}{}^\lambda C^d_{b]e}$$

(Eqn. 2-51)

$${}^\lambda R_{ac} = {}^0 R_{ac} - 2{}^0\nabla_{[a}{}^\lambda C^d_{b]c} + 2{}^\lambda C^e_{c[a}{}^\lambda C^d_{b]e}$$

(Eqn. 2-52)

If ${}^0 g_{ac}$ is a solution of the "vacuum Einstein equation" (pg. 185 Wald [20]), then ${}^0 R_{ac} = 0$. However, in the analysis this simply provides us with a convenient reference configuration. With such a reference configuration:

$${}^\lambda R_{ac} = 2{}^0\nabla_{[a}{}^\lambda C^d_{b]c} + 2{}^\lambda C^e_{c[a}{}^\lambda C^d_{b]e}\,.$$

(Eqn. 2-53)

If we define

$$\delta R_{ac}(\lambda) = \left(\frac{dR_{ac}}{d\lambda}\right)_{\lambda=0},$$

(Eqn. 2-54)

then:

$$\delta R_{ac} = -2\,{}^{o}\nabla_{[a}\delta C^{b}_{b]c}\,, \quad where \quad \delta C^{c}_{ab} = \frac{1}{2}\,{}^{o}g^{cd}\{{}^{o}\nabla_{a}\delta g_{bd} + {}^{o}\nabla_{b}\delta g_{ad} - {}^{o}\nabla_{d}\delta g_{ab}\}.$$

(Eqn. 2-55)

Thus,

$$\delta R_{ac} = -\frac{1}{2}\,{}^{o}g^{bd\,o}\nabla_{a}{}^{o}\nabla_{c}(\delta g_{bd}) - \frac{1}{2}\,{}^{o}g^{bd\,o}\nabla_{b}{}^{o}\nabla_{d}(\delta g_{ac}) + {}^{o}g^{bd\,o}\nabla_{b}{}^{o}\nabla_{(c}(\delta g)_{a)d}$$

(Eqn. 2-56)

So, dropping the "o" superscript, this simplifies to:

$$g^{ab}\delta R_{ab} = \nabla^{a}\nabla^{b}(\delta g_{ab}) - \nabla^{a}\left(g^{cd}\nabla_{a}(\delta g_{cd})\right) = \nabla^{a}\mathrm{v}_{a} \quad where\ \mathrm{v}_{a} = \nabla^{b}(\delta g_{ab}) - g^{cd}\nabla_{a}(\delta g_{cd}).$$

(Eqn. 2-57)

This will be useful in Sec. 3.2.2.

Example 2.1. Show that any two covariant derivatives, ∇_a and $\widetilde{\nabla}_a$ differ by a tensor C^{b}_{ca}:

$$(\nabla_a - \widetilde{\nabla}_a)V^b = C^{b}_{ca}V^c\,,$$

with C symmetric on its covariant indices: $C^{b}_{ca} = C^{b}_{ac}$. (Use abstract indices.) Then, use Leibnitz to show the analogous formula holds for any tensor:

$$(\nabla_a - \widetilde{\nabla}_a)T^{b\ldots c}_{d\ldots e} = C^{b}_{fa}T^{f\ldots c}_{d\ldots e} + \cdots + C^{c}_{fa}T^{b\ldots f}_{d\ldots e} - C^{f}_{da}T^{b\ldots c}_{f\ldots e} - \cdots - C^{f}_{ea}T^{b\ldots c}_{d\ldots e}\,.$$

Solution

Consider $(\nabla_a - \widetilde{\nabla}_a)(V^c w_c)$, since a covariant derivative is a map between $\binom{m}{n}$ and $\binom{n}{m}$ tensor fields which satisfies, among other things: $\nabla_a f = \text{grad } f$ for scalars. Then, since $V^c w_c$ is a scalar we have $(\nabla_a - \widetilde{\nabla}_a)(V^c w_c) = 0$. Now, using Leibnitz: $w_c(\nabla_a - \widetilde{\nabla}_a)V^c + V^c(\nabla_a - \widetilde{\nabla}_a)w_c = 0$. Thus we find that

$(\nabla_a - \widetilde{\nabla}_a)$ defines a map of the 1-form w_c to a $\binom{0}{2}$ tensor; furthermore, the map is linear. Consequently $(\nabla_a - \widetilde{\nabla}_a)$ must be a tensor of type $\binom{1}{2}$ as this is the type which maps 1-forms to $\binom{0}{2}$ tensors. Denote the tensor field by $-C^d_{ac}$:

$$(\nabla_a - \widetilde{\nabla}_a)w_c = -C^d_{ac}w_d$$

C^d_{ac} is also symmetric on its covariant indices as can be shown by considering $w_c\nabla_c f$:

$$(\nabla_a - \widetilde{\nabla}_a)\nabla_c f = \nabla_a\nabla_c f - \widetilde{\nabla}_a\nabla_c f = (\nabla_a\nabla_c - \widetilde{\nabla}_a\widetilde{\nabla}_c)f,$$

which is symmetric on a and c. So,

$$w_d(\nabla_a - \widetilde{\nabla}_a)V^d + V^c C^d_{ac}w_d = 0,$$

and $(\nabla_a - \widetilde{\nabla}_a)V^b = C^b_{ca}V^c$, so $C^b_{ca} = C^b_{ac}$.

Now, individually the covariant derivatives satisfy Leibnitz' rule as will any linear combination of them. So, the application of $(\nabla_a - \widetilde{\nabla}_a)$ to any tensor can be expressed in terms of its action on the Vector and Dual spaces of the tensor product space. Thus for contravariant indices we get a contraction as with a a dual vector: $(\nabla_a - \widetilde{\nabla}_a)w_c = -C^d_{ac}w_d$, while for covariant indices we get that of a vector: $(\nabla_a - \widetilde{\nabla}_a)V^b = C^b_{ca}V^c$. Thus,

$$(\nabla_a - \widetilde{\nabla}_a)T^{b\ldots c}_{d\ldots e} = C^b_{fa}T^{f\ldots c}_{d\ldots e} + \cdots + C^c_{fa}T^{b\ldots f}_{d\ldots e} - C^f_{da}T^{b\ldots c}_{f\ldots e} - \cdots - C^f_{ea}T^{b\ldots c}_{d\ldots e}$$

2.5.2 Riemann tensor

Let's re-derive the Riemann tensor by way of the Cartan structure equations (a complete description of the method is given in Ch. 4) using abstract indices. The Riemann tensor is defined as follows:

$$\frac{1}{2}R^a_{bcd}V^b = \nabla_{[c}\nabla_{d]}V^a$$

(Eqn. 2-58)

Using an orthonormal frame and Cartan's formalism:

$$\nabla_k e_j = \Gamma^i_{jk}e_j$$

(Eqn. 2-59)

$$R^i_{jk\ell} = e_k\Gamma^i_{j\ell} - e_\ell\Gamma^i_{jk} + \Gamma^i_{mk}\Gamma^m_{j\ell} - \Gamma^i_{m\ell}\Gamma^m_{jk}$$

(Eqn. 2-60)

Cartan's structure equations, using the notation ω^i dual basis vectors and $[u,v]^i = u^j \frac{\partial}{\partial x^j} v^i - v^j \frac{\partial u^i}{\partial x^j}$, and have $\omega^i(e_k) = \delta^i_k = constant$ for an orthonormal frame:

$$\begin{aligned}(d\omega^i)_{jk} &= d\omega^i(e_j, e_k) = 2e^a_b \nabla_{[a} W^i_{b]} e^a_j e^b_k = (\nabla_a \omega^i_b e^a_j e^b_k - \nabla_b \omega^i_a e^a_j e^b_k) \\ &= [\nabla_j(\omega^i_b e^b_k) - \omega^i_b \nabla_j e^b_k] - [\nabla_k(\omega^i_a e^b_j) - \omega^i_a \nabla_k e^b_j] \\ &= \nabla_j\{\omega^i(e_k)\} - \nabla_k\{\omega^i(e_j)\} + \omega^i(\nabla_k e_j) - \omega^i(\nabla_j e_k) \\ &= -\omega^i(\nabla_j e_k - \nabla_k e_j) = -\omega^i(e^a_j \nabla_a e_k - e^a_k \nabla_a e_j) \\ &= -\omega^i([e_j, e_k]) = -\omega^i(c^\ell_{jk} e_\ell) = -c^i_{jk}\end{aligned}$$

Thus

$$(d\omega^i)_{jk} = -c^i_{jk}$$

(Eqn. 2-61)

If we next define connection forms: $\omega^i_j = \Gamma^i_{jk}\omega^k$ we expect $\omega_{ij} = -\omega_{ji}$ in an orthonormal frame (or any frame where the metric has constant components). Let's show this. Start with

$$\nabla_k e_j = \Gamma^i_{jk} e_j \quad \rightarrow \quad e^a_k \nabla_a e^b_j = \Gamma^i_{jk} e^b_i (e^a_k \omega^k_a) \quad \rightarrow \quad \nabla_a e^b_j = \Gamma^i_{jk} \omega^k_a e^b_i$$
$$\rightarrow \quad \nabla_a e^b_j = \omega^i_{ja} e^b_i \ .$$

Now have:

$$e_i \cdot \nabla_a e_j = \nabla_a(g_{ij}) - e_j \cdot \nabla_a e_i = -e_j \cdot \nabla_a e_i \ \Rightarrow \ \omega_{ij} = -\omega_{ji}$$

Notice that:

$$(d\omega^i)_{k\ell} = -c^i_{k\ell} = \Gamma^i_{k\ell} - \Gamma^i_{\ell k} = (\omega^i_j)_\ell (\omega^j)_k - (\omega^i_j)_k (\omega^j)_\ell$$

Consider

$$\omega^i_j = \omega^i_{ja} = (\Gamma^i_{jk}\omega^k)_a \ \ and \ \ (\omega^i_j)_\ell = (\Gamma^i_{jk}\omega^k)_a e^a_\ell = \Gamma^i_{jk}\omega^k_a e^a_\ell$$

So, $(\omega^i_j)_\ell (\omega^j)_k = \Gamma^i_{j\ell}\delta^j_k = \Gamma^i_{k\ell}$ and we can now write:

$$(d\omega^i)_{k\ell} = -(\omega^i_j \wedge \omega^j)_{k\ell} \ \rightarrow \ d\omega^i = -\omega^i_j \wedge \omega^j$$

(Eqn. 2-62)

Recall $\Gamma_{ijk} = \frac{1}{2}\Big(e_k g_{ij} + e_j g_{ik} - e_i g_{jk} + \underbrace{c_{kij} + c_{jik} - c_{ijk}}\Big)$, where we have $(\omega_{ij})_k = \omega_{ija} e^a_k = \Gamma_{ijk}\omega^k_a e^a_k = \Gamma_{ijk}$ and $(d\omega_j)_{k\ell} = c_{ik\ell}$. So, In an orthonormal basis, or one where the metric is constant:

$$(\omega_{ij})_k = \frac{1}{2}\Big((d\omega_i)_{jk} + (d\omega_j)_{ki} - (d\omega_k)_{ij}\Big).$$

(Eqn. 2-63)

Since the Ricci tensor

$$R^i_j = d\omega^i_j + \omega^i_k \wedge \omega^k_j$$

(Eqn. 2-64)

Implying

$$\left(R^i_j\right)_{ab} = \left(d\omega^i_j\right)_{ab} + \left(\omega^i_k \wedge k_j\right)_{ab}$$

(Eqn. 2-65)

Note that $\omega^i_j = \Gamma^i_{jk}\omega^k$, $\omega^i(e_j) = \delta^i_j$, $\left(\omega^i_j\right)_\ell = \Gamma^i_{j\ell}$, $\left(\omega^i_j\right)_a = \left(\Gamma^i_{jk}\omega^k_a\right)$ will be useful in what follows.

We have:

$$\left(d\omega^i_j\right)_{k\ell} = \left(d\omega^i_j\right)_{ab} e^a_k e^b_\ell = \nabla_{[a}\left(\omega^i_j\right)_{b]} e^a_k e^b_\ell$$

$$= e^a_k \nabla_a \left(\omega^i_j\right)_b e^b_\ell - e^b_\ell \nabla_b \left(\omega^i_j\right)_a e^a_k$$

$= \nabla_k\{\Gamma^i_{jm}\omega^m_b\}e^b_\ell - \nabla_\ell\{\Gamma^i_{jm}\omega^m_a\}e^a_k = e_k\Gamma^i_{j\ell} - e_\ell\Gamma^i_{jk} - \Gamma^i_{jm}\Gamma^m_{\ell k} + \Gamma^i_{jm}\Gamma^m_{k\ell} = e_k\left(\Gamma^i_{j\ell}\right) - e_\ell\left(\Gamma^i_{jk}\right) + 2\Gamma^i_{jm}\Gamma^m_{[k\ell]}$.

Thus,

$$\left(d\omega^i_j\right)_{k\ell} = e_k\left(\Gamma^i_{j\ell}\right) - e_\ell\left(\Gamma^i_{jk}\right) + 2\Gamma^i_{jm}\Gamma^m_{[k\ell]}$$

(Eqn. 2-66)

We also have:

$$\left(\omega^i_m \wedge \omega^m_j\right)_{k\ell} = \left(\omega^i_m \wedge \omega^m_j\right)_{ab} e^a_k e^b_\ell = \left(\omega^i_m\right)_{[a}\left(\omega^m_j\right)_{b]} e^a_k e^b_\ell$$

$$= \left(\omega^i_m\right)_k\left(\omega^m_j\right)_\ell - \left(\omega^i_m\right)_\ell\left(\omega^m_j\right)_k$$

Thus,

$$\left(\omega^i_m \wedge \omega^m_j\right)_{k\ell} = \Gamma^i_{mk}\Gamma^m_{j\ell} - \Gamma^i_{m\ell}\Gamma^m_{jk}$$

(Eqn. 2-67)

Since

$$R^i_{jk\ell} = \left(R^i_j\right)_{ab} e^a_k e^b_\ell$$

(Eqn. 2-68)

We have:

$$R^i_{jk\ell} = e_k\left(\Gamma^i_{j\ell}\right) - e_\ell\left(\Gamma^i_{jk}\right) + \Gamma^i_{mk}\Gamma^m_{j\ell} - \Gamma^i_{m\ell}\Gamma^m_{jk} + 2\Gamma^i_{jm}\Gamma^m_{[k\ell]},$$

(Eqn. 2-69)

which is the Riemann tensor.

Example 2.2. Show that $\Gamma^i_{[jk]} = -\frac{1}{2}c^i_{jk}$.

Solution

We have $e^a_k\nabla_a e^b_j = \Gamma^i_{jk}e^b_i$, so

$$c^i_{jk} e^b_i = [\vec{e}_j, \vec{e}_k] = e^a_j \nabla_a e^b_k - e^b_k \nabla_b e^a_j = \Gamma^i_{jk} e^b_i - \Gamma^i_{jk} e^b_i = -2\Gamma^i_{[jk]} e^b_i$$

Thus,

$$\Gamma^i_{[jk]} = -\frac{1}{2} c^i_{jk} \, .$$

Example 2.3. Use the previous relation and $\nabla_a g_{bc} = 0$ to show that

$$\Gamma_{ijk} = \frac{1}{2}\left(e_k g_{ij} + e_j g_{ik} - e_i g_{jk} + c_{kij} + c_{jik} - c_{ijk}\right) \, .$$

Solution

$\nabla_a g_{bc} = 0 \;\rightarrow\; e^a_i \nabla_a \left(e^b_j e^c_k g_{bc}\right)$ and we have:

$$e^a_i \nabla_a \left(e^b_j e^c_k g_{bc}\right) = e^a_i \left(\nabla_a e^b_j\right) e^c_k g_{bc} + e^a_i (\nabla_a e^c_k) e^b_j g_{bc} + e^a_i e^b_j e^c_k (\nabla_a g_{bc})$$

$$e^a_i \nabla_a g_{jk} = \left(e^a_i \nabla_a e^b_j\right) e^c_k g_{bc} + (e^a_i \nabla_a e^c_k) e^b_j g_{bc}$$

$$e_i g_{jk} = \Gamma^n_{ji} e^b_n e^c_k g_{bc} + \Gamma^n_{ki} e^c_n e^b_j g_{bc}$$

$$e_i g_{jk} = \Gamma^n_{ji} g_{nk} + \Gamma^n_{ki} g_{jn} = \Gamma_{kji} + \Gamma_{jki}$$

$$\begin{cases} e_i g_{jk} = \Gamma_{kji} + \Gamma_{jki} \text{ now permute indices:} \\ e_j g_{ik} = \Gamma_{kij} + \Gamma_{ikj} \\ e_k g_{ji} = \Gamma_{ijk} + \Gamma_{jik} \end{cases}$$

$$e_k g_{ij} + e_j g_{ik} - e_i g_{jk}$$
$$= \Gamma_{ijk} + \Gamma_{jik} + \Gamma_{kij} + \Gamma_{ikj} - \Gamma_{kji} - \Gamma_{jki} + \left(+\Gamma_{ijk} - \Gamma_{ijk}\right)$$
$$= 2\Gamma_{ijk} + \left(\Gamma_{ikj} - \Gamma_{ijk}\right) + \left(\Gamma_{jik} - \Gamma_{jki}\right) +$$
$$\left(\Gamma_{kij} - \Gamma_{kji}\right)$$
$$= 2\Gamma_{ijk} - 2\Gamma_{i[jk]} + 2\Gamma_{j[ki]} - 2\Gamma_{k[ji]} = 2\Gamma_{ijk} +$$
$$c_{ijk} + c_{jki} + c_{kji}$$
$$= 2\Gamma_{ijk} - \left(c_{kij} + c_{jik} - c_{ijk}\right)$$

So,

$$\Gamma_{ijk} = \frac{1}{2}\left(e_k g_{ij} + e_j g_{ik} - e_i g_{jk} + c_{kij} + c_{jik} - c_{ijk}\right).$$

Example 2.4. Prove the Gauss-Codacci equation:

$${}^{(3)}R_{\alpha\beta\gamma\delta} = h^{\alpha'}_{\alpha} h^{\beta'}_{\beta} h^{\gamma'}_{\gamma} h^{\delta'}_{\delta} R_{\alpha'\beta'\gamma'\delta'} + K_{\alpha\delta} K_{\beta\gamma} - K_{\alpha\gamma} K_{\beta\delta} \, .$$

Solution

A tensor over the tangent space of the 3-surface at a given point p is simply

$$T^{a\dots b}_{c\dots d} = h^a_e \cdots h^b_f h^g_c \cdots h^j_d T^{e\dots f}_{g\dots j} \,,$$

where h_{ab} is the spatial metric associated with a spacetime metric g_{ab} induced by the choice of foliation Σ_t . (t^a a vector field on the manifold for which $t^a \nabla_a t = 1$, t the global time function that labels Σ_t .) Denote $D_K T^{a\dots b}_{c\dots d} = h^a_e \cdots h^b_f h^g_c \cdots h^j_d h^n_k \nabla_n T^{e\dots f}_{g\dots j}$, then:

$$^{(3)}R^d_{abc} w_d = D_a D_b w_c - D_b D_a w_c \,.$$

Now we project our what is meant by $D_a D_b w_c$:

$$D_a D_b w_c = D_a\left(h^d_b h^e_c \nabla_d w_e\right)$$

$$= h^f_a h^g_b h^k_c \nabla_f \left(h^d_g h^e_k \nabla_d w_e\right)$$

$$= h^f_a h^g_b h^k_c \left\{\left(\nabla_f h^d_g\right) h^e_k \nabla_d w_e + h^d_g \left(\nabla_f h^e_k\right)\nabla_d w_e + h^d_g h^e_k \nabla_f \nabla_d w_e\right\}$$

$$= \left(h^f_a h^g_b \nabla_f h^d_g\right)\left(h^k_c h^e_k\right)\nabla_d w_e + \left(h^f_a h^k_c \nabla_f h^e_k\right)\left(h^g_b h^d_g\right)\nabla_d w_e +$$

$$h^f_a \left(h^g_b h^d_g\right)\left(h^k_c h^e_k\right)\nabla_f \nabla_d w_e$$

$$h_{ab} = g_{ab} + n_a n_b \quad \rightarrow \quad h^g_b h^d_g = \left(g^g_b + n_b n^g\right)\left(g^d_g + n_g n^d\right)$$

$$= g^d_b + n_b n^d + n_b n^d + n_b n^d \underbrace{\left(n_g n^j\right)}_{-1}$$

$$= g^t_b + n_b n^d = \underline{h^d_b}$$

Thus,

$$\nabla_f h^d_g = \nabla_f\left(g^d_g + n_g n^d\right) = \left(\nabla_f n_g\right) n^d + n_g\left(\nabla_f n^d\right)$$

and we get:

$$h^f_a h^g_b \nabla_f h^t_g = g^f_a h^g_b \left(\nabla_f n_g\right) n^d_t h^f_a \underbrace{h^g_b n_g}_{0} \nabla_f n^d$$

$$= h^f_a\left(g^g_b + n_b n^g\right)\left(\nabla_f n_g\right) n^d =$$

$$h^f_a \nabla_f \left(g^g_b n_g\right) n^d_t h^f_a n_b n^d \left(\tfrac{1}{2}\right) \nabla_f n_g$$

$$h^f_a h^g_b \nabla_f h^t_g = h^f_a \left(\nabla_f n_b\right) n^d + h^f_a n_b d^t \left(\tfrac{1}{2}\nabla_f\left(n^g n_g\right)\right) = \underbrace{\left(h^f_a \nabla_f n_b\right)}_{K_{ab}} n^d =$$

$$K_{ab} n^d.$$

So,

$$D_a D_b w_c = h^e_c K_{ab} n^d \nabla_d w_e + h^d_b K_{ac} n^e \nabla_d w_e + h^f_a h^d_b h^e_c \nabla_f \nabla_d w_e$$

$$h^d_b n^e \nabla_d w_e = h^d_b \nabla_d\left(n^e w_e\right) - h^d_b w_e \nabla_d n^e$$

$$= -K_b^e w_e$$

Now,

$$^{(3)}R_{abc}^d w_d = h_c^e(K_{ab} - K_{ba})n^d \nabla_d w_e - K_{ac}K_b^e w_e + K_{bc}K_a^e w_e + \underbrace{h_a^f h_b^d}_{symmetric\ in\ L,d.} h_c^e(\nabla_f \nabla_d - \nabla_d \nabla_f)w_e$$

So

$$^{(3)}R_{abc}^d w_d = h_a^f h_b^d h_c^e R_{fde}^g w_g - K_{ac}K_b^e w_e + K_{bc}K_a^e w_e$$

and

$$w_g = h_g^j w_j \text{ since } w_g \text{ is on } \Sigma.$$

Thus,

$$^{(3)}R_{abc}^d = h_a^f h_b^d h_c^e h_j^d R_{fgk}^j - K_{ac}K_b^d + K_{bc}K_a^d$$

In a given coordinate system we thus have:

$$^{(3)}R_{\alpha\beta\gamma\delta} = h_\alpha^{\alpha'} h_\beta^{\beta'} h_\gamma^{\gamma'} h_\delta^{\delta'} R_{\alpha'\beta'\gamma'\delta'} + K_{\alpha\delta}K_{\beta\gamma} - K_{\alpha\gamma}K_{\beta\delta}.$$

Example 2.5. Show that

$$\dot{\Gamma}_{bc}^a = \frac{1}{2}g^{ad}(\nabla_{bjcd} + \nabla_{cjbd} - \nabla_{djbc})$$

$$\dot{R}_{ab} = 2\nabla_{[c}\dot{\Gamma}_{b]a}^c$$

and

$$\dot{R} = \nabla^a \nabla^b j_{ab} - \nabla_a \nabla^a j - j^{ab}R_{ab}.$$

Solution

We have:

$$\dot{R}_{ab} = 2\nabla_{[c}\dot{\Gamma}_{b]a}^c = \nabla_c \dot{\Gamma}_{ba}^c - \nabla_b \dot{\Gamma}_{ca}^c$$

where $R = g^{ab}R_{ab}$ we have: $\dot{R} = \dot{g}^{ab}R_{ab} + g^{ab}\dot{R}_{ab}$, thus

$$\dot{\Gamma}_{bc}^a = \frac{1}{2}g^{ad}(\nabla_b j_{cd} + \nabla_c j_{bd} - \nabla_d j_{bc})$$

Using this in $\dot{R}_{ab} = 2\nabla_{[c}\dot{\Gamma}_{b]a}^c$:

$$\dot{R} = \nabla^a \nabla^b j_{ab} - \nabla_a \nabla^a j + (\dot{g}^{\dot{a}b})R_{ab}$$

Thus,

$$-j_{ab} = -\dot{g}_{ab} \Rightarrow (\dot{g}^{\dot{a}b}) = -j^{cd}$$

and we have the desired:

$$\dot{R} = \nabla^a \nabla^b j_{ab} - \nabla_a \nabla^a j - j^{ab}R_{ab}.$$

2.6 Lie Groups: Groups that are also manifolds

Before we can describe Yang-Mills Fields and *Gauge* Covariant Derivatives we must learn the math language of Lie Groups.

2.6.1 Introduction

A Lie group is a group that is also a manifold, in which the operation of group multiplication is a diffeomorphism.

<u>Left multiplication</u>: $L_g h = gh \quad (g, h \in G)$. Then $L_g : G \to G$ is a diffeomorphism.

<u>Right multiplication</u>: $R_g h = hg$ is also a diffeomorphism.

The "Adjoint" map A_g is defined by $A_g h = g^{-1}hg$, which is also a diffeomorphism:

$$A_g = R_g L_{g^{-1}}$$

(Eqn. 2-70)

Given a vector ξ^a at the identity *e* of *G* the diffeomorphism L_g produces a vector $L_{g*}\xi^a$ at $L_g(e) = ge = g$. In other words, there is a vector field on *G*:

$$\xi^a(g) = L_{g*}[\xi^a(e)] \, ,$$

(Eqn. 2-71)

which is obtained from $\xi^a(e)$ by left multiplication.

So,

$$L_{g*}\xi^a(h) = L_{g*}L_n\xi^a(e) = \left(L_g L_n\right)_* \xi^a(e) = L_{gh*}\xi^a(e) = \xi^a(gh)$$

(Eqn. 2-72)

Thus, to each tangent vector in $T_e G$ correspnds a left invariant vector field on G (i.e., left invariant $L_{g*}(\xi^a)_{field} = (\xi^a)_{field}$). It can be proven that the Lie bracket $[\vec{\xi}, \vec{\eta}]$ of two left invariant vector fields $\vec{\xi}, \vec{\eta}$ is itself a left invariant vector field. Also, if Ψ_λ is the family diffeomorphisms generated by $\vec{\xi}$ and ϕ is another 'diffeo', then $\phi\Psi_\lambda\phi^{-1}$ is the diffeo generated by $\phi_*\vec{\xi}$.

$$\vec{\xi}(x) = \frac{d}{d\lambda}\Psi_\lambda(x)\bigg|_{\lambda=0} = \frac{d}{d\lambda}\Psi_{\lambda+\alpha}(0)\bigg|_{\lambda=0} \quad where \quad \Psi_\alpha(0) = \Psi_0(x) = x$$

(Eqn. 2-73)

and

$$\phi_*[\vec{\xi}(x)] = \frac{d}{d\lambda}[\phi \circ \Psi_\lambda(x)]\Big|_{\lambda=0}$$

(Eqn. 2-74)

At $\phi(x), \phi_*[\vec{\xi}(x)]$ is tangent to the path $\lambda \to \phi \circ \Psi_\lambda(x)$. So, $\phi_*[\vec{\xi}(\phi^{-1}(x))]$ is tangent to $\lambda \to \phi \circ \Psi_\lambda \circ \phi^{-1}(x)$ at x. Thus, $\phi_*\xi$ is tangent at x to $\lambda \to \phi \circ \Psi_\lambda \circ \phi^{-1}(x)$. Therefore, $\phi_*\vec{\xi}$ generates the diffeomorphism $\phi \circ \Psi_\lambda \circ \phi^{-1}$. Now, let $\phi = L_g$:

$$[\vec{\xi}, \vec{\eta}] = \mathcal{L}_{\vec{\xi}}\eta = -\frac{d}{d\lambda}(\Psi_{\lambda*}\vec{\eta})\Big|_{\lambda=0} \to \quad \phi_*[\vec{\xi}, \vec{\eta}] = \phi_*\left\{\frac{d}{d\lambda}(\Psi_{\lambda*}\vec{\eta})\right\}$$
$$= -\frac{d}{d\lambda}(\phi_*\Psi_{\lambda*}\vec{\eta})\,.$$

(Eqn. 2-75)

Thus,

$$\phi_*[\vec{\xi}, \vec{\eta}] = -\frac{d}{d\lambda}[(\phi\Psi_\lambda\phi^{-1})_*\phi_*\vec{\eta}] = \mathcal{L}_{\phi_*\vec{\xi}}\phi_*\vec{\eta}$$

(Eqn. 2-76)

Since $\vec{\xi}$ and $\vec{\eta}$ are left inv. $\phi_*\vec{\xi} = \vec{\xi}, \phi_*\vec{\eta} = \vec{\eta}$ (for fields). So,

$$\phi_*[\vec{\xi}, \vec{\eta}] = \mathcal{L}_\varepsilon\vec{\eta} = [\vec{\xi}, \vec{\eta}]$$

(Eqn. 2-77)

and the bracket is proven to be left invariant as well.

The vector space T_eG can be identified with the vector space of left inv. vector fields on G. T_eG is closed under the Lie bracket justifying the name Lie *algebra*.

If $\xi^a(e)$ and $\eta^a(e)$ are vectors at $e \in G$, then $[\xi, \eta]^a(e)$ is a vector at e. Since $[\xi, \eta]$ is linear in ξ and η it defines a bilinear map $T_eG \times T_eG \to T_eG$. Thus there is an antisymmetric tensor C^a_{bc} on bc at e with

$$[\xi, \eta]^a(e) = C^a_{bc}\xi^b(e)\eta^c(e)$$

(Eqn. 2-78)

Define $C^a_{bc}(g) = L_{g*}C^a_{bc}(e)$ then C^a_{bc} is left invariant, and

$$[\xi, \eta]^a = C^a_{bc}\xi^b\eta^c\ (everywhere).$$

(Eqn. 2-79)

C^a_{bc} is known as the structure tensor of the Lie algebra, and the concrete values C^k_{ij} are the structure constants of T_eG.

Example 2.6.
a. Prove that $w^a = [u, v]^a$ is a vector field by proving that the Leibnitz rule is satisfied.
b. Show that in any chart: $[u, v]^k = (u^i \partial_i v^k - v^i \partial_i u^k)$.

Solution
(a) Start with:
$w(fg) = [u, v](fg) = u(v(fg)) - v(u(fg)) = u(v(f)g + fv(g)) - v(u(f)g + fu(g))$
$= u(v(f)g) + u(fv(g)) - v(u(f)g) + v(fu(g))$
$= u(v(f))g - v(u(f))g + fu(v(g)) - fv(u(g))$
$= g[u, v] + f[u, v](g)$
$w(fg) = gw(f) + fw(g)$, and the Leibnitz rule is satisfied.

(b) In a chart x: $v(f) = v^i \partial_i f$, thus :
$[u, v](f) = u(v(f)) - v(u(f)) \rightarrow \ [u, v]^k \partial_k f = u^i \partial_i (v^j \partial_j f) - v^j \partial_j (u^i \partial_i f)$
So,
$[u, v]^k \partial_k f = u^i (\partial_i v^j) \partial_j f + u^i \left(\partial_i (\partial_j f)\right) v^j - v^j (\partial_j u^i) \partial_i f - v^j u^i \partial_j (\partial_i f)$
$= (u^i \partial_i v^j) \partial_j f - (v^j \partial_j u^i) \partial_i f + (u^i v^j \partial_i \partial_j f - u^i v^j \partial_j \partial_i f) = (u^i \partial_i v^k - v^i \partial_i u^k) \partial_k f$
Thus,

$$[u, v]^k = (u^i \partial_i v^k - v^i \partial_i u^k).$$

Example 2.7. A vector lies in a submanifold M of N if it is the tangent to the curve $c(\lambda)$ that lies in M.
(a) Prove that a vector V^a at a point $P \in M$ lies in M if and only if $V^a \nabla_a f = 0$, for all functions f on N that are constant on M.
(b) Show that if u^a and v^a are vector fields on N for which $u^a(P)$ and $v^a(P)$ lie in M when $P \in M$, then $[u, v]^a$ similarly lies in M. (Show that $[u, v]^a \nabla_a f = 0$, f constant on M.)

Solution
(a) In a given chart x^i we have $V^a \nabla_a f = V^i \partial_i f$. If the dimension of M is m and N is n then we can arrange $i = 1 \ldots m \ldots n$ such that the first m coordinates are a basis for the submanifold. Now,

$$V^i \partial_i f = V^i \partial_i f + \cdots + V^m \partial_m f + V^{m+1} \partial_{m+1} f + \cdots V^n \partial_n f$$

Now, if we specify all functions which are constant on M under f then all the derivatives $\partial_i f, \cdots, \partial_m f$ are zero, so

$$V^i \partial_i f = V^{m+1} \partial_{m+1} f + \cdots V^n \partial_n f \,.$$

If for all f this vanishes then each of the $V^{m+1}, \cdots, V^n$ must be zero, thus if $V^a \nabla_a f = 0$ for all f on N that are constant on M then V^i can only have nonzero compoents can only have nonzero components $i = l \ldots m$, i.e., V^a lies in M. the converse, if V^a lies in M then the $i = m + 1 \ldots n$ components must be zero and if the function f is constant on M, i.e., $\partial_i f = 0 \; \forall : \in l \ldots m$, the $V^i \partial_i f = 0 \Rightarrow V^a \nabla_a f = 0$.

(b) We have $V^i \partial_i f = 0$ and $u^i \partial_i f = 0$ for the f as specified:
$[u, v]^a \nabla_a f = [u, v]^i \partial_i f$

$$= \left(u^j \frac{\partial v^i}{\partial x^j} - v^j \frac{\partial u^i}{\partial x^j} \right) \frac{\partial f}{\partial x^i} \begin{cases} \text{since } \dfrac{\partial f}{\partial x^i} = 0 \text{ for } i = l..m \\ \text{also } u^i \text{ and } v^i \text{ are zero for } i = m + 1 \ldots n \end{cases}$$

Let $i_M = l \ldots m, i_N = m + 1 \ldots n$, then

$$[u, v]^a \nabla_a f = \left(u^{jM} \frac{\partial v^{iN}}{\partial x^{jM}} - v^{jM} \frac{\partial u^{iN}}{\partial x^{jM}} \right) \frac{\partial f}{\partial x^{iN}} = 0 \,.$$

2.6.2 The general linear group $GL_n(R)$

The Lie group might be very general, such as in $GL_n(R)$. The general linear group $GL_n(R)$ is the group of inventible $n \times n$ matrices $a = \|a^i_j\|$. Any other group of matrices is a subgroup of some GL_n. Natural coordinates on $G = GL_n$ and the n^2matrix elements a^i_j of $a \in G$. The associated coordinate basis: $\partial / \partial a^i_j$ consists of vectors tangent to curves:

$$\lambda \to a(\lambda) = \|a^m_n + \lambda \delta^m_i \delta^j_n\| \,.$$

(Eqn. 2-80)

If V is tangent at e to the path $g^i_j(\lambda)$ its components along the natural coordinate basis are

$$V^i_j(e) = \frac{d}{d\lambda} g^i_j(\lambda)|_{\lambda=0}$$

(Eqn. 2-81)

where

$$V^i_j(a) = L_{a*} V^i_j(e) = \frac{d}{d\lambda} [a^i_k g^k_j(\lambda)]\big|_{\lambda=0} = a^i_k V^k_j(e)$$

(Eqn. 2-82)

Thus, a left invariant vector field V on G has components $V_j^i(a)$ at $a \in G$ obtained from its compoennts at the origin by left-multiplication of V_j^i by a_j^i.

Now, consider the *1* – parameter subgroup defined by

$$g(\lambda) = e^{\lambda v}$$

(Eqn. 2-83)

where v is the matrix $\|v_n^m\|$. Then,

$$\Psi_\lambda(a) = ag(\lambda)$$

(Eqn. 2-84)

is the 1 – parameter family of diffeo's generated by the left-inv. vector field $\vec{v}$:

$$\frac{d}{d\lambda}ag(\lambda)\bigg|_{\lambda=0} = a_k^i v_j^k(e) = v_j^i(a)\,.$$

(Eqn. 2-85)

Thus, the left-inv. vector fields $\vec{v}$ generate right multiplication by $e^{\lambda x}$.

For another interesting property, consider $[u, v]$:

$$[u,v] = \mathcal{L}_u v = \frac{d}{d\lambda}\Psi_{-\lambda}V\bigg|_{\lambda=0}\,,$$

(Eqn. 2-86)

where

$$(\Psi_{-\lambda}V)_j^i = v_n^i[\Psi_\lambda(e)]e^{-\lambda x} = v_n^i[g(\lambda)]g_j^n(-\lambda) = g_m^i(\lambda)v_n^m(e)g_j^n(-\lambda).$$

(Eqn. 2-87)

Since

$$\frac{d}{d\lambda}\Psi_{-\lambda}V_j^i = u_m^i(\lambda)v_n^m(e)g_j^n(-\lambda) - g_m^i(\lambda)v_n^m(e)u_j^n(-\lambda)$$

(Eqn. 2-88)

we have

$$[u,v]_j^i = u_m^i v_n^m \delta_j^n - \delta_m^i v_n^m u_j^n = u_m^i v_j^m - v_m^i u_j^m\,,$$

(Eqn. 2-89)

which is a simple relation between the Lie bracket $[u, v]$ and the commutator of two matrices.

Example 2.8. In the group GL_n find the left-invariant vector fields e_j^i that agree with the vector $\frac{\partial}{\partial a_j^i}$ at the identity. Compute the structure constants $C_{jkm}^{i\ell n}$ defined by $\left[e_k^\ell, e_m^n\right] = C_{jkm}^{i\ell n} e_i^j$.

Partial Solution

Start with $\frac{\partial}{\partial a_j^i}$ is targent to $a_j^i(\lambda) = \left\| a_n^m + \lambda \delta_i^m \delta_n^j \right\|$. Rename e_j^i by v_j^i for notational convenience. So have that v_j^i is left unvariant if $v_j^i(g) = \left[L_g^* V(e)\right]_j^i$. For GL_n left multiplication corresponds to matrix multiplication. Let $v_j^i(e) = \frac{\partial}{\partial a_j^i}$, i.e., $v_j^i(e)$ is tangent to $a_j^i(\lambda)$ at $\lambda = 0$. The left-invariant vector fields v_j^i that agree with $\frac{\partial}{\partial a_j^i}$ at the identity are:

$$v_j^i(g) = \frac{d}{d\lambda} g_k^i a_j^k(\lambda)\bigg|_{\lambda=0} = g_k^i v_j^k(e) = g_k^i \left(\frac{\partial}{\partial a_j^k}\right)$$

Now consider $\left[v_k^\ell, v_m^n\right] = C_{jkm}^{i\ell n} v_i^j$ to proceed. The rest is left to the reader.

Example 2.9. Prove the Jacobi identity by

(a) Showing for any three vector fields u, v, w that

$$\left[u, [v, w]\right] + \left[v, [w, u]\right] + \left[w, [u, v]\right] = 0.$$

(b) Showing that for any Lie group, the structure constant tensor satisfies: $C_{d[a}^e C_{bc]}^d = 0$.

Solution

We consider a local coordinate system, then

$$\begin{aligned}
&\left[u, [v, w]\right] + \left[v, [w, u]\right] + \left[w, [u, v]\right] \\
&= \partial_u(\partial_v \partial_w - \partial_w \partial_v) = (\partial_v \partial_w - \partial_w \partial_v)\partial_u + \partial_v(\partial_w \partial_u - \partial_u \partial_w) \\
&\quad -(\partial_w \partial_u - \partial_u \partial_w)\partial_v + \partial_w(\partial_u \partial_v - \partial_v \partial_u) - (\partial_u \partial_v - \partial_v \partial_u)\partial_w \\
&= 0
\end{aligned}$$

Using the Jacobi identity and the definition of the structure tensor:

$$\left[u^a, [v^b, w^c]\right] + \left[v^b, [w^c, u^a]\right] + \left[w^c, [u^a, v^b]\right] = 0$$

With $[X^a, Y^a]^c = C_{ab}^c X^a Y^b$ we have:

$$[u^a, C^d_{ab}v^b w^c] + [v^b, C^d_{ca}w^c u^a] + [w^c, C^d_{ab}u^a v^b] = 0$$

or

$$C^e_{ad}C^d_{bc}u^a v^b w^c + C^e_{bd}C^d_{ca}v^b w^c u^a + C^e_{cd}C^d_{ab}w^c u^a v^b = 0$$

If we add up cyclic permutations we get:

$$C^e_{ad}C^d_{bc}(u^a v^b w^c + v^b w^c u^a + w^c u^a v^b) + C^e_{bd}C^d_{ca}(v^b w^c u^a + w^c u^a v^b + u^a v^b w^c) + C^e_{cd}C^d_{ab}(w^c u^a v^b + u^a v^b w^c + v^b w^c u^a) = 0$$

$$= (C^e_{ad}C^d_{bc} + C^e_{bd}C^d_{ca} + C^e_{cd}C^d_{ab})(u^a v^b w^c + v^b w^c u^a + w^c u^a v^b) = 0$$

For general u, v, w we then get: $(C^e_{ad}C^d_{bc} + C^e_{bd}C^d_{ca} + C^e_{cd}C^d_{ab}) = 0$. Consider the lower three free indices and ask if we have antisymmetric on $a \leftrightarrow b$ (yes). Same for a and c , and b and c, thus fully antisymmetric, thus have:

$$C^e_{d[a}C^d_{bc]} = 0\,.$$

2.6.3 Rotations

Rotations in n dimensions

The rotation group, *SO* (*n*), is the group of orientation-preserving linear maps of R^n to itself that preserve the flat metric δ_{ab}. Thus a rotation can be regarded as a tensor R^a_b on R^n with one up and one down index, mapping vectors v^a to rotated vectors $\overline{v}^a = R^a_b v^b$. In terms of the natural orthonormal frame for R^n, the components R^i_j form an $n \times n$ a matrix satisfying

$$R^T R = I,$$

where $R^{Ti}_j = R^j_i$. A matrix satisfying the above is called an *orthogonal* matrix, and the "*SO*(*n*)" means the group of special orthogonal transformations of R^n – special means having determinant one. The full orthogonal group, *O(n)* is twice as large, including all reflections.

A rotated vector has components $\hat{v}^i = R^i_j v^j$ along the original basis, $\{e_i\}$. The rotated vector is then

$$\overline{v} = e_i \overline{v}^i = e_i R^i_j v^j = \tilde{e}_i v^i.$$

The last expression is what one would naturally write down if one began with the action of R on the basis vectors given by:

$$\overline{e}_i = e_j R^j_i,$$

to obtain, by linearity, its action on v.

Rotations in 2 dimensions

The counterclockwise rotation by angle θ of a vector v in R^2 is described by the above equation with the matrix

$$[R^i_j(\theta)] = \begin{bmatrix} \cos\theta & -\sin\theta \\ \sin\theta & \cos\theta \end{bmatrix}.$$

(Eqn. 2-90)

That is,

$$\tilde{v}^i = R^i_j v^j \text{ and } \tilde{e}_i = e_j R^j_i.$$

(Eqn. 2-91)

The group has a single parameter, θ, which runs from 0 to 2π, and the same point is labeled by θ and 2π. Thus, the manifold of the group $SO(2)$ is a circle. Up to reparameterization, there is one path through the identity, namely $\theta \to R^i_j(\theta)$, and the corresponding tangent vector can be identified with the matrix $v^i_j = \frac{d}{d\theta} R^i_j \Big|_{\theta=0} = -\epsilon^i_j$, thus

$$v^i_j = \begin{bmatrix} 0 & -1 \\ 1 & 0 \end{bmatrix}.$$

(Eqn. 2-92)

The Lie algebra is then the 1-dimensional vector {rv}, r real, with the single Lie bracket, $[v, v] = 0$. (Equivalently, the structure constant tensor $C^a_{bc} = 0$.) Using the relation, $\epsilon^i_k \epsilon^k_j = \delta^i_j$, we can directly verify: exponentiating a matrix in the Lie algebra gives back a 1-parameter subgroup (here there is only one parameter, and the subgroup is the whole group).

$$e^{\theta v} = e^{-\theta\epsilon} = 1 - \theta\epsilon + \frac{1}{2!}\theta^2 + \frac{1}{3!}\theta^3\epsilon - \cdots$$
$$= \cos\theta - \epsilon\sin\theta = R(\theta)$$

(Eqn. 2-93)

The group SO(2) is isomorphic to the group U(l) of 1×1 unitary matrices: that is, U(l) is the set $\{e^{i\theta}\}$ under ordinary multiplication. The isomorphism is simply $R^i_j(\theta) \to e^{i\theta}$. There is another, larger group, that has an identical Lie algebra to that of SO(2) = U(1), namely the group R of real numbers under addition. Again, because R is a 1-dimensional group, its Lie algebra consists of the single element v tangent to the curve $\lambda \to \lambda$ through the identity at $\lambda = 0$. Two Lie groups with the same Lie algebra are locally isomorphic: a neighborhood of the identity of each group is isomorphic, but globally they are different. In this case R is a *covering group* of SO(2) = U(1): it is a simply-connected group and there is a homomorphism ψ of R to SO(2) = U(1)

$$\psi(\lambda) = \lambda \bmod 2\pi$$

(Eqn. 2-94)

The homomorphism wraps the line infinitely many times around the circle.

Rotations in 3 dimensions

The counterclockwise rotation by angle θ about an axis along the unit vector n is described with matrix

$$\left[R^i_j(\theta n)\right] = \exp\left[-\epsilon^i_{jk} n^k\right]$$

(Eqn. 2-95)

Here n is a vector from the origin to any point on the unit sphere, and θ is any number between 0 and π. We thus have a map from a ball of radius π onto the rotation group, given by

$$\theta n \to R(\theta n).$$

(Eqn. 2-96)

Each point in the interior of the ball is mapped to a distinct point of the rotation group. Because a rotation by π about n is the same as a rotation by π about –n, diametrically opposite points on the surface of the ball are mapped to the same rotation:

$$R(\pi n) = R(-\pi n).$$

(Eqn. 2-97)

The rotation group is therefore the manifold obtained from a 3-ball by identifying diametrically opposite points of its surface; this space is called RP^3 , or real projective 3-space.

Example 2.10. Consider the vector fields L^a_x, L^a_y, and L^a_z that generate rotations about the axes of $\mathbb{R}^3$:

$$L_z = x\partial_y - y\partial_x, \qquad L_y = z\partial_x - x\partial_z, \qquad L_x = y\partial_z - z\partial_y\,.$$

Find their commutation relations and check that L_x commutes with $L^2 = L^2_x + L^2_y + L^2_z$.

Solution

$$\left[L_x, L_y\right] = y\partial_x + \left(yz\partial_z\partial_x - yx\partial_z\partial_z - z^2\partial_x\partial_y + xz\partial_z\partial_y\right)$$
$$-x\partial_y - \left(zy\partial_x\partial_z - xy\partial_z\partial_z - z^2\partial_x\partial_y + xz\partial_z\partial_y\right) = y\partial_x -$$
$$x\partial_y = -L_z$$

$$[L_z, L_x] = x\partial_z - z\partial_x = -L_y$$

$[L_y, L_z] = z\partial_y - y\partial_z = -L_x$

If we use $x = x_1, y = x_2, z = x_3$ then $\left[L_{x_i}, L_{x_j}\right] = -L_{x_k}\ \ for\ \ (ijk) = (123)$ and all cyclic permutations. Thus,

$$[L_x, L^2] = [L_x, L_x^2] + [L_x, L_y^2] + [L_x, L_z^2]$$
$$= L_xL_y^2 - L_y^2L_x + L_xL_z^2 - L_z^2L_x = (-L_z + L_yL_x)L_y - L_y(L_xL_y + L_z) + L_xL_z^2 - L_z^2L_x$$
$$= -L_zL_y - L_yL_z + (L_zL_x + L_y)L_z - L_z(L_xL_z - L_y) = -\{L_zL_y + L_yL_z + L_yL_z - L_zL_y\}$$
$$= 0$$

2.6.4 SO(3) and SU(2)

Euler angles

We need Euler angles primarily to describe the relation between SO(3) and SU(2) that underlies spinors; the relation belies the deeper meaning of Hamilton's discovery that multiplication in the rotation group can be written as multiplication of quaternions. We follow Goldstein's notation, but Goldstein likes to rotate coordinates instead of vectors. We will use the convention that a rotation is active, the motion of a vector. The components of a counterclockwise-rotated vector relative to fixed coordinate axes are the same as the components of Goldstein's fixed (passive) vector relative to clockwise-rotated axes. So Goldstein's clockwise matrices are our counterclockwise matrices. To get our formulas, change the sign of each angle in the formulas.

A rotation can be specified by what it does to an orthonormal frame $\{e_i\}$ that starts aligned with the fixed frame i, j, k. The final location of e_3 can be specified by saying that it connects the origin to a point on the unit sphere with coordinates (θ, ϕ). The other two frame vectors are then determined up to a rotation about e_3. Then an arbitrary frame orientation can be obtained in three steps (it can of course also be gotten by a single rotation about some axis):

(i) Rotate through an angle ψ about the z-axis $R(\psi\, k)$.
(ii) Rotate through angle θ about the x-axis $R(\theta\, i)$
(iii) Rotate through angle ϕ about the z-axis $R(\phi\, k)$

Step (i) leaves e_3 fixed. Step (ii) rotates e_3 to a position in the y-z plane at an angle θ from the z-axis.

Step (iii) rotates e_3 to a final angle ϕ from the y-axis. The final frame $\bar{e}_i$ is then given by

$$\bar{e}_i = e_m R_i^m(\psi,\theta,\phi) = e_m R_k^m(\psi k) R_j^k(\theta i) R_i^j(\psi k),$$

(Eqn. 2-98)

and any vector rotated by $R_j^i(\psi,\theta,\phi)$ has components (along the fixed frame i, j, k)

$$\bar{v}^i = R_j^i(\psi,\theta,\phi) v^j.$$

(Eqn. 2-99)

Example 2.11. Examine the Lie algebra SO(3). (see [21] for further details)

(a) Show that the matrices $(K_i)_{jk} = -\epsilon_{ijk}$ satisfy the commutation relations $[K_j, K_k] = \epsilon^i_{jk} K_i$.

(b) Show that:

$$e^{K_1\theta} = \begin{pmatrix} 1 & 0 & 0 \\ 0 & \cos\theta & -\sin\theta \\ 0 & \sin\theta & \cos\theta \end{pmatrix},$$

and find the analogous

$$e^{K_2\theta} = \begin{pmatrix} \cos\theta & 0 & \sin\theta \\ 0 & 1 & 0 \\ -\sin\theta & 0 & \cos\theta \end{pmatrix} \text{ and } e^{K_3\theta} = \begin{pmatrix} \cos\theta & -\sin\theta & 0 \\ \sin\theta & \cos\theta & 0 \\ 0 & 0 & 1 \end{pmatrix}.$$

(c) Find the left-invariant vector fields $\hat{e}_1$, $\hat{e}_2$, and $\hat{e}_3$ at a point $R(\phi,\theta,\Psi)$ of SO(3) in terms of the vectors $\vec{\partial}_\theta$, etc.

Solution

(a) If $(K_i)_{jk} = -\epsilon_{ijk}$ then:

$$K_1 = \begin{pmatrix} 0 & 0 & 0 \\ 0 & 0 & -1 \\ 0 & 1 & 0 \end{pmatrix},\quad K_2 = \begin{pmatrix} 0 & 0 & 1 \\ 0 & 0 & 0 \\ -1 & 0 & 0 \end{pmatrix},\quad K_3 = \begin{pmatrix} 0 & -1 & 0 \\ 1 & 0 & 0 \\ 0 & 0 & 0 \end{pmatrix}$$

and

$$[K_1, K_2] = \begin{pmatrix} 0 & -1 & 0 \\ 1 & 0 & 0 \\ 0 & 0 & 0 \end{pmatrix} = K_3, [K_1, K_3] = \begin{pmatrix} 0 & 0 & -1 \\ 0 & 0 & 0 \\ 1 & 0 & 0 \end{pmatrix} = -K_2, [K_2, K_3]$$
$$= \begin{pmatrix} 0 & 0 & 0 \\ 0 & 0 & -1 \\ 0 & 1 & 0 \end{pmatrix} = K_1$$

Since $\epsilon^3_{12} = 1,\ \epsilon^2_{13} = -1,\ and\ \epsilon^1_{23} = 1$, we see that:

$$[K_j, K_k] = \epsilon^i_{jk} K_i\,.$$

(b) $e^{K_1\theta} = \sum_{n=0}^{\infty}\frac{\theta^n}{n!}K_1^n = 1 + \theta K_1 + \frac{\theta^2 K_1^2}{2} + \frac{\theta^3 K_1^3}{3!} + \frac{\theta^4 K_1^4}{4!} + \frac{\theta^5 K_1^5}{5!} + \frac{\theta^6 K_1^6}{6!}\ldots$
and since:

$K_1^2 = \begin{pmatrix} 0 & 0 & 0 \\ 0 & -1 & 0 \\ 0 & 0 & -1 \end{pmatrix}, K_1^3 = \begin{pmatrix} 0 & 0 & 0 \\ 0 & 0 & 1 \\ 0 & -1 & 0 \end{pmatrix}, K_1^4 =$
$\begin{pmatrix} 0 & 0 & 0 \\ 0 & 1 & 0 \\ 0 & 0 & 1 \end{pmatrix}$ $and \quad K_1 = K_1^{(1+4n)}$, we have:

$$e^{K_1\theta} = 1 + K_1\left(\theta - \frac{1}{3!}\theta^3 + \frac{1}{5!}\theta^5 - \cdots\right)$$
$$+ K_1^2\left(\frac{1}{2}\theta^2 - \frac{1}{4!}\theta^4 + \frac{1}{6!}\theta^6 - \cdots\right)$$
$$= K_1(\sin\theta) + \begin{pmatrix} 1 & 0 & 0 \\ 0 & (\cos\theta) & 0 \\ 0 & 0 & (\cos\theta) \end{pmatrix} = \begin{pmatrix} 1 & 0 & 0 \\ 0 & \cos\theta & -\sin\theta \\ 0 & \sin\theta & \cos\theta \end{pmatrix}$$

Similarly for $e^{K_2\theta} = \begin{pmatrix} \cos\theta & 0 & \sin\theta \\ 0 & 1 & 0 \\ -\sin\theta & 0 & \cos\theta \end{pmatrix}$ and $e^{K_3\theta} =$
$\begin{pmatrix} \cos\theta & -\sin\theta & 0 \\ \sin\theta & \cos\theta & 0 \\ 0 & 0 & 1 \end{pmatrix}$.

(c) $R(\phi,\theta,\Psi) = R(\phi K_3)R(\theta K_1)R(\Psi K_3)$ (regarding a vector as a matrix)
$\vec{\partial}_\theta = \frac{d}{d\lambda}R(\phi,\theta+\lambda,\Psi)|_{\lambda=0} = R(\phi K_3)\frac{d}{d\lambda}R[(\theta+\lambda)K_1]_{\lambda=0}R(\Psi K_3) =$
$R(\phi K_3)R(\theta K_1)K_1 R(\Psi K_3)$
So,

$K_1 R(\Psi K_3) = \begin{pmatrix} 0 & 0 & 0 \\ 0 & 0 & -1 \\ 0 & 1 & 0 \end{pmatrix}\begin{pmatrix} \cos\Psi & -\sin\Psi & 0 \\ \sin\Psi & \cos\Psi & 0 \\ 0 & 0 & 1 \end{pmatrix} =$
$\begin{pmatrix} 0 & 0 & 0 \\ 0 & 0 & -1 \\ \sin\Psi & \cos\Psi & 0 \end{pmatrix}$
and

$R^{-1}(\Psi K_3)K_1 R(\Psi K_3) = \begin{pmatrix} 0 & 0 & -\sin\Psi \\ 0 & 0 & -\cos\Psi \\ \sin\Psi & \cos\Psi & 0 \end{pmatrix} = \cos\Psi\, K_1 -$
$\sin\Psi\, K_2$.

So,

$$\vec{\partial}_\theta = R(\phi K_3)R(\theta K_1)R(\Psi K_3)[K_1 \cos\Psi - K_2 \sin\Psi]$$
$$= R(\phi,\theta,\Psi)K_1 \cos\Psi - R(\phi,\theta,\Psi)K_2 \sin\Psi$$

Thus,

$$\vec{\partial}_\theta = \hat{e}_1(R)\cos\Psi - \hat{e}_2(R)\sin\Psi\,,$$

where $\hat{e}_i(R)$ is the left invariant vector field associated with $\hat{e}_i(e) = K_i$.

Similarly, we have:

$$\vec{\partial}_\phi = R(\phi K_3)K_3 R(\theta K_1)R(\Psi K_3)$$

and

$R^{-1}(\theta K_3)K_3 R(\theta K_1)$

$$= \begin{pmatrix} 1 & 0 & 0 \\ 0 & \cos\theta & \sin\theta \\ 0 & -\sin\theta & \cos\theta \end{pmatrix}\begin{pmatrix} 0 & -1 & 0 \\ 1 & 0 & 0 \\ 0 & 0 & 0 \end{pmatrix}\begin{pmatrix} 1 & 0 & 0 \\ 0 & \cos\theta & -\sin\theta \\ 0 & \sin\theta & \cos\theta \end{pmatrix}$$
$$= K_3\cos\theta + K_2\sin\theta$$

$$\vec{\partial}_\phi = R(\phi K_3)R(\theta K_1)(K_3\cos\theta + K_2\sin\theta)R(\Psi K_3)$$
$$= R(\phi,\theta,\Psi)K_3\cos\theta + R(\phi K_3)R(\theta K_1)K_2 R(\Psi K_3)\sin\theta$$

$$R^{-1}(\Psi K_3)K_2 R(\Psi K_3) = \begin{pmatrix} \cos\Psi & \sin\Psi & 0 \\ -\sin\Psi & \cos\Psi & 0 \\ 0 & 0 & 1 \end{pmatrix}\begin{pmatrix} 0 & 0 & 1 \\ 0 & 0 & 0 \\ -\cos\Psi & -\sin\Psi & 0 \end{pmatrix}$$
$$= \begin{pmatrix} 0 & 0 & \cos\Psi \\ 0 & 0 & -\sin\Psi \\ -\cos\Psi & \sin\Psi & 0 \end{pmatrix} = K_2\cos\Psi + K_1\sin\Psi$$

$$\vec{\partial}_\phi = R(\phi,\theta,\Psi)K_3\cos\theta + R(\phi,\theta,\Psi)(K_2\cos\Psi + K_1\sin\Psi)\sin\theta$$
$$= \hat{e}_3(R)\cos\theta - \hat{e}_2(R)\cos\Psi\sin\Psi + \hat{e}_1(R)\sin\Psi\sin\theta$$

(c) $\vec{\partial}_\Psi = R(\phi K_3)R(\theta K_1)R(\Psi K_3)K_3 = R(\phi,\theta,\Psi)K_3 = \hat{e}_3$

So,

$\vec{\partial}_\Psi = \hat{e}_3$:

$\vec{\partial}_\phi = \hat{e}_3\cos\theta + \hat{e}_2\cos\Psi\sin\theta + \hat{e}_1\sin\Psi\sin\theta$

$\vec{\partial}_\theta = \hat{e}_1\cos\Psi - \hat{e}_2\sin\Psi$

are grouped to give:

$$\frac{1}{\sin\theta}\vec{\partial}_\phi = \cot\theta\,\vec{\partial}_\Psi = \hat{e}_2\cos\Psi + \hat{e}_1\sin\Psi$$
$$\hat{e}_2 = \cos\Psi\left(\frac{1}{\sin\theta}\vec{\partial}_\phi = \cot\theta\,\vec{\partial}_\Psi\right) - \sin\Psi\,\vec{\partial}_\theta$$

$$\hat{e}_1 = \sin\Psi\left(\frac{1}{\sin\theta}\vec{\partial}_\phi = \cot\theta\,\vec{\partial}_\Psi\right) - \cos\Psi\,\vec{\partial}_\theta$$

Thus:

$$\hat{e}_1 + i\hat{e}_2 = e^{-i\Psi}\vec{\partial}_\theta + \frac{ie^{-i\Psi}}{\sin\theta}\left(\vec{\partial}_\phi - \cos\theta\,\vec{\partial}_\Psi\right) \text{ and } \hat{e}_3 = \partial_\Psi.$$

Example 2.12. Check that the vector fields above reproduce the Lie algebra of SO(3), e.g., verify that $[\hat{e}_j, \hat{e}_k] = \epsilon^i_{jk}\hat{e}_i$. (See [21] for further details.)

Solution

Starting with

$$\hat{e}_1 = \frac{\sin\Psi}{\sin\theta}\left(\vec{\partial}_\phi - \cos\theta\,\vec{\partial}_\Psi\right) + \cos\Psi\,\vec{\partial}_\theta$$

$$\hat{e}_2 = \frac{\cos\Psi}{\sin\theta}\left(\vec{\partial}_\phi - \cos\theta\,\vec{\partial}_\Psi\right) + \sin\Psi\,\vec{\partial}_\theta$$

$$\hat{e}_3 = \vec{\partial}_\Psi$$

We get:

$$\hat{e}_x = \hat{e}_1 + i\hat{e}_2 = e^{-1^{-\Psi}}\left[\partial_\theta + \frac{i}{\sin\theta}\left(\partial_\phi - \cos\theta\,\partial_\Psi\right)\right]$$

$$\hat{e}_y = \hat{e}_1 - i\hat{e}_2 = e^{i\Psi}\left[\partial_\theta + \frac{i}{\sin\theta}\left(\partial_\phi - \cos\theta\,\partial_\Psi\right)\right]$$

$$[\hat{e}_x, \hat{e}_y] = \left\{-1\frac{\cos\theta}{\sin\theta}(-i)e^{-i\Psi}\hat{e}_y - i\partial_\theta\left\{\frac{i}{\sin\theta}\left(\partial_\phi - \cos\theta\,\partial_\Psi\right)\right\}\right\}$$
$$-\left\{1\frac{\cos\theta}{\sin\theta}(-i)e^{i\Psi}\hat{e}_x - i\partial_\theta\left\{\frac{i}{\sin\theta}\left(\partial_\phi - \cos\theta\,\partial_\Psi\right)\right\}\right\}$$

$$= \frac{\cos\theta}{\sin\theta}\left(\frac{2i}{\sin\theta}\left(\partial_\phi - \cos\theta\,\partial_\Psi\right)\right) - 2i\frac{(-\cos\theta)}{\sin^2\theta}\left(\partial_\phi - \cos\theta\,\partial_\Psi\right) +$$
$$\frac{2i}{\sin\theta}(-\sin\theta)\partial_\Psi = -2i\vec{\partial}_\Psi$$

$$[\hat{e}_1 + i\hat{e}_2, \hat{e}_1 - i\hat{e}_2] = -i[\hat{e}_1, \hat{e}_2] + i[\hat{e}_2, \hat{e}_1] = -2i[\hat{e}_1, \hat{e}_2] - 2i\vec{\partial}_\Psi$$
$$[\hat{e}_1, \hat{e}_2] = \hat{e}_3$$

$$[\hat{e}_1, \hat{e}_3] = -\left[\frac{\cos\Psi}{\sin\theta}\left(\vec{\partial}_\phi - \cos\theta\,\vec{\partial}_\Psi\right) - \sin\Psi\,\vec{\partial}_\theta\right] = -\hat{e}_2$$

$$[\hat{e}_2, \hat{e}_3] = -\left[\frac{-\sin\Psi}{\sin\theta}\left(\vec{\partial}_\phi - \cos\theta\,\vec{\partial}_\Psi\right) - \cos\Psi\,\vec{\partial}_\theta\right] = -\hat{e}_1$$

So, $[\hat{o}_1, \hat{o}_2] - \hat{o}_3, [\hat{o}_2, o_3] - \hat{o}_1, [\hat{o}_3, \hat{o}_1] - \hat{o}_2$

Thus, $[\hat{e}_j, \hat{e}_k] = \epsilon^i_{jk}\hat{e}_i$.

2.6.5 Homogeneous Spacetimes

Action of a Lie group

A Lie group *G* acts on itself by left multiplication and by right multiplication. There are similarly two ways that G can act on any manifold M.

Definition. A left action of a Lie group *G* on a manifold M assigns to each $a \in G$ a diffco $a : M \rightarrow M$, satisfying

$$a(bp) = (ab)p,$$

all $a, b \in G, p \in M$.

Definition. A right action of a Lie group *G* on a manifold *M* assigns to each $a \in G$ a diffeo $a : M \rightarrow M$, satisfying

$$(pa)b = p(ab)$$

all $a, b \in G, p \in M$.

An action of *G* on *M* is *transitive* if every point of *M* can be reached from any other point of *M* by the action of an element of G:

$$\forall p, q \in M, \ a \in G$$
$$q = pa.$$

(written for a right action). An action is *simply transitive* if every element of *G* except *e* moves every element of *M*: $pa = p \Rightarrow a = e$. If *G* acts simply transitively on *M,* then, as a manifold, *G* is diffeomorphic to *M.* Proof: Pick a point *p* in *M.* Define a diffeo $\psi_p : G \longrightarrow M$ by $\psi_{(p)}(a) = pa$ map is onto because the action is transitive; it is 1–1 because the action is *simply* transitive.

Note that the arbitrariness in the diffeo lies in choosing what point *p* to assign to the identity *e* of *G.* Once one has chosen an origin, *p,* the rest of the diffeo is determined by the group action.

Theorem A simply transitive action of a Lie group *G* on a manifold *M* maps the Lie algebra $\mathfrak{g}$ of *G* to an isomorphic Lie algebra of vector fields on *M.* In fact, given a vector field ξ^a in $\mathfrak{g}$, each diffeo $\psi_{(p)}$ drags ξ^a to the *same* vector field ξ^a on *M,* independent of the choice of origin, *p.*

Proof. We have shown previously that diffeos commute with Lie derivatives: let $\bar{\xi}^a = \psi^{\alpha}_{(p)a}\xi^a$,

$$[\bar{\xi},\bar{\eta}] = [\overline{\xi,\eta}],$$

We must show that $\bar{\xi}^a$ is independent of the choice of diffeo $\psi_{(p)}$. Let $g(\lambda)$ be the 1-parameter subgroup generated by ξ^a. Then at a point $q = pa$, $\bar{\xi}^a$ is tangent to $p(ag(\lambda)) = (pa)g(\lambda) = qg(\lambda)$. In the final expression, p does not appear: the curve $\lambda \to qg(\lambda)$ is independent of *p*, so its tangent, $\bar{\xi}^a$ is independent of *p*.

Spacetimes foliated by with homogenous spacelike slices

Let *M with* $g_{\alpha\beta}$ be a spacetime on which a group G acts on the right and for which each $a \in G$ corresponds to an isometry of M – each diffeo a : .*M* → *M* leaves the metric of *M* invariant. Homogenous universes are, in addition, foliated by a family $\sum_t$ of spacelike hypersurfaces on each of which *G* acts simply transitively. Then each point of a slice $\sum_t$ is identical to every other point of a slice. Homogeneous does not imply *isotropic:* a uniform magnetic field and a macroscopic crystal are homogeneous but not isotropic. In a homogeneous but anisotropic gravitational field, for example, tidal forces are direction-dependent.

The transitive slices of a homogeneous spacetime are orthogonal to a family of timelike geodesics. The proof uses the fact that the component of momentum along a Killing vector is conserved by a free particle.

Theorem. Let u^{α} be tangent to a geodesic $\gamma(\lambda)$ with affine parameter λ, and let ξ^a be a Killing vector. Then $u_{\alpha}\xi^a$ is constant along γ. Proof:

$$\frac{d}{d\lambda}(u_{\alpha}\xi^a) = u^{\beta}\nabla_{\beta}(u_{\alpha}\xi^a) = u^{\alpha}u^{\beta}\nabla_{\beta}\xi_{\alpha} + \xi_{\alpha}u^{\beta}\nabla_{\beta}u_{\alpha} = 0.$$

(Eqn. 2-100)

where each term vanishes separately, the first by Killing's equation ($\nabla_{\beta}\xi_{\alpha}$ antisymmetric), the second by the geodesic equation, $u^{\beta}\nabla_{\beta}u^{\alpha} = 0$. For a timelike geodesic, $p_{\alpha}\xi^a$ is a conserved component of the 4-momentum $p^{\alpha} = mu^{\alpha}$.

Consider the family of geodesics normal to a homogenous hypersurface Σ_0, with unit normal t^{α}. We want to show that each geodesic γ in the family is normal to every hypersurface $\sum_t$. The images $\bar{\xi}^a$ of the vectors in the Lie algebra of G span the 3-dimensional tangent space of $\sum_t$ at each point of $\sum_t$. Hence γ is normal to $\sum_t$ if and only if its tangent t^α is normal to each vector ξ^α. But, by construction, on $\sum_0$, $t_\alpha \xi^\alpha = 0$ and by the theorem, $t_\alpha \xi^\alpha = 0$ is conserved along γ so $t_\alpha \xi^\alpha = 0$ on $\sum_t$. If we choose the parameter *t* to be proper time along each geodesic γ then $t_\alpha = -\nabla_\alpha t$ (the sign makes t^α future pointing, $t^\alpha \nabla_\alpha t = 1$).

We can write the metric in a simple form in terms of a basis of 1-forms w^μ with $w^0 = \nabla_\alpha t$ and with w^i the image on each $\sum_t$ of a right-invariant basis on *G*. Unlike the left-invariant vector fields that generate the right action of *G*, these right-invariant forms depend on the choice of diffeo $\psi_{(p)}$. One must pick a basis, w^i, orthogonal to w^0, at a point *p* of $\sum_0$, and extend the basis to every pomt *q* of $\sum_0$ by the diffeos *a*. That is,

$$w^i(q) = a^*[w^i(p)],$$

(Eqn. 2-101)

where $q = pa$. Finally, extend w^i to all of M by requiring that $\mathcal{L}_t w^i = 0$, or, equivalently, $w^i(t) = \phi_t w^i(0)$, where $\{\phi_t\}$ is the family of diffeos generated by t^α. Because the Lie algebras of left- and right-invariant vector fields of a group are isomorphic, we have the relation,

$$d\omega^i = -\frac{1}{2} c^i_{jk} \omega^j \wedge \omega^k,$$

(Eqn. 2-102)

where c^i_{jk} are the structure constants of the Lie algebra corresponding to the basis ω^i. Abstractly, the
images ξ, η under $\psi_{(p)}$ of right-invariant vector fields on *G* satisfy

$$[\xi, \eta]^\alpha = c^\alpha_{\beta\gamma} \xi^\beta \eta^\gamma,$$

(Eqn. 2-103)

where $c^\alpha_{\beta\gamma}$ is the image of the right-invariant structure constant tensor on *G*, and we can write

$$\nabla_{[\beta} \omega^i_{\gamma]} = -\frac{1}{2} c^\alpha_{\beta\gamma} \omega^i_\alpha .$$

(Eqn. 2-104)

By construction. the basis ω^μ is invariant under the action of *G*. Because the metric $g_{\alpha\beta}$ is invariant under *G*. Its components along ω^μ are constants on each $\sum_t$. We have

$$g_{\alpha\beta} = -\nabla_\alpha t \nabla_\beta t + h_{ij}(t) \omega^i_\alpha \omega^j_\beta.$$

(Eqn. 2-105)

Because the metric components depend only on t, and the structure constants c^i_{jk} *are* constants, the Einstein equations are now a set of ordinary differential equations, easily solved by computer.

Bianchi's classification of 3-dimensional Lie groups
With only one exception ($SO(3) \times R$, acting on $S^2 \times R$), every group that acts transitively on a 3-manifold Σ has a subgroup that acts simply transitively on Σ. When the action is simply transitive, Σ is a copy of G, so one can classify homogeneous spaces by classifying the 3-dirnensional Lie groups. Ignoring the global topology of the space, this is equivalent to classifying the 3-dimensional Lie algebras. The classification is due to Bianchi, first applied to homogenous spacetimes by Taub (Ann. Math. 53, 472-90, 1951) [22].

We want to classify the possible structure constant tensors C^a_{bc} of a 3-dimensional Lie group G. First note that an isomorphism $\psi: G \to G$ maps C^a_{bc} to the new structure constant tensor, $\psi^a_d \psi^{-1}_b{}^e \psi^{-1}_c{}^f C^d_{ef}$. Any invertible linear transformation ψ^a_b can be obtained in this way from some isomorphism in this way. Because we want to regard as identical two Lie groups that are related by a diffeo, we need to find the classes of tensors C^a_{bc} invariant under similarity transformations (or, from a passive viewpoint, we want classes of structure constants C^i_{jk} invariant under a change of basis).

What does it mean to find invariant classes of tensors? For a vector A^α, there are only two classes – the class consisting of the vector 0, and the class consisting of all nonzero vectors: Given any two nonzero vectors, A^α, B^α, there is always an invertible linear map that takes $A^\alpha$to B^α. In 3-dimensions, any nonzero antisymmetric tensor A^{ab} is similarly equivalent to any other antisymmetric tensor. Pick a metric ξ^{ab} and a corresponding totally antisymmetric tensor ϵ_{abc} (The nonnalization, $\frac{1}{2}\epsilon_{acd}\epsilon_b^{cd} = \delta_{ab}$, fixes ϵ_{abc} up to an overall sign.) Then one can uniquely associate each antisymmetric tensor with a vector by taking the dual:

$$^*A^a \overset{\text{def}}{=} \frac{1}{2}\epsilon^a_{bc}A^{bc},$$

(Eqn. 2-106)

and the operation is invertible:

$$A^{ab} = \epsilon^{ab*}_c A^c.$$

(Eqn. 2-107)

(It is, of course, this duality relation for tensors on R^3 that allows one to replace the wedge product, $u \wedge v$, of two vectors by the cross product, $u \times v = {}^*(u \wedge v)$.) Then a Lie transformation that takes ${}^*A^a$ to ${}^*B^a$ also takes A^{ab} to B^{ab}.

Finally, symmetric tensors, M^{ab} are classified by their rank and signature. The rank of M is the number of nonzero eigenvalues, the dimension of the space of vectors for which $M^{ab}v_b \neq 0$. One can always diagonalize a symmetric tensor, i.e., write a symmetric matrix M^{ab} of rank n in terms of n orthogonal normalized eigenvectors, $u, \dots, v$; $M^{ab} = \pm u^a u^b \pm \cdots \pm v^a v^b$. Any two tensors M, M' with the same number of nonzero eigenvectors and the same signature (set of signs) are equivalent, because there is a similarity transformation that maps $u \to u', \dots, v \to v'$. Tensors M^{ab} with the same rank but different signatures cannot be in the same invariance class, because a similarity transformation maps $M^{ab} \to M'^{ab} = \pm u'^a u'^b \pm \cdots \pm v'^a v'^b$, thereby maintaining the set of signs.

The Bianchi classification proceeds as follows. The tensor c^a_{bc} antisymmetric on b, c, is equivalent to its dual,

$$ {}^*c^{ab} = \frac{1}{2}\epsilon^{bcd} c^a_{cd}. $$

(Eqn. 2-108)

Write $*C$ as the sum of its symmetric and antisymmetric parts: Let

$$ M^{ab} = {}^*c^{ab}, \quad A_a = \epsilon_{abc}{}^*c^{bc}. $$

(Eqn. 2-109)

Then

$$ {}^*c^{ab} = M^{ab} + \frac{1}{2}\epsilon_c^{ab} A^c, $$

(Eqn. 2-110)

and A is the trace of c,

$$ A_a = \epsilon_{abc}\left(\frac{1}{2}\epsilon^{cde} c^b_{de}\right) = \delta_a^{[d}\delta_b^{e]} c^b_{de} = c^b_{ab}. $$

(Eqn. 2-111)

Finally, c can be expressed in terms of A and M:

$$ c^a_{bc} = \epsilon_{bcd}{}^*c^{ad} = \epsilon_{bcd}(M^{ad} - \epsilon_e^{ad} A^e) = \epsilon_{bcd}M^{ad} + \delta^a_{[b}A_{c]}\,. $$

(Eqn. 2-112)

A Lie algebra is determined by a structure constant tensor satisfying the Jacobi identity,

$$ c^a_{e[b}c^e_{cd]} = 0. $$

(Eqn. 2-113)

This turns out to be simply the relation,

$$M^{ab}A_b = 0.$$

(Eqn. 2-114)

Conclusion: A 3-dimensional Lie algebra is specified by a symmetric tensor M^{ab} and a vector A^a satisfying $M^{ab}A_b = 0$. The 3-dimensional Lie algebras are, thus, classified by the rank and signature of M^{ab} and by whether or not A^a is zero:

Bianchi Type	A^a	Classification of M
I	0	0 0 0
II	0	+ 0 0
VII	0	+ + 0
VI	0	+ – 0
IX	0	+ + +
VIII	0	+ + –
V	+	0 0 0
IV	+	0 0 +
VII′	+	0 + +
VI′ (III if α = –1/2)	+	0 + –

Note that an orientation-reversing transformation changes the sign of c^{abc} and hence of the sign of the tensor M^{ab} that is constructed froin the new c^{abc} by using a *fixed* ϵ^{abc}. Thus the signatwe of M^{ab} is determined only up to an overall sign. (This is tricky, because a similarity transformation of M^{ab} cannot change its signature – under a parity transformation, both ϵ_{abc} and c^a_{bc} change sign, and M^{ab} is unchanged. The arbitrariness in the signature of M^{ab} arises from the arbitrariness in the choice of ϵ_{abc}). A final subtlety in the classification arises from the fact that classes of equivalent tensors c^a_{bc} correspond to classes of equivalent pairs (A^a, M^{ab}) – One cannot make independent similarity transformations of A^a and M^{ab}. When $A^a = 0$, or $M^{ab} = 0$, we are done. When $A^a \neq 0$, and $M^{ab} \neq 0$, rank M is 1 or 2. For rank one, there turns out to be nothing new. For M of rank 2 (types VII′ and VI′ above), however, one can construct an invariant scalar, α:

$$\alpha \epsilon_{acd}\epsilon_{bef} M^{ce} M^{df} = A_a A_b,$$

(Eqn. 2-115)

or, equivalently, expressing M and A in terms of c^a_{bc} :

$$c_{ad}^{c} c_{bc}^{d} = \frac{1+\alpha}{\alpha} c_{ac}^{c} c_{bd}^{d} .$$

(Eqn. 2-116)

There are, then, two continuous families of Lie groups, parameterized by a, with $A^a \neq 0$, rank $M = 2$, signature ++ and +–.

The 3-metric of a homogeneous spacetime has constant components with respect to the group-invariant basis, ω^i. That is, $h_{ab} = h_{ij}\omega_a^i \omega_b^j$. with h_{ij} constant. Because the metric is invariant under the action of the group, so are the Riemann and Ricci tensors. Their components can depend only on h_{ij} and on the structure constants c_{ijk} of the group. In fact, because derivatives of the connection coefficients vanish, the tensors must be quadratic in c_{jk}^{i} with the metric entering only to raise or lower indices. The computation of the Ricci tensor is as follows.

$$R_{ij} = \Gamma_{lk}^{k}\Gamma_{ij}^{l} - \Gamma_{lj}^{k}\Gamma_{ik}^{l} - \Gamma_{il}^{k} c_{kj}^{l}$$

(Eqn. 2-117)

$$R_{ij} = \frac{1}{4}\left(c_{kl}^{k} - ck_{lk} - c_{lk}^{k}\right)\left(c_{j\,i}^{l} - c_{ij}^{l} - c_{ij}^{l}\right)$$
$$- \frac{1}{4}\left(c_{lj}^{k} - c_{lj}^{k} - c_{j\,l}^{k}\right)\left(c_{k\,i}^{l} - c_{ik}^{l} - c_{ik}^{l}\right)$$
$$- \frac{1}{2}\left(c_{l\,i}^{k} - c_{il}^{k} - c_{il}^{k}\right) c_{kj}^{l}$$

(Eqn. 2-118)

$$R_{ij} = -c_{(ij)}^{l} c_{k\,l}^{k} - c_{kli} c_{j}^{(kl)} + \frac{1}{4} c_{ikl} c_{j}^{kl}$$

(Eqn. 2-119)

2.7 Geometry of Yang-Mills Fields

Let B be a manifold and G a group that acts freely on B (if $L_B x = x$ for any $x \in B$ then $g = e$, then 'acts freely'). Let M be the set of orbits of G, $M = B/G$. If B looks locally like a Cartesian product, $B_{local} \approx M \times G$, then B is called a principle bundle with group G. In a principle bundle (defined below) the fiber space is isomorphic to the group, $F \approx G$, and he action of the group preserves the fibers. A principle fiber bundle B admits a cross section ($B = M \times G$, a Cartesian product, globally) iff (if and only if) B is 'trivial'.

Consider tangent bundles $\pi: B \to M$, where π is smooth and surjective, and for every $x \in M$ $\exists$ $nghb. U_x \ni \pi^{-1}(U_x)$ diffeomorphic to $U \times F$ (F is the fiber) via a diffeo that takes $\pi^{-1}(x)$ to $\{x\} \times F \ \ \forall_x \in U$ and $F = \pi^{-1}(x)$, is the fiber.

Def. Fiber Bundle
Def. Bundle:
A Bundle is defined as a tuple $(\mathcal{E}, M, \pi)$, where $\mathcal{E}$ is the total space, M is the base space, and π is a surjective map from $\mathcal{E}$ to M.
Def. Product Bundle:
A tuple $(M \times a, M, \pi)$.
Def. Fiber Bundle:
A bundle that is locally a product bundle.

Example: The Mobius strip is a fiber bundle: it is locally a product bundle of R^1 (the fiber) over S^1.
Example: The cylinder is a product bundle (global) of an R^1 fiber over S^1.

2.7.1 Connections

Consider the meaning of parallel transport in terms of a bundle of frames over M. Given x_0 and $\vec{e}_\alpha(x_0)$ a covariant derivative ∇ tells one how to parallel transport $\vec{e}_\alpha(x_0)$ along a given curve $x(\lambda)$, such that it is the unique solution to

$$\vec{x} \cdot \nabla \vec{e}_\alpha = 0$$

(Eqn. 2-120)

In effect ∇ tells us how to lift a path $x(\lambda)$ in M to a path $\{\vec{e}_\mu(\lambda), x(\lambda)\}$ in B through $\vec{e}_\mu(0)$ at x_0. Lifted paths of parallel transported bases are called horizontal paths in B. Without a connection, i.e., ∇, we don't know what horizontal means, but there is a natural definition of vertical paths. A vertical path satisfies $\pi[b(\lambda)] = x,\ \ x \text{ indep.} of\ \lambda$., and since $\vec{e}_\mu(\lambda) = \vec{e}_\nu(0) a^\nu_\mu(\lambda)$ this is a path in GL_n. Note: a vector ξ^c on B is vertical if $\pi_* \xi = 0$. If B is a principal bundle, then we have $b(\lambda) = bg(\lambda)$, say, where $g(\lambda)$ is a path in G. So $\dot{b}^c$ can be identified with a vector $\dot{g}^a$ at the identity in G.

Consider G to act on B on the right for definiteness, define:

$$\Psi(\text{b})\text{: G} \to \text{B},\ \ g \longmapsto bg$$

(Eqn. 2-121)

Then $\Psi(\text{b})_*$ maps $\xi^a \in \mathcal{G}$ to $\xi^c(b) = \Psi(b)^c_a \xi^a(e)$ (where $\mathcal{G}$ is the Lie algebra of G). $\Psi(\text{b})_*$ is a diffeo generating right action, this then generates left inv. vector field on g which is mapped to a left inv. vector field on B, i.e., the Lie algebra structure is preserved.

2.7.2 Connection on bundle of Lorentz frames

Let's now consider a bundle, E, of Lorentz frames over a manifold M:

Vector on E, use:
ξ^c *with* $c, d, \ldots.$ $(I, J, \ldots$ concrete),
Vector on M, use:
ξ^α *with* $\alpha, \beta, \ldots.$ $(\mu, v, \ldots$ concrete),

If ξ^c is vertical then it is tangent to $\vec{e}_v \Lambda^v_\mu(\lambda)$ and can be associated with $\frac{d}{d\lambda}\Lambda^v_\mu(\lambda)\Big|_{\lambda=0} = \dot{\Lambda}^v_\mu$ an element of the Lie algebra of the Lorentz group.

Define a "Lie-algebra valued *1*–form" $\overline{\omega}^v_{\mu c}$ on E by

$$\overline{\omega}^v_{\mu c}\xi^c = \dot{\Lambda}^v_\mu\,, \quad \xi^c \text{ vertical}$$

(Eqn. 2-122)

$$\overline{\omega}^v_{\mu c}\xi^c = 0\,, \quad \xi^c \text{ horizontal}$$

(Eqn. 2-123)

Any ξ^c can be decompeted into vertical and horizontal parts: $\xi^c = h^c + v^c$. We could like to derive a relation between $\overline{\omega}$ ... and the Cartan 1–forms $\omega^v_{\mu\alpha}$ on M. So, what is the relation between $\omega^v_{\mu\alpha}$ and $\overline{\omega}^v_{\mu c}$?

Consider a mapping $e: M \to E$ with $x \longmapsto \left(x, \vec{e}_\mu(x)\right)$. We have a pullback: $e^*\overline{\omega}^v_\mu(\dot{x}) = \overline{\omega}^v_\mu(e_*\dot{x})$, where $\dot{x}^\alpha$ is a vector in M tangent to $x(\lambda)$, now $e_*\dot{x}$ is the vector tangent to $\vec{e}_\mu\big(x(\lambda)\big)$ in E:

$$\vec{e}_\mu\big(x(\lambda)\big) = \vec{E}_v(\lambda)\Lambda^v_\mu(\lambda)$$

(Eqn. 2-124)

where $\vec{E}_v(\lambda)$ describes parallel-transport of $\vec{e}_\mu(0)$ along $x(\lambda)$, and $\Lambda^v_\mu(\lambda)$ describes a Lorentz Transform.

So,

$$e_*\dot{x} = \vec{E}_v + \vec{e}_v(0)\dot{\Lambda}^v_\mu\,,$$

(Eqn. 2-125)

where $\vec{E}_v$ is the horizontal part and $\vec{e}_v(0)\dot{\Lambda}^v_\mu$ is the vertical part. Thus, $\overline{\omega}^v_\mu(e_*\dot{x}) = \dot{\Lambda}^v_\mu$ (from definition of $\overline{\omega}$).

Now consider $\omega^v_\mu(\dot{x}) = \Gamma^v_{\mu\alpha}\dot{x}^\sigma$:

$$e_\nu \omega^\nu_\mu(\dot{x}) = \dot{x}^\sigma \nabla_\sigma e_\mu = \frac{D}{D\lambda} e_\mu(x(\lambda)) = \frac{D}{D\lambda}[E_\nu(\lambda)\Lambda^\nu_\mu(\lambda)]$$
$$= \left(\frac{D}{D\lambda} E_\nu(\lambda)\right) \Lambda^\nu_\mu(\lambda)\Big|_{x=0} + E_\nu(\lambda)|_{\lambda=0} \frac{d}{d\lambda} \Lambda^\nu_\mu$$
(Eqn. 2-126)

and

$$\omega^\nu_\mu(\dot{x}) = \dot{\Lambda}^\nu_\mu = \overline{\omega}^\nu_\mu(e_* \dot{x}) = e^* \overline{\omega}^\nu_\mu(\dot{x}) \, .$$
(Eqn. 2-127)

So, $\omega^\nu_\mu(\dot{x})$ is the pull-back of $\overline{\omega}^\nu_\mu$.

2.7.3 Connections on a Principle Fiber Bundle

Let a, b, c be Lie algebra indices

$$B \overset{\pi}{\rightarrow} M \text{ principle bundle with group G.}$$

Write a vertical path as $b(\lambda) = bg(\lambda)$, we can associate the tangent vector $\dot{b}^c$ with an element of the Lie algebra $\dot{g}^a$ in $\mathcal{G}$ (where $\mathcal{G}$ is the Lie algebra of G). The definition of connection for a Principal Fiber B essentially repeats the properties found for the Lorentz analysis:

$$\text{Denote} \qquad R_g : B \rightarrow B \qquad \forall g \in G$$
$$b \longmapsto b_g$$

Definition of a connection $\overline{A}^a_c$ on B: a Lie-algebra valued *1*–form on B for which:

(i) $\overline{A}^a_c \xi^c = \dot{g}^a$ If ξ^c is a vertical vector tangent to $b_g(\lambda)$ at b.

(ii) The subspace of horizontal vectors at each b in B, $\left\{\xi^c | \overline{A}^a_c \xi^c = 0\right\}$, is isomorphic to $T_x M$ $\left(x = \pi(b)\right)$.

(iii) If ξ^c is horizontal $R_{g*}\xi^c$ is horizontal.

Now, given a cross-section $b(x)$ of B we will get a "vector potential" A^a_α on the spacetime M:

$$A^a_\alpha = b^*_\alpha \overline{A}^a_c \, .$$

Example 2.13. Let B be a principal G-bundle over M. Let ε^a be an isovector on M, corresponding to a g-valued scalar on $\bar{\varepsilon}^a$ that satisfies:

$$\bar{\varepsilon}(bg) = g^{-1}\bar{\varepsilon}(b) g \, .$$

(a) Show that $D\bar{\varepsilon} = \text{hor}(d\bar{\varepsilon})$ is the gauge covariant derivative:

$$D_c \bar{\varepsilon}^a = d_c \bar{\varepsilon}^a + C^a_{bd} A^b_c \bar{\varepsilon}^d \, .$$

(b) Verify that $D\bar{\varepsilon}$ satisfies the defining relation for an isovector:

$$D\bar{\varepsilon}(bg) = g^{-1}D\bar{\varepsilon}(b)g\,.$$

(c) Let $b\colon M \to B$ be the cross-section of B for which $\bar{\varepsilon}^a\big(b(x)\big) = \varepsilon^a(x)$. Define the gauge covariant derivative $D_\alpha\varepsilon^a(x)$ by $D_\alpha\varepsilon^a(x) \equiv b_\alpha^{c*}D_c\bar{\varepsilon}^a\big(b(x)\big)$, show that:

$$D_\alpha\varepsilon^a(x) = d_\alpha\varepsilon^a(x) + C_{bd}^a A_\alpha^b \varepsilon^d(x)\,.$$

Solution

(a) Consider a path $b(\lambda)$ in the bundle with tangent vector $\dot{b}^c$ at $\lambda = 0$. $b(\lambda)$ can be related to a horizontal path $b_{||}(\lambda)$ through $b(o)$ using $g(\lambda)$ a path through $e \in G$:

$$b(\lambda) = b_{||}(\lambda)g(\lambda)$$

If $D\bar{\varepsilon} = \text{hor}\,(d\bar{\varepsilon})$ then $\dot{b}^c D_c\bar{\varepsilon} = \frac{d}{d\lambda}\Big(\bar{\varepsilon}\big(b_{11}(\lambda)\big)\Big)\Big|_{\lambda=0}$

So,

$\dot{b}^c D_c\bar{\varepsilon} = \frac{d}{d\lambda}\Big(\bar{\varepsilon}\big(b_{11}(\lambda)\big)\Big)\Big|_{\lambda=0}$ and using $\bar{\varepsilon}(bg) = g^{-1}\bar{\varepsilon}(b)g$

$$= \frac{d}{d\lambda}\big(g(\lambda)\bar{\varepsilon}\big(b(\lambda)\big)g(\lambda)^{-1}\big)\Big|_{\lambda=0}$$

$$gg^{-1} = 1$$

$$\dot{g}g^{-1} + g(\dot{g}^{-1}) = 0$$

$$= g\frac{d}{d\lambda}\bar{\varepsilon}\big(b(\lambda)\big)\Big|_{\lambda=0} g^{-1} + \dot{g}\bar{\varepsilon}\big(b(o)\big)e - e\bar{\varepsilon}\big(b(o)\big)\dot{g}$$

$$(\dot{g}^{-1})|_{\lambda=0} = -g^{-1}\dot{g}g^{-1}|_{\lambda=0}$$

$$= -\dot{g}$$

$$= \dot{b}^c d_c\bar{\varepsilon} + \dot{g}\vec{\varepsilon} - \vec{\varepsilon}\dot{g}$$

Using the definition of connection on B: $\dot{g}^b = A_c^b\dot{b}^c$ and $[\dot{g},\varepsilon] = C_{bd}^a\dot{g}^b\varepsilon^d$

$$\dot{b}^c D_c\bar{\varepsilon}^a = \dot{b}_c^c d_c\bar{\varepsilon}^a + C_{bd}^a\dot{g}^b\varepsilon^d = \dot{b}^c d_c\bar{\varepsilon}^a = C_{bd}^a A_c^b\dot{g}^b\bar{\varepsilon}^d$$

$$= \dot{b}^c(d_c\bar{\varepsilon}^a + C_{bd}^a A_c^b\bar{\varepsilon}^d)$$

Thus,

$$D_c\bar{\varepsilon}^a = d_c\bar{\varepsilon}^a + C_{bd}^a A_c^b\bar{\varepsilon}^d\,.$$

(b) Does $D\bar{\varepsilon}(bg) = g^{-1}D\bar{\varepsilon}(b)g$? We have:

$$\bar{\varepsilon}(bg) = g^{-1}\bar{\varepsilon}(b)g$$

So,

$D\bar{\varepsilon}(bg) = D(g^{-1})\bar{\varepsilon}(b)g + g^{-1}D\bar{\varepsilon}(b)g + g^{-1}\bar{\varepsilon}(b)Dg$, but, since either $g^{-1}(\lambda)$ or $g(\lambda)$ is a vertical path $D(g^{-1}) = 0, D(g) = 0$, so

$$D\bar{\varepsilon}(bg) = g^{-1}D\bar{\varepsilon}(b)g$$

(c) $D_\alpha \varepsilon^a(x) \equiv b_\alpha^{c*} D_c \bar{\varepsilon}^a\big(b(x)\big)$

$$\begin{aligned}(b_\alpha^c)^* D_c \bar{\varepsilon}^a &= (b_\alpha^c)^*[d_c \bar{\varepsilon}^a + C_{bd}^a A_c^b \bar{\varepsilon}^d] \\ &= (b_\alpha^c)^* d_c \bar{\varepsilon}^a + C_{bd}^a (b_\alpha^c)^* A_c^b \bar{\varepsilon}^d \\ &= d_\alpha \bar{\varepsilon}^a\big(b(x)\big) + C_{bd}^a A_\alpha^b \bar{\varepsilon}^d\big(b(x)\big)\end{aligned}$$

But $\bar{\varepsilon}^a\big(b(x)\big) = \varepsilon^a(x)$, so

$$D_\alpha \varepsilon^a(x) = d_\alpha \varepsilon^a(x) + C_{bd}^a A_\alpha^b \varepsilon^d(x).$$

2.8 Gauge transformations

Gauge transformations can be associated with change of cross section, i.e., a bundle automorphism. Given a connection A_c^a on a principle bundle $B \overset{\pi}{\rightarrow} M$, we want the change in A_c^a arising from an automorphism:

$$\hat{g}: B \rightarrow B$$

where

$$\hat{g}(b) = bg(b) = bg(x), \qquad x \subset \pi(b)$$

(Eqn. 2-128)

and where the latter leaves the base space fixed, assumes base $\subset B$.

For spacetime independent

$$\hat{g} = R_g \;\; and \;\; \big(R_g^* A_c\big)\dot{b}^c = A_c^a\big(R_{g*}\dot{b}^c\big)\,.$$

(Eqn. 2-139)

$\dot{b}^c$ horizontal $\rightarrow R_{g*}\dot{b}^c$ horizontal $\rightarrow A_c R_{g*}\dot{b}^c = 0$

$\dot{b}^c$ vertical $\rightarrow$ a tangent to $b(\lambda) = bh(\lambda)$ with $h(\lambda)$ a path through $e \in G$.

So, $R_{g*}\dot{b}^c$ tangent at bg to $bh(\lambda)g = b_g g^{-1}h(\lambda)g$. Thus $R_{g*}\dot{b}^c$ can be associated with tangent at $e \in G$ to path $g^{-1}h(\lambda)g$:

$$\frac{d}{d\lambda}[g^{-1}h(\lambda)g] = g^{-1}\dot{h}g$$

(Eqn. 2-130)

Since $g^{-1}\dot{h}g$ is a Lie algebra valued *1*-form, we need only regard the associated vector in G:

$$A_c^a\big(R_{g*}\dot{b}^c\big) = g^{-1}\dot{h}g \;\;\rightarrow\;\; R_g^* A_c = g^{-1}A_c g$$

(Eqn. 2-131)

For spacetime dependent

$\hat{g}_*$ will no longer take horizontal vectors to horizontal, so the analysis becomes more involved:

$$\hat{g}^*\overline{A}_c\dot{b}^c = \overline{A}_c\hat{g}_*\dot{b}^c$$

(Eqn. 2-132)

For $\dot{b}^c$ tangent at b_0 to $b(\lambda) \to \hat{g}_*\dot{b}^c$ tangent at $bg\ to\ b(\lambda)g[b(\lambda)]$:

$$\frac{d}{d\lambda}b(\lambda)g[b(\lambda)]|_{\lambda=0} = \frac{d}{d\lambda}b(\lambda)g + \frac{d}{d\lambda}b_g\big(b(\lambda)\big)\Big|_{\lambda=0}$$

(Eqn. 2-133)

where $\frac{d}{d\lambda}b(\lambda)g$ is $R_{g*}\dot{b}$ and $\frac{d}{d\lambda}bg\big(b(\lambda)\big)\Big|_{\lambda=0} = \frac{d}{d\lambda}bg\left(g^{-1}g\big(b(\lambda)\big)\right)$ and $g^{-1}g\big(b(\lambda)\big)$ is a path through the identity in G.

Since $\frac{d}{d\lambda}\left(g^{-1}g\big(b(\lambda)\big)\right) = g^{-1}d_c g\dot{b}^c$, we have

$$\overline{A}_c\big(\hat{g}_*\dot{b}^c\big) = \overline{A}_c\big(R_{g*}\dot{b}^c\big) + g^{-1}d_c g\dot{b}^c$$

(Eqn. 2-134)

And we get

$$\hat{g}^*\overline{A}_c = g^{-1}\overline{A}_c g + g^{-1}d_c g\ .$$

(Eqn. 2-135)

If we apply to bundle of frames with $\vec{e}'_\mu = \vec{e}_\nu\Lambda^\nu_\mu(x),\ \omega^\nu_\mu = e^*\overline{\omega}^\nu_\mu,\ \ \omega^\nu_\mu(\dot{x}) = \Gamma^\nu_{\mu\alpha}\dot{x}^\alpha$, we get:

$$e'^*\overline{\omega}^\nu_{\mu c} = \Lambda^{-1}\overline{\omega}_c\Lambda + \Lambda^{-1}d_c\Lambda$$

(Eqn. 2-136)

$$\hat{\Lambda}^*\overline{\omega}^\mu_\nu = \Lambda^\mu_\sigma\overline{\omega}^\sigma_\tau\Lambda^\tau_\nu + \Lambda^\mu_\sigma d\Lambda^\sigma_\nu$$

(Eqn. 2-137)

$$\Gamma^{\mu r}_{\nu\lambda}\vec{e}^\lambda = \Lambda^\mu_\sigma\Gamma^\sigma_{\tau\varphi}\vec{e}^\varphi\Lambda^\tau_\nu + \Lambda^\mu_\sigma d_\tau\Lambda^\sigma_\nu\vec{e}^\tau \text{ (using } \vec{e}^\lambda = e^\tau\Lambda^\lambda_\tau\text{)}$$

(Eqn. 2-138)

Thus,

$$\Gamma^\mu_{\nu\lambda} = \Lambda^\mu_\sigma\Lambda^\tau_\nu\Lambda^\varphi_\lambda\Gamma^\sigma_{\tau\varphi} + \Lambda^\mu_\sigma\Lambda^\varphi_\lambda d_\varphi\Lambda^\sigma_\nu$$

(Eqn. 2-139)

which is the law for change of Γ's under a change of orthonormal frame.

Example 2.14. (a) Let h_{ab} be a tensor on g (equivalently, a left-invariant tensor field on G). Show that $h_{ab}\varepsilon^a\eta^b$ is invariant under maps A_g. (b) Deduce that $g_{bc}D_\alpha\varepsilon^b D^\alpha\varepsilon^c$ is invariant under gauge transformations. (c)

Extend the definition of gauge-covariant derivative to an arbitrary field, say,
$T^{a...b\,\alpha...\beta}_{c...d\,\gamma...\delta}$ with space-time and Lie-algebra indices.

Solution

(a) $A_g\colon \varepsilon(x) \to g(x)^{-1}\varepsilon(x)g(x)$

So $A_g\colon h_{ab}\varepsilon^a\eta^b \to h_{ab}g(x)^{-1}\varepsilon^a(x)g(x)g^{-1}(x)\eta^b(x)g(x)$

$$= h_{ab}g(x)^{-1}\varepsilon^a(x)\eta^b(x)g(x)$$

$$= g(x)^{-1}\left[\underbrace{h_{ab}\varepsilon^a(x)\eta^b(x)}_{group\ scalar}\right]g(x) =$$

$$[h_{ab}\varepsilon^a(x)\eta^b(x)]\underbrace{g(x)^{-1}g(x)}_{1}$$

Thus, $A_g[h_{ab}\varepsilon^a\eta^b] = h_{ab}\varepsilon^a\eta^b$

(b) $g_{bc}D_\alpha\varepsilon^b D^\alpha\varepsilon^c = g_{bc}\left(d_\alpha\varepsilon^b + C^b_{ed}A^e_\alpha\varepsilon^d\right)\left(d^\alpha\varepsilon^c + C^c_{ed}A^{e\alpha}\varepsilon^d\right)$

$$= g_{bc}d_\alpha\varepsilon^b d^\alpha\varepsilon^c + g_{bc}\left[C^b_{ed}A^e_\alpha\varepsilon^d d^\alpha\varepsilon^c + C^c_{ed}A^{e\alpha}\varepsilon^d d_\alpha\varepsilon^b\right] +$$

$$g_{bc}C^b_{ed}A^e_\alpha\varepsilon^d C^c_{fg}A^{f\alpha}\varepsilon^g$$

$$= g_{bc}d_\alpha\varepsilon^b d^\alpha\varepsilon^c + [C_{ced}A^e_\alpha\varepsilon^d d^\alpha\varepsilon^c + C_{ced}A^e_\alpha\varepsilon^d d^\alpha\varepsilon^c] +$$

$$C_{ced}C^c_{fg}A^e_\alpha A^{f\alpha}\varepsilon^d\varepsilon^g$$

$$= g_{bc}d_\alpha\varepsilon^b d^\alpha\varepsilon^c + \underbrace{(2C_{ced}A^e_\alpha)}_{B^\alpha_{cd}}(d^\alpha\varepsilon^c)\varepsilon^d + \underbrace{\left(C_{ced}C^c_{fg}A^e_\alpha A^{f\alpha}\right)}_{D_{dg}}\varepsilon^d\varepsilon^g$$

↑ $F_{bc} = g_{bc}X(d_\alpha, d^\alpha)$ ↑ ↑

↑ a tensor on $\mathbb{G}$ an α-valued tensor on $\mathbb{G}$ a tensor on $\mathbb{G}$

Thus the expression is invariant under gauge transformation from previous problem.

(c) $D_\mu T^{a...b\,\alpha...\beta}_{c...d\,\gamma...\delta} =$

$$\underbrace{\nabla_\mu\left(T^{a...b\,\alpha...\beta}_{c...d\,\gamma...\delta}\right)}_{\substack{the\ usual\ cov.spacetime\\ derivative\ acting\ on\\ \alpha\to\delta\ indices}} + C^a_{ef}A^e_\mu T^{f...b}_{c...d} + \cdots + C^b_{ef}A^e_\mu T^{a...f}_{c...d}$$

$$-C^f_{ec}A^e_\mu T^{a...b}_{f...d} \cdots - C^f_{ed}A^e_\mu T^{a...b}_{c...f}$$

↗

Making use of $w^i = -C^i_{jk}w^j w^k$ for Cartesian basis, this just requires a change in the appropriate sign in the $D_c\overline{w}_a$ analysis.

2.9 Curvature

Let's now consider the Curvature.

Definition: Horizontal projection h. The tensor h^C_D is defined by $h^C_D\xi^D = \xi^C\,, if\ \xi$ horizontal, and $h^C_D\xi^D = 0,\ if\ \xi$ vertical.

If σ^a_C is a l-form, its horizontal part is hor $\sigma^a_C = h^D_C\sigma^a_D$. A horizontal form thus kills vertical vectors:

$$(\text{hor}\ \sigma^a_C)\xi^C = 0\,, if\ \ \xi \text{ vertical}$$

Exterior derivative

Let's now define the exterior derivative of Lie algebra valued forms: Let f^a be a Lie algebra valued scalar on B and let ξ^C be a vector field on B tangent to $b(\lambda)$.

Definition: $d_C f^a$ is the l-form given by

$$\xi^C d_C f^a = \frac{d}{d\lambda} f^a : b(\lambda)]_{\lambda=0}$$

(Eqn. 2-140)

(Since $f^a(b)$ is a vector in G, then $f^a(b_2)\cdot f^a(b_1)$ is a vector in G and the RHS is a well defined vector in G.)

Definition: If w^a_C is a l-form $(dw)^a_{CD}$ is the two form satisfying

$$(dw)^a_{CD}\xi^C\eta^D = d_C(w^a_D\eta^D)\xi^C - d_D(w^a_C\xi^C)\eta^D - w^a_C[\xi,\eta]^C$$

(Eqn. 2-141)

In a chart b^I (and basis $\vec{e}_i$ for G), taking $\vec{\xi} = \frac{\partial}{\partial b^{I'}}\ \vec{\eta} = \frac{\partial}{\partial b^{J'}}$ we have:

$$(dw)^i_{IJ} = \partial_I w^i_J - \partial_J w^i_J.$$

(Eqn. 2-142)

One extends to p-forms by Leibnitz: in a chart b^I, a basis $\vec{e}_i$ for the exterior derivative of a p-form is then

$$(d\sigma)^{i\ldots j}_{IJ\ldots K} = (p+1)\partial_{[I}\partial^{i\ldots j}_{J\ldots K]}$$

(Eqn. 2-143)

(you simply ignore the Lie algebra indices).

Definition: Let A^a_C be a connection on B. **The curvature** of A is a Lie algebra valued 2-form Ω^a_{CD} defined by

$$\Omega = \text{hor}\ dA$$

TM:

$$\Omega^a_{CD} = d_C A^a_D + C^a_{bc} A^b_C A^c_D$$

Proof: Look at $\Omega^a_{CD}\xi^C\eta^D$ for the 3 cases $\begin{cases} \xi, \eta \text{ vertical} \\ \xi, \eta \text{ horizontal} \\ \xi \text{ vertical}, \eta \text{ horizontal} \end{cases}$

Both vertical. Since $\Omega^a_{CD}\xi^C(b)\eta^D(b)$ depends only on ξ and η at b, we can pick them arbitrarily elsewhere. $\xi^C(b)$ is vertical and so corresponds to a Lie algebra vector $\overline{\xi}^a = A^a_C\xi^C$. Extend ξ^C so that it is vertical everywhere and corresponds everywhere to the same vector $\overline{\xi}^a$. Similarly let η^C correspond to $\overline{\eta}^a$ everywhere. Then$(dA)^a_{CD}\xi^C\eta^D = d_C\left(\overline{\xi}^a\right)\eta^C - d_D(\overline{\eta}^a)\xi^D - C^a_{bc}\overline{\xi}^b\overline{\eta}^c$. But because $\overline{\xi}^a$ and $\overline{\eta}^a$ are constant Lie-algebra valued scalars,

$$d\overline{\xi}^a = 0, \quad d\overline{\eta}^a = 0.$$

Thus $(dA)^a_{CD}\xi^C\eta^D = C^a_{bc}\overline{\xi}^b\overline{\eta}^c$, and we get:

$$\Omega^a_{CD}\xi^C\eta^D = -C^a_{bc}\overline{\xi}^b\overline{\eta}^c + C^a_{bc}\overline{\xi}^b\overline{\eta}^c = 0$$

Both horizontal. We arrive at the same result, namely,

$$\Omega^a_{CD}\xi^C\eta^D = (dA)^a_{CD}\xi^C\eta^D\,,$$

but now with ξ vertical, η horizontal. We will again get $\Omega(\xi,\eta) = 0$. To see this, let ξ be the vertical vector field corresponding to a fixed $\overline{\xi}^a$ in G. Let η^C be horizontal. Then (ξ,η) is horizontal.

Proof: ξ^C generates the diffeos R_{a_λ} of B, where a_λ is the 1-parameter group tangent to $\overline{\xi}^a$ at $e \in G$. By the def of a connection, $R_{a_\lambda *}\eta$ is horizontal. Thus

$$\Omega^a_{CD} = (dA)^a_{CD}\xi^C\eta^D = 0$$

For an electromagnetic field, the curvature $\Omega_{CD} = d_C\overline{A}_D$ is a tensor on B whose pullback to M is the field tensor $F_{\mu\nu}$. [The 4 space-time dimensional $\overline{A}_C$ is, in a chart (η, X^μ), $\overline{A}_I = \left(1, -A_\mu\frac{Q}{\hbar c}\right)$]. $F_{\mu\nu}$ is gauge invariant or equivalently Ω_{CD} is invariant under $A_C \to \hat{g}^*A_C$ where, as before, $\hat{g}$ is the bundle automorphism corresponding to g(x). In the general case, Ω^a_{CD} is not gauge invariant, but it transforms simply under $\hat{g}$: Write Ω' for the curvature of $\hat{g}^* A_C$.

Theorem. $\Omega'_{CD} = g^{-1}\Omega_{CD}g$

Proof. $\Omega_{IJ} = \partial_I \bar{A}_J - \partial_J \bar{A}_I + [A_I, A_J]$, where $[A_I, A_J]^a = C^a_{bc}A^b_I A^c_J$.

$$\hat{g}^* A_I = g^{-1}A_I g + g^{-1}\partial_I g$$

Thus (using $\partial_I g^{-1} = -g^{-1}\partial_I g g^{-1}$):

$$\Omega'_{IJ} = \partial_I(g^{-1}A_J g + g^{-1}\partial_J g) - \partial_J(g^{-1}A_I g + g^{-1}\partial_I g) + [g^{-1}A_I g + g^{-1}\partial_I g,\ \ g^{-1}A_J g + g^{-1}\partial_J g]$$

$$\Omega'_{IJ} = g^{-1}(\partial_I A_J - \partial_J A_I + [A_I, A_J])g - g^{-1}\Omega_{IJ}g.$$

The pullback $F^a_{\alpha\beta} = \psi^*(\Omega^a_{CD})$ of Ω by a cross section ψ is the "Yang-Mills" field associated with A^a_α.

Example 2.15. Let G be a Lie group with Lie algebra g. Recall that given an element $g\epsilon G$, there is an automorphism, $A_g: G \to G$ that leaves the identity fixed: $A_g(a) \mapsto g^{-1}ag$.
(a) Regard G as a group of matrices and find the image under A_{g*} of an element ε^a of g.
(b) Let ζ be an element of g and let $g(\lambda) = \exp(\lambda\zeta)$. What is

$$\frac{d}{d\lambda}A_{g(\lambda)*}\varepsilon^a\ ?$$

Solution

We have $\begin{array}{l} A_g: G \to G \\ A_g(a) \mapsto g^{-1}ag \end{array}$ where the Lie group is G and the Lie algebra is $\mathbb{G}$.

Consider $\varepsilon^a \in \mathbb{G}$ which is tangent to the curve on the group manifold $a(\lambda), a \in G$, at the identity, i.e., $a(0) = e$. Then $A_{g*}\varepsilon^a$ is $\in \mathbb{G}$ which is tangent at $g^{-1}a(\lambda)g|_{\lambda=0} = g^{-1}eg = e$ to the path $\lambda \to g^{-1}a(\lambda)g$:

$$A_{g*}\varepsilon^a = \frac{d}{d\lambda}(g^{-1}a(\lambda)g)\Big|_{\lambda=0} = g^{-1}\dot{a}g$$

And, for G a group of matrices, where $\varepsilon^a \to \dot{a}^k_\ell = \frac{d}{d\lambda}g^{-1}a(\lambda)^k_\ell\Big|_{\lambda=0}$, we have:

$$\left(A_{g*}\varepsilon^a\right)^i_j = (g^{-1})^i_k \dot{a}^k_\ell g^\ell_j\ .$$

(b) $\zeta \in \mathbb{G}$ and $g(\lambda) = \exp(\lambda\zeta) \leftarrow$ different λ param. than above.

$$A_{g(\lambda)*}\varepsilon^a = g^{-1}\dot{a}g = \exp(-\lambda\zeta)\,\dot{a}\exp(\lambda\zeta)$$

$$\frac{d}{d\lambda}A_{g(\lambda)*}\varepsilon^a = -\zeta\exp(-\lambda\zeta)\,\dot{a}\exp(\lambda\zeta) + \exp(-\lambda\zeta)\,\dot{a}\exp(\lambda\zeta)\,\zeta$$

$$= \exp(-\lambda\zeta)\,[\dot{a},\zeta]\exp(\lambda\zeta)$$

$$= g^{-1}[\dot{a},\zeta]g$$

and since $\zeta = \frac{d}{d\lambda}g(\lambda)\Big|_{\lambda=0} = \dot{g}$, thus, $\frac{d}{d\lambda}A_{g(\lambda)*}\varepsilon^a = g^{-1}[\dot{a},\dot{g}]g$, and we have:

$$\left(\frac{d}{d\lambda}A_{g(\lambda)*}\varepsilon^a\right)^i_{\ j} = (g^{-1})^i_{\ k}[\dot{a},\dot{g}]^k_{\ \ell}\, g^\ell_{\ j}$$

Example 2.16. Let C^a_{bc} be the structure constant tensor of a Lie algebra g.
(a) Define a left- and right- invariant metric (the Killing metric) $h_{ab} = C^d_{ae}C^e_{bd}$. Use the Jacobi identity to show that C_{abc} is totally antisymmetric, where $C_{abc} = h_{ag}C^g_{bc}$.
(b) Find h_{ab} for so(3).

Solution
(a) We have $C_{abc} = h_{ag}C^g_{bc} = C^d_{ae}C^e_{gd}C^g_{bc}$. We also have the Jacobi identity:
$C^a_{e[b}C^e_{cd]} = 0 \ \rightarrow\ C^a_{eb}C^e_{cd} = C^a_{ec}C^e_{bd} - C^a_{ed}C^e_{bc} = C^a_{e[d}C^e_{c]b}$
So,

$$C_{abc} = C^d_{ae}C^e_{g[c}C^g_{b]d} = C^d_{ae}C^e_{gc}C^g_{bd} - C^d_{ae}C^e_{gb}C^g_{cd}$$

Thus,
$C_{abc} - C_{bac} = C^e_{gc}\left(C^d_{ae}C^g_{bd} - C^d_{be}C^g_{ad}\right) - C^g_{cd}\left(C^d_{ae}C^e_{gb} - C^d_{be}C^e_{ga}\right) = 2C_{abc}$

So, $C_{abc} = -C_{bac}$, now we merely generalize this result: $C_{abc} = -C_{acb} = C_{cab} = -C_{cba}$, thus C_{abc} is antisymmetric under exchange $a \leftrightarrow b, a \leftrightarrow c, b \leftrightarrow c$, thus C_{abc} is completely antisymmetric.

(b) $SO(3)$ *has* $C^a_{bc} = \epsilon^a_{bc}$ (basis in adjoint. rep), thus $h_{ab} = C^d_{ae}C^e_{bd} = \epsilon^d_{ae}\epsilon^e_{bd} = -2\delta_{ab}$.

2.10 Parallel transport: the gauge-covariant derivative.

The physical motivation for regarding $\vec{A}$ as a connection, as specifying a kind of parallel transport, came from quantum mechanics. Similar clarity regarding gauge transformations in general becomes apparent in quantum theory as well. Consider the Schrodinger equation for a charge Q in a magnetic field $B = \nabla \times A$ and potential well V is

$$\frac{1}{2m}\left(\frac{\hbar}{i}\nabla - \frac{Q}{c}\vec{A}\right)^2 \Psi + V\Psi = i\hbar\partial_t\Psi.$$

(Eqn. 2-144)

Suppose that V represents the potential of a box confining the particle to a region small enough that $QBR^2 \ll \hbar c$, where R is the radius of the region. If one picks an origin x_0 and writes

$$\Psi = \psi \exp\left(i\int_{x_0}^{x} A - dx\frac{Q}{\hbar c}\right),$$

(Eqn. 2-145)

then we get the corresponding "uncharged" equation:

$$-\frac{\hbar^2}{2m}\nabla^2\psi + V\psi = i\hbar\partial_t\psi.$$

(Eqn. 2-146)

Although the line integral above depends on the path from x_0 to x, for R small its value is independent of the (short) piece of the path that lies within the box. Let $\psi_E(x,t)$ be an energy eigenstate when the box is centered at $x = 0$. Suppose the box is slowly moved from x_1 to x_2, along a path $x(t)$, $V(x,t) = V_0[x - x(t)]$ with $\dot{x} \ll \left[\langle\frac{p^2}{m^2}\rangle\right]^{1/2}$. Then to first order in $\dot{x}$, $\psi_E[x - x(t), t]$ satisfies the above and when the box reaches x_2 this physical parallel transport has changed the uncharged wave function ψ from

$$\psi_E(x - x_1, t) \text{ to } \psi_E(x - x_2, t).$$

The integral equation above then implies that for the charge particle, described by Ψ, physical parallel-transport takes

$$\Psi(x - x_1, t) \text{ to } \Psi(x - x_2, t)\exp\left(i\int_{x_1}^{x_2} A - dx\frac{Q}{\hbar c}\right).$$

(Eqn. 2-147)

The line integral depends on the path from x_1 to x_2, and the fact that the box is small means that the "path" traveled by the box is well-defined – the trajectories of different points of the box are close enough to give the same change of phase. Then, under parallel-transport along a path $x(\lambda)$, the phase of a charged particle's wave function changes by

$$\exp\left(i \int A - dx \frac{Q}{\hbar c}\right).$$

(Eqn. 2-148)

The possible phases of Ψ at a point x are just the complex numbers of magniture 1, elements $e^{i\eta}$ of the unit complex circle. The space M, together with a circle at each point x of M is a circle bundle over M; so the space of possible phases of Ψ is a circle bundle B, and the phase $\eta(x)$, defined by

$$\Psi(x) = |\Psi(x)|e^{i\eta(x)}$$

(Eqn. 2-149)

is a cross section. A gauge transformation, $A \rightarrow A + \nabla f,\ \ \eta \rightarrow \eta + \frac{Q}{\hbar c} f$, can, insofar as η is concerned, be regarded as a change of cross section. In which case the change in phase above becomes:

$$\frac{d\eta}{d\lambda} - \frac{Q}{\hbar c}\vec{A} - \frac{d\vec{x}}{d\lambda} = 0$$

(Eqn. 2-150)

If we define a vector field $\overline{A}_C$ on B by $\overline{A}_\eta = 1, \quad \overline{A}_i = -\frac{Q}{\hbar c} A_i$, or

$$\overline{A} = d\eta - \frac{Q}{\hbar c} A\,,$$

(Eqn. 2-151)

then parallel transport on the bundle has

$$\overline{A}_C \dot{x}^C = 0,$$

(Eqn. 2-152)

where the horizontal vector $\dot{x}^C$ has components $(\dot{\eta}, \dot{x}^i)$.

To summarize: When a charge particle is physically transported in a background magnetic field, the change in phase of the particle's wave function can be described as parallel transport on a principal U(1) bundle with connection given by the above equation.

Finally consider the result of a bundle automorphism given by $\eta'(x) = \eta(x) + \alpha(x)$. With respect to the new coordinates $(\eta', x'^i) = (\eta + \alpha, x^i)$, the components of A are given by

$$\overline{A}_{i'} = \overline{A}_i - \partial_i \alpha$$
$$\overline{A}_{\eta'} = \overline{A}_\eta.$$

The change in the physical potential A is given by

$$A' = A + \frac{\hbar c}{Q}\nabla_{\alpha},$$

(Eqn. 2-153)

which is the gauge transformation associated with a change of phase α in the wave function of a charge particle.

2.11 EM and Y-M Examples

Let's now consider some examples that help to clarify the theoretical framework, for both EM and Y-M seen as gauge fields, that has been developed and adopted thus far.

Example 2.17. Show Maxwell's equations in empty space can be written in the form:

$$dF = 0 \quad and \quad d^{*}F = 0,$$

(Eqn. 2-154)

where

$$^{*}F^{ab} = \frac{1}{2}\epsilon_{abcd}F^{cd}.$$

(Eqn. 2-155)

Solution

$F_{\alpha\beta}$ is defined by:

$$F_{\alpha\beta} = \begin{array}{c} \\ t \\ x \\ y \\ z \end{array} \begin{array}{c} \begin{array}{cccc} t & x & y & z \end{array} \\ \begin{bmatrix} 0 & -E_x & -E_y & -E_z \\ E_x & 0 & B_z & B_y \\ E_y & -B_z & 0 & B_x \\ E_z & B_y & -B_x & 0 \end{bmatrix} \end{array}$$

Also,

$$\bar{F} = \frac{1}{2}F_{\alpha\beta}dx^{\alpha} \wedge dx^{\beta} = E_x dx \wedge dt + E_y dy \wedge dt + E_z dz \wedge dt + B_x dy \wedge dz + B_y dz \wedge dx + B_z dx \wedge dy$$

and

$$d\bar{F} = d(E_x dx \wedge dt + \cdots + B_z dx \wedge dy)$$
$$= \left(\frac{\partial E_x}{\partial t}dt + \frac{\partial E_x}{\partial x}dx + \frac{\partial E_x}{\partial y}dy + \frac{\partial E_z}{\partial z}dz\right) \wedge dx \wedge dt + \cdots$$
$$= \frac{\partial E_x}{\partial y}dy \wedge dx \wedge dt + \frac{\partial E_x}{\partial z}dz \wedge dx \wedge dt + \text{other E's similarly}$$
$$+ \left(\frac{\partial B_x}{dt}dt + \frac{\partial B_x}{dx}dx\right) \wedge dy \wedge dz + \text{other B's similarly}$$

$$= dy \wedge dx \wedge dt\left(\frac{\partial E_x}{\partial y} + \frac{\partial E_y}{\partial x} - \frac{\partial B_z}{\partial t}\right) + \text{cycles}$$

$$+dx \wedge dy \wedge dz\left(\underbrace{\frac{\partial B_x}{\partial x} + \frac{\partial B_y}{\partial y} + \frac{\partial B_z}{\partial z}}_{\nabla \cdot B}\right)$$

$$= (\nabla \cdot B)(dx \wedge dy \wedge dz) + \left(\frac{\partial E_y}{\partial x} - \frac{\partial E_x}{\partial y} + \frac{\partial B_z}{\partial t}\right)(dt \wedge dx \wedge dy)$$

$$+\left(\frac{\partial E_z}{\partial y} - \frac{\partial E_y}{\partial z} + \frac{\partial B_x}{\partial t}\right)(dt \wedge dy \wedge dz) + \left(\frac{\partial E_x}{\partial z} - \frac{\partial E_z}{\partial x} + \frac{\partial B_y}{\partial t}\right)(dt \wedge dz \wedge dx)$$

$$= (\nabla \cdot B)(dx \wedge dy \wedge dz) + \left(\nabla \times \vec{E} + \dot{\vec{B}}\right)_z (dt \wedge dx \wedge dy) + \left(\nabla \times \vec{E} + \vec{B}\right)_x (dt \wedge dy \wedge dz)$$

$$+\left(\nabla \times \vec{E} + \vec{B}\right)_y (dt \wedge dz \wedge dzx)$$

If we make use of Maxwell's equation's:

$$\nabla \cdot B = 0$$

$$\nabla \times \vec{E} = -\dot{\vec{B}}\ ,$$

we then see that:

$$d\bar{F} = 0.$$

${}^*F_{\alpha\beta}$ is defined by: ${}^*F^{ab} = \frac{1}{2}\epsilon_{abcd}F^{cd}$:

$$ {}^*F_{\alpha\beta} = \begin{bmatrix} 0 & B_x & B_y & B_z \\ -B_x & 0 & +E_z & -E_y \\ -B_y & -E_z & 0 & +E_x \\ -B_z & +E_y & -E_x & 0 \end{bmatrix},$$

Thus,

$$ {}^*\bar{F} = -B_x dx \wedge dt - B_y dy \wedge dt - B_z dz \wedge dt + E_x dz \wedge dy + E_y dx \wedge dz + E_z dy \wedge dx$$

and we have:

$$d{}^*\bar{F} = -\left(\frac{\partial B_x}{\partial y} dy + \frac{\partial B_x}{\partial z} dz\right) \wedge dx \wedge dt + \cdots$$

$$d{}^*\bar{F} = (\nabla \cdot E)(dx \wedge dy \wedge dz) + \left(\nabla \times \vec{B} + \dot{\vec{E}}\right)_z (dt \wedge dx \wedge dy)$$

$$+\left(\nabla \times \vec{B} + \dot{\vec{E}}\right)_x (dt \wedge dy \wedge dz) + \left(\nabla \times \vec{B} + \dot{\vec{E}}\right)_y (dt \wedge dz \wedge dzx)$$

Now the other Maxwell relations are $\nabla \cdot E = 4\pi\rho$ and $\frac{\partial \vec{E}}{\partial t} - \nabla \times B = 4\pi J$.
So,

$$d^*F = 4\pi\rho(dx \wedge dy \wedge dz) + 4\pi J_z(dt \wedge dx \wedge dy) - 4\pi J_x(dt \wedge dy \wedge dz) - 4\pi J_y(dt \wedge dz \wedge dzx)$$

and

$$d^*F = 4\pi^*J$$

Notice $d^{2*}F = 0 \Rightarrow \frac{\partial\rho}{\partial t} - \Delta \cdot J = 0$ the continuity equation. Inn empty space $\rho = 0, \vec{J} = 0$ and we get

$$d^*F = 0.$$

Example 2.18. Write Maxwell's equations in a background Schwarzschild space-time with coordinates (t, r, θ, φ) and metric:

$$g_{ab} = -e^{2v}dt^2 + e^{2\lambda}dr^2 + r^2(d\theta^2 + \sin^2\theta\, d\phi^2).$$

Solution

Instead of Minkowski: $g_{ab} = -dt^2 + dx^2 + dy^2 + dz^2$ and a two form $\bar{F}$ with components:

$$F_{\alpha\beta} = \begin{bmatrix} 0 & -E_x & -E_y & -E_z \\ E_x & 0 & B_z & -B_y \\ E_y & -B_z & 0 & B_x \\ E_z & B_y & -B_x & 0 \end{bmatrix},$$

we now have metric

$$\begin{aligned} g_{ab} &= -e^{2v}dt^2 + e^{2\lambda}dr^2 + r^2(d\theta^2 + \sin^2\theta\, d\phi^2) \\ &= -d\tau^2 + dR^2 + d\Theta^2 + d\Phi^2, \end{aligned}$$

where $d\tau = e^v dt, dR - e^\lambda dr, d\Theta = rd\theta, d\Phi = r\sin\theta\, d\phi$ and the components of our two form are now written

$$F_{\alpha\beta} = \begin{bmatrix} 0 & -E_r & -E_\theta & -E_\phi \\ E_r & 0 & B_\phi & -B_\theta \\ E_\theta & -B_\phi & 0 & B_r \\ E_\phi & B_\theta & -B_r & 0 \end{bmatrix}$$

Now $d\bar{F} = 0$ is expressed by:

$$d\bar{F} = 0 \Rightarrow \begin{cases} \nabla \cdot B = 0 \\ \nabla \times \vec{E} = -\dot{\vec{B}} \end{cases}$$

modified by e^λ and e^v terms from usual spherical coordinate form. We have to modify Maxwell's eqn's for spherical coordinates $\{\hat{e}_r, \hat{e}_\theta, \hat{e}_\phi\} \Rightarrow \{e^\lambda\hat{e}_r, \hat{e}_\theta, \hat{e}_\phi\}, \frac{\partial}{\partial r} \Rightarrow e^{-\lambda}\frac{\partial}{\partial r}, \frac{\partial}{\partial \tau} = e^{-v}\frac{\partial}{\partial t}$, so:

$$d\bar{F} = 0 \Rightarrow \begin{cases} "\nabla \cdot B = 0" \\ "\nabla \times \vec{E} = -\dot{\vec{B}}" \end{cases} \Rightarrow \quad \frac{1}{r^2} e^{-\lambda} \frac{\partial}{\partial r}(r^2 B_r)$$

$$+ \frac{1}{r \sin\theta} \frac{\partial}{\partial \theta}(\sin\theta\, B_\theta) + \frac{1}{r \sin\theta} \frac{\partial}{\partial \phi}(B_\phi) = 0$$

So,

$$\hat{e}_r(e^{\lambda}) \frac{1}{r \sin\theta}\left[\frac{\partial}{\partial \theta}(\sin\theta\, E_\phi) - \frac{\partial E_\theta}{\partial \phi}\right]$$

$$+ \hat{e}_\theta \left[\frac{1}{r \sin\theta} \frac{\partial E_r}{\partial \phi} - \frac{e^{-\lambda}}{r} \frac{\partial}{\partial r}(rE_\phi)\right]$$

$$+ \hat{e}_\phi \frac{1}{r}\left[e^{-\lambda} \frac{\partial}{\partial r}(rE_\theta) - \frac{\partial E_r}{\partial \theta}\right]$$

$$= -e^{-v} \frac{\partial \vec{B}}{\partial t}$$

$d^{*}F = 0$ ∇·E=0

$$: \quad \frac{1}{r^2} e^{-\lambda} \frac{\partial}{\partial r}(r^2 E_r) + \frac{1}{r \sin\theta} \frac{\partial}{\partial \theta}(\sin\theta\, E_\theta)$$

$$+ \frac{1}{r \sin\theta} \frac{\partial}{\partial \phi}(E_\phi) = 0$$

While for "$\nabla \times \vec{E} = -\dot{\vec{B}}$" $\Rightarrow$ same as above with $E \to B$ and $B \to -E$. We get a wave equation:

$$\frac{\partial^2 \vec{E}}{\partial \tau^2} = \nabla^2 E$$

or

$$\left(e^{-v} \frac{\partial}{\partial t}\right)\left(e^{-v} \frac{\partial}{\partial t}\right)\vec{E}$$

$$= \left[\frac{1}{r^2} e^{-\lambda} \frac{\partial}{\partial r}\left(r^2 e^{-\lambda} \frac{\partial}{\partial r}\right) + \frac{1}{r^2 \sin\theta} \frac{\partial}{\partial \theta}\left(\sin\theta \frac{\partial}{\partial \theta}\right)\right.$$

$$\left. + \frac{1}{r^2 \sin^2\theta} \frac{\partial^2}{\partial \phi^2}\right]\vec{E}.$$

Example 2.19. Find the angular momentum of a dyon: an electric charge e and a magnetic charge g at rest on the z-axis. Using the relation

$$L = \frac{1}{4\pi} \int dV\, \vec{\phi} \cdot \vec{E} \times \vec{B} = \frac{1}{4\pi} \int dV\, \varepsilon_{\alpha\beta\gamma} B^{\alpha} \phi^{\beta} E^{\gamma}\,,$$

show that $\varepsilon_{\alpha\beta\gamma}\phi^\beta E^\gamma = -e\nabla_\alpha \cos\theta$, and use Gauss' theorem to obtain $L = \pm eg$, where he sign depends on whether e is above or below g.

Solution

We have $\vec{B} = g\frac{\vec{n}'}{(r')^2}$ $\vec{E} = e\frac{\vec{n}}{r^2}$, for starters. The electromagnetic momentum density is: $\vec{P} = \frac{(\vec{E}\times\vec{B})}{4\pi}$. Total momentum $= \int dV\,\vec{P}$. The only vector available is $\vec{R}$ so $\vec{P} \propto \vec{R}$, so total momentum $= \frac{\vec{R}}{|R|}\int dV \frac{\vec{R}}{|R|}\cdot$ $(\vec{E}\times\vec{B})$ where $\vec{R}\cdot(\vec{n}\times\vec{n}') = 0$!, so the total momentum = 0. With the total vanishing this considerably simplifies matters, the angular momentum of the e-m field is then independent of choice of origin. So the situation to be studied above will give an unambiguous result.

Angular momentum in e-m field: $\vec{L} = \frac{1}{4\pi}\int \vec{x}\times(\vec{E}\times\vec{B})d^3x$

Once again, the only vector valued quantity from the geometric configuration above is $\vec{R}$, so $\vec{L}\propto\vec{R}$, thus we need only consider the value of $L_z = |\vec{L}| = L$. Also, in terms of the rotational K.V. field ϕ.

$$L = \frac{1}{4\pi}\int dV\,\vec{\phi}\cdot\vec{E}\times\vec{B} = \frac{1}{4\pi}\int dV\,\varepsilon_{\alpha\beta\gamma}B^\alpha\dot{\phi}^\beta E^\gamma$$

So, we need to simplify $\varepsilon_{\alpha\beta\gamma}\phi^\beta E^\gamma$, and for the configuration chosen $\vec{E} = \frac{e}{r^2}\hat{r}$, so lets use spherical coordinates, for which the volume element is $r^2\sin\theta$. Also, ϕ^β has only a $\beta = \phi$ component: $\phi^\phi = \varepsilon_{\alpha\beta\gamma}\phi^\beta E^\gamma =$ $\varepsilon_{\alpha\beta\gamma}\phi^\beta\left(\frac{e}{r^2}\right) = \varepsilon_{\alpha\phi r}\left(\frac{1}{r}\right)\left(\frac{e}{r^2}\right)$ (and in the latter, only the $\alpha = \theta$ component is nonzero).

$\varepsilon_{\theta\phi r}(r^{-1})\left(\frac{e}{r^2}\right) = r^{-1}e\sin\theta = -e\nabla_\alpha\cos\theta$ for $\alpha = \theta$.

So, $\varepsilon_{\alpha\beta\gamma}\phi^\beta E^\gamma = -e\nabla_\alpha\cos\theta$ for the geometry shown.

Now consider:

$$L = \frac{1}{4\pi}\int dV\,B^\alpha(-e\nabla_\alpha\cos\theta) = \frac{1}{4\pi}\int dV\,\nabla_\alpha B^\alpha\cos\theta \quad where \quad \nabla_\alpha B^\alpha = \nabla\cdot B = 4\pi g\delta(\vec{x}-\vec{R}).$$

If g is above e on the z-axis then L = eg; if g is below then L = –eg since $\cos(\pi) = -1$. Since we knew beforehand that $\vec{L} \propto \vec{R}$ this then allows us to generalize the result to:

$$\vec{L}_{em} = eg\frac{\vec{R}}{|R|}$$

where $\vec{R}$ is the vector from e to g.

Example 2.20. Define the Poisson Bracket for electromagnetism, and show that the canonical transformations generated by the Gauss constraint,

$$\int_\Sigma dx\, f\nabla_\alpha E^\alpha = 0,$$

are a family of gauge transformations. Hence Σ is a spatial plane of Minkowski space and the smearing function f is smooth and vanishes at spatial infinity.

Solution

The Lagrangian for EM is:

$$L_{EM} = -\frac{1}{4}F_{\alpha\beta}F^{\alpha\beta} = -\frac{1}{4}(\partial_\alpha A_\beta - \partial_\beta A_\alpha)(\partial^\alpha A^\beta - \partial^\beta A^\alpha)$$
$$= -\frac{1}{2}(\partial_\alpha A_\beta)(\partial^\alpha A^\beta) + \frac{1}{2}(\partial_\alpha A_\beta)(\partial^\beta A^\alpha).$$

We take our canonical coordinates to be the vector potential A^α evaluated on the hypersurface Σ. For this reason we decompose it into normal and tangential parts:

$$V = -A_\alpha n^\alpha \;\; and \;\; {}^{(3)}A_\alpha = h_\alpha^\beta A_\beta = \vec{A}$$

where $h_{\alpha\beta} = \eta_{\alpha\beta} + n_\alpha n_\beta$.

$$\partial_\alpha A_\beta = \partial_\alpha\left(\eta_\beta^\gamma A_\gamma\right) = \partial_\alpha\left(h_\beta^\gamma - n_\beta n^\gamma\right)A_\gamma = \partial_\alpha(\vec{A} + \vec{n}V)_\beta$$

Thus

$$\eta_\alpha^\sigma \partial_\sigma = (h_\alpha^\sigma - n_\alpha n^\sigma)\partial_\sigma = h_\alpha^\sigma \partial_\sigma - n_\alpha n^\sigma \partial_\sigma = \left(\vec{\nabla} - \vec{n}\partial_{\vec{n}}\right)_\alpha$$

and

$$\partial_\alpha A_\beta = ({}^{(3)}\nabla_\alpha - n_\alpha n^\sigma \partial_\sigma)({}^{(3)}A_\alpha + n_\beta V).$$

The Lagrangian density is thus:

$$L_{EM} = -\frac{1}{4}\left(\nabla_\alpha A_\beta - \nabla_\beta A_\alpha + (n_\beta \nabla_\alpha - n_\alpha \nabla_\beta)V - n_\alpha \dot{A}_\beta + n_\beta \dot{A}_\alpha\right)^2$$

$$\left(\vec{\nabla}\times\vec{A}\right)^i = \varepsilon_{ijk}\nabla^j A^k = \left(\nabla^j A^k - \nabla^k A^j\right)^i$$
$$(\nabla\times A)\cdot(\nabla\times A) = \varepsilon_{ijk}\varepsilon^{\ell}\left(\nabla^j A^k - \nabla^k A^j\right)\left(\nabla_\ell \mathrm{A}_m - \nabla_m A_\ell\right)$$
$$= 2\left(\nabla^j A^k - \nabla^k A^j\right)\cdot\left(\nabla^j A^k - \nabla^k A^j\right)$$

So,

$$L_{EM} = -\frac{1}{2}\left(\vec{\nabla}\times\vec{A}\right)\cdot\left(\vec{\nabla}\times\vec{A}\right) = \frac{1}{4}(\cdots)^2$$

where

$$\left(n_\beta\dot{A}_\alpha - n_\alpha\dot{A}_\beta\right)^2 = -2\dot{\vec{A}}^2 - 2n_\beta\dot{A}_\alpha n^\alpha\dot{A}^\beta$$
$$\leftarrow\begin{cases}\dot{A}_\alpha = \dot{h}_\alpha^\beta A_\beta + h_\alpha^\beta\dot{A}_\beta\\ so\; n^\alpha\dot{A}_\alpha = \left(n^\alpha h_\alpha^\beta\right)\dot{A}_\beta = 0\end{cases}$$

And

$$\left(n_\beta\dot{A}_\alpha - n_\alpha\dot{A}_\beta\right)\left(n^\beta\nabla^\alpha V - n^\alpha\nabla^\beta V\right) = -2\dot{A}_\alpha\nabla^\alpha V$$

Thus,

$$L_{EM} = -\frac{1}{2}\left(\vec{\nabla}\times\vec{A}\right)\cdot\left(\vec{\nabla}\times\vec{A}\right) + \frac{1}{2}\left(\dot{\vec{A}} + \vec{\nabla}\mathrm{V}\right)\cdot\left(\dot{\vec{A}} + \vec{\nabla}\mathrm{V}\right).$$

So,

$$\vec{\pi} = \frac{\partial L}{\partial\dot{\vec{A}}} = \left(\dot{\vec{A}} + \vec{\nabla}\mathrm{V}\right)$$

Recall that with $\vec{\mathrm{B}} = \vec{\nabla}\times\vec{A}$, thus $\vec{\nabla}\times\vec{E} + \frac{\partial\vec{B}}{\partial t} = 0$ becomes $\vec{\nabla}\times\left(E + \frac{\partial\vec{A}}{\partial t}\right) = 0$, thus $E + \frac{\partial\vec{A}}{\partial t} = -\vec{\nabla}\mathrm{V}$, for some scalar potential V, but this is the same V given here (aside from gauge trans.) arrived at from the covariant Lagrangian. So,

$$\dot{\vec{A}} + \vec{\nabla}\mathrm{V} = -\vec{E}$$

So, the canonically conjugate momenta conjugate to $\vec{A}$ is

$$\pi_{\vec{A}} = -\vec{E}$$

Trivially $\pi_V = 0$.

Notice,

$$H_{EM} = \vec{\pi}\cdot\dot{\vec{A}} - L_{EM}$$
$$= \vec{\pi}\cdot\vec{\pi} - \vec{\pi} - \vec{\nabla}\mathrm{V} + \frac{1}{2}\vec{B}\cdot\vec{B} - \frac{1}{2}\vec{\pi}\cdot\vec{\pi}$$
$$= \frac{1}{2}\vec{\pi}\cdot\vec{\pi} + \frac{1}{2}\vec{B}\cdot\vec{B} + \mathrm{V}\vec{\nabla}\cdot\vec{\pi} - \vec{\nabla}\cdot(V\vec{\pi})$$

Now consider

$$\{f, g\}_{EM} = \int_S d^3x \left(\frac{\delta f}{\delta \pi^\alpha(x)} \frac{\delta g}{\delta A_\alpha(x)} - \frac{\delta g}{\delta A_\alpha(x)} \frac{\delta f}{\delta \pi^\alpha(x)} \right)$$

$$\{f, g\}_{EM} = \int_S d^3x \left(\frac{\delta f}{\delta A_\alpha} \frac{\delta g}{\delta E^\alpha} - \frac{\delta f}{\delta E^\alpha} \frac{\delta g}{\delta A_\alpha} \right)$$

Now,

$$\left\{ \int_\Sigma d^3x\, f \nabla_\alpha E^\alpha, E^\alpha(x) \right\} = \frac{\delta}{\delta A_\alpha} \left\{ \int_\Sigma^{\beta x} f \nabla_\alpha E^\alpha \right\} = 0$$

and

$$\{\textstyle\int_\Sigma d^3x\, f \nabla_\alpha E^\alpha, A_\alpha(x)\} = -\frac{\delta}{\delta E^\alpha} \{\int_\Sigma d^3x f \nabla_\alpha E^\alpha\} =$$

$$-\textstyle\int_\Sigma dx\, f \frac{\delta}{\delta E^\alpha} (\nabla_\alpha E^\alpha) = -\int_\Sigma dx\, f \nabla_\alpha \left(\frac{\delta E^\alpha}{\delta E^\alpha} \right)$$

$$= +\textstyle\int_\Sigma dx\, \nabla_\alpha f \delta(x) = \nabla_\alpha f(x)$$

So,

$$\{\textstyle\int_\Sigma d^3x f \nabla_\alpha E^\alpha, F(A, E)\} = \int dy \left[\frac{\delta P_f}{\delta A_\alpha} \frac{\delta F}{\delta E^\alpha} - \frac{\delta P_f}{\delta E_\alpha} \frac{\delta F}{\delta A} \right] =$$

$$\textstyle\int dy \left[+\nabla_\alpha f(x) \frac{\delta F}{\delta A_\alpha} \right]$$

$$\{\textstyle\int_\Sigma d^3x f \nabla_\alpha E^\alpha, F(A, E)\} = -\frac{d}{d\lambda} F(\Psi_\lambda A_\alpha E^\alpha)|_{\lambda=0}$$

and using $\Psi_\lambda A_\alpha = A_\alpha + \nabla_\alpha f_\lambda$. Since $\int_\Sigma d^3x f \nabla_\alpha E^\alpha = 0 \Rightarrow$ $\frac{d}{d\lambda} F(\Psi_\lambda A, E) = 0$, we have:

$$F(A_\alpha, E^\alpha) = F(A_\alpha + \nabla_\alpha f, E^\alpha)$$

where

$$A_\alpha \to A_\alpha + \nabla_\alpha f$$

is a family of gauge transformations leaves F invariant.

Example 2.21. Begin with electromagnetism. Let the gauge-covariant derivative of Ψ be given by

$$D_\alpha \Psi = (\partial_\alpha - ieA_\alpha)\Psi\,.$$

Show that the result of parallel transport of Ψ around γ is given by

$$\exp\left(ie \oint_\gamma A_\alpha\, dx^\alpha \right) = \exp\left(ie \int_S F_{\alpha\beta} dS^{\alpha\beta} \right),$$

where S is any surface with boundary γ. Hint: Define $U(\lambda)\Psi(\gamma(0)) \equiv \Psi(\gamma(\lambda))$, where $\Psi(\lambda)$ is the result of parallel transporting Ψ along γ by a parameter distance λ, $\dot{\gamma}^\alpha D_\alpha \Psi = 0$. Then $U(\lambda)$ satisfies

$$\frac{dU}{d\lambda} = ie\dot{\gamma}^\alpha A_\alpha U,$$

from which it can be shown that $U(\lambda) = \exp\left(ie \int_0^\lambda A_\alpha \dot{\gamma}^\alpha(\lambda)d\lambda\right)$.

Solution

We have $D_\alpha \Psi = (\partial_\alpha - ieA_\alpha)\Psi$ (the form for electromagnetism), and defining $U(\lambda)\Psi(\gamma(0)) \equiv \Psi(\gamma(\lambda))$, where $\dot{\gamma}^\alpha D_\alpha \Psi = 0$. We have to start:

$$\frac{d}{d\lambda}\left(U(\lambda)\Psi(\gamma(0))\right) = \frac{dU}{d\lambda}\Psi(\gamma(0)) = \frac{d\Psi(\lambda)}{d\lambda} = \dot{\gamma}^\alpha \partial_\alpha \Psi(\gamma(\lambda))$$
$$= \dot{\gamma}^\alpha (ieA_\alpha)\Psi(\gamma(\lambda))$$

Thus,

$$\frac{dU}{d\lambda} = ie\dot{\gamma}^\alpha A_\alpha U \Psi(\gamma(0)) = ie\dot{\gamma}^\alpha A_\alpha U$$

So,

$$\frac{dU}{U} = ie\dot{\gamma}^\alpha(\lambda) A_\alpha d\lambda$$

and

$$\ln U(\lambda) = ie \int_0^\lambda A_\alpha \dot{\gamma}^\alpha(\lambda) d\lambda + C$$

If we use $U(\lambda) = \exp\left(ie \int_0^\lambda A_\alpha \dot{\gamma}^\alpha(\lambda)d\lambda\right) \Rightarrow U(0) = 1$ and $C = 0$. So, $\Psi(\gamma(\lambda))$ transported around loop. $\Psi(\gamma)$ is:

$$\Psi(\gamma) = \exp\left(ie \oint_\gamma A_\alpha \, dx^\alpha\right)\Psi(\gamma(0))$$

Furthermore, using Stokes theorem, the phase factor may be expressed as

$$\exp\left(ie \oint_\gamma A_\alpha \, dx^\alpha\right) = \exp\left(ie \int_S F_{\alpha\beta} dS^{\alpha\beta}\right).$$

Electromagnetism: charged scalar fields.

In electromagnetism, we have seen that a charge scalar field ψ on M can be regarded as a field $\overline{\psi}$ on a U(1) bundle B over M, with

$$\mathcal{L}_\eta \overline{\psi} = im\overline{\psi},$$

for a field of charge m. A cross section b: M → B corresponds to a chart (η, z) on B for which $b(x) = (0, x)$ is a cross section. The value of ψ corresponding to the $\eta = 0$ cross section is

$$\psi(x) = \overline{\psi} \circ b(x) = \overline{\psi}(0, x).$$

A vector potential A_α of the electromagnetic field is (up to a constant) the pullback to M of a connection $\overline{A}_C$ on B:

$$A_\alpha = \frac{\hbar c}{e} \pi_\alpha^C \overline{A}_C\,.$$

The gauge-covariant derivative of a field ψ is along a path $x(\lambda)$ in M is

$$\dot{x}^\alpha D_\alpha \psi = \dot{x}^\alpha \left(d_\alpha - i \frac{me}{\hbar c} A_\alpha \right) \psi,$$

and this is equivalent to the derivative of $\overline{\psi}$ along a horizontal lift of $x(\lambda)$, i.e., along a horizontal path $b_{||}(\lambda)$ over $x(\lambda)$:

$$\dot{x}^\alpha D_\alpha \psi = \frac{d}{d\lambda} \left(\overline{\psi} \left(b_{||}(\lambda) \right) \right) \Big|_{\lambda=0} = 0.$$

Proof: A horizontal lift $b_{||}(\lambda)$ safisfies $\overline{A}_C \dot{b}^C = 0$. Writing $\dot{b} = (\dot{\eta}, \dot{x})$, we have

$$\dot{\eta} + \dot{x}^\alpha \overline{A}_\alpha = 0,$$

for a horizontal path. Then the horizontal derivative of $\overline{\psi}$ is

$$\frac{d}{d\lambda} \left(\overline{\psi} \left(b_{||}(\lambda) \right) \right) \Big|_{\lambda=0} = \frac{d}{d\lambda} [\overline{\psi}(\eta(\lambda), x(\lambda))] = \dot{x}^\alpha d_\alpha \overline{\psi} + \dot{\eta} \partial_\eta \overline{\psi}$$

$$= \dot{x}^\alpha \left(d_\alpha - i \frac{me}{\hbar c} A_\alpha \right) \overline{\psi}\,.$$

In other words, the gauge-covariant derivative,

$$D_\alpha \psi = \left(d_\alpha - i \frac{Q}{\hbar c} A_\alpha \right) \psi,$$

corresponds to the horizontal derivative of $\overline{v}$,

$$D_C \psi = h_C^D (d_D \overline{v}), \quad or \ D\overline{\psi} = \text{hor } d\overline{v},$$
$$D_a v = b_a^{*D} D_C \overline{v}.$$

2.12 Isovectors and the Y-M Lagrangian

Let G be a Lie group with Lie algebra g. An isovector, ξ^a, is a g-valued scalar on a manifold M. A gauge transformation $g(x)$ acts on an isovector field ξ^i by the adjoint representation:

$$g : \xi(x + g(x)^{-1} \xi x) g(x)$$

(Eqn. 2-156)

This transformation law is equivalent to saying that ξ is the pullback, $\xi(x) = \overline{\xi}(b(x))$, to M of a g-valued scalar $\overline{\xi}^a$ on B, safistying $\overline{\xi}(bg) = g^{-1} \overline{\xi} g$. What is the gauge-covariant derivative? In analogy with electromagnetism, write

$$D_C\bar{\xi} = h_C^D \xi_D \bar{\xi} \quad \text{or} \quad D\bar{\xi} = \text{hor } d\bar{\xi},$$

(Eqn. 2-157)

Thus

$$D_a\xi = b_a^{*C} D_C\bar{\xi}$$

(Eqn. 2-158)

More explicitly, let $\dot{b}^C$ be a vector tangent to $b(\lambda)$. Its horizontal projection, $b_{||}^C$ is tangent to the horizontal path $b_{||}(\lambda)$ through $b(0)$, related to $b(\lambda)$ by $b(\lambda) = b_{||}(\lambda)g(\lambda)$ with $g(\lambda)$ a path through $e \in G$. Then $\dot{g}^a = A_C^a \dot{b}^C$ and $\dot{b}^C D_C\xi$ is given by

$$\dot{b}^C D_C\bar{\xi} = \frac{d}{d\lambda}\left(\bar{\xi}\left(b_{||}(\lambda)\right)\right)\Big|_{\lambda=0} = \frac{d}{d\lambda}[\bar{\xi}(b(\lambda), g(\lambda))]$$
$$= \frac{d}{d\lambda}[g(\lambda)\bar{\xi}(b(\lambda), g^{-1}(\lambda))] = \dot{b}^C d_C\bar{\xi} + \dot{g}\bar{\xi} - \bar{\xi}\dot{g}$$

(Eqn. 2-159)

Thus,

$$\dot{b}^C D_C\bar{\xi}^a = \xi^C\left(d_C\bar{\xi}^a - c_{bd}^a A_C^d \xi^d\right),$$

(Eqn. 2-160)

using $[\dot{g}, \xi^{-1}] = c_{bd}^a \dot{g}^d \xi^i$ In other words, the horizontal derivative of an isovector on B is

$$D_C\bar{\xi}^a = d_C\bar{\xi}^a + c_{bd}^a A_C^d \bar{\xi}^d,$$

(Eqn. 2-161)

and the corresponding gauge-covariant derivative on M is then

$$D_a\xi^a = d_a\xi^a + c_{bd}^a A_C^d \xi^d.$$

(Eqn. 2-162)

Here $A_\alpha^a = b_a^{*C} A_C^a$ is (up to a constant) the Yang-Mills vector potential corresponding to the connection A_C^a.

The gauge-covariant derivative of an isovector is an isovector, because its definition looks at the value of the cross section $b(x)$ at the point x_0 where the derivative is evaluated. In other words, if one changes cross section from $b(x)$ to $\bar{b}(x) = b(z)g(x)$, the ordinary derivative of $\xi(x) = \bar{\xi}[b(x)g(x)] =$ involves the derivative dg of the gauge transformation. The gauge-covariant derivative of $\bar{\xi}$, however, is (along the direction $\dot{x}(\lambda)$) the derivative of $\bar{\xi}\left[b_{||}\left(x(\lambda)\right)g(x_0)\right]$.

Example 2.22. Show directly from the definition that $D_a\xi^a$ is an isovector, that

$$D\overline{\xi}(x) = g(x)^{-1} d\xi(x) g(x).$$
(Note that the definition of covariant derivative given above can be used for fields that transform according to arbitrary representations of the group.)

Y-M Lagrangian Solution

Since $F^a_{\alpha\beta}$ is an isovector, $\mathcal{L} = F^a_{\alpha\beta} F_a^{\alpha\beta}$ is an isoscalar – and also a space scalar- and is therefore the natural generalization of the electromagnetic Lagrangian density $F_{\alpha\beta}F^{\alpha\beta}$:
$$I_{Y-M} - \frac{1}{4}\int dr\, F^a_{\alpha\beta} F_a^{\alpha\beta}.$$
To find the resulting field equations, note that
$$\delta F^a_{\mu\nu} = \delta\left[\partial_\mu A^a_\nu - \partial_\nu A^a_\mu + C^a_{bc} A^b_\mu A^c_\nu\right]$$
$$= \partial_\mu \delta A^a_\nu - \partial_\nu \delta A^a_\mu + C^a_{bc} A^b_\mu \delta A^c_\nu - C^a_{bc} A^b_\nu \delta A^c_\mu$$
Thus
$$\delta F^a_{\mu\nu} = D_\mu \delta A^a_\nu - D_\nu \delta A^a_\mu$$
Note that just as $\delta\Gamma^\alpha_{\beta\gamma}$ is a tensor, δA^a_α is an isovector: The inhomogeneous term in the gauge transformation law drops out if one takes the difference
$$\hat{A}^a_\alpha - A^a_\alpha,$$
of two vector potentials, so $\hat{A}^a_\alpha - A^a_\alpha$ is an isovector and
$$\delta A^a_\alpha = \frac{d}{d\lambda} A^a_\alpha(\lambda) = \lim_{\lambda\to 0}\frac{1}{\lambda}[A^a_\alpha(\lambda) - A^a_\alpha(0)],$$
is therefore an isovector. Thus
$$\delta I_{Y-M} = \int_\Omega dr\, D_\alpha \delta A^a_\beta F_a^{\alpha\beta} = \int_\Omega dr\left[D_\alpha\left(\delta A^a_\beta F_a^{\alpha\beta}\right) - \delta A^a_\beta D_\alpha F_a^{\alpha\beta}\right],$$
by Leibnitz, and since $D_\alpha = \nabla_\alpha$ when acting on an isoscalar:
$$\delta I_{Y-M} = \int_\Omega dr\, \nabla_\alpha\left(\delta A^a_\beta F_a^{\alpha\beta}\right) - \int dr\, \delta A^a_\beta D_\alpha F_a^{\alpha\beta} =$$
$$-\int_\Omega dr\, \delta A^a_\beta D_\alpha F_a^{\alpha\beta},$$
for variations δA^a_ν that vanish at the boundary $\partial\Omega$.Thus
$$\frac{\delta I_{YM}}{\delta A^a_\alpha(x)} = D_\beta F_a^{\alpha\beta}(x),$$
and the source free field equation is
$$D_\beta F_a^{\alpha\beta} = 0.$$
This is the analogue of half of Maxwell's equations. The other half, $\nabla_{[\alpha}F_{\beta\gamma]} = 0$ has the obvious analog
$$D_{[\alpha}F^a_{\beta\gamma]} = 0$$

Which is also true. For gravity, $F^a_{\alpha\beta}$ is the Riemann tensor and $D_\beta F_a^{\alpha\beta} = 0$ is the Bianchi identity.

Proof of the identity $D_{[\alpha}F^a_{\beta\gamma]} = 0$. Start with $F^a_{\beta\gamma} = 2\nabla_{[\beta}A^a_{\gamma]} + C^a_{bc}A^b_\beta A^c_\gamma$:

$$
\begin{aligned}
D_\alpha F^a_{\beta\gamma} = \nabla_\alpha F^a_{\beta\gamma} + C^a_{bc}A^b_\alpha F^c_{\beta\gamma} \\
&= 2\nabla_\alpha \nabla_{[\beta}A^a_{\gamma]} + C^a_{bc}\nabla_\alpha\left(A^b_\beta A^c_\gamma\right) \\
&+ C^a_{bc}A^b_\alpha\left(2\nabla_{[\beta}A^c_{\gamma]} + C^c_{de}A^d_\beta A^e_\gamma\right)
\end{aligned}
$$

Then

$$
\begin{aligned}
D_{[\alpha}F^a_{\beta\gamma]} &= 2\nabla_{[\alpha}\nabla_\beta A^a_{\gamma]} + C^a_{bc}\left(\nabla_{[\alpha}A^b_\beta A^c_{\gamma]} + A^b_{[\beta}\nabla_\alpha A^c_{\gamma]}\right) + C^a_{bc}A^b_{[\alpha}2\nabla_\beta A^c_{\gamma]} \\
&+ C^a_{bc}C^c_{de}A^b_{[\alpha}A^d_\beta A^e_{\gamma]}
\end{aligned}
$$

and since $\nabla_{[\alpha}\nabla_\beta A^a_{\gamma]} = 0$, even in curved spacetime, and with cancellations:

$$D_{[\alpha}F^a_{\beta\gamma]} = C^c_{[de}C^a_{b]c}A^b_\alpha A^d_\beta A^e_\gamma = 0,$$

where the last step follows from the Jacobi Identity.

Example 2.23. For Yang-Mills Theory, $D_\alpha\Psi = (\partial_\alpha - igA_\alpha)\Psi$, where $A(\lambda)$ is matrix $A^a_{b\alpha}$, and Ψ is an isovector, Ψ^α. For different λ, the matrices $A(\lambda)$ do not commute, and we need to show that $U(\lambda)$ is given by

$$U(\lambda) = P\exp\left(ig\int_0^\lambda A_\alpha\,\dot{\gamma}^\alpha(\lambda)d\lambda\right).$$

So, in a power series expansion of the exponential, the matrices are to be ordered as encountered along the path, the largest values of the parameter λ being to the left.

Solution

For Yang-Mills: $D_\alpha\Psi^\alpha = \left(\partial_\alpha - igA^a_{b\alpha}(\lambda)\right)\Psi^\alpha$. Note: for different λ the matrices $A(\lambda)$ do not commute (non-Abelian theory), so we introduce the "path" ordering operator P. We have

$$\frac{d}{d\lambda}U = ig\dot{\gamma}^\alpha A_\alpha U \quad\Rightarrow\quad U(\lambda) = \exp\left(ig\int_0^\lambda A_\alpha\,\dot{\gamma}^\alpha(\lambda)d\lambda\right),$$

but if we wish to expand the exponential we must contend with:

$$\int_0^\lambda A_\alpha \dot{\gamma}^\alpha(\lambda)d\lambda = \int_0^{\lambda_1} A_\alpha \dot{\gamma}^\alpha(\lambda)d\lambda + \cdots + \int_{\lambda_N}^{\lambda} A_\alpha \dot{\gamma}^\alpha(\lambda)d\lambda$$
$$\simeq A_\alpha^1\Gamma_1^\alpha + \cdots A_\alpha^N\Gamma_N^\alpha$$
where $\Gamma_1^\alpha = \int_0^{\lambda_1} \dot{\gamma}^\alpha\,(\lambda)d\lambda$, etc., and $A_\alpha^1 = A_\alpha(\lambda_1)$, etc. Then
$$U(\lambda) \simeq 1 + ig\int_d^\lambda A_\alpha \dot{\gamma}^\alpha(\lambda)d\lambda + \frac{1}{2}\left(ig\int_0^\lambda A_\alpha \dot{\gamma}^\alpha(\lambda)d\lambda\right)^2 + (\cdots)^3$$
where for the square and higher order terms an ordering property enters. Algebraically we have
$$\left(\int_0^\lambda A_\alpha \dot{\gamma}^\alpha(\lambda)d\lambda\right)^2 \cong (A_\alpha^1\Gamma_1^\alpha \cdots A_\alpha^N\Gamma_N^\alpha)^2$$
with no ordering prescribed. so it could be
$$= A_\alpha^1(A_\alpha^1\Gamma_1^\alpha \cdots \Gamma_N^\alpha)\Gamma_1^\alpha + A_\alpha^2(A_\alpha^1\Gamma_1^\alpha \cdots \Gamma_\alpha^N\Gamma_N^\alpha)\Gamma_2^\alpha + \cdots$$
But this ordering doesn't account for the non-abelian character of the variable A_α. This can be resolved easily:
$$\frac{d}{d\lambda}U = ig\dot{\gamma}^\alpha A_\alpha U \quad \Rightarrow \quad U(\lambda) = \exp\left(ig\int_0^\lambda A_\alpha \dot{\gamma}^\alpha(\lambda)d\lambda\right)$$
and $U(\lambda_1) \simeq \exp(igA_\alpha^1\Gamma_1^\alpha)$ for λ_1 infinitesimal. So
$$\Psi(\gamma(\lambda_1)) = U(\lambda_1)\Psi(\gamma(0)) = \exp(igA_\alpha^1\Gamma_1^\alpha)\,\Psi(\gamma(0)).$$
Now go another infinitesimal increment from $\gamma(\lambda_1) \to \gamma(\lambda_2)$ where $\lambda_2 - \lambda_1$ is infinitesinal:
$$\Psi(\gamma(\lambda_2)) = U(\lambda_2 - \lambda_1)\Psi(\gamma(\lambda_1)) = \exp(igA_\alpha^2\Gamma_2^\alpha)\exp(igA_\alpha^1\Gamma_1^\alpha)\,\Psi(\gamma)$$
Continuing in this fashion we see that
$$P\exp\left(ig\int_0^\lambda A_\alpha \dot{\gamma}^\alpha(\lambda)d\lambda\right) = \exp(igA_\alpha^N\Gamma_N^\alpha)\cdots\exp(igA_\alpha^1\Gamma_1^\alpha)\,\Psi(\gamma(0)),$$
where this merely prescribes the proper ordering on the path integral under a power series expansion such matrices A_α are ordered with largest parameter values to the left. So, we have
$$U(\lambda) = \underline{P}\exp\left(ig\int_0^\lambda A_\alpha \dot{\gamma}^\alpha(\lambda)d\lambda\right).$$
Now, in quantum field theory [9], the mathematical tools to solve this already exist, using perturbation theory methods. Since

$$D_\alpha \Psi = (\partial_\alpha - igA_\alpha)\Psi = 0$$

with parallel transport along $\gamma(\lambda)$:

$$\dot{\gamma}^\alpha \partial_\alpha \Psi = ig\dot{\gamma}^\alpha A_\alpha \Psi \qquad \dot{\gamma}^\alpha \partial_\alpha \to \partial_t, \quad -gA_\alpha \dot{\gamma}^\alpha \to H_I$$

Thus

$$\partial_t \Psi = -\mathrm{i}H_I \Psi \;\; or \;\; i\frac{\partial}{\partial t}\Psi = H_I \Psi \; with \; \Psi = U(t)\Psi(\lambda(0)).$$

Thus,

$$i\frac{\partial U}{\partial t} = H_I(\mathrm{t})\mathrm{U}$$

and we can write:

$U(t) = I - i\int_{-\alpha}^{t} dt_1\, H_I(\mathrm{t}_1)\mathrm{U}(\mathrm{t}_1)$, now iterating on $\mathrm{U}(\mathrm{t}_1)$ ...

$$U(t) = I - i\int_{-\infty}^{t} dt_1\, H_I(\mathrm{t}_1) + (-\mathrm{i})^2 \int_{-\infty}^{t} dt_1 \int_{-\infty}^{t_1} dt_2\, H_I(\mathrm{t}_1)H_I(\mathrm{t}_2) + \cdots$$
$$+(-\mathrm{i})^{\mathrm{n}} \int_{-\infty}^{t} dt_1 \int_{-\infty}^{t_1} dt_2 \cdots \int_{-\infty}^{t_{n-1}} dt_n\, H_I(\mathrm{t}_1)\cdots H_I(\mathrm{t}_\mathrm{n}) + \cdots$$

Let's introduce the time ordered product:

$$U(t) = \sum_{n=0}^{\infty} \frac{(-1)^2}{n!} \int_{-\infty}^{t} dl_1 \int_{-\infty}^{t} dt_2 \cdots \int_{-\infty}^{t} dt_n \cdots T[H_I(\mathrm{t}_1)\cdots H_I(\mathrm{t}_\mathrm{n})]$$
$$= T\exp\left[-i\int_{-\infty}^{t} dt'\, H_I(t')\right]$$

Transforming back (with bounds of integral appropriately modified for this case):

$$U(\lambda) = P\exp\left(-i\int_{0}^{\lambda} A_\alpha\, \dot{\gamma}^\alpha(\lambda)d\lambda\right).$$

2.13 Exercises

(Ex. 2.1) Derive Eqn. 2-7.
(Ex. 2.2) Derive Eqn. 2-30.
(Ex. 2.3) Derive Eqn. 2-33.
(Ex. 2.4) Derive Eqn. 2-37.
(Ex. 2.5) Derive Eqn. 2-50.
(Ex. 2.6) Derive Eqn. 2-57.
(Ex. 2.7) Derive Eqn. 2-61.

(Ex. 2.8) Derive Eqn. 2-62.
(Ex. 2.9) Derive Eqn. 2-69.
(Ex. 2.10) Derive Eqn. 2-76.
(Ex. 2.11) Derive Eqn. 2-89.
(Ex. 2.12) Derive Eqn. 2-119.
(Ex. 2.13) Derive Eqn. 2-127.
(Ex. 2.14) Derive Eqn. 2-131.
(Ex. 2.15) Derive Eqn. 2-135.
(Ex. 2.16) Derive Eqn. 2-139.
(Ex. 2.17) Derive Eqn. 2-146.
(Ex. 2.18) Derive Eqn. 2-153.
(Ex. 2.19) Derive Eqn. 2-162.

3 Einstein's geometrodynamic equations and Action formulation

3.1 Einstein's Equations and the General Theory of Relativity (GR)

In this chapter we describe Einstein's geometrodynamics equations and the Action formulation, starting from Einstein's differential formulation linking geometric curvature to matter/field stress-energy:

$$G_{\mu\nu} + \Lambda g_{\mu\nu} = \frac{8\pi G}{c^4} T_{\mu\nu}\,,$$

(Eqn. 3-1)

where coupling constant G is now seen directly in its role linking a geometrodynamic tensor and the stress-energy tensor. As mentioned in Ch. 2, the constant G has dimensionful units (unlike alpha in electrodynamics, which is dimensionless). G describes a coupling to a manifold, while alpha describes a coupling to a field.

Starting from the formulation of Einstein's relativity in differential form, as with Newton's equations, we then seek (in Sec. 3.2) a variational formulation (the Lagrangian) and then a variational integral (functional) formulation (the Action). See Book 1 [11] for a review of these topics. In Sec. 3.3 the focus shifts to global issues, with discussion of algebraic topology consequences of a boundary of a boundary being zero. A consequence of this, with minimal assumptions, is the form of the fields equations indicated above. In Sec. 3.4 the focus is on local issues, with discussion of classical gravitational waves. Wave solutions *with fields* are usually an indicator of a quantizable solution (at second quantization at least, see Book 4 [3]), but here we know that the QFT for Gravitation is not renormalizable due to the dimensionality of the coupling, and due to the positive energy constraint. Furthermore, it is difficult enough to observe classical gravitational waves, so to see a purported quantum gravitational wave would only be possible via cosmological effects. The modern-day LIGO observatory has begun mapping the Universe in terms of such (classical) wave signals much as previous astronomers mapped the skies using optical and EM wave signals.

3.2 Action formulation of GR

In this section we review the various Action formulations for general relativity. We begin by examining the Hawking Action [23] $S_{Hawking} = S_H$ and the Landau Action $S_{Landau} = S_L$. We begin by showing that $S_H \neq S_L$ but $\delta S_H = \delta S_L$ with the usual choice of boundary conditions and variation (Hilbert).

3.2.1 The Hawking and Landau Actions

The variant of the GR Action that will be referred to as the 'Hawking' Action has the form:

$$S_{Hawking} = \int_{u} \sqrt{-g} R dV + 2 \int_{\partial u} K dS$$
$$= \int_{u} \left(\sqrt{-g} G_L + \partial_\alpha w^\alpha\right) dV + 2 \int_{\partial u} K dS$$

(Eqn. 3-2)

where,

$$w^\alpha = \left[\sqrt{-g}\left(g^{\beta\gamma}\Gamma^{\alpha}{}_{\beta\gamma} - g^{\beta\alpha}\Gamma^{\gamma}{}_{\beta\gamma}\right)\right] \; and \; G_L = g^{ab}\left(\Gamma^{c}{}_{ad}\Gamma^{d}{}_{bc} - \Gamma^{c}{}_{ab}\Gamma^{d}{}_{cd}\right),$$

(Eqn. 3-3)

and the (non-tensor) Γ's are defined as usual by:

$$\Gamma^{\alpha}{}_{\beta\gamma} = \frac{1}{2} g^{\alpha\delta}\left(g_{\delta\beta,\gamma} + g_{\delta\gamma,\beta} - g_{\beta\gamma,\delta}\right).$$

(Eqn. 3-4)

The split $\sqrt{-g}R = \sqrt{-g}G_L + \partial_\alpha w^\alpha$ is not covariant and is due to Landau, for which the action is simply:

$$S_{Landau} = \int_u \sqrt{-g}G_L \, dV.$$

(Eqn. 3-5)

Note that w^α is not a vector (it is merely a four-component object) and G_L is not a scalar.

The difference in the actions involves boundary terms, for which different variation conventions are applicable, each yielding the Einstein equations upon appropriate variation. Let's examine this more closely:

$$S_H - S_L = \int_u \partial_\alpha w^\alpha \, dV + 2\int_{\partial u} K dS = \int_{\partial u} X dS$$

(Eqn. 3-6)

where $K = h^\alpha{}_\beta \nabla_\alpha n^\beta = \nabla_\beta n^\beta + n^\alpha \nabla_\alpha \left(\frac{1}{2} n_a n^a\right) = 2n^\alpha{}_{,\alpha}$ (since $n_a n^a = -1$), and:

$$X = g^{\beta\gamma} n_\alpha \Gamma^\alpha{}_{\beta\gamma} - g^{\beta\alpha} n_\alpha \Gamma^\gamma{}_{\beta\gamma} + 2\nabla_\alpha n^\alpha.$$

(Eqn. 3-7)

Note that $2\nabla_\alpha n^\alpha = 2n^\alpha \Gamma^\gamma{}_{\alpha\gamma} + 2\frac{du^\alpha}{dx^\alpha}$, so:

$$X = n^\alpha \left(g^{\beta\gamma} \Gamma_{\alpha\beta\gamma} + \Gamma^\gamma{}_{\alpha\gamma}\right) + 2n^\alpha{}_{,\alpha} \; .$$

(Eqn. 3-8)

Substituting the $\Gamma^\alpha{}_{\beta\gamma}$ definitions, we then get:

$$X = g^{\alpha\beta} n_{\alpha,\beta} + g^\gamma{}_\beta n^\beta{}_{,\gamma} \; .$$

(Eqn. 3-9)

In case the shortcut above isn't satisfying, we can do the calculation inserting the $\Gamma^\alpha{}_{\beta\gamma}$ definitions in the first equation for X:

$$X = \left(g^{bc} g^{ad} \frac{1}{2} - g^{ba} g^{cd} \frac{1}{2}\right)\left(g_{db,c} + g_{dc,b} - g_{bc,d}\right) n_a + 2h^a{}_b n^b{}_{,a} + 2h^a{}_b \Gamma^b{}_{ca} n^c$$

Thus,

$$X = -n^d g^{bc} g_{dc,b} - n^d n^c n^b g_{bd,c} + 2h^{ac} n_{c,a}$$

(Eqn. 3-10)

and

$$X = g^{ab}n_{a,b} + g^{c}{}_{b}n^{b}{}_{,c} \,,$$

(Eqn. 3-11)

as before. Thus, $S_H \neq S_L$ but $\delta S_H = \delta S_L$, in general.

3.2.2 The Einstein-Hilbert Action

There are two main types of variations to consider:
(i) Hilbert variation, also Hawking, have $\delta g^{ab} = 0$ on the boundary.
(ii) Palatini variation, also Landau, have $\delta g^{ab} = 0$ and $\nabla_a(\delta g_{bc}) = 0$, on the boundary.
We will mostly use (i) in what follows.
Note that we can set up Gaussian Normal coordinates on the various members of a suitable partition of the surface integral. We can choose, locally, a coordinate patch such that:

$$g_{\alpha\beta} = h_{\alpha\beta} - n_\alpha n_\beta \text{ and } n_\alpha n^\alpha = -1.$$

(Eqn. 3-12)

Eventually we will convert back to the volume integral form for the surface term and see if the 2nd normal derivative terms have cancelled out, as they should. (This was the purpose for Hawking's choice of S_H in the first place).

If we consider the Hawking Action and simplify to no boundary curvature contributions (from K *or* δK) we get the Einstein-Hilbert Action:

$$S[g^{ab}] = \int \sqrt{-g}\, R dV$$

(Eqn. 3-13)

Variation gives:

$$\delta S = \int \{\sqrt{-g}(\delta R_{ab})g^{ab} + \sqrt{-g}R_{ab}\delta g^{ab} + R\delta(\sqrt{-g})\}\, dV$$

(Eqn. 3-14)

For $\delta\sqrt{-g}$:
$Tr\left[\frac{dA}{d\tau}A^{-1}\right] = \frac{1}{detA}\frac{d(detA)}{d\tau} \Rightarrow \delta g_{ab}g^{ab} = \frac{1}{detg}\delta(detg),$
(note $(g^{ab})^{-1} = g_{ab}$),

$$\delta\sqrt{-detg} = \frac{1}{2}(-detg)^{-1/2}\delta(-detg)$$
$$= \frac{1}{2}(-detg)^{-1/2}(-1)(-detg)\delta g_{ab}g^{ab}$$
$$\delta\sqrt{-g} = \frac{1}{2}\sqrt{-g}g_{ab}\delta g^{ab}$$
So,
$$\delta S = \underbrace{\int \nabla^a\nabla_a\sqrt{-g}dV}_{\text{boundary term}} + \int \left(R_{ab} - \frac{1}{2}Rg_{ab}\right)\delta g^{ab}\sqrt{-g}dV$$
(Eqn. 3-15)

Let's consider the boundary terms
$$\int_u \nabla_a V^a dV = \int_{\partial u} V_a n^a dS$$
(Eqn. 3-16)

On the left side, we have the spacetime between hypersurfaces describing the three-geometry at early and late times. On the right side, we have the boundary that is time-like if space-time finite, and spacelike hypersurfaces chosen (achronal). n^a is the unit normal to the boundary, assumed non-null, $h^{ab} = g^{ab} \pm n^a n^b$, and $n^a h^{ab} = 0$ is chosen to agree. Note that MTW convention (-, +, +, +) has time-like negative, so the plus sign above is chosen accordingly:

$$V_a n^a = n^a g^{bc}[\nabla_c(\delta g_{ab}) - \nabla_a(\delta g_{bc})] = n^a h^{bc}[\nabla_c(\delta g_{ab}) - \nabla_a(\delta_{bc})].$$
(Eqn. 3-17)

Since δg^{ab} is required to vanish on the boundary with the various conventions described, so its variation on the boundary also vanishes: $h^{bc}\nabla_c(\delta g_{ab}) = 0$. So,

$$V_a n^a = -n^a h^{bc}\nabla_a(\delta g_{bc})$$
(Eqn. 3-18)

Now, $K \equiv K^a_a = h^a_b \nabla_a n^b$, so $\delta K = h^a_b(\delta C)^b_{ac} n^c$
$$= \frac{1}{2}n^c h^a_b g^{bd}[\nabla_a(\delta g_{cd}) + \nabla_c(\delta g_{ad}) - \nabla_d(\delta g_{ac})]$$
$$= \frac{1}{2}n^c h^{ad}\nabla_c(\delta g_{ab})$$
(Eqn. 3-19)

note that in the last expression, δg_{ab} can be replaced by δh_{ab}. So, with the variational constraint,

$$V_a n^a = -2\delta K$$

(Eqn. 3-20)

and we have,

$$\delta S = -2\int_{\partial u} \delta K\, dS + \int_{u} G_{ab}\delta g^{ab}\, dV \qquad G_{ab} = \sqrt{-g}\left(R_{ab} - \frac{1}{2}Rg_{ab}\right)$$

(Eqn. 3-21)

Since we assume δK and K are zero on the boundary, we get $G_{ab} = 0$, which is Einstein's equation with no matter and zero cosmological constant.

The Hawking action is then described by including the $2\int_{\partial u} K$:

$$S' = S + 2\int_{\partial u} K$$

(Eqn. 3-22)

which yields Einstein's equations upon variation. For the Wald/Hawking action a surface contributions is added such that the action retains manifestly scalar properties and so that there are no surface terms (a useful form for path integral analysis).

3.2.3 The Einstein-Hilbert Action variation using Gaussian Normal Coordinates and Lie Notation

Let's now consider the Hilbert action for gravitation, S_G, where the Lagrangian density is $\mathcal{L}_G = \sqrt{-g}R$:

$$\frac{dS_G}{d\lambda} = \int \frac{d\mathcal{L}_G}{d\lambda} d^4x$$

$$= \int \nabla^a v_a \sqrt{-g} d^4x + \int \underbrace{\left(R_{ab} - \frac{1}{2}Rg_{ab}\right)}_{G_{ab}} \delta g^{ab} \sqrt{-g} d^4x.$$

(Eqn. 3-23)

If we satisfy $\delta g^{ab} = 0$ on the boundary and Einstein's Equations, $G_{ab} = 0$ (for no matter), then (in what follows the unit normal vector n^a is written upper case N^a):

$$\frac{dS_G}{d\lambda} = \int_u \nabla^a v_a \sqrt{-g} d^4x = \int_{\partial u} \nabla_a N^a dx$$

$$= \int_{\partial u} N^a \{(h^{bc} - N^b N^c)[\nabla_c(\delta g_{ab}) - \nabla_a(\delta g_{bc})]$$

(Eqn. 3-24)

$$\frac{dS_G}{d\lambda} = \int_{\partial u} N^a h^{bc}[\nabla_c(\delta g_{ab}) - \nabla_a(\delta g_{bc})] = \int_{\partial u} -N^a h^{bc} \nabla_a(\delta g_{bc})$$

(Eqn. 3-25)

The latter term is the variation of K, which is easily shown with the one-parameter Lie notation:

$${}^{\lambda}K = h^a_{\ b}\ {}^{\lambda}\nabla_a N^b = h^a_b({}^{o}\nabla_a N^b + C^b_{ac} N^c) \quad \rightarrow \quad \delta^{\lambda} K = h^a_b(\delta C^b_{ac}) N^c.$$

(Eqn. 3-26)

Thus,

$$\delta^{\lambda} K = h^a_b \frac{1}{2} g^{bd}[\nabla_a(\delta g_{cd}) + \nabla_c(\delta g_{ad}) - \nabla_d(\delta g_{ac})] N^c$$

$$= \frac{1}{2} N^c h^{ad} \nabla_c(\delta g_{ad}).$$

(Eqn. 3-27)

So,

$$\frac{dS_G}{d\lambda} = -2 \int_{\partial u} \delta K + \int_u G_{ab}\, \delta g^{ab} \sqrt{-g} d^4x$$

(Eqn. 3-28)

Thus, redefining $S'_G = S_G + 2 \int_{\partial u} K$ we get an action that is a variational extremum when Einstein's equations are satisfied (and matter Lagrangian could be added with no difficulty.) Thus, the Hawking form of the Action can be written

$$S_H = \int_u \sqrt{-g} R d^4x + 2 \int_{\partial u} K.$$

(Eqn. 3-29)

The Landan action does not restrict itself to a manifestly scalar description. In a given coordinate system the total divergence term from the Hilbert action is immediately discarded, this is done by a non-tensorial break-up of the Ricci scalar and discarding any surface terms.

$$\sqrt{-g}R = \sqrt{-g}\,\underbrace{g^{iK}\left(\Gamma^{m}_{i\ell}\Gamma^{\ell}_{Km} - \Gamma^{\ell}_{iK}\Gamma^{m}_{\ell m}\right)}_{\text{Landau "G" term}}$$
$$+\frac{\partial}{\partial x^m}\underbrace{\left[\sqrt{-g}g^{iK}\Gamma^{m}_{iK} - \sqrt{-g}g^{im}\Gamma^{\ell}_{i\ell}\right]}_{w^m}$$
$$= \sqrt{-g}G + \frac{\partial}{\partial x^m}[----]$$

(Eqn. 3-30)

$$S_{Hawking} - S_{Landau} = \int_u \partial_m w^m dV + 2\int_{\partial u} K\,dS$$
$$= \int_{\partial u}\left\{\left(g^{iK}\Gamma^{m}_{iK} - g^{im}\Gamma^{\ell}_{i\ell}\right)n_m + 2\nabla_m n^m\right\}dS$$
$$= \int_{\partial u} g^{ab}\left(n_{a,b} + g_{bc}n^{c}_{,a}\right)\sqrt{-g}d^3x \neq \underline{0}$$

(Eqn. 3-31)

Thus, the actions are inequivalent. However, it can be shown that $\delta(S_H - S_E) = 0$ with $\delta g^{ab} = 0$ on ∂u:

$$\delta(S_H - S_L) = \int_{\partial u} g^{ab}\left((\delta n_a)_{,b} + g_{bc}(\delta n^c)_{,a}\right)\sqrt{-g}d^3$$
$$= \int_{\partial u}\left[n^a n^b(\delta n_a)_{,b} + g_{bc}n^a n^b(\delta n^c)_{,a}\right]\sqrt{-g}d^3x$$

$$\delta(S_H - S_L) = \int_{\partial u} n^b\sqrt{-g}\,\underbrace{\left[n^a(\delta n_a)_{,b} + n_a(\delta n^a)_{,b}\right]}_{\substack{0 \\ \text{using the } n_a n^a = -1 \text{ constraint}}}d^3x = 0$$

3.2.4 The Landau Action

Let's now consider the Landau formulation:

$$S_L = \int_u \sqrt{-g}G_L\,d^4x,\ \ where\ \ G_L = g^{ab}\left(\Gamma^{c}_{ad}\Gamma^{d}_{bc} - \Gamma^{c}_{ab}\Gamma^{d}_{cd}\right).$$

(Eqn. 3-32)

We have:

$$\delta S_L = \int_u \left(\delta(\sqrt{-g})G_L + \sqrt{-g}\delta G_L\right)d^4x\,,$$

(Eqn. 3-33)

where

$$\delta\left(\sqrt{-g}\right) = \frac{1}{2}\sqrt{-g}g^{ab}\delta g_{ab} = -\frac{1}{2}\sqrt{-g}g_{ab}\delta g^{ab}$$

(Eqn. 3-34)

For the variation on the Landau form we can reuse prior calculations if we make use of the relation:

$$\sqrt{-g}G_L = \sqrt{-g}R - \left[\frac{\partial}{\partial x^\ell}\left(\sqrt{-g}g^{ik}\Gamma^\ell_{ik}\right) - \frac{\partial}{\partial x^k}\left(\sqrt{-g}g^{ik}\Gamma^\ell_{i\ell}\right)\right].$$

(Eqn. 3-35)

Thus,

$$\delta\int_u \sqrt{-g}G_L d^4x$$
$$= \int_u G_{ab}\,\delta g^{ab}\sqrt{-g}d^4x - 2\int_{\partial u}\delta K\,dS$$
$$- \delta\int_{\partial u}\left(n_j\sqrt{-g}\left[g^{ik}\Gamma^j_{ik} - g^{ij}\Gamma^\ell_{i\ell}\right]\right)d^3x$$

(Eqn. 3-36)

In the Landau formulation we have Palatini-type variation, thus need two conditions at the boundary: (i) $\delta g^{ab} = 0$ and (ii) $\nabla_a(\delta g_{bc}) = 0$. The second is <u>not</u> assumed in the Hawking analysis, and we now have:

$$\delta\int_u \sqrt{-g}G_L d^4x = \int_u G_{ab}\delta g^{ab}\sqrt{-g}\,d^4x.$$

(Eqn. 3-37)

Thus zero variation with the chosen boundary variation constraints also gives $G_{ab} = 0$ for (Einstein's) equations of motion.

Note that the intial form chosen by Landau was motivated by dropping all divergence terms. Thus, starting from:

$$R = g^{\alpha\beta}\left[\underline{\Gamma^\gamma_{\alpha\beta,\gamma}} - \Gamma^\gamma_{\alpha\gamma,\beta} + \Gamma^\gamma_{\alpha\beta}\Gamma^\delta_{\gamma\delta} - \Gamma^\gamma_{\alpha\delta}\Gamma^\delta_{\beta\gamma}\right]$$

(Eqn. 3-38)

and using:

$\sqrt{-g}g^{\alpha\beta}\Gamma^\gamma_{\alpha\beta,\gamma} = \left(\sqrt{-g}g^{\alpha\beta}\Gamma^\gamma_{\alpha\beta}\right)_{,\gamma} - \Gamma^\gamma_{\alpha\beta}\left(\sqrt{-g}g^{\alpha\beta}\right)_{,\gamma}$ and

$\left(\sqrt{-g}g^{\alpha\beta}\right)_{,\gamma} = g^{\alpha\beta}\left(\sqrt{-g}\right)_{,\gamma}\sqrt{-g}g^{\alpha\beta}_{,\gamma}$

Also have relations: $\Gamma^{\delta}_{\gamma\delta} = \left(\ln\sqrt{-g}\right)_{,\gamma} = \frac{1}{\sqrt{-g}}\left(\sqrt{-g}\right)_{,\gamma}$ and $g^{\alpha\beta}_{,\gamma} = -\Gamma^{\delta}_{\delta\gamma}g^{\delta\beta} - \Gamma^{\beta}_{\delta\gamma}g^{\alpha\delta}$,

So:

$$\left(\sqrt{-g}g^{\alpha\beta}\right)_{,\gamma} = g^{\alpha\beta}\left(\sqrt{-g}\Gamma^{\delta}_{\gamma\delta}\right) - \sqrt{-g}\left(\Gamma^{\delta}_{\delta\gamma}g^{\delta\beta} + \Gamma^{\beta}_{\delta\gamma}g^{\alpha\delta}\right)$$

Similarly

$$\sqrt{-g}g^{\alpha\beta}\Gamma^{\gamma}_{\alpha\beta,\gamma} = \left(\sqrt{-g}g^{\alpha\beta}\Gamma^{\gamma}_{\alpha\gamma}\right)_{,\beta} - \Gamma^{\gamma}_{\alpha\gamma}\left(\sqrt{-g}g^{\alpha\beta}\right)_{,\beta}$$

and relation: $g^{\gamma\delta}\Gamma^{\alpha}_{\gamma\delta} = \frac{1}{\sqrt{-g}}\left(\sqrt{-g}g^{\alpha\beta}\right)_{,\beta}$ gives:

$$\begin{aligned}\sqrt{-g}R &= \left(\sqrt{-g}g^{\alpha\beta}\Gamma^{\gamma}_{\alpha\beta}\right)_{,\gamma} - \left(\sqrt{-g}g^{\alpha\beta}\Gamma^{\gamma}_{\alpha\gamma}\right)_{,\beta} \\ &+\sqrt{-g}\left[\Gamma^{\gamma}_{\alpha\alpha\beta}\left(\Gamma^{\alpha}_{\delta\gamma}g^{\delta\beta} + \Gamma^{\beta}_{\delta\gamma}g^{\alpha\delta}\right) - g^{\gamma\delta}\Gamma^{\alpha}_{\gamma\delta}\Gamma^{\gamma}_{\alpha\gamma} - g^{a\delta}\Gamma^{\delta}_{\gamma\delta}\Gamma^{\gamma}_{\alpha\beta}\right] \\ &+\sqrt{-g}g^{\alpha\beta}\left(\Gamma^{\gamma}_{\alpha\beta}\Gamma^{\delta}_{\gamma\delta} - \Gamma^{\gamma}_{\gamma\delta}\Gamma^{\delta}_{\beta\gamma}\right) \\ &= \underbrace{\left\{\sqrt{-g}\left(g^{\alpha\beta}\Gamma^{\gamma}_{\alpha\beta} - g^{\alpha\gamma}\Gamma^{\delta}_{\gamma\delta}\right)\right\}_{,\gamma}}_{\equiv w^{\gamma}_{,\gamma}} + \sqrt{-g}\underbrace{g^{\alpha\beta}\left(\Gamma^{\gamma}_{\gamma\delta}\Gamma^{\delta}_{\beta\gamma} - \Gamma^{\gamma}_{\gamma\beta}\Gamma^{\delta}_{\gamma\delta}\right)}_{\equiv G_L}\end{aligned}$$

(Eqn. 3-39)

And dropping the divergence part:

$$S = \int \sqrt{-g}\, G_L d^4x.$$

(Eqn. 3-40)

3.2.5 Relations between Actions

How are the different actions related?

$$L_{Hilbert} = \sqrt{-g}R = L_{Landau} + \partial F\left(g, \partial g, \sqrt{-g}\right)$$

(Eqn. 3-41)

where L_{Landau} has only 1st derivatives but is nontensorial. The ∂F part has 1st and 2nd derivatives and is nontensorial. $L_{Hilbert}$ is tensorial but has 2nd derivatives. Depending on application, one formulation or the other might be optimal.

We've seen a direct relation between the Landau and Hilbert forms, what of between the Landau and Hawking forms? Consider a variation on $g_{ab}(\lambda)$ that carries ${}^0\nabla \rightarrow {}^\lambda\nabla = \partial$ in the coordinates chosen (also, recall that ${}^\lambda\nabla_a V^a = \partial_a V^a = {}^0\nabla_a V^a + C^a_{ca}(\lambda)V^c$ and ${}^\lambda\nabla_a = {}^0\nabla_a + C_a$:

$$\mathcal{L}_H({}^0\nabla) = \mathcal{L}_L({}^0\nabla) + \partial F \Rightarrow \mathcal{L}_H({}^\lambda\nabla) = \mathcal{L}_L({}^\lambda\nabla) + {}^\lambda\nabla F$$

(Eqn. 3-42)

The new form matches the tensorial division $\mathcal{L}_H = \mathcal{L}_{Hawk}(\nabla) + \nabla J$, thus must correspond uniquely and we have

$$\mathcal{L}_L({}^0\nabla_\mathrm{a}) + \partial F = \mathcal{L}_L\left({}^0\nabla_\mathrm{a} = {}^\lambda\nabla_a - C^a_{ca}\right) + {}^\lambda\nabla_a F$$
$$= \mathcal{L}_{Hawking}\left({}^\lambda\nabla_a\right) + {}^\lambda\nabla_a F$$

Thus

$$\mathcal{L}_L({}^0\nabla_\mathrm{a}) = \mathcal{L}_{Hawking}({}^0\nabla_\mathrm{a} + C^a_{ca}).$$

(Eqn. 3-43)

The Actions possible:

$$Hilbert\colon S[g^{ab}] = \int_u \sqrt{-g}\, R dV$$

(Eqn. 3-44)

yields Einstein equation when no boundary exists or when boundary is in asymptotically flat region, or where $\nabla(\delta g) = 0$ in general somehow.

$$Hawking\colon\ S[g^{ab}] = \int_u \sqrt{-g}\, R dV + 2\int_{\partial u} K dS$$

(Eqn. 3-45)

yields Einstein's vacuum equation and accounts for boundary effects. And,

$$Landau\colon S[g^{ab}, \partial g] = \int_u \sqrt{-g}\, G_L dV$$

(Eqn. 3-46)

which uses Palatini variation, and again yields Einstein's equations.

Note that using the Hilbert Action requires care if surface terms involved:

$$S_{Hilbert} = \int_u \left(\sqrt{-g} G + w^m_{,m}\right) dV$$

(Eqn. 3-47)

where

$$G = g^{iK}\left(\Gamma^m_{i\ell}\Gamma^\ell_{Km} - \Gamma^\ell_{iK}\Gamma^m_{\ell m}\right)$$

(Eqn. 3-48)

and

$$w^m = \sqrt{-g} g^{iK}\Gamma^m_{iK} - \sqrt{-g} g^{im}\Gamma^\ell_{i\ell}$$

(Eqn. 3-49)

So does

$$S_{Hilbert} = \int_u \sqrt{-g}\, G dV \ ?$$

It depends. Care must be taken to ensure that the $\frac{\partial}{\partial x^m} w^m$ terms are true boundary terms and not artifacts of the coordinates chosen. Such problems can arise due to the use of angular coordinates and will do so in general if angular coordinates are used. Consider the Kruskal-Szekeres Coordinates:

$$\frac{\partial}{\partial x^m} w^m = \underbrace{\frac{\partial}{\partial \theta} w^\theta}_{\substack{\uparrow \\ \text{not a boundary term!}}} + \frac{\partial}{\partial \varphi} w^\varphi + \underbrace{\frac{\partial}{\partial u} w^u + \frac{\partial}{\partial V} w^V}_{\substack{\text{for kruskal variables,whatever the case} \\ \text{they represent true boundary terms}}}$$

(Eqn. 3-50)

So, modify G to include the $\frac{\partial}{\partial \theta}$ term: $G' = G + \frac{\partial}{\partial \theta} w^\theta$,and we get

$$S_{Hilbert} = \int_u \sqrt{-g}\left(G + \frac{\partial}{\partial \theta} w^\theta\right) dV$$

(Eqn. 3-51)

3.2.6 Initial Value Equations

The idea that physics predicts the future (at least the future values of a wave function) can be regarded as the existence of an initial value formulation. For a Newtonian particle, one freely specifies its position q^i and momemtum q_i at an initial time t_o. Its subsequent behavior is then governed by

$$\dot{q}^i = \frac{\partial H}{\partial P_i} \quad and \quad \dot{p}_i = \frac{\partial H}{\partial q_i},$$

(Eqn. 3-52)

where H depends only on q and p at t, not on their time derivatives, e.g.:

$$\dot{q}^i = p^i \quad and \quad \dot{p}_i = \frac{\partial V(g)}{\partial q^i}.$$

(Eqn. 3-53)

Similarly, for a scalar field in flat space, one freely specifies its initial value ϕ_S on a $t = t_o$ hypersurface S and the initial value of its conjugate momemtum, $p = \dot{\phi}$. Its subsequent evolution is then given by:

$$\dot{\phi} = p \quad and \quad \dot{p} = \nabla^2 \phi,$$

(Eqn. 3-54)

where again the RHS depends only on p, ϕ and their spatial derivatives – only on p and ϕ on S. If we set

$$H = \frac{1}{2}\int_{S_t} [p^2 + (\nabla\phi)^2]\, dS,$$

(Eqn. 3-55)

then:

$$\dot{\phi} = \frac{\partial H}{\partial p} \quad and \quad \dot{p_i} = \frac{\partial H}{\partial \phi}.$$

(Eqn. 3-56)

In electromagnetism a new feature arises. In flat space, with no sources (for example), one specifies B^a and E^a on S, but these spatial vectors are not freely specifiable. They must satisfy the constraints

$$\nabla \cdot \mathrm{B} = 0 \quad \text{and} \quad \nabla \cdot \mathrm{E} = 0.$$

(Eqn. 3-57)

Equations which involve only E and B on S. The evolution equations

$$\dot{B} = -\nabla \times \mathrm{E} \quad \text{and} \quad \dot{\mathrm{E}} = \nabla \times \mathrm{B}$$

(Eqn. 3-58)

are the time derivatives of E and B in terms of the values of E and B on S. They evolve the constraints.

For a scalar field one specifies two scalars ϕ, p on S, and for EM two spatial vectors B^a, E^a, while for gravity one specifies two spatial tensors, h_{ab} (the spatial metric) and p_{ab} which can be regarded as the first time derivative of h_{ab}.

Let S be a spacelike hypersurface of a spacetime M, $g_{\alpha\beta}$ and let $h_{\alpha\beta} = g_{\alpha\beta} + n_\alpha n_\beta$, where n_α is the unit normal to S. let t be a scalar field for which S is a $t = t_o$ surface and with $\nabla_\alpha t \neq 0$ on S so that the future pointing normal is

$$n_\alpha = -N\, \nabla_\alpha t,$$

(Eqn. 3-59)

where

$$N = \left(-\nabla_\beta t\, \nabla^\beta t\right)^{-1/2}.$$

(Eqn. 3-60)

In a chart $\left(x^o = t, x^i\right), N = (-g^{oo})^{-1/2}$, and $n_\mu = -\delta^o_\mu |g^{oo}|^{-1/2}$.

If one regards the t = constant surface, S, as a manifold by itself, a vector v^a on S is the tangent to a path $c(\lambda)$ on S. When S is regarded as a submanifold of M, the curve $c(\lambda)$ lies in M and its tangent is a vector v^α

in M for which $v^\alpha n_\alpha = 0$. Formally, if i: S = M is the embedding map, v^a is dragged by i to v^α: $v^\alpha = i^\alpha_a v^a$. With respect to the chart x^i on S, v^a has components v^i, and with respect to the chart (t, x^i) on M, v^α has components $v^\mu = (0, v^i)$. Any contravariant tensor $T^{a...b}$ on S can be similarly idendified with a tensor $T^{\alpha...\beta}$ on M, orthogonal in all indices to n_α: $0 = T^{\alpha...\beta} n_\alpha = \cdots = T^{\alpha...\beta} n_\beta$. Formally $T^{\alpha...\beta} := \iota^\alpha_a \dots \iota^\beta_b T^{a...b}$, and in the chart (t, x^i) only purely spatial components of $T^{\alpha...\beta}$ are nonzero, and they are identical to the components $T^{i...j}$ of $T^{a...b}$.

Given a metric $g_{\alpha\beta}$ on M, one can extend the identification to arbitrary tensors by associating a covariant vector σ_a on S with the unique covariant vector σ_α in M for which $\sigma_\alpha n^\alpha = 0$ and $\sigma_a = \iota^\alpha_a \sigma_\alpha$. Again with respect to any chart (t, x^i) for which S is a t = constant surface, the spatial components of σ_α are identical to the components σ_i of σ_a. However $\sigma_o = \sigma_\alpha n^\alpha$ must vanish only if one picks the scalars x^i to be constant along curves tangent to n^α. Then $n^\alpha \nabla_\alpha x^i = 0$, so n^α is parallel to ∂_t:

$$n^\mu = \frac{1}{N} \delta^\mu_0 ,$$

and $\sigma_\alpha n^\alpha = 0 \rightarrow \sigma_o = 0$.

To summarize: Tensors $T^{a...}_{c...}$ on S can be naturally identified with tensors $T^{\alpha...}_{\gamma...}$ on M for which

$$0 = T^{\alpha...}_{\beta...} n_\alpha = \cdots = T^{\alpha...}_{\beta...} n^\beta .$$

In particular, the spatial metric ${}^3g_{ab} = \iota^\alpha_a \iota^\beta_b g_{\alpha\beta}$ on S is identified with the tensor $h_{\alpha\beta} = g_{\alpha\beta} + n_\alpha n_\beta$ on M. In any chart (t, x^i) for which S is a t = constant surface,

$$h_{ij} = {}^3g_{ij} .$$

In a chart of which n^α is along ∂_t, these are the only nonzero components of $h_{\alpha\beta}$. If $T^{\beta...}_{...\gamma}$ is a tensor on S, its covariant derivative with respect to the spatial metric $h_{\alpha\beta}$ is thus:

$$D_\alpha T^{\beta...}_{...\gamma} = h^\delta_\alpha h^\beta_\epsilon \dots h^\zeta_\gamma \nabla_\delta T^{\epsilon...}_{...\zeta} .$$

(Eqn. 3-61)

Example 3.1. Show that D_α is the unique derivative operator that maps tensors on S to tensors on S and satisfies: (i) Leibnitz; (ii) $D_{[\alpha}D_{\beta]}f = 0$; (iii) $D_\alpha h\beta\gamma = 0$.

Solution

Consider a chart in which n^α is along ∂_t it is natural to take h_{ij} and $\dot{h}_{ij} = \partial_t h_{ij}$ as initial data. Now $\dot{h}_{ij}$ are the components of a tensor on S. Let

$$K_{\alpha\beta} = h_\alpha^\gamma h_\beta^\delta \nabla_\gamma n_\delta\,,$$

then, in a chart (t, x^i) with $g^{oi} = 0$, we have $K_{\alpha o} = 0$ and

$$g^{oo} = 1 \to N = 1 = (-g^{oo})^{\frac{1}{2}}, \qquad K_{ij} = \frac{1}{2}N^{-1}\dot{h}_{ij}$$

Proof:

$K_{ij} = \nabla_i n_j = \partial_i n_j - \Gamma_{ji}^k n_k = -n^o r_{oij} \quad (n^\mu = g^{\mu o} = \delta_o^\mu\ )$

So,

$$K_{ij} = \frac{1}{2}n^\alpha \partial_\alpha h_{ij} = \frac{1}{2N}\dot{h}_{ij}$$

Now, $K_{\alpha o} = 0$ is immediate from $h_{\alpha o} = 0$. Thus, as initial data on S, we take the two tensor fields

$h_{\alpha\beta}$ and $K_{\alpha\beta}$. Note that $K_{\alpha\beta}$ is symmetric (if it's symmetric in a chart, it's symmetric); and, since $n^\beta \nabla_\alpha n_\beta = 0\ (from\ n^\beta n_\beta = -1)$, we have:

$$K_{\alpha\beta} = h_\alpha^\gamma \nabla_\gamma n_\beta.$$

In electromagnetism, the constraint equations were

$$n_\alpha(\nabla_\beta F^{\alpha\beta} - 4\pi j^\alpha) = 0$$

$$\nabla \cdot E = 4\pi\rho$$

and

$$n_\alpha \nabla_\beta^* F^{\alpha\beta} = 0.$$

$$\nabla \cdot B = 0$$

Here, they will be the equations:

$$n_\alpha(G^{\alpha\beta} - 8\pi\, T^{\alpha\beta}) = 0,$$

from which we get the momentum and Hamiltonian constraints that follow.

Momentum Constraint

$$h_\beta^\alpha G^{\beta\gamma} n_\gamma = -8\pi\, J^\alpha, \tag{Eqn. 3-62}$$

where J^α is the momentum density seen by an observer with velocity n^α:

$$J^\alpha = h^\alpha_\beta T^{\beta\gamma} n_\gamma \,.$$

(Eqn. 3-63)

Claim: We can show that

$$D_\beta\left(K^{\alpha\beta} - h^{\alpha\beta} K\right) = h^\alpha_\beta G^{\beta\gamma} n_\gamma.$$

(Eqn. 3-64)

By definition of the derivative D_α and of $K_{\alpha\beta}$,

$$D_\alpha\left(K^{\alpha\beta} - h^{\alpha\beta} K\right) = h^\alpha_\gamma h^\beta_\delta \nabla_\beta \left[K^{\gamma^\delta} - h^{\gamma^\delta} K\right]$$
$$= h^\alpha_\gamma h^\beta_\delta \nabla_\beta\left[\left(h^{\gamma\epsilon} h^{\delta\zeta} - h^{\gamma\delta} h^{\epsilon\zeta}\right)\nabla_\epsilon n_\zeta\right]$$

$$= \underbrace{\left(h^{\alpha\gamma} h^{\beta\delta} - h^{\alpha\beta} h^{\gamma\delta}\right)\nabla_\beta \nabla_\gamma n_\delta}_{I} +$$
$$\underbrace{h^\alpha_\gamma h^\beta_\delta \nabla_\epsilon n_\zeta \nabla_\beta\left(h^{\gamma\epsilon} h^{\delta\zeta} - h^{\gamma\delta} h^{\epsilon\zeta}\right)}_{II}$$

$I = h^{\alpha\gamma} h^{\beta\delta}\left(\nabla_\beta \nabla_\gamma - \nabla_\gamma \nabla_\beta\right) n_\delta = h^{\alpha\gamma} h^{\beta\delta} R_{\beta\gamma\delta\epsilon} n^\epsilon = h^{\alpha\gamma}\left(g^{\beta\delta} + n^\beta n^\delta\right) R_{\beta\gamma\delta\epsilon} n^\epsilon = h^{\alpha\gamma} R_{\gamma\epsilon} n^\epsilon$

Thus,

$I = h^\alpha_\beta G^{\beta\gamma} n_\gamma$

Using $n^\beta \nabla_\alpha n_\beta = 0, h_{\alpha\beta} n^\beta = 0, \nabla_\beta\left(h^{\gamma\delta}\right) = \nabla_\beta\left(n^\gamma n^\delta\right), and\ \ h^\gamma_\alpha \nabla_\gamma n_\beta = K_{\alpha\beta}$:

$$II = h^\alpha_\gamma h^\beta_\delta \nabla_\epsilon n_\zeta\left[\nabla_\beta(n^\gamma n^\epsilon) h^{\delta\zeta} + h^{\zeta\epsilon} \nabla_\beta\left(n^\delta n^\zeta\right) - \nabla_\beta\left(n^\delta n^\zeta\right) h^{\epsilon\zeta} - h^{\gamma\delta} \nabla_\beta\left(n^\epsilon n^\zeta\right)\right]$$
$= h^\alpha_\gamma h^\beta_\delta \nabla_\epsilon n_\zeta\left(\nabla_\beta n^\gamma n^\epsilon h^{\delta\zeta} - h^{\gamma\delta} n^\epsilon \nabla_\beta n^\zeta\right) = K^{\alpha\zeta} n^\epsilon \nabla_\epsilon n_\zeta - K^{\alpha\zeta} n^\epsilon \nabla_\epsilon n_\zeta = 0.$

Thus, $D_\alpha\left(K^{\alpha\beta} - h^{\alpha\beta} K\right) = I + II = h^\alpha_\beta G^{\beta\gamma} n_\gamma = 0.$

Hamiltonian Constraint

$$G^{\alpha\beta} n_\alpha n_\beta = 8\pi\rho,$$

(Eqn. 3-65)

where $\rho = T^{\alpha\beta} n_\alpha n_\beta$ is the energy density seen by an observer with 4-velocity n^α.

Claim:

$$^3R - K^{\alpha\beta}K_{\alpha\beta} + K^2 = 16\pi\rho,$$

(Eqn. 3-66)

where $R = {}^3R$ is the Ricci scalar of the spatial metric. To prove this, we must show that

$$^3R - K^{\alpha\beta}K_{\alpha\beta} + K^2 = 2G^{\alpha\beta}n_\alpha n_\beta.$$

(Eqn. 3-67)

We need the following lemma:

The Gauss-Codazzi equation

$$^3R_{\alpha\beta\gamma\delta} = h^\mu_\alpha h^\nu_\beta h^\sigma_\gamma h^\tau_\delta R_{\mu\nu\sigma\tau} + K_{\alpha\delta}K_{\beta\gamma} - K_{\alpha\gamma}K_{\beta\delta}\,.$$

(Eqn. 3-68)

(^4R projected to $\delta = {}^3R - KK + KK =$ intrinsic curv.)

By writing

$$\frac{1}{2}\,{}^3R_{\alpha\beta\gamma\delta}\xi^\delta = D_{[\alpha}D_{\beta]}\xi_\gamma = h^\lambda_{[\alpha}h^\mu_{\beta]}h^\nu_\gamma\nabla_\lambda\underbrace{\left(h^\sigma_\mu h^\tau_\nu\nabla_\sigma\xi_\tau\right)}_{D_\mu\xi_\nu},$$

(Eqn. 3-69)

where ξ^k is a vector on S, and proceeding in analogy with the momentum constraint's algebra. Contracting on β and δ (and relabeling indices), we get

$$^3R_{\alpha\beta} = h^\mu_\alpha h^\nu_\beta h^{\sigma\tau}R_{\mu\sigma\nu\tau} + K_{\alpha\gamma}K^\gamma_\beta - K_{\alpha\beta}K,$$

and

$$^3R = h^{\alpha\beta}h^{\gamma\delta}R_{\alpha\gamma\beta\delta} + K_{\alpha\beta}K^{\alpha\beta} - K^2.$$

Finally

$$h^{\alpha\beta}h^{\gamma\delta}R_{\alpha\gamma\beta\delta} = \left(g^{\alpha\beta} + n^\alpha n^\beta\right)\left(g^{\gamma\delta} + n^\gamma n^\delta\right)R_{\alpha\gamma\beta\delta} = R + 2R_{\alpha\beta}n^\alpha n^\beta$$
$$= 2G_{\alpha\beta}n^\alpha n^\beta,$$

Summary:

$$D_\beta\left(K^{\alpha\beta} - h^{\alpha\beta}K\right) = -8\pi\,J^\alpha$$

(Eqn. 3-70)

$$^3R - K^{\alpha\beta}K_{\alpha\beta} + K^2 = 16\pi\rho$$

(Eqn. 3-71)

The remainder of the Einstein equations (the space-space components) involving $G^{\gamma\delta}h^\alpha_\gamma h^\beta_\delta$ are dynamical equations for $K_{\alpha\beta}$: they involve first time derivatives of $K_{\alpha\beta}$.

In writing the dynamical equations, we will retain the freedom to pick an arbitrary gauge – to allow, in particular, the metric to have nonzero g_{oi} components. One fixes a slicing (foliation) of the spacetime manifold M by 3-dimensional hypersurfaces S_t diffeomorphic to a fixed 3-manifold S,

$$i_t : S \to S_t.$$

The image of a point p of S is then a curve $t \to i_t(p)$, whose tangent we will call t^α. If $\{x^i\}$ is a chart on S, $\{t, x^i\}$ is a chart on M and $t^\alpha = \partial_t$. Specifying metric $g_{\alpha\beta}$ on M is equivalent (see ADM formalism in Ch. 5) to giving (i) a metric $h_{\alpha\beta}$ on each S_t, and (ii) a decomposition

$$t^\alpha = Nn^\alpha + S^\alpha,$$

(Eqn. 3-72)

of t^α into parts along S_t and perpendicular to S_t. The vector S^α represents the shift in your spatial position as you move along t^α, and it is called the shift vector. In a chart (t, x^i) with $t^\alpha = \partial_t$, we have $h_{ij} = g_{ij}$ and $N = (-g^{oo})^{-1/2}$ as before; and

$$S^o = 0, \qquad S^i = \frac{g^{oi}}{g^{oo}}$$

(Eqn. 3-73)

Thus one can recover $g_{\alpha\beta}$ from $h_{\alpha\beta}, N, S^\alpha$ and the fixed vector field t^α. Equivalently, given a family ${}^3g_{ab}(t), N(t)$, and $S^a(t)$ of 3-metrics, lapses and shifts on S one has a unique metric $g_{\alpha\beta}$ on M given in its contravariant form by

$$g^{\alpha\beta} = h^{\alpha\beta} - n^\alpha n^\beta.$$

(Eqn. 3-74)

where on each surface S_t ,

$$n^\alpha = \frac{t^\alpha - S^\alpha}{N}.$$

(Eqn. 3-75)

with

$$h^{\alpha\beta} = i_a^\alpha i_a^\beta g^{ab}.$$

(Eqn. 3-76)

and

$$S^\alpha = i_a^\alpha S^a.$$

(Eqn. 3-77)

Using Lie derivatives, one can show that:

$$\boldsymbol{K_{\alpha\beta} = \frac{1}{2}\pounds_n h_{\alpha\beta}}.$$

(Eqn. 3-78)

Proof: In a chart with $n^\mu = \delta^\mu_o$ and $g^{oi} = 0$, we have $\pounds_n h_{\mu v} = n^\sigma \partial_\sigma h_{\mu v} = \dot{h}_{\mu v}$. So,

$$K_{\alpha\beta} = \frac{1}{2N}(\pounds_t - \pounds_s)h_{\alpha\beta} = \frac{1}{2N}\dot{h}_{\alpha\beta} - \frac{1}{N}D_{(\alpha}S_{\beta)},$$

where one writes $\dot{h}_{\alpha\beta} = \pounds_t\, h_{\alpha\beta}$, because in a chart with $t^\alpha = \partial_t$, $\partial_t h_{\mu v} = \pounds_t\, h_{\mu v}$.

3.2.7 Hamiltonian Formulation

We will obtain the dynamical equations by casting the Einstein action in a form that explicitly displays both the Hamiltonian and the constraints of the theory. The Lagrangian of a system subject to a constraint $C(q,\dot{q}) = 0$ can be written in the manner

$$L(q,\dot{q},\lambda) = L_o(q,\dot{q}) + \lambda C(q,\dot{q}).$$

(Eqn. 3-79)

If λ is regarded as an independent configuration space variable, then (because λ does not occur in L) the Euler-Lagrange equation for λ is just the constraint

$$C(q,\dot{q}) = 0.$$

(Eqn. 3-80)

If the Legendre transformation $p_i = \frac{\partial L_o}{\partial \dot{q}_i}(q,\dot{q})$ is invertible to give $\dot{q} = \dot{q}(p,q)$, we can define as usual

$$H(p,q,\lambda) = p_i\dot{q}^i(p,q) - L(p,\dot{q}(p,q),\lambda)$$
$$= \left(p_i\dot{q}^i_i - L_o\right)(p,q) - \lambda C(p,q).$$

(Eqn. 3-81)

Now, Hamilton's equations with this Hamiltonian are simply the Euler-Lagrange equations for the Lagrangian

$$L(p,q,\dot{q},\lambda) = \int [p\dot{q} - H_o(p,q) + \lambda C(p,q)]\, dt$$
$$= \int [p\dot{q} - H(p,q,\lambda)]\, dt.$$

(Eqn. 3-82)

Because $\dot{p}$ and $\dot{\lambda}$ do not occur in L, we have two constraints, one of which is Hamilton's equation for $\dot{q}$:

$$0 = \frac{\partial L}{\partial \lambda} = C(p,q)$$

And

$$0 = \frac{\partial L}{\partial p_i} = \dot{q}^i - \frac{\partial H}{\partial p_i} \rightarrow \dot{q}^i = \frac{\partial H}{\partial p_i}.$$

Hamilton's equation for $\dot{p}$ is the Euler-Lagrange equation for q:

$$0 = \frac{d}{dt}\left(\frac{\partial L}{\partial \dot{q}}\right) - \frac{\partial L}{\partial q} = \dot{p}_i - \frac{\partial H}{\partial q^i}.$$

(Eqn. 3-83)

We now write the Maxwell and Einstein Lagrangian in the form above as an example (further with GR is done in the ADM description in Ch. 5).

Maxwell: Let $n^\alpha = -\nabla^\alpha t$ be the unit normal to a t = constant plane of Minkowski space. Let (Φ, A_α) be a vector potential, with A_α spatial $(A_\alpha n^\alpha = 0)$. Then, writing

$$\epsilon_{\alpha\beta\gamma} = n^\delta \epsilon_{\delta\alpha\beta\gamma}.$$

(Eqn. 3-84)

For the spatial antisymmetric tensor, we have

$$E_\alpha = F_{\alpha\beta} n^\beta = -\partial_t \dot{A}_\alpha + D_\alpha \phi = -\dot{A}_\alpha + D_\alpha \Phi,$$

(Eqn. 3-85)

And

$$B^\alpha = \frac{1}{2} \epsilon^{\alpha\beta\gamma} F_{\beta\gamma}.$$

(Eqn. 3-86)

The Lagrangian is

$$L = \int_{S_t} -\frac{1}{4} F^{\alpha\beta} F_{\alpha\beta} \, dV,$$

(Eqn. 3-87)

where $-\frac{1}{4} F^{\alpha\beta} F_{\alpha\beta} = \frac{1}{2}(E^2 - B^2)$.

The momentum conjugate to A_α is

$$\pi^\alpha(x) = \frac{\delta L}{\delta \dot{A}_\alpha(x)} = -E^\alpha(x),$$

(Eqn. 3-88)

And thus

$$E^2 = \pi^\alpha \pi_\alpha = \pi^\alpha (\dot{A}_\alpha + D_\alpha \Phi),$$

(Eqn. 3-89)

Finally

$$L = \int dx \left[\pi^\alpha(\dot{A}_\alpha + D_\alpha \Phi) - \frac{1}{2}(E^2 - B^2)\right]$$
$$= \int dx \left[\pi^\alpha \dot{A}_\alpha - \frac{1}{2}(\pi^2 - B^2) + \Phi D_\alpha \pi^\alpha\right].$$

(Eqn. 3-90)

Equivalently,

$$L = \int dx \left[E^{\alpha} \dot{A}_{\alpha} - \frac{1}{2}(E^2 - B^2) - \Phi D_{\alpha} E^{\alpha} \right].$$

(Eqn. 3-91)

The second term is the Hamiltonian and the last has the form of a Lagrange multiplier, Φ, multiplying the Gauss constraint

$$\pi = 2Kh^{1/2} \Rightarrow h^{1/2} p_{\alpha\beta} = \pi_{\alpha\beta} - \frac{1}{2} h_{\alpha\beta} \pi$$

(Eqn. 3-92)

$$\Rightarrow \mathcal{L} = N \left[h^{1/2}\, {}^3R + h^{-1/2} \left(\pi^{ab} \pi_{ab} - \frac{1}{2} \pi^2 \right) \right]$$

(Eqn. 3-93)

And

$$\mathcal{H} = -h^{1/2}\, {}^3R + \left(\pi^{ab} \pi_{ab} - \frac{1}{2} \pi^2 \right) h^{-1/2}$$

(Eqn. 3-94)

Claim:

$$I = \int dt \int dx \left[\pi^{ab} \dot{h}_{ab} - N\mathcal{H} - S^{\alpha} \mathcal{H}_{\alpha} \right]$$

Proof: $\int dx \left[\pi^{\alpha\beta} \dot{h}_{\alpha\beta} - S^{\alpha} \mathcal{H}_{\alpha} \right] = \int dx \left[\pi^{\alpha\beta} \dot{h}_{\alpha\beta} + 2S^{\alpha} D_{\beta} \pi^{\beta}_{\alpha} \right],$

$= \int dx\, \pi^{\alpha\beta} \left(\dot{h}_{\alpha\beta} - 2D_{\beta} S_{\alpha} \right)$ (after integ. by parts)

$= \int dx\, \pi^{\alpha\beta} (2NK_{\alpha\beta})$

$= \int dx \left(\pi^{\alpha\beta} \pi_{\alpha\beta} - \frac{1}{2} \pi^2 \right) 2N.$

Then

$$\int dt\, dx \left[\pi^{\alpha\beta} \dot{h}_{\alpha\beta} - N\mathcal{H} - S^{\alpha} \mathcal{H}_{\alpha} \right]$$

$$= \int dt\, dx \left\{ \left(\pi^{\alpha\beta} \pi_{\alpha\beta} - \frac{1}{2} \pi^2 \right) 2Nh^{-1} - N \left[-\, {}^3R + \left(\pi^{\alpha\beta} \pi_{\alpha\beta} - \frac{1}{2} \pi^2 \right) h^{-1} \right] \right\}$$

$= I.$

The momentum constraint

$$-\mathcal{H}_{\alpha} = D_{\beta} \pi^{\beta}_{\alpha} = 0$$

is like the Gauss constraint. The Hamiltonian is

$$H = \int dS N\, \mathcal{H}.$$

3.3 Boundary of a Boundary is zero [24]

In Algebraic Topology (review in Appendix) there is the identity that the exterior derivative of the exterior derivative of a 1-form ($\boldsymbol{A}$) is zero:

$$\boldsymbol{ddA} = 0.$$

(Eqn. 3-95)

Note that a 1-form is interchangeable with a contragradient vector in tensor notation. In component form, the exterior derivative on $\boldsymbol{A} = A_\alpha$ is:

$$\boldsymbol{dA} = d_\beta A_\alpha = A_{\beta,\alpha} - A_{\alpha,\beta}.$$

(Eqn. 3-96)

This is exactly the form of potential field used in the EM Field tensor 'F' when 'F' is written as a 2-form:

$$F_{\alpha\beta} = \boldsymbol{dA} = A_{\beta,\alpha} - A_{\alpha,\beta},$$

(Eqn. 3-97)

thus we automatically have, from topological identity, that:

$$\boldsymbol{dF} = 0.$$

(Eqn. 3-98)

This gives us the homogenous Maxwell relations, which is half-way, what of the equations with source? The other Maxwell relations, when cast in the 4-vector notation describing the current 4-vector, leads to:

$$F^{\alpha\beta}{}_{,\beta} = 4\pi j^\alpha, \;\; j^\alpha = \rho u^\alpha.$$

(Eqn. 3-99)

Making use of the dual transform we have:

$${}^*\boldsymbol{F}_{\mu\nu} = \frac{1}{2}\epsilon_{\alpha\beta\mu\nu}F^{\alpha\beta} \;\; and \;\; {}^*\boldsymbol{j}_{\beta\mu\nu} = \frac{1}{2}\epsilon_{\alpha\beta\mu\nu}j^\alpha.$$

(Eqn. 3-100)

So we can rewrite the source relations in dual form with greater clarity:

$$\boldsymbol{d}(\,{}^*\boldsymbol{F}) = 4\pi\;{}^*\boldsymbol{j}$$

(Eqn. 3-101)

where it is now trivial to have:

$$\boldsymbol{dd}\;{}^*\boldsymbol{F} = 0.$$

(Eqn. 3-102)

So, the full set of Maxwell's equations are he precise form required from topological identity alone, once a 1-form potential is posited to exist as the basis of the EM field tensor as indicated.

Let's now consider the boundary of a boundary identity in the context of General Relativity, does it uniquely specify the Einstein equations in this situation in a similar way? Yes, but significantly more work, here is an outline. To begin the homogenous equation are easy to show as before as they relate to the Bianchi identities:

First Bianchi Identity:

$$R^{\alpha}{}_{[\beta\gamma\delta]} = 0.$$

(Eqn. 3-103)

Second Bianchi Identity:

$$R^{\alpha}{}_{[\gamma\delta;\mu]} = 0.$$

(Eqn. 3-104)

Thus, $G^{\alpha}{}_{\beta,\alpha} = 0$, the sourceless Einstein Tensor equation is simply a form of the contracted second Bianchi identity.

To get the full form of the Einstein relation requires examination of the key forms that occur in a forms-based analysis of GR. This is precisely the Cartan formalism from Ch. 4. The key forms involved now are a connection 1-forms, the torsion 2-forms, and the curvature 2-forms. This analysis is shown in [24] and is left as an exercise.

Any gauge field with a principle fiber bundle formulation (with cross-section) will have equations of motion from topological identity alone. All the relevant fields of matter are gauge fields and Gravitation can be expressed in terms of a gauge field (albeit non-renormalizable) so here it also found to have topologically defined equations of motion.

3.4 Waves [24]

As with electromagnetic waves from whatever source, we want to consider simple, weak field, solution to the Einstein field equations (which will consist of wave solutions on a Minkowski background). Let's write our weak field $g_{\alpha\beta}$ in terms of the Minkowski metric $\eta_{\alpha\beta}$:

$$g_{\alpha\beta} = \eta_{\alpha\beta} + h_{\alpha\beta}.$$

(Eqn. 3-105)

In what follows we follow the succinct description given in [24], but even more succinctly, so for further details see [24]. What follows is a first-order expansion in the theory with perturbation parameter $|h_{\alpha\beta}| \ll 1$. Thus, to first order in $|h_{\alpha\beta}|$ we simply have:

$$g^{\alpha\beta} = \eta^{\alpha\beta} - h^{\alpha\beta}.$$

(Eqn. 3-106)

The definition of the Ricci tensor at first order is simply:

$$R^{(1)}_{\alpha\beta} = \frac{1}{2}\left(-\Box\, h_{\alpha\beta} + h^{\sigma}{}_{\beta,\sigma\alpha} + h^{\sigma}{}_{\alpha,\sigma\beta} - h_{,\alpha\beta}\right),$$

(Eqn. 3-107)

and $\Box = \eta^{\alpha\beta}\frac{\partial}{\partial x^\alpha}\frac{\partial}{\partial x^\beta}$ is the d'Alembertian operator. The Einstein Field equations are thus:

$$-\Box\, h_{\alpha\beta} + h^\sigma{}_{\beta,\sigma\alpha} + h^\sigma{}_{\alpha,\sigma\beta} - h_{,\alpha\beta} = 16\pi\left(T_{\alpha\beta} - \frac{1}{2}\eta_{\alpha\beta}T\right).$$

(Eqn. 3-108)

Consider the gauge transformation:

$$x'^\alpha = x^\alpha + \xi^\alpha.$$

(Eqn. 3-109)

At first order this means that:

$$h'_{\alpha\beta} = h_{\alpha\beta} - \xi_{\alpha,\beta} - \xi_{\beta,\alpha}.$$

(Eqn. 3-110)

If we choose

$$\Box\, \xi_\alpha = h^\sigma{}_{\alpha,\sigma} - \frac{1}{2}h^\sigma{}_{\alpha,\alpha}$$

(Eqn. 3-111)

which is known as Lorentz gauge, we get:

$$\Box\, h_{\alpha\beta} = -16\pi\left(T_{\alpha\beta} - \frac{1}{2}\eta_{\alpha\beta}T\right).$$

(Eqn. 3-112)

At component level, as in electromagnetism, the solution is the retarded potential. If free-space propagation ($T_{\alpha\beta} = 0$) we get the standard wave equation:

$$\Box\, h_{\alpha\beta} = 0.$$

(Eqn. 3-113)

Since we have set c=1, we see that this is a wave equation for perturbations that move at the speed of light. For plane-wave propagation along the z-axis and use of a transverse traceless gauge, we arrive at nonzero tensor components (upon taking real parts):

$$h_{xx} = -h_{yy} = A_+ e^{-i\omega(t-z)}$$

(Eqn. 3-114)

and

$$h_{xy} = h_{yx} = A_\times e^{-i\omega(t-z)}.$$

(Eqn. 3-115)

(Notation indicates quadrupole moments of two types.)

3.5 Exercises

(Ex. 3.1) Derive Eqn. 3-9.
(Ex. 3.2) Derive Eqn. 3-27.
(Ex. 3.3) Derive Eqn. 3-39.

(Ex. 3.4) Derive Eqn. 3-61.
(Ex. 3.5) Derive Eqn. 3-70.
(Ex. 3.6) Derive Eqn. 3-71.
(Ex. 3.7) Derive Eqn. 3-78.
(Ex. 3.8) Derive Eqn. 3-101.
(Ex. 3.9) Derive Eqn. 3-106.
(Ex. 3.10) Derive Eqn. 3-107.
(Ex. 3.11) Derive the relation described at the end of Sec. 3.3.

Chapter 4. Exterior Products and the Cartan Method

General Relativity analysis usually requires computation of the Riemann tensor for the space-time of interest. The most economical way to do this is by use of exterior products and the Cartan Method (a.k.a. Cartan Calculus).

4.1 Derivation of the Cartan calculus

Beginning with the conventional definition of exterior derivative, we add:

(1) d produces a $(p+1)$ from $d\sigma$ from a p-form σ.
(2) $d(\alpha \wedge \beta) = d\alpha \wedge \beta + (-1)^p \alpha \wedge d\beta$ and $d^2 = 0$
(3) $df = \nabla_a f$ with ∇_a the covariant derivative.

The exterior (antisymmetric) differentiation of tensors is defined by:

$$\langle df, \vec{u} \rangle = \partial_\mu f \qquad \langle d\vec{v}, \vec{u} \rangle = \nabla_u v$$

(Eqn. 4-1)

where

$$d(\bar{\alpha} \wedge \bar{\beta}) = (d\bar{\alpha}) \wedge \bar{\beta} + (-1)^P \bar{\alpha} \wedge d\bar{\beta}$$

(Eqn. 4-2)

where α is a p-form and β is a q–form. We extend the definition to exterior derivative on exterior product of tensor valued p-form $\vec{S}$ with ordinary q-form β thus:

$$d(\bar{S} \wedge \bar{\beta}) = d\grave{S} \wedge \bar{\beta} + (-1)^P \bar{S} \wedge d\bar{\beta}.$$

(Eqn. 4-3)

So, if $S = S^{\alpha\beta}_{|\gamma\delta|} e_\alpha e_\beta dx^\gamma \wedge dx^\delta$ then

$$dS = d\left(e_\alpha e_\beta S^{\alpha\beta}_{|\gamma\delta|}\right) \wedge \left(dx^\gamma \wedge dx^\delta\right)$$

(Eqn. 4-4)

Now, lets apply the formalism: $e^a_\mu w^\mu_b = \delta^a_b \Rightarrow d(\delta^a_b) = 0$, so:

$$de^a_\mu \wedge w^\mu_b + e^a_\mu dw^\mu_b = 0 \text{ (here } e^a_\mu \text{ is a vector valued 0-form)}$$

(Eqn. 4-5)

Let $d\left(e_\mu\right)^a_b = (e_\nu)^a \left(w^\nu_\mu\right)_b$, i. e., $de_\mu = e_\nu w^\nu_\mu$ in less explicit notation, where w^ν_μ is a 1-form. Since $\nabla_\beta e_\alpha = e_\mu \Gamma^\mu_{\alpha\beta}$ we find that $\left(w^\nu_\mu\right)_b = \Gamma^\nu_{\mu\gamma}(w^\gamma)_b$, thus $w^\nu_\mu = \Gamma^\mu_{\mu\lambda} w^\lambda$. Now, consider $V^a = \left(e_\mu\right)^a V^\mu$ lets take dV^a:

$$dv = e_\mu\left(dv^\mu + w^\mu_\nu v^\mu\right) \Rightarrow d^2 v \text{ follows.}$$

(Eqn. 4-6)

Now the exterior derivative for a 1-form contracted on a bivector $u \wedge v$:

$$\langle d\alpha, u \wedge v\rangle = \partial_u\langle \alpha, v\rangle - \partial_v\langle \alpha, u\rangle - \langle \alpha, [u, v]\rangle$$

(Eqn. 4-7)

Generalizing to a tensor valued 1-form:

$$\langle dS, u \wedge v\rangle = \partial_u\langle S, v\rangle - \partial_v\langle S, u\rangle - \langle S, [u, v]\rangle$$

(Eqn. 4-8)

So, for $S = dw$ a vector valued 1-form:

$$\langle d^2 w, u \wedge v\rangle = \nabla_u \nabla_v w - \nabla_v \nabla_u w - \nabla_{[u,v]} w = R(u, v) w$$

(Eqn. 4-9)

Note that the general p-form is a completely antisymmetric tensor of rank $\binom{0}{P}$:

$$\alpha = \frac{1}{P!} \alpha_{i_1 i_2 \dots i_P} \omega^{i_1} \wedge \omega^{i_2} \wedge \dots \wedge \omega^{i_P} \equiv \alpha_{|i_1 i_2 \dots i_P|} \omega^{i_1} \dots \omega^{i_P}$$

(Eqn. 4-10)

where the vertical bars denote summation only over $i_1 < i_2 < \cdots < i_P$.
Example, suppose you have:

$\vec{u}$ is a vector-valued 0-form (vector)
$\bar{\sigma}$ is a scalar valued 1-form (1 form)

Then:

$$d(u \otimes \sigma) = du \wedge \sigma + u \otimes d\sigma \qquad \textit{shortened } \; d(u\sigma)$$
$$= du \wedge \sigma + ud\sigma.$$

(Eqn. 4-11)

Cartan differential forms (1928, 1946 Lecons sur la Geometrie des Espares de Riemann [25,26]) then start by considering

$$\vec{v} = \vec{e}_\mu V^\mu$$

(Eqn. 4-12)

for which

$$d\vec{v} = d\vec{e}_\mu V^\mu + e_\mu dv^\mu.$$

(Eqn. 4-13)

Let's now write the differential of the typical vector-valued 1-form e_μ as $de_\mu = e_v \omega_\mu^v$ there the "components" ω_μ^v in the expansion of de_μ are 1-forms. Furthermore, note that $\omega_\mu^v = \Gamma_{\mu\lambda}^v \omega^\lambda$. So, the expansion of the vector valued one-form is:

$$dv = e_\mu \left(dV^\mu + \omega_v^\mu v^v\right)$$

(Eqn. 4-14)

Similarly:

$$\begin{aligned} d^2 v &= de_\alpha \wedge (dV^\alpha + \omega_v^\alpha v^v) + e_\mu\left(d^2 v^\mu + d\omega_v^\mu v^v - \omega_v^\mu \wedge dv^v\right) \\ &= e_\mu\left(\omega_\alpha^\mu \wedge dv^\alpha + \omega_\alpha^\mu \wedge \omega_v^\alpha v^v + d^2 v^\mu + d\omega_v^\mu v^v - \omega_v^\mu \wedge dv^\alpha\right) \\ &= e_\mu\left(d\omega_v^\mu + \omega_\alpha^\mu \wedge \omega_v^\alpha\right) v^v \\ &= e_\mu R_v^\mu v^v \end{aligned}$$

(Eqn. 4-15)

Now,

$$\langle d\alpha, u \wedge v\rangle = \partial_u \langle \alpha, v\rangle - \partial_v \langle \alpha, u\rangle - \langle \alpha, [u, v]\rangle$$

(Eqn. 4-16)

Generalizes to

$$\langle dS, u \wedge v\rangle = \nabla_u \langle S, v\rangle - \nabla_v \langle S, u\rangle - \langle S, [u, v]\rangle$$

(Eqn. 4-17)

So,

$$\langle d^2\omega, u \wedge v\rangle = \nabla_u \nabla_v \omega - \nabla_v \nabla_u \omega - \nabla_{[u,v]}\omega = R'(u, v)\omega$$

(Eqn. 4-18)

Thus, the $\binom{1}{1}$-tensor valued 2-form R evaluated on the bivector (parallelogram) $u \wedge v$, is identical with curvature operator $R'(u, v)$ above:

$$\langle R, u \wedge v\rangle = R(u, v).$$

(Eqn. 4-19)

So,

$$e_\mu \otimes \omega^v \langle R_v^\mu u \wedge v\rangle = e_\mu \otimes \omega^v R_{v\alpha\beta}^\mu u^\alpha u^\beta$$

(Eqn. 4-20)

Thus,

$$R_v^\mu = R_{v|\alpha\beta|}^\mu \omega^\alpha \omega^\beta \; with\; sum\; over\; \alpha, \beta\; restricted\; to\; \alpha < \beta,$$

(Eqn. 4-21)

provides packaging of 21 curvature components into six curvature 2-forms.

The $\Gamma^{\mu}_{\alpha\beta} \to R^{\mu}_{\nu\alpha\beta}$ calculation becomes a $\omega^{\mu}_{\nu} \to R^{\mu}_{\nu}$ calculation.

The $g_{\mu\nu} \to \Gamma^{\mu}_{\alpha\beta}$ calculation becomes a $g_{\mu\nu} \to \omega^{\mu}_{\nu}$ calculation.

The $g_{\mu\nu} \to \omega^{\mu}_{\nu}$ calculation proceeds from two features:

(1) The symmetry of the covariant derivatives

(2) And its compatibility with the metric

Condition (1) appears hidden in $d^2P = 0 \;\; where \; \langle dP, v\rangle = V,$, i.e. dP is a special form of the generic $\binom{1}{1}$ tensor $T = e_{\mu}T^{\mu}_{\nu}\omega^{\nu}$ with $T^{\mu}_{\nu} = \delta^{\mu}_{\nu}$:

$$dP = e_{\mu}\omega^{\mu}$$

(Eqn. 4-22)

Putting this in $\langle dS, u \wedge v\rangle = \nabla_u\langle S, v\rangle - \nabla_v\langle S, u\rangle - \langle S, [u, v]\rangle$ we get:

$$\langle d^2P, u \wedge v\rangle = \nabla_u v - \nabla_v u - [u, v] = 0.$$

(Eqn. 4-23)

Note that ∇ is said to be symmetric (torsion free) when $\nabla_u v - \nabla_v u - [u, v]$. So, $d^2P = 0$ demands the symmetry of the covariant derivative and vice-versa:

$$0 = d^2P = d(e_{\mu}\omega^{\mu}) = de_{\mu} \wedge \omega^{\mu} + e_{\mu}d\omega^{\mu} = e_{\mu}(\omega^{\mu}_{\nu} \wedge \omega^{\nu} + d\omega^{\mu})$$

(Eqn. 4-24)

$$0 = d\omega^{\mu} + \omega^{\mu}_{\nu} \wedge \omega^{\nu} \;\text{ from symmetry.}$$

(Eqn. 4-25)

Consider (2) can be expressed as:

$$d(u \cdot v) = (du) \cdot v + u \cdot (dv) \quad \text{thus} \;\; d(\cdot) = 0$$

(Eqn. 4-26)

Let $u = e_{\mu}$ and $v = e_{\nu}$:

$$d(g_{\mu\nu}) = (de_{\mu}) \cdot e_{\nu} + e_{\mu} \cdot (de_{\nu}) = (e_{\alpha}\omega^{\alpha}_{\mu}) \cdot e_{\nu} + e_{\mu} \cdot (e_{\alpha}\omega^{\alpha}_{\mu})$$

(Eqn. 4-27)

then

$$dg_{\mu\nu} = \omega_{\nu\mu} + \omega_{\mu\nu} \;\text{ from compatability } \; and \quad \omega_{\mu\nu} = g_{\mu\alpha}\omega^{\alpha}_{\nu} = \Gamma_{\mu\nu\alpha}\omega^{\alpha}.$$

(Eqn. 4-28)

Returning to the expression

$$\langle d^2\omega, u \wedge v\rangle = \nabla_u\nabla_v\omega - \nabla_v\nabla_u\omega - \nabla_{[u,v]}\omega = R(u,v)\omega,$$

(Eqn. 4-29)

let's consider evaluation of the Riemann tensor from the Cartan forms. Now $R(u,v)$ above is the curvature operator which is related to Riemann by:

$$\text{Riemann:} \qquad (\vec{w}^\alpha, \vec{e}_\beta, \vec{e}_\gamma, \vec{e}_\delta) \equiv \langle \vec{w}^\alpha, R(\vec{e}_\gamma, \vec{e}_\delta)\vec{e}_\beta\rangle$$

(Eqn. 4-30)

So, in our notation:

$$R^i_{jk\ell} = \left(R^i_j\right)_{ab} e^a_k e^b_\ell$$

(Eqn. 4-31)

Recall $de_\mu \wedge w^\mu + e_\mu dw^\mu = 0$

$$e_\mu\left(w^\mu_\nu w^\nu + dw^\mu\right) = 0 \Rightarrow \underline{dw^\mu + w^\mu_\nu \wedge w^\mu = 0}$$

(Eqn. 4-32)

Now compatibility of covariant derivative with the metric means:

$$\left.\begin{aligned} d(u^a g_{ab} v^b) &= (du^a)g_{ab}v^b + u^a g_{ab} dv^b \\ d(u,v) &= (du)\cdot v + u\cdot(dv) \end{aligned}\right\} \underline{d(g_{ab}) = 0}$$ (for orthonomal coord frame)

(Eqn. 4-33)

Let $u = e_\mu, v = e_\nu$

$$(d\vec{e}_\mu)\bar{g}^{\mu\nu}\vec{e}_\nu + \vec{e}_\mu d(g^{\mu\nu})\vec{e}_\nu + \vec{e}_\mu g^{\mu\nu}(d\vec{e}_\nu) = 0$$

$$\vec{e}_\mu w^\mu_\gamma g^{\gamma\nu}\vec{e}_\nu + \vec{e}_\mu d(g^{\mu\nu})\vec{e}_\nu + e_\mu g^{\mu\delta} e_\nu w^\nu_\delta = 0$$

$$-d(g^{\mu\nu}) = w^\mu_\gamma g^{\gamma\nu} + w^\nu_\delta g^{\mu\delta} \Rightarrow \underline{w_{\mu\nu} + w_{\nu\mu} = 0}$$

(Eqn. 4-34)

$$v = e_\mu v^\mu$$

$$dv = e_\mu\left(dv^\mu + w^\mu_\nu v^\nu\right) \Rightarrow d^2v = e_\mu w^\mu_\alpha \wedge (dv^\alpha + w^\alpha_\nu v^\nu) + e_\mu w^\mu_\nu v^\nu$$

$$d^2v = e_\mu\left(dw^\mu_\nu + w^\mu_\alpha \wedge w^\alpha_\nu\right)v^\nu = \vec{e}_\mu R^\mu_\nu v^\nu$$

$$\underline{R^\mu_\nu \equiv dw^\mu_\nu + w^\mu_\alpha \wedge w^\alpha_\nu}$$

(Eqn. 4-35)

Synopsis of Computation of Curvature using exterior differential forms:

(1) Symmetry of covariant differentiation (i.e., torsion free):

$$\nabla_U V - \nabla_V U = [U,V]$$

(Eqn. 4-36)

Consider a tensor valued one-form S of any rank, it satisfies:

$$\langle dS, U \wedge V\rangle = \nabla_U\langle S, V\rangle - \nabla_V\langle S, U\rangle - \langle S, [U, V]\rangle$$

(Eqn. 4-37)

If $S = dw$, a vector valued 1-form, then using $\langle dw, U\rangle = \nabla_U w$ we get:

$$\langle d^2w, U \wedge V\rangle = \nabla_U\nabla_V w - \nabla_V\nabla_U w - \nabla_{[U,V]}w = \mathcal{R}(U, V)w$$

(Eqn. 4-38)

Where $\mathcal{R}^\mu_\nu = R^\mu_{\nu|\propto\beta|}w^\propto \wedge w^\beta$. Now consider $S = e_\mu w^\mu = dP$ in Cartan's notation:

$$\langle d^2P, U \wedge V\rangle = \nabla_U V - \nabla_V U - [U, V] = 0 \Longrightarrow \underline{d^2P = 0}$$

(Eqn. 4-39)

$0 = d(e_\mu w^\mu) = de_\mu \wedge w^\mu + e_\mu dw^\mu \rightarrow\ de_\nu = e_\mu w^\mu_\nu \rightarrow\ 0 = dw^\mu + w^\mu_\nu \wedge w^\mu$

(2) Compatibility :

$$d(u \cdot v) = (du) \cdot v + u \cdot (dv)$$
$$d(e_\mu \cdot e_\nu) = de_\mu \cdot e_\nu + e_\mu \cdot (de_\nu)$$
$$dg_{\mu\nu} = w_{\mu\nu} + w_{\nu\mu}$$

(Eqn. 4-40)

For an orthonormal frame $dg_{\mu\nu} = 0$ and $w_{\mu\nu} = -w_{\nu\mu}$, leaving only six $w_{\mu\nu}$ to solve for in four dimensions: $\mathcal{R}^\mu_\nu = dw^\mu_\nu + w^\mu_\alpha \wedge w^\alpha_\nu$.

4.2 Riemann tensor for 2-sphere

Let's use Cartan calculus to write down the independent components of the Riemann tensor in an orthonormal frame for the 2-sphere with metric:

$$ds^2 = a^2 d\Omega^2 = a^2(d\theta^2 + \sin^2\theta\, d\phi^2)\,,$$

(Eqn. 4-41)

and find the Ricci scalar.

Solution

We have:

$$R^i_{jk\ell} = (R^i_j)_{ab} e^a_k e^b_\ell = \left[(dw^i_j)_{ab} + (w^i_k \wedge w^k_j)\right] e^a_k e^b_\ell\,,$$

where $dw^i = -w^i_j \wedge w^j$. For $ds^2 = (w^\theta)^2 + (w^\phi)^2$ we have:

$$w^\theta = a d\theta \quad and \quad w^\phi = a\sin\theta\, d\phi\,.$$

We have: $\begin{cases} dw^\mu = -w^\mu_\nu \wedge w^\nu \quad (w_{\mu\nu} + w_{\nu\mu} = 0) \\ R^\mu_\nu = dw^\mu_\nu + w^\mu_\alpha \wedge w^\alpha_\nu \end{cases}$

Thus,

$dw^\theta = ad^2\theta = 0 \quad \rightarrow \quad w_\phi^\theta = 0 \ or \ w_\phi^\theta \propto w^\phi$
$dw^\phi = a\cos\theta\, d\theta \wedge d\phi = ad\theta \wedge \cos\theta\, d\phi = w^\theta \wedge \cos\theta\, d\phi$
$= -\cos\theta\, d\phi \wedge w^\theta$

Thus,

$$w_\theta^\phi = \cos\theta\, d\phi.$$

And we get:
$R_\mu^\mu = 0$
$R_\theta^\phi = dw_\alpha^\phi + w_\alpha^\phi \wedge w_\theta^\alpha = -\sin\theta\, d\theta \wedge d\phi$
$R_\theta^\phi = \frac{-1}{a^2} w^\theta \wedge w^\phi$
$R_{\theta k\ell}^\phi = -\frac{1}{a^2}\left(w^\theta \wedge w^\phi\right)_{ab} e_k^a e_\ell^b$
Thus,
$R_{\theta\phi\theta}^\phi = \frac{1}{a^2}$ and the rest zero. The Ricci scalar is: $R = R_\mu^\mu = R_\phi^\phi = \frac{1}{a^2}$.

4.3 Riemann tensor for 3-sphere
Let's use the Cartan calculus to write down the independent components of the Riemann tensor in an orthonormal frame for the 3-sphere with metric:

$$ds^2 = a^2(dx^2 + \sin^2 x\,[d\theta^2 + \sin^2\theta\, d\phi^2])$$

(Eqn. 4-42)

and verify that $R_{abcd} = a^{-2}(g_{ac}g_{bd} - g_{ad}g_{bc})$.

Solution
We have $ds^2 = a^2(dx^2 + \sin^2 x\,[d\theta^2 + \sin^2\theta\, d\phi^2]) = (w^x)^2 + \left(w^\theta\right)^2 + \left(w^\phi\right)^2$, thus:
$w^x = adx \qquad dw^x = 0$
$w^\theta = a\sin x\, d\theta \qquad dw^\theta = a\cos x\, dx \wedge d\theta$
$= -\cos x\, d\theta \wedge w^x$
$w^\phi = a\sin x \sin\theta\, d\phi \qquad dw^\phi$
$= a\cos x\sin\theta\, dx \wedge d\phi + a\sin x\cos\theta\, d\theta \wedge d\phi$
$= -\cos x\sin\theta\, d\phi \wedge w^x +$
$\cos\theta\, d\phi \wedge w^\theta$
Using $dw^\mu = -w_\nu^\mu \wedge w^\nu$:

$$dw^x : \begin{cases} w_\theta^x = 0 \text{ or } \propto w^\theta \\ w_\phi^x = 0 \text{ or } \propto w^\phi \end{cases}$$

$$dw^\theta : \begin{cases} w_x^\theta = \cos x \, d\theta \propto w^\theta \\ w_\phi^\theta = 0 \text{ or } \propto w^\phi \end{cases}$$

$$dw^\phi : \begin{cases} w_x^\phi = \cos x \sin\theta \, d\phi \propto w^\phi \\ w_\theta^\phi = \cos\theta \, d\phi \propto w^\phi \end{cases}$$

These choices are consistent:

$$w_x^\theta = \cos x \, d\theta \qquad \Rightarrow \qquad dw_\theta^x = \sin x \, dx \wedge d\theta = \frac{1}{a^2} w^x \wedge w^\theta$$

$$w_x^\phi = \cos x \sin\theta \, d\phi \Rightarrow$$
$$dw_\phi^x = \sin x \sin\theta \, dx \wedge d\phi - \cos x \cos\theta \, d\theta \wedge d\phi$$

$$w_\theta^\phi = \cos\theta \, d\phi \qquad \Rightarrow \qquad dw_\phi^\theta = \sin\theta \, d\theta \wedge d\phi = \frac{1}{a^2} \frac{1}{\sin^2 x} w^\theta \wedge w^\phi$$

Using $R_\nu^\mu = dw_\nu^\mu + w_\alpha^\mu \wedge w_\nu^\alpha$ *and* $dw_\phi^x = \frac{1}{a^2} w^x \wedge w^\phi - \frac{1}{a^2} \frac{\cos x \cos\theta}{\sin^2 x \sin\theta} w^\theta \wedge w^\phi$, we get:

$$R_\theta^x = \frac{1}{a^2} w^x \wedge w^\theta + w_\alpha^x \wedge w_\theta^\alpha$$
$$= \frac{1}{a^2} w^x \wedge w^\theta + \cos x \sin\theta \, d\phi \wedge \cos\theta \, d\phi$$

$$R_\theta^x = \frac{1}{a^2} w^x \wedge w^\theta \Rightarrow R_{x\theta x\theta} = \frac{1}{a^2} = \frac{1}{a^2}\left(\underbrace{\eta_{xx}\eta_{\theta\theta}}_{1} - \underbrace{\eta_{x\theta}\eta_{\theta x}}_{0}\right)$$

$$R_\phi^x = dw_\phi^x + w_\theta^x \wedge w_\phi^\theta = dw_\phi^x + \cos x \, d\theta \wedge \cos\theta \, d\phi = \frac{1}{a^2} w^x \wedge w^\phi$$

$$R_\phi^x = \frac{1}{a^2} w^x \wedge w^\phi \Rightarrow R_{x\phi x\phi} = \frac{1}{a^2} = \frac{1}{a^2}\left(\underbrace{\eta_{xx}\eta_{\phi\phi}}_{1} - \underbrace{\eta_{x\phi}\eta_{\phi x}}_{0}\right)$$

$$R_\phi^\theta = \frac{1}{a^2} \frac{1}{\sin^2 x} w^\theta \wedge w^\phi + w_x^\theta \wedge w_\phi^x$$
$$= \frac{1}{a^2} \frac{w^\theta \wedge w^\phi}{\sin^2 x} + \cos x \, d\theta \wedge \cos x \sin\theta \, d\phi$$
$$= \frac{1}{a^2} w^\theta \wedge w^\phi \left(\frac{1}{\sin^2 x} - \frac{\cos^2 x}{\sin^2 x}\right) = \frac{1}{a^2} w^\theta \wedge w^\phi$$

$$R_\phi^\theta = \frac{1}{a^2} w^\theta \wedge w^\phi \Rightarrow R_{\theta\phi\theta\phi} = \frac{1}{a^2} = \frac{1}{a^2}(\eta_{\theta\theta}\eta_{\phi\phi} - \eta_{\theta\phi}\eta_{\phi\theta})$$

So, $R_{ijk\ell} = \frac{1}{a^2}(\eta_{ij}\eta_{k\ell} - \eta_{i\ell}\eta_{jk})$ follows from above, then $R_{ijk\ell} w_a^i w_b^j w_c^k w_d^\ell$ gives:

$$R_{abcd} = a^{-2}(g_{ac}g_{bd} - g_{ad}g_{bc}).$$

4.4 Riemann tensor for the FRW space-times

Let's use Cartan calculus to find the independent components of the Riemann and Ricci tensor in an orthonormal frame for the FRW spacetime metric

$$ds^2 = -dt^2 + a(t)^2(dx^2 + \sin^2 x\, d\Omega^2).$$

(Eqn. 4-43)

Solution

We have $ds^2 = -dt^2 + a(t)^2(dx^2 + \sin^2 x\, d\Omega^2) = -(w^t)^2 + (w^x)^2 + (w^\theta)^2 + (w^\phi)^2$, thus:

$$w^t = dt \qquad dw^t = 0$$

$$w^x = a(t)dx \qquad dw^x = \dot{a}dt \wedge dx = -\frac{\dot{a}}{a} w^x \wedge w^t$$

$$w^\theta = a(t)\sin x\, d\theta \qquad dw^\theta = \dot{a}\sin x\, dt \wedge d\theta + a\cos x\, dx \wedge d\theta$$

$$= \frac{\dot{a}}{a} w^\theta \wedge w^t - \frac{\cot x}{a} w^\theta \wedge w^x$$

$$w^\phi = a(t)\sin x \sin\theta\, d\phi$$

$$dw^\phi = \dot{a}\sin x \sin\theta\, dt \wedge d\phi + a\cos x \sin\theta\, dx \wedge d\phi + a\sin x\cos\theta\, d\theta \wedge d\phi$$

$$= \frac{-\dot{a}}{a} w^\phi \wedge w^t - \frac{\cot x}{a} w^\phi \wedge w^x - \frac{\cot x}{a\sin x} w^\phi \wedge w^\theta .$$

Thus,

$$w_t^\phi = \frac{-\dot{a}}{a} w^\phi = \dot{a}\sin x \sin\theta\, d\phi = \frac{\dot{a}}{a} w^\phi \to dw_t^\phi$$

$$= \left(\frac{\ddot{a}}{a} - \left(\frac{\dot{a}}{a}\right)^2\right) w^t \wedge w^\phi + \frac{\dot{a}}{a} dw^\phi$$

$$w_x^\phi = \frac{\cot x}{a} w^\phi = \cos x \sin\theta\, d\phi$$

$$w_\theta^\phi = \frac{\cot\theta}{a\sin x} w^\phi = \cos\theta\, d\phi$$

$$w_t^\theta = \frac{\dot{a}}{a} w^\theta = \dot{a}\sin x\, d\theta = \frac{\dot{a}}{a} w^\theta \to dw_t^\theta$$

$$= \left(\frac{\ddot{a}}{a} - \left(\frac{\dot{a}}{a}\right)^2\right) w^t \wedge w^\theta + \frac{\dot{a}}{a} dw^\theta$$

$$w_x^\theta = \frac{\cot x}{a} w^\theta = \cos x\, d\theta$$

$$w_t^x = \frac{\dot{a}}{a} w^x = \dot{a}dx = \frac{\dot{a}}{a} w^x \to dw_t^x = \left(\frac{\ddot{a}}{a} - \left(\frac{\dot{a}}{a}\right)^2\right) w^t \wedge w^x + \frac{\dot{a}}{a} dw^x$$

We can use the results from the 3-sphere to help now:

${}^{(4)}R_j^i = {}^{(3)}R_j^i + w_t^i \wedge w_j^t$ for $i, j \neq t$

$= {}^{(3)}R^i_j + w^i_t \wedge w^j_t$ (since lowering t introduces an additional negative)

${}^{(4)}R_{ijk\ell} = {}^{(3)}R_{ijk\ell} + \left(w_{it} \wedge w_{jt}\right)_{ab} e^a_k e^b_\ell$ (since $w_{it} = \frac{\dot{a}}{a} w_i$)

$$= {}^{(3)}R_{ijk\ell} + \left(\frac{\dot{a}}{a}\right)^2 \left(w^i_a \wedge w^j_b\right) e^a_k e^b_\ell$$

$$= {}^{(3)}R_{ijk\ell} + \left(\frac{\dot{a}}{a}\right)^2 \left(\eta_{ij}\eta_{k\ell} - \eta_{i\ell}\eta_{jk}\right)$$

This simplifies to:

$${}^{(4)}R_{ijk\ell} = \left(\frac{1+\dot{a}^2}{a^2}\right)\left(\eta_{ik}\eta_{j\ell} - \eta_{i\ell}\eta_{jk}\right) \quad \text{for} \quad i,j \neq t$$

Now, for $j = t$, $(i \neq t)$, and using $A = \left(\frac{\ddot{a}}{a} - \left(\frac{\dot{a}}{a}\right)^2\right)$:

$${}^{(4)}R^i_t = A w^t \wedge w^i + \left(\frac{\dot{a}}{a}\right) dw^i + w^i_\alpha \wedge w^\alpha_t$$

$$= A w^t \wedge w^i + \left(\frac{\dot{a}}{a}\right) dw^i + w^i_k \wedge \left(\left(\frac{\dot{a}}{a}\right) w^k\right)$$

$$= A w^t \wedge w^i + \left(\frac{\dot{a}}{a}\right)\left[\underbrace{dw^i + w^i_\alpha \wedge w^\alpha}_{=0} - \underbrace{w^i_t}_{\left(\frac{\dot{a}}{a}\right) w^i} \wedge w^t\right] = \left[\frac{\ddot{a}}{a} - \left(\frac{\dot{a}}{a}\right)^2 + \left(\frac{\dot{a}}{a}\right)^2\right] w^t \wedge w^i$$

${}^{(4)}R_{itti} = \left(\frac{\ddot{a}}{a}\right)$ so:

$${}^{(4)}R^t_{iti} = \left(\frac{\ddot{a}}{a}\right)$$

Consider $i,j \neq t$ comp. of Ricci:

$$R_{ij} = R^\alpha_{i\alpha j} = R^t_{itj} + R^k_{ikj} = \left(\frac{\ddot{a}}{a}\right)\eta_{ij} + \left(\frac{1+\dot{a}^2}{a^2}\right)\left(\eta_{ij}\eta^k_k - \eta^k_j\eta_{ik}\right)$$

Thus,

$R_{ij} = \left[\left(\frac{\ddot{a}}{a}\right) + 2\left(\frac{1+\dot{a}^2}{a^2}\right)\right]\eta_{ij}$ and $R_{tt} = R^k_{tkt} = -R^t_{ktk} = -3\left(\frac{\ddot{a}}{a}\right) \Rightarrow R_{tt} = 3\left(\frac{\ddot{a}}{a}\right)$.

The Ricci scalar is thus:

$R = R^\mu_\mu = 3\left[\left(\frac{\ddot{a}}{a}\right) + 2\left(\frac{1+\dot{a}^2}{a^2}\right)\right] + 3\left(\frac{\ddot{a}}{a}\right) = 6\left[\frac{\ddot{a}}{a} + \frac{(1+\dot{a}^2)}{a^2}\right]$ and we get:

$$R = 6\left[\frac{\ddot{a}}{a} + \frac{(1+\dot{a}^2)}{a^2}\right].$$

(Eqn. 4-44)

4.5 Riemann tensor for the Kasner space-times

Let's use Cartan calculus to find the independent components of the Riemann and Ricci tensor for the Kasner spacetime with metric

$$ds^2 = -dt^2 + t^{(2p_1)}(dx^1)^2 + t^{(2p_2)}(dx^2)^2 + t^{(2p_3)}(dx^3)^2 .$$

(Eqn. 4-45)

Solution

We have to start:

$$w^t = dt \qquad dw^t = 0$$

$$\left.\begin{aligned} w^1 = t^{P_1}dx^1 \quad & dw^1 = \frac{P_1}{t}w^1 \wedge w^t \Rightarrow w_t^1 = -\frac{P_1}{t}w^1 \\ w^2 = t^{P_2}dx^2 \quad & dw^2 = \frac{P_2}{t}w^2 \wedge w^t \Rightarrow w_t^2 = -\frac{P_2}{t}w^2 \\ w^3 = t^{P_3}dx^3 \quad & dw^3 = \frac{P_3}{t}w^3 \wedge w^t \Rightarrow w_t^3 = -\frac{P_3}{t}w^3 \end{aligned}\right\} w_t^i = -\frac{P_i}{t}w^i$$

Thus,

$$dw_t^i = \frac{P_i}{t^2}w^t \wedge w^1 + \frac{P_i^2}{t^2}w^1 \wedge w^t = \frac{P_i(1-P_i)}{t^2}w^t \wedge w^i$$

$$R_v^\mu = dw_v^\mu + w_\alpha^\mu \wedge w_v^\alpha$$

For $\mu, v \neq$ t

$$R_j^i = w_\alpha^i \wedge w_j^\alpha = w_t^i \wedge w_j^t = \frac{P_iP_j}{t^2}w^i \wedge w^j \Rightarrow R_{jij}^i = \frac{P_iP_j}{t^2} ,$$

where the repeated indices in the last equation are not summed.

For $\mu = i, v = t$:

$$R_t^i = dw_t^i + w_\alpha^i \wedge w_t^\alpha = dw_t^i = \frac{P_i(1-P_i)}{t^2}w^i \wedge w^i \Rightarrow R_{tti}^i = \frac{P_i(1-P_i)}{t^2}$$

$R_{\mu v} = R_{\mu\alpha v}^\alpha$ for $\mu, v \neq t$:

$$R_{j\alpha j}^\alpha = \frac{1}{t^2}[(P_1 + P_2 + P_3 - P_j)P_j + P_j(1-P_j)].$$

so

$$R_{jj} = \frac{1}{t^2}[P_j(P_1 + P_2 + P_3 + 1) - 2P_j^2].$$

$$R_{tat}^\alpha = \sum_i -\frac{P_i(1-P_i)}{t^2} = R_{tt} = \frac{1}{t^o}[P_1^2 + P_2^2 + P_3^2 - P_1 - P_2 - P_3]$$

To satisfy the vacuum Einstein equation.: $R_{\mu v} = 0$.

$R_{tt} = 0$ gives $P_1^2 + P_2^2 + P_3^2 = P_1 + P_2 + P_3$
$R_j^j = 0$ gives $(P_1 + P_2 + P_3)^2 + (P_1 + P_2 + P_3) = 2(P_1^2 + P_2^2 + P_3^2)$
Using 1st eqn. $\Rightarrow (P_1 + P_2 + P_3)^2 = +(P_1 + P_2 + P_3) \Rightarrow (P_1 + P_2 + P_3) = 0$ or 1.

So, $\begin{cases} P_1 + P_2 + P_3 = 1 \\ P_1^2 + P_2^2 + P_3^2 = 1 \end{cases}$

(Eqn. 4-46)

4.6 Riemann tensor for the Mixmaster space-time [21]
If the symmetry of the 3-sphere is broken by replacing δ_{ij} in the previous problem by a general diagonal metric, the corresponding homogeneous space-time is the Mixmaster universe:

$$ds^2 = -dt^2 + e^{2\alpha}\{e^{2\beta_1}(w^1)^2 + e^{2\beta_2}(w^2)^2 + e^{2\beta_3}(w^3)^2\},$$

(Eqn. 4-47)

where α and β_i are functions only of t and $\beta_1 + \beta_2 + \beta_3 = 0$. Find the Einstein equations for a vacuum Mixmaster universe.

Solution
Mixmaster has metric:

$$ds^2 = -dt^2 + e^{2\alpha}\left\{e^{2\beta_1}\left(w^{1'}\right)^2 + e^{2\beta_2}\left(w^{2'}\right)^2 + e^{2\beta_3}\left(w^{3'}\right)^2\right\}$$

where $\alpha = \alpha(t), \beta_i = \beta_i(t), \beta_1 + \beta_2 + \beta_3 = 0$. Recall $dw^{i'} \frac{1}{2} C^{i'}_{j'k'} w^{j'} \wedge w^{k'}$ and for the adjoint rep $C^{i'}_{j'k'} = \epsilon^{i'}_{j'k'}$. So, $dw^{i'} = w^{3'} \wedge w^{2'}$ and cyclic permutations for the others. Let $w^t = dt$ and $w^i = e^{\alpha+\beta_i} w^{i'}$. We're interested in $\frac{1}{2} R^{\mu}_{\nu\gamma\delta} w^\gamma \wedge w^\delta = dw^\mu_\nu + w^\mu_\rho \wedge w^\rho_\nu$. Due to symmetries notational and otherwise we need only consider $R^1_{2\gamma\delta}$ and $R^1_{t\gamma\delta}$, from which we can then get the others by symmetry arguments.

$$\frac{1}{2} R^1_{2\gamma\delta} w^\gamma \wedge w^\delta = dw^1_2 + w^1_3 \wedge w^3_2 + w^1_t \wedge w^t_2$$

And

$$\frac{1}{2} R^1_{t\gamma\delta} w^\gamma \wedge w^\delta = dw^1_t + w^1_2 \wedge w^2_t + w^1_3 \wedge w^3_t$$

So,
$dw^1 = (\dot{\alpha} + \dot{\beta}_1) w^t \wedge w^1 - e^{-\alpha+2\beta_i} w^2 \wedge w^3 = w^1_t \wedge w^t - w^1_2 \wedge w^2 - w^1_3 \wedge w^3$

Guess: $w_t^1 = (\dot{\alpha} + \dot{\beta}_1)w^1$ or $w_3^1 = \frac{1}{2}e^{-\alpha+2\beta_1}w^2$ or $w_2^1 = -\frac{1}{2}e^{-\alpha+2\beta_1}w^3$.

Similarly,

$dw^2 = (\dot{\alpha} + \dot{\beta}_2)w^t \wedge w^2 - e^{-\alpha+2\beta_2}w^3 \wedge w^1$ so guess:

$w_t^2 = (\dot{\alpha} + \dot{\beta}_2)w^2$ or $w_1^2 = \frac{1}{2}e^{-\alpha+2\beta_2}w^3$ or $w_3^2 = -\frac{1}{2}e^{-\alpha+2\beta_2}w^1$.

A consistent choice is then apparent: $\begin{cases} w_2^1 = -e^{-\alpha+2\beta_1}w^3 \\ w_3^2 = -e^{-\alpha+2\beta_2}w^1 \\ w_1^3 = -e^{-\alpha+2\beta_3}w^2 \end{cases}$

Now,

$dw_2^1 = (\dot{\alpha} - 2\dot{\beta}_2)w^t \wedge w^3 - e^{-\alpha+2\beta_1}dw^3$, where $dw^3 = (\dot{\alpha} + \dot{\beta}_3)w^t \wedge w^3 - e^{-\alpha+2\beta_3}w^1 \wedge w^2$.

Thus,

$$\frac{1}{2}R^1_{2\gamma\delta}w^\gamma \wedge w^\delta$$
$$= (\dot{\alpha} - 2\dot{\beta}_1)w^t \wedge w^3 - e^{-\alpha+2\beta_1}\left[(\dot{\alpha} + \dot{\beta}_3)w^t \wedge w^3 - e^{-\alpha+2\beta_3}w^1 \wedge w^3\right] + \left(e^{-\alpha+2\beta_3}w^2\right) \wedge \left(e^{-\alpha+2\beta_2}w^1\right) + (\dot{\alpha} + \dot{\beta}_1)w^1 \wedge (\dot{\alpha} + \dot{\beta}_2)w^2$$

and we get specifically:

$$R^1_{2t3} = 2\left[\dot{\alpha} - 2\dot{\beta}_1 - (\dot{\alpha} + \dot{\beta}_3)e^{-\alpha+2\beta_1}\right]$$
$$R^1_{212} = 2\left[(\dot{\alpha} + \dot{\beta}_1)(\dot{\alpha} + \dot{\beta}_2) + e^{-2\alpha}\left(e^{2(\beta_1+\beta_3)} - e^{2(\beta_2+\beta_3)}\right)\right]$$
$$R^1_{212} = 2\left[(\dot{\alpha} + \dot{\beta}_1)(\dot{\alpha} + \dot{\beta}_2) + e^{-2\alpha}\left(e^{-2\beta_2} - e^{-2\beta_1}\right)\right]$$

Similarly,

$dw_t^1 = (\ddot{\alpha} + \ddot{\beta}_1)w^t \wedge w^1 + (\dot{\alpha} + \dot{\beta}_1)^2 w^t \wedge w^1 - (\dot{\alpha} + \dot{\beta}_1)e^{-\alpha+2\beta_1}w^2 \wedge w^3$

Thus,

$$\frac{1}{2}R^1_{t\gamma\delta}w^\gamma \wedge w^\delta$$
$$= dw_t^1 + \left(e^{-\alpha+2\beta_2}w^3\right) \wedge (\dot{\alpha} + \dot{\beta}_2)w^2 + \left(e^{-\alpha+2\beta_3}w^2\right) \wedge (\dot{\alpha} + \dot{\beta}_3)w^3$$

and we get:

$$R^1_{tt1} = 2\left[(\ddot{\alpha} + \ddot{\beta}_1) + (\dot{\alpha} + \dot{\beta}_1)^2\right]$$
$$R^1_{t23} = 2\left[e^{-\alpha+2\beta_3}(\dot{\alpha} + \dot{\beta}_3) + e^{-\alpha+2\beta_1}(\dot{\alpha} + \dot{\beta}_2) - e^{-\alpha+2\beta_1}(\dot{\alpha} + \dot{\beta}_1)\right]$$

Grouping:

$$R_{11} = R^2_{121} + R^3_{131} + R^t_{1t1}$$
$$= 2\left[(\dot{\alpha} + \dot{\beta}_1)(\dot{\alpha} + \dot{\beta}_2) + (\dot{\alpha} + \dot{\beta}_1)(\dot{\alpha} + \dot{\beta}_3) - (\dot{\alpha} + \dot{\beta}_1)^2 - (\ddot{\alpha} + \ddot{\beta}_1)\right]$$
$$+2e^{-2\alpha}\left[\left(e^{-2\beta_2} - e^{-2\beta_1}\right) + \left(e^{-2\beta_3} - e^{-2\beta_1}\right)\right]$$

$$R_{tt} = R^1_{t1t} + R^2_{t2t} + R^3_{t3t}$$
$$= 6\ddot{\alpha} + 2\left[(\dot{\alpha} + \dot{\beta}_1)^2 + (\dot{\alpha} + \dot{\beta}_2)^2 + (\dot{\alpha} + \dot{\beta}_3)^2\right]$$

So,

$$R_{11} + R_{22} + R_{33} + R_{tt} = 6\dot{\alpha}^2 - \dot{\beta}_1^2 - \dot{\beta}_2^2 - \dot{\beta}_3^2 + 2e^{-2\alpha}[0]$$
$$= 6\dot{\alpha}^2 - \dot{\beta}_1^2 - \dot{\beta}_2^2 - \dot{\beta}_3^2.$$

The Vacuum Einstein equation is $R_{\mu\nu} = 0$, so we get:

$$6\dot{\alpha}^2 = \dot{\beta}_1^2 + \dot{\beta}_2^2 + \dot{\beta}_3^2 .$$

(Eqn. 4-48)

Other applications of the Cartan method are used to solve the most complicated Riemann tensor calculations, these include CTC spacetime analysis in Ch. 7, BH spacetime analysis in Ch. 8, and the RNaDS and Lovelock space-times in Book 6 [4].

4.7 Exercises

(Ex. 4.1) Derive Eqn. 4-4.
(Ex. 4.2) Derive Eqn. 4-6.
(Ex. 4.3) Derive Eqn. 4-7.
(Ex. 4.4) Derive Eqn. 4-21.
(Ex. 4.5) Derive Eqn. 4-35.

Chapter 5. The ADM formalism

In this chapter we analyze space-times that have metric that can be written in the form:

$$ds^2 = -N^2(r,t)dt^2 + L^2(r,t)(dr + N^r(r,t)dt)^2 + R^2(r,t)d\Omega^2.$$

(Eqn. 5-1)

This is often referred to as the lapse and shift formalism due to ADM [2], or simply, the ADM formalism. Doing the Riemann tensor calculation for a particular ADM metric (or local ADM metric) is often very difficult, such that the solution approach is not only important to get a solution, but to get a solution with useful groupings of variables for the analysis that follows. For this reason four separate approaches are described: (i) the Cartan approach); (ii) the extremal path method; (iii) the 3+1 split method with geodesic path method; (iv) the change of basis method. The methods can be used in combination, especially method (iv). Some examples are shown for each of the approaches.

5.1 The ADM Lapse and Shift formalism

Consider

$$ds^2 = -N(t,r)^2dt^2 + L(t,r)^2(dr + N^r(t,r)dt)^2 + R(t,r)^2(d\theta^2 + \sin^2\theta\, d\varphi).$$

(Eqn. 5-2)

For a general action we have:

$$S = \frac{1}{16\pi}\int d^4x\sqrt{-^4g}\left(^4R + 6\ell^{-2} - F^{\mu v}F_{\mu v}\right) + boundary\ terms$$

(Eqn. 5-3)

for which we must calculate ^4R. Calculation of ^4R is done in three ways:

(1) Via exterior differential forms: $w^\alpha \to w^\alpha_\beta \to R^\alpha_\beta \to R^\alpha_{\alpha\delta\gamma}$ (Sec. 5.2)
(2) Via extremal path method (Sec. 5.3)
(3) Via 3 + 1 split: ${}^4R = {}^3R + KK, {}^3R = {}^3\Gamma{}^3\Gamma + $ 3-divergence. Since the Γs in the ${}^3R$ are based on a coordinate frame it is easiest to calculate them using the geodesic path method (Sec. 5.4).

In addition to the above three ways to evaluate ${}^4R$, it is often advantageous to perform a change of basis first, and these methods are described in Sec.'s 5.5-5.7.

Sometimes it is easier to calculate the Landau form of the Action, which uses (see Ch. 2):

$$G_L = g^{\alpha\beta}\left(\tilde{\Gamma}^\gamma_{\alpha\delta}\tilde{\Gamma}^\delta_{\beta\gamma} - \tilde{\Gamma}^\delta_{\alpha\beta}\Gamma^\gamma_{\delta\gamma}\right),$$

(Eqn. 5-4)

where $\tilde{\Gamma}^\alpha_{\beta\gamma}$ is the tensor obtained by incorporating the added structure of a reference basis. The reference basis is chosen so that the connections fall-off sufficiently at the boundaries so as to yield no boundary contributions to the Action.

$$S = \int \sqrt{-g}\, G_L d^4x.$$

(Eqn. 5-5)

The choice of basis is indicated by the choice of boundary conditions. For asymptotically flat sections at spatial infinity, the Minkowski reference basis is appropriate. This reference basis is chosen not just at the boundary, as in approaches that retain 2nd order terms in the integrand of the action; rather, the reference basis is chosen throughout the spacetime.

Calculation of the $\tilde{\Gamma}$'s is done by first calculating the Γ's for a coordinate frame, necessary to avoid structure constant terms in the calculation, via the extremal path method if not in coordinate frame, then see Sec. 5.6). The Γ's are then evaluated in the basis of the Minkowski reference fluid, and interpreted as a tensor $\tilde{\Gamma}$.

5.2 Evaluation of ${}^4R$ using exterior differential forms (Cartan Method)

So, starting with the metric written in "lapse and shift" form due to ADM [2]:

$$ds^2 = -N^2(r,t)dt^2 + L^2(r,t)(dr + N^r(r,t)dt)^2 + R^2(r,t)d\Omega^2$$

where

$$I_G = \frac{1}{16\pi}\int_M d^4 x\sqrt{-g}\,{}^4R = \int_M dt\, dr \frac{{}^4R\sqrt{-g}}{4\sin\theta} = \int_M dt\, dr\mathcal{L} + \int_{\partial M} \alpha \cdot ds$$

(Eqn. 5-6)

Choose for basis 1-forms as follows:

$w^o = Ndt, w^1 = L(dr + N^r dt), w^2 = Rd\theta, w^3 = R\sin\theta\, d\varphi$
$D = \partial_t - N^r_{,}\partial_r$
$dw^o = -N'(LN)^{-1}w^o \wedge w^1$
$dw^1 = (DL - (N^r)'L)(LN)^{-1}w^o \wedge w^1$
$dw^2 = (DR)/(NR)w^o \wedge w^2 + R^1/(RL)w^1 \wedge w^2$
$dw^3 = (DR)/(NR)w^o \wedge w^2 + R^1/(RL)w^1 \wedge w^3 + \cot\theta\, R^{-1}w^2 \wedge w^3$

$\boldsymbol{dw^\alpha = -w^\alpha_\beta \wedge w^\beta}$**:** $\left(w^\mu_v = \Gamma^\mu_{v\lambda}w^\lambda, \text{but non} - \text{coord. frame}\right)$
$w^0_1 = w^1_0 = N'(LN)^{-1}w^0 + (DL - (N^r)'L)(LN)^{-1}w^1$
$w^2_0 = w^0_2 = (DR)/(NR)w^2$
$w^3_0 = w^0_3 = (DR)/(NR)w^3$
$w^2_1 = w^1_2 = R'(RL)^{-1}w^2$
$w^3_1 = w^1_3 = R'(RL)^{-1}w^3$
$w^3_2 = w^2_3 = R^{-1}\cot\theta\, w^3$

$\boldsymbol{R^\mu_v = dw^\mu_v + w^\mu_\alpha \wedge w^\alpha_v}$**:**
$R^0_1 = dw^0_1$
$w^0_1 = N'L^{-1}dt + (L'(LN^r)')N^{-1}(dr + N^r dt)$

$$dw^0_1 = \left\{\frac{N'}{L} + \frac{N^r}{N}(L' - (LN^r)')\right\}' dr \wedge dt + \left\{\frac{(L' - (LN^r)')}{N}\right\}' dt \wedge dr$$

$$R^0_1 = (LN)^{-1}\left\{\left(\frac{N'}{L} + \frac{N^r}{N}(L' - (LN^r)')\right) - \left(\frac{(L' - (LN^r)')}{N}\right)'\right\}' w^1 \wedge w^0$$

$R^0_2 = dw^0_2 + w^0_\alpha \wedge w^\alpha_2 = dw^0_2 + w^0_1 \wedge w^1_2$
$w^0_2 = (DR)/(NR)w^2 = (N^{-1}DR)d\theta$
$dw^0_2 = (DR)d(N^{-1}) \wedge d\theta + N^{-1}d(\acute{R} - N^r R') \wedge d\theta$

$$= -N^{-1}(DR)\left(\frac{N'}{N}\right) dr \wedge d\theta + N^{-1}(DR)'dr \wedge d\theta$$

$$= -N^{-1}(DR)\left(\frac{\acute{N}}{N}\right) dt \wedge d\theta + N^{-1}(DR)'dt \wedge d\theta$$

$dr = L^{-1}w^1 - N^r dt \qquad d\theta = R^{-1}w^2$

$$dw_2^0 = (LNR)^{-1}\left[DR' - \left(\frac{N'}{N}\right)(DR)\right]w^1 \wedge w^2$$
$$+ (N^2R)^{-1}[D^2R - N^{-1}(DR)(DN)]w^0 \wedge w^2$$
$$w_1^0 \wedge w_2^1 = \{-(L^2NR)^{-1}R'N'\}w^0 \wedge w^2$$
$$+ \{-(L^2NR)^{-1}R'(DL - (N^r)'L)\}w^1 \wedge w^2$$
$$R_2^0 = \frac{1}{NR}\left\{\frac{D^2R}{N} - N^{-2}(DR)(DN) - L^2R'N'\right\}w^0 \wedge w^2$$
$$+ \frac{1}{LNR}\left\{DR' - \left(\frac{N'}{N}\right)DR - L^{-1}R'(DL - (N^r)'L)\right\}w^1 \wedge w^2$$

$$R_3^0 = dw_3^0 + w_1^0 \wedge w_3^1 + w_2^0 \wedge w_3^2$$

$$w_3^0 = (DR)(NR)^{-1}w^3 = (DR)N^{-1}\sin\theta\, d\varphi$$
$$dw_3^0 = \left(\frac{DR}{N}\right)'\sin\theta\, dt \wedge d\varphi + \left(\frac{DR}{N}\right)'\sin\theta\, dr \wedge d\varphi + \left(\frac{DR}{N}\right)\cos\theta\, d\theta$$
$$\wedge\, d\varphi$$
$$= (NR)^{-1}\left(\frac{DR}{N}\right)' w^0 \wedge w^3 + (LR)^{-1}\left(\frac{DR}{N}\right)' w^1 \wedge w^3 -$$
$$(NR)^{-1}\left(\frac{DR}{N}\right)' N^r w^0 \wedge w^3$$

$$+R^2\left(\frac{DR}{N}\right)\cot\theta\, w^2 \wedge w^3$$
$$= (NR)^{-1}D\left(\frac{DR}{N}\right)' w^0 \wedge w^3 + (LR)^{-1}\left(\frac{DR}{N}\right)' w^1 \wedge w^3 +$$
$$R^2\left(\frac{DR}{N}\right)\cot\theta\, w^2 \wedge w^3$$

$w_1^0 \wedge w_3^1 = \{w_1^0 \wedge w_2^1$ with $w^2 \to w^3$ in result$\}$
$$= \{-(L^2NR)^{-1}R'N'\}w^0 \wedge w^2 + \left\{-(L^2NR)^{-1}R'\big(DL -$$
$$N^{r\prime}L\big)\right\}w^1 \wedge w^2$$

$$w_2^0 \wedge w_3^2 = -(NR^2)^{-1}(DR)\cot\theta\, w^2 \wedge w^3$$

$R_3^0 = \{R_2^0$ with $w^2 \to w^3$ because the $w^2 \wedge w^3$ terms cancel$\}$
$$= (NR)^{-1}\left\{\frac{D^2R}{N} - N^2(DR)(DN)L^{-2}R'N'\right\}w^0 \wedge w^3$$
$$+(LNR)^{-1}\left\{DR' - \left(\frac{N'}{N}\right)DR - L^{-1}R'(DL - (N^r)'L)\right\}w^1 \wedge w^3$$

$$R_1^1 = dw_1^1 + w_\alpha^1 \wedge w_1^\alpha = dw_1^1 + w_0^1 \wedge w_1^0$$

$$w_2^1 = -R'(RL)^{-1}w^2 = -L^{-1}R'd\theta$$

$$dw_2^1 = \left(-\frac{R'}{L}\right)' dr \wedge d\theta + \left(-\frac{R'}{L}\right)' dt \wedge d\theta$$
$$= (RN)^{-1}D\left(-\frac{R'}{L}\right)w^0 \wedge w^2 + (RL)^{-1}\left(-\frac{R'}{L}\right)' w^1 \wedge w^2$$

$$w_0^1 \wedge w_2^0 = (LN^2R)^{-1}N'(DR)w^0 \wedge w^2$$
$$+ (LN^2R)^{-1}(DL - (N^r)'L)(DR)w^1 \wedge w^2$$

$$R_2^1 = (RN)^{-1}\left\{(LN)^{-1}N'(DR) - D\left(\frac{R'}{L}\right)\right\}w^0 \wedge w^2$$
$$+(RN)^{-1}\left\{N^{-2}(DL - N^{r\prime}L)(DR) - \left(\frac{R'}{L}\right)'\right\}w^1 \wedge w^2$$

$$R_3^1 = dw_3^1 + w_0^1 \wedge w_3^0 + w_2^1 \wedge w_3^2$$

$$R_3^1 = \{R_2^1 \text{ with } w^2 \to w^3 \text{ because the } w^2 \wedge w^3 \text{ terms cancel}\}$$

$$R_3^2 = dw_3^2 + w_\alpha^2 \wedge w_3^\alpha = dw_3^2 + w_0^2 \wedge w_3^0 + w_1^2 \wedge w_3^1$$

$$w_3^2 = R^{-1}\cot\theta\, w^3 = -\cos\theta\, d\varphi$$

$$dw_3^2 = \sin\theta\, d\theta \wedge d\varphi = R^{-2}w^2 \wedge w^3$$

$$w_0^2 \wedge w_3^0 + w_1^2 \wedge w_3^1 = (NR)^{-2}(DR)^2 w^2 \wedge w^3 - (RL)^{-2}(R')^2 w^2 \wedge w^3$$

$$R_3^2 = R^{-2}\{1 + N^{-2}(DR)^2 - L^{-2}(R')^2\}w^2 \wedge w^3$$

$R_\nu^\mu = R_{\nu|\alpha\beta|}^\mu w^\alpha \wedge w^\beta$ and $R = g^{\mu\nu}R_{\mu\nu} = g^{\alpha\beta}g^{\gamma\delta}R_{\alpha\gamma\beta\delta}$:

Thus,
$$R = -2R_{oioi} + R_{ijij} = 2R_{ioi}^o + R_{jij}^i$$
$$= 2(R_{101}^0 + R_{202}^0 + R_{303}^0) + 2(R_{212}^1 + R_{313}^1 + R_{323}^2)$$
$$= 2(R_{101}^0 + 2R_{202}^0) + 2(2R_{212}^1 + R_{323}^2) \text{ (due to spherical symmetry)}$$

$$R^0_{101} = (LN)^{-1}\left(\frac{(L' - (LN^r)')}{N}\right)' - (LN)^{-1}\left(\frac{N'}{L} + \frac{N^r}{N}(L' - (LN^r)')\right)$$

$$R^0_{202} = (RN)^{-1}\left(\frac{D^2R}{N} - \frac{(DR)(DN)}{N^2} - \frac{R'N'}{L^2}\right)$$

$$R^1_{212} = (RL)^{-1}\left(\frac{\left(DL - N^{r'}L\right)}{N^2}(DR) - \left(\frac{R'}{L}\right)'\right)$$

$$R^2_{323} = R^2\left(1 + \frac{(DR)^2}{N^2} - \frac{\left(R'\right)^2}{L^2}\right)$$

$$\boldsymbol{\mathcal{L}} = \frac{\left(\sqrt{-\boldsymbol{g}}\,{}^4\boldsymbol{R}\right)}{(\mathbf{4}\sin\boldsymbol{\theta})} \ :$$

$$\mathcal{L} = \frac{1}{2}R^2\left\{\left(\frac{(L' - (LN^r)')}{N}\right)' - \left(\frac{N'}{L} + \frac{N^r}{N}(L' - (LN^r)')\right)'\right\}$$
$$+(RL)\left\{\frac{D^2R}{N} - \frac{(DR)(DN)}{N^2} - \frac{R'N'}{L^2}\right\} + (RN)\left\{\frac{\left(DL - N^{r'}L\right)(DR)}{N^2} - \left(\frac{R'}{L}\right)'\right\}$$
$$+\frac{1}{2}(LN)\left\{1 + \frac{(DR)^2}{N^2} - \frac{\left(R'\right)^2}{L^2}\right\}$$

$$D^2R = (\partial_t - N^r\partial r)\left(\acute{R} - N^rR'\right) = \left(\acute{R} - N^rR'\right)^{\cdot} - N^r\left(\acute{R} - N^rR'\right)'$$

$$\mathcal{L} = \frac{R\dot{R}}{N}(L' - (LN^r)') + RR'\left(\frac{N'}{L} + \frac{N^r}{N}(L' - (LN^r)')\right)$$

$$-\left(\frac{RL}{N}\right)'\left(\acute{R} - N^rR'\right) + \left(\frac{RLN^r}{N}\right)'\left(\acute{R} - N^rR'\right) - \frac{RL}{N^2}(DR)(DN) - \frac{R}{L}R'N'$$

$$+\frac{R}{N}\left(DL - N^{r'}L\right)(DR) + (RN)'\left(\frac{R'}{N}\right) + \frac{1}{2}(LN)\left\{1 + \frac{(DR)^2}{N^2} - \frac{\left(R'\right)^2}{L^2}\right\}$$

$$+\left[\frac{1}{2}R^2\frac{(L' - (LN^r)')}{N}\right]' - \left[\frac{1}{2}R^2\left(\frac{N'}{L} + \frac{N^r}{N}(L' - (LN^r)')\right)\right]'$$

$$+\left[\frac{RL}{N}(\acute{R}-N^rR')\right]'-\left[\frac{RLN^r}{N}(\acute{R}-N^rR')\right]'-\left[\frac{RN}{L}R'\right]'$$

$$\mathcal{L}=\mathcal{L}^{(1)}+\partial_\mu\alpha^\mu$$
$$\alpha^t=\frac{1}{2}R^2N^{-1}(L'-(LN^r)')+RLN^{-1}DR$$
$$\alpha^r=\frac{1}{2}R^2\left(N'L^{-1}+N^rN^{-1}(L'-(LN^r)')\right)-RLN^rN^{-1}DR-RNL^{-1}R'$$

$$\mathcal{L}^{(1)}=-RRN^{-1}\left(DL-N^{r'}\right)+RR'N'L^{-1}+RR'N^{-1}N^r\left(DL-N^{r'}\right)$$
$$-(RLN^{-1})^0DR+(RLN^rN^{-1})'DR-RLN^{-2}(DR)(DN)-$$
$$RL^{-1}R'N'$$
$$+RN^{-1}\left(DL-N^{r'}L\right)(DR)+(RN)'R'L^{-1}+\tfrac{1}{2}LN+\tfrac{1}{2}LN^{-1}(DR)^2-$$
$$\tfrac{1}{2}NL^{-1}(R')^2$$
$$=RR'N'L^{-1}-D(RLN^{-1})DR+(RLN^{-1})N^{r'}(DR)$$
$$-RLN^{-2}(DR)(DN)-RN'R'L^{-1}+(RN)'R'L^{-1}+\tfrac{1}{2}LN+$$
$$\tfrac{1}{2}LN^{-1}(DR)^2-\tfrac{1}{2}NL^{-1}(R')^2$$
$$=\tfrac{1}{2}LN-\tfrac{1}{2}LN^{-1}(DR)^2-RL(-N^{-2}DN)DR-\left\{RN^{-1}DLDR-\right.$$
$$\left.(RLN^{-1})N^{r'}(DR)\right\}$$
$$-RLN^{-2}(DR)(DN)+\tfrac{1}{2}NL^{-1}(R')^2+RR'N'L^{-1}$$

$$\mathcal{L}^{(1)}=\frac{1}{2}LN-\frac{1}{2}LN^{-1}(DR)^2+\frac{1}{2}NL^{-1}(R')^2+RR'(N')L^{-1}$$
$$-RN^{-1}(DR)(L'-(LN^r)')$$

Take $\mathcal{L}^{(1)}$ as the volume contribution to the Action, then

$$\pi_R=\frac{\delta L}{\partial R}=N^{-1}(N^rLR)'-N^{-1}(LR)'$$

(Eqn. 5-7)

$$\pi_L=\frac{\delta L}{\delta L'}=RN^{-1}DR$$

(Eqn. 5-8)

Inverting:

$$\dot{R}=NR^{-1}\pi_L+N^rR'$$

(Eqn. 5-9)

$$\dot{L}=-N^t\pi_RR^{-1}+LN\pi_LR^{-2}+(N^rL)'$$

(Eqn. 5-10)

Substituting back into the Lagrangian, and separating in the usual Hamiltonian form given by the Legendre transformation:

$$I_G = \int dr\, dt\left(\pi_L \dot{L} + \pi_R \dot{R} - NH - N^r H_r\right)$$

(Eqn. 5-11)

$$H = \frac{L\pi_L^2}{2R^2} - \frac{\pi_L \pi_R}{R} - \frac{1}{2}\left\{\left(\frac{2RR'}{L}\right)' - \frac{(R')^2}{L} - L\right\}$$

(Eqn. 5-12)

$$H_r = R'\pi_R - L\pi_L'$$

(Eqn. 5-13)

H has regained a second-order term through elimination of N' by integration by parts. This was required to separate the N as a Lagrange multiplier. Surface terms arising from N' are not a problem if ${}^4R$ is evaluated in terms of the Γ's, with (non-tensorial) divergences dropped. This eliminates second order terms at the outset, as well as N' terms. When taken for change of basis to that for which the derivatives in the Γ's can be taken as covariant, a tensorial interpretation results. This approach may not need surface term corrections as in the usual ADM.

5.3 The Extremal Path Method for evaluating the Connection Coefficients

A geodesic is a (parametrized) curve that extremizes the integral:

$$I = \frac{1}{2}\int g_{\mu\nu}\, \dot{x}^\mu \dot{x}^\nu d\lambda,$$

(affine parametrization).

(Eqn. 5-14)

Variation leads to the "geodesic equation":

$$\ddot{x}^\mu + \Gamma^\mu_{\alpha\beta}\dot{x}^\alpha \dot{x}^\beta = 0,$$

(Eqn. 5-15)

where the Γ's are the connection coefficients needed in the G_L calculation:

$$ds^2 = -N(t,r)^2 dt^2 + L(t,r)^2(dr + N^r(t,r)dt)^2 + R(t,r)^2 d\Omega^2$$

(Eqn. 5-16)

Let's start with path integral to be extremized:

$$I = \int \frac{1}{2}\left(\frac{ds}{d\lambda}\right)^2 d\lambda$$

(Eqn. 5-17)

So,

$$(\delta I)_t = \frac{1}{2}\int \Bigg\{ -2N^2\left(\frac{dt}{d\lambda}\right)\delta\left(\frac{dt}{d\lambda}\right) - 2N\frac{\partial N}{\partial t}\delta t\left(\frac{dt}{d\lambda}\right)^2$$

$$+ 2L\frac{\partial L}{\partial t}\delta t\left(\frac{dr}{d\lambda} + N^r\frac{dt}{d\lambda}\right)^2$$

$$+ 2L^2\left(\frac{dr}{d\lambda} + N^r\frac{dt}{d\lambda}\right)\left(\frac{\partial N^r}{\partial t}\right)\left(\frac{dt}{d\lambda}\right)\delta t + N^r\delta\left(\frac{dt}{d\lambda}\right)$$

$$+ 2R\left(\frac{\partial R}{\partial t}\right)\delta t\left(\frac{d\Omega}{d\lambda}\right)^2\Bigg\} d\lambda$$

(1) $\frac{d}{d\lambda}\left(N^2\left(\frac{dt}{d\lambda}\right) - L^2\left(\frac{dr}{d\lambda} + N^r\frac{dt}{d\lambda}\right)N^r\right) - N\frac{\partial N}{\partial t}\left(\frac{dt}{d\lambda}\right)^2 + L\frac{\partial L}{\partial t}\left(\frac{dr}{d\lambda} +\right.$ $\left.N^r\frac{dt}{d\lambda}\right)^2 + L^2\left(\frac{dr}{d\lambda} + N^r\frac{dt}{d\lambda}\right)\left(\frac{\partial N^r}{\partial t}\right)\left(\frac{dt}{d\lambda}\right) + R\left(\frac{\partial R}{\partial t}\right)\left(\frac{d\Omega}{d\lambda}\right)^2 = 0$

N^r term leads to coupled geodesic equations as there are $\frac{d^2t}{d\lambda^2}$ and $\frac{d^2r}{d\lambda^2}$ terms in the above. Calculation of $(\delta I)_r$ needed next:

$$(\delta I)_r = \frac{1}{2}\int \Bigg\{ -2N\frac{\partial N}{\partial r}\delta r\left(\frac{dt}{d\lambda}\right)^2 + 2L\frac{\partial L}{\partial t}\delta t\left(\frac{dr}{d\lambda} + N^r\frac{dt}{d\lambda}\right)^2$$

$$+ 2R\left(\frac{\partial R}{\partial r}\right)\delta r\left(\frac{d\Omega}{d\lambda}\right)^2$$

$$+ 2L^2\left(\frac{dr}{d\lambda} + N^r\frac{dt}{d\lambda}\right)\left(\delta\left(\frac{dr}{d\lambda}\right) + \frac{\partial N^r}{\partial r}\delta r\left(\frac{dt}{d\lambda}\right)\right)\Bigg\} d\lambda$$

(2) $-\frac{d}{d\lambda}\left(L^2\left[\frac{dr}{d\lambda} + N^r\frac{dt}{d\lambda}\right]\right) - N\frac{\partial N}{\partial r}\left(\frac{dt}{d\lambda}\right)^2 + L\frac{\partial L}{\partial r}\left(\frac{dr}{d\lambda} + N^r\frac{dt}{d\lambda}\right)^2 +$ $R\frac{\partial R}{\partial r}\left(\frac{d\Omega}{d\lambda}\right)^2$

$+L^2\left(\frac{dr}{d\lambda} + N^r\frac{dt}{d\lambda}\right)\left(\frac{\partial N^r}{\partial r}\right)\left(\frac{dt}{d\lambda}\right) = 0$

Equations (1) and (2) are grouped to have no $\frac{d^2r}{d\lambda^2}$ term:

$$\frac{d^2t}{d\lambda^2}+\left(\frac{\partial t}{\partial\lambda}\right)^2\left(-\frac{DN}{N}+\frac{LDL}{N^2}(N^r)^2+\frac{L^2DN^r}{N^2}(N^r)-\frac{L^2N^{r\prime}N^r}{N^2}+\frac{2\dot{N}}{N}\right)$$
$$+\left(\frac{\partial t}{\partial\lambda}\right)\left(\frac{\partial t}{\partial\lambda}\right)\left(\frac{LDL}{N^2}2N^r+\frac{L^2DN^r}{N^2}-\frac{L^2}{N^2}[\dot{N}^r+N^r(N^r)']\right.$$
$$\left.+\frac{2N'}{N}\right)+\left(\frac{\partial r}{\partial\lambda}\right)^2\left(\frac{LDL}{N^2}-\frac{L^2(N^r)'}{N^2}\right)$$
$$+\left(\left(\frac{d\theta}{d\lambda}\right)^2+\sin^2\theta\left(\frac{d\varphi}{d\lambda}\right)^2\right)\frac{RDR}{N^2}=0$$

Relating to the form:

$$\frac{d^2r}{d\lambda^2}+\Gamma^t_{\alpha\beta}\left(\frac{dx^\alpha}{d\lambda}\right)\left(\frac{dx^\beta}{d\lambda}\right)=0$$

We get:

$$\Gamma^t_{tt}=\frac{2\dot{N}}{N}-\frac{DN}{N}+N^2D\left(\frac{1}{2}(N^r)^2L^2\right)-N^{-2}L^2N^r(N^r)$$

(Eqn. 5-18)

$$\Gamma^t_{rt}=\frac{N'}{N}+\frac{1}{2}N^{-2}(DL^2)N^r+N^{-2}L^2(-N^r(N^r)')$$

(Eqn. 5-19)

$$\Gamma^t_{rr}=N^{-2}\left(\frac{1}{2}DL^2-L^2(N^r)'\right)$$

(Eqn. 5-20)

$$\Gamma^t_{\theta\theta}=\sin^{-2}\theta\,\Gamma^t_{\theta\theta}=N^{-2}\frac{1}{2}DR^2$$

(Eqn. 5-21)

If $N^r=0$ and there is no "t" dependence (D operation yields zero), then only Γ^t_{rt} above is nonzero with $(N'N^{-1})$, this agrees with a previous calc.

Now to group with no $\frac{d^2t}{d\lambda^2}$ term:

$$-2L\frac{dL}{d\lambda}\left(\frac{dr}{d\lambda}+N^r\frac{dt}{d\lambda}\right)-L^2\left(\frac{d^2r}{d\lambda^2}+N^r\frac{d^2t}{d\lambda^2}+\frac{dN^r}{d\lambda}\frac{dt}{d\lambda}\right)$$
$$-NN\frac{\partial N}{\partial r}\left(\frac{dt}{d\lambda}\right)^2+L\frac{\partial L}{\partial r}\left(\frac{dr}{d\lambda}+N^r\frac{dt}{d\lambda}\right)^2+R\frac{\partial R}{\partial r}\left(\frac{d\Omega}{d\lambda}\right)^2$$
$$+L^2\left(\frac{dr}{d\lambda}+N^r\frac{dt}{d\lambda}\right)\left(\frac{\partial N^r}{\partial t}\right)\left(\frac{dt}{d\lambda}\right)$$

$$\frac{d^2r}{d\lambda^2}+N^r\left(-\Gamma^t_{\alpha\beta}\left(\frac{dx^\alpha}{d\lambda}\right)\left(\frac{dx^\beta}{d\lambda}\right)\right)+\dot{N}^r\left(\frac{dt}{d\lambda}\right)^2+N^{r'}\left(\frac{dr}{d\lambda}\right)\left(\frac{dt}{d\lambda}\right)$$
$$+\frac{2}{L}\left(\dot{L}\left(\frac{dt}{d\lambda}\right)+L'\left(\frac{dr}{d\lambda}\right)\right)\left(\frac{dr}{d\lambda}+N^r\frac{dt}{d\lambda}\right)+\frac{N}{L^2}N'\left(\frac{\partial t}{\partial\lambda}\right)^2$$
$$-\frac{L'}{L}\left(\frac{dr}{d\lambda}+N^r\frac{dt}{d\lambda}\right)^2$$
$$-RL^{-2}R'\left(\frac{d\Omega}{d\lambda}\right)^2-\left(\frac{dr}{d\lambda}+N^r\frac{dt}{d\lambda}\right)(N^r)'\left(\frac{dt}{d\lambda}\right)=0$$

Thus,

$$\frac{d^2r}{d\lambda^2}+\left(\frac{\partial t}{\partial\lambda}\right)^2\left(-N^r\Gamma^t_{tt}+\dot{N}^r+2N^rL'L^{-1}+NN'L^{-2}-L'L^{-1}(N^r)^2\right.$$
$$\left.-N^r(N^r)'\right)$$
$$+\left(\frac{\partial t}{\partial\lambda}\right)\left(\frac{\partial r}{\partial\lambda}\right)\left(-N^r\Gamma^t_{tt}+\frac{2L'}{L}+\frac{2L'}{L}N^r-\frac{2L'}{L}N^r\right)$$
$$+\left(\frac{\partial r}{\partial\lambda}\right)^2\left(-N^r\Gamma^t_{rr}+\frac{2L'}{L}-\frac{L'}{L}\right)$$
$$+\left(\frac{\partial\theta}{\partial\lambda}\right)^2(-N^r\Gamma^t_{\theta\theta}-L^{-2}RR')++\left(\frac{\partial\varphi}{\partial\lambda}\right)^2\left(-N^r\Gamma^t_{\varphi\varphi}-L^{-2}RR'\sin^{-2}\theta\right)$$
$$=0$$

Thus,

$$\Gamma^r_{tt}=-N^r\Gamma^t_{tt}+\dot{N}^r+2N^rL'L^{-1}-(N^r)^2L'L^{-1}+NN'L^{-2}-N^r(N^r)'$$
(Eqn. 5-22)

$$\Gamma^r_{tr}=-N^r\Gamma^t_{tr}+2L'L^{-1}$$
(Eqn. 5-23)

$$\Gamma^r_{rr}=-N^r\Gamma^t_{rr}+L'L^{-1}$$
(Eqn. 5-24)

$$\Gamma^r_{\theta\theta}=-N^r\Gamma^t_{\theta\theta}-L^{-2}RR'=\sin^{-2}\theta\,\Gamma^r_{\varphi\varphi}$$
(Eqn. 5-25)

These also agree with results for $N^r=0$ and no "t" dependence. Now consider:

$$(\delta I)_\theta=\frac{1}{2}\int\left\{2R^2\left(\frac{d\theta}{d\lambda}\right)\delta\left(\frac{d\theta}{d\lambda}\right)+2R^2\sin\theta\cos\theta\,\delta\theta\left(\frac{d\varphi}{d\lambda}\right)^2\right\}d\lambda$$
$$-\frac{d}{d\lambda}\left(R^2\left(\frac{d\theta}{d\lambda}\right)\right)+R^2\sin\theta\cos\theta\left(\frac{d\varphi}{d\lambda}\right)^2=0$$

Thus,

$$\frac{d^2\theta}{d\lambda^2} + 2R^{-1}\left(R'\frac{dr}{d\lambda} + \dot{R}\frac{dt}{d\lambda}\right)\left(\frac{d\theta}{d\lambda}\right) - \sin\theta\cos\theta\left(\frac{d\varphi}{d\lambda}\right)^2 = 0$$

And the coefficients are:

$$\Gamma^{\theta}_{\varphi\varphi} = -\sin\theta\cos\theta$$

(Eqn. 5-26)

$$\Gamma^{\theta}_{r\theta} = \left(\frac{R'}{R}\right)$$

(Eqn. 5-27)

$$\Gamma^{\theta}_{t\theta} = \left(\frac{\dot{R}}{R}\right)$$

(Eqn. 5-28)

For the $\varphi(\lambda)$ variation $\frac{d}{d\lambda}\left(R^2\sin^2\theta\left(\frac{d\varphi}{d\lambda}\right)\right) = 0 \Rightarrow \frac{d^2\varphi}{d\lambda^2} +$
$\underbrace{2R}_{R^2\sin^2\theta}\frac{dR}{d\lambda}\sin^2\theta\frac{d\varphi}{d\lambda} + \underbrace{2R^2}_{R^2\sin^2\theta}\sin\theta\cos\theta\frac{d\theta}{d\lambda}$

And the coefficients are:

$$\Gamma^{\varphi}_{\theta\varphi} = \cot\theta, \Gamma^{\varphi}_{r\varphi} = \left(\frac{R'}{R}\right), \Gamma^{\varphi}_{t\varphi} = \left(\frac{\dot{R}}{R}\right)$$

(Eqn. 5-29)

Let's apply the geodesic method to BH spacetimes in coordinates that cross the Horizon.

Kruskal coordinates

The Kruskal coordinates are used to arrive at an action.

$$ds^2 = \frac{32M^3}{r}e^{-r/2M}(dv^2 - du^2) - r^2(d\theta^2 + \sin^2\theta\, d\varphi^2)$$
$$= e^{2A}(dv^2 - du^2) - r^2 d\Omega^2$$

(Eqn. 5-30)

The geodesics are variational minimal of the distance measures between events, (we consider here only converse spaces). Using affine parametrization, extremizing

$$I = \frac{1}{2}\int g_{\mu\nu}\dot{x}^{\mu}\dot{x}^{\nu}d\lambda$$

(Eqn. 5-31)

Is known to yield $\ddot{x}^{\mu} + \Gamma^{\mu}_{\alpha\beta}\dot{x}^{\alpha}\dot{x}^{\beta} = 0$, where $\Gamma^{\mu}_{\alpha\beta}$'s are the needed connection coefficients.

Proof that carries of extremal length are geodesics on pg 316 MTW [21]. Discussion given in Box 13.3 of MTW of how a "geometric" principle of extremal length may be replaced by a "dynamic" extremal principle based on the simplest coord. inv. equalization of $\frac{1}{2}\dot{x}^2$, i.e. $\frac{1}{2}g_{\mu\nu}\frac{dx^\mu}{d\lambda}\frac{dx^\nu}{d\lambda}$ which in the form $I = \frac{1}{2}\int g_{\mu\nu}\frac{dx^\mu}{d\lambda}\frac{dx^\nu}{d\lambda}d\lambda = \int L\left(\dot{x},\frac{dr}{d\lambda}\right)d\lambda$, with affine parameter, yields the equation $\ddot{x}^\mu + \Gamma^\mu_{\alpha\beta}\dot{x}^\alpha\dot{x}^\beta = 0$.

Geodesic Lagrangian Method for calculation of the connection coefficient is then done.

$$ds^2 = e^{2A(u,v)}(dv^2 - du^2) - r^2(d\theta^2 + \sin^2\theta\, d\varphi^2)$$

(Eqn. 5-32)

$$x = \frac{\partial x}{\partial v}, x' = \frac{\partial x}{\partial u}, \dot{x} = \frac{\partial x}{\partial \lambda}$$

(Eqn. 5-33)

$$I = \int g_{\mu\nu}\dot{x}^\mu\dot{x}^\nu d\lambda$$

(Eqn. 5-34)

Vary $v(\lambda)$ first:

$$\delta I = 0 = \int d\lambda\{2\dot{A}\delta v e^{2A}(\dot{v}^2 - \dot{u}^2) + e^{2A}2\dot{v}\delta\dot{v}$$
$$- 2r\dot{r}\delta v(\dot{\theta}^2 + \sin^2\theta\,\dot{\varphi}^2)\}$$
$$0 = \dot{A}e^{2A}(\dot{v}^2 - \dot{u}^2) - (e^{2A}\dot{v})^0 - r\dot{r}(\dot{\theta}^2 + \sin^2\theta\,\dot{\varphi}^2)$$
$$\ddot{v} + \dot{A}\dot{v}^2 + \dot{A}\dot{u}^2 + e^{-2A}r\dot{r}\dot{\theta}^2 + e^{-2A}r\dot{r}\sin^2\theta\,\dot{\varphi}^2 + 2A'\dot{u}\dot{v} = 0$$

(Eqn. 5-35)

$$\Gamma^v_{vv} = \dot{A}, \Gamma^v_{uu} = \dot{A}, \Gamma^v_{\theta\theta} = \mathrm{e}^{-2A}r\dot{r}, \Gamma^v_{\varphi\varphi} = e^{-2A}r\dot{r}\sin^2\theta, \Gamma^v_{uv} = A'$$

(Eqn. 5-36)

Vary $u(\lambda)$ to get:

$$0 = A'e^{2A}(\dot{v}^2 - \dot{u}^2) - (e^{2A}\dot{v})^0 - rr'(\dot{\theta}^2 + \sin^2\theta\,\dot{\varphi}^2)$$
$$\ddot{u} + A'\dot{u}^2 + A'\dot{v}^2 - rr'(\dot{\theta}^2 + \sin^2\theta\,\dot{\varphi}^2)e^{-2A} + 2\dot{A}\dot{u}\dot{v} = 0$$

(Eqn. 5-37)

$$\Gamma^u_{uu} = A', \Gamma^u_{vv} = A', \Gamma^u_{\theta\theta} = -rr'\mathrm{e}^{-2A}, \Gamma^u_{\varphi\varphi} = rr'e^{-2A}\sin^2\theta, \Gamma^u_{vu} = \dot{A}$$

(Eqn. 5-38)

Vary $\theta(\lambda)$ to get:

$$\delta I = 0 = \int d\lambda\{-2r^2\dot{\theta}\delta\dot{\theta} - 2r^2\sin\theta\cos\theta\,\delta\theta\dot{\varphi}^2\}$$

$$0 = \ddot{\theta} + \frac{2}{r}\dot{r}\dot{\theta} - \sin\theta\cos\theta\,\delta\theta\dot{\varphi}^2$$

(Eqn. 5-39)

$$\Gamma^{\theta}_{u\theta} = \frac{r'}{r}, \Gamma^{\theta}_{v\theta} = \frac{\dot{r}}{r}, \Gamma^{\theta}_{\varphi\varphi} = -\sin\theta\cos\theta$$

(Eqn. 5-40)

Vary $\varphi(\lambda)$ to get:

$$\delta I = 0 = \int d\lambda\{(r\sin\theta)^2\dot{\varphi}2\delta\dot{\varphi}\}$$

$$0 = \ddot{\varphi} + \frac{2}{r}(\dot{r}\dot{v} + r'\dot{u})\dot{\varphi} + 2\cot\theta\,\dot{\theta}\dot{\varphi}$$

(Eqn. 5-41)

$$\Gamma^{\varphi}_{v\varphi} = \frac{\dot{r}}{r}, \Gamma^{\varphi}_{u\varphi} = \frac{r'}{r}, \Gamma^{\varphi}_{\theta\varphi} = -\cot\theta$$

(Eqn. 5-42)

The Γ'^{s} have been verified on the computer and

$$R = g^{iK}R_{iK} = g^{iK}\left[\Gamma^{\ell}_{iK,\ell} - \Gamma^{\ell}_{i\ell,K} + \Gamma^{\ell}_{iK}\Gamma^{m}_{\ell m} - \Gamma^{m}_{i\ell}\Gamma^{\ell}_{Km}\right]$$

(Eqn. 5-43)

$$= 2e^{-2A}\left[\frac{2}{r}(r'' - \ddot{r}) + A''\dot{A}' + \frac{1}{r^2}((r')^2 - (\dot{r})^2)\right] - \frac{2}{r^2}$$

(in a coord. system).

(Eqn. 5-44)

Now, let's compute G:

We have

$$G' = G + \frac{1}{\sqrt{-g}}\frac{\partial}{\partial\theta}\left[\sqrt{-g}\left(g^{iK}\Gamma^{\theta}_{iK} - g^{i\theta}\Gamma^{\ell}_{i\ell}\right)\right]$$

(Eqn. 5-45)

$$G = -2e^{-2A}\left[\frac{1}{r^2}((r')^2 - (\dot{r})^2) + \frac{2}{r}\left(r'A' - \dot{r}\dot{A}\right)\right]$$

(Eqn. 5-46)

$$\sqrt{-g} = e^{2A}r^2\sin\theta$$

So,

$$S = \int\sqrt{-g}\,G'dV = \int e^{2A}r^2\sin\theta\left(G - \frac{2}{r^2}\right)d\theta d\varphi du dv$$

$$= -8\pi\int\left\{(r')^2 - (\dot{r})^2 + 2r\left(r'A' - \dot{r}\dot{A}\right) + e^{2A}\right\}du dv$$

(Eqn. 5-47)

where

$$A' = -\frac{1}{4M}\left(1-\frac{2M}{r}\right)r', A = -\frac{1}{4M}\left(1-\frac{2M}{r}\right)\dot{r}, e^{2A} = \frac{32M^3}{r}e^{-r/2M}$$

So,

$$\underline{S = \frac{4\pi}{M}\int\left\{r[(r')^2-(\dot{r})^2] - \frac{64M^4}{r}e^{-r/2M}\right\}dudv}$$

(Eqn. 5-48)

Thus

$$\delta S = 0 = \frac{4\pi}{M}\int\left[[(r')^2-(\dot{r})^2]\delta r + r[2r'\delta r' - 2\dot{r}\delta\dot{r}]\right.$$
$$\left. - 64M^4\left(-\frac{1}{r^2}-\frac{1}{\partial\mu r}\right)e^{-r/2M}\delta r\right]dudv$$

$$(r')^2-(\dot{r})^2-2(r')^2-2rr''+2\dot{r}^2+2r\ddot{r}$$
$$+64M^4\left(-\frac{1}{r^2}-\frac{1}{\partial\mu r}\right)e^{-r/2M} = 0$$

$$-\frac{r}{M}(r''-\ddot{r}) - \frac{1}{2M}((r')^2-(\dot{r})^2) + \frac{32M^3}{r}\left(\frac{1}{r}-\frac{1}{\partial\mu}\right)e^{-r/2M} = 0$$

$$\boxed{r(r''-\ddot{r}) + \frac{1}{2}\left(r'^2-\dot{r}^2\right) - \frac{32M^4}{r}\left(\frac{1}{r}-\frac{1}{2M}\right)e^{-r/2M} = 0}$$

(Eqn. 5-49)

Novikov

$R^* = 0$ observer in Novikov:

$r_{\max} = 2M \qquad R^x = (r_{\max}/2M - 1)^{1/2} = 0$

$r = \frac{1}{2}r_{\max}(1+\cos\eta) = M(1+\cos\eta)$

$\tau = (r_{\max}^3/8M)^{1/2}(\eta+\sin\eta) = M(\eta+\cos\eta)$

Analysis of radial geodesic motion:

$$ds^2 = -\left(1-\frac{2M}{r}\right)dt^2 + \left(1-\frac{2M}{r}\right)^{-1}dr^2 + r^2d\Omega^2$$
$$g^{\alpha\beta}P_\alpha P_\beta + m^2 = 0 \quad , \quad P_0 = -E, \qquad P_\varphi = \pm L$$
$$\frac{-E^2}{(l-{}^{2M}/_r)} + \left(1-\frac{2M}{r}\right)^{-1}(P^r)^2 + \frac{L^2}{r^2} + m^2 = 0 \qquad P^r = m\frac{dr}{d\tau}$$

$$\left(\frac{dr}{d\tau}\right)^2 = \tilde{E}^2 - \left(1 - {}^{2M}/_{r}\right)\left(1 + {}^{\tilde{L}^2}/_{r^2}\right) \qquad \tilde{E} = {}^{E}/_{m}, \qquad \tilde{L} = {}^{L}/_{m}$$

(Eqn. 5-50)

Radial motion $\tilde{L} \cdot= 0$, $R \equiv {}^{2M}\Big/(1 - \tilde{E}^2)$

$\left(\frac{dr}{d\tau}\right)^2 = 2M\left(\frac{1}{r} - \frac{1}{R}\right) \leftarrow$ if $L = \left(g^{\alpha\beta}P_\alpha P_\beta + m^2\right)$ then this expression would be in a Lagrangian, ready to be quantized:

$$L = \left(\frac{dr}{d\tau}\right)^2 - \frac{2M}{r} + \frac{2M}{R}$$

(Eqn. 5-51)

Alternate derivation of radial dynamical equation:

$$ds^2 = -d\tau^2 = \frac{32M^3}{r} e^{-r/2M}(du^2 - dv^2) + r^2 d\Omega^2$$

$$\left(\frac{r}{2M} - 1\right) e^{r/2M} = u^2 - v^2$$

$$\frac{\partial r}{\partial \tau} = \frac{\partial r}{\partial v}\frac{\partial v}{\partial \tau} + \frac{\partial r}{\partial u}\frac{du}{d\tau}$$

And since $\left(\frac{\partial \tau}{\partial v}\right)^2 = \frac{32M^3}{r} e^{-r/2M} = -\left(\frac{\partial \tau}{\partial u}\right)^2$. Also $\frac{\partial r}{\partial v} = -v\left(\frac{8M^2}{r} e^{-r/2M}\right)$ and $\frac{\partial r}{\partial u} = u\left(\frac{8M^2}{r} e^{-r/2M}\right)$. So,

$$\left(\frac{\partial r}{\partial \tau}\right)^2 = (v^2 - u^2)\frac{\left(\frac{8M^2}{r} e^{-r/2M}\right)^2}{\left(\frac{32M^3}{r} e^{-r/2M}\right)} - \frac{2uv\left(\frac{8M^2}{r} e^{-r/2M}\right)^2}{i\left(\frac{32M^3}{r} e^{-r/2M}\right)}$$

$$= -\left(1 - \frac{2M}{r}\right) + 2iuv\left(\frac{2M}{r}\right) e^{-r/2M}$$

$$\left(\frac{\partial r}{\partial \tau}\right)^2\Bigg|_{u=0} = -\left(1 - \frac{2M}{r}\right)\Bigg|_{u=0} = \frac{\partial r_{\min}}{\partial \tau} = \frac{\partial \theta}{\partial \tau}$$

$$\left(\frac{\partial \theta}{\partial \tau}\right)^2 = -\left(1 - \frac{2M}{\theta}\right) \qquad \underline{\theta = r_{\min} = throat.}$$

(Eqn. 5-52)

Note about comparison with prior deviation, there Schwarzschild coordinates are used which break down at $r = 2M$, however if $E \to 0$ as $r \to 2M$ the result is also clear there too. So, $\left(\frac{\partial \theta}{\partial \tau}\right)^2 - \frac{2M}{\theta} = -1$ (when $\dot{\theta} = 0 \quad \theta = 2M$). Comparing with the usual coulomb system in 1-D.

$$K.E. = \frac{1}{2}mv^2 = \frac{1}{2}m\left(\frac{\partial r}{\partial t}\right)^2$$

(Eqn. 5-53)

$$P.E. = \frac{-q^2}{r}$$

(Eqn. 5-54)

$$Energy = K.E. + P.E. = \text{const}$$

(Eqn. 5-55)

$$E = \frac{1}{2}m\left(\frac{\partial r}{\partial t}\right)^2 - \frac{q^2}{r}$$

(Eqn. 5-56)

$$\left(\frac{\partial r}{\partial t}\right)^2 - \left(\frac{2q^2}{m}\right)\left(\frac{1}{r}\right) = \frac{2E}{m}$$

(Eqn. 5-57)

Let $q^2 = Mm$ and $E = \frac{-m}{2}$

$\left(\frac{\partial r}{\partial t}\right)^2 - \frac{2M}{r} = -1$ … exactly like the throat dynamics.

(Eqn. 5-58)

So, there is a direct relationship to a coulomb system, however, instead of $0 < r < \infty$ we have $0 < \theta < 2M$. If $\theta = 2M$ is simply regarded as the classical turning point the q.m. the system will extend to $\theta > 2M$ which can't be interpreted within the classical geometric description .Classically there is infall, collapse of (modified) hydrogen atom. Q.M. there are quantized energy states, including a non-collapse minimum.

5.4 Evaluation of ^4R using 3 + 1 split and geodesic path method

We have:

$G_{\mu\nu} = R_{\mu\nu} - \frac{1}{2}g_{\mu\nu}R$, $g_{\mu\nu} = -n_\mu n_\nu + h_{\mu\nu}$, $n_\mu n^\mu = -1$, $n^\mu h_{\mu\nu} = 1$

(Eqn. 5-59)

$R = 2(G_{ab}n^a n^b - R_{ab}n^a n^b)$

(Eqn. 5-60)

$$2G_{ab}n^a n^b = R + 2R_{ab}n^a n^b = R_{abcd}(g^{ac} + n^a n^c)(g^{bd} + n^b n^d)$$
$$= R_{abcd}h^{ac}h^{db}$$

(Eqn. 5-61)

(Gauss-Codacci) $\quad = {}^{(3)}R + (K_a^a)^2 - K_{ab}K^{ab}$

(Eqn. 5-62)

$$
\begin{aligned}
R_{ab}n^a n^b = R^C_{acb}n^a n^b &= n^a(\nabla_c\nabla_a - \nabla_a\nabla_c)n^C \\
&= (\nabla_a n^a)^2 - (\nabla_c n^a)(\nabla_a n^c) - \nabla_a(n^a\nabla_c n^c) + \nabla_c(n^a\nabla_a n^c) \\
&= (K^a_a)^2 - K_{ac}K^{ac} - \nabla_a(n^a\nabla_c n^c) + \nabla_c(n^a\nabla_a n^c)
\end{aligned}
$$

(Eqn. 5-63)

$$\underline{{}^{(4)}R = {}^{(3)}R + K_{ab}K^{ab} - (K^a_a)^2 + \text{ divergences (tensorial)}}$$

(Eqn. 5-64)

$$K_{ab} = -\frac{1}{2}N^{-1}\left[\dot{h}_{ab} - D_a N_b - D_b N_a\right]$$

(Eqn. 5-65)

$$ds^2 = -N^2 dt^2 + L^2(dr + N^r dt)^2 + R^2 d\Omega^2$$

(Eqn. 5-66)

$$N^r = g_{rk}N^k = g_{rr}N^r = L^2 N^r$$

(Eqn. 5-67)

In the $\{t, r, \theta, \varphi\}$ coordinate frame:

$$\begin{pmatrix} {}^{(4)}g_{oo} & {}^{(4)}g_{ok} \\ {}^{(4)}g_{io} & {}^{(4)}g_{ik} \end{pmatrix} = \begin{pmatrix} (N_r N^r - N^2) & N^r & o & o \\ N^r & & & \\ o & & ({}^{(3)}g_{ik}) & \\ o & & & \end{pmatrix}$$

(Eqn. 5-68)

And

$$\begin{pmatrix} {}^{(4)}g^{oo} & {}^{(4)}g^{om} \\ {}^{(4)}g^{ko} & {}^{(4)}g^{km} \end{pmatrix} = \begin{pmatrix} -1/N^2 & N_r/N^2 & o & o \\ N_r/N^2 & ({}^{(3)}g^{km}(N_r/N)^2) & o & o \\ o & o & ({}^{(3)}g^{km}) & \\ o & o & & \end{pmatrix}$$

(Eqn. 5-69)

${}^{(3)}R$ is based on ${}^{(3)}g_{ik} = g_{ik}$: $d\ell^2 = L^2 dr^2 + R^2 d\Omega^2$, so ${}^{(3)}g \rightarrow {}^{(3)}\Gamma \rightarrow {}^{(3)}R$ is the calculation, and Γ's easiest to evaluate by extremal path method.

$K_{ik} \equiv -\frac{1}{2}£_n g_{ik}$ (note the sign convention)

(Eqn. 5-70)

or

$(dn)_i \equiv -K_{ik}dx^k$

(Eqn. 5-71)

Either of which yields:

$$K_{ik} = \frac{1}{2N}\left\{-\dot{g}_{ik} + N_{(i|k)}\right\}$$

(Eqn. 5-72)

(the sign convention makes no difference in ${}^4R \sim \dot{K}^2$)

$$N_{i|k} = \frac{\partial N_i}{\partial x^k} - {}^3\Gamma^m_{ik} N_m \ , \qquad N_m = {}^3g_{m\ell} N^\ell$$

(Eqn. 5-73)

$$K_{rr} = \frac{1}{2N}\{-\dot{g}_{rr} + N_{(r|r)}\} = \frac{1}{2N}\left(-(2LL') + 2\left[(N_r)_{,r} - \Gamma^r_{rr} N_r\right]\right)$$

(Eqn. 5-74)

$$N_r = L^2 N^r \Rightarrow (L^2 N^r)_{,r} - \left(\frac{L'}{L}\right) L^2 N^r = LL'N^r + L^2 N^r_{,r}$$

(Eqn. 5-75)

$$K_{rr} = \frac{1}{2N}\{-(2LL') + 2L(LN^r)'\} = -N^{-1} L\{\dot{L} - (LN^r)'\}$$

(Eqn. 5-76)

$$K_{\theta\theta} = \frac{1}{2N}\left\{-\dot{g}_{\theta\theta} + N_{(\theta|\theta)}\right\} = \frac{1}{2N}\{-\dot{g}_{\theta\theta} - 2\Gamma^r_{\theta\theta} N_r\}$$
$$= \frac{1}{2N}\left\{-2R\acute{R} - 2\left(-\frac{RR'}{L^2}\right)(L^2 N^r)\right\}$$
$$= -\frac{R}{N}\left\{\acute{R} - R'N^r\right\}$$

(Eqn. 5-77)

$$K_{\varphi\varphi} = \sin^2\theta\, K_{\theta\theta}$$

(Eqn. 5-78)

Now for the Γ's, using $d\ell^2 = L^2 dr^2 + R^2 d\Omega^2 \ and \ \ddot{x}^\mu + \Gamma^\mu_{\alpha\beta}\dot{x}^\alpha \dot{x}^\beta = 0$:

$$I = \int \frac{1}{2}\left(\frac{d\ell}{d\lambda}\right)^2 d\lambda = \int \frac{1}{2}\left\{L^2\left(\frac{dr}{d\lambda}\right)^2 + R^2\left(\left(\frac{d\theta}{d\lambda}\right)^2 + \sin^2\theta\left(\frac{d\varphi}{d\lambda}\right)^2\right)\right\} d\lambda$$

$$(\delta I)_r = \int \frac{1}{2}\left\{2LL'\left(\frac{dr}{d\lambda}\right)^2 \delta r + 2L^2 \frac{dr}{d\lambda}\frac{d}{d\lambda}(\delta r)\right.$$
$$\left. + 2RR'\left(\left(\frac{d\theta}{d\lambda}\right)^2 + \sin^2\theta\left(\frac{d\varphi}{d\lambda}\right)^2\right)\right\}$$
$$= \int \left\{LL'\left(\frac{dr}{d\lambda}\right)^2 - \frac{d}{d\lambda}\left(L^2 \frac{dr}{d\lambda}\right) + RR'\left(\frac{d\Omega}{d\lambda}\right)^2\right\} \delta r d\lambda$$

So,

$$-L^2\frac{d^2 r}{d\lambda^2} - 2L\frac{dL}{d\lambda}\frac{dr}{d\lambda} + LL'\left(\frac{dr}{d\lambda}\right)^2 + RR'\left(\left(\frac{d\theta}{d\lambda}\right)^2 + \sin^2\theta\left(\frac{d\varphi}{d\lambda}\right)^2\right) = 0$$

(Eqn. 5-79)

and we have:

$$\Gamma^r_{rr} = \left(\frac{L'}{L}\right)$$

(Eqn. 5-80)

$$\Gamma^r_{tr} = \left(\frac{2\dot{L}}{L}\right)$$

(Eqn. 5-81)

$$\Gamma^r_{\theta\theta} = \frac{RR'}{L^2}$$

(Eqn. 5-82)

$$\Gamma^r_{\varphi\varphi} = \frac{RR'}{L^2}\sin^2\theta$$

(Eqn. 5-83)

$$(\delta I)_\theta = \int \left\{ R^2 \left(\frac{d\theta}{d\lambda}\right) \frac{d}{d\lambda}(\delta\theta) + R^2 \sin\theta\cos\theta \left(\frac{d\varphi}{d\lambda}\right)^2 \right\} d\lambda$$

So,

$$-\frac{d}{d\lambda}\left(R^2\left(\frac{d\theta}{d\lambda}\right)\right) + R^2 \cos\theta\sin\theta\left(\frac{d\varphi}{d\lambda}\right)^2 = 0$$

and we have:

$$\Gamma^\theta_{\varphi\varphi} = -\cos\theta\sin\theta$$

(Eqn. 5-84)

$$\Gamma^\theta_{r\theta} = \left(\frac{R'}{R}\right)$$

(Eqn. 5-85)

$$(\delta I)_\varphi = \int \left\{ R^2 \sin^2\theta \frac{d\varphi}{d\lambda}\frac{d}{d\lambda}(\delta\varphi) \right\} d\lambda$$

So,

$$\frac{d}{d\lambda}\int \left\{ R^2 \sin^2\theta \frac{d\varphi}{d\lambda} \right\} = 0$$

and we have:

$$\Gamma^\varphi_{\theta\varphi} = \cot\theta$$

(Eqn. 5-86)

$$\Gamma^\varphi_{r\varphi} = \frac{R'}{R}$$

(Eqn. 5-87)

We can now calculate ${}^3R = g^{ik}\left[\Gamma^j_{ik,j} - \Gamma^j_{ij,k} + \Gamma^j_{ik}\Gamma^\ell_{j\ell} - \Gamma^j_{i\ell}\Gamma^\ell_{kj}\right]$:

$$\Gamma_{rr}^{r} = \left(\frac{L'}{L}\right) \qquad\qquad\qquad\qquad g^{rr} = L^{-2}$$

$$\Gamma_{\theta\theta}^{r} = \sin^{-2}\theta\,\Gamma_{\varphi\varphi}^{r} = \left(-\frac{RR'}{L^2}\right) \qquad g^{\theta\theta} = \sin^2\theta\,g^{\varphi\varphi} =$$
$$R^{-2}$$

$$\Gamma_{r\theta}^{\theta} = \Gamma_{r\varphi}^{\varphi} = \left(\frac{R'}{R}\right) \qquad\qquad \sqrt{^3g} = LR^2\sin\theta$$

$$\Gamma_{\varphi\varphi}^{\theta} = \cos\theta\sin\theta$$

$$\Gamma_{\theta\varphi}^{\varphi} = \cot\theta$$

$$^3R = g^{ik}\big[\Gamma_{i\ell}^{j}\Gamma_{kj}^{\ell} - \Gamma_{ik}^{j}\Gamma_{j\ell}^{\ell}\big] + \frac{1}{\sqrt{^3g}}\big\{\sqrt{^3g}\big(g^{ik}\Gamma_{ik}^{\ell} - g^{j\ell}\Gamma_{jk}^{k}\big)\big\}_{,\ell}$$

$$= {}^3G_L + \frac{1}{\sqrt{^3g}}\left(\sqrt{^3g}\,w^{\ell}\right)_{,\ell}$$

(Eqn. 5-88)

where

$$^3G_L = L^{-2}\left[(\Gamma_{rr}^{r})^2 + 2\big(\Gamma_{r\theta}^{\theta}\big)^2 - (\Gamma_{rr}^{r})^2 - \Gamma_{rr}^{r}\big(2\Gamma_{r\theta}^{\theta}\big)\right]$$

$$+ R^{-2}\left[2\Gamma_{\theta\theta}^{r}\Gamma_{r\theta}^{\theta} + \big(\Gamma_{\theta\varphi}^{\varphi}\big)^2 - \Gamma_{\theta\theta}^{r}\big(\Gamma_{rr}^{r} + 2\Gamma_{r\theta}^{\theta}\big)\right]$$

$$+ R^{-2}\sin^{-2}\theta\left[2\Gamma_{\varphi\varphi}^{r}\Gamma_{r\varphi}^{\varphi} + 2\Gamma_{\varphi\varphi}^{\theta}\Gamma_{\theta\varphi}^{\varphi} - \Gamma_{\varphi\varphi}^{r}\big(\Gamma_{rr}^{r} + 2\Gamma_{r\theta}^{\theta}\big) - \Gamma_{\varphi\varphi}^{\theta}\big(\Gamma_{\theta\varphi}^{\varphi}\big)\right]$$

$$= 2L^{-2}\big(\Gamma_{r\theta}^{\theta}\big)^2 - 2L^{-2}\Gamma_{r\theta}^{\theta}\Gamma_{rr}^{r} - 2R^{-2}\Gamma_{\theta\theta}^{r}\Gamma_{rr}^{r} + R^{-2}\big(\Gamma_{\theta\varphi}^{\varphi}\big)^2 +$$
$$R^{-2}\sin^{-2}\theta\left(\Gamma_{\varphi\varphi}^{\theta}\Gamma_{\theta\varphi}^{\varphi}\right)$$

$$= 2L^{-2}R^{-2}(R')^2 + 2R^{-1}L^{-3}L'R' + R^{-2}\cot^2\theta - R^{-2}\cot^2\theta -$$
$$2L^{-3}R^{-1}R'L'$$

$$= 2L^{-2}R^{-2}(R')^2$$

(Eqn. 5-89)

$$\sqrt{^3g}\,w^{r} = \sqrt{^3g}\big(g^{ik}\Gamma_{ik}^{r} - g^{jr}\Gamma_{jk}^{k}\big)$$

$$= \sqrt{^3g}\left(g^{rr}\Gamma_{rr}^{r} + g^{\theta\theta}\Gamma_{\theta\theta}^{r} + g^{\varphi\varphi}\Gamma_{\varphi\varphi}^{r} - g^{rr}\big(\Gamma_{rr}^{r} + 2\Gamma_{r\theta}^{\theta}\big)\right)$$

$$= \sqrt{^3g}(L^{-3}L' - 2R^{-1}L^{-2}R' - L^{-3}L' -$$
$$2L^{-2}R^{-1}R')$$

$$= -4RL^{-1}\sin\theta\,R'$$

(Eqn. 5-90)

$$\left(\sqrt{^3g}\,w^{r}\right)_{,r} = 4L^{-1}\sin\theta\left[(R')^2 + RR''\right] + 4RL^{-2}L'\sin\theta\,R'$$

$$\sqrt{{}^3g}w^\theta = \sqrt{{}^3g}\left(g^{\varphi\varphi}\Gamma^\theta_{\varphi\varphi} - g^{\theta\theta}\Gamma^\varphi_{\theta\varphi}\right)$$
$$= (LR^2\sin\theta)\left(-R^{-2}\left(\frac{\cos\theta}{\sin\theta}\right) - R^{-2}\cot\theta\right) = -2L\cos\theta$$

$$\left(\sqrt{{}^3g}w^\theta\right)_{,\theta} = 2L\sin\theta$$

$$\left(\sqrt{{}^3g}w^\varphi\right)_{,\varphi} = 0$$

So,

$$\frac{1}{\sqrt{{}^3g}}\left(\sqrt{{}^3g}w^\ell\right)_{,\ell} = 2R^{-2} - 4L^{-2}R^{-2}[(R')^2 + RR''] + 4R^{-1}L^{-3}L'R'$$

(Eqn. 5-91)

$${}^3R = 4L^{-2}R^{-1}R'' + 4L^{-3}R^{-1}L'R' - 2L^{-2}R^{-2}(R')^2 + 2R^{-2}$$

(Eqn. 5-92)

$$K^{ab}K_{ab} - K^2 = -\left(2K^r_r\left(K^\theta_\theta + K^\varphi_\varphi\right) + 2K^\theta_\theta K^\varphi_\varphi\right)$$
$$= -2\left(K_{rr}\left(R^{-2}K_{\theta\theta} + (R\sin\theta)^{-2}K_{\varphi\varphi}\right) + R^{-2}(R\sin\theta)^{-2}K_{\varphi\varphi}\right)$$
$$= -2R^{-2}(2K_{rr}K_{\theta\theta}L^{-2} + R^{-2}(K_{\theta\theta})^2)$$
$$= -2R^{-2}N^{-2}\left((2L^{-1}R(LN^r)')\left(\dot{R} - R'N^r\right) + \left(\dot{R} - R'N^r\right)^2\right)$$

(Eqn. 5-93)

$$N\sqrt{{}^3g} = NLR^2\sin\theta$$

(Eqn. 5-94)

$$N\sqrt{{}^3g}(K^{ab}K_{ab} - K^2)$$
$$= -2N^{-1}\sin\theta\left[2R\left(\dot{L} - (LN^r)'\right)\left(\dot{R} - R'N^r\right) + L\left(\dot{R} - R'N^r\right)^2\right]$$

(Eqn. 5-95)

$$\sqrt{-{}^4g}\,{}^{(4)}R = N\sqrt{{}^3g}\{{}^3R + K^{ab}K_{ab} - (K^a_a)^2\}$$
$$= [-4L^{-1}NRR'' + 4L^{-2}RNL'R' - 2L^{-1}N(R')^2 + 2NL]\sin\theta - 2N^{-1}\sin\theta\,[\cdots]$$

(Eqn. 5-96)

Thus,

$$S = \frac{1}{16\pi}\int_{\infty}\sqrt{-^4g}\,{}^{(4)}R dr dt d\theta d\varphi$$

$$= \int dt \int_{-\infty}^{\infty} dr \begin{bmatrix} -N^{-1}\left\{\left(R\left(\dot{L} - (LN^r)'\right)\left(\dot{R} - R'N^r\right) + \frac{1}{2}L\left(\dot{R} - R'N^r\right)^2\right)\right\} \\ +N\left\{-L^{-1}RR'' + L^{-2}RR'L' - \frac{1}{2}L^{-1}(R')^2 + \frac{1}{2}L\right\} \end{bmatrix}$$

(Eqn. 5-97)

How does this compare with the ${}^{(4)}L$ obtained earlier? Many surface terms are already eliminated:

$${}^4L_{old} - {}^4L_{new} = \left[\frac{1}{2}R^2\left(\frac{\left(\dot{L} - (LN^r)'\right)}{N}\right)\right]^{\cdot}$$
$$- \left[\frac{1}{2}R^2\left(\frac{N'}{L} + \frac{N^r}{N}\left(\dot{L} - (LN^r)'\right)\right)\right]'$$
$$-RDR\frac{\left(\dot{L} - (LN^r)'\right)}{N} + RR'N'L^{-1} + \left[\left(\frac{RL}{N}\right)DR\right]^{\cdot} - \left(\frac{RL}{N}\right)^{\cdot}DR$$
$$-\left[\left(\frac{RLN^r}{N}\right)DR\right]' + \left(\frac{RLN^r}{N}\right)'DR - \frac{RL}{N^2}(DR)(DN) - \frac{R}{N}R'N'$$
$$+RN^{-1}\left(DL - N^{r\prime}L\right)(DR) + RN^{-1}\left(DL - N^{r\prime}L\right)(DR) + \frac{L}{N}(DR)^2$$

which reduces to:

$${}^4L_{old} - {}^4L_{new} = \left[\frac{1}{2}R^2\left(\frac{\left(\dot{L}-(LN^r)'\right)}{N}\right) + \left(\frac{RL}{N}\right)DR\right]^{\cdot} - \left[\frac{1}{2}R^2\left(\frac{N'}{L} + \frac{N^r}{N}\left(\dot{L} - (LN^r)'\right)\right) + \left(\frac{RLN^r}{N}\right)DR\right]'$$

(Eqn. 5-98)

Working with the 3 + 1 split has greatly simplified the analysis – surface terms need not arise, only to be dropped again.

5.5 Change of basis analysis, starting with 2-sphere example

Let's consider a change of basis: $\tilde{e}_\alpha = e_\beta A^\beta_\alpha$. This results in a gauge transformation to get the new connections (non-tensorial), and have:

$$\tilde{S}^\alpha_{\beta\gamma} = (A^{-1})^\alpha_\delta A^\sigma_\beta A^x_\gamma S^\delta_{\sigma x}$$

(Eqn. 5-99)

$$\tilde{\Gamma}^\alpha_{\beta\gamma} = (A^{-1})^\alpha_\delta A^\sigma_\beta A^x_\gamma \Gamma^\delta_{\sigma x} + (A^{-1})^\alpha_\mu \left(A^\mu_\beta\right)_{,\gamma}$$

(Eqn. 5-100)

Consider a 2-sphere:

$$ds^2 = R^2(d\theta^2 + \sin^2\theta\, d\varphi^2)$$

$$-d\left(2R^2\frac{d\theta}{d\lambda}\right) + 2R^2\sin\theta\cos\theta\left(\frac{d\varphi}{d\lambda}\right)^2 = 0$$

(Eqn. 5-101)

$$\frac{d^2\theta}{d\lambda^2} - \sin\theta\cos\theta\left(\frac{d\varphi}{d\lambda}\right)^2 = 0 \quad\Rightarrow\quad \boxed{\Gamma^{\theta}_{\varphi\varphi} = -\sin\theta\cos\theta}$$

(Eqn. 5-102)

$$-d\left(2R^2\sin^2\theta\frac{d\varphi}{d\lambda}\right) = 0$$

$$\frac{d^2\varphi}{d\lambda^2} + \cot\theta\left(\frac{d\varphi}{d\lambda}\right)\left(\frac{d\theta}{d\lambda}\right) = 0 \quad\Rightarrow\quad \boxed{\Gamma^{\varphi}_{\theta\varphi} = -\cot\theta}$$

(Eqn. 5-103)

So, can now compute ${}^2R$:

$${}^2R = g^{\mu\nu}\left\{\Gamma^{\alpha}_{\mu\nu,\alpha} - \Gamma^{\alpha}_{\mu\alpha,\nu} + \Gamma^{\alpha}_{\beta\alpha}\Gamma^{\beta}_{\mu\nu} - \Gamma^{\alpha}_{\beta\nu}\Gamma^{\beta}_{\mu\alpha}\right\} \qquad \text{(in coord. Frame)}$$

(Eqn. 5-104)

$$= g^{\theta\theta}\left\{-\Gamma^{\varphi}_{\theta\varphi,\theta} - \left(\Gamma^{\varphi}_{\theta\varphi}\right)^2\right\} + g^{\varphi\varphi}\left(\Gamma^{\theta}_{\varphi\varphi,\theta} - \Gamma^{\varphi}_{\theta\varphi}\Gamma^{\theta}_{\varphi\varphi}\right)$$

$$= R^{-2}\{(1+\cot^2\theta) - \cot^2\theta\} + (R\sin\theta)^{-2}\{(\sin^2\theta - \cos^2\theta) + \cos^2\theta\}$$

$$= 2R^{-2}$$

(Eqn. 5-105)

Using the Landau expression:

$${}^2R = g^{\mu\nu}\left\{\Gamma^{\alpha}_{\beta\nu} - \Gamma^{\beta}_{\mu\alpha} + \Gamma^{\alpha}_{\beta\alpha}\Gamma^{\beta}_{\mu\nu}\right\} + \frac{1}{\sqrt{{}^2g}}\left\{\sqrt{{}^2g}\left(g^{\mu\nu}\Gamma^{\alpha}_{\mu\nu} - g^{\mu\alpha}\Gamma^{\nu}_{\mu\nu}\right)\right\}_{,\alpha}$$

(Eqn. 5-106)

$$= R^{-2}\begin{bmatrix} (\cot\theta)^2 + (\sin\theta)^{-2}(-\sin\theta\cos\theta)\cot\theta \\ +(\sin\theta)^{-1}\{\sin\theta\,[(\sin\theta)^{-2}(-\sin\theta\cos\theta)\cot\theta]\}_{,\theta} \end{bmatrix} = 2R^{-2}$$

(Eqn. 5-107)

Rather than the coordinate basis: $e_\theta = \partial_\theta, e_\varphi = \partial_\varphi$, let's transform to a Lorentzian basis:

$$\tilde{e}_\theta = r^{-1}\partial_\theta \quad and \quad \tilde{e}_\varphi = (r\sin\theta)^{-1}\partial_\varphi$$

So,

$$\tilde{e}_\alpha = e_\beta A^{\beta}_{\alpha} \Rightarrow \left|A^{\beta}_{\alpha}\right| = r^{-1}\begin{pmatrix}1 & 0\\ 0 & (\sin\theta)^{-1}\end{pmatrix}; \left|(A^{-1})^{\gamma}_{\delta}\right| = r\begin{pmatrix}1 & 0\\ 0 & \sin\theta\end{pmatrix}$$

$$\tilde{g}^{\mu\nu} = (A^{-1})^{\mu}_{\gamma}(A^{-1})^{\nu}_{\delta}g^{\gamma\delta} \quad ; \quad \tilde{g}_{\mu\nu} = A^{\gamma}_{\mu}A^{\delta}_{\nu}g_{\gamma\delta}$$

$$\tilde{g}^{\theta\theta} = r^2 g^{\theta\theta} = \left(\frac{r}{R}\right)^2 \qquad \tilde{g}_{\theta\theta} = \left(\frac{R}{r}\right)^2 \quad \sqrt{{}^2\tilde{g}} = \left(\frac{R}{r}\right)^2$$

$$\tilde{g}^{\varphi\varphi} = \left(\frac{r}{R}\right)^2 \qquad \tilde{g}_{\varphi\varphi} = \left(\frac{R}{r}\right)^2$$

$$\tilde{\Gamma}^{\theta}_{\varphi\varphi} = r(r\sin\theta)^{-2}(-\sin\theta\cos\theta) = -r^{-1}\cot\theta$$
$$\tilde{\Gamma}^{\varphi}_{\theta\varphi} = r^{-1}\cot\theta$$
$$\Gamma^{\varphi}_{\varphi\theta} = r^{-1}\cot\theta + (A^{-1})^{\varphi}_{\varphi} r^{-1}\partial_{\theta}(A^{\varphi}_{\varphi}) = 0$$

Note that this result is not symmetric due to incompatibility of connection with metric. So, putting this together:

$$^{2}R = \tilde{g}^{\mu\nu}\left\{\left(\tilde{\Gamma}^{\alpha}_{\mu\nu,\alpha} - \tilde{\Gamma}^{\alpha}_{\mu\alpha,\nu}\right) + \left(\tilde{\Gamma}^{\alpha}_{\beta\alpha}\tilde{\Gamma}^{\beta}_{\mu\nu} - \tilde{\Gamma}^{\alpha}_{\beta\nu}\tilde{\Gamma}^{\beta}_{\mu\alpha}\right)\right\}$$

(Eqn. 5-108)

For the right group we simply have $-R^{-2}(2\cot^2\theta)$. For the left grouping we have:

$$\tilde{g}^{\theta\theta}\left\{-(r^{-1}\partial_{\theta})\Gamma^{\varphi}_{\theta\varphi}\right\} + \tilde{g}^{\varphi\varphi}\left\{(r^{-1}\partial_{\theta})\Gamma^{\theta}_{\varphi\varphi}\right\} = 2R^{-2}(1+\cot^2\theta)$$

(Eqn. 5-109)

Thus,

$$^{2}R = 2R^{-2}.$$

5.6 Change of basis analysis for Landau form

We want to take the Landau form $R = G + \frac{1}{\sqrt{g}}\left(\sqrt{g}w^{\ell}\right)_{,\ell}$ over to a new basis, but this is non-trivial because w^{ℓ} is not a true tensor. This requires a review of the Landau derivation. Starting with:

$$R = \tilde{g}^{\mu\nu}\left\{\tilde{\Gamma}^{\alpha}_{\mu\nu,\delta}A^{\delta}_{\alpha} - \tilde{\Gamma}^{\alpha}_{\mu\alpha,\delta}A^{\delta}_{\nu} + \tilde{\Gamma}^{\alpha}_{\beta\alpha}\tilde{\Gamma}^{\beta}_{\mu\nu} - \tilde{\Gamma}^{\alpha}_{\nu\beta}\tilde{\Gamma}^{\beta}_{\mu\alpha}\right\}$$

(Eqn. 5-110)

and the gauge relation:

$$\sqrt{-\tilde{g}}\,\tilde{g}^{\alpha\beta}\tilde{\Gamma}^{\gamma}_{\alpha\beta,\delta}A^{\delta}_{\gamma} = \left(\sqrt{-\tilde{g}}\,\tilde{g}^{\alpha\beta}\tilde{\Gamma}^{\gamma}_{\alpha\beta}A^{\delta}_{\gamma}\right)_{,\delta} - \tilde{\Gamma}^{\gamma}_{\alpha\beta}\left(\sqrt{-\tilde{g}}\,\tilde{g}^{\alpha\beta}A^{\delta}_{\gamma}\right)_{,\delta}$$

(Eqn. 5-111)

Also recall:

$\Gamma^{\delta}_{\gamma\delta} = \left(\ln\sqrt{-g}\right)_{,\gamma}$ (in coord. frame)

(Eqn. 5-112)

$g^{\alpha\beta}_{,\gamma} = -\Gamma^{\alpha}_{\delta\gamma}g^{\delta\beta} - \Gamma^{\beta}_{\delta\gamma}g^{\alpha\delta}$ (from $g^{\alpha\beta};\gamma = 0$)

(Eqn. 5-113)

$\Gamma^{\gamma\delta}\Gamma^{\nu}_{\gamma\delta} = -\frac{1}{\sqrt{-g}}\left(\sqrt{-g}g^{\alpha\beta}\right)_{,\beta}$ (both the above)

(Eqn. 5-114)

The new basis is not a coordinate frame, so modifications are needed.

$$\tilde{\Gamma}^{\gamma}_{\alpha\beta}\left(\sqrt{-\tilde{g}}\tilde{g}^{\alpha\beta}A^{\delta}_{\gamma}\right)_{,\delta} = \sqrt{-\tilde{g}}A^{\delta}_{\gamma}\tilde{\Gamma}^{\gamma}_{\alpha\beta}\left(\tilde{g}^{\alpha\beta}_{,\delta}\right) + \tilde{\Gamma}^{\gamma}_{\alpha\beta}\tilde{g}^{\alpha\beta}\left(\sqrt{-\tilde{g}}A^{\delta}_{\gamma}\right)_{,}$$

$$= \sqrt{-\tilde{g}}\tilde{\Gamma}^{\gamma}_{\alpha\beta}\left[-\tilde{\Gamma}^{\alpha}_{\delta\gamma}\tilde{g}^{\delta\beta} - \tilde{\Gamma}^{\beta}_{\delta\gamma}\tilde{g}^{\alpha\delta}\right] + \tilde{\Gamma}^{\gamma}_{\alpha\beta}\tilde{g}^{\alpha\beta}\left(\sqrt{-g}A^{\delta}_{\gamma}\right)$$

$$= \sqrt{-\tilde{g}}\tilde{\Gamma}^{\gamma}_{\alpha\beta}\left[-\tilde{\Gamma}^{\alpha}_{\delta\gamma}\tilde{g}^{\delta\beta} + \tilde{\Gamma}^{\beta}_{\delta\gamma}\tilde{g}^{\alpha\delta}\right] + \tilde{\Gamma}^{\gamma}_{\alpha\beta}\tilde{g}^{\alpha\beta}\left(\sqrt{-g}A^{\delta}_{\gamma}\right)$$

Similarly for $\sqrt{-\tilde{g}}\tilde{g}^{\alpha\beta}\tilde{\Gamma}^{\gamma}_{\alpha\gamma,\delta}A^{\delta}_{\beta} = \left(\sqrt{-\tilde{g}}\tilde{g}^{\alpha\beta}\tilde{\Gamma}^{\gamma}_{\alpha\gamma}A^{\delta}_{\beta}\right)_{,\delta} -$ $\tilde{\Gamma}^{\gamma}_{\alpha\gamma}\left(\sqrt{-\tilde{g}}\tilde{g}^{\alpha\beta}A^{\delta}_{\beta}\right)_{,\delta}$

$$\tilde{\Gamma}^{\gamma}_{\alpha\gamma}\left(\sqrt{-\tilde{g}}\tilde{g}^{\alpha\beta}A^{\delta}_{\beta}\right)_{,\delta} = \sqrt{-\tilde{g}}\,\tilde{\Gamma}^{\gamma}_{\alpha\gamma}\,\underline{A^{\delta}_{\beta}\tilde{g}^{\alpha\beta}_{,\delta}} + \tilde{\Gamma}^{\gamma}_{\alpha\gamma}\tilde{g}^{\alpha\beta}\left(\sqrt{-\tilde{g}}A^{\delta}_{\beta}\right)_{,\delta}$$

$$= -\sqrt{-\tilde{g}}\,\tilde{\Gamma}^{\gamma}_{\alpha\gamma}\left[\tilde{\Gamma}^{\alpha}_{\delta\beta}\tilde{g}^{\delta\beta} + \tilde{\Gamma}^{\beta}_{\delta\beta}\tilde{g}^{\alpha\delta}\right] + \tilde{\Gamma}^{\gamma}_{\alpha\gamma}\tilde{g}^{\alpha\beta}\left(\sqrt{-\tilde{g}}A^{\delta}_{\beta}\right)_{,\delta}$$

using

$$A^{\delta}_{\beta}\tilde{g}^{\alpha\beta}_{,\delta} = -\tilde{\Gamma}^{\alpha}_{\delta\beta}\tilde{g}^{\delta\beta} - \tilde{\Gamma}^{\beta}_{\delta\beta}\tilde{g}^{\alpha\delta}$$

Thus,

$$\sqrt{-\tilde{g}}R = \tilde{g}^{\alpha\beta}\left\{\sqrt{-\tilde{g}}\left(\tilde{\Gamma}^{\gamma}_{\delta\beta}\tilde{\Gamma}^{\delta}_{\alpha\gamma} + \underline{\tilde{\Gamma}^{\gamma}_{\alpha\delta}\tilde{\Gamma}^{\delta}_{\beta\gamma}}\right) - \sqrt{-\tilde{g}}\left(\tilde{\Gamma}^{\gamma}_{\delta\gamma}\tilde{\Gamma}^{\delta}_{\alpha\beta} + \tilde{\Gamma}^{\gamma}_{\beta\gamma}\tilde{\Gamma}^{\delta}_{\alpha\delta}\right)\right.$$
$$\left. + \sqrt{-\tilde{g}}\left(\underline{\tilde{\Gamma}^{\mu}_{\nu\mu}\tilde{\Gamma}^{\nu}_{\alpha\beta}} - \underline{\tilde{\Gamma}^{\mu}_{\alpha\nu}\tilde{\Gamma}^{\nu}_{\beta\mu}}\right)\right\}$$
$$+ \left(\sqrt{-\tilde{g}}\tilde{g}^{\alpha\beta}\left[\tilde{\Gamma}^{\gamma}_{\alpha\beta}A^{\delta}_{\gamma} - \tilde{\Gamma}^{\gamma}_{\alpha\gamma}A^{\delta}_{\beta}\right]\right)_{,\delta} - \tilde{\Gamma}^{\gamma}_{\alpha\beta}\tilde{g}^{\alpha\beta}\left(\sqrt{-\tilde{g}}A^{\delta}_{\gamma}\right)_{,\delta}$$
$$+ \tilde{\Gamma}^{\gamma}_{\alpha\gamma}\tilde{g}^{\alpha\beta}\left(\sqrt{-\tilde{g}}A^{\delta}_{\beta}\right)_{,\delta}$$

$$\sqrt{-\tilde{g}}R = \sqrt{-\tilde{g}}\tilde{g}^{\alpha\beta}\left\{\tilde{\Gamma}^{\gamma}_{\delta\beta}\tilde{\Gamma}^{\delta}_{\alpha\gamma} - \tilde{\Gamma}^{\gamma}_{\beta\gamma}\tilde{\Gamma}^{\delta}_{\alpha\delta}\right\}$$
$$- \tilde{g}^{\alpha\beta}\left\{\tilde{\Gamma}^{\gamma}_{\alpha\beta}\left(\sqrt{-\tilde{g}}A^{\delta}_{\gamma}\right)_{,\delta} - \tilde{\Gamma}^{\gamma}_{\alpha\gamma}\left(\sqrt{-\tilde{g}}A^{\delta}_{\beta}\right)_{,\delta}\right\}$$
$$+ \left(\sqrt{-\tilde{g}}\tilde{g}^{\alpha\beta}\left[\tilde{\Gamma}^{\gamma}_{\alpha\beta}A^{\delta}_{\gamma} - \tilde{\Gamma}^{\gamma}_{\alpha\gamma}A^{\delta}_{\beta}\right]\right)_{,\delta}$$

When $A^{\mu}_{\mu} = \delta^{\mu}_{\mu}$ and $\tilde{\Gamma}^{\delta}_{\gamma\delta} = \frac{1}{\sqrt{-g}}\left(\sqrt{-g}\right)_{,\gamma}$ for coord frame, the familiar equation results, so

$$\sqrt{-\tilde{g}}R = \sqrt{-\tilde{g}}\tilde{g}^{\alpha\beta}\left\{\left(\tilde{\Gamma}^{\gamma}_{\delta\beta}\tilde{\Gamma}^{\delta}_{\alpha\gamma} - \tilde{\Gamma}^{\gamma}_{\alpha\beta}\frac{1}{\sqrt{-\tilde{g}}}\left(\sqrt{-\tilde{g}}A^{\delta}_{\gamma}\right)_{,\delta}\right)\right.$$
$$\left. + \left(\tilde{\Gamma}^{\gamma}_{\alpha\gamma}\frac{1}{\sqrt{-\tilde{g}}}\left(\sqrt{-\tilde{g}}A^{\delta}_{\beta}\right)_{,\delta} - \tilde{\Gamma}^{\gamma}_{\beta\gamma}\tilde{\Gamma}^{\delta}_{\alpha\delta}\right)\right\}$$
$$+ \left(\sqrt{-\tilde{g}}\tilde{g}^{\alpha\beta}\left[\tilde{\Gamma}^{\gamma}_{\alpha\beta}A^{\delta}_{\gamma} - \tilde{\Gamma}^{\gamma}_{\alpha\gamma}A^{\delta}_{\beta}\right]\right)_{,\delta}$$

Returning to the ${}^2R$ calculation as a check:

$$ {}^2R = \tilde{g}^{\theta\theta}\left\{\Gamma^{\varphi}_{\theta\varphi}\frac{1}{\sqrt{\tilde{g}}}\left(\sqrt{{}^2\tilde{g}}A^{\theta}_{\theta}\right)_{,\theta} - \left(\Gamma^{\varphi}_{\theta\varphi}\right)^2\right\} $$

$$ + \tilde{g}^{\varphi\varphi}\left\{\Gamma^{\varphi}_{\theta\varphi}\Gamma^{\theta}_{\varphi\varphi} - \Gamma^{\theta}_{\varphi\varphi}\frac{1}{\sqrt{{}^2\tilde{g}}}\left(\sqrt{{}^2\tilde{g}}A^{\theta}_{\theta}\right)_{,\theta}\right\} $$

$$ +\frac{1}{\sqrt{{}^2\tilde{g}}}\left(\sqrt{{}^2\tilde{g}}\left\{\tilde{g}^{\theta\theta}\left[\Gamma^{\varphi}_{\theta\varphi}\,A^{\theta}_{\theta}\right] + \tilde{g}^{\varphi\varphi}\left[\Gamma^{\theta}_{\varphi\varphi}\,A^{\theta}_{\theta}\right]\right\}\right)_{,\theta} $$

$$ = \left(\frac{r}{R}\right)^2\left\{-\left(\Gamma^{\varphi}_{\theta\varphi}\right)^2 + \Gamma^{\varphi}_{\theta\varphi}\Gamma^{\theta}_{\varphi\varphi} + \left(\Gamma^{\theta}_{\varphi\varphi,\theta} - \Gamma^{\varphi}_{\theta\varphi,\theta}\right)r^{-1}\right\} $$
$$ = \left(\frac{r}{R}\right)^2\{-2r^{-2}\cot^2\theta + (r^{-1}[1+\cot^2\theta] + r^{-1}[1+\cot^2\theta])r^{-1}\} $$
$$ = 2R^2 $$

5.7 The spatial ADM metric evaluated in coordinate frame and orthonormal frame

The spatial metric, $d\ell^2 = L^2dr^2 + R^2d\Omega^2$, has been examined in the coordinate frame:

$$ e_r = \partial_r\,, e_\theta = \partial_\theta\,, e_\varphi = \partial_\varphi. $$

The connection coefficients are:

$$ \Gamma^r_{rr} = L^{-1}L' $$
$$ \Gamma^r_{\theta\theta} = \sin^{-2}\theta\,\Gamma^r_{\varphi\varphi} = $$
$$ -L^{-2}RR' $$
$$ \Gamma^\theta_{r\theta} = \Gamma^\varphi_{r\varphi} = R^{-1}R' $$
$$ \Gamma^\theta_{\varphi\varphi} = \cos\theta\sin\theta $$
$$ \Gamma^\varphi_{\theta\varphi} = \cot\theta $$

(and the Γ^i_{jk}'s are symmetric on the $(jk)'^s$ since this is a coord. frame with metric compatible with connection). Now to switch to the orthonormal frame:

$$ \tilde{e}_r = \partial_r\,, \tilde{e}_\theta = r^{-1}\partial_\theta\,, \tilde{e}_\varphi = (r\sin\theta)^{-1}\partial_\varphi $$

($\tilde{e}_\alpha\tilde{e}_\beta\eta^{\alpha\beta} = 1$ where $\eta_{\alpha\beta} = dr\times dr + r^2d\theta\times d\theta + r^2\sin^2\theta\,d\varphi\times d\varphi$)

So, for $\tilde{e}_\alpha = e_\beta A^\beta_\alpha$ defining A^β_α:

$$ \left|A^\beta_\alpha\right| = \begin{pmatrix} 1 & 0 & 0 \\ 0 & r^{-1} & 0 \\ 0 & 0 & (r\sin\theta)^{-1} \end{pmatrix};\ \left|(A^{-1})^\gamma_\delta\right| = \begin{pmatrix} 1 & 0 & 0 \\ 0 & r & 0 \\ 0 & 0 & r\sin\theta \end{pmatrix} $$

$$\tilde{g}^{\mu\nu} = (A^{-1})^{\mu}_{\gamma}(A^{-1})^{\nu}_{\delta}g^{\gamma\delta} \qquad ; \qquad \tilde{g}_{\mu\nu} = A^{\gamma}_{\mu}A^{\delta}_{\nu}g_{\gamma\delta}$$

$$\tilde{g}^{rr}_{\theta\theta} = g^{rr} = L^{-2} \qquad \tilde{g}_{rr} = L^2$$

$$\tilde{g}^{\theta\theta} = r^2 g^{\theta\theta} = \left(\frac{r}{R}\right)^2 \qquad \tilde{g}_{\theta\theta} = \left(\frac{R}{r}\right)^2 \qquad \underline{\sqrt{{}^3\tilde{g}} = \left(\frac{R}{r}\right)^2}$$

$$\tilde{g}^{\varphi\varphi} = (r\sin\theta)^2\tilde{g}^{\varphi\varphi} = \left(\frac{r}{R}\right)^2 \qquad \tilde{g}_{\varphi\varphi} = \left(\frac{R}{r}\right)^2$$

$$\tilde{\Gamma}^{\alpha}_{\beta\gamma} = (A^{-1})^{\alpha}_{\delta}A^{\sigma}_{\beta}A^{x}_{\gamma}\Gamma^{\delta}_{\sigma x} + (A^{-1})^{\alpha}_{\mu}\left(A^{\mu}_{\beta}\right)_{,\gamma}$$

$$\begin{cases} \tilde{\Gamma}^{\theta}_{r\theta} = \Gamma^{\theta}_{r\theta} = R^{-1}R' \\ \tilde{\Gamma}^{\theta}_{\theta r} = \Gamma^{\theta}_{\theta r} + (A^{-1})^{\theta}_{\theta}\left(A^{\theta}_{\theta}\right)_{,r} = R^{-1}R' - r^{-1} \end{cases}$$

$$\begin{cases} \tilde{\Gamma}^{\varphi}_{r\varphi} = \Gamma^{\varphi}_{r\varphi} = R^{-1}R'\left(= \tilde{\Gamma}^{\theta}_{r\theta}\right) \\ \tilde{\Gamma}^{\varphi}_{\varphi r} = R^{-1}R' - r^{-1}\left(= \tilde{\Gamma}^{\theta}_{\theta r}\right) \end{cases}$$

$$\tilde{\Gamma}^{\theta}_{\varphi\varphi} = r(r\sin\theta)^{-2}\Gamma^{\theta}_{\varphi\varphi} = -r^{-1}\cot\theta$$

$$\begin{cases} \tilde{\Gamma}^{\varphi}_{\theta\varphi} = r^{-1}\tilde{\Gamma}^{\varphi}_{\theta\varphi} = r^{-1}\cot\theta \\ \tilde{\Gamma}^{\varphi}_{\varphi\theta} = (r^{-1}\cot\theta) + (A^{-1})^{\varphi}_{\varphi}\left(A^{\varphi}_{\varphi}\right)_{,\theta} = (r^{-1}\cot\theta) - r^{-1}\cot\theta = 0 \end{cases}$$

From the calculation previously:

$$\begin{aligned} {}^3R = \tilde{g}^{\alpha\beta}\Bigg\{&\left(\tilde{\Gamma}^{\gamma}_{\delta\beta}\,\tilde{\Gamma}^{\delta}_{\alpha\gamma} - \tilde{\Gamma}^{\gamma}_{\alpha\delta}\frac{1}{\sqrt{\tilde{g}}}\left(\sqrt{\tilde{g}}A^{\delta}_{\gamma}\right)_{,\delta}\right) \\ &+\left(\tilde{\Gamma}^{\gamma}_{\alpha\gamma}\frac{1}{\sqrt{\tilde{g}}}\left(\sqrt{\tilde{g}}A^{\delta}_{\beta}\right)_{,\delta} - \tilde{\Gamma}^{\gamma}_{\beta\gamma}\,\tilde{\Gamma}^{\delta}_{\alpha\delta}\right)\Bigg\} + \frac{1}{\sqrt{\tilde{g}}}\left(\sqrt{\tilde{g}}w^{\delta}\right) \end{aligned}$$

Where $w^{\delta} = \tilde{g}^{\alpha\beta}\left[\tilde{\Gamma}^{\gamma}_{\alpha\beta}A^{\delta}_{\gamma} - \tilde{\Gamma}^{\gamma}_{\alpha\gamma}A^{\delta}_{\beta}\right]$ and where all <u>explicit</u> derivatives are coordinate derivatives:

$$^{3}R = \tilde{g}^{rr}\left\{\left(\left(\tilde{\Gamma}^{r}_{rr}\right)^{2} + 2\tilde{\Gamma}^{\theta}_{\theta r}\tilde{\Gamma}^{\theta}_{\theta r} - \tilde{\Gamma}^{r}_{rr}\frac{1}{\sqrt{\tilde{g}}}\frac{\partial}{\partial r}\left(\sqrt{\tilde{g}}\right)\right)\right.$$

$$\left. + \left(\tilde{\Gamma}^{r}_{rr} + 2\tilde{\Gamma}^{\theta}_{r\theta}\right)\frac{1}{\sqrt{\tilde{g}}}\frac{\partial}{\partial r}\left(\sqrt{\tilde{g}}\right) - \left(\tilde{\Gamma}^{r}_{rr} + 2\tilde{\Gamma}^{\theta}_{\theta r}\right)\right\}$$

$$+\tilde{g}^{\theta\theta}\left\{\left(\tilde{\Gamma}^{r}_{\theta\theta}\tilde{\Gamma}^{\theta}_{\theta r} + \tilde{\Gamma}^{\theta}_{r\theta}\tilde{\Gamma}^{r}_{\theta\theta} + \tilde{\Gamma}^{\varphi}_{\varphi\theta}\tilde{\Gamma}^{\varphi}_{\theta\varphi} - \tilde{\Gamma}^{r}_{\theta\theta}\frac{1}{\sqrt{\tilde{g}}}\left(\sqrt{\tilde{g}}\right)_{,r}\right)\right.$$

$$\left. + \tilde{\Gamma}^{\varphi}_{\theta\varphi}\frac{1}{\sqrt{\tilde{g}}}\left(\sqrt{\tilde{g}}A^{\theta}_{\theta}\right)_{,\theta} - \left(\tilde{\Gamma}^{\varphi}_{\theta\varphi}\right)^{2}\right\}$$

$$+\tilde{g}^{\varphi\varphi}\left(\tilde{\Gamma}^{r}_{\varphi\varphi}\tilde{\Gamma}^{\varphi}_{\varphi r} + \tilde{\Gamma}^{\varphi}_{r\varphi}\tilde{\Gamma}^{r}_{\varphi\varphi} + \tilde{\Gamma}^{\varphi}_{\theta\varphi}\tilde{\Gamma}^{\theta}_{\varphi\varphi} + \tilde{\Gamma}^{\theta}_{\varphi\varphi}\tilde{\Gamma}^{\varphi}_{\varphi\theta} - \tilde{\Gamma}^{r}_{\varphi\varphi}\tfrac{1}{\sqrt{\tilde{g}}}\left(\sqrt{\tilde{g}}\right)_{,r} -\right.$$

$$\left.\tilde{\Gamma}^{\theta}_{\varphi\varphi}\frac{\left(\sqrt{\tilde{g}}A^{\theta}_{\theta}\right)}{\sqrt{\tilde{g}}}\right)$$

$$+\tfrac{1}{\sqrt{\tilde{g}}}\left\{\left(\sqrt{\tilde{g}}\left[\tilde{g}^{rr}\tilde{\Gamma}^{r}_{rr} + \tilde{g}^{\theta\theta}\tilde{\Gamma}^{r}_{\theta\theta} + \tilde{g}^{\varphi\varphi}\tilde{\Gamma}^{r}_{\varphi\varphi} - \tilde{g}^{rr}\left(\tilde{\Gamma}^{r}_{rr} + 2\tilde{\Gamma}^{\theta}_{r\theta}\right)\right]\right)_{,r} +\right.$$

$$\left.\left(\sqrt{\tilde{g}}A^{\theta}_{\theta}\left[\tilde{g}^{\varphi\varphi}\tilde{\Gamma}^{r}_{\varphi\varphi} - \tilde{g}^{\theta\theta}\tilde{\Gamma}^{\varphi}_{\theta\varphi}\right]\right)_{,\theta}\right\}$$

$$^{3}R = \tilde{g}^{rr}\left\{2\tilde{\Gamma}^{\theta}_{\theta r}\tilde{\Gamma}^{\theta}_{r\theta} + 2\tilde{\Gamma}^{\theta}_{r\theta}\frac{1}{\sqrt{\tilde{g}}}\frac{\partial}{\partial r}\left(\sqrt{\tilde{g}}\right) - 4\tilde{\Gamma}^{r}_{rr}\tilde{\Gamma}^{\theta}_{r\theta} - 4\left(\tilde{\Gamma}^{\theta}_{r\theta}\right)^{2}\right\}$$

$$+\tilde{g}^{\theta\theta}\left\{\begin{array}{c}2\tilde{\Gamma}^{r}_{\theta\theta}\left(\tilde{\Gamma}^{\theta}_{\theta r} + \tilde{\Gamma}^{\theta}_{r\theta} - \tfrac{1}{\sqrt{\tilde{g}}}\left(\sqrt{\tilde{g}}\right)_{,r}\right) + \tilde{\Gamma}^{\varphi}_{\theta\varphi}\left(\tilde{\Gamma}^{\varphi}_{\varphi\theta} + \tfrac{1}{\sqrt{\tilde{g}}}\left(\sqrt{\tilde{g}}A^{\theta}_{\theta}\right)_{,\theta} - \tilde{\Gamma}^{\varphi}_{\theta\varphi} + \tilde{\Gamma}^{\theta}_{\varphi\varphi}\right. \\ +\tilde{\Gamma}^{\theta}_{\varphi\varphi}\left(\tilde{\Gamma}^{\varphi}_{\varphi\theta} + \tfrac{1}{\sqrt{\tilde{g}}}\left(\sqrt{\tilde{g}}A^{\theta}_{\theta}\right)_{,\theta}\right)\end{array}\right.$$

$$+\tfrac{1}{\sqrt{\tilde{g}}}\left\{\left(\sqrt{\tilde{g}}\left[2\tilde{g}^{\theta\theta}\tilde{\Gamma}^{r}_{\theta\theta} - 2g^{rr}\tilde{\Gamma}^{\theta}_{r\theta}\right]\right)_{,r} + \left(\sqrt{\tilde{g}}A^{\theta}_{\theta}\tilde{g}^{\theta\theta}\left[\tilde{\Gamma}^{\theta}_{\varphi\varphi} - \tilde{\Gamma}^{\varphi}_{\theta\varphi}\right]_{,\theta}\right)\right\}$$

$$^{3}R = L^{-2}\left\{2\left(\frac{R'}{R} - \frac{1}{r}\right)\left(\frac{R'}{R}\right) + 2\left(\frac{R'}{R}\right)\frac{1}{\sqrt{\tilde{g}}}\frac{\partial}{\partial r}\left(\sqrt{\tilde{g}}\right) - 4\left(\frac{L'}{L}\right)\left(\frac{R'}{R}\right)\right.$$

$$\left. - 4\left(\frac{R'}{R}\right)^{2}\right\}$$

$$+\left(\frac{r}{R}\right)^2\left\{\begin{array}{c} 2(-r^{-2}L^{-2}RR')\left(\left(\frac{R'}{R}-\frac{1}{r}\right)+\frac{R'}{R}-\frac{1}{\sqrt{\tilde{g}}}\frac{\partial}{\partial r}\left(\sqrt{\tilde{g}}\right)\right) \\ +(r^{-1}\cot\theta)\left(\frac{1}{\sqrt{\tilde{g}}}\left(\sqrt{\tilde{g}}A_\theta^\theta\right)\right)_{,\theta} \\ -r^{-1}\cot\theta-r^{-1}\cot\theta+(r^{-1}\cot\theta)\left(-\frac{1}{\sqrt{\tilde{g}}}\left(\sqrt{\tilde{g}}A_\theta^\theta\right)_{,\theta}\right) \end{array}\right\}$$

$$+\frac{1}{\sqrt{\tilde{g}}}\left\{\left(\sqrt{\tilde{g}}\left[2\left(\frac{r}{R}\right)^2\right](-r^{-2}L^{-2}RR')-2L^{-2}R^{-1}R'\right)_{,r}+\right.$$
$$\left.\left(\sqrt{\tilde{g}}A_\theta^\theta\left(\frac{r}{R}\right)^2[-r^{-1}\cot\theta-r^{-1}\cot\theta]\right)_{,\theta}\right\}$$

$$=4L^{-2}\left(\frac{R'}{R}\right)\frac{1}{\sqrt{\tilde{g}}}\frac{\partial}{\partial r}\left(\sqrt{\tilde{g}}\right)-4L^{-3}L'\left(\frac{R'}{R}\right)-6L^{-2}\left(\frac{R'}{R}\right)^2$$
$$+2rR^{-2}\cot\theta\left(\frac{1}{\sqrt{\tilde{g}}}\left(\sqrt{\tilde{g}}A_\theta^\theta\right)_{,\theta}-r^{-1}\cot\theta\right)$$
$$+\frac{1}{\sqrt{\tilde{g}}}\left\{\left(-4\sqrt{\tilde{g}}L^{-2}R^{-1}R'\right)_{,r}-2R^{-2}\left(\sqrt{\tilde{g}}\cot\theta\right)_{,\theta}\right\}$$

$${}^3R=-4L^{-2}R^{-1}R''+4L^{-3}L'R^{-1}R'-2L^{-2}R^{-2}(R')^2+2R^{-2}$$

Let's now consider a change of basis from spherical coordinate frame $e_\alpha=\left(e_r,e_\theta,e_\varphi\right)$ to Cartesian coordinate frame $\tilde{e}_\beta=\left(\tilde{e}_x,\tilde{e}_y,\tilde{e}_z\right)$ and $\tilde{e}_\alpha=A_\alpha^\beta e_\beta$, and where $d\ell^2=L^2dr^2+R^2d\Omega^2=\tilde{g}_{\mu\nu}dx^\mu dx^\nu$ with $dx^1=dx, dx^2=dy, dx^3=dz$. We can determine A_α^β from coordinate relations:

$$x=r\sin\theta\cos\varphi$$
$$y=r\sin\theta\sin\varphi$$
$$z=r\cos\theta$$

Thus,
$$\frac{\partial}{\partial x}=\sin\theta\cos\varphi\frac{\partial}{\partial r}+\cos\theta\cos\varphi\frac{1}{r}\frac{\partial}{\partial r}-\frac{\sin\varphi}{r\sin\theta}\frac{\partial}{\partial\varphi}$$
$$\frac{\partial}{\partial y}=\sin\theta\sin\varphi\frac{\partial}{\partial r}+\cos\theta\sin\varphi\frac{1}{r}\frac{\partial}{\partial r}+\frac{\cos\varphi}{r\sin\varphi}\frac{\partial}{\partial\varphi}$$

$$\frac{\partial}{\partial y} = \cos\theta \frac{\partial}{\partial r} - \sin\theta \frac{1}{r}\frac{\partial}{\partial \theta}$$

$$\begin{pmatrix} \tilde{e}_x \\ \tilde{e}_y \\ \tilde{e}_z \end{pmatrix} = \begin{matrix} \\ x \\ y \\ z \end{matrix} \begin{pmatrix} \mathrm{r} & \theta & \varphi \\ \sin\theta\cos\varphi & r^{-1}\cos\theta\cos\varphi & -(r\sin\theta)^{-1}\sin\varphi \\ \sin\theta\sin\varphi & r^{-1}\cos\theta\sin\varphi & (r\sin\theta)^{-1}\cos\varphi \\ \cos\theta & -r^{-1}\sin\theta & 0 \end{pmatrix} \begin{pmatrix} e_r \\ e_\theta \\ e_\varphi \end{pmatrix}$$

and

$$\begin{pmatrix} e_r \\ e_\theta \\ e_\varphi \end{pmatrix} = \begin{pmatrix} \sin\theta\cos\varphi & \sin\theta\sin\varphi & \cos\theta \\ r\cos\theta\cos\varphi & r\cos\theta\sin\varphi & -r\sin\theta \\ -r\sin\theta\sin\varphi & r\sin\theta\cos\varphi & 0 \end{pmatrix} \begin{pmatrix} \tilde{e}_x \\ \tilde{e}_y \\ \tilde{e}_z \end{pmatrix}$$

$$|\tilde{g}_{\mu\nu}| = |A_\mu^\gamma||g_{\gamma\delta}||A_\nu^\delta|^t$$

$$|\tilde{g}_{\mu\nu}| = |A_\mu^r||g_{rr}||A_\nu^r|^t + |A_\mu^\theta||(g_{\theta\theta})||A_\nu^\theta|^t + |A_\mu^\varphi||(g_{\varphi\varphi})||A_\nu^\varphi|^t$$

$$= \begin{pmatrix} \sin\theta\cos\varphi \\ \sin\theta\sin\varphi \\ \cos\theta \end{pmatrix} L^2(\sin\theta\cos\varphi \ \sin\theta\sin\varphi \ \cos\theta)$$

$$+ \begin{pmatrix} r^{-1}\cos\theta\cos\varphi \\ r^{-1}\cos\theta\sin\varphi \\ -r^{-1}\sin\theta \end{pmatrix} R^2(\cdots) +$$

$$\begin{pmatrix} -(r\sin\theta)^{-1}\sin\varphi \\ (r\sin\theta)^{-1}\cos\varphi \\ 0 \end{pmatrix} (R\sin\theta)^2(\cdots)$$

$$\det|\tilde{g}_{\mu\nu}| = \left(\det|A_\mu^\gamma|\right)^2 \det|g_{\gamma\delta}|$$

$$\det(A^{-1}) = r^2 \det \begin{vmatrix} \sin\theta\cos\varphi & \sin\theta\sin\varphi & \cos\theta \\ \cos\theta\cos\varphi & \cos\varphi\sin\varphi & -\sin\theta \\ -\sin\theta\sin\varphi & \sin\theta\cos\varphi & 0 \end{vmatrix}$$

$= r^2\{\sin^3\theta\,[\cos^2\varphi + \sin^2\varphi] + \sin\theta\cos^2\theta\,[\cos^2\varphi +$
$\sin^2\varphi]\}$ $\quad (r\sin\theta)^2 = x^2 + y^2$
$= r^2\{\sin\theta\,[\cos^2\varphi + \sin^2\varphi]\} = r^2\sin\theta$

$$\det|\tilde{g}_{\mu\nu}| = (r^2\sin\theta)^{-2}LR^2\sin\theta = r^{-2}(r\sin\theta)^{-1}LR^2$$
$$= r^{-3}(x^2+y^2)^{-1/2}LR^2$$

In MTW notation:
$d\ell^2 = L^2dr^2 + R^2d\Omega^2 = \tilde{g}_{\mu\nu}dx^\mu dx^\nu = g^{\mu\nu}dx_\mu dx_\nu$

$$= g^{\mu\nu}\frac{\partial}{\partial x^\mu}\otimes\frac{\partial}{\partial x^\nu}$$
$$= L^{-2}\left(\sin\theta\cos\varphi\,\tilde{e}_x + \sin\theta\sin\varphi\,\tilde{e}_y + \cos\theta\,\tilde{e}_z\right)^2$$
$$+R^{-2}\left(r\cos\theta\cos\varphi\,\tilde{e}_x + r\cos\theta\sin\varphi\,\tilde{e}_y + r\sin\theta\,\tilde{e}_z\right)^2$$
$$+(R\sin\theta)^{-2}\left(-r\sin\theta\sin\varphi\,\tilde{e}_x + r\sin\theta\cos\varphi\,\tilde{e}_y\right)^2$$
$$= \tilde{e}_x\otimes\tilde{e}_x\left\{\sin^2\theta\cos^2\varphi\,L^{-2} + \left(\frac{r}{R}\right)^2(\cos^2\theta\cos^2\varphi + \sin^2\varphi)\right\}$$
$$+2\tilde{e}_x\otimes\tilde{e}_y\left\{\begin{matrix}\sin^2\theta\sin\varphi\cos\varphi\,L^{-2}\\ +\left(\frac{r}{R}\right)^2(\cos^2\theta\sin\varphi\cos\varphi - \sin\varphi\cos\varphi)\end{matrix}\right\}$$
$$+2\tilde{e}_x\otimes\tilde{e}_z\left\{\sin\theta\cos\theta\cos\varphi\,L^{-2} - \left(\frac{r}{R}\right)^2\sin\theta\cos\theta\cos\varphi\right\}$$
$$+\tilde{e}_y\otimes\tilde{e}_y\left\{\sin^2\theta\sin^2\varphi\,L^{-2} + \left(\frac{r}{R}\right)^2(\cos^2\theta\sin^2\varphi + \cos^2\varphi)\right\}$$
$$+2\tilde{e}_y\otimes\tilde{e}_z\left\{\sin\theta\cos\theta\sin\varphi\,L^{-2} + \left(\frac{r}{R}\right)^2\sin\theta\cos\theta\sin\varphi\right\}$$
$$+\tilde{e}_z\otimes\tilde{e}_z\left\{\cos^2\theta\,L^{-2} + \left(\frac{r}{R}\right)^2\sin^2\theta\right\}$$

(Eqn. 5-115)

5.8 Exercises

(Ex. 5.1) Derive Eqn. 5-7.
(Ex. 5.2) Derive Eqn. 5-8.
(Ex. 5.3) Derive Eqn. 5-9.
(Ex. 5.4) Derive Eqn. 5-10.
(Ex. 5.5) Derive Eqn.s 5-11, 5-12, and 5-13.
(Ex. 5.6) Derive Eqn. 5-18.
(Ex. 5.7) Derive Eqn. 5-22.
(Ex. 5.8) Derive Eqn. 5-26.
(Ex. 5.9) Derive Eqn. 5-36.
(Ex. 5.10) Derive Eqn. 5-46.
(Ex. 5.11) Derive Eqn. 5-48.
(Ex. 5.12) Derive Eqn. 5-49.
(Ex. 5.13) Derive Eqn. 5-50.
(Ex. 5.14) Derive Eqn. 5-51.
(Ex. 5.15) Derive Eqn. 5-52.
(Ex. 5.16) Derive Eqn. 5-72.
(Ex. 5.17) Derive Eqn. 5-76.
(Ex. 5.18) Derive Eqn. 5-83.
(Ex. 5.19) Derive Eqn. 5-115.

Chapter 6. The FNC formalism

In Ch. 5 we saw that a space-like Cauchy surface could be described aa part of a foliation with local metric given in terms of "Lapse" and "Shift" between sheaves of the foliation. We now consider, instead of a foliation of space-like hypersurface, a congruence of time-like geodesics. For our reference geodesic in particular, we have the "Observer" frame. As the name suggests, this could be the actual perspective of an observer in the formalism. By 'frame' the name also suggests a coordinate system but this we haven't done yet. Let's now consider a central observer (on a geodesic) and consider the maximal extension of a coordinate frame to second order in perturbations (e.g., Fermi Normal Coordinates). For this observer, can the metric along their time-like path (parameterized by t) be written in terms of the Riemann tensor for the indicated coordinate frame? The answer is yes and this will be shown in three different types of derivations in Section 6.2-6.4.

6.1 Introduction

Since Einstein space-time solutions are locally flat, for a geodesic observer we expect a local metric description to be flat up to 2nd order derivatives. The contributions at 2nd order will be found to satisfy the following metric form for choice of coordinates known as Fermi Normal Coordinates (FNCs) (Manasse and Misner (1963). J. Math. Phys. 4, 735-745 [27]):

$$ds^2 = \left(-1 - R_{\hat{0}\hat{\ell}\hat{0}\hat{m}}\, x^{\hat{\ell}} x^{\hat{m}}\right)dt^2 - \left(\frac{4}{3} R_{\hat{0}\hat{\ell}\hat{\jmath}\hat{m}}\, x^{\hat{\ell}} x^{\hat{m}}\right) dt\, dx^{\hat{\jmath}}$$
$$+ \left(\delta_{\hat{\imath}\hat{\jmath}} - \frac{1}{3} R_{\hat{\imath}\hat{\ell}\hat{\jmath}\hat{m}}\, x^{\hat{\ell}} x^{\hat{m}}\right) dx^{\hat{\imath}} dx^{\hat{\jmath}} + O\left(|x^{\hat{\jmath}}|^3\right) dx^{\hat{\alpha}} dx^{\hat{\beta}}$$

(Eqn. 6-1)

In an approach to geodesic analysis based on a geodesic observer we will eventually make use of a Jacobi Field analysis to extend the domain of application. In essence, we will be returning to the congruence of geodesics indicated at the outset. Since we will be analyzing a congruence of geodesics there is the congruence analysis method according to the Raychaudhuri Equation formalism. Let's examine this before proceeding.

6.1.1 Raychaudhuri Formalism

Consider a smooth congruence of (time-like) geodesics → a smooth vector field ξ^a, and normalize to $\xi^a\xi_a = -1$. Then define

$$B_{ab} = \nabla_b \xi_a \rightarrow \text{purely spatial since } B_{ab}\xi^a = 0$$

(Eqn. 6-2)

Consider $\gamma_s(\tau)$: a smooth <u>one</u>-parameter subfamily of geodesics in the congruence, let η^a be the orthogonal deviation from γ_0. η^a represents an infinitesential spatial displacement from γ_0 to a nearby geodesic in the subfamily, and:

$$\mathcal{L}_\xi \eta^a = 0 \text{ thus } \xi^b \nabla_b \eta^a = \eta^b \nabla_b \eta^a \xi^a = B^a_b \eta^b$$

(Eqn. 6-3)

So, B^a_b measures the failure of η^a to be parallely transported.

Define

$$h_{ab} = g_{ab} + \xi_a \xi_b$$

(Eqn. 6-4)

$$\theta = B^{ab} h_{ab}$$

(Eqn. 6-5)

$$\sigma_{ab} = B_{(ab)} - \frac{1}{3}\theta h_{ab}$$

(Eqn. 6-6)

$$\omega_{ab} = B_{[ab]}$$

(Eqn. 6-7)

Then:

$$B_{ab} = \frac{1}{3}\theta h_{ab} + \sigma_{ab} + \omega_{ab}$$

(Eqn. 6-8)

Along any geodesic in the congruence, θ measures the avg. expansion of the infinitesimally nearby surrounding geodesics, ω_{ab}, being the antisymmetric part of the linear map B_{ab}, measures their rotation; and σ_{ab} measures their shear, $\xi^c \nabla_c B_{ab} = -B^c_b B_{ac} + R_{cba}{}^d \xi^c \xi^d$. From

$$\xi^c \nabla_c B_{ab} = -B^c_a B_{ac} + R_{cba}{}^d \xi^c \xi_d$$

(Eqn. 6-9)

The trace gives:

$$\xi^c \nabla_c \theta = \frac{d\theta}{d\tau} = -\frac{1}{3}\theta^2 - \sigma_{ab}\sigma^{ab} + \omega_{ab}\,\omega^{ab} - R_{cd}\,\xi^c \xi^d$$

(Eqn. 6-10)

and the trace free symmetric point:

$$\xi^c \nabla_c \sigma_{ab} = -\frac{2}{3}\theta\sigma_{ab} - \sigma_{ac}\sigma^c_b - \omega_{ac}\omega^c_b + \frac{1}{3}h_{ab}\,(\sigma_{cd}\sigma^{cd} - \omega_{cd}\omega^{cd})$$
$$+C_c b_{ad}\xi^c \xi^d + \tfrac{1}{2}\left(h_{ac}h_{bd}R^{cd} - \tfrac{1}{3}h_{ab}h_{cd}R^{cd}\right)$$

(Eqn. 6-11)

and the antisymmetric point:

$$\xi^c \nabla_c \omega_{ab} = -\frac{2}{3}\theta\omega_{ab} - 2\sigma^c_{[b}\omega_{a]c}$$

(Eqn. 6-12)

Can choose $\omega_{ab} = 0$ and it would always be so, not true for the others. So to determine the evolution of θ along the geodesic we have to contend with two coupled <u>nonlinear</u> ODE'S.

If we specify a θ and σ_{ab} initially ($\omega_{ab} = 0$) it is conceivable that the eqn's could be solved but the evolution through the inevitable caustics for θ_0 negative pose a technical difficulty since we then have to patch through to a time-reversed, inverted θ evolution. In effect, a geodesic with Jacobi field analysis is needed, so that is what will be done.

6.1.2 FNC analysis for CTC metric to be studied in Ch. 7

Let τ be the proper time as measured by the closed null geodesic (cng) observer's clock. Let $P = P_0(\tau)$ be that observer's world line. The observer carries an orthonormal tetrad $\{e_{\hat{\alpha}}\}$ with

$$e_{\hat{0}} = \vec{u} = dP_0/d\tau \quad \text{and} \quad e_{\hat{\alpha}} \cdot e_{\hat{\beta}} = \eta_{\alpha\beta}$$

(Eqn. 6-13)

We know the world line to be a geodesic, so no acceleration, is there any rotation?

$$\Gamma_{\hat{j}\hat{k}\hat{0}} = -w^{\hat{\imath}} e_{\hat{0}\hat{\imath}\hat{j}\hat{k}}$$

(Eqn. 6-14)

Proper time $\tau = \rho_0 \phi$ $\qquad e_{\hat{0}} = \vec{u} = \frac{1}{\rho_0}\frac{\partial}{\partial e}$ $\qquad x^{\hat{0}} = \tau = \rho_0 \phi$

$x^{\hat{3}} = (\rho - \rho_0) \quad x^{\hat{2}} = z \quad x^{\hat{\imath}} = t$

$$e_{\hat{3}} = \frac{\partial}{\partial \rho}\,;\; e_{\hat{2}} = \frac{\partial}{\partial z}\,;\; e_{\hat{1}} = \frac{\partial}{dt}$$
If rotation is zero we get FNCs.

What is the geodesic associated with a curve:

$$e(\phi(\lambda), t = -2\alpha, \rho = \rho_0, z = 0)$$

(Eqn. 6-15)

Or simply regard the vector $\frac{\partial}{\partial \phi}$ at the instant $\phi = 0, t = -2\alpha, \rho = \rho_{0,} z = 0$: let $\vec{u}(\lambda = 0) = \frac{\partial}{\rho_0 \partial \phi}$ then (note: $u^\alpha(\lambda = 0) =$ only $u^\phi = \frac{1}{\rho_0}$):

$$\nabla_{\vec{u}} \vec{u} = 0 \Rightarrow u^\alpha_{;\beta} u^\beta = \left(u^\alpha_{,\beta} + \Gamma^\alpha_{\gamma\beta} u^\gamma\right) u^\beta = 0$$

(Eqn. 6-16)

Thus

$$\frac{dx^\beta}{d\lambda} \frac{\partial}{\partial x^\beta} \left(\frac{dx^\alpha}{d\lambda}\right) + \Gamma^\alpha_{\gamma\beta} \frac{dx^\gamma}{d\lambda} \frac{dx^\beta}{d\lambda} = 0$$

leads to:

$$\frac{\partial u^\phi}{\partial \phi} + \Gamma^\phi_{\phi\phi}\, u^\phi = 0\,.$$

(Eqn. 6-17)

In Ch. 7 we show that the CTC metric:

$$ds^2 = f[dt^2 - (pd\phi)^2] + (1 - f^2) dtpd\phi - d\rho^2 - dz^2$$

(Eqn. 6-18)

has

$$\Gamma^\phi_{\phi\phi} = \frac{(1-f^2)}{(1-f^2)} \left(\frac{\partial f}{\partial t}\right) \text{ and } f|_G = 0 \rightarrow \Gamma^\phi_{\phi\phi}|_G = -4\rho_0 \alpha \tanh(3\alpha^2).$$

Thus,

$$u^\phi = \frac{1}{\rho_0} \exp\{4\rho_0 \alpha \tanh(3\alpha^2)\, \phi\} \;\; \textit{and} \text{ other } \;\; u^\alpha = 0.$$

(Eqn. 6-19)

$\vec{u} = u^\phi \frac{\partial}{\partial \phi}$ *is then:*

$$\vec{u} = \exp\{4\rho_0 \alpha \tanh(3\alpha^2)\, \phi\}\, (1/\rho_0)\, \frac{\partial}{\partial \phi}$$

(Eqn. 6-20)

which describes an exponential blue shift. So, CTC's are evidently not allowed, and anything that would allow a CTC is not allowed (restatement of Hawking's CPC [28]). Since violation of WEC allows for CTC, violation of WEC not allowed.

Now consider a change the signature of the metric such that:

$$ds^2 = f[dt^2 - (pd\phi)^2] + (1 - f^2)dtpd\phi + d\rho^2 + dz^2$$

Now to consider the pseudo-orthonormal frame along the geodesic:

$$e_0 = \frac{\partial}{\partial x^0}\bigg|_G = \exp\left[-\frac{df}{dt}\bigg|_G \phi\right]\frac{1}{\rho_0}\frac{\partial}{\partial\phi} = \exp[4\rho_0\alpha\tanh(3\alpha^2)\,\phi]\frac{1}{\rho_0}\frac{\partial}{\partial\phi}$$

Want $e_0 \cdot e_1 = -1$ at G :

$$e_1 = \exp[4\rho_0\alpha\tanh(3\alpha^2)\,\phi]\ \frac{\partial}{\partial t}$$

$$e_2 = \frac{\partial}{\partial\rho}$$

$$e_3 = \frac{\partial}{\partial z}$$

$$\frac{De_\alpha}{dx_0} = (e_\alpha)^{\mu'}_{;\upsilon'}(e_0)^{\upsilon'} = (e_\alpha)^{\mu'}_{;\phi}\left[\underbrace{\exp\{4\rho_0\alpha\tanh(3\alpha^2)\,\phi\}\frac{1}{\rho_0}}_{\gamma}\right]$$

(Eqn. 6-21)

What is $(e_\alpha)^{\mu'}_{;\phi}$?

$(e_\alpha)^{\mu'}_{;\phi}$: $\quad (e_\alpha)^{\mu'}_{;\phi} = (e_\alpha)^{\phi}_{,\phi} + \Gamma^{\phi}_{\alpha\phi}\,(e_0)^{\alpha} = \frac{\gamma}{\rho_0}\exp[\gamma\phi] + \frac{\Gamma^{\phi}_{\phi\phi}}{-\gamma}\,\frac{\exp[\phi]}{\rho_0}$

So, $(e_\alpha)^{\mu}_{;\phi} = 0$

$(e_\alpha)^{\mu'}_{;\phi}$: $\quad (e_1)^{t}_{;\phi} = \exp[-\gamma\phi]\,(-\gamma) + \Gamma^{t}_{t\phi}\exp[-\gamma\phi]$

$$\Gamma^{t}_{t\phi} = g^{t\mu'}\Gamma_{\mu t\phi} = g^{t\mu'}\tfrac{1}{2}\left(g_{\mu t,\phi} + g_{\mu\phi,t} - g_{t\phi,\mu'}\right)$$

$$= g^{t\phi}\,\tfrac{1}{2}\,g_{\phi\phi,t} + g^{tt}\tfrac{1}{2}\left(g_{t\phi,t} - g_{t\phi,t}\right) = -\frac{(1-f^2)}{(1-f^2)}\left(\frac{\partial f}{\partial t}\right)$$

$$= \gamma$$

So, $(e_\alpha)^{\mu}_{;\phi} = 0$

$(e_2)^{\mu'}_{;\phi}$: $\quad (e_2)^{\rho}_{;\phi} = \Gamma^{\rho}_{;\phi}(e_2)^{\mu} = \Gamma^{\rho}_{;\phi} = 0$

So, $(e_2)^{\mu}_{;\phi} = 0$

Similarly $(e_3)^{\mu'}_{;\phi} = 0$. Thus, $(e_\alpha)^{\mu'}_{;\phi} = 0$ and $\frac{De_\alpha}{dx_0} = 0 \Rightarrow$ The tetrad e_α is parallel transported.

In the coordinates of this space-time $x^{\mu'}: t, \phi, \rho, z$ we have

$R_{\alpha\beta\gamma\delta} = R_{\mu'v'\sigma'\tau'}\,(e_\alpha)^{\mu'}(e_\beta)^{v'}\;(e_\gamma)^{\sigma'}\;(e_\sigma)^{\tau'}\}$

$(e_0)^{\mu'} = \frac{1}{\rho_0}\,e^{\gamma\phi}\delta_\phi^{\mu'}$

and

$$\left.\begin{aligned} R_{0\beta\gamma\delta} &= \frac{1}{\rho_0}\,e^{\gamma\phi}\,R_{\phi v'\sigma'\tau'}(e_\beta)^{v'}(e_\gamma)^{\sigma'}(e_\delta)^{\tau'} \\ R_{1\beta\gamma\delta} &= e^{-\gamma\phi}\,R_{tv'\sigma'\tau'}(e_\beta)^{v'}(e_\gamma)^{\sigma'}(e_\delta)^{\tau'} \\ R_{2\beta\gamma\delta} &= R_{\rho v'\sigma'\tau'}(e_\beta)^{v'}(e_\gamma)^{\sigma'}(e_\delta)^{\tau'} \\ R_{3\beta\gamma\delta} &= R_{zv'\sigma'\tau'}(e_\beta)^{v'}(e_\gamma)^{\sigma'}(e_\delta)^{\tau'} \end{aligned}\right\}\text{ same for remaining indices}$$

$$R_{\alpha\beta\gamma\delta} = \left(\frac{1}{\rho_0}\,e^{\gamma\phi}\right)^{\left[\delta_0^\alpha+\delta_0^\beta+\delta_0^\gamma+\delta_0^\delta\right]}\left(e^{-\gamma\phi}\right)^{\left[\delta_1^\alpha+\delta_1^\beta+\delta_1^\gamma+\delta_1^\delta\right]}R_{\mu'v'\sigma'\tau'};$$

where $\alpha = \begin{pmatrix}0\\1\\2\\3\end{pmatrix} \rightarrow \mu' = \begin{pmatrix}\phi\\t\\\rho\\z\end{pmatrix}$ *etc.*

Now

$g_{00} = -1 + R_{0\ell 0m}|_G\,x^\ell x^m + \cdots$ $\qquad \ell = 1,2,3$ *only*

(Eqn. 6-22)

$g_{0i} = -\frac{2}{3}\,R_{0\ell im}|_G\,x^\ell x^m + \cdots$ $\qquad x_1 = e^{\gamma\phi}t\,;\, x_1 = \rho;\;\; x_3 = z$

(Eqn. 6-23)

$g_{ij} = \delta_{ij} + \frac{1}{3}\,R_{i\ell jm}|_G\,x^\ell x^m + \cdots$

(Eqn. 6-24)

So, the Fermi normal metric is:

$ds^2 = \;g_{00}e^{2\gamma\phi}\,\left(\frac{1}{\rho_0}\frac{\partial}{\partial\phi}\right)^2 + \cdots$

$R_{i\ell jm}|_G = F(\phi)$ *only*

What is $R_{ijk\ell}$? ($i = 1,2,3$ *not* 0)

What is $R_{i'j'k'\ell'}$? ($i' = \rho, z, t$ *not* 0)

Let's use MTW conventions for $R_{\alpha\beta\gamma\delta}$:

$g_{00} = -1 + R_{0\ell 0m}|_G\,x^\ell x^m + \cdots$

(Eqn. 6-25)

$g_{0i} = -\frac{2}{3}\,R_{0\ell im}|_G\,x^\ell x^m + \cdots$

(Eqn. 6-26)

$$g_{ij} = \delta_{ij} - \frac{1}{3} R_{i\ell jm}|_G \, x^\ell x^m + \cdots$$

(Eqn. 6-27)

And

$$R_{i'j'k'\ell'} = g_{i'\mu'} R^{\mu'}_{j'k'\ell'}$$
$$R^{\mu}_{v\alpha\beta} = \partial_\alpha \Gamma^{\mu}_{v\beta} - \partial_\beta \Gamma^{\mu}_{v\alpha} + \Gamma^{\mu}_{\sigma\alpha}\Gamma^{\sigma}_{v\beta} - \Gamma^{\mu}_{\sigma\beta}\Gamma^{\sigma}_{v\alpha}$$
$$\Gamma^{\mu}_{v\beta} = g^{\mu r\gamma}\Gamma_{\gamma v\beta} = g^{\mu\gamma}\frac{1}{2}\left(g_{\gamma v,\beta} + g_{\gamma\beta,v} - g_{v\beta,\gamma}\right)$$
$$g_{\mu v} = \begin{pmatrix} -f & -\frac{1}{2}\rho(1-f^2) & 0 & 0 \\ \frac{1}{2}\rho(1-f^2) & f\rho^2 & 0 & 0 \\ 0 & 0 & 1 & 0 \\ 0 & 0 & 0 & 1 \end{pmatrix}$$

(Eqn. 6-28)

and

$$-g^{\mu v} = \begin{pmatrix} \frac{-1}{\rho^2 \frac{1}{4}(1-f^2)^2} & \begin{Bmatrix} -f\rho^2 & \frac{1}{2}(1-f^2)\rho \\ -\frac{1}{2}(1-f^2)\rho & f \end{Bmatrix} & 0 & 0 \\ 0 & 0 & -1 & 0 \\ 0 & 0 & 0 & -1 \end{pmatrix}$$

(Eqn. 6-29)

$$g_{\mu v,i} = \begin{pmatrix} -\frac{\partial f}{\partial x^i} & \left(-\frac{1}{2}\frac{\partial \rho}{\partial x^i}(1-f^2) + \rho f\frac{\partial f}{\partial x^i}\right) & 0 & 0 \\ \left(-\frac{1}{2}\frac{\partial \rho}{\partial x^i}(1-f^2) + \rho f\frac{\partial f}{\partial x^i}\right) & \left(\rho^2\frac{\partial f}{\partial x^i} + 2f\rho\frac{\partial \rho}{\partial x^i}\right) & 0 & 0 \\ 0 & 0 & 0 & 0 \end{pmatrix}$$

(Eqn. 6-30)

Thus

$$\left.\frac{\partial f}{\partial x^t}\right|_G = \left.\frac{\partial f}{\partial t}\right|_G = 4\alpha \tanh(3\alpha^2)$$
$$f = 1 - \frac{\operatorname{sech}[(\rho-\rho_0)^2 + z^2 + t^2 - \alpha^2]}{\operatorname{sech}(3\alpha^2)}$$

$$\left.\frac{\partial f}{\partial \rho}\right|_G = 0 \quad \left.\frac{\partial f}{\partial z}\right|_G = 0$$

$$\left.\frac{\partial f}{\partial u}\right|_G = \frac{\operatorname{sech}[(\rho-\rho_0)^2+z^2+t^2-\alpha^2]}{\operatorname{sech}(3\alpha^2)} \tanh[(\rho-\rho_0)^2+z^2+t^2 - \alpha^2]\frac{\partial}{\partial u}[(\rho-\rho_0)^2+z^2+t^2]\Big|_G$$

$$= \tanh(3\alpha^2)\, 2(\rho-\rho_0)\frac{\partial \rho}{\partial u} + 2z\frac{\partial z}{\partial u} + 2t\frac{\partial t}{\partial u}\Big|_G$$

$$= \tanh(3\alpha^2) 2(-2\alpha)\frac{\partial t}{\partial u}\Big|_G \qquad \text{only } u = t \neq 0.$$

$$\left.\frac{\partial^2 f}{\partial u \partial v}\right|_G = [2(-2\alpha)]^2\left(\frac{\partial t}{\partial u}\right)\left(\frac{\partial t}{\partial v}\right) + \tanh(3\alpha^2)2\left[\left(\frac{\partial t}{\partial u}\right)\left(\frac{\partial t}{\partial v}\right) + \left(\frac{\partial z}{\partial u}\right)\left(\frac{\partial z}{\partial v}\right) + \left(\frac{\partial \rho}{\partial v}\right)\left(\frac{\partial \rho}{\partial u}\right)\right]$$

$f_{j\alpha}\big|_G = -4\alpha \tanh(3\alpha^2)$ for $\alpha = t$ the rest are zero

$f_{j\alpha\beta}\big|_G = (4\alpha)^2 + 2\tanh(3\alpha^2)$ for $\alpha = \beta = t$

$= 2\tanh(3\alpha^2)$ for $\alpha = \beta$ either z or ρ

So, only $f_{,t}$; $f_{,tt}$; $f_{,\rho\rho}$; $f_{,zz}$ cases.

$$g_{\mu\nu,i}\big|_G = \begin{pmatrix} -f_{,t} & 0 & \phi \\ 0 & \rho_0^2 f_{,t} & \\ & \phi & \phi \end{pmatrix} \text{for } i = t \qquad g_{\mu\nu,\phi} = 0$$

$$= \begin{pmatrix} 0 & -\frac{1}{2} & \phi \\ -\frac{1}{2} & 0 & \\ & \phi & \phi \end{pmatrix} \text{ for } i = \rho$$

$= (\phi)$ for $i = z$

$$g_{\mu\nu,ij}\big|_G = \begin{pmatrix} -f_{,ij} & \rho_0 f_{ij} f_{ji} & \phi \\ \rho_0 f_{ij} f_{ji} & \{2\rho_0(\rho_{jj} f_{ji} + \rho_{ji} f_{ij}) + \rho_0^2 f_{,ij}\} & \\ & \phi & \phi \end{pmatrix}\Bigg|_G$$

$$= \begin{pmatrix} -f_{,tt} & \rho_0 (f_{,t})^2 & \phi \\ \rho_0 (f_{,t})^2 & \rho_0^2 f_{,tt} & \\ & \phi & \phi \end{pmatrix}\Bigg|_G \quad \text{for } i = j = t$$

$$= \left.\begin{pmatrix} -f_{,ii} & 0 & \phi \\ 0 & \rho_0^2 f_{,ii} & \\ & \phi & \phi \end{pmatrix}\right|_G \quad \text{for } i = j \text{ either } \rho \text{ or } z$$

$$= \left.\begin{pmatrix} 0 & 0 & \phi \\ 0 & 2\rho_0 f_{,t} & \\ & \phi & \phi \end{pmatrix}\right|_G \quad \text{for either } i = t\ ,\ j = \rho \text{ or switched}$$

$\Gamma_{\gamma v\beta} = \frac{1}{2}\left(g_{\gamma v,\beta} + g_{\gamma\beta,v} - g_{v\beta,\gamma}\right)$ $\qquad g_{tt,t}\ ;\ g_{\phi\phi,t}\ ;\ g_{\phi t,\rho}$

$\Gamma_{\gamma v\beta}\big|_G$ has only $\Gamma_{ttt}|_G = \frac{1}{2} f_{,t} = 2\alpha\tanh(3\alpha^2)$

And

$\Gamma_{\phi\phi t}\big|_G = \Gamma_{\phi t\phi}\big|_G = -\Gamma_{t\phi\phi}\big|_G = \frac{1}{2}\ g_{\phi\phi,t}\big|_G = -2\alpha\tanh(3\alpha^2)\ \rho_0^2$

And

$\Gamma_{\phi t\rho}\big|_G = \Gamma_{\phi\rho t}\big|_G = -\Gamma_{\rho\phi t}\big|_G = \frac{1}{2}\ g_{\phi t\rho}\big|_G = -\frac{1}{4}$

$\partial_\alpha\Gamma_{\gamma v\beta} = \frac{1}{2}\left(g_{\gamma v,\beta\alpha} + g_{\gamma\beta,v\alpha} - g_{v\beta,\gamma\alpha}\right)$

$\partial_t\Gamma_{\gamma vt} = \frac{1}{2}\left(g_{\gamma v,tt} + g_{\gamma t,vt} - g_{vt,\gamma t}\right)$

$\rightarrow \partial_t\Gamma_{ttt} = \frac{1}{2} g_{tt,tt} = -f_{,tt}$ $\qquad \partial_t\Gamma_{tv\beta} = \frac{1}{2}\left(g_{tv,\beta t} + g_{t\beta,vt} - g_{v\beta,tt}\right)$

$\rightarrow \partial_t\Gamma_{\phi tt} = g_{\phi t,tt} = \rho_0\left(f_{,t}\right)^2$ $\qquad \rightarrow \partial_t\Gamma_{t\phi\phi} = -\frac{1}{2} g_{\phi\phi,tt}$

$\partial_t\Gamma_{t\phi t} = 0$

$\rightarrow \partial_t\Gamma_{\phi\phi t} = \frac{1}{2} g_{\phi\phi,tt} = \rho_0^2 f_{,tt}$

$\partial_k\Gamma_{\gamma v\beta} = \frac{1}{2}\left(g_{\gamma v,\beta k} + g_{\gamma\beta,vk} - g_{v\beta,\gamma k}\right)$

$\partial_k\Gamma_{\gamma v\beta} = \frac{1}{2}\left(g_{\gamma v,kk} + g_{\gamma k,vk} - g_{vk,\gamma k}\right) = \frac{1}{2} g_{\gamma v,kk}$

$\rightarrow \partial_k\Gamma_{ttk} = -f_{,ii}\left(\frac{1}{2}\right)$

$\rightarrow \partial_k\Gamma_{\phi\phi k} = \rho_0^2 f_{,ii}\left(\frac{1}{2}\right)$

$\partial_k\Gamma_{kv\beta} = \frac{1}{2}\left(g_{kv,\beta k} + g_{k\beta,vk} - g_{v\beta,kk}\right) = \frac{1}{2} g_{v\beta,kk}$

$\rightarrow \partial_k \Gamma_{ktt} = \left(-\frac{1}{2}\right)\left(-f_{,ii}\right)$
$\rightarrow \partial_k \Gamma_{k\phi\phi} = \left(-\frac{1}{2}\right)\left(\rho_0^2 f_{,ii}\right)$

$\partial_t \Gamma_{\gamma\nu\beta} = \frac{1}{2}\left(g_{\gamma\nu,\beta t} + g_{\gamma\beta,\nu t} - g_{\nu\beta,\gamma t}\right)$
$\rightarrow \partial_t \Gamma_{\phi\phi\rho} = \partial_t \Gamma_{\phi\rho\phi} = \partial_t \Gamma_{\rho\phi\phi} = \frac{1}{2} g_{\phi\phi,\rho t} = \rho_0 f_{,t}$

$\partial_\rho \Gamma_{\gamma\nu\beta} = \frac{1}{2}\left(g_{\gamma\nu,\beta\rho} + g_{\gamma\beta,\nu\rho} - g_{\nu\beta,\gamma\rho}\right)$
$\rightarrow \partial_\rho \Gamma_{\phi\phi t} = \partial_\rho \Gamma_{\phi t\phi} = \partial_\rho \Gamma_{t\phi\phi} = \frac{1}{2} g_{\phi\phi,t\rho} = \rho_0 f_{,t}$

The rest of the derivation of this section will be left as an exercise.

6.2 FNC Derivation via construction (a leading order analysis)
Now consider neighboring geodesics. Eqn. for geodesic deviation in the neighborhood of the cng. can be studied in FNC using MTW convention: with vectors e_α and with metric 2nd order in "spatial" displacement. Thus, in FNC coord system all $g_{\alpha\beta,\gamma}$ terms are zero. So have:

$$\frac{D^2\eta^\mu}{d\tau^2} + R^\mu_{\beta\gamma\delta}\,(e_0)^\beta \eta^\gamma (e_0)^\delta = 0$$

(Eqn. 6-31)

Consider

$$ds^2 = \left(-1 - F(x^0)_{\ell m}\, x^\ell x^m\right)(dx^0)^2 - G(x^0)_{\ell i m}\, x^\ell x^m dx^i dx^0 + \left(\delta_{ij} - H(x^0)_{i\ell j m}\, x^\ell x^m\right) dx^i dx^j$$

(Eqn. 6-32)

Geodesic eqn:

$k^\alpha_{;\beta} k^\beta = 0$

In FNC coord's this is $k^\alpha_{,\beta} k^\beta = 0$

$k^\beta \frac{\partial k^\alpha}{\partial x^\beta} = 0$ and also have $k^\beta k_\beta = 0$

So,

$K^\alpha = k^\alpha + \delta k^\alpha$

$$k^\beta \frac{\partial \delta k^\alpha}{\partial x^\beta} + \delta k^\beta \frac{\partial k^\alpha}{\partial x^\beta} = 0$$

$$k^\beta\, \delta\, \frac{\partial k^\alpha}{\partial x^\beta} + \cdots = 0$$

$$k^\phi \frac{\partial(\delta k^\alpha)}{\partial\psi} + \delta k^\phi \frac{\partial k^\alpha}{\partial\psi} = 0$$

$$\alpha = \phi \rightarrow \frac{\partial \ln(\delta k^{\phi})}{\partial \phi} \rightarrow \delta k^{\phi} = \frac{c}{k^{\phi}}$$

$$\alpha \neq \phi \rightarrow \frac{\partial(\delta k^{\alpha'})}{\partial \phi} = 0 \rightarrow \delta k^{\alpha'} = F(\rho, z, t) \text{ not } \phi$$

Let's now consider correspondence of appropriate orders to get constructive derivation:

Fermi Normal Coordinates:

$$ds^2 = (-1 - R_{0\ell 0m}|_G x^{\ell} x^m) dt^2 - \left(\tfrac{3}{4} R_{0\ell jm}\big|_G x^{\ell} x^m\right) dt dx^j$$
$$+ \left(\delta_{ij} - \tfrac{1}{3} R_{i\ell jm}\big|_G x^{\ell} x^m\right) dx^i dx^j + O\left(|x^j|^3\right) dx^{\alpha} dx^{\beta}$$

(Eqn. 6-33)

$R_{\alpha\beta\gamma\delta} = R_{[\alpha\beta][\gamma\delta]} = R_{[\gamma\delta][\alpha\beta]}$

What are $\Gamma_{\mu\beta\gamma}$ and $R^{\alpha}_{\beta\gamma\delta} = \Gamma^{\alpha}_{\beta\delta,\gamma} - \Gamma^{\alpha}_{\beta\gamma,\delta} + \Gamma^{\alpha}_{\mu\gamma}\Gamma^{k}_{\beta\delta} - \Gamma^{\alpha}_{\mu\delta}\Gamma^{k}_{\beta\gamma}$

$$\Gamma_{\mu\beta\gamma} = \tfrac{1}{2}\left(g_{\mu\beta,\gamma} + g_{\mu\gamma,\beta} - g_{\beta\gamma,\mu}\right)$$

$\ddot{g}_{00} = -1 - R_{0\ell 0m} x^{\ell} x^m + O(|x|^3)$ *Here* $R_{\alpha\beta\gamma\delta}\big|_G$ is the tensor evaluated.

$g_{0i} = -\tfrac{2}{3} R_{0\ell jm} x^{\ell} x^m +$
$O(|x|^3)$ along the geodesic, so usually it has a "t".

$g_{ij} = \delta_{ij} - \tfrac{1}{3} R_{i\ell jm} x^{\ell} x^m + O(|x|^3)$ dependence in the F.N.C system.
Here, due to the axial symmetry the cng that we are regarding as G must have $R_{\alpha\beta\gamma\delta}\big|_G = const.$

For the spatial case of cng as G the F.N.C's will have no time dependence to all orders since $R_{\alpha\beta\gamma\delta}\big|_G$ is terms independent.

$\Gamma_{000} = 0$

$\Gamma_{00i} = \tfrac{1}{2} g_{00,i} = -R_{0i0m}|_G x^m + O(|x|^2) = O(|x|)$

$\Gamma_{i00} = -\Gamma_{00i}$

$$\Gamma_{0ij} = \frac{1}{2}\left(g_{0i,j} + g_{0j,i}\right)$$
$$= -\frac{1}{3}\left(R_{0jim}x^m + R_{0\ell ij}x^\ell + R_{0ijm}x^m + R_{0\ell ji}x^\ell\right) + O(|x|^2)$$
$$= -\tfrac{2}{3}\left(R_{0(ji)m}x^m + R_{0\ell(ij)}x^\ell\right) + O(|x|^2) = O(|x|)$$

$$\Gamma_{i0j} = \frac{1}{2}\left(g_{0i,j} + g_{0j,i}\right)$$
$$= -\frac{1}{3}\left(R_{0jim}x^m + R_{0\ell ij}x^\ell - R_{0ijm}x^m - R_{0\ell ji}x^\ell\right) + O(|x|^2)$$
$$= -\tfrac{2}{3}\left(R_{0[ji]m}x^m + R_{0\ell[ij]}x^\ell\right) + O(|x|^2) = O(|x|)$$

$$\Gamma_{ijk} = \frac{1}{2}\left(g_{ij,k} + g_{ik,j} - g_{jk,i}\right)$$
$$= -\tfrac{1}{6}\left(R_{ikjm}x^m + R_{i\ell jk}x^\ell + R_{ijkm}x^m + R_{i\ell kj}x^\ell - R_{jikm}x^m - R_{j\ell ki}x^\ell\right) + O(|x|^2)$$
$$= \tfrac{1}{3}\left(R_{j\ell ki}x^\ell + R_{jikm}x^m\right) + O(|x|^2) = O(|x|)$$

$$\boxed{\begin{aligned} &\Gamma_{000} = 0 \\ &\Gamma_{00i} = -\Gamma_{00i} = -R_{0i0m}|_G x^m + O(|x|^2) = O(|x|) \\ &\Gamma_{0ij} = -\tfrac{2}{3}\left(R_{0(ji)m}x^m + R_{0\ell(ij)}x^\ell\right) + O(|x|^2) = O(|x|) \\ &\Gamma_{i0j} = -\tfrac{2}{3}\left(R_{0[ji]m}x^m + R_{0\ell[ij]}x^\ell\right) + O(|x|^2) = O(|x|) \\ &\Gamma_{ijk} = \tfrac{1}{3}\left(R_{j\ell ki}x^\ell + R_{jikm}x^m\right) + O(|x|^2) = O(|x|) \end{aligned}} \quad R's \to R|_G$$

(Eqn. 6-34)

$$\tilde{R}_{\mu\beta\gamma\delta} = g_{\mu\alpha}\,\tilde{R}^{\alpha}_{\beta\gamma\delta} \qquad \begin{aligned} g^{00} &= -1 + R_{0\ell 0m}x^\ell x^m + O(|x|^3) \\ g^{0i} &= O(|x|^2) \\ g^{ij} &= \delta^{ij} + \tfrac{1}{3}R_{i\ell jm}x^\ell x^m + O(|x|^3) \end{aligned}$$

$$\Gamma^0_{0i} = +R_{0i0m}|_G x^m + O(|x|^2)$$
$$\Gamma^i_{00} = \delta^{ij}\left(-R_{0j0m}\big|_G x^m\right) = -R_{0i0m}|_G x^m + O(|x|^2)$$

$$\Gamma^0_{ij} = \tfrac{2}{3}\left(R_{0(ji)m}\, x^m + R_{0\ell(ij)}\, x^\ell\right) + O(|x|^2)$$
$$\Gamma^i_{j0} = \Gamma^i_{0j} = -\tfrac{2}{3}\left(R_{0[ji]m}\, x^m + R_{0\ell[ij]}\, x^\ell\right) + O(|x|^2)$$
$$\Gamma^i{}_{jk} = \frac{1}{3}\left(R_{j\ell ki}\, x^\ell + R_{jikm}\, x^m\right) + O(|x|^2)$$

$$\begin{aligned}\tilde{R}^\alpha_{\beta\gamma\delta} &= \Gamma^\alpha_{\beta\delta,\gamma} - \Gamma^\alpha_{\beta\gamma,\delta} + \Gamma^\alpha_{\mu\gamma}\Gamma^\mu_{\beta\delta} - \Gamma^\alpha_{\mu\delta}\Gamma^\mu_{\beta\gamma} \\ &= \Gamma^\alpha_{\beta\delta,\gamma} - \Gamma^\alpha_{\beta\gamma,\delta} + O(|x|^2)\end{aligned}$$

$R^\alpha_{\beta\gamma 0} = \Gamma^\alpha_{\beta 0,\gamma}$ only $\tilde{R}^\alpha_{\beta i0}$ and $\tilde{R}^\alpha_{\beta ij}$ due to axial symmetry and to $O(|x|)$ otherwise

$$\tilde{R}^\alpha_{\beta i0}: \quad \begin{aligned}\tilde{R}^0_{ji0} &= +R_{0j0i}\big|_G + O(|x|) \rightarrow \tilde{R}_{0j0i} = +R_{0j0i}\big|_G \\ \tilde{R}^j_{0i0} &= -R_{0j0i}\big|_G + O(|x|) \\ \tilde{R}^j_{ki0} &= -\tfrac{2}{3}\left(R_{0[kj]i} + R_{0i[jk]} +\right)\big|_G + O(|x|) \\ &= -\tfrac{2}{3}\left(\tfrac{3}{2} R_{0ijk}\right)\Big|_G \\ &= -R_{0ijk}\big|_G \\ \tilde{R}_{0ijk} &= R_{0ijk}\big|_G + O(|x|)\end{aligned}$$

So,

$$\tilde{R}^\alpha_{\beta i0} \rightarrow \begin{cases}\tilde{R}_{0ijk} - R_{0ijk}\big|_G + O([x]) \\ \tilde{R}_{0j0i} - R_{0j0i}\big|_G + O([x])\end{cases} \textit{as expected}$$

The following was useful in the above:

$$\begin{aligned}\Gamma^0_{0i} &= g^{0\mu}\Gamma_{\mu 0i} \\ &= \underline{g^{00}\Gamma_{00i}} + \Gamma_{\mu 0i}\left(O(|x|^2)\right)\end{aligned}$$
$$\Gamma^i_{00} = g^{i\alpha}\Gamma_{\alpha 00} = g^{i0}\Gamma_{000} + \underline{g^{ij}\Gamma_{j00}}$$
$$\begin{aligned}\Gamma^0_{ij} &= g^{00}\Gamma_{0ij} + g^{0k}\Gamma_{kij} \\ &= \underline{g^{00}\Gamma_{0ij}} + O(|x|^2)\end{aligned}$$
$$\begin{aligned}\Gamma^i_{0j} &= g^{i0}\Gamma_{00j} + g^{ik}\Gamma_{k0j} \\ &= \underline{g^{ik}\Gamma_{k0j}} + O(|x|^2)\end{aligned}$$

Now,

$$\swarrow \tilde{R}^0_{0ij} = R_{0j0i} - R_{0i0j} + O(|x|) = O(|x|)$$
$$\tilde{R}^\alpha_{\beta ij} = \Gamma^\alpha_{\beta j,i} - \Gamma^\alpha_{\beta i,j} + O(|x|)^2$$
$$\nwarrow \tilde{R}^0_{kij} = \Gamma^0_{kj,i} - \Gamma^0_{ki,j} + O(|x|^2)$$

$$\tilde{R}^{\alpha}_{\beta ij} = \frac{3}{3}\left[R_{0(jk)i} + R_{0i(kj)} - R_{0(ik)j} - R_{0j(ki)}\right] + O(|x|)$$
$$\tilde{R}_{0kij} = R_{0kij}\big|_G + O(|x|)$$
$$\hat{R}^{n}_{mij} = \Gamma^{n}_{mj,i} - \Gamma^{n}_{mi,j} + O(|x|^2) = \frac{1}{3}\left(R_{mijn} + R_{mnji} - R_{mjin} - R_{mnij}\right)$$
$$R_{mijn} = R_{mnji} - R_{mjni} + O(|x|) = R_{nmij}\big|_G + O(|x|)$$
$$\tilde{R}_{nmij} = R_{nmij}\big|_G + O(|x|)$$

Thus,

$$\tilde{R}^{\alpha}_{\beta ij} \rightarrow \begin{cases} \tilde{R}_{00ij} = O(|x|) = 0 \text{ by antisymmetry} \\ \tilde{R}_{0kij} = R_{0kij}\big|_G + O(|x|) \\ \tilde{R}_{nmij} = R_{nmij}\big|_G + O(|x|) \end{cases}$$
$$\tilde{R}^{\alpha}_{\beta i0} \rightarrow \begin{cases} \tilde{R}_{0ijk} = R_{0ijk}\big|_G + O(|x|) \\ \tilde{R}_{0j0i} = R_{0j0i}\big|_G + O(|x|) \end{cases}$$

And $\tilde{R}_{\alpha\beta 00} = O(|x|) = 0$ by antisymmetry

(Eqn. 6-35)

6.3 Full derivation of FNC and Geodesic Deviation from geodesic Equation (Levi-Civita)

Consider the geodesic equation ($\nabla_{\vec{k}}\,\vec{k} = 0$):

$$\frac{dk^{\alpha}}{d\tau} + \Gamma^{\alpha}_{\beta\mu}\, k^{\beta} k^{\mu} = 0$$

(Eqn. 6-36)

Consider a one – parameter family of geodesics $e(s,\tau)$ with tangent vectors $K^{\alpha}(s)$ and associated connections $\Gamma(e)$, where τ is the affine parameter and s is a continuous index over the family of geodesics.

$$k^{\alpha}(s) = k^{\alpha}_{s} + \underbrace{\frac{dk^{\alpha}_{s}}{ds}\bigg|_{s=0}}_{\delta k^{\alpha}}(s) + O(s^2)$$

(Eqn. 6-37)

So, $\frac{dk^{\alpha}_{s}}{d\tau} + \Gamma^{\alpha}_{\beta\mu}\{e(\tau,s)\}k^{\beta}_{s}k^{\mu}_{s} = 0$ is a geodesic eqn. for any of the geodesics.

$$\frac{d}{ds}\left[\frac{dk^{\alpha}_{s}}{d\tau} + \Gamma^{\alpha}_{\beta\mu}\, k^{\beta}_{s} k^{\mu}_{s}\right] = 0 \qquad \leftarrow \text{evaluate at} \quad s = 0$$

(Eqn. 6-38)

$$\frac{d(\delta k^\alpha)}{d\tau} + \frac{d\Gamma^\alpha_{\beta\mu}}{ds} k_0^\beta k_0^\mu + 2\Gamma^\alpha_{\beta\mu}(\delta k^\beta)\, k_0^\mu = 0$$

(Eqn. 6-39)

$$\frac{d(\delta k^\alpha)}{d\tau} + \frac{d\Gamma^\alpha_{00}}{ds} + 2\Gamma^\alpha_{\beta 0}(\delta k^\beta) = 0$$

(Eqn. 6-40)

Only have $\alpha = i$ non zero $\quad \frac{d\Gamma^i_{00}}{ds} = R_{0i0m}|_G \overset{\downarrow \eta^m}{\frac{dx^m}{ds}} + O(|x|) =$
$-R_{0i0m}|_G\, \eta^m + O(|x|)$
To lowest order:

$$\frac{d(\delta k^\alpha)}{d\tau} + \frac{d\Gamma^\alpha_{00}}{ds} = 0$$

(Eqn. 6-41)

$$\alpha = i \quad : \quad \frac{d(\delta k^i)}{d\tau} + \left.\frac{d\Gamma^i_{00}}{ds}\right|_{s=0} = 0$$

(Eqn. 6-42)

$$\frac{d(\delta k^i)}{d\tau} + \left(R^i_{00m}\Big|_G \left.\frac{dx^m}{ds}\right|_{s=0}\right) = 0$$
$$\frac{d(\delta k^i)}{d\tau} + R^i_{\beta\mu m}\Big|_G\, \eta^m|_{s=0}\, k_0^\beta k_0^\mu = 0$$
$$\frac{d(\delta k^i)}{d\tau} + R^i_{\beta\mu\alpha}\Big|_G \eta^m k_0^\beta k_0^\mu = 0$$

(Eqn. 6-43)

Recall $\qquad \delta k^\alpha = \left.\frac{dk_s^\alpha}{ds}\right|_{s=0} = \frac{dk_s^\alpha}{dx^\mu}\frac{dx^\mu}{ds} = \frac{dk_s^\alpha}{dx^\mu}\,\eta^\mu$

While $\qquad \frac{d\eta^\alpha}{d\tau} = \frac{d\eta^\alpha}{dx^m}\frac{dx^\mu}{d\tau} = \frac{d\eta^\alpha}{dx^\mu}\, k^\mu$

If we restrict our 1-parameter family such that we have an η^μ that is lie dragged by k^μ we then have:
$[\vec{\eta}, \vec{k}] = 0$ gives $\eta^\alpha k^\beta_{,\alpha} - k^\alpha \eta^\beta_{,\alpha} = 0$, so

$$\eta^\alpha \frac{dk^\beta}{dx^\alpha} = k^\alpha \frac{d\eta^\alpha}{dx^\alpha}$$

(Eqn. 6-44)

This would then give $\delta k^\alpha = \frac{d\eta\eta^\alpha}{d\tau}$ so that we would have:

$$\frac{d^2\eta^i}{d\tau^2} + R^i_{\beta\mu\alpha}\Big|_G \eta^\alpha\, k_0^\beta k_0^\mu = 0$$

(Eqn. 6-45)

Considering the $\alpha = 0$ case: $\frac{d(\delta k^0)}{d\tau} = 0$ to lowest order. So, in tensional form we have to lowest order:

$$\frac{d^2\eta^\alpha}{d\tau^2} + R^\alpha_{\beta\mu\nu}\big|_G \eta^\nu k_0^\beta k_0^\mu = 0$$

(Eqn. 6-46)

the geodesic deviation equation with the restriction on our 1-parameter family such that η^μ is lie dragged. Generalizing the 1-parameter family using the tensional nature, to an arbitrary coordinate system, we have the geodesic deviation eqn., undoubtedly in a manner very much like Levi-Civita's original deviation as we are working in Fermi Normal Coordinates as he did in his paper.

If we do not restrict ourselves to 1-parameter families that follow from a lie dragged displacement vector we have the desired general equations:

$$\frac{d(\delta k^0)}{d\tau} = 0$$

(Eqn. 6-47)

$$\frac{d(\delta k^i)}{d\tau} = R^i_{00m}\big|_G \eta^m$$

(Eqn. 6-48)

6.4 Derivation of Deviation From Wald [20]

Consider a 2-d submanifold spanned by $\gamma_s(t)$, denoted Σ. γ_s, a one parameter family of geodesics, tangent T^a, deviation X^a map $(t, s) \rightarrow \gamma_s(t)$ is smooth, $1-1$, with smooth inverse.

"Gauge Freedom" in X^a under change of affine parameterization, by rescaling t by an s-dependent factor X^a can always be chosen orthogonal to T^a. Thus, the Lie dragging relation can always be solved (coordinate vector fields) and we get the geodesic deviation equation.

So, we have geodesic eqn. as the simplest parameterization to study:

$$\frac{d^2\eta^i}{d\tau} + R^i_{\beta\mu j}\big|_G \eta^j k_0^\beta k_0^\mu = 0$$

(Eqn. 6-49)

($k_0^\mu \eta_\mu = 0$ with param. that yields geod. dev. Eqn.)

$$\frac{d^2\eta^i}{d\tau^2} + R^i_{00j}\big|_G \eta^j = 0$$

(Eqn. 6 50)

$$\ddot{\eta}^i + \lambda^i_j \, \eta^j = 0$$

(Eqn. 6-51)

where $\lambda^i_j = g^{ik} \, \lambda_{kj}$ etc.

The "Gauge Freedom" used to specify a coordinatization of Σ results in a specific vector field for the tangent vectors which is one case among many, certainly of measure zero. A different gauge of rescaling will yield some other vector field.

General equations:

$$\frac{d(\delta k^0)}{d\tau} = 0$$

(Eqn. 6-52)

$$\frac{d(\delta k^i)}{d\tau} = R^i_{00\mu}\big|_G \eta^\mu(\tau) \xrightarrow{|\eta^m| \to 0} 0$$

(Eqn. 6-53)

Now, $\eta^m|_{s=0} = \left.\frac{dx^m}{ds}\right|_{s=0}$ = deviation vector from G

So, η^m is a function of τ : $\eta^m = \eta^m(\tau)$ and can be found by the geodesic deviation eqn.:

$$\frac{d^2\eta^i}{d\tau^2} + R^i_{00j}\big|_G \eta^j = 0$$

Thus, $\frac{d(\delta k^i)}{d\tau} = \frac{d^2\eta^i}{d\tau^2}$

$\delta k^i = \frac{d\eta^i}{d\tau} + const$ so we have $[\vec{\eta}, \vec{k}] = \vec{a}$ to lowest order.

(Eqn. 6-54)

<u>So, to lowest order</u>:

$$\frac{d^2\eta^i}{d\tau^2} + R^i_{00j}\big|_G \eta^j = 0 \text{ with I.C.'s } \eta^j(\tau = 0) \text{ and } \frac{d\eta^j(\tau = 0)}{d\tau}$$

(Eqn. 6-55)

Solves for η^i and,

$\delta k^i = \frac{d\eta^i(\tau)}{d\tau} + a^i$ then yields $\delta k^i(\tau)$ for a given $\delta k^i(0)$, where a^i is determined by $\delta k^i(0)$ or vic

(Eqn. 6-56)

Difficulties using a 1-parameter family of the geodesics when we're partly interested in that instance when the analysis fails, i.e., the formation of caustics.

Suppose we start with

$$\frac{d^2\eta^\mu}{d\tau^2} + R^\mu_{\alpha\beta\gamma}\eta^r k^\alpha k^\beta = 0 \qquad \text{Geodesics deviation}$$

(Eqn. 6-57)

$$\frac{dk^\alpha}{d\tau} + \Gamma^\alpha_{\beta\mu}k^\beta k^\mu = 0 \qquad \text{Geodesic equation}$$

(Eqn. 6-58)

Fermi Normal Coordinates:

$$ds^2 = \left(-1 - R_{0\ell 0m}\, x^\ell x^m\right)dt^2 - \left(\tfrac{4}{3}\, R_{0\ell jm}\, x^\ell x^m\right)dtdx^j$$
$$+ \left(\delta_{ij} - \tfrac{1}{3}R_{i\ell jm}\, x^\ell x^m\right)dx^i dx^j + 0\left(\left|x^j\right|^3\right)dx^\alpha dx^\beta$$

(Eqn. 6-59)

Where $R_{\alpha\beta\gamma\delta}$ are the components of the Riemann tensor along the world line $x^j = 0$. Assuming I've evaluated $R_{\alpha\beta\gamma\delta}\big|_G$.

$g_{00} = -1 - R_{0\ell 0m}\, x^\ell x^m$ $\qquad$ $R_{\alpha\beta\gamma\delta} = R_{[\alpha\beta][\gamma\delta]} = R_{[\gamma\delta][\alpha\beta]}$

$g_{0i} = -\frac{2}{3}R_{0\ell jm}\, x^\ell x^m$ $\qquad$ $R^\alpha_{\beta\gamma\delta} = \Gamma^\alpha_{\beta\delta,\gamma} - \Gamma^\alpha_{\beta\gamma,\delta} + \Gamma^\alpha_{\mu\gamma}\Gamma^\mu_{\beta\delta} -$
$\Gamma^\alpha_{\beta\delta}\Gamma^\mu_{\beta\gamma}$

$g_{ij} = \delta_{ij} - \frac{1}{3}R_{i\ell jm}\, x^\ell x^m$

$\Gamma_{\mu\beta\gamma} = \frac{1}{2}\left(g_{\mu\beta,\gamma} + g_{\mu\gamma,\beta} - g_{\beta\gamma,\mu}\right)$ $\qquad$ $\Gamma_{\mu\beta\gamma} = \Gamma_{\mu\gamma\beta}$

$\Gamma_{000} = 0$

$\Gamma_{00i} = \frac{1}{2}\left(g_{00,i} + g_{0i,0} - g_{0i,0}\right) = \frac{1}{2}g_{00,i} = -\frac{1}{2}\left(R_{0i0m}x^m + R_{0\ell 0i}x^\ell\right) = 0$

$\Gamma_{i00} = \frac{1}{2}\left(g_{i0,0} + g_{i0,0} - g_{00,i}\right) = -\Gamma_{00i}$

$\Gamma_{0ij} = \frac{1}{2}\left(g_{i0,j} + g_{0j,i} - g_{ij,0}\right) = \frac{1}{2}\left(g_{0i,j} + g_{0j,i}\right) = -\frac{1}{3}\big[R_{0jim}\, x^m +$
$R_{0\ell ij}\, x^\ell + R_{0ijm}x^m + R_{0\ell ji}x^\ell\big]$

$\Gamma_{i0j} = \frac{1}{2}\left(g_{i0,j} + g_{ij,0} - g_{0j,i}\right) = -\frac{1}{3}\big(R_{0jim}\, x^m + R_{0\ell ij}\, x^\ell + R_{0jim}x^m +$
$R_{0\ell ji}x^\ell\big) = 0$

$\Gamma_{ijk} = \frac{1}{2}\left(g_{ij,k} + g_{ik,j} - g_{jk,i}\right) = \frac{1}{3}\left(R_{j\ell ki}\, x^\ell + R_{jikm}\, x^m\right)$

In Misner paper $\Gamma_v{}^\sigma{}_\alpha = \Gamma^\sigma_{v\alpha}$

$\Gamma_{0ij} = -\frac{1}{3}\left[R_{0jim}\, x^m + R_{0\ell ij}\, x^\ell + R_{0ijm}x^m + R_{0\ell ji}x^\ell\right]$

$\Gamma_{ijk} = \frac{1}{3}\left[R_{j\ell ki}\, x^\ell + R_{jikm}x^m\right]$

$R^\alpha_{\beta\gamma\delta} = \Gamma^\alpha_{\beta\delta,\gamma} - \Gamma^\alpha_{\beta\gamma,\delta} + \Gamma^\alpha_{\mu\gamma}\Gamma^\mu_{\beta\delta} - \Gamma^\alpha_{\beta\delta}\Gamma^\mu_{\beta\gamma}$

Since the

$\Gamma's$ are $O(|x^j|)$ and $(g^{\alpha\sigma})_{,\mu}$ is $O(|x^j|)$ while $g^{\alpha\sigma}(\Gamma_{\sigma\beta\delta\gamma})$ is $O(const)$ we need only consider such.

$R^{\alpha}_{\beta\gamma\delta} \cong \eta^{\alpha\sigma}\Gamma_{\sigma\beta\delta\gamma} - \eta^{\alpha\sigma}\Gamma_{\sigma\beta\gamma,\delta}$

$R_{\mu\beta\gamma\delta} \simeq g_{\mu\alpha}\eta^{\alpha\sigma}\Gamma_{\sigma\beta\delta,\gamma} - g_{\mu\alpha}\eta^{\alpha\sigma}\Gamma_{\sigma\beta\gamma,\delta} \simeq \Gamma_{\mu\beta\delta,\gamma} - \Gamma_{\mu\beta\gamma,\delta}$

$\tilde{R}_{0ijk} \simeq \Gamma_{0ik,j} - \Gamma_{0ij,k} \simeq \frac{1}{3}\left[R_{0kij} + R_{0jik} + R_{0ikj} + R_{0jik}\right] - R_{0jik} - R_{0kij} - R_{0ijk} + R_{0kij}$

$\simeq -\frac{1}{3}\left[2R_{0ikj} + R_{0jik} - R_{0kij}\right]$

$\simeq \Gamma_{0ik,j} - \Gamma_{0ij,k} \simeq \frac{1}{2}\left(g_{0i,k} + g_{0k,i}\right)_{,j} - \frac{1}{2}\left(g_{0i,j} + g_{0j,i}\right)_{,k}$

$\simeq \frac{1}{2}\left(g_{0i,kj} + g_{0k,ij} - g_{0i,jk} - g_{0j,ik}\right)$

$\simeq \frac{2}{3}R_{0ijk} + \frac{1}{3}\left(R_{0kji} - R_{0jki}\right)$

$R_{\alpha[\beta\gamma\delta]} = 0 \rightarrow R_{\alpha\beta\gamma\delta} - R_{\alpha\gamma\beta\delta} + R_{\alpha\delta\beta\gamma} = 0$

$R_{0kji} = R_{0jki} - R_{0ikj}$

$\tilde{R}_{0ijk} = R_{0ijk}$

$\tilde{R}_{ijk\ell} \cong \Gamma_{ij\ell,k} - \Gamma_{ijk,\ell} \cong \frac{1}{3}\left[R_{jk\ell i} + R_{ji\ell k} - R_{j\ell ki} - R_{jik\ell}\right]$

$\simeq \frac{1}{3}\left[R_{jk\ell i} - R_{j\ell ki}\right] + \frac{2}{3}R_{ji\ell k}$

Have $R_{\alpha[\beta\gamma\delta]} = 0$ and $R_{[\alpha\beta\gamma\delta]} = 0$

$\tilde{R}_{ijk\ell} \simeq \frac{1}{3}\left[R_{ikj\ell} + R_{i\ell kj}\right] + \frac{2}{3}R_{ji\ell k}$

$R_{ikj\ell} - R_{i\ell kj} + R_{ij\ell k} = 0 \quad R_{i\ell kj} - R_{ij\ell k} + R_{ikj\ell} = 0$

$\tilde{R}_{ijk\ell} \simeq \frac{1}{3}\left[R_{ijk\ell}\right] + \frac{2}{3}R_{ijk\ell} = R_{ijk\ell}$

Perhaps a FNC system can be generalized to any geodesic or non-geodesic (it already has been to some extent) by merely writing

$ds^2 = \left(-1 - A_{0\ell 0m}x^{\ell}x^{m}\right)dt^2 - B_{0\ell jm}\,x^{\ell}x^{m}\,dt dx^j + C_{i\ell jm}\,x^{\ell}x^{m}\,dx^i dx^j + 0\left(|x^j|^3\right)dx^{\alpha}dx^{\beta}$

(Eqn. 6-60)

and then derive A, B, C, by calc. $R_{\alpha\beta\gamma\delta}$ along chosen geodesic and equating. See Levi-Civita, Math. Ann 97, 291 (1926) [29] or J. L. Synge; "Relativity, The General Theory" [30] to verify.

Riemann normal coordinates gives: (satisfy $\Gamma^{\sigma}_{\mu\nu} = 0$ at a single point)

$$ds^2 = \left\{\eta_{\mu\nu} + \frac{1}{3} R_{\alpha\mu\beta\nu} x^\alpha x^\beta + O(|x|)^3\right\} dx^\mu dx^\nu$$

(Eqn. 6-61)

See L. P. Eisenhart Riemannian Geometry [31].

Levi-Civita, in the paper above, developed for the first time the equation of geodesic deviation and used Fermi coord. as a technique for simplifying this eqn. to display its properties more clearly. We will now retrace a summary of that effort. Consider to start:

$$\Gamma_{0ij} = -\tfrac{1}{3}\left[R_{0jim} x^m + R_{0\ell ij} x^\ell + R_{0ijm} x^m + R_{0\ell ij}\, x^\ell\right]$$
$$\Gamma_{ijk} = \tfrac{1}{3}\left[R_{j\ell ki} x^\ell + R_{jikm} x^m\right]$$

$$g_{00} = -1 - R_{0\ell 0m}\, x^\ell x^m$$
$$g_{0i} = -\tfrac{2}{3} R_{0\ell jm}\, x^\ell x^m$$
$$g_{ij} = \delta_{ij} - \tfrac{1}{3} R_{i\ell jm}\, x^\ell x^m$$

g^{00} to $O\left(\left|x^\ell\right|^2\right)$ is $g^{00} \simeq -1 + R_{0\ell 0m}\, x^\ell x^m$

$$g^{0i} \simeq 0\left(\left|x^\ell\right|^2\right)$$
$$g^{ij} \simeq \delta^{ij} + \tfrac{1}{3} R_{i\ell jm}\, x^\ell x^m$$

$$\frac{dk^\alpha}{d\tau} + \Gamma^\alpha_{\beta\mu} k^\beta k^\mu = 0$$
$$k^\alpha = \frac{dx^\alpha}{d\tau}$$

$$k^\alpha = \dot{k}^\alpha + \delta k^\alpha$$
$$\frac{dk^\alpha}{d\tau} + \Gamma^\alpha_{\beta\mu}\left(\dot{k}^\beta + \delta k^\beta\right)\left(\dot{k}^\mu + \delta k^\mu\right) = 0$$
$$\frac{dk^\alpha}{d\tau} + \Gamma^\alpha_{\beta\mu}\left(\dot{k}^\beta \delta k^\beta + \delta k^\mu \dot{k}^\mu\right) = 0$$
$$\frac{dk^\alpha}{d\tau} + 2\Gamma^\alpha_{\beta\mu} \dot{k}^\beta \delta k^\mu = 0$$

In FNC

$$\frac{dk^\alpha}{d\tau} + \Gamma^\alpha_{\beta\mu} \dot{k}^\beta \dot{k}^\mu = 0$$

$$\frac{d\dot{k}^\alpha}{dx^\mu}\frac{dx^\mu}{d\tau} + \Gamma^\alpha_{\beta\mu}\dot{k}^\beta\dot{k}^\mu = 0$$
$$\frac{d\dot{k}^\alpha}{dx^\mu}\dot{k}^\mu + \Gamma^\alpha_{\beta\mu}\dot{k}^\beta\dot{k}^\mu = 0$$
$$\frac{d\dot{k}^\alpha}{dx^0}\dot{k}^\mu + \Gamma^\alpha_{00}\dot{k}^0\dot{k}^0 = 0$$

So, $\frac{dk^\alpha}{d\tau} + 2\Gamma^\alpha_{0\mu}\dot{k}^0\delta k^\mu = 0$. Since $\frac{d(\dot{k}^0)}{d\tau} = 0$ we get
$$\frac{d(\delta k^\alpha)}{d\tau} + 2\Gamma^\alpha_{0\mu}\dot{k}^0\delta k^\mu = 0$$

Thus,
$$\frac{d(\delta k^\alpha)}{dx^0} + 2\Gamma^\alpha_{0\mu}(\delta k^\mu) = 0$$
$$\Gamma^\alpha_{0\mu} = g^{\alpha\beta}\Gamma_{\beta 0\mu} \to zero$$
$\frac{d(\delta k^\alpha)}{dx^0} = 0$ to 1st <u>order</u>.

So, we must consider the other factors

0th order have $\frac{dk^\alpha}{d\tau} = 0$

(Eqn. 6-62)

1st order have (two infinitesimals δk^α and δk^μ)
$$(\dot{k}^0)\frac{d(\delta k^\alpha)}{dx^0} + \Gamma^\alpha_{\beta\mu}\dot{k}^\beta\dot{k}^\mu 2\Gamma^\alpha_{\beta\mu}\dot{k}^\beta\delta k^\mu = 0$$
$$\frac{d(\delta k^\alpha)}{dx^0} + \Gamma^\alpha_{00}(\dot{k}^0) = 0$$

(Eqn. 6-63)

2nd order
$$(\delta k^i)\frac{d(\delta k^\alpha)}{dx^i} + \Gamma^\alpha_{\beta\mu}\delta k^\beta\delta k^\mu = 0$$
$$\frac{d(\delta k^\alpha)}{dx^i} + \Gamma^\alpha_{ji}\delta k^j = 0$$

(Eqn. 6-64)

Consider a one-parameter family of geodesics C(s) (where s is not affine param. but an "index") with tangent vectors k^α(s) and associated conventions $\Gamma(C(s))$.

$$k^\alpha_s = k^\alpha_0 + \left.\frac{dk^\alpha}{ds}\right|_{s=0}(s) + O(s^2)$$

(Eqn. 6-65)

For geodesic eqn we have

$$\frac{d}{d\tau}k^{\alpha}(s) + \Gamma^{\alpha}_{\beta\mu}|(s)k^{\beta}(s)k^{\mu}(s) = 0$$
$$\frac{d}{ds}\left[\frac{d}{d\tau}k^{\alpha}_{s} + \Gamma^{\alpha}_{\beta\mu}\big(C(s)\big)k^{\beta}_{s}k^{\mu}_{s}\right]\Bigg|_{s=0} = 0$$
↑ or any s for that matter

We have:

$\frac{dk^{\alpha}}{d\tau} + \Gamma^{\alpha}_{\beta\mu}k^{\beta}k^{\mu} = 0$; $k^{\alpha} = \dot{k}^{\alpha} + \delta k^{\alpha}$; $\dot{k}^{\alpha}$ has only $\alpha = 0$ *comp*

↓

$\frac{d}{d\tau}\left(\dot{k}^{\alpha} + \delta k^{\alpha}\right) + \Gamma^{\alpha}_{\beta\mu}\left(\dot{k}^{\alpha} + \delta k^{\beta}\right)\left(\dot{k}^{\mu} + \delta k^{\mu}\right) = 0$

(Only have Γ_{0ij} and Γ_{ijk} nonzero $\Rightarrow \Gamma^{\alpha}_{\beta\mu}\dot{k}^{\beta} = 0$ *and* $\Gamma^{\alpha}_{\beta\mu}\dot{k}^{\mu} = 0$).

$$\frac{d\dot{k}^{\alpha}}{d\tau} + \frac{d(\delta k^{\alpha})}{d\tau} + \Gamma^{\alpha}_{\beta\mu}\left(\delta k^{\beta}\right)\left(\delta k^{\mu}\right) = 0$$

$$\boxed{\frac{d(\delta k^{\alpha})}{dx^{i}} + \Gamma^{\alpha}_{ji}\left(\delta k^{j}\right) = 0}$$

(Eqn. 6-66)

$$\frac{d(\delta k^{\alpha})}{\delta k^{j}} = \Gamma^{\alpha}_{ji}dx^{i}$$

$\alpha = 0$:

$$\frac{d(\delta k^{0})}{\delta k^{j}} = -\Gamma^{0}_{ji}\,dx^{i} = g^{00}\,\Gamma_{0ji}\,dx^{i}$$
$$= +\left[R_{0ijm}x^{m} + R_{0\ell ji}x^{\ell} + R_{0jim}x^{m} + R_{0\ell ji}x^{\ell}\right]dx^{i}$$

$$d(\delta k^{0}) = \left(-\delta k^{j}\right)\left[2R_{0jim}x^{m} + 3R_{0mji}x^{m}\right]dx^{i}$$
$$\delta k^{0} = -\delta k^{j}\Big[2R_{0jia}x^{a}x^{i} + 3R_{0aji}x^{a}x^{i} + 2R_{0jii}\tfrac{1}{2}\left(x^{i}\right)\left(x^{i}\right) +$$
$$3R_{0iji}\frac{x^{i}}{2}x^{i} + const$$
$$\delta k^{0} = -\delta k^{j}\left[2R_{0jia}x^{a}x^{i} + 3R_{0aji}x^{a}x^{i} + \tfrac{3}{2}3R_{0iji}\,x^{i}x^{i}\right] + const$$

$\alpha = k$:

$$\frac{d(\delta k^{k})}{\delta k^{j}} = \Gamma^{k}_{ji}dx^{i}$$

Recap

$\frac{dk^{\alpha}}{d\tau} + \Gamma^{\alpha}_{\beta\mu}k^{\beta}k^{\mu} = 0$ $\qquad\qquad$ $k^{\alpha} = \dot{k}^{\alpha} + \delta k^{\alpha}$

= tangent to curve $\tau(x)$

$\Gamma^{\alpha}_{0\beta} = 0$ only have Γ_{0ij} and Γ_{ijk} nonzero on fundamental geodesic.

$$\frac{d}{d\tau}\left(\dot{k}^{\alpha} + \delta k^{\alpha}\right) + \Gamma^{\alpha}_{\beta\mu}\left(\tau(x)\right)\left(\dot{k}^{\alpha} + \delta k^{\beta}\right)\left(\dot{k}^{\alpha} + \delta k^{\mu}\right) = 0$$

$$\frac{d\dot{k}^{\alpha}}{d\tau} + \frac{d(\delta k^{\alpha})}{d\tau} + \Gamma^{\alpha}_{\beta\mu}\dot{k}^{\beta}\dot{k}^{\mu} + 2\Gamma^{\alpha}_{\beta\mu}\dot{k}^{\beta}\delta k^{\mu} + \Gamma^{\alpha}_{\beta\mu}\left(\delta k^{\beta}\delta k^{\mu}\right) = 0$$

$\dot{k}^{\alpha} = const$, thus $\frac{d}{d\tau}\dot{k}^{\alpha} = 0$

$\Gamma^{\alpha}_{\beta\mu}\dot{k}^{\beta} = 0$ since $\Gamma^{\alpha}_{0\mu} = 0$

So, to first order in δk^{α} :

$$\frac{d(\delta k^{\alpha})}{d\tau} = 0$$

$$\boxed{\delta k^{\alpha} = const}$$

$$\frac{d^2\eta^{\mu}}{d\tau^2} + R^{\mu}_{\alpha\beta\gamma}\Big|_G \eta^{\gamma}\dot{k}^{\alpha}\dot{k}^{\beta} = 0$$

(Eqn. 6-67)

$\eta^{\mu}| = \eta^{\mu}$ projected orthogonal to $\dot{k}^{\alpha}$ geodesic should exhibit the same growth with affine parameter as that for the displacement of a geodesic with $\delta k^{\alpha} = const.$ along that affine parameter-thus linear growth.

Geodesic equation gives lowest order behavior: linear.
Geodesic deviation equation gives up to 2nd order ODE behavior: so can see oscillatory or exponential.

What is $R^{\mu}_{\alpha\beta\gamma}\Big|_G \dot{k}^{\alpha}\dot{k}^{\beta} = R^{\mu}_{00\gamma}\Big|_G =?$

$R^{\mu}_{00\gamma}\Big|_G$ has $R^{0}_{\ell 0m}$ term as expected.

To lowest order have $R_{0\ell 0m} = R_{0\ell 0m}|_G$; $R_{0\ell jm} = R_{0\ell jm}\Big|_G$

$R_{i\ell jm} = R_{i\ell jm}\Big|_G$ rest have displacement factor (higher order).

So, $R^{\mu}_{00\gamma} = g^{\mu v}R_{v00\gamma} \simeq \eta^{\mu v}\, R_{0v0\gamma}$

$$\frac{d^2\eta^\mu}{d\tau^2} + R^\mu_{\alpha\beta\gamma}\Big|_G \eta^\gamma\,(\dot{k}^\alpha + \delta k^\alpha)(\dot{k}^\beta + \delta k^\beta) = 0$$

To lowest order

$$\frac{d^2\eta^\mu}{d\tau^2} + R^\mu_{00\gamma}\Big|_G \eta^\gamma \dot{k}^0 \dot{k}^0 = 0$$

(Eqn. 6-68)

$$\frac{d^2\eta^\mu}{d\tau^2} + \lambda^\mu_\nu\,\eta^\nu = 0 \qquad \lambda^\mu_\nu = R^\mu_{00\gamma}\Big|_G \dot{k}^0 \dot{k}^0 = \underline{\text{constant}}$$

(Eqn. 6-69)

Can have exponential and oscillatory, η^μ terms. Use of FNC implies no rotation of geodesic so λ^μ_ν has only expansion, contraction, and shear.

6.5 Exercises

(Ex. 6.1) Derive Eqn. 6-8.
(Ex. 6.2) Derive Eqn. 6-9.
(Ex. 6.3) Derive Eqn. 6-10.
(Ex. 6.4) Derive Eqn. 6-17.
(Ex. 6.5) Derive Eqn. 6-20.
(Ex. 6.6) Derive Eqn. 6-21.
(Ex. 6.7) Derive Eqn.'s 6-22, 6.23, and 6.24..
(Ex. 6.8) Derive Eqn. 6-34.
(Ex. 6.9) Derive Eqn. 6-35.
(Ex. 6.10) Derive Eqn. 6-37.
(Ex. 6.11) Derive Eqn. 6-41.
(Ex. 6.12) Derive Eqn. 6-46.
(Ex. 6.13) Derive Eqn. 6-47 and 6.48.
(Ex. 6.14) Derive Eqn. 6-62, 6.63, and 6.64.
(Ex. 6.15) Derive Eqn. 6-66.
(Ex. 6.16) Derive Eqn. 6-67.
(Ex. 6.17) Derive Eqn. 6-68 and 6-69.
(Ex. 6.18) Complete the derivation at the end of Sec. 6.1 using any of the methods shown..

Chapter 7. Closed Time-like Curve (CTC) Analysis

In Ch. 7 and 8 we explore pathologies of geometrodynamics, starting with closed time-like curves (CTCs) in Ch. 7 and then Black Holes in Ch. 8. In Ch. 7 we see that in order for CTCs and other pathologies to NOT occur, the Weak Energy Condition (WEC) must be satisfied. We also see that WEC must be satisfied as part of the basis for the positive energy theorem [12] (critical in Book 5 [9] on QFT).

7.1 Space-time analysis involving CTCs

We will find in what follows that WEC is indicated (generically) to prevent CTC infinite self-energy boost processes. Thus we arrive at GR with the WEC constraint. As with the nonrenormalizability of the GR gauge field representation, we see here another barrier to GR quantization. There is no barrier to quantization-based *thermalization* of GR, however, and this is examined in Book 6 [4].

The Positive Energy Theorem (PET) [2,12,24,33-38] is applicable when in a spacetime where:
(i) the Einstein Field equations apply;
(ii) the weak energy condition (WEC) applies; and
(iii) the spacetime has an asymptotically Euclidean ('flat') spacelike hypersurface.
Under these conditions, use of the ADM formalism will show that the ADM four-momentum is time-like and future-pointing (so positive energy, unless ADM four-momentum is precisely zero, which would indicate empty Minkowski spacetime). The 'future-pointing' aspect requires orientability on the manifold (to be globally consistent). If there is orientability, then the Manifold is compatible with a spinor field on that Manifold (matter fields are spinor fields). One simplified variant of the PET proof assumes the existence of such a spinor structure [12].

The ADM Total energy is, thus, positive in the indicated circumstances (in Shell analysis, Ch. 10, this indicates the allowed mass spectrum is constrained to be positive definite). The constraint for 'positive energy' is really a constraint for the weak energy condition to be valid (since we are happy to say the Einstein Equations apply, and observationally we see a Universe that is, almost freakishly, 'flat'). We will find that if we do not enforce WEC, we get CTCs, and when we get CTCs we have infinite repeated boosting (if not going 'one way', then go 'the other') resulting in blue-shift infinite energy instability on any geodesic approaching a CTC. Such problematic CTCs can be defined in a variety of ways, including for a variety of families of solutions, which are explored in what follows, and the only universal way to block their appearance is to constrain such that the WEC is true (or Average WEC, AWEC, when working with QFT).

7.2 Taub-NUT space [39]

Before discussing (4D) Taub space, and its extension to NUT space, let's first understand the underlying 2D Misner space:

7.2.1 Misner Space

The topology on the Misner manifold M is $R \times S^1$. The metric, g, on the manifold is:

$$ds^2 = -t^{-1}dt^2 + td\psi^2, \;\; 0 \le \psi \le 2\pi, \;\; 0 < t < \infty, \quad \text{(Eqn. 7-1)}$$

which is singular at $t = 0$.

Misner space, (M, g), can be extended by defining: $\psi' = \psi - log(t)$, now denoted (M, g'). Note that the coordinate transformation that is needed in any such extension must itself be singular at the coordinate singularity to cancel the divergence. Let's now consider the extension indicated under ψ':

$$ds^2 = 2d\psi' dt + t(d\psi')^2, \;\; 0 \le \psi' \le 2\pi, \;\; 0 < t < \infty.$$

Note that the singularity at $t = 0$ has been removed, so we can trivially extend the domain of t for the larger manifold M', which we denote (M', g'): $-\infty < t < \infty$. Can we do any further extension on the manifold? In essence, this is asking if the manifold is locally extendible.

Def.: Locally inextendible.[39]

A manifold is locally inextendible if there is no open set $U \subset M$ with non-compact closure in M, such that the pair $(U, g|_U)$ has an extension (U, g') in which the closure of the image of U is compact.

Notice that locally inextendible can be related to geodesic completeness (see below for details).

Since the extended Misner space-time (M', g') has the topology of the surface of an infinite cylinder, there ae no open sets with non-compact closure. Thus the extended Misner space-time is inextendible. This does not mean that the analysis is done, however, as more than one extension may be possible. If we re-examine the extension definition we see a trivial alternate extension: $\psi'' = \psi + log(t)$, which gives the space-time (M'', g''):

$$ds^2 = -2d\psi'' dt + t(d\psi'')^2, \;\; 0 \le \psi'' \le 2\pi, \;\; -\infty < t < \infty. \quad \text{(Eqn. 7-2)}$$

Misner space-time has the two inextendible space-times indicated and no more. The next question becomes, are these space-times equivalent? To show they are not equivalent will take some effort, but the same diagrams and analysis will need to be examined in analyzing the CTC's that appear, so following the derivation indicated in [39] we now address the issue of geodesic completeness in these space-times, and to do this we must analyze the geodesics of the space-time.

7.2.2 Geodesics of Extended Misner Space-time

Working with the (M'', g'') extension, but dropping the double prime to single prime notationally, we have:

$$ds^2 = -2d\Psi' dt + t(d\Psi')^2 = d\Psi'(-2dt + td\Psi')$$

(Eqn. 7-3)

We have for geodesics: $d\Psi' = 0$ and $-2dt + td\Psi' = 0$. Let's begin by showing that the latter is a null geodesic.

Shifting to an indexed notation let have: $x^1 = t\,, x^2 = \Psi' \Rightarrow g_{12} = -1\,,$ $g_{22} = t,\, g_{22,1} = 1.$
Thus:

$$\Gamma_{221} = \Gamma_{212} = -\Gamma_{122} = \frac{1}{2} g_{22,1} = \frac{1}{2}$$

Thus:

$$\frac{d^2x^a}{d\lambda^2} + \Gamma^a_{bc} \frac{dx^b}{d\lambda}\frac{dx^c}{d\lambda} = 0$$

$$\rightarrow \quad \frac{d^2x^2}{d\lambda^2} + 2g^{22}\Gamma_{221}\frac{dx^2}{d\lambda}\frac{dx^1}{d\lambda} + g^{21}\Gamma_{122}\left(\frac{dx^2}{d\lambda}\right)^2 = 0$$

This generates the coordinate geodesic equations:

(1) $\frac{d^2\Psi'}{d\lambda^2} + \frac{1}{t}\frac{dt}{d\lambda}\frac{d\Psi'}{d\lambda} + \left(+\frac{1}{2}\right)\left(\frac{d\Psi'}{d\lambda}\right)^2 = 0 \quad \rightarrow \quad \frac{d^2x^1}{dx^2} +$
$2g^{12}\Gamma_{221}\frac{dx^2}{d\lambda}\frac{dx^1}{d\lambda} = 0$

(2) $\frac{d^2t}{d\lambda^2} + \frac{d\Psi'}{d\lambda}\frac{dt}{d\lambda} = 0.$

Let's confirm that $-2dt + td\Psi' = 0$ is a geodesic, we have with another derivative:

$$-2\frac{dt}{d\lambda} + t\frac{d\Psi'}{d\lambda} = 0 \quad and \quad -2\frac{d^2t}{d\lambda^2} + \frac{dt}{d\lambda}\frac{d\Psi'}{d\lambda} + t\frac{d^2\Psi'}{d\lambda^2} = 0$$

(Eqn. 7-4)

Are (1) and (2) consistent with the above? The answer is yes, but the proof is left as an exercise.

So, $-2dt + td\Psi' = 0$ does indeed describe a geodesic, but is it a null geodesic? (dropping the prime in what follows) Let's approach this using abstract index notation and consider the tangent vector to the geodesic denoted K^a:

$$\text{null: } K^a K_a|_{\lambda=0} = 0$$
$$\text{geodesic: } K^b \nabla_b K^a = 0$$

Thus,

$$0 = K_a K^b \nabla_b K^a = \frac{1}{2} K \cdot \nabla(K^2) = \frac{1}{2}\frac{d}{d\lambda}K^2$$

Thus

$$\left. K^2\right|_{\lambda=0} = 0 \text{ and } \left.\frac{d}{d\lambda}K^2\right|_{\lambda=0} = 0 \Big\} \quad \Rightarrow \quad \underline{K^2(\lambda) = 0}$$

Let's use the notation where the components of K^a are K^t, K^Ψ ($K^\Psi = \frac{d\Psi}{d\lambda}$) and

$$g_{t\Psi} = -1, g_{tt} = 0, g_{\Psi\Psi} = t.$$

We therefore have:

$0 = g_{ij}K^i K^j = -2K^t K^\Psi + t(K^\Psi)^2 = K^\Psi(tK^\Psi - 2K^t)$

and

$$K^b \nabla_b K^2 = K^b \nabla_b [K^\Psi(tK^\Psi - 2K^t)] = 2K_a K^b \nabla_b K^a$$
$$= 2K_\Psi(1) + 2K_t(2)$$

Thus, $2K_\Psi(1) + 2K_t(2) = 0 \quad \rightarrow \quad -2dt + td\Psi = 0$, and the second geodesic is indeed a null geodesic.

Now that the geodesics are identified, are they complete? To show geodesic incompleteness we need to show geodesics that reach the boundary of the space-time in finite affine parameter. A consequence of incompleteness is lack of being simply connected, which is typically associated with a non-trivial relation to a covering space. As usual, further clarity is sought by examination of the covering space. To do this for a covering space on both extensions we will consider a new coordinate transformation.

Recall

$$(M,g): \quad ds^2 = -t^{-1}dt^2 + td\Psi^2 \quad t > 0 \qquad 0 \le \Psi \le 2\pi$$

where Ψ has identification at boundaries. Consider the coordinate transformation

$$t = \frac{1}{4}(\hat{t}^2 - \hat{x}^2), \Psi = 2\tanh^{-1}\left(\frac{\hat{x}}{\hat{t}}\right) \Rightarrow ds^2 = -d\hat{t}^2 + d\hat{x}^2$$

(Eqn. 7-5)

Since $t > 0 \Rightarrow \hat{t} > 0$, also, since $\hat{t}^2 = 4 + t\hat{x}^2 \Rightarrow |\hat{x}| < \hat{t}$. Thus, without regard to the identifications in Ψ just yet, we see that (M, g) is covered by the future light cone of Minkowski space.

<u>Identification</u>: points with $\Psi = \alpha + n2\pi$ are identified with $\Psi = \alpha$ where $0 \leq \alpha \leq 2\pi$ under the identification.

$$\hat{t} = 2t\cosh\left({}^{\Psi}/_{2}\right)$$
$$\hat{x} = 2t\sinh\left({}^{\Psi}/_{2}\right)$$
$$\left(\hat{t}(\alpha), \hat{x}(\alpha)\right) = \left(\hat{t}(\alpha + n2\pi), \hat{x}(\alpha + n2\pi)\right)$$
$$ds^2 = -d\hat{t}^2 + d\hat{x}^2$$

(Eqn. 7-6)

In other words, one identifies all points $(\hat{t}\cosh n\pi + \hat{x}\sinh n\pi, \hat{x}\cosh n\pi + \hat{t}\sinh n\pi)$ for integer values of *n*. The space (M, g) is thus the quotient of $(I, \tilde{\eta})$ by the discrete subgroup G of the Lorentz group consisting of boosts by multiples of a given factor, where *I* is the upper Minkowski quadrant common to both extended Misner spaces:

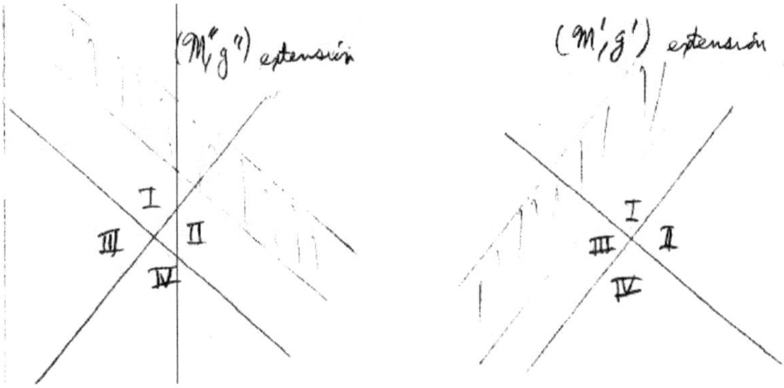

Fig. 7.1. Misner space

As for the identification under (possibly repeated) constant boost factor, let's specifically consider the (M'', g'') case:

$$ds^2 = -2d\Psi''dt + t(d\Psi'')^2 = d\Psi'(-2dt + td\Psi'') \text{ and } \Psi = \Psi'' - \log t$$

(Eqn. 7-7)

For the null geodesics we have $d\Psi' = 0$, $-2dt + td\Psi' = 0$, but in the new coordinates it is the usual Minkowski form: $d\hat{t} = \pm d\hat{x}$, where we use the $d\hat{t} = -d\hat{x}$ solution for quadrants I and II. We, thus, have

$$\hat{t} = -\hat{x} + \alpha,$$

(Eqn. 7-8)

as the null geodesic that is complete in the extension (M'', g'') shown in the Figure below:

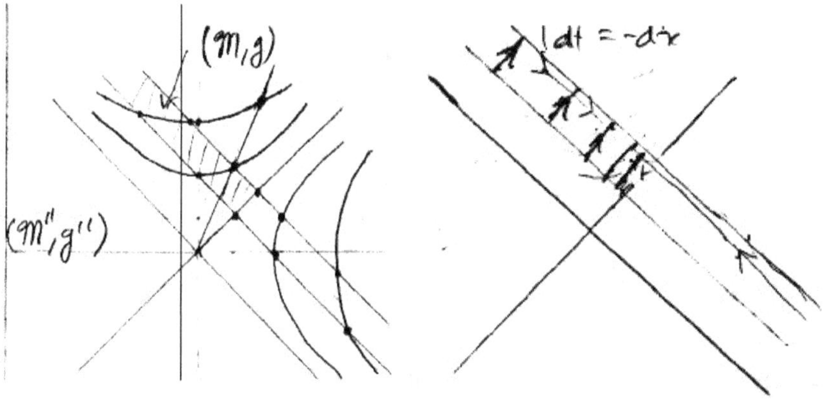

Fig. 7.2. Misner Space with null geodesic

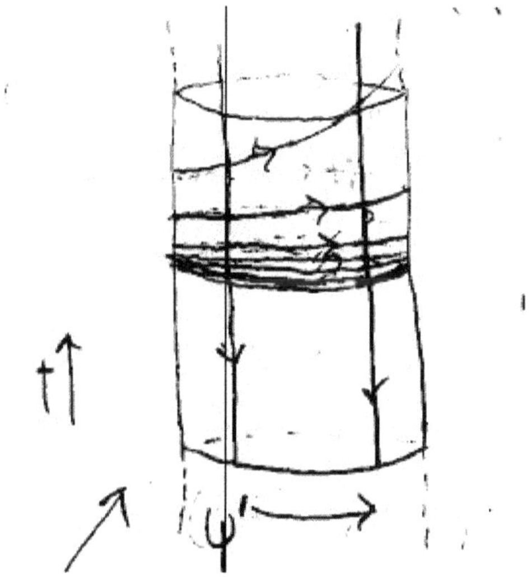

Fig. 7.3. Misner Space with null geodesic upon identification map.

(M'', g'') is covered by (I,II) of Minkowski, or $(I + II, \eta)/G$ is (M'', g'').
(M', g') is covered by (I,III) of Minkowski, or $(I + III, \eta)/G$ is (M', g').

Consider a future oriented right moving null geodesic in (M'', g'') (regarded from the larger Minkowski covering space it is obviously not complete) $\hat{t} = \hat{x} + \alpha$ and $\alpha > 0$ for the upper quadrant:

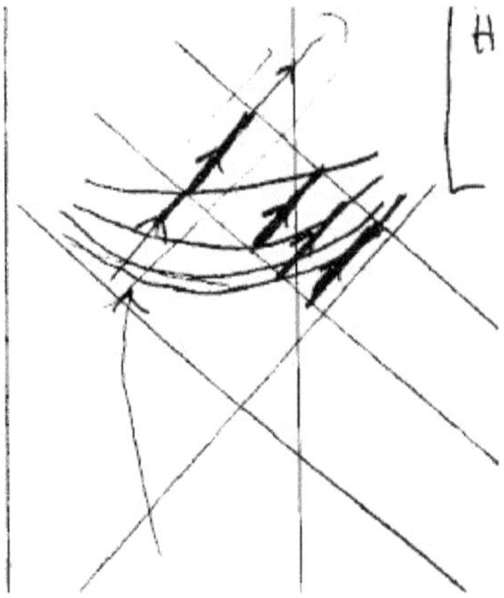

Fig. 7.4. Misner Space null geodesic affine distance finite to boundary.

For this trajectory we have intercepts at higher and higher hyperbolae in the upper quadrant, e.g.,

$$t = \frac{1}{4}(\hat{t}^2 - \hat{x}^2)$$

(Eqn. 7-9)

increases. While

$$\Psi = 2\tanh^{-1}\left(\frac{\hat{x}}{\hat{t}}\right) \quad \rightarrow \quad \Psi = 2\tanh^{-1}\left(1 - \frac{\alpha}{\hat{t}}\right), \qquad \hat{t} > 0.$$

(Eqn. 7-10)

For a given geodesic the least value of $\hat{t}$ is:

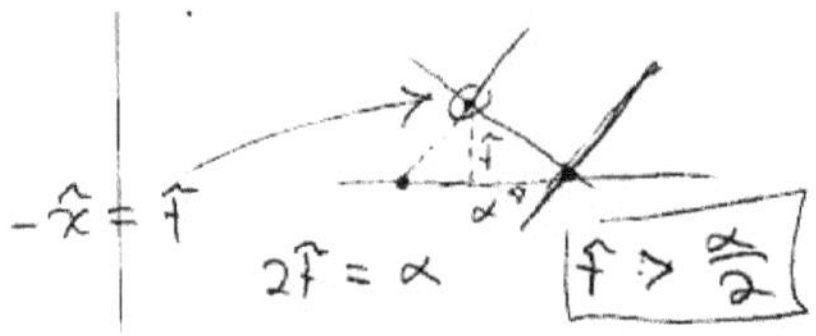

Fig. 7.5.

So,

$$\Psi|_{min} = 2\tanh^{-1}(1-2) \rightarrow -\infty,$$

(Eqn. 7-11)

at the 'start' of the future oriented geodesic going to $\hat{t} \rightarrow \infty$. So, Ψ increases on this geodesic.

Consider $ds^2 = 2d\Psi'dt + t(d\Psi')^2 = d\Psi'(2dt + td\Psi')$ and $\Psi = \Psi' + \log t$. We have

$$\text{Null geodesics} : d\Psi' = 0 \ , \ 2dt + td\Psi' = 0$$

and

$$d\Psi' = -2\frac{dt}{t} \Rightarrow t = A\exp\left(-\frac{1}{2}\Psi'\right), \qquad \text{(Eqn. 7-12)}$$

and we have:

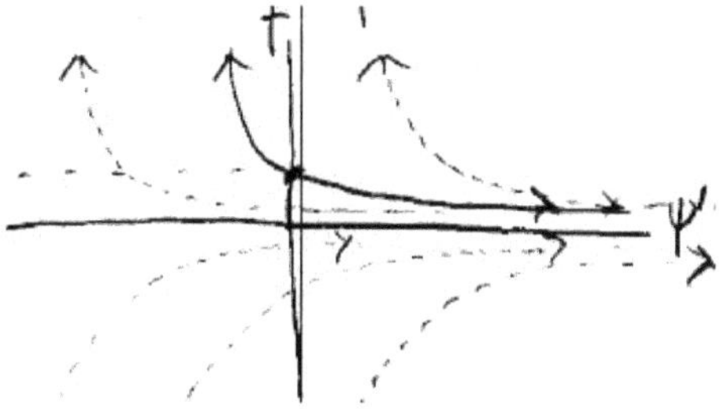

Fig. 7.6. Ψ' Characteristics.

For null: $\Psi = \Psi' + \log t = \Psi' - \left(\frac{1}{2}\Psi'\right) + \log A$

$$\Psi = \frac{1}{2}\Psi' + const \rightarrow \ \underline{\Psi \propto \Psi' \text{ on null}}$$

So, in the future light cone region, as affine parameter increases t increases and Ψ' increases, this does not show agreement between the $(I + II, \tilde{\eta})/G$ Misner space and $ds^2 = 2d\Psi'dt + t(d\Psi')$, but with the other extension. So, consider $ds^2 = -2d\Psi''dt + t(d\Psi'')^2 = d\Psi''(-2dt + td\Psi'')$, again we have

$$d\Psi'' = 2\frac{dt}{t} \Rightarrow t = B\exp\left(\frac{1}{2}\Psi''\right)$$

and we have the figures for $\tilde{t} = -\tilde{x} + \beta, \beta > 0, \tilde{t} > \frac{\beta}{2}$:

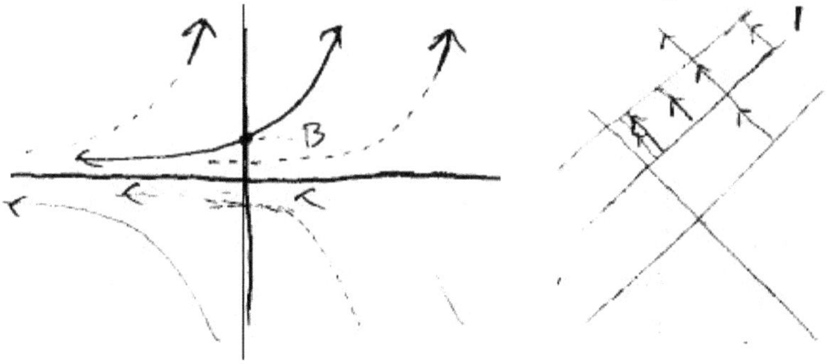

Fig. 7.7.

So $\Psi \to \infty$ at start, and goes to $-\infty$ at future infinity.So, for $(I + III, \tilde{\eta})/G$ Misner, as affine parameter increases t increases and Ψ decreases. So $(I + II, \tilde{\eta})/G$ corresponds to (M'', g'') with

$$ds^2 = -2d\Psi''dt + t(d\Psi'')^2$$

(Eqn. 7-13)

The action of the identification group G in the region I is properly discontinuous, as it is in region I+II and I+III. To be properly discontinuous (39]) requires that we satisfy a manifold condition and a Hausdorff condition for the Quotient space [39]. Trying to merge the extensions in results in a space that is not Hausdorff [39]. If a space with identification boundary is considered that the lines 'cross', forming an upper and lower wedge is not Hausdorff, but if restricted to one wedge, then it is Hausdorff and can be embedded in Misner space [39]:

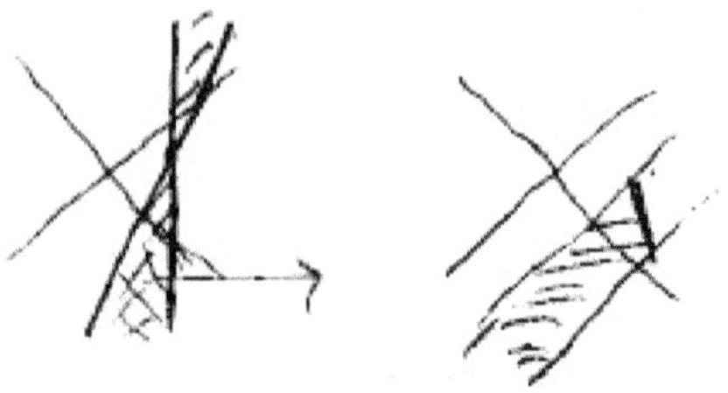

Fig. 7.8. Other Misner-like spaces.

Now that we've considered a 2D case with behavior of interest, let's move to a 4D formulation due to Taub to explore this further in a full 4D manifold scenario.

7.2.3 Taub Space

$$ds^2 = -U^{-1}dt^2 + (2l)^2U(t)(d\Psi + \cos d\varphi)^2 + (t^2 + l^2)(d\theta^2 + \sin^2\theta\, d\theta),$$

(Eqn. 7-14)

where $U(t) = -1 + 2\frac{(mt+l^2)}{(t^2+l^2)}$, $0 \leq \Psi \leq 4\pi$, $0 \leq \theta \leq \pi$, $0 \leq \varphi \leq 4\pi$. This space satisfies the empty-space Einstein equations: $R_{\mu\nu} = 0$. Taub gave the metric for $U(t) > 0$, where t=constant hypersurfaces are space-like. Newman, Unti, and Tamburino (NUT) gave the region where $U(t) < 0$. As noted by Misner [27], there are some unusual properties: (1) The Taub region evolves into NUT space which (i) is lacking interpretation as a process in itself, and (ii) because it arrives at NUT space.

(2) NUT space has CTC's.
(3) NUT space has no ('decent') space-like hypersurface.
(4) Even though the curvature tensor vanishes at infinity, the metric is not flat at infinity.

To start the examination of Taub first consider the spatial metric on the hypersurface t=constant:

$$dl^2 = (2l)^2 U(t)(d\Psi + \cos d\varphi)^2 + (t^2 + l^2)(d\theta^2 + \sin^2\theta\, d\theta),$$

or

$$dl^2 = b^2(d\Psi + \cos d\varphi)^2 + a^2(d\theta^2 + \sin^2\theta\, d\theta),$$

(Eqn. 7-15)

where $b^2 = (2l)^2 U(t)$ and $a^2 = (t^2 + l^2)$. If $a^2 = b^2$ then dl^2 would be a Killing metric for SO(3) or SU(2). We considered SO(3) and SU(2) in Book 2 [13] discussion on Lie Algebras. Let's revisit that analysis here but consistent with the notation of [39]:

7.2.4 Review of SO(3) and SU(2)

$$R_x(\theta) = \begin{pmatrix} 1 & 0 & 0 \\ 0 & \cos\theta & \sin\theta \\ 0 & -\sin\theta & \cos\theta \end{pmatrix} \quad R_y(x) = \begin{pmatrix} \cos x & 0 & -\sin x \\ 0 & 1 & 0 \\ \sin x & 0 & \cos x \end{pmatrix} \quad R_z(\phi) = \begin{pmatrix} \cos\phi & \sin\phi & 0 \\ -\sin\phi & \cos\phi & 0 \\ 0 & 0 & 1 \end{pmatrix}$$

Euler angles for a pt. (x convention): $P(\Psi, \theta, \phi) = R_z(\Psi)R_x(\theta)R_z(\phi)$.
$P(0,0,0) = I$ is the identity.

For small t:

$$R_z(t)P(\Psi,\theta,\phi) = P(\Psi + \mathrm{t}, \theta, \phi)$$

$$R_x(t)P(\Psi,\theta,\phi) = P\left(\Psi + t\sin\Psi\cot\theta, \theta + t\cos\Psi, \phi + t\frac{\sin\Psi}{\sin\theta}\right)$$

$$R_y(t)P(\Psi,\theta,\phi) = P\left(\Psi - t\cos\Psi\cot\theta, \theta - t\sin\Psi, \phi + t\frac{\cos\Psi}{\sin\theta}\right)$$

(Eqn. 7-16)

Now, a diffeomorphism $\Psi\colon \mathrm{M} \to \mathrm{M}$ maps a vector ξ^a at x_0 to a vector $\Psi_*\xi^a$ at $\Psi(x_0)$. So, the diffeo R_p (R for right multiplication) produces a vector:

$$\mathrm{R}_*\xi^a \;\; at \;\; R_p(I) = IR_p = R_p$$

(Eqn. 7-17)

So, the vector $V_z = \frac{d}{dt}[R_z(t)I]|_{t=0}$ becomes:

$$R_{p^*}V_z = \frac{d}{dt}[R_z(t)P]|_{t=0} = \frac{\partial}{\partial\Psi}$$

(Eqn. 7-18)

We thus arrive at a Right invariant basis of vector fields for $SO(3)$:

$\vec{V}_z: \vec{e}_3 = {\partial}/_{\partial\Psi}$

$\vec{V}_y: \vec{e}_2 = -\sin\Psi\,{\partial}/_{\partial\theta} - \cos\Psi\left(\cot\theta\,{\partial}/_{\partial\Psi} - \frac{1}{\sin\theta}{\partial}/_{\partial\phi}\right)$

$\vec{V}_x: \vec{e}_1 = -\cos\Psi\,{\partial}/_{\partial\theta} - \sin\Psi\left(\cot\theta\,{\partial}/_{\partial\Psi} - \frac{1}{\sin\theta}{\partial}/_{\partial\phi}\right)$

(Eqn. 7-19)

Now at I on the group manifold there is a linear mapping on the vector space there that gives the dual vector space, that of the 1-forms:

$$e^a\sigma_b(I) = \delta^a_b(I) = const$$

(Eqn. 7-20)

This relation is extended to the whole group by requiring

$$\delta^a_b(P) = R_{p*}\delta^a_b(I)$$

(Eqn. 7-21)

such that δ^a_b is right invariant, then $e^a\sigma_b = \delta^a_b$ everywhere, if e_i and σ^j are the corresponding bases:

$$(e^a\sigma_b)(e_i)_a(\sigma^i)^b = \delta^a_b(e_i)_a(\sigma^i)^b = \delta^j_i$$

(Eqn. 7-22)

$e_i\sigma^j = \delta^j_i$, as desired and with $R_{P^*}(e^a\sigma_b) = e^a\sigma_b$, $R_{P^*}e^a = e^a$ (right inv), we have:

$$R_{P^*}(e^a\sigma_b) = R_{P^*}e^a R_{P^*}\sigma_b = e^a R_{P^*}\sigma_b = e^a\sigma_b$$

(Eqn. 7-23)

So, $R_{P^*}\sigma_b = \sigma_b$, thus the form field basis that corresponds to the right inv. basis of vector field is a right inv. form basis:

$$\begin{pmatrix} \sigma^1 = \sin\Psi\sin\theta\, d\phi + \cos\Psi\, d\theta \\ \sigma^2 = \cos\Psi\sin\theta\, d\phi - \sin\Psi\, d\theta \\ \sigma^3 = \cos\theta\, d\phi + d\Psi \end{pmatrix} \text{derived considering } \langle\sigma^j, e_i\rangle = \delta_{ij}$$

(Eqn. 7-24)

A similar calculation for the diffeo's corresponding to left multiplication gives:

Left inv. vector fields on SO(3):

$$\vec{V}_2: \vec{e}_3 = \partial/\partial\phi$$

$$\vec{V}_y: \vec{e}_2 = -\sin\phi\,\partial/\partial\phi - \cos\phi\left(\cot\theta\,\partial/\partial\phi - \frac{1}{\sin\theta}\partial/\partial\Psi\right)$$

$$\vec{V}_x: \vec{e}_1 = \cos\phi\,\partial/\partial\theta - \sin\phi\left(\cot\theta\,\partial/\partial\phi - \frac{1}{\sin\theta}\partial/\partial\Psi\right)$$

(Eqn. 7-25)

Left inv. form fields:

$$\sigma^1 = \sin\phi\sin\theta\, d\Psi + \cos\phi\, d\theta$$

$$\sigma^2 = \cos\phi\sin\theta\, d\Psi - \sin\phi\, d\theta$$

$$\sigma^3 = \cos\theta\, d\Psi + d\phi$$

(Eqn. 7-26)

Let's now, return to the Taub spatial metric and make use of the above diffeo. notation:

$$d\ell^2 = a^2(d\theta^2 + \sin^2\theta\, d\phi^2) + b^2(d\Psi + \cos\theta\, d\phi)^2$$
$$= a^2((\sigma_R^1)^2 + (\sigma_R^2)^2) + b^2(\sigma_R^3)^2$$

(Eqn. 7-27)

Also, note $(\sigma_L^1)^2 + (\sigma_L^2)^2 + (\sigma_L^3)^2 = (\sigma_R^1)^2 + (\sigma_R^2)^2 + (\sigma_R^3)^2$

Define the Killing metric: $K_{ab} = \delta_{ij}\sigma_a^i\sigma_b^j$. K_{ab} is both left and right inv. under SO(3), so it is convenient to rewrite:

$$d\ell^2 \Rightarrow g_{ab} = a^2(K_{ab} - (\sigma_R^3)_a(\sigma_R^3)_b) + b^2(\sigma_R^3)_a(\sigma_R^3)_b$$

(Eqn. 7-28)

Thus g_{ab} is right inv. under SU(2), g_{ab} is also left inv. under the U(1) subgroup of SU(2) that fixes $\Psi_a = (\sigma_R^3)_a$. The symmetry group of Taub is thus $\underline{SU(2)_R \times U(1)_L}$.

The vector fields generated by the <u>right action of SU(2)</u> are the <u>left inv. vector fields of SU(2)</u>, (and vice versa). So, the KVs generating the right action of SU(2) are

$$(\vec{e}_i)_{left}$$

from before, and the additional KV generating the left U(1) action is

$$(\vec{e}_3)_R = \partial/\partial\Psi.$$

(Eqn. 7-29)

The 4-D group of isometries from these 4 KV's are transitive on the hypersurfaces $t = const$. Minkowski, by comparison, has the 10D isometries of the Poincare Group, or the Lorentz group plus translations.

7.2.5 Taub Fiber analysis (part 1)

Locally Taub space-time is S^2 with fiber (Def. in Sec. 2.7) and the isometries map fibers into fibers. Unlike S^2, the fiber is not simply connected. So, using the fiber terminology, Taub is locally a simple S^2 structure with fiber. Although Taub space is simply connected, we will arrive at results similar to the Misner space analysis by considering M as a fiber over S^2 with fiber $R^1 \times S^1$. The bundle projection is then:

$$\pi: M \to S^2$$

is defined by $(t, \Psi, \theta, \phi) \to (\theta, \phi)$, which is the product with the t-axis of the Hopf fibering. Now we know that Taub space has a 4D group of isometries transitive on the hypersurfaces t=constant, thus the isometries map fibers of the bundle into fibers. So, consider a fiber, $\mathcal{F}$ and induced metric $\tilde{g}$. If the fiber is the (t, Ψ) plane, then the induced metric is:

$$\tilde{g}: ds^2 = -U^{-1}dt^2 + 4l^2 U d\psi^2.$$

(Eqn. 7-30)

Note this is precisely the form of 3D Misner space with $t \to U(t)$ and introduction of the constant $4l^2$.This means that we can expect much the same result as before, extendability ("locally inextendible"), but with geodesic incompleteness. Arriving at that result without switching to the more abstract fiber formalism would be far more work. This application is worked out in the notation of [39], but is also shown in (Penrose) abstract index notation.

Suppose the fiber is spanned by $\left\{\frac{\partial}{\partial t}, \frac{\partial}{\partial \psi}\right\}$ and the horizontal sub-space is spanned by $\left\{\frac{\partial}{\partial \theta}, \frac{\partial}{\partial \varphi} - \cos\theta \frac{\partial}{\partial \psi}\right\}$. Other than the obvious orthonormal relations, We then have

$$\left(\frac{\partial}{\partial \psi}\right) \cdot \left(\frac{\partial}{\partial \varphi} - \cos\theta \frac{\partial}{\partial \psi}\right) = g_{\psi\varphi} - \cos\theta\, g_{\psi\psi} = 4l^2 U \cos\theta - \cos\theta\, 4l^2 U$$
$$= 0,$$

as well. So, dividing any vector at the tangent space of q into a vertical vector and a horizontal vector can be written:

$$g(\vec{X}, \vec{Y}) = g_V(\vec{X}_V, \vec{Y}_V) + (t^2 + l^2) g_H(\pi_* \vec{X}, \pi_* \vec{Y}),$$

(Eqn. 7-31)

where $g_V = \tilde{g}$ and g_H is the metric of a 2-sphere.

Although g is not a direct sum of g_V and $(t^2 + l^2)g_H$ it can be regarded as such, *locally*. With such a decomposition we then have much the same analysis as before, with extensions to the space-time via direct analytic extension. So, let's carefully consider

$$g_V = \tilde{g}: ds^2 = -U^{-1}dt^2 + 4l^2 U d\psi^2, \;\; where \;\; U(t) = -1 + 2\frac{(mt + l^2)}{(t^2 + l^2)}.$$

(Eqn. 7-32)

We need to know the zero's and asymptotes of $U(t)$:

$$U|_{t=-\infty} = -1 \;\; and \;\; U|_{t=\infty} = -1$$

and

$$U|_{t=t\pm} = 0 \;\; where \;\; t_\pm = m \pm (m^2 + l^2)^{1/2}$$

and

$$\left.\frac{dU}{dt}\right|_{t=t1,2} = 0 \;\; where \;\; t_{1,2} = -\left(\frac{l^2}{m}\right) \pm \left(\left(\frac{l^2}{m}\right)^2 + l^2\right)^{1/2}$$

So:

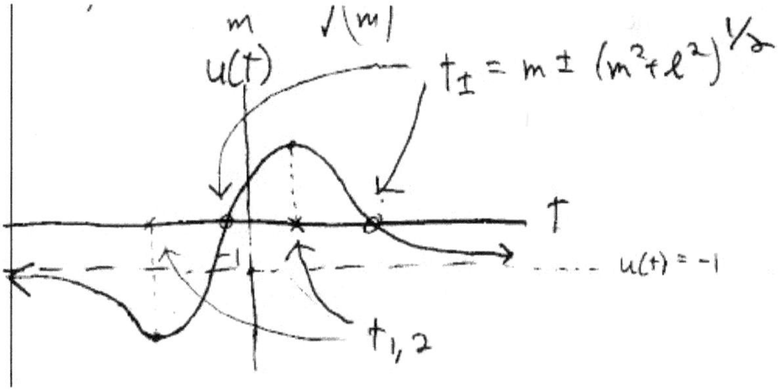

Fig. 7.9. Analysis of U(t)

As with Misner, consider the characteristics of the null geodesic for the induced metric of the fiber:

$$ds^2 = -U^{-1}dt^2 + 4l^2 U d\Psi^2, \quad 0 \le \Psi \le 4\pi, \quad t_- < t < t_+.$$

Similar to Misner, substitute

$$\Psi' = \Psi + \frac{1}{2l}\int \frac{dt}{U(t)}.$$

(Eqn. 7-33)

Then we get:

$$ds^2 = 4ld\Psi'(lU(t)d\Psi' - dt),$$

(Eqn. 7-34)

and we've extended to $-\infty < t < \infty$. Regarding the $(lU(t)d\Psi' - dt) = 0$ null geodesics:

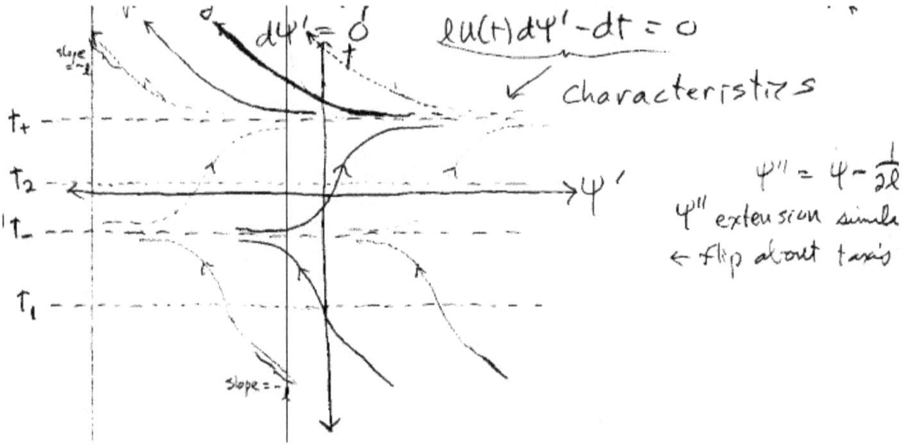

Fig. 7.10. Analysis of Characteristics.

Thus, we find the behavior is as with Misner space, one family of null geodesics in a given extension crosses both horizons $t = t_+$ and $t = t_-$, and one family spirals around these surfaces and is incomplete.

7.2.6 Taub Fiber analysis (part 2) [22]

The metric has Topology $R \times S^3$:

$$ds^2 = -U^{-1}dt^2 + (2\ell)^2 U(d\Psi + \cos\theta \, d\phi)^2 + (t^2 + \ell^2)(d\theta^2 + \sin^2\theta \, d\phi^2)$$

(Eqn. 7-35)

$$U(t) = -1 + \frac{2(mt + \ell^2)}{t^2 + \ell^2} \qquad 0 \le \Psi \le 4\pi, 0 \le \theta \le \pi, 0 \le \phi = 2\pi$$

(Eqn. 7-36)

Metric singular at $t = t_\pm = m \pm (m^2 + \ell^2)^{1/2}$ extending across these surfaces one arrives at NUT space. Before discussing extension consider Misner space:

$(\mathfrak{M})$ topology $R \times S^1$
(g) $ds^2 = -t^{-1}dt^2 + td\Psi^2$
$0 \le \Psi \le 2\pi$, metric singular at $t = 0$, and $0 < t < \infty$.

$(\mathfrak{M}, g)$ can be extended by defining: $\Psi' = \Psi - \log t$

(The coord. trans. utilized in any extension must itself be singular at the point in question in order that it may cancel the appropriate term in the metric.)

The metric becomes:

$(\mathfrak{M}', g') \qquad ds^2 = 2d\Psi' dt + t(d\Psi')^2 \qquad \underline{new - \infty < t < \infty}$

(Eqn. 7-37)

Region $t > 0$ is isometric to $(\mathfrak{M}, g)$

Similarly $\Psi' = \Psi + \log t$ yields $(\mathfrak{M}'', g'')$ with $ds^2 = -2d\Psi'' dt + t(d\Psi'')^2$

We have two inequivalent locally inextendable analytic extensions of $(\mathfrak{M}, g)$ both of which are geodesically incomplete.

(locally inextendible: there is no open set $\mathfrak{U} < M$ with non-compact closure in $\mathfrak{M}$, such that the pair $(\mathfrak{U}, g|_{\mathfrak{U}})$ has an extension $(\mathfrak{U}', g')$ in which the closure of the image of $\mathfrak{U}$ is compact)

Let our manifold $\mathfrak{M}$ be $\mathbb{R}^2$ with a closed disk removed: (metric Lorentz). Consider an open set which "borders" that disk (almost touching). The closure of $\mathcal{U}$ is non-compact (by construction it adjuts the edge of the manifold). The pair $(\mathcal{U}, g|_{\mathcal{U}})$ has the trivial extension $(\mathcal{U}', g')$ in wich $\mathcal{U}'$ includes the edge of the disk (thus extending on $\mathfrak{W}$) and g' is simply g of $\mathbb{R}^2$. The closure of $\mathcal{U}'$ is now compact. Such is an example of a space that is <u>not</u> locally inextendible.

Local inextendibility is a stronger condition of extendibility, it allows for a fibre bundle type extension of the manifold. Regular extension of $(\mathfrak{W}, g)$ simply asks that there be an isometric C^r embedding $\mu: \mathfrak{W} \to \mathfrak{W}'$ for $(\mathfrak{W}', g')$.

Now Misner space is not simply connected so let's study $(\mathfrak{M}, g), (\mathfrak{M}', g'), (\mathfrak{M}'', g'')$ in the covering space and see what is found. The metric for $(\mathfrak{M}, g)$:

$$ds^2 = -t^{-1}dt^2 + td\Psi^2$$

(Eqn. 7-38)

which has Lorentz signature and is restricted to $t > 0$. We only have 2 dimensions, so conformally flat. What is the coordinate trans. that yields $ds^2 = -d\tilde{t}^2 + d\tilde{x}^2$? Consider

$$t = \frac{1}{4}(\tilde{t}^2 - \tilde{x}^2) \;\; and \;\; \Psi = 2\tanh^{-1}\left(\frac{\tilde{x}}{\tilde{t}}\right)$$

(Eqn. 7-39)

Then

$$dt = \frac{1}{2}(\tilde{t}d\tilde{t} - \tilde{x}d\tilde{x})$$

(Eqn. 7-40)

and

$$d\Psi = \frac{2}{1-\left(\frac{\tilde{x}}{\tilde{t}}\right)^2} d\left(\frac{\tilde{x}}{\tilde{t}}\right) = \frac{2}{1-\left(\frac{\tilde{x}}{\tilde{t}}\right)^2}\left\{\frac{d\tilde{x}}{\tilde{t}} - \frac{\tilde{x}d\tilde{t}}{\tilde{t}^2}\right\} = \frac{2(\tilde{t}d\tilde{x} - \tilde{x}d\tilde{t})}{\tilde{t}^2 - \tilde{x}^2}$$

(Eqn. 7-41)

$$(dt)^2 = \frac{1}{2}(\tilde{t}^2 d\tilde{t}^2 - 2\tilde{t}\tilde{x}d\tilde{t}d\tilde{x} + \tilde{x}^2 d\tilde{x}^2)$$

$$(ds)^2 = -\frac{(\tilde{t}^2 d\tilde{t}^2 + \tilde{x}^2 d\tilde{x}^2 - 2\tilde{t}\tilde{x}d\tilde{t}d\tilde{x})}{\tilde{t}^2 - \tilde{x}^2} + \frac{(\tilde{t}^2 d\tilde{x}^2 - 2\tilde{t}\tilde{x}d\tilde{x}d\tilde{t} + \tilde{x}^2 d\tilde{t}^2)}{\tilde{t}^2 - \tilde{x}^2} = -d\tilde{t}^2 + d\tilde{x}^2$$

So we have found the coordinate transformation that shows that locally Misner space is simply Minkowski space, i.e., Misner space is covered by a portion of Mink., i.e., Misner is flat with identifications. What is the rule for identifications?

The identification involve the coordinate Ψ, where

$$t = \frac{1}{4}(\tilde{t}^2 - \tilde{x}^2) \;\; and \;\; \Psi = 2\tanh^{-1}\left(\frac{\tilde{x}}{\tilde{t}}\right).$$

(Eqn. 7-42)

Switching to

$$\tilde{t} = 2t\cosh\left(\Psi/_2\right) \quad and \quad \tilde{x} = 2t\sinh\left(\Psi/_2\right)$$

(Eqn. 7-43)

we are able to work the domain $0 \le \Psi \le 2\pi$ more easily, and with the identification of $\Psi = 0$ with $\Psi = 2\pi$ or $\Psi = \alpha + n2\pi$ with $\Psi = \alpha$: $\left(\tilde{t}(\alpha), \tilde{x}(\alpha)\right) = \left(\tilde{t}(\alpha + 2\pi), \tilde{x}(\alpha + 2\pi)\right)$. Thus:

$$\begin{aligned}\tilde{t}(\alpha + 2\pi) &= 2t\cosh\left(\frac{\alpha}{2} + \pi\right) \\ &= 2t\cosh\left(\frac{\alpha}{2}\right)\cosh\pi + 2t\sinh\left(\frac{\alpha}{2}\right)\sinh\pi \\ &= \tilde{t}(\alpha)\cosh\pi + \tilde{x}(\alpha)\sinh\pi\end{aligned}$$

and

$$\tilde{x}(\alpha + 2\pi) = 2t \sinh\left(\frac{\alpha}{2} + \pi\right)$$
$$= 2t \sinh\left(\frac{\alpha}{2}\right)\cosh(\pi) + 2t \cosh\left(\frac{\alpha}{2}\right)\sinh(\pi)$$
$$= \tilde{x}(\alpha)\cosh(\pi) + \tilde{t}(\alpha)\sinh(\pi)$$

Thus, $(\tilde{t}, \tilde{x})$ is identified with $(\tilde{t} \cosh \pi + \tilde{x} \sinh \pi, \tilde{x} \cosh \pi + \tilde{t} \sinh \pi)$. So, one identifies all points

$$(\tilde{t} \cosh n\pi + \tilde{x} \sinh n\pi, \tilde{x} \cosh n\pi + \tilde{t} \sinh n\pi)$$
(Eqn. 7-44)

for all integer values of *n*. Since $t > 0 \Rightarrow \tilde{t} > 0$, and since $\tilde{t}^2 = 4t + \tilde{x}^2$ then $|\tilde{x}| < \tilde{t}$. Thus, $(\mathfrak{M}, g)$ is the region of Minkowski space in the future lightcone (I) of the origin with identifications as given. The space $(\mathfrak{M}, g)$ is thus the quotient of $(I, \tilde{\eta})$ by the discrete subgroup G of the Lorentz group consisting of boosts by multiples of a given factor.

The action of G in region I is properly discontinuous, as it is in regions I + II and I + III.
$(I + II, \tilde{\eta})/G$ is $(\mathfrak{W}', g')$ $(\tilde{t} > -\tilde{x})$ (from cylinder, and null geodesic picture)
$(I + III, \tilde{\eta})/G$ is $(\mathfrak{W}'', g'')$ $(\tilde{t} > \tilde{x})$
$(I + II + III, \tilde{\eta})/G$ is not Hausdorff.
$(\tilde{m} - \{p\}, \tilde{\eta})/G$ is the maximal non-hausdorff extension of $(\mathfrak{M}, g)$.

$L(\tilde{m})/G$ is a Hausdorff manifold!

The bundle of linear frames, i.e., the collection of all pairs $(X, Y), X, Y \in T_q$ at all $q \in \tilde{m}$. Action of $A \in G$ on $\tilde{m}$ induces an action A_* *on* $\hat{L}(\mathfrak{W})$ which takes (X, Y) to (A_*X, A_*Y), etc. The fact that the bundle of frames can be well behaved even though the space is not, suggests that it is useful to look at singularities by using the bundle of linear frames.

Back to 4-d Taub space $(\mathfrak{M}, g)$. As $\mathfrak{M}$ is simply connected, one cannot simplify matters by going to a covering space as was done with Misner space. However a similar result can be achieved by considering $\mathfrak{M}$ as a fibre bundle over S^2 with fibre $R^1 \times S^1$. Bundle projection is defined by:

$$\pi: \mathfrak{M} \to S^2 \text{ such that } (t, \Psi, \theta, \phi) \to (\theta, \phi)$$

The product with the t-axis of the *Hopf* fibering $S^3 \to S^2$ which has fibre S^1.

Recall that Taub space has a 4-dim group of isometries whose surfaces of transitivity are the three-spheres $\{t = const.\}$. The group of isometrics maps fibres of the bundle defined by π into fibres, and so the pairs $(\mathcal{F}, \tilde{g})$ are all isometric, where $\mathcal{F}$ is a fibre $(\mathcal{F} \cong R^1 \times S^1)$ and $\tilde{g}$ is the induced metric.

The fibre can be regarded as the (t, Ψ) plane, and the metric $\tilde{g}$ on $\mathcal{F}$ is obtained from g by dropping terms in $d\theta$ and $d\phi$:

$$ds^2 = -U^{-1}dt^2 + 4\ell^2 U(d\Psi)^2$$

(Eqn. 7-45)

So we have described $\mathfrak{M}$ as a fibre bundle, i.e., a bundle which is locally a product bundle $(\mathfrak{N} \times a, \mathfrak{N}, \pi)$, the usefulness of this is that while $\mathfrak{M}$ cannot be decomposed into the product manifold $(R^1 \times S^1) \times S^2$, it can be locally in terms of fibre bundles. An example of this: $S^3 \to S^2$, where the S^1 fibre is not indep. over S^2, it has continuity requirements. Thus, not a global product, but locally anything can be chosen and the rest of S^1 over S^2 adjusted to fit.

There are four Killing Vectors for NUT space:

$$\xi_t = {}^{\partial}/_{\partial t}$$
$$\xi_x = -\sin\theta\frac{\partial}{\partial\theta} - \cos\phi\left[\cot\theta\,{}^{\partial}/_{\partial\phi} + \lambda\tan\frac{1}{2}\theta\,{}^{\partial}/_{\partial t}\right]$$
$$\xi_y = \cos\phi\,{}^{\partial}/_{\partial\theta} - \sin\theta\left[\cot\theta\,{}^{\partial}/_{\partial\phi} + \lambda\tan\frac{1}{2}\theta\,{}^{\partial}/_{\partial t}\right]$$
$$\xi_z = {}^{\partial}/_{\partial\phi} - \lambda\,{}^{\partial}/_{\partial t} \qquad \lambda = 2\ell$$

(Eqn. 7-46)

No Ψ in eqn's so every Killing vectors lies in a hypersurface of constant Ψ.

a, b, c are each one of (xyz):

$[\xi_a, \xi_b] = -\epsilon_{abc}\xi_c$ (Lie Bracket's)

$[\xi_a, \xi_\Psi] = 0$

(Eqn. 7-47)

The usefulness of the fiber bundle decomposition is clear. Just as $R^1 \times S^3$ is not a direct product of $R^1 \times S^1$ with S^2 and yet can be considered so locally, we find that: the metric g is not the direct sum of g_v and

$(t^2 + \ell^2)g_H$ but it can nevertheless be regarded as such a sum locally, where g_v and g_H are defined according to the following. The tangent space T_q at the point $q \in \mathfrak{M}$ can be decomposed into a vertical subspace V_q which is tangent to the fibre and is spanned by ${\partial}/{\partial t}$ and ${\partial}/{\partial \Psi}$, and a horizontal subspace H_q which is spanned by ${\partial}/{\partial \theta}$ and ${\partial}/{\partial \phi} - \cos\theta\, {\partial}/{\partial \Psi}$. The Taub metric respects this local form:

$$ds^2 = -U^{-1}dt^2 + (2\ell)^2 U(d\Psi + \cos\theta\, d\phi)^2 + (t^2 + \ell^2)(d\theta^2 + \sin^2\theta\, d\phi^2)$$

(Eqn. 7-48)

$$d\omega^0 = U^{-1/2}dt \Rightarrow {\partial}/{\partial t}$$

$$d\omega^1 = 2\ell U^{1/2}(d\Psi + \cos\theta\, d\phi) \Rightarrow \langle d\Psi + \cos\theta\, d\phi, \frac{\partial}{\partial z}\rangle = 0$$

$$d\omega^2 = (t^2 + \ell^2)^{1/2} d\theta \Rightarrow {\partial}/{\partial \theta}$$

$$d\omega^3 = (t^2 + \ell^2)^{1/2} \sin\theta\, d\phi \Rightarrow {\partial}/{\partial \phi}$$

(Eqn. 7-49)

$${\partial}/{\partial z} = {\partial}/{\partial \phi} - \cos\theta\, {\partial}/{\partial \Psi}$$

(Eqn. 7-50)

$$ds^2 = -U^{-1}dt^2 + (2\ell)^2 U d\Psi^2 + (2\ell)^2 U(2\cos\theta\, d\Psi d\phi + \cos^2\theta\, d\phi^2) + (t^2 + \ell^2)(d\theta^2 + \sin^2\theta\, d\phi^2)$$

(Eqn. 7-51)

Let's analyze the fiber bundle for Taub space-time carefully, and switch to notation by [21]:

$$ds^2 = -U^{-1}dt^2 + (2\ell)^2 U(d\Psi + \cos\theta\, d\phi)^2 + (t^2 + \ell^2)(d\theta^2 + \sin^2\theta\, d\phi^2)$$

becomes

$$\bar{g} = -U^{-1}dt \otimes dt + (2\ell)^2 U(d\Psi + \cos\theta\, d\phi) \otimes (d\Psi + \cos\theta\, d\phi) + (t^2 + \ell^2)(d\theta \otimes d\theta + \sin^2\theta\, d\phi \otimes d\phi)$$

(Eqn. 7-52)

Decompose all vectors in the tangent space T_q at $q \in \mathfrak{M}$ such that locally we have a direct sum of metrics for the corresponding direct product from the fibre bundle decomposition of the space.

$$\bar{g} = \bar{g}_V + \bar{g}_H$$

(Eqn. 7-53)

$$\bar{g}_V = -U^{-1}dt \otimes dt + (2\ell)^2 U(d\Psi + \cos\theta\, d\phi) \otimes (d\Psi + \cos\theta\, d\phi)$$

(Eqn. 7-54)

$$\bar{g}_H = (t^2 + \ell^2) d\theta \otimes d\theta + (t^2 + \ell^2) \sin^2\theta \, d\phi \otimes d\phi$$

(Eqn. 7-55)

Decompose the vectors of T_q into two groups: V_q and H_q

The metric induced on the fiber is

$$\overline{g}'_V = -U^{-1} dt \otimes dt + 4\ell^2 U d\Psi \otimes d\Psi$$

(Eqn. 7-56)

The vectors which span the fiber comprise V_q: $\left\{\frac{\partial}{\partial t}, \frac{\partial}{\partial \Psi}\right\}$ and satisfy $g_H(V_q, V_q) = 0$. The vectors which span H_q should be such that $g_V(H_q, H_q) = 0$, so H_q: $\left\{\frac{\partial}{\partial \theta}, \frac{\partial}{\partial \phi} - \cos\theta \frac{\partial}{\partial \Psi}\right\}$. For such a decomposition of T_q we have:

$$g(T_q, T_q) = g'_V(V_q, V_q) + g'_V(H_q, H_q) + q_H(V_q, V_q) + q_H(H_q, H_q)$$

(Eqn. 7-57)

or, with re-labeling:

$$g(\vec{X}, \vec{Y}) = g_V(\vec{X}_V, \vec{Y}_V) + (t^2 + \ell^2) g_H(\pi_* X_H, \pi_* Y_H)$$

(Eqn. 7-58)

So, given T_q for $q \in \mathfrak{M}$, we want to decompose T_q into two subspaces: V_q, the vertical subspace, which is tangent to the fibre; and H_q, the horizontal subspace (whose projection spans the base space), and consists of those vectors which are orthogonal to $\{\mathfrak{M} - \pi^{-1}(S^2)\}$ (which is approx. the fiber). The metric on the fiber is $ds^2 = -U^{-1} dt^2 + 4\ell^2 U (d\Psi)^2$, so the vectors which span V_q are $\frac{\partial}{\partial t}$ and $\frac{\partial}{\partial \Psi}$. The metric on $\{\mathfrak{M} - \pi^{-1}(S^2)\}$ is

$$\overline{g}'_V = -U^{-1} dt \otimes dt + (2\ell)^2 U (d\Psi + \cos\theta \, d\phi) \otimes (d\Psi + \cos\theta \, d\phi)$$

(Eqn. 7-59)

Those vectors orthogonal to $\overline{g}_V$ are:

$$H_q: \left\{ {}^{\partial}/_{\partial\theta}, {}^{\partial}/_{\partial\phi} - \cos\theta \, {}^{\partial}/_{\partial\Psi} \right\}$$

(Eqn. 7-60)

Clearly $\pi_*\left(\frac{\partial}{\partial\theta}\right) = \frac{\partial}{\partial\theta}$ and $\pi_*\left\{ {}^{\partial}/_{\partial\phi} - \cos\theta \, {}^{\partial}/_{\partial\Psi} \right\} = {}^{\partial}/_{\partial\phi}$ span the base space. So, if every vector can be decomposed into these two spaces then

$$g(\vec{X}, \vec{Y}) = g_V(\vec{X}_V, \vec{Y}_V) + (t^2 + \ell^2) g_H(\pi_* \vec{X}_H, \pi_* \vec{Y}_H)$$

(Eqn. 7-61)

And we have a local direct sum. Although not orthogonal, $\partial / \partial\phi - \cos\theta \; \partial / \partial\Psi$ and $\partial / \partial\Psi$ do span the subspace spanned by $\partial / \partial\phi$ and $\partial / \partial\Psi$, so every vector can be decomposed.

Now, the interesting part of the metric g is contained in g_V since it corresponds to the part of $\mathfrak{M}$ which is <u>not</u> locally <u>in</u>extendible. Taub gave the metric for $U(t) > 0$ in which $t = const$ hypersurfaces are spacelike. Newman, Unti and Tamburino gave the region where $U(t) < 0$.

$$U(t) = -1 + \frac{2(mt + \ell^2)}{t^2 + \ell^2}$$

$$U(t) > 0 \rightarrow \quad \frac{2(mt + \ell^2)}{t^2 + \ell^2} > 1 \rightarrow \quad 2mt + \ell^2 > t^2 + \ell^2 \quad \rightarrow \quad 2mt + \ell^2 > t^2$$

Define $y(t) = t^2 - 2mt - \ell^2$, then $y'(t) = 2t - 2m$ and $y''(t) = 2$, with zero-crossings at:

$$t_{\pm} = m \pm \sqrt{m^2 + \ell^2}$$

and

$$U'(t) = 0 \;\text{ at }\; t_{1,2} = -\frac{\ell^2}{m} \pm \sqrt{\left(\frac{\ell^2}{m}\right)^2 + \ell^2}.$$

Thus,
$U(t) > 0 \Rightarrow y(t) < 0 \Rightarrow$ region between t_- and t_+.
$U(t) < 0 \Rightarrow y(t) > 0 \Rightarrow$ region to left of t_- and right of t_+.

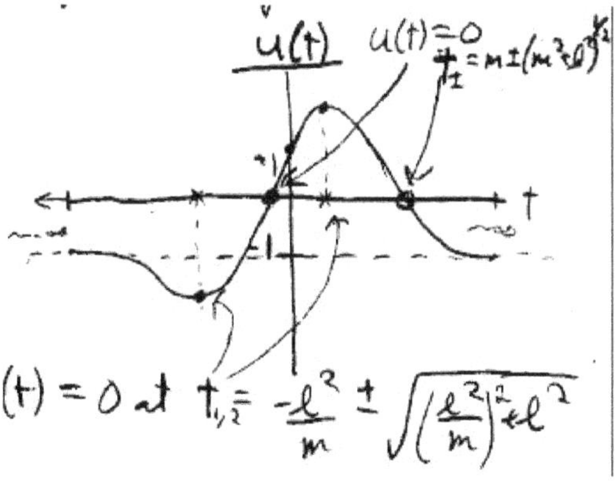

Fig. 7.11. Analysis of U(t).

Consider analytic extensions of $(\mathcal{F}, g_v)$.

g_v: $ds^2 = -U^{-1}dt^2 + 4\ell^2 U(d\Psi)^2$
g_v has singularities at $t = t_{\pm}$ where $U = 0$

Consider the manifold defined by Ψ and by $t_- < t < t_+$, $(\mathcal{F}_0, g_v)$ can be extended by defining

$$\Psi' = \Psi + \frac{1}{2\ell}\int \frac{dt}{U(t)}$$

(similar to Misner case where $\log t = \int \frac{dt}{t}$). The metric becomes:

$$g_v' : ds^2 = 4\ell d\Psi'(\ell U(t)d\Psi' - dt)$$

(Eqn. 7-62)

This is analytic on $\mathcal{F}'$ (with topology $S' \times R'$) defined by Ψ' and $-\infty < t < \infty$, as expected $t_- < t < t_+$ of $(\mathcal{F}', g_v')$ is isometric with $(\mathcal{F}_0, g_v)$.

We have ctc's in the extended regions. One family of null geodesics crosses both horizons $t = t_-$ and $t = t_+$ (two horizons unlike Misner's one) but the other family spirals round near these surfaces and is <u>incomplete</u>. Another extension follows from: $\Psi'' = \Psi - \frac{1}{2\ell}\int \frac{dt}{U(t)}$.

$$U(t) = -1 + \frac{2(mt + \ell^2)}{t^2 + \ell^2} \quad and \quad \frac{dU(t)}{dt} = \frac{2m}{t^2 + \ell^2} - \frac{2(mt + \ell^2)}{(t^2 + \ell^2)^2}(2t)$$

Ψ' extension: $ds^2 = 4\ell d\Psi'(\ell U(t)d\Psi' - \mathrm{dt})$, and null geodesics: $d\Psi' = 0 \rightarrow \Psi' = \text{const.}$, and have $\ell U(t)d\Psi' - \mathrm{dt} = 0 \rightarrow \frac{dt}{d\Psi'} = \ell U(t)$:

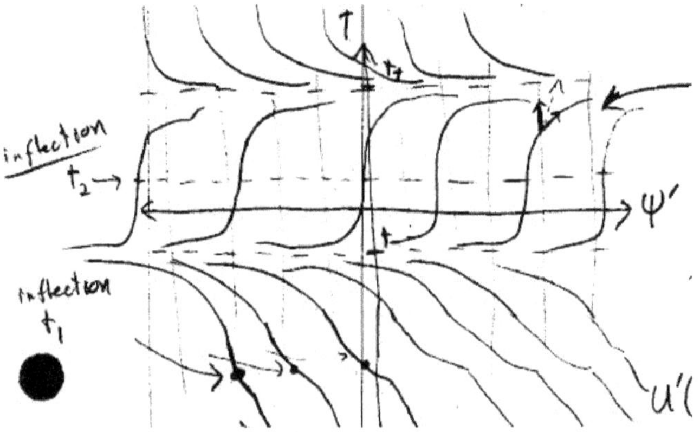

Fig. 7.12. Analysis of Characteristics

$$\frac{dt}{d\Psi'} < 0 \ for \ t < t_- \text{ and } t > t_+$$

$$\frac{dt}{d\Psi'} = 0 \text{ for } t = t_\pm$$

$$\frac{dt}{d\Psi'} > 0 \ for \ t_- < t < t_+$$

For Ψ'' extension: $ds^2 = 4\ell d\Psi''(\ell U(t)d\Psi'' + \mathrm{dt})$, flip above picture about t-axis.

The behavior is very much as with Misner space, one family of null geodesics in a given extension crosses both horizons $t = t_-$ and $t = t_+$ but the other family spirals round near these surfaces and is incomplete. To show incompleteness:

$$ds^2 = -U^{-1}dt^2 + 4\ell^2 U(d\Psi)^2$$

$$\Psi' = \Psi + \frac{1}{2\ell}\int \frac{dt}{U(t)} \qquad \Psi'' = \Psi - \frac{1}{2\ell}\int \frac{dt}{U(t)}$$

$$d\Psi' = d\Psi + \frac{1}{2\ell}\frac{1}{U(t)}dt \qquad d\Psi'' = d\Psi - \frac{1}{2\ell}\frac{1}{U(t)}dt$$

$$d\Psi' d\Psi'' = (d\Psi)^2 - \frac{1}{4\ell^2}\frac{1}{U(t)^2}(dt)^2$$

$$ds^2 = 4\ell^2 U(t) d\Psi' d\Psi''$$

$(\mathcal{F}_0, g_v)$ is $0 \le \Psi \le 4\pi$ and $t_- < t < t_+$ with $ds^2 = -U^{-1}dt^2 + 4\ell^2 U(d\Psi)^2$

Covering space of $\mathcal{F}_0$ is $\widetilde{\mathcal{F}_0}$ defined by

$$-\infty < \Psi < \infty \quad \text{and} \quad t_- < t < t_+$$

Using the above double null form we have the domain

$$-\infty < \Psi' < \infty \quad \text{and} \quad -\infty < \Psi'' < \infty.$$

Define new coordinates:

$$U_\pm = \tan^{-1}(\exp(\Psi'/\alpha_\pm)) \;\; and \;\; V_\pm = \tan^{-1}(-\exp(\Psi''/\alpha_\pm))$$

where

$$-\infty < \Psi' < \infty \Rightarrow 0 < U_\pm < {}^{\pi}/_{2} \;\; and \;\; -\infty < \Psi'' < \infty \Rightarrow -\frac{\pi}{2} < V_\pm < 0.$$

Thus

$$dU_\pm = \frac{1}{1 + (\exp(\Psi'/\alpha_\pm))^2}\exp(\Psi'/\alpha_\pm)\left\{\frac{d\Psi'}{\alpha_\pm}\right\}$$

$$dV_\pm = \frac{1}{1 + (\exp(-\Psi''/\alpha_\pm))^2}(-\exp(-\Psi''/\alpha_\pm))\left\{\frac{-d\Psi''}{\alpha_\pm}\right\}$$

So,

$$\tan(U_\pm) = \exp(\Psi'/\alpha_\pm) \to \Psi' = \alpha_\pm \ln(\tan(U_\pm)) \to d\Psi' = \alpha_\pm \frac{\sec^2(U_\pm)}{\tan(U_\pm)} dU_\pm$$

Thus

$$d\Psi' = 2\alpha_\pm \csc(2U_\pm) dU_\pm$$

Similarly:

$$\tan(V_\pm) = -\exp(-\Psi''/\alpha_\pm) \to \Psi'' = -\alpha_\pm \ln(-\tan(V_\pm)) \to d\Psi'' = -\alpha_\pm \frac{\sec^2(V_\pm)}{\tan(V_\pm)} dV_\pm$$

and

$$d\Psi'' = -2\alpha_\pm \mathrm{cosec}(2V_\pm) dV_\pm.$$

$$ds^2 = 4\ell^2 U(t) d\Psi' d\Psi'' = 4\ell^2 U(t)(-4\alpha_\pm^2)\mathrm{cosec}(2U_\pm)\mathrm{cosec}(2V_\pm) dU_\pm dV_\pm$$

with

$$U(t) = -1 + \frac{2(mt + \ell^2)}{t^2 + \ell^2}$$

and

$$\alpha_+^2 = \left(\frac{t_+ - t_-}{4\ell(mt + \ell^2)}\right)^2 \qquad \alpha_-^2 = \left(\frac{t_+ - t_-}{n4\ell(mt + \ell^2)}\right)$$

$$U(t_\pm) = 0 \Rightarrow t_\pm^2 + \ell^2 = 2(mt_\pm + \ell^2)$$

$$\alpha_+^2(t_+) = \left(\frac{t_+ - t_-}{2\ell(t_+^2 + \ell^2)}\right)$$

The transformation with $U_\pm, V_\pm$ is to study the structure at infinity. In such a Penrose representation the extension is fairly clear. Covering space of $\mathcal{F}_0$ is $\widetilde{\mathcal{F}_0}$ defined by

$$-\infty < \Psi < \infty \quad \text{and} \quad t_- < t < t_+$$

On $\widetilde{\mathcal{F}_0}$ the metric is written in double null form with

$$-\infty < \Psi' < \infty \quad \text{and} \quad -\infty < \Psi'' < \infty$$

Penrose transformation:

$$-\infty < \Psi' < \infty \Rightarrow 0 < U_\pm < {}^{\pi}/_{2}$$

$$-\infty < \Psi'' < \infty \Rightarrow -\frac{\pi}{2} < V_\pm < 0$$

Consider $U(t)$ *near* t_-:

$$U(t) \simeq \left\{\frac{2m}{t_-^2 + \ell^2} + \frac{2(2t_-)}{(t_-^2 + \ell^2)}\right\}(t - t_-) \simeq \frac{(m + 2t_-)}{(mt_- + \ell^2)}(t - t_-)$$

$$\int_{t_-+\xi}^{t} \frac{dt}{U(t)} \simeq \frac{(mt_- + \ell^2)}{(m + 2t_-)} \int_{t_-+\xi}^{t} \frac{dt}{(t - t_-)} \simeq \beta[\ln(t - t_-)]|_{t_-+\xi}^{t} \cong \beta \ln\left(\frac{\alpha}{\xi}\right)$$

as $\xi \to 0, t \to t_-$, we see that $\int \frac{dt}{U(t)} \Rightarrow \infty$. Thus:

$\Psi'' \to -\infty$ follows from $t \to t_-$, thus $V_\pm \to -\frac{\pi}{2}$

$\Psi' \to +\infty$ follows from $t \to t_-$, thus $U_\pm \to \frac{\pi}{2}$

Similarly for $t \to t_+$ we see that $\int \frac{dt}{U(t)} \Rightarrow -\infty$

$\Psi'' \to +\infty$ follows from $t \to t_+$, thus $V_\pm \to 0$

$\Psi' \to -\infty$ follows from $t \to t_+$, thus $U_\pm \to 0$

Now consider the $\mathcal{F}_0$ domain:

$$\mathcal{F}_0: \quad 0 \le \Psi \le 4\pi, \qquad t_- < t < t_+$$

with $\Psi = 0$ and $\Psi = 4\pi$ identified $\Rightarrow$ so not simply connected, with covering space is given by $-\infty < \Psi < \infty$. In the region $t_- < t < t_+$ we can use both sets of null coordinates, thus arriving at the double null form for the metric, also, $-\infty < \Psi' < \infty$, $-\infty < \Psi'' < \infty$. Transforming to a Penrose type representation with coordinates (U_+, V_+) and (U_-, V_-) on $\widetilde{\mathcal{F}_0}$ we get the metric $\tilde{g}_v$:

$$ds^2 = 4\ell^2 U(t)(-4\alpha_\pm^2)\csc(2U_\pm)\csc(2V_\pm)dU_\pm dV_\pm$$

As $t \to t_-$: $\left\{\frac{1}{2\ell}\int_t^{t'} \frac{dt}{U(t)}\right\} \to +\infty \quad (t' > t_-)$

$$\Rightarrow \quad \left.\begin{matrix} \Psi' \to \infty \\ \Psi'' \to -\infty \end{matrix}\right\} \Rightarrow \begin{matrix} U_\pm \to \frac{\pi}{2} \\ V_\pm \to -\frac{\pi}{2} \end{matrix}$$

As $t \to t_+$: $\left\{\frac{1}{2\ell}\int_{t'}^{t} \frac{dt}{U(t)}\right\} \to -\infty \quad (t' > t_+)$

$$\Rightarrow \quad \left.\begin{matrix} \Psi' \to -\infty \\ \Psi'' \to +\infty \end{matrix}\right\} \Rightarrow \begin{matrix} U_\pm \to 0 \\ V_\pm \to 0 \end{matrix}$$

In the region $t_- < t < t_+$ $\quad t = const$ hypersurfaces are spacelike. From the extension $g_v'(11)$ it is also clear that $t = t_\pm$ hypersurfaces are null.

Thus,

$t = t_+$ hypersurface hase $\underbrace{U_\pm = 0}_{\Psi' \to -\infty}$, $-\frac{\pi}{2} \leq V_\pm \leq 0$

$\uparrow$ $\qquad\qquad$ $\uparrow \Psi$ finite to $\Psi = +\infty$

$\Psi'' \to -\infty$ while $-\left\{\frac{1}{2\ell}\int \cdots\right\} \to +\infty$

Thus $\Psi \to -\infty$ limit

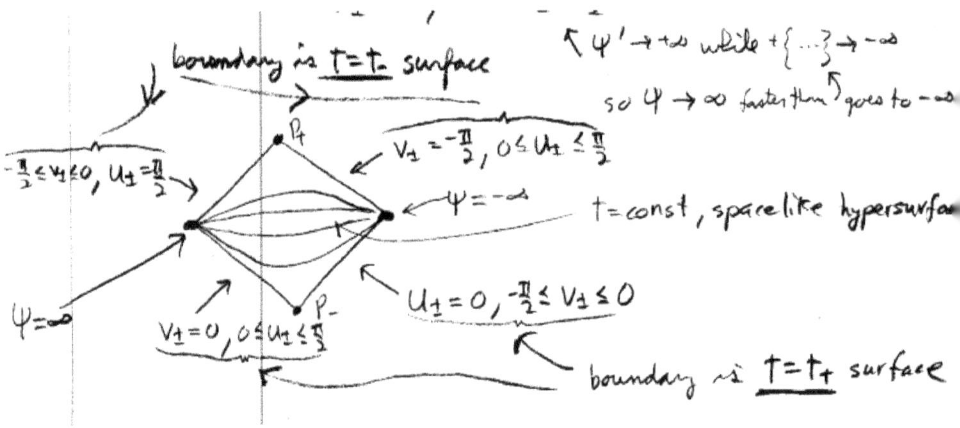

Fig. 7.13. Penrose Diagram.

On a given spacelike surface of the covering space we have $-\infty < \Psi < \infty$, which represents the orbit of a one-dimentional group of isometries, i.e., $\Psi = \Psi + n4\pi$.

Near P_+, P_- the action of the group is similar to that of the Lorentz group in 2-d Minkowski space:

$$ds^2 = 4\ell^2 U(t)(-4 \propto_\pm) \csc(2U_\pm) \csc(2V_\pm)\, dU_\pm dV_\pm$$

(Eqn. 7-63)

Recall our two inequivalent extensions, first consider $\mathcal{F}'$ which followed from Ψ' and letting $-\infty < t < \infty$. If we regard $\Psi' = \Psi + \frac{1}{2\ell}\int \frac{dt}{U(t)}$ once again but now consider the continuation to $t < t_-$ and $t > t_+$ we find for $t < t_-$, we've labled $U_\pm(t_-) = \frac{\pi}{2}$, consider the values of Ψ' for $t < t_-$:

$$\Psi'(t)|_{t=(t_-)^-} = \Psi + \{-\infty\}$$

$$g(\partial_\Psi, \partial_\phi - \cos\theta\, \partial_\Psi) = ?$$

$$\{(z_0, z_1) \mid |z_0|^2 + |z_1|^2 = 1\}$$

$U = \begin{pmatrix} z_0 & -\overline{z_1} \\ z_1 & \overline{z_0} \end{pmatrix} \rightarrow$ works as $SU(2)$

$S^3, SU(2)$ isomorphic

$z_0 = \cos\frac{\theta}{2} e^{i\frac{\Psi-\phi}{2}}$

$z_1 = \sin\frac{\theta}{2} e^{i\frac{\Psi+\phi}{2}}$

$(z_0, z_1) \Rightarrow (\theta, \phi)$

Action of $U(1)$ on (z_0, z_1):

$(z_0, z_1) \longrightarrow (Uz_0, Uz_1)$ with $U = e^{i\eta}$

still projects down to (θ, ϕ)! $\begin{cases} z_0 \longrightarrow \cos\theta\, e^{i^{(\Psi+2\eta)}/_2 - {}^{\phi}/_2} \\ z_1 \longrightarrow \sin\theta\, e^{i^{(\Psi+2\eta)}/_2 + {}^{\phi}/_2} \end{cases}$ grouping forced

So η ranging from $0 \longrightarrow 2\pi$ sweeps "Ψ" through all values.

Misner is covered by $t < -x$ region of Minkowski $\left(\sim {}^1/_2\right)$. Covering space of X is a pair $\left(\tilde{X}, f\right)$ where $\tilde{X}$ is a connected and locally connected space, and f a continuous mapping of $\tilde{X}$ onto X, such that, each point $x \in X$ has a neighborhood $v(x)$ such that the restriction of f to each connected component $C_\propto$ of $f^{-1}\big(v(x)\big)$ is a homeomorphism from $C_\propto$ onto $v(x)$. So, as long as the boost identification is finite each point $x \in X$ has a neighborhood $v(x)$ such that the restriction of f to each connected component $C_\propto$ of $f^{-1}\big(v(x)\big)$ is a homeomorphism for $C_\propto$ onto $v(x)$. Thus Misner space is covered by half of Minkowski.

7.2.7 Killing Vector Structure of Taub

Metric g_{ab}:

$$d\ell^2 = a^2(d\theta^2 + \sin^2\theta\, d\phi^2) + b^2(d\Psi + \cos\theta\, d\phi)^2$$
$$= a^2[(\sigma^1)^2 + (\sigma^2)^2] + b^2(\sigma^3)^2$$

(Eqn. 7-64)

With

$\sigma^1 = \cos\Psi\, d\theta + \sin\Psi \sin\theta\, d\phi$

$\sigma^2 = \sin\Psi\, d\theta - \cos\Psi \sin\theta\, d\phi \quad or \quad \sigma^\pm = \sigma^1 \pm \sigma^2$

$\sigma^3 = d\Psi + \cos\theta\, d\phi$

$$= e^{\pm i\Psi}(d\theta \mp i \sin\theta\, d\phi)$$

(Eqn. 7-65)

If K_{ab} is the Killing metric on SU(2), $K_{ab} = \delta_{ij}\sigma_a^i\sigma_b^j$, and have

$$g_{ab} = a^2(K_{ab} - \sigma_a^3\sigma_b^3) + b^2\sigma_a^3\sigma_b^3$$

(Eqn. 7-66)

If we write $\Psi^a = \frac{\partial}{\partial\Psi}$, $\Psi_a = K_{ab}\Psi^b$, then $\Psi_a = \sigma_a^3$, then

$$g_{ab} = a^2(K_{ab} - \Psi_a\Psi_b) + b^2\Psi_a\Psi_b.$$

(Eqn. 7-67)

Now $\{\sigma_a^i\}$ is a right invariant basis for SU(2); and K_{ab} is both left and right invariant under SU(2), so g_{ab} is left invariant under the U(1) subgroup of SU(2) that fixes Ψ_a. Symmetry group of Taub is thus SU(2) × U(1).

7.3 CTC Preliminary analysis

7.3.1 Initial Taub generalization

In formulating the closed time-like curve (CTC) problem we seek a simple space-time that in a bounded region admits CTC's. A cylindrical symmetry is desired throughout. Also, it is important that the metric is C^2 (differentiable to second order) so that subsequent calculations with the Riemann tensor are simple. Having decided on such space-time we want to know if the weak energy condition is satisfied on the CTC or throughout the space-time in general. We are also interested in the behavior of the stress energy tensor at the Cauchy Horizon associated with the CTC (especially if the weak energy condition is satisfied,).

Before examining the stress energy problem for CTC's as described above, it is instructive to first consider the Misner space-time as example. In the 4D Misner example there are competing processes to determine whether or not the stress-energy is divergent near the CTC (classically). There is a blue-shifting factor associated with the CTC and a cross-sectional area expansion associated with its repeated projections. The necessary stress-energy divergence is then an indication of positive energy violation.

2D-Misner space-time instability

2D-Misner space-time is a manifold with topology $\mathbb{R} \times S$, which is a metric space. It will demonstrate some of the distinctive behavior that we must contend with in full 4D space-time CTC analysis., in particular, closed-null-geodesic (cng) instabilities. Here's 2D-Misner:

$$ds^2 = -t^{-1}dt^2 + td\psi^2, \quad 0 \le \psi \le 2\pi \quad , \quad 0 < t < \infty.$$

Picture a cylinder to see this:

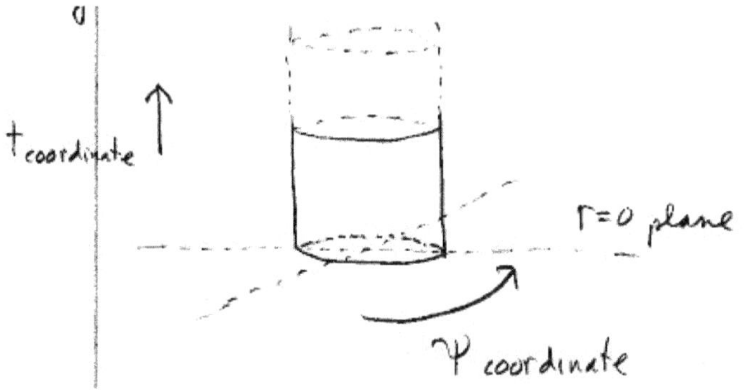

Fig. 7.14. Misner embed in 2D-Minkowski.

Let's consider how Misner embeds in 2D-Minkowski space-time. Consider the coordinate transformation:

$$t = \frac{1}{4}(\tilde{t}^2 - \tilde{x}^2), \quad \psi = 2\tanh^{-1}\frac{\tilde{x}}{\tilde{t}} .$$

We now have 2D-Minkowski:

$$ds^2 = -d\tilde{t}^2 + d\tilde{x}^2$$

The domain $0 \le \psi \le 2\pi,\ 0 < t < \infty$ becomes $\tilde{t} > |\tilde{x}|$. In other words, Misner is covered by the future light cone of 2D-Minkowski space-time. But we have an identification of points due to the periodicity in $\psi = \psi + n2\pi$, thus, the points related by boost $\beta = \tanh \pi$ are identified:

$$(\tilde{t}, \tilde{x}) = (\tilde{t}\cosh n\pi + \tilde{x}\sinh n\pi, \quad \tilde{x}\cosh n\pi + \tilde{t}\sinh n\pi).$$

So, 2D-Misner is the quotient space of the future light cone of 2D-Minkowski by the discrete subgroup of the Lorentz group consisting of boosts by $\beta = \tanh \pi$.

Misner Space-time is geodesically incomplete (meaning I can 'walk' right out of this place):

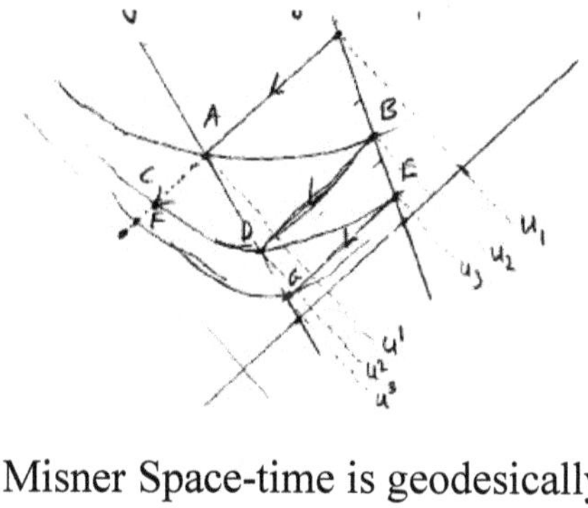

Fig. 7.15. Misner Space-time is geodesically incomplete.

$$u_{total} = (u_1 - u^1) + \frac{1}{\beta}(u_2 - u^2) + \frac{1}{\beta^2}(u_3 - u^3) + \cdots \ , \ \beta > 1.$$

(Eqn. 7-68)

We have the set of relations: $u_n = \beta u^{n-1}$, so:

$$u_{total} = u_1$$

(Eqn. 7-69)

by rearrangement, which can be done if u_{total} is finite. Now to check if finite:

$$u_{total} < (u_1 - u^1)(1 + \frac{1}{\beta} + \frac{1}{\beta^2} + \cdots).$$

Since $\beta > 1$, we have:

$$u_{total} < \frac{(u_1 - u^1)}{1 - \frac{1}{\beta}} \ ,$$

(Eqn. 7-70)

thus u_{total} is finite and $u_{total} = u_1$.

Note, as indicated in the sketch, there is another way to perform the sum by using the boost relations on variables u and v:

$$u = t + x \to u(\cosh\psi + \sinh\psi) = u\beta \ ,$$
$$v = t - x \to u(\cosh\psi - \sinh\psi) = v\alpha \ .$$

(Eqn. 7-71)

Let's shift all of the segments to the same line as shown in the sketch:
(i) Boost A to B: $u_2 = \beta u^1$, $v_2 = \alpha v^1$.
(ii) Let the ull ray propagate from B to D: $v_2 = v^2$.
(iii) Boost back to original frame: $v_C = \frac{1}{\alpha} v^2 = v^1$.
Thus,

$$u_1 - u_C = \frac{1}{\beta}(u_2 - u^2)$$

and

$$u_C - u_F = \frac{1}{\beta^2}(u_3 - u^3), \ldots$$

Again summing to get

$$u_{total} = u_1 \ .$$

Now consider the implications of an incomplete geodesic, where we reach the boundary of our space-time, when this is accomplished with an infinite number of boosts. To be specific, consider a packet of energy, a photon say, moving along the incomplete geodesic described. With each identification its energy will be boosted by the β factor. Thus, in finite parameter distance the energy diverges, and this takes place near the edge of the manifold (the light cone) where the identification is practically that of a closed null geodesic (cng). This describes the classical instability that is inherent to a space-time with cng's.

Unlike the 2D-Misner space-time, however, in a full four dimensional theory we can have defocusing which may conspire to remove the instability. The resulting divergence of null rays, however, is indicative of a weak energy violation.

In 4D, does a first CTC, a cng, have the same blue-shifting instability as described in Misner space-time? Consider the metric

$$ds^2 = f(-dt^2 + (\rho d\phi)^2) - (1 - f^2)dt(\rho d\phi) + d\rho^2 + dz^2.$$

(Eqn. 7-72)

At $f = 1$: $ds^2 = -dt^2 + (\rho d\phi)^2 + d\rho^2 + dz^2$ (the Minkowski metric in cylindrical coordinates).
At $f = 0$: $ds^2 = -dt\rho d\phi + d\rho^2 + dz^2$ (here the light-cone has tilted so much that it admits a cng).
If $f < 0$. then CTC occur.

Choose f such that for a given observer at rest in the asymptotically flat region the first cng is simply:

$$C(\varphi(\lambda), t = t_0, \rho = \rho_0, z = z_0).$$

Now, $\nabla_u \vec{u} = 0$ for affinely parameterized C, $\vec{u} = \frac{dC}{d\lambda}$, which can be written in index notation as $u^\alpha{}_{;\beta}u^\beta = 0$ and only $u^\varphi \neq 0$. So

$$u^\varphi \frac{\partial}{\partial\varphi} u^\varphi = -\Gamma^\varphi{}_{\varphi\varphi}\big|_G u^\varphi u^\varphi, \quad where\ \Gamma^\varphi{}_{\varphi\varphi}\big|_G = const.\ due\ to\ cylindrical\ sym.$$

Thus,

$$u^\varphi = Aexp\left(\left[-\Gamma^\varphi{}_{\varphi\varphi}\big|_G\right]\varphi\right) = \exp\left[-\rho_0\left(\frac{\partial f}{\partial t}\right)\varphi\right] \quad (choosing\ A = 1).$$

(Eqn. 7-73)

So, in general we have $(\partial f/\partial t) \neq 0$ which will then yield the familiar blue-shifting instability depending on which way the cng is traversed

(with respect to the sign of $(\partial f/\partial t)$. The relation to Misner becomes clear if one notes that for certain choices of f the neighborhood of the cng is
$\mathbb{R}^2 \times Misner$. Thus, the cng in such a space-time has a similar blue-shift divergence to that of Misner space-time. A divergence of null rays can counteract this but this usually implies a divergence of the weak energy condition. Note, however, that from

$$\frac{d\theta}{dv} = -R_{ab}k^a k^b - 2\sigma^2 - \frac{1}{2}\theta^2 + \omega^2,$$

(Eqn. 7-74)

and in the instance where the vorticity is large, the blue-shift may be counteracted without necessitating weak energy violation at the cng. This is not to say that there isn't still weak energy violation somewhere in the space-time.

To recap where we are so far, the space-time studied has metric:

$$ds^2 = f(-dt^2 + (\rho d\phi)^2) - (1 - f^2)dt(\rho d\phi) + d\rho^2 + dz^2$$

For which at $f = 1$: $ds^2 = -dt^2 + (\rho d\phi)^2 + d\rho^2 + dz^2 \rightarrow$ Minkowski spacetime in cylindrical coordinates. And, at $f = 0$ we have: $ds^2 = -dt\rho d\phi + d\rho^2 + dz^2$: The lightcone for $f = 0$ has tipped so much that it admits a closed null geodesic. If f goes < 0 for some values of its argument, as it does in general, we have closed time-like curves.

For $ds^2 = -dt(\rho d\phi) + d\rho^2 + dz^2$ we have for the future oriented, pure time-like vector:

$$t^a = \frac{1}{\sqrt{2}}\left(\frac{\partial}{\partial t} + \frac{1}{\rho}\frac{\partial}{\partial \phi}\right) \Rightarrow (t^a, t^a) = -1$$

And the mutually orthogonal spacelike vectors are

$$z^a = \frac{1}{\sqrt{2}}\left(\frac{\partial}{\partial t} - \frac{1}{\rho}\frac{\partial}{\partial \phi}\right)$$

$$x^a = \frac{\partial}{\partial \rho}$$

$$y^a = \frac{\partial}{\partial z}$$

Let us consider the general null vector: $k^a = t^a + z^a \cos\alpha + (\cos\beta\, x^a \sin\beta\, y^a)\sin\alpha$, in regards to the null convergence condition $R_{ab}K^aK^b \geq 0$. From the Einstein equations, $R_{ab} - \frac{1}{2}g_{ab}R + \Lambda g_{ab} = 8\pi T_{ab}$, this condition implies the weak energy condition.

Let $x_1 = \alpha,\quad x_2 = \sin\alpha\cos\beta,\quad x_3 = \sin\alpha\sin\beta,\quad (x_1^2 + x_2^2 + x_3^2 = 1)$ then

$$R_{\mu\nu}K^\mu K^\nu = \underbrace{\left(\frac{1}{2}R_{tt} + \frac{1}{\rho}R_{t\phi} + \frac{1}{2\rho^2}R_{\phi\phi}\right)}_{A} + \underbrace{\left(R_{tt} - \frac{1}{\rho^2}R_{\phi\phi}\right)}_{B}x_1$$

$$+ \underbrace{\left(\frac{1}{2}R_{tt} - \frac{1}{\rho}R_{t\phi} + \frac{1}{2\rho^2}R_{\phi\phi}\right)x_1^2}_{H}$$

$$+ \underbrace{\left(\sqrt{2}R_{tp} + \frac{1}{\rho}\sqrt{2}R_{\phi\rho}\right)}_{C}x_2 + \underbrace{\left(\sqrt{2}R_{tz} + \frac{1}{\rho}\sqrt{2}R_{\phi z}\right)}_{D}x_3$$

$$+ \underbrace{\left(\sqrt{2}R_{t\rho} - \frac{1}{\rho}\sqrt{2}R_{\phi\rho}\right)}_{E}x_1x_2$$

$$+ \underbrace{\left(\sqrt{2}R_{tz} - \frac{1}{\rho}\sqrt{2}R_{\phi z}\right)}_{F}x_1x_3 + \underbrace{2R_{\rho z}}_{G}x_2x_3 + \underbrace{R_{\rho\rho}}_{I}x_2^2 + \underbrace{R_{zz}}_{J}x_3^2$$

$$= A + Bx_1 + Cx_2 + Dx_3 + Ex_1x_2 + Fx_1x_3 + Gx_2x_3 + Hx_1^2 + Ix_2^2 + Jx_3^2$$

(Eqn. 7-75)

Consider $x_2 = x_3 = 0$ then we have two inequalities corresponding to the $x_1 = \pm 1$ cases:

$$\left.\begin{array}{l} A + B + H \geq 0 \\ A - B + H \geq 0 \end{array}\right\} \underline{A + K \geq |B|}$$

(Eqn. 7-76)

Thus, even in case $B = 0$ we must still satisfy $\underline{A + H \geq 0}$.

$$A + H = R_{tt} + \frac{1}{\rho^2}R_{\phi\phi} \geq 0$$

(Eqn. 7-77)

After a wee bit of work this is:

$$R_{tt} + \frac{1}{\rho^2} R_{\phi\phi} = \frac{-2}{\rho(f^2+1)} \left(\frac{\partial f}{\partial \rho}\right) \geq 0 \quad \text{thus} \quad \boxed{-\left(\frac{\partial f}{\partial \rho}\right) \geq 0}$$

(Eqn. 7-78)

In order for our space-time to be asymptotically flat we must have $f(\rho \to \infty) = 1$, for well-definedness at $\rho = 0$ we must also have $f(\rho \to 0) = 1$. To get a cng or ctc's we must have $f(\rho_c) = 0$ at some critical ρ. It is thus clear from these arguments that condition on $(\partial f / \partial \rho)$ above cannot be generically satisfied for this class of spacetimes. If we were just interested in satisfying the $(\partial f / \partial \rho)$ condition at the cng which generates the Cauchy Horizon we would have to adjust f such that only one cng generates the Cauchy Horizon (of the later cng's, that might have been part of the Cauchy Horizon, generally half don't satisfy $-(\partial f / \partial \rho) \geq 0$). The one cng that then defines the Cauchy Horizon then has $(\partial f / \partial \rho) = 0$ this being a special limiting case. There is a further restriction in this instance:

On the first cng we clearly have $f = 0$, for which

$$\left.\begin{aligned} A + B + H = 2R_{tt} &= \left(\frac{\partial^2 f}{\partial \rho^2}\right) + \left(\frac{\partial^2 f}{\partial z^2}\right) - \left(\frac{\partial f}{\partial \rho}\right) \geq 0 \\ A - B + H = 2\frac{1}{\rho^2} R_{\phi\phi} &= \left(\frac{\partial^2 f}{\partial \rho^2}\right) - \left(\frac{\partial^2 f}{\partial z^2}\right) - 3\rho \left(\frac{\partial f}{\partial \rho}\right) \geq 0 \end{aligned}\right\} \text{new condition: } \left(\frac{\partial^2 f}{\partial \rho^2}\right) + \left(\frac{\partial^2 f}{\partial z^2}\right) =$$

(Eqn. 7-79)

Thus, in order to get a minima, $f = 0$, at ρ_c we musthave $\frac{\partial^{2n} f}{\partial \rho^{2n}} > 0 \;\; n \geq 2$, $\frac{\partial^{2m} f}{\partial z^{2m}} > 0 \;\; m \geq 2$, and $\frac{\partial^2 f}{\partial z^2} = 0$.

Thus, weak energy is satisfied on the single cng which defines the Cauchy Horizon for a very restricted class of $f(\rho, z, t)$. But, in general, the weak energy condition is not satisfied throughout the space-time for the entire class of space-times represented by the metric.

Preliminary Conclusion

A careful analysis will be done on the above, but some preliminary conclusions are now apparent: macroscopic CTC's require weak energy violation, agreeing with the Chronology Protection Conjecture [28].

7.3.2 tanh3 redo

Consider a metric that admits CTC's only at a particular instant in a small toroidal region; also want the metric to be C^2so that subsequent calculations with Riemann are possible. So, want a function $f(\rho, z)$ such that at a particular z we have:

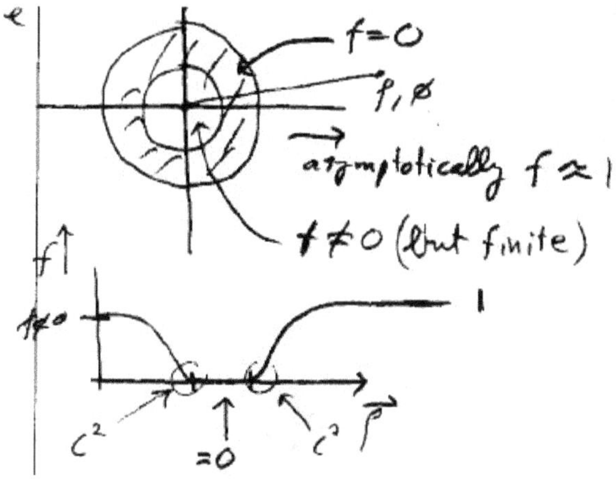

Fig. 7.16. A toroidal CTC region.

Since the metric is toroidal we have a $\rho^2 + z^2$ factor, so consider:

$$f(\rho, z) = \begin{cases} 0 & for \;\; [(\rho - \rho_0)^2 + z^2] < \alpha^2 \\ \tanh^3([(\rho - \rho_0)^2 + z^2] - \alpha^2) & for \;\; [(\rho - \rho_0)^2 + z^2] \geq \alpha^2 \end{cases}$$

(Eqn. 7-80)

This function is C^2. Only want CTC's at a particular instant and want the transition to be at least C^2. Let's compare a metric that admits toroidal CTC's with Minkowski:

Normal Minkowski: $ds^2 = dt^2 - dp^2 - dz^2 - (\rho d\phi)^2$

(Eqn. 7-81)

Toroidal: $ds^2 = dt(\rho d\phi) - dp^2 - dz^2$ (increasing ϕ null geodesic)

(Eqn. 7-82)

Given the above, for the region to exist for only a finite time, we need to let α be a function of t. There is the following possibility:

$$ds^2 = f(\rho, z, t)\{dt^2 - dp^2 - dz^2 - (\rho d\phi)^2\} + [1 - f(\rho, z, t)]\{dt(\rho d\phi) - dp^2 - dz^2\}$$
$$f(\rho, z, t) = \tanh^3([\rho - \rho_0]^2 + z^2)$$

which simplifies to:

$$ds^2 = f(\rho, z, t)\{dt^2 - (\rho d\phi)^2\} + [1 - f(\rho, z, t)]\{dt(\rho d\phi)\}$$
$$f(\rho, z, t) = \tanh^3([\rho - \rho_0]^2 + z^2)$$

(Eqn. 7-83)

Here's a sketch of the light cones:

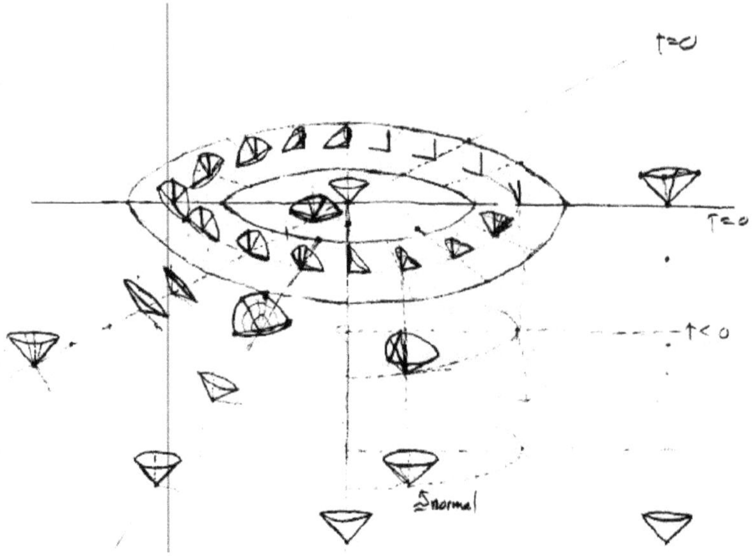

Fig. 7.17. Light cones in the tanh3 CTC spacetime.

The Cauchy Horizon exists for $|t| < t_0$ in the region $[\rho - \rho_0]^2 + z^2 < \alpha^2$. The first CTC occurs at $t = -t_0$, $\rho = \rho_0$, and $z = 0$:

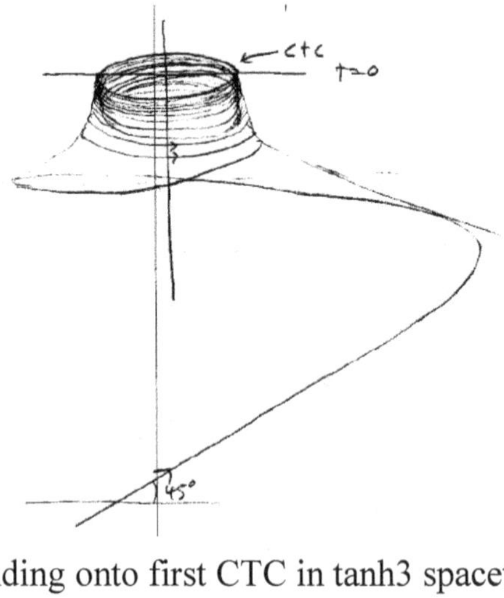

Fig. 7.18. Winding onto first CTC in tanh3 spacetime.

7.3.3 Blue Shift Analysis for tanh3

Let's start from the cylindrical space-time:

$$ds^2 = f(\rho, z, t)\{dt^2 - (\rho d\phi)^2\} + [1 - f(\rho, z, t)]\{dt(\rho d\phi)\}$$

(Eqn. 7-84)

where

$$f(\rho, z, t) = \begin{cases} 0 & [(\rho - \rho_0)^2 + z^2 + t^2] < \alpha^2 \\ \tanh^3((\rho - \rho_0)^2 + z^2 + t^2 - \alpha^2) & [(\rho - \rho_0)^2 + z^2 + t^2 \geq \alpha^2 \end{cases}$$

(Eqn. 7-85)

Normal Minkowski in Cylindrical coordinates has: $ds^2 = dt^2 - d\rho^2 - dz^2 - (\rho d\phi)^2$, while the above metric admits toroidal cng's (actually null): $ds^2 = dt(\rho d\phi) - d\rho^2 - dz^2$. The first CTC (cng) occurs at $t = -\alpha, \rho = \rho_0, z = 0$.

Now that we've identified the curve of interest, we would like to calculate the tangent vector of the CTC as it is parallel transported about the CTC to its starting point (so as to extract the blue shift factor on one loop). Let's call the curve of interest $e(\phi)$, where $t = -\alpha, \rho = \rho_0, z = 0$. Let's derive the $\Gamma's$ and $R's$ to proceed. Note $\vec{u} = \frac{de(\phi)}{d\phi}$ and $\nabla_{\vec{u}}\vec{u} = 0$:

$$0 = u^\alpha_{;\beta}u^\beta = \left(u^\alpha_{,\beta} + \Gamma^\alpha_\gamma u^\gamma\right)u^\beta = u^\alpha_{,\phi} + \Gamma^\alpha_\gamma u^\gamma = u^\alpha_{,\phi} + \Gamma^\phi_{\phi\phi}u^\phi$$

(Eqn. 7-86)

Only nonzero is u^ϕ. (Recall in coordinates t, ρ, z, ϕ: $u^\phi = \frac{\partial}{\partial\phi}$ and $u^\beta = \frac{\partial e(\phi)}{\partial x^\beta}$).

We now have the curve $e(\lambda)$ of interest, and it is a geodesic. So consider a coordinate system $\{x^\alpha(e)\}$ then the geodesic eqn. gives

$$\frac{d^2x^\alpha\{e(\lambda)\}}{d\lambda^2} + \Gamma^\alpha_{\gamma\beta}\frac{dx^\gamma}{d\lambda}\frac{dx^\beta}{d\lambda} = 0.$$

(Eqn. 7-87)

As we move along $e(\lambda)$, for various values of λ, only the $x^\phi\big(e(\lambda)\big)$ components are changing in the ϕ, t, ρ, z system. So,

$$\frac{dx^\phi}{d\lambda} = \frac{d\phi}{d\lambda}\frac{d}{d\phi}x\{e(\lambda)\}, \quad where \quad e(\lambda)$$
$$= \delta(\rho - \rho_0)\delta(z)\delta(t + \alpha) \times \{a\phi(\lambda) + b\},$$
$$where\ a = 1\ and\ b = 0\ chosen.$$

(Eqn. 7-88)

The geodesic equation:

$$\frac{d^2x^\phi}{d\lambda^2} + \Gamma^\phi_{\phi\phi}\frac{dx^\phi}{d\lambda}\frac{dx^\phi}{d\lambda} = 0$$

(Eqn. 7-89)

then becomes:

$$\frac{du^\phi}{d\phi} + \Gamma^\phi_{\phi\phi} = 0.$$

(Eqn. 7-90)

So, what is $\Gamma^\phi_{\phi\phi}$? Consider:

$$ds^2 = fdt^2 - f\rho^2 d\phi^2 + (1-f)\rho dt d\phi - d\rho^2 - dz^2, and\ note\ that\ ds^2 = g_{\mu\nu}dx^\mu dx^\nu,$$

where $f = f(\rho, z, t)$. We thereby obtain:

$g_{tt} = f$
$g_{t\phi} = g_{\phi t} = \frac{1}{2}(1-f)\rho$

$g_{\rho\rho} = -1, g_{zz} = -1$
$g_{\phi\phi} = -f\rho^2$
and the rest are zero.

Now using $\Gamma_{\mu\beta\gamma} = \frac{1}{2}\left(g_{\mu\beta,\gamma} + g_{\mu\gamma,\beta} - g_{\beta\gamma,\mu}\right)$, we obtain:

$$\Gamma^\phi_{\phi\phi} = g^{\phi\mu}\Gamma_{\mu\phi\phi} = g^{\phi\mu}\frac{1}{2}\left(-g_{\phi\phi,\mu}\right).$$

(Eqn. 7-91)

To proceed we need $g^{\mu\nu}g_{\nu\gamma} = \delta^\mu_\gamma$, So, let

$$g^{\mu\nu} = \begin{pmatrix} A & B & 0 \\ C & D & \\ 0 & & -1 \end{pmatrix}$$

Need to invert:

$$\begin{pmatrix} f & \frac{1}{2}(1-f)\rho \\ \frac{1}{2}(1-f)\rho & -f\rho \end{pmatrix} = \begin{pmatrix} a & b \\ b & c \end{pmatrix}$$

$$\begin{pmatrix} A & B \\ C & D \end{pmatrix}\begin{pmatrix} a & b \\ b & c \end{pmatrix} = \begin{pmatrix} Aa + Bb & Ab + Bc \\ Ca + Db & Cb + Dc \end{pmatrix}$$

$$Aa + Bb = 1 \quad Ab + Bc = 0 \qquad A = -\frac{c}{b}B$$
$$Cb + Dc = 1 \quad Ca + Db = 0 \qquad C = -\frac{b}{a}D$$

$$\left(-\frac{ac}{b} + b\right)B = 1 \quad B = \frac{b}{b^2 - ac}$$
$$\left(-\frac{b^2}{c} + c\right)D = 1 \quad D = \frac{a}{b^2 - ac}$$

Thus,

$$\begin{pmatrix} A & B \\ C & D \end{pmatrix} = \frac{1}{\det\begin{pmatrix} a & b \\ b & c \end{pmatrix}} \begin{pmatrix} c & -b \\ -b & a \end{pmatrix}$$

and we have:

$$g^{\mu\nu} = \begin{pmatrix} \frac{-f\rho^2}{-\rho^2\left(f^2+\frac{1}{4}(1-f)^2\right)} & \frac{-\frac{1}{2}(1-f)\rho}{-\rho^2\left(f^2+\frac{1}{2}(1-f)^2\right)} & 0 & 0 \\ \frac{-\frac{1}{2}(1-f)\rho}{-\rho^2\left(f^2+\frac{1}{2}(1-f)^2\right)} & \frac{f}{-\rho^2\left(f^2+\frac{1}{4}(1-f)^2\right)} & 0 & 0 \\ 0 & 0 & -1 & 0 \\ 0 & 0 & 0 & -1 \end{pmatrix}$$

The only $g^{\phi\mu}$ is $\mu = t$, so $\Gamma^{\phi}_{\phi\phi} = g^{\phi t}\frac{1}{2}\left(-g_{\phi\phi,t}\right)$ and:

$$\Gamma^{\phi}_{\phi\phi} = \frac{\frac{1}{2}(1-f)}{\rho\left(f^2+\frac{1}{2}(1-f)^2\right)}\frac{1}{2}\left(-\frac{\partial}{\partial t}(-f\rho^2)\right) = \frac{\frac{1}{4}(1-f)\rho}{\left(f^2+\frac{1}{2}(1-f)^2\right)}\left(\frac{\partial f}{\partial t}\right)$$

(Eqn. 7-92)

Consider, specifically, the null geodesic that winds onto the CTC of interest. There exists such a geodesic, $e(\lambda)$, such that for (affine) parameter $\lambda > \lambda_0$, $\frac{dx^{\alpha}\{e(\lambda)\}}{d\lambda}$ has in the coord. ϕ, t, ρ, z:

$$\frac{d\phi\{e(\lambda)\}}{d\lambda} \gg \frac{dt\{e(\lambda)\}}{d\lambda} \text{ or } \frac{d\rho\{e(\lambda)\}}{d\lambda} etc.,$$

such that for any $\epsilon > 0$, $\left|\frac{dt[e(\lambda)]}{d\lambda}\right| < \epsilon$, same for ρ and z.

This follows by definition of the curve that has the CTC as limit curve. (That curve being sliced along a given ϕ and the broken loops arrived at labeled C_n, such that for every $P\epsilon$ CTC, C_n converges to P, [39]). So, in such a case we have

$$\frac{du^{\phi}}{d\phi} + \Gamma^{\phi}_{\phi\phi} = 0$$

where $\Gamma^{\phi}_{\phi\phi}$ is implicitly evaluated along the $e(\lambda)$ what has

$$[(\rho - \rho_0)^2 + z^2 + t^2] \approx \alpha^2 + \delta \quad and\ \delta \text{ infinitesimal}$$

(Eqn. 7-93)

So, $f(\rho, z, t) = \tanh^3((\rho - \rho_0)^2 + z^2 + t^2 - \alpha^2)$ and on the latter marked part, call it x, $has\ x \ll 1$. Now have:

$$\frac{\partial f}{\partial t} = 3\tanh^2(x)\,\mathrm{sech}^2(x)\,(2t) \simeq 3x^2(2t)$$

(Eqn. 7-94)

$$\left.\Gamma^{\phi}_{\phi\phi}\right|_e \simeq \frac{\frac{1}{4}(1 - x^3)\rho}{\left(x^6 + \frac{1}{2}(1 - x^3)^2\right)}(3x^2)(2t) \simeq \frac{1}{2}\rho_0 \frac{1}{(1 - x^3)} 6x^2 t$$
$$\simeq 3\rho_0(2\alpha t')^2(-\alpha) = -3\rho_0\alpha\delta^2$$

(Eqn. 7-95)

So,

$\left.\Gamma^{\phi}_{\phi\phi}\right|_e \simeq -12\alpha^3\rho_0(t')^2$, thus

$$\frac{du^{\phi}}{d\phi} = 12\alpha^3\rho_0(t')^2 = 3\rho_0\alpha\delta^2$$

(Eqn. 7-96)

Let $\quad \left.U^{\phi}\right|_{\phi=\theta} = 3\rho_0\alpha\delta^2\theta + const.$

$\left.U^{\phi}\right|_{\phi=\theta+2\pi} = 3\rho_0\alpha\delta^2(\theta + 2\pi\pi) + const.$

Let $\quad \left.U^{\phi}\right|_{\phi=0} = 1 = 3\rho_0\alpha\delta^2(0) + C \;\rightarrow\; C = 1$

So,

$$\left.U^{\phi}\right|_{\phi=2\pi} = 1 + 3\rho_0\alpha\delta^2(2\pi) = \left.U^{\phi}\right|_{\phi=0} = (1 + 6\pi\rho_0\alpha\delta^2).$$

(Eqn. 7-97)

So, we have a blue shift factor that goes as δ^2 for δ small. So, need to evaluate

$$\prod_{n=0}^{\infty}(1 + 6\pi\rho_0\alpha\delta_n^2)$$

(Eqn. 7-98)

where $\delta_n = \delta\big(e(\phi + n2\pi, \rho \approx \rho_0, z \approx 0, t \approx -\alpha)\big)$, with $\delta_\rho = \delta(e|_{start})$:

$$\prod_{n=0}^{\infty}(1+6\pi\rho_0\alpha\delta_n^2) = \exp\left\{\sum_{n=0}^{\infty} 6\pi\rho_0\alpha\delta_n^2\right\} = \exp\left\{6\pi\rho_0\alpha\sum_{n=0}^{\infty}\delta_n^2\right\}$$

(Eqn. 7-99)

How is δ_{n+1} related to δ_n? At some time t such that $t \lesssim -\alpha \ : \ \delta \ll 1$, we have:

$$ds^2 = \tanh^3\delta\,[dt^2 - (\rho d\phi)^2] + [1-\tanh^3\delta](dt\rho d\phi) - d\rho^2 - dz^2$$

(Eqn. 7-100)

We know $|d\phi| \gg |dt|, |d\rho|, |dz|$ in region, also $|dt| > |d\rho|$ or $|dz|$ (speed of light < c). So,

$$ds^2 \approx \tanh^3\delta\,[-\rho_0^2(d\phi)^2] + \rho_0(dt)(d\phi)$$

(Eqn. 7-101)

We want the null geodesic of interest so…

$$0 = \delta^3(-\rho_0^2(d\phi)^2) + \rho_0(dt)(d\phi) = d\phi[dt - \delta^3\rho_0 d\phi],$$

thus the dominant contribution to the null geodesic curve from $dt, d\rho$ or dz comes from dt as $e \to CTC$. $\underline{So\ \delta \approx (t-\alpha)^2}$. Thus,

$$\phi = \frac{t-\alpha}{\rho_0\delta_n^3} + C',$$

(Eqn. 7-102)

and can choose $C' = 0$. Thus $n2\pi = \frac{1}{\rho_0\delta_n^{5/2}}$ or $\delta_n^{5/2} = \frac{1}{\rho_0 n2\pi} \to \delta_n = \frac{1}{(2\pi\rho_0)^{2/5}n^{2/5}}$. If we consider the sum on δ_n^2 we get a divergent result

$$\sum_{n=0}^{\infty}\delta_n^2 \cong \int_0^{\infty} dn\left(\frac{1}{(2\pi\rho_0)^{4/5}}n^{-4/5}\right) \quad \to \quad \text{divergent}$$

(Eqn. 7-103)

So, the blue shift is divergent.

7.4 CTC analysis for sech space

7.4.1 Blue Shift and Jacobi Analysis for sech

The program is to find a simple space-time that in a bounded space-time region admits CTC's. A cylindrical symmetry is desired throughout. It is important that the metric be C^2 so that subsequent calculations with the Riemann tensor are possible. Having decided upon such a space-time, the Cauchy Horizon due to the CTC's will be studied. Is the stress energy tensor divergent at the Cauchy Horizon? To address this, the blue-shifting of the CTC will be calculated and then the divergence of the Jacobi field will be studied via Raychaudhuri's equation. Having calculated both of these values, the energy in an infinitesimal space-time region should go as $[(blue)^2/\theta]$ for one loop. Is this value greater than one? If so we have a divergence. Once a few (more) specific cases have been studied a more general analysis will be done.

Recall:
Normal Minkowski: $ds^2 = dt^2 - (\rho d\phi)^2 - d\rho^2 - dz^2$.
Toroidal Space-time: $ds^2 = dt(\rho d\phi) - d\rho^2 - dz^2$ (increasing ϕ null geodesic)

Before we considered:

$$ds^2 = f(\rho, z, t)\{dt^2 - (\rho d\phi)^2\} + [1 - f(\rho, z, t)]\{dt(\rho d\phi)\} - d\rho^2 - dz^2$$

$$f(\rho, z, t) = \tanh^3([\rho - \rho_0]^2 + z^2)$$

Let's now consider the slightly different form:

$$ds^2 = f(\rho, z, t)\{dt^2 - (\rho d\phi)^2\} + \left[1 - \left(f(\rho, z, t)\right)^2\right]\{dt(\rho d\phi)\} - d\rho^2 - dz^2$$

$$f(\rho, z, t) = 1 - \frac{\operatorname{sech}(x^2 - \alpha^2)}{\operatorname{sech}(3\alpha^2)}$$

$$f|_{x=\pm 2\alpha} = 0.$$

(Eqn. 7-104)

The first CTC occurs at $t = -2\alpha, \rho = \rho_0, z = 0$ and is in fact null, as it should be. Now, calculate the tangent vector of the CTC as it is parallel propagated about the null geodesic back to its starting point, so as to extract the blue shift factor. Denote the curve $e(\phi(\lambda), t = -t\alpha, \rho = \rho_0, z = 0)$ as $e(\lambda)$. Notice that $\frac{de(\lambda)}{d\lambda} = \frac{\partial e(\lambda)}{\partial \lambda}\frac{d\phi}{d\lambda}$ in the ϕ, t, ρ, z coord. system.Using

$\nabla_{\vec{u}}\vec{u} = 0$ for a geodesic eqn., where $u = \frac{d}{d\lambda} = \frac{d\phi_i}{d\lambda}\frac{\partial}{\partial\phi}$ and

$$e(\lambda) = \delta(\rho - \rho_0)\delta(z)\delta(t\alpha)\delta(\phi - \lambda).$$

(Eqn. 7-105)

The components of u are $u^\phi = \frac{d\phi}{d\lambda}$, with the rest zero. Using this we can compute the following:

$$0 = u^\alpha_{;\beta}u^\beta = \left(u^\alpha_{,\beta} + \Gamma^\alpha_{\gamma\beta}u^\gamma\right)u^\beta$$

or long form:

$$0 = \frac{\partial}{\partial x^\beta}\left(\frac{dx^\alpha}{d\lambda}\right)\frac{dx^\beta}{d\lambda} + \Gamma^\alpha_{\gamma\beta}\frac{dx^\gamma}{d\lambda}\frac{dx^\beta}{d\lambda} \quad \textit{with}\text{ only} \quad \begin{matrix} u^\phi \neq 0 \\ \alpha = \phi \end{matrix}$$

Thus,

$$0 = u^\phi \frac{\partial}{\partial\phi}u^\phi + \Gamma^\phi_{\phi\phi}u^\phi u^\phi \;\; \textit{and}\text{ let } u^\phi = v$$

and we now have:

$$0 = \frac{\partial v}{\partial\phi} + \Gamma^\phi{}_{\phi\phi}v.$$

(Eqn. 7-106)

So, need to know $\Gamma^\phi{}_{\phi\phi}$:

Let's go back to:

$ds^2 + fdt^2 - f(\rho d\phi)^2 + (1-f^2)dt(\rho d\phi) - d\rho^2 - dz^2$

Using $ds^2 = g_{\mu v}dx^\mu dx^v$, this can be written in matrix form:

$$g_{\mu v} = \begin{pmatrix} f & \frac{1}{2}(1-f^2)\rho & 0 & 0 \\ \frac{1}{2}(1-f^2)\rho & -f\rho^2 & 0 & 0 \\ 0 & 0 & -1 & 0 \\ 0 & 0 & 0 & -1 \end{pmatrix}$$

Inverting:

$$g^{\mu v} = \begin{pmatrix} \frac{-1}{\rho^2\left(f^2+\frac{1}{4}(1+f^2)^2\right)}\begin{pmatrix} -f\rho^2 & -\frac{1}{2}(1+f^2)\rho \\ -\frac{1}{2}(1+f^2)\rho & f \end{pmatrix} & (0) \\ (0) & \begin{pmatrix} -1 & 0 \\ 0 & 1 \end{pmatrix} \end{pmatrix}$$

Now, since $\Gamma_{\mu\beta\gamma} = \frac{1}{2}\left(g_{\mu\beta,\gamma} + g_{\mu\gamma,\beta} - g_{\beta\gamma,\mu}\right)$ we can then write:

$$\Gamma^\phi{}_{\phi\phi} = g^{\phi\mu}\Gamma_{\mu\phi\phi}$$

and since the metric has no ϕ dependence we have

$$\Gamma^\phi{}_{\phi\phi} = g^{\phi\mu}\frac{1}{2}\left(-g_{\phi\phi,\mu}\right)$$

where the only $g^{\phi\mu}$ that is nonzero is $g^{\phi t}$, so:

$$\Gamma^{\phi}{}_{\phi\phi} = g^{\phi t}\frac{1}{2}\left(-g_{\phi\phi,t}\right) = \frac{\frac{1}{2}(1-f^2)\rho}{\rho^2\left(f^2+\frac{1}{4}(1-f^2)^2\right)}\cdot\frac{1}{2}\frac{\partial}{\partial t}(+f\rho^2)$$
$$= \frac{(1-f^2)}{(1+f^2)^2}\left(\frac{\partial f}{\partial t}\right)$$

To determine $\Gamma^{\phi}_{\phi\phi}\Big|_{\substack{x=-2\alpha \\ (t=-2\alpha)}}$ we need to know $\frac{\partial f}{\partial t}\Big|_{x=-2\alpha}$ (note: $f|_{x=-2\alpha} = 0$ and $\rho|_{\substack{x=-2\alpha \\ (t=-2\alpha)}} = \rho_0$):

$$\frac{\partial f}{\partial t} = \frac{-1}{\operatorname{sech}(3\alpha^2)}\{-\operatorname{sech}(x^2-\alpha^2)\tanh(x^2-\alpha^2)\,(2x)\}$$

and

$$\frac{\partial f}{\partial t}\Big|_{x=-2\alpha} = \tanh(3\alpha^2)\,(-4\alpha).$$

So,

$$\Gamma^{\phi}_{\phi\phi}\Big|_{x=-2\alpha} = \rho_0(-4\alpha)\tanh(3\alpha^2) = -4\rho_0\alpha\tanh(3\alpha^2)$$

(Eqn. 7-107)

thus:

$$\frac{dv}{v} = +4\rho_0\alpha\tanh(3\alpha^2)\,d\phi \quad and \quad \ln v$$
$$= +4\rho_0\alpha\tanh(3\alpha^2)\,\phi + \text{const.}$$

(Eqn. 7-108)

Thus

$$u^{\phi} = A\exp\{4\rho_0\alpha\tanh(3\alpha^2)\,\phi\}.$$

(Eqn. 7-109)

Normalize $u^{\phi}\Big|_{\phi=0} = 1$, then $A = 1$ and we have:

$$u^{\phi} = \exp\{4\rho_0\alpha\tanh(3\alpha^2)\,\phi\} \rightarrow \left(u^{\phi}\Big|_{\phi=n2\pi}\right) = \left(u^{\phi}\Big|_{\phi=2\pi}\right)^n, \quad n \geq 1.$$

(Eqn. 7-110)

So, the blueshift factor on one loop is

$$\exp\{4\rho_0\alpha\tanh(3\alpha^2)\,2\pi\} > 1.$$

(Eqn. 7-111)

Now, to examine the Jacobi field of the null closed geodesic and find the behavior of the expansion factor. Consider the congruence of null geodesics with tangent vector K, g(K,K)=0, and:

$$K\big|_{\rho=\rho_0,\ t=-2\alpha,\ z=0} = \frac{\partial}{\partial\phi}.$$

Let's begin by rewriting:
$$ds^2 = fdt^2 - f(\rho d\phi)^2 + (1-f^2)dt(\rho d\phi) - d\rho^2 - dz^2$$
$$= \big(fdt + (\rho d\phi)\big)\big(dt - f(\rho d\phi)\big) - d\rho^2 - dz^2$$
$$= 2E^3 \otimes E^4 - E^1 \otimes E^1 - E^2 \otimes E^2 \rightarrow g(E^3, E^4) = 1, g(E^1, E^1) = g(E^2, E^2) = -1$$
where
$$E^1 = d\rho\,, E^2 = dz\,, E^3 = (fdt + \rho d\phi)\,, E^4 = \frac{1}{2}\big(dt - f(\rho d\phi)\big)$$

The $E^{i'}s$ are the basis forms, let's calculate their duals (and use component notation (t, ϕ, ρ, z):
$$(E^1)_a = (0{,}0{,}1{,}0) \quad (E^2)_a = (0{,}0{,}0{,}1) \quad (E^4)_a = \frac{1}{2}(1, -\rho f, 0{,}0) \quad (E^3)_a = (f, \rho, 0{,}0)$$
Denote $(E_i)^a = \left(E^i\right)_b g^{ba}$:
$$(E_1)^a = \begin{pmatrix} 0 \\ 0 \\ -1 \\ 0 \end{pmatrix} \Rightarrow E_1 = -\frac{\partial}{\partial \rho} \quad and \quad (E_2)^a = \begin{pmatrix} 0 \\ 0 \\ 0 \\ -1 \end{pmatrix} \Rightarrow E_2 = -\frac{\partial}{\partial z}$$
while:
$$(E_4)^a = \tfrac{1}{2}(1, -\rho f, 0{,}0)\|g\| = \frac{-(1/2)}{\rho^2 + \frac{1}{4}(1-f^2)^2}\begin{pmatrix} [-f\rho^2 + \rho^2 f(1-f^2)] \\ \left[-\frac{1}{2}\rho(1-f^2) - \rho f^2\right] \\ 0 \\ 0 \end{pmatrix}$$
$$= \frac{-(1/2)}{\left[\frac{1}{2}\rho(1-f^2)\right]^2}\begin{pmatrix} \rho^2\left(-f + \frac{1}{2}f - \frac{1}{2}f^2\right) & = & \rho^2\left(-\frac{1}{2}f(1+f^2)\right) \\ \rho\left(-f^2 - \frac{1}{2} + \frac{1}{2}f^2\right) & = & \rho\left(-\frac{1}{2}(1+f^2)\right) \\ 0 & & 0 \\ 0 & & 0 \end{pmatrix}$$
$$= \frac{1}{\rho(1+f^2)}\begin{pmatrix} \rho f \\ 1 \\ 0 \\ 0 \end{pmatrix}$$
Thus,
$$E_4 = \frac{1}{\rho(1+f^2)}\left\{\rho f\frac{\partial}{\partial t} + \frac{\partial}{\partial \phi}\right\}.$$

Does this check?

$$(E_4)^a(E_4)_a = \frac{2}{\rho(1+f^2)}(\rho f - \rho f) = 0, which\ is\ correct.$$

Likewise

$$(E_3)^a = (f, \rho, 0, 0)\|g\| = \frac{-1}{\left[\frac{1}{2}\rho(1+f^2)\right]^2}\begin{pmatrix} \left(-f^2\rho^2 - \frac{1}{2}(1-f^2)\rho^2\right) \\ \left(-\frac{1}{2}(1-f^2)\rho f + \rho f\right) \\ 0 \\ 0 \end{pmatrix}$$

$$= \frac{-1}{\left[\frac{1}{2}\rho(1+f^2)\right]^2}\begin{pmatrix} \rho^2\left[-f^2 - \frac{1}{2} + \frac{1}{2}f^2\right] & = & -\rho^2\frac{1}{2}(1+f^2) \\ \rho f\left[1 - \frac{1}{2} + \frac{1}{2}f^2\right] & = & \rho f\frac{1}{2}(1+f^2) \\ 0 & & 0 \\ 0 & & 0 \end{pmatrix}$$

$$= \frac{2}{\rho(1+f^2)}\begin{pmatrix} \rho \\ -f \\ 0 \\ 0 \end{pmatrix}$$

Thus,

$$E_3 = \frac{2}{\rho(1+f^2)}\begin{pmatrix} \rho \\ -f \\ 0 \\ 0 \end{pmatrix} = \frac{2}{\rho(1+f^2)}\left\{\rho\frac{\partial}{\partial t} - f\frac{\partial}{\partial\phi}\right\}$$

Does this check?

$$(E_3)^a(E_3)_a = (stuff)(f\rho - \rho f) = 0$$

$$(E_4)_a(E_3)^a = \frac{1}{2}(1, -\rho f, 0, 0) + \frac{2}{\rho(1+f^2)}\begin{pmatrix} \rho \\ -f \\ 0 \\ 0 \end{pmatrix}$$

$$= \frac{2}{\rho(1+f^2)}(\rho + \rho f^2) = 1$$

The rest zero, so checks. Also, as $f \to 0$, $E_4 \backsimeq \frac{1}{\rho}\frac{\partial}{\partial\phi}$ and $E_3 \backsimeq 2\frac{\partial}{\partial t}$.

Let $E_4 = K$, K is the tangent vector to the congruence of null geodesics being studied. Consider Z to be the vector representing the separation of corresponding points on neighboring curves. The integral curves of z are lie dragged along the null congruence:

$$L_K Z = 0 = Z^a_{;b}K^b - Z^b K^a_{;b}$$

(Eqn. 7-112)

where

$$\frac{D}{d\lambda}Z^a = K^a_{;b}Z^b$$

(Eqn. 7-113)

Since K is a geodesic: $\frac{D}{d\lambda}K^a = K^a_{;b}K^b = 0$ (affine param.). Thus,

$$\left.\begin{matrix}\text{In the } E_i \text{ basis} \\ (E_1, E_2, E_3, E_4)\end{matrix}\right\} K_b = (0,0,0,1)$$

$\underline{K^a_{;4} = 0}$ follows from geod. eqn.

So $\frac{D}{d\lambda}Z^a = K^a_{;\alpha}Z^\alpha$ *and* $\frac{DZ^4}{d\lambda} = K^4_{;\alpha}Z^\alpha$. Thus, for the 1, 2, 3 comp.s we have a system of ordinary diff eq.'s:

$$\frac{D}{d\lambda}Z^\alpha = K^\alpha_{;\beta}Z^\beta$$

(Eqn. 7-114)

Projection of Z into Q_q (the quotient of tangent vector space by K) obeys a propagation eqn. which involves <u>only this projection</u>.

$K_a = (0,0,0,1) \qquad (K^a K_a) = 0 = K^a g_{ab} K^b$
$(K^a K_a)_{;C} = 0 \ , K^a_{;C} = 0$

$$K^3 = g^{3a}K_a = g^{34}K_4 = K_4 \qquad \begin{matrix} K_{4;C} = 0 \\ \underline{K^3_{;C} = 0} \end{matrix}$$

This implies $\frac{dZ^3}{d\lambda} = 0 \quad Z^3 = \text{const.}$ along null geodesic.

$Z^3 = \text{const.}$ can be interpreted as saying that light rays emitted from the same source at different times maintain a constant separation in time. One is interested in the behavior of neighboring null geodesics which have purely spatial separations, i.e., vectors Z for which $Z^3 = 0$. So we are now considering the projection of vectors onto the S_q subspace (consisting of equivalence classes of vectors in the subspace of the tangent space that ae multiples of K), and they obey the eqn.:

$$\frac{d}{dv}Z^m = K^m_{;n}Z^n \text{ where } m, n = 1,2 \text{ only.}$$

(Eqn. 7-115)

Since the Z^m obey 1st order linear ordinary diff. eq., they can be expressed in terms of their values at some point q by:

$$Z^m(V) = \hat{A}_{mn}(V)Z^n|_q$$

(Eqn. 7-116)

where $\frac{d}{dv}\hat{A}_{mn}(V) = K_{m;p}\hat{A}_{pn}(V)$

The matrix A_{mn} can be regarded at representing the shape and orientation of the cross-sections of null geodesics which is circular at q. We need a projection operator onto the 1, 2 subspace:

$$h^a_b Z^b = Z^a \;\; for \;\; a = 1{,}2 \;, = 0 \;\; for \;\; a = 3{,}4$$

$$Z^b = a_1(E_1)^b + a_2(E_2)^b + a_3(E_3)^b + a_4(E_4)^b$$

$$\boxed{h^a_b = \delta^a_b - (E_3)^b(E_4)_b - (E_4)^b(E_3)_b}$$

From the Raychaudhuri eqn. (optical scalar):

$$\frac{d}{dv}\hat{\theta} = -R_{ab}K^aK^b + 2\hat{\omega}^2 - 2\hat{\sigma}^2 - \frac{1}{2}\hat{\theta}^2$$

(Eqn. 7-117)

and

$2\hat{\omega}^2 = \hat{\omega}_{ab}\hat{\omega}^{ab}$ $\qquad$ $\hat{\omega}_{ab} = h^c_a h^d_b K_{[c;d]}$ antisymmetric

$2\hat{\sigma}^2 = \hat{\sigma}_{ab}\hat{\sigma}^{ab}$ $\qquad$ $\hat{\theta}_{ab} = h^c_a h^d_b K_{(c;d)}$

$\hat{\sigma}_{ab} = \hat{\theta}_{ab} - \frac{1}{2}h_{ab}\theta$ $\qquad$ $\hat{\theta} = \hat{\theta}_{ab}h^{ab}$

$$h^b_a = \delta^a_b - L^aK_b - K^aL_b$$

Need to calculate $K_{c;d}$ and R_{ab} to proceed, where we have the following:

$$K_{c;d} = K_{c,d} - \Gamma^a_{dc}K_a$$
$$R^a_{bcd} = \partial\,\Gamma^a_{db}/\partial x^c - \partial\,\Gamma/\partial x^d + \Gamma^a_{cf}\Gamma^f_{db} - \Gamma^a_{df}\Gamma^f_{cb}$$
$$K_{\alpha;\beta} = K_{\alpha,\beta} - \Gamma^\mu_{\alpha\beta}K_\mu$$
$$R_{\mu v} = \Gamma^\alpha_{\mu v,\alpha} - \Gamma^\alpha_{\mu\alpha,v} + \Gamma^\alpha_{\beta\alpha}\Gamma^\beta_{\mu v} - \Gamma^\alpha_{\beta v}\Gamma^\beta_{\mu\alpha}$$
$$\Gamma^\alpha_{\beta\gamma} = g^{\alpha\mu}\Gamma_{\mu\beta\gamma}$$
$$\Gamma_{\mu\beta\gamma} = \frac{1}{2}\left(g_{\mu\beta,\gamma} + g_{\mu\gamma,\beta} - g_{\beta\gamma,\mu} + C_{\mu\beta\gamma} + C_{\mu\gamma\beta} - C_{\beta\gamma\mu}\right)$$
$$\left[e_\alpha, e_\beta\right] \equiv C^\gamma_{\alpha\beta}e_\gamma$$
$$[u, v] = \left(u^\beta v^\alpha_{,\beta} - V^\beta u^\alpha_{,\beta}\right)\underbrace{\left(\frac{\partial}{\partial x^\alpha}\right)}_{i.e.\,e_\alpha}$$

(Eqn. 7-118)

These equation will be evaluated in the next section.

7.4.2 Focusing Theorem [21]

Let's consider the focusing theorem in this context. The cross-sectional area A evolves according to:

$$\frac{d^2 A^{1/2}}{d\lambda^2} = -\left(|\sigma|^2 + \frac{1}{2} R_{\alpha\beta} K^{\alpha} K^{\beta}\right) A^{1/2}$$

(Eqn. 7-119)

where λ is affine parameter along the central ray $\vec{K} = {}^{d}/_{d\lambda}$ and where:

$$|\sigma|^2 = \frac{1}{2} K_{a;\beta} K^{a;\beta} - \frac{1}{4} \left(K^{\mu}_{;\mu}\right)^2.$$

(Eqn. 7-120)

Recall the form of the metric under study:

$$ds^2 = f dt^2 - f(\rho d\phi)^2 + (1 - f^2) dt(\rho d\phi) - d\rho^3 - dz^2, \ f = f(t, \rho, z),$$

and

$$\vec{K}\Big|_{start} = {}^{d}/_{d\lambda}$$

or, in (t, ϕ, ρ, z) *notaton*: $K = (0,1,0,0)$.

From blue shift analysis:

$$\vec{K} = \exp[4\rho_0 \alpha \tanh(3\alpha^2)\, \phi] \frac{\partial}{\partial \phi} = (0, \exp[4\rho_0 \alpha \tanh(3\alpha^2)\, \phi], 0,0)$$

Recall

$$g_{\mu\nu} = \begin{pmatrix} f & \frac{1}{2}(1 - f^2)\rho & 0 & 0 \\ \frac{1}{2}(1 - f^2)\rho & -f\rho^2 & 0 & 0 \\ 0 & 0 & -1 & 0 \\ 0 & 0 & 0 & -1 \end{pmatrix}$$

and

$$g^{\mu\nu} = \begin{pmatrix} \frac{-1}{\rho^2\left(f^2 + \frac{1}{4}(1-f^2)^2\right)} \begin{pmatrix} -f\rho^2 & -\frac{1}{2}(1 - f^2)\rho \\ -\frac{1}{2}(1 - f^2)\rho & f \\ 0 & 0 \\ 0 & 0 \end{pmatrix} & \begin{matrix} 0 & 0 \\ 0 & 0 \\ -1 & 0 \\ 0 & -1 \end{matrix} \end{pmatrix}$$

$$R_{\mu\nu} = \Gamma^{\alpha}_{\mu\nu,\alpha} - \Gamma^{\alpha}_{\mu\alpha,\nu} + \Gamma^{\alpha}_{\beta\alpha} \Gamma^{\beta}_{\mu\nu} - \Gamma^{\alpha}_{\beta\nu} \Gamma^{\beta}_{\mu\alpha}$$

and

$$\Gamma^{\alpha}_{\beta\gamma} = g^{\alpha\mu}\Gamma_{\mu\beta\gamma} \quad where \quad \Gamma_{\mu\beta\gamma} = \frac{1}{2}\left(g_{\mu\beta,\gamma} + g_{\mu\gamma,\beta} - g_{\beta\gamma,\mu}\right)$$

Since

$$K_{\alpha;\beta} = K_{\alpha,\beta} - \Gamma^{\mu}_{\alpha\beta}K_{\mu}$$

where components are $(0, \exp[\dots], 0,0)$, only $K_{\phi} \neq 0$, and it has ϕ dependence. So,

$K_{\alpha;\beta} = K_{\phi,\beta} - \Gamma^{\phi}_{\phi\beta}K_{\phi}$ and $\Gamma^{\phi}_{\phi\beta} = g^{\phi\gamma}\Gamma_{\gamma\phi\beta} = g^{\phi t}\Gamma_{t\phi\beta} + g^{\phi\phi}\Gamma_{\phi\phi\beta}$.

Using the relation

$$\Gamma_{\gamma\phi\beta} = \frac{1}{2}\left(g_{\gamma\phi,\beta} + g_{\gamma\beta,\phi} - g_{\phi\beta,\gamma}\right)$$

we then have $\Gamma_{t\phi\beta} = \frac{1}{2}\left(g_{t\phi,\beta} - g_{\phi\beta,t}\right)$, $\Gamma_{\phi\phi\beta} = \frac{1}{2}\left(g_{\phi\phi,\beta} - g_{\phi\beta,\phi}\right)$, thus

$$K_{\alpha;\beta} = K_{\phi,\beta} - \left\{g^{\phi t}\frac{1}{2}\left(g_{t\phi,\beta} - g_{\phi\beta,t}\right) + g^{\phi\phi}\frac{1}{2}g_{\phi\phi,\beta}\right\}K_{\phi}.$$

We can now evaluate the various terms:

$$K_{\phi;\phi} = K_{\phi,\phi} - \frac{1}{2}g^{\phi t}g_{\phi\phi,t}K_{\phi}$$

$$K_{\phi;t} = -\left\{\frac{1}{2}g^{\phi\phi}g_{\phi\phi,t}\right\}K_{\phi}$$

$$K_{\phi;\rho} = \left(\frac{1}{2}g^{\phi t}g_{t\phi,\rho} - \frac{1}{2}g^{\phi\phi}g_{\phi\phi,\rho}\right)K_{\phi}$$

$$K_{\phi;z} = \left(-\frac{1}{2}g^{\phi t}g_{t\phi,z} - \frac{1}{2}g^{\phi\phi}g_{\phi\phi,z}\right)K_{\phi}$$

$$K^{\mu}_{;\mu} = g^{\mu\gamma}K_{\gamma;\mu} = g^{\mu\phi}K_{\phi;\mu} = g^{\phi\phi}K_{\phi;\phi} + g^{t\phi}K_{\phi;t}$$

$$K_{\alpha;\beta}K^{\alpha;\beta} = K_{\alpha;\beta}g^{\alpha\gamma}g^{\beta\delta}K_{\gamma;\delta} = K_{\phi;\beta}g^{\phi\phi}g^{\beta\delta}K_{\phi;\delta}$$

$$= g^{\phi\phi}\left\{\begin{matrix} K_{\phi;t}\left(g^{t\phi}K_{\phi;\phi} + g^{tt}K_{\phi;t}\right) + K_{\phi;\phi}\left(g^{\phi t}K_{\phi;t} + g^{tt}K_{\phi;\phi}\right) \\ +K_{\phi;\rho}\left(g^{\rho\rho}K_{\phi;\phi}\right) + K_{\phi;z}\left(g^{zz}K_{\phi;z}\right) \end{matrix}\right\}$$

(Eqn. 7-121)

Thus,

$$|\sigma|^2 = \frac{1}{2}g^{\phi\phi}\left\{2g^{\phi t}K_{\phi;t}K_{\phi;\phi} + g^{tt}\left(K_{\phi;t}\right) + g^{\phi\phi}\left(K_{\phi;\phi}\right)^2 - \left[\left(K_{\phi;\rho}\right)^2 + \left(K_{\phi;z}\right)^2\right]\right\}$$
$$-\frac{1}{4}\left\{\left(g^{\phi\phi}\right)^2\left(K_{\phi;\phi}\right)^2 + \left(g^{t\phi}\right)^2\left(K_{\phi;t}\right)^2 + 2g^{\phi\phi}g^{\phi t}K_{\phi;t}K_{\phi\phi}\right\}$$

$$|\sigma|^2 = \frac{1}{4}\left(g^{\phi\phi}\right)^2\left(K_{\phi;\phi}\right)^2 + \left[\frac{1}{2}g^{\phi\phi}g^{tt} - \frac{1}{4}\left(g^{\phi t}\right)^2\right]\left(K_{\phi;t}\right)^2 + \frac{1}{2}g^{\phi\phi}g^{\phi t}K_{\phi;t}K_{\phi\phi} - \frac{1}{2}g^{\phi\phi}[\cdots].$$

(Eqn. 7-122)

The null geodesic along which we study the area A is $C(\lambda)$, which in the coord. system chosen has $t = -2\alpha, \rho = \rho_0, z = 0$, where we have chosen f to be:

$$f(\rho, z, t) = 1 - \frac{\operatorname{sech}(\{[\rho - \rho_0]^2 + z^2 + t^2\} - \alpha^2)}{\operatorname{sech}(3\alpha^2)}$$

and

$$f(\rho_0, 0, -2\alpha) = 0 = f|_{C(\lambda)}.$$

Computing the derivatives along the curve:

$$\left.\frac{\partial f}{\partial \rho}\right|_{C(\lambda)} = \frac{\operatorname{sech}(\{[\rho - \rho_0]^2 + z^2 + t^2\} - \alpha^2)}{\operatorname{sech}(3\alpha^2)} \tanh(\{\cdots\}$$
$$- \alpha^2)\, 2(\rho - \rho_0)|_{C(\lambda)} = 0$$

$$\left.\frac{\partial f}{\partial z}\right|_{C(\lambda)} = \{\cdots\} 2z|_{C(\lambda)} = 0$$

$$\left.\frac{\partial f}{\partial t}\right|_{C(\lambda)} = \tanh(3\alpha^2)\, 2(-2\alpha) = -4\alpha \tanh(3\alpha^2)$$

and

$$g_{\mu\nu}|_{C(\lambda)} = \begin{pmatrix} 0 & \frac{1}{2}\rho_0 & & \\ \frac{1}{2}\rho_0 & 0 & -1 & \\ & & & -1 \end{pmatrix} \qquad g^{\mu\nu}|_{C(\lambda)}$$
$$= \begin{pmatrix} 0 & \frac{1}{\rho_0} & & \\ \frac{1}{\rho_0} & 0 & -1 & \\ & & & -1 \end{pmatrix}$$

So,

$$g_{\mu\nu,t}|_{C(\lambda)} = \begin{pmatrix} -4\alpha \tanh(3\alpha^2) & 0 & 0 & 0 \\ 0 & +4\alpha\rho_0^2 \tanh(3\alpha^2) & 0 & 0 \\ 0 & 0 & 0 & 0 \\ 0 & 0 & 0 & 0 \end{pmatrix}$$

and

$$g_{\mu\nu,\phi} = 0\,, \quad g_{\mu\nu,\rho}|_C = \begin{pmatrix} 0 & \frac{1}{2} & \\ \frac{1}{2} & 0 & [0] \\ [0] & & [0] \end{pmatrix}, \quad g_{\mu\nu,z}|_C = \begin{pmatrix} [0] & [0] \\ [0] & [0] \end{pmatrix}.$$

Thus,

$|\sigma|^2 = 0$ since $g^{\phi\phi}\big|_{C(\lambda)} = 0.$

(Eqn. 7-123)

Let's now evaluate $R_{\alpha\beta}K^\alpha K^\beta$:

Since

$R_{\alpha\beta}K^\alpha K^\beta\big|_C = R_{\phi\phi}K^\phi K^\phi$ for $\vec{K}$ chosen

$R_{\phi\phi} = \Gamma^\alpha_{\phi\phi,\alpha} - \Gamma^\alpha_{\phi\alpha,\phi} + \Gamma^\alpha_{\beta\alpha}\Gamma^\beta_{\phi\phi} - \Gamma^\alpha_{\beta\phi}\Gamma^\beta_{\phi\alpha}$

$\Gamma^\alpha_{\phi\phi} = g^{\alpha\mu}\frac{1}{2}\left(g_{\mu\phi,\phi} + g_{\mu\phi,\phi} - g_{\phi\phi,\mu}\right) = g^{\alpha\mu}\frac{1}{2}\left(-g_{\phi\phi,\mu}\right)$

and

$\Gamma^\alpha_{\phi\phi}\big|_C = \left(\frac{1}{\rho_0}\right)(-4\alpha\rho_0^2\tanh(3\alpha^2))\rho_0^2$ for $\alpha = \phi$ only

---

$\Gamma^\alpha_{\phi\phi,\alpha} = g^{\alpha\mu}_{,\alpha}\frac{1}{2}\left(-g_{\phi\phi,\mu}\right) + \frac{1}{2}g^{\alpha\mu}\left(-g_{\phi\phi,\mu\alpha}\right)$

$\Gamma^\alpha_{\phi\phi,\alpha}\big|_C = g^{tt}_{,t}\frac{1}{2}\left(-g_{\phi\phi,t}\right) + \frac{1}{2}g^{\rho\rho}\left(-g_{\phi\phi,\rho\rho}\right) + \frac{1}{2}g^{zz}\left(-g_{\phi\phi,zz}\right)$

$= 4\left(\frac{\partial f}{\partial t}\right)\frac{1}{2}(-4\alpha\rho_0^2\tanh(3\alpha^2)) + \frac{1}{2}(\rho_0^2 2) + \frac{1}{2}(\rho_0^2 2)$

$= [-2(4\alpha\tanh(3\alpha^2))^2 + 2]\rho_0^2$

$\Gamma^\beta_{\phi\alpha} = g^{\beta\mu}\frac{1}{2}\left(g_{\mu\phi,\alpha} + g_{\mu\alpha,\phi} - g_{\phi\alpha,\mu}\right)$

$\Gamma^\beta_{\phi\alpha}\big|_C = g^{\phi t}\{\ \} + g^{t\phi}\{\ \} + g^{\rho\rho}\{\ \} + g^{zz}\{\ \}$

$\downarrow \quad \downarrow \quad \downarrow \quad \downarrow$

$\frac{1}{2}\left(g_{t\phi,\alpha} - g_{\phi\alpha,t}\right)\quad \frac{1}{2}\left(g_{\phi\phi,\alpha}\right)\quad \frac{1}{2}\left(-g_{\phi\alpha,\rho}\right)\quad \frac{1}{2}\left(-g_{\phi\alpha,z}\right)$

$\downarrow \quad \downarrow \quad \downarrow \quad \downarrow \quad \downarrow$

$g_{t\phi,\rho} = \frac{1}{2}\quad g_{\phi\phi,t}\quad g_{\phi\phi,t}\quad g_{\phi t,\rho} = \frac{1}{2}\quad 0$

$\Gamma^\beta_{\phi\alpha}\big|_C = \left(\frac{1}{\rho_0}\right)\left(\frac{1}{2}\right)\left(\frac{1}{2}\right)\quad \beta = \phi, \alpha = \rho\quad \frac{1}{2\rho_0}$

$\left(\frac{1}{\rho_0}\right)\left(\frac{1}{2}\right)(-4\alpha\rho_0^2\tanh(3\alpha^2))\quad \beta = \phi, \alpha = \phi\quad +\rho_0(4\alpha\tanh(3\alpha^2))$

$\left(\frac{2}{\rho_0}\right)\left(\frac{1}{2}\right)(-4\alpha\rho_0^2\tanh(3\alpha^2))\quad \beta = t, \alpha = t\quad -\rho_0(4\alpha\tanh(3\alpha^2))$

$(-1)\left(\frac{1}{2}\right)\left(-\frac{1}{2}\right)\quad \beta = \rho, \alpha = t\quad \frac{1}{4}$

Thus,

$$
\begin{aligned}
R_{\phi\phi} &= [2-2(4\alpha\tanh(3\alpha^2))^2]\rho_0^2 + \Gamma^{\alpha}_{\phi\alpha}\Gamma^{\phi}_{\phi\phi} - \Gamma^{\alpha}_{\beta\phi}\Gamma^{\beta}_{\phi\alpha} \\
&= [2-2(4\alpha\tanh(3\alpha^2))^2]\rho_0^2 + \left(\Gamma^{t}_{\phi t}\right)\left[\Gamma^{\phi}_{\phi\phi} - \Gamma^{t}_{\phi t}\right] \\
&= [2-2(4\alpha\tanh(3\alpha^2))^2]\rho_0^2 + 2\rho_0^2(4\alpha\tanh(3\alpha^2))^2 \\
&= [2-4(4\alpha\tanh(3\alpha^2))^2]\rho_0^2
\end{aligned}
$$

(Eqn. 7-124)

$$
\frac{d^2A^{1/2}}{d\lambda^2} = \{1-2(4\alpha\tanh(3\alpha^2))^2\}\rho_0^2\left(K^{\phi}\right)^2 A^{1/2}
$$

$$\downarrow$$

$$
\frac{d}{d\lambda}\left(\frac{\partial A^{1/2}}{\partial x^{\beta}}\frac{dx^{\beta}}{d\lambda}\right) = \frac{d}{d\lambda}\left(\frac{\partial A^{1/2}}{\partial x^{\beta}}K^{\beta}\right) = \frac{\partial}{\partial x^{\alpha}}\left[\frac{\partial A^{1/2}}{\partial\phi}K^{\phi}\right]\frac{dx^{\alpha}}{d\lambda}
$$

$$
= \left[\frac{\partial^2 A^{1/2}}{\partial\phi^2} + \frac{\partial K^{\phi}}{\partial\phi}\frac{\partial A^{1/2}}{\partial\phi}\right]K^{\phi}
$$

From geodesic eqn., affinely parametrized:

$$
\frac{\partial K^{\phi}}{\partial\phi} = -\Gamma^{\phi}_{\phi\phi}K^{\phi}
$$

(Eqn. 7-125)

Now $\Gamma^{\phi}_{\phi\phi} = \rho_0(4\alpha\tanh(3\alpha^2))$, so:

$$
\begin{aligned}
&K^{\phi}\frac{\partial^2 A^{1/2}}{\partial\phi^2} - [\rho_0(4\alpha\tanh(3\alpha^2))]\left(\frac{\partial A^{1/2}}{\partial\phi}\right)\left(K^{\phi}\right)^2 \\
&= (1-2(4\alpha\tanh(3\alpha^2))^2)\rho_0^2 A^{1/2}(t)
\end{aligned}
$$

$$
\begin{aligned}
&\frac{\partial^2 A^{1/2}}{\partial\phi^2} - \underbrace{[\rho_0(4\alpha\tanh(3\alpha^2))]}_{a}K^{\phi}\frac{\partial A^{1/2}}{\partial\phi} \\
&- \underbrace{(1-2(4\alpha\tanh(3\alpha^2))^2)\rho_0^2}_{b}K^{\phi}A^{1/2} = 0
\end{aligned}
$$

$$
K^{\phi} = \exp(a\phi).
$$

(Eqn. 7-126)

What is the vector field for the congruence of null geodesics that contain the 1st ctc formed and are parametrized by the coordinates (ρ, z, t), a three parameter family, about that 1st ctc?

$$ds^2 = \big(fdt + (\rho d\phi)\big)\big(dt - f(\rho d\phi)\big) - d\rho^2 - dz^2$$

so

$$du = \frac{1}{2}\big(dt - f(\rho d\phi)\big) \Rightarrow \frac{\partial}{\partial u} = \frac{1}{\rho(1+f^2)}\left\{\rho f \frac{\partial}{\partial t} + \frac{\partial}{\partial \phi}\right\}$$

and

$$\left.\frac{\partial}{\partial u}\right|_{1^{st}\,ctc} = \frac{1}{\rho_0}\frac{\partial}{\partial \phi}$$

which is tangent to the 1st ctc as desired, but the vector can't simply be constant if the ctc is to be affinely parametrized. In fact $u^\phi = A\exp\{4\rho_0\alpha\tanh(3\alpha^2)\,\phi\}$ thus the vector has a ϕ dependence. Suppose we choose vector such that in the coordinate system chosen $u^\phi\big|_{\phi=0} = 1$, then

$$\left.\frac{\partial}{\partial u}\right|_{1^{st}\,ctc} = \exp\underbrace{\{4\rho_0\alpha\tanh(3\alpha^2)\,\phi\}}_{a}\frac{1}{\rho_0}\frac{\partial}{\partial \phi}$$

The choice of $\frac{\partial}{\partial u}$ that agrees with this is then

$$\frac{\partial}{\partial u} = \frac{\exp\{a\phi\}}{\rho(1+f^2)}\left\{\rho f\frac{\partial}{\partial t} + \frac{\partial}{\partial \phi}\right\}.$$

(Eqn. 7-127)

7.4.3 Transverse oscillations on 1st CTC

Let's evaluate the oscillations in the transverse z-direction as we wind around in the ϕ coordinate. To do this let's make some slight notational shifts to be ready to consider the geodesic deviation equation in a convenient way.

CTC metric:

$$ds^2 = f(-dt^2 + (\rho d\phi)^2) - (1 - f^2)dt\rho d\phi + d\rho^2 + dz^2$$

(Eqn. 7-128)

where *f* is of the form:

$$f|_{x\ asymptotic} = 1 \qquad (x^2 = [\rho - \rho_0] + z^2 + t^2)$$

(Eqn. 7-129)

thus,

$$f|_{x=0} = 0 \quad , \quad f|_{\substack{x>0\\ \rho<\rho_0}} < 0$$

(Eqn. 7-130)

So, basically want:

$$f|_{some\ geodesic} = 0.$$

(Eqn. 7-131)

Choose

$$f(\rho,z,t) = 1 - \frac{\mathrm{sech}(x^2-\alpha^2)}{\mathrm{sech}\,3\alpha^2} \Rightarrow f|_{x=\pm 2\alpha} = 0$$

(Eqn. 7-132)

The first CTC (a closed null geodesic) occurs at $t = -2\alpha, \rho = \rho_0, z = 0$. The CTC curve can be written:

$$\begin{aligned} \mathcal{C}(\phi(\lambda), t = -2\alpha, \rho = \rho_0, z = 0) &= \tau(\lambda) \\ &= \delta(\rho-\rho_0)\delta(z)\delta(t+2\alpha)\delta(\phi - f(\lambda)) \end{aligned}$$

(Eqn. 7-133)

Using the geodesic equation we can find the coordinate parameterization of the affinely parametrized geodesic $\mathcal{C}$:

$$\nabla_{\vec{u}}\vec{u} = 0 \Rightarrow u^{\alpha}_{;\beta}u^{\beta} = \left(u^{\alpha}_{,\beta} + \Gamma^{\alpha}_{\gamma\beta}u^{\gamma}\right)u^{\beta} = 0 \ \ (\vec{u}\text{ tangent vector})$$

Only $u^{\phi} \neq 0$, so:

$$0 = u^{\phi}\frac{\partial}{\partial\phi}u^{\phi} + \Gamma^{\phi}_{\phi\phi}\Big|_G u^{\phi}u^{\phi}$$

Thus,

$$\frac{du^{\phi}}{u^{\phi}} = -\Gamma^{\phi}_{\phi\phi}\Big|_G d\phi \ \Rightarrow \ \ln u^{\phi} = \left(-\Gamma^{\phi}_{\phi\phi}\right)\phi + const$$

Solving:

$$u^{\phi} = A\exp\left\{\left(-\Gamma^{\phi}_{\phi\phi}\Big|_G\right)\phi\right\} \quad \left(\text{let } u^{\phi} = 1 \text{ at } \phi = 0 \ \Rightarrow A = 1\right)$$

So we have:

$$u^{\phi} = \exp\left\{\left(-\Gamma^{\phi}_{\phi\phi}\Big|_G\right)\phi\right\}.$$

(Eqn. 7-134)

Since f does not have any ϕ dependence:

$$\begin{aligned} \Gamma^{\phi}_{\phi\phi} = g^{\phi t}\frac{1}{2}\left(-g_{\phi\phi,t}\right) &= \frac{\frac{1}{2}(1-f^2)\rho}{-\frac{1}{4}(1-f^2)^2\rho^2}\,\frac{1}{2}\left(-\rho^2\frac{\partial f}{\partial t}\right) \\ &= \frac{(1-f^2)}{(1-f^2)^2}\rho\left(\frac{\partial f}{\partial t}\right) \end{aligned}$$

On the geodesic:

$$\Gamma^{\phi}_{\phi\phi}\Big|_G = \rho_0\left(\frac{\partial f}{\partial t}\right)\Big|_G.$$

For $\partial f/\partial t$ we have:

$$\frac{\partial f}{\partial t} = \frac{\mathrm{sech}(x^2-\alpha^2)\tanh(x^2-\alpha^2)\,(2x)}{\mathrm{sech}(3\alpha^2)}$$

and

$$\left.\left(\frac{\partial f}{\partial t}\right)\right|_G = \tanh(3\alpha^2)\,(-4\alpha).$$

So,

$$\left.\Gamma^{\phi}_{\phi\phi}\right|_{x=-2\alpha} = -4\alpha\rho_0\tanh(3\alpha^2)$$

and

$$u^{\phi} = \exp[(4\alpha\rho_0\tanh(3\alpha^2))\phi],$$

which has an exponentially divergent blue shift, where:

$$\left(u^{\phi}\big|_{\phi=n2\pi}\right) = \left(u^{\phi}\big|_{\phi=2\pi}\right)^n$$

(Eqn. 7-135)

Geodesic deviation equation:

$$\nabla_{K^a}\nabla_{K^b}\eta^c + R^c_{abd}K^aK^b\eta^d = 0 \Rightarrow \frac{d^2\eta^\mu}{d\tau^2} + R^\mu_{\alpha\beta\gamma}K^\alpha K^\beta\eta^\gamma = 0,$$

where $K^v = \frac{\partial x^v}{\partial\tau}$ and $\frac{d}{d\tau}K^v = 0$ (parallel transport) we have:

$$\frac{d}{d\tau}\left(\frac{d\eta^\mu}{dx^v}\frac{\partial x^v}{\partial\tau}\right) = \frac{\partial x^v}{\partial\tau}\frac{d}{d\tau}\left(\frac{d\eta^\mu}{dx^v}\right) = K^vK^\delta\left(\frac{d^2\eta^\mu}{dx^\delta dx^v}\right).$$

Since there is only a K^ϕ component in K^a we get:

$$\frac{d^2\eta^\mu}{d\phi^2} + \left.\mathrm{R}^\mu_{\phi\phi\gamma}\right|_G\eta^\gamma = 0$$

(Eqn. 7-136)

and $\left.\mathrm{R}^\mu_{\phi\phi\gamma}\right|_G$ is a constant.

From:

$$ds^2 = -fdt^2 + f(\rho d\phi)^2 - (1-f^2)dt\rho d\phi + d\rho^2 + dz^2$$

we have:

$$g_{\mu v} = \begin{pmatrix} -f & -\frac{1}{2}\rho(1-f^2) & 0 & 0 \\ -\frac{1}{2}\rho(1-f^2) & f\rho^2 & 0 & 0 \\ 0 & 0 & 1 & 0 \\ 0 & 0 & 0 & 1 \end{pmatrix}$$

and

$$g^{\mu v} = \begin{pmatrix} \frac{1}{\left[-\frac{1}{4}\rho^2(1-f^2)^2\right]} \begin{pmatrix} +f\rho^2 + \frac{1}{2}(1-f^2)\rho \\ +\frac{1}{2}(1-f^2)\rho - f \end{pmatrix} & & 0 & 0 \\ & & 0 & 0 \\ 0 & 0 & 1 & 0 \\ 0 & 0 & 0 & 1 \end{pmatrix}$$

So,

$g_{\mu v,\phi} = 0$ and $g_{\mu v,i}\ for\ \ i = \rho, z, t$:

$$g_{\mu v,i} =$$

$$\begin{pmatrix} \begin{bmatrix} -\frac{\partial f}{\partial x^i} & \left(-\frac{1}{2}\frac{\partial \rho}{\partial x^i}(1-f^2) + \rho f\frac{\partial f}{\partial x^i}\right) \\ \left(-\frac{1}{2}\frac{\partial \rho}{\partial x^i}(1-f^2) + \rho f\frac{\partial f}{\partial x^i}\right) & \left(\rho^2\frac{\partial f}{\partial x^i} + 2f\rho\frac{\partial \rho}{\partial x^i}\right) \end{bmatrix}_1 & [0] \\ [0] & [0] \end{pmatrix}$$

Just studying the bracket $[\quad]_1$ for $g_{\mu v,ij}$:

$$g_{\mu v,ij} =$$

$$\begin{bmatrix} -\frac{\partial^2 f}{\partial x^i \partial x^j} & \left(f\frac{\partial f}{\partial x^i}\frac{\partial \rho}{\partial x^i} + \frac{\partial \rho}{\partial x^i}; f\frac{\partial f}{\partial x^i} + \rho\frac{\partial f}{\partial x^j}\frac{\partial f}{\partial x^i} + \rho f\frac{\partial^2 f}{\partial x^i \partial x^j}\right) \\ (\dots\dots same\quad) & \left(2\rho\frac{\partial f}{\partial x^j}\frac{\partial f}{\partial x^i} + \rho^2\frac{\partial^2 f}{\partial x^i \partial x^j} + 2\frac{\partial f}{\partial x^j}\rho\frac{\partial \rho}{\partial x^i} + 2f\frac{\partial \rho}{\partial x^j}\frac{\partial \rho}{\partial x^i}\right) \end{bmatrix}_2$$

$$[\quad]_1|_G = \begin{bmatrix} -\frac{\partial f}{\partial x^i} & -\frac{1}{2}\frac{\partial \rho}{\partial x^i} \\ -\frac{1}{2}\frac{\partial \rho}{\partial x^i} & \rho^2\frac{\partial f}{\partial x^i} \end{bmatrix}\Bigg|_G \qquad \underline{\text{just setting } f = 0}$$

$$[\quad]_2|_G = \begin{bmatrix} -\frac{\partial^2 f}{\partial x^i \partial x^j} & \rho\frac{\partial f}{\partial x^j}\frac{\partial f}{\partial x^i} \\ \rho\frac{\partial f}{\partial x^j}\frac{\partial f}{\partial x^i} & \left(2\rho\frac{\partial \rho}{\partial x^j}\frac{\partial f}{\partial x^i} + \rho^2\frac{\partial^2 f}{\partial x^i \partial x^j} + 2\frac{\partial f}{\partial x^j}\rho\frac{\partial \rho}{\partial x^i}\right) \end{bmatrix}\Bigg|_G$$

Thus,

$$g_{\mu v}\big|_G = \begin{pmatrix} 0 & -\frac{1}{2}\rho_0 & 0 & 0 \\ -\frac{1}{2}\rho_0 & 0 & 0 & 0 \\ 0 & 0 & 1 & 0 \\ 0 & 0 & 0 & 1 \end{pmatrix}$$

and

$$g^{\mu v}|_G = \begin{pmatrix} 0 & -\frac{2}{\rho_0} & 0 & 0 \\ -\frac{2}{\rho_0} & 0 & 0 & 0 \\ 0 & 0 & 1 & 0 \\ 0 & 0 & 0 & 1 \end{pmatrix} \quad (have\ let\ \rho|_G = \rho_0)$$

For our choice of *f*:

$$[\ \ ]_1|_G = \begin{bmatrix} -f_{,t}|_G & 0 \\ 0 & -\rho_0^2 f_{,t}|_G \end{bmatrix}_{i=t}, \begin{bmatrix} 0 & -\frac{1}{2} \\ -\frac{1}{2} & 0 \end{bmatrix}_{i=\rho}, [\phi]_{i=z}$$

and

$$\left.\left(\frac{\partial f}{\partial x^i}\right)\right|_G = 4\alpha\tanh(3\alpha^2)\left.\left(\frac{\partial t}{\partial x^i}\right)\right|_G$$
$$= 4\alpha\tanh(3\alpha^2)\ \ (for\ i = t;\ = 0\ otherwise).$$

Using this we can now compute the Riemann tensor:

$$R^\mu_{\nu\alpha\beta} = \Gamma^\mu_{\nu\beta,\alpha} - \Gamma^\mu_{\nu\alpha,\beta} + \Gamma^\mu_{\sigma\alpha}\Gamma^\sigma_{\nu\beta} - \Gamma^\mu_{\sigma\beta}\Gamma^\sigma_{\nu\alpha}$$
$$R^\mu_{\phi\phi\gamma} = \Gamma^\mu_{\phi\gamma,\phi} - \Gamma^\mu_{\phi\phi,\gamma} + \Gamma^\mu_{\sigma\phi}\Gamma^\sigma_{\phi\gamma} - \Gamma^\mu_{\sigma\gamma}\Gamma^\sigma_{\phi\phi}$$

Need $\Gamma^\mu_{\phi\gamma} = g^{\mu\delta}\frac{1}{2}\left(g_{\delta\phi,\gamma} + g_{\delta\gamma,\phi} - g_{\phi\gamma,\delta}\right) = \frac{1}{2}g^{\mu\delta}\left(g_{\phi\delta,\gamma} - g_{\phi\gamma,\delta}\right) = g^{\mu\delta}g_{\phi[\delta,\gamma]}$

We have:

$$\Gamma^\mu_{\phi\gamma}\Big|_G = \frac{1}{2}g^{\mu\delta}\Big|_G\left(g_{\phi\delta,\gamma}\big|_G - g_{\phi\gamma,\delta}\big|_G\right)$$

and

$$g_{\phi\phi,i}\big|_G = -\rho_0[4\alpha\tanh(3\alpha^2)]_{(i=t)}\ \ and\ \ g_{\phi t,i}\big|_G = 0\ \ and\ others\ zero.$$

So,

$$\Gamma^\mu_{\phi i} = \frac{1}{2}g^{\mu\phi}\Big|_G g_{\phi\phi,i}\big|_G \Rightarrow \begin{matrix}\mu = t \\ i = t\end{matrix},\ \text{otherwise zero} \rightarrow \Gamma^t_{\phi i} = [4\alpha\tanh(3\alpha^2)]$$

and

$$\Gamma^\mu_{\phi\phi} = \frac{-1}{2}g^{\mu i}\Big|_G g_{\phi\phi,i} \Rightarrow \begin{matrix}\mu = \phi \\ i = t\end{matrix} \rightarrow -[4\alpha\tanh(3\alpha^2)]$$

So,

$$\left\{\begin{matrix}\Gamma^\phi_{\phi\phi} = -[4\alpha\tanh(3\alpha^2)] \\ \Gamma^t_{\phi t} = [4\alpha\tanh(3\alpha^2)]\end{matrix}\right\}\ \text{other } \Gamma^\mu_{\phi\gamma}\text{'s are zero}$$

Since $\Gamma^\mu_{\sigma\phi}\Gamma^\sigma_{\phi\gamma} = \Gamma^\mu_{\phi\phi}\Gamma^\phi_{\phi\gamma} + \Gamma^\mu_{\phi t}\Gamma^t_{\phi\gamma}$, we have:

$$\Gamma^\phi_{\sigma\phi}\Gamma^\sigma_{\phi\phi} = \left(\Gamma^\phi_{\phi\phi}\right)^2_{\mu=\gamma=\phi}\ \text{and}\ \Gamma^t_{\sigma\phi}\Gamma^\sigma_{\phi t} = \left(\Gamma^t_{\phi t}\right)^2_{\mu=\gamma=\phi}$$

are the only nonzero terms. Also

$$\Gamma^\mu_{\sigma\gamma}\Gamma^\sigma_{\phi\phi} = \Gamma^\mu_{\phi\gamma}\Gamma^\phi_{\phi\phi} \Rightarrow \begin{cases}\left(\Gamma^\phi_{\phi\phi}\right)^2 & \mu = \gamma = \phi \\ \left(\Gamma^t_{\phi t}\right)\left(\Gamma^\phi_{\phi\phi}\right) & \mu = \gamma = t\end{cases}$$

So,

$$\Gamma^{\mu}_{\phi\phi} = g^{\mu v}\frac{1}{2}\left(g_{v\phi,\phi} + g_{v\phi,\phi} - g_{\phi\phi,v}\right) = -\frac{1}{2}g^{\mu v}g_{\phi\phi,v}$$

and

$$\Gamma^{\mu}_{\phi\phi,\gamma} = \tfrac{1}{2}g^{\mu v}_{,\gamma}g_{\phi\phi,v} \sim \tfrac{1}{2}g^{\mu v}g_{\phi\phi,v\gamma}.$$

We need $g^{\mu v}_{,i}$:

$$g^{\mu v}_{,i} = [\quad]_1|_G = \frac{\partial}{\partial x^i}\left[\frac{1}{4}\rho^2(1+f^2)^2\right]^{-1}\begin{bmatrix} -f\rho^2 & -\frac{1}{2}(1+f^2)\rho \\ -\frac{1}{2}(1+f^2)\rho & f \end{bmatrix}$$

$$+\frac{1}{\left[\frac{1}{4}\rho^2(1+f^2)^2\right]}\left[\begin{matrix} \left(-\frac{\partial f}{\partial x^i}\rho^2 + 2\rho\frac{\partial \rho}{\partial x^i}f\right) & \left(f\frac{\partial f}{\partial x^i}\rho - \frac{1}{2}(1+f^2)\frac{\partial \rho}{\partial x^i}\right) \\ (\qquad\qquad) & \frac{\partial f}{\partial x^i} \end{matrix}\right]\Bigg|_G$$

$$= \frac{1}{\left(\rho^2/_4\right)}\begin{bmatrix} -\rho_0^2\left(\frac{\partial f}{\partial x^i}\right) & \frac{1}{2}\frac{\partial \rho}{\partial x^i} \\ \frac{1}{2}\frac{\partial \rho}{\partial x^i} & \left(\frac{\partial f}{\partial x^i}\right) \end{bmatrix}\Bigg|_G$$

So,

$$g^{\mu v}_{,\rho} = \frac{1}{\left(\frac{\rho_0^2}{4}\right)}\begin{bmatrix} 0 & \frac{1}{2} \\ \frac{1}{2} & 0 \end{bmatrix}; \qquad g^{\mu v}_{,t}$$

$$= \frac{1}{\left(\frac{\rho_0^2}{4}\right)}\begin{bmatrix} -\rho_0^2 & 0 \\ 0 & 1 \end{bmatrix}[4\alpha\tanh(3\alpha^2)]; \qquad and \quad g^{\mu v}_{,z} = 0.$$

Now $\Gamma^{\mu}_{\phi\phi\gamma}\Big|_G = -\frac{1}{2}g^{\mu v}_{,\gamma}g_{\phi\phi,v} - \frac{1}{2}g^{\mu v}g_{\phi\phi,v\gamma}\Big|_G = -\frac{1}{2}g^{\mu t}_{,\gamma}g_{\phi\phi,t} - \frac{1}{2}g^{\mu v}g_{\phi\phi,v\gamma}$, so we also need $g_{\phi\phi,v\alpha}$:

$$g_{\phi\phi,v\alpha} = g_{\phi\phi,ij} = \left(2\rho\frac{\partial \rho}{\partial x^j}\frac{\partial f}{\partial x^i} + \rho^2\frac{\partial^2 f}{\partial x^i \partial x^j} + 2\rho\frac{\partial f}{\partial x^j}\frac{\partial \rho}{\partial x^i}\right)$$

where

$$\frac{\partial^2 f}{\partial x^i \partial x^j} = \frac{\partial}{\partial x^j}\left\{\frac{\operatorname{sech}(x^2-\alpha^2)}{\operatorname{sech}(3\alpha^2)}\tanh(x^2-\alpha^2)\,(2x)\frac{\partial x}{\partial x^i}\right\}$$

or

$$\frac{\partial^2 f}{\partial x^i \partial x^j} = [\mathrm{sech}^2(3\alpha^2) - \tanh^2(3\alpha^2)](2x)^2\left(\frac{\partial x}{\partial x^i}\right)\left(\frac{\partial x}{\partial x^j}\right) + \tanh(3\alpha^2)\left[2\left\{\frac{\partial \rho}{\partial x^j}\frac{\partial \rho}{\partial x^i} + \cdots\right\}\right]$$

together with $x = \sqrt{[\rho - \rho_0]^2 + z^2 + t^2}$ we then have:

$$\frac{\partial x}{\partial x^j} = \frac{1/2}{x}\left\{2(\rho - \rho_0)\frac{\partial \rho}{\partial x^j} + 2z\frac{\partial z}{\partial x^j} + 2t\frac{\partial t}{\partial x^j}\right\}$$

$$\frac{\partial^2 x}{\partial x^j \partial x^i} = -\frac{1}{x^2}\left\{2(\rho - \rho_0)\frac{\partial \rho}{\partial x^j} + \cdots\right\}\left\{2(\rho - \rho_0)\frac{\partial \rho}{\partial x^i} + \cdots\right\} + \frac{1}{x}\left\{2\left(\frac{\partial \rho}{\partial x^j}\right)\left(\frac{\partial \rho}{\partial x^i}\right) + \cdots\right\}$$

$$x|_G = 2\alpha$$

$$\left.\frac{\partial x}{\partial x^j}\right|_G = \frac{\frac{1}{2}}{2\alpha}\left\{-(4\alpha)\frac{\partial t}{\partial x^j}\right\} = -\frac{\partial t}{\partial x^j} \quad \text{and}$$
$$\frac{\partial}{\partial x^j}\left(2x\frac{\partial x}{\partial x^j}\right) = 2\left[\left(\frac{\partial \rho}{\partial x^j}\right)^2 + \cdots\right]$$

So, we have

$$\begin{cases} f_{,tt} = [\mathrm{sech}^2(3\alpha^2) - \tanh^2(3\alpha^2)](4\alpha)^2 + 2\tanh(3\alpha^2) \\ f_{,\rho\rho} = f_{,zz} = \frac{1}{\alpha}(4\alpha)\tanh(3\alpha^2) = 2\tanh(3\alpha^2) \end{cases} \quad \text{others zero}$$

Thus,

$$g_{\phi\phi,\rho t} = g_{\phi\phi,t\rho} = -2\rho_0(4\alpha\tanh(3\alpha^2))$$
$$g_{\phi\phi,tt} = \rho_0^2[(\mathrm{sech}^2(3\alpha^2) - \tanh^2(3\alpha^2))(4\alpha)^2 + 2\tanh(3\alpha^2)]$$
$$g_{\phi\phi,\rho\rho} = g_{\phi\phi,zz} = \rho_0^2 2\tanh(3\alpha^2).$$

Let's now return to $\left.\Gamma^{\mu}_{\phi\phi,\gamma}\right|_G = -\frac{1}{2}g^{\mu t}_{,\gamma}g_{\phi\phi,t} - \frac{1}{2}g^{\mu v}g_{\phi\phi,v\gamma}$:

$g^{\mu t}_{,\gamma}$ has two terms $g^{\phi t}_{,\rho}$ and $g^{tt}_{,t}$

and

$g^{\mu v}$ has terms $g^{\phi t}, g^{\rho\rho}, g^{zz}$.

So, for $\mu = \phi$ and $\gamma = \rho$:

$$\Gamma^{\phi}_{\phi\phi,\rho}\Big|_G = \tfrac{1}{2} g^{\phi t}_{,\rho} g_{\phi\phi,t} - \tfrac{1}{2} g^{\phi v} g_{\phi\phi,v\rho} = -\tfrac{1}{2} \underbrace{g^{\phi t}_{,\rho}}\ \underbrace{g_{\phi\phi,t}} - \tfrac{1}{2} g^{\phi t} g_{\phi\phi,t\rho} =$$
$$-[4\alpha \tanh(3\alpha^2)]$$

and for $\mu = t$ and $\gamma = t$:

$$\Gamma^{t}_{\phi\phi,t}\Big|_G = -\frac{1}{2} g^{tt}_{,t} g_{\phi\phi,t} - \frac{1}{2} g^{tv} g_{\phi\phi,vt} = 2\rho_0^2 \left(\frac{\partial f}{\partial t}\right)_G.$$

Similarly:

$$\Gamma^{\phi}_{\phi\phi,t}\Big|_G = -\frac{1}{2}\left(-\frac{2}{\rho_0}\right) g_{\phi\phi,tt}$$
$$= \rho_0[(\mathrm{sech}^2(3\alpha^2) - \tanh^2(3\alpha^2))(4\alpha)^2 + 2\tanh(3\alpha^2)]$$
$$\Gamma^{\rho}_{\phi\phi,t}\Big|_G = -\frac{1}{2}[-2\rho_0(4\alpha \tanh(3\alpha^2))] = +\rho_0(4\alpha \tanh(3\alpha^2))$$
$$\Gamma^{\rho}_{\phi\phi,\rho}\Big|_G = -\frac{1}{2}[2\rho_0^2 \tanh(3\alpha^2)] = -\rho_0^2(\tanh(3\alpha^2))$$
$$\Gamma^{z}_{\phi\phi,z}\Big|_G = -\frac{1}{2}[2\rho_0^2 \tanh(3\alpha^2)] = -\rho_0^2(\tanh(3\alpha^2))$$

So,

$$\Gamma^{\mu}_{\phi\phi,\gamma}\Big|_G = -\frac{1}{2} g^{\mu t}_{,\gamma} g_{\phi\phi,t} - \frac{1}{2} g^{\mu v} g_{\phi\phi,v\gamma}$$
$= -[4\alpha \tanh(3\alpha^2)] \quad \rightarrow \quad \mu = \phi, \gamma = \rho$

$= \rho_0[(\mathrm{sech}^2(3\alpha^2) - \tanh^2(3\alpha^2))(4\alpha)^2 + 2\tanh(3\alpha^2)] \quad \rightarrow$
$\mu = \phi, \gamma = t$

$= +2\rho_0^2[4\alpha \tanh(3\alpha^2)]^2 \quad \rightarrow \quad \mu = t, \gamma = t$

$= +\rho_0(4\alpha\tanh(3\alpha^2)] \quad \rightarrow \quad \mu = \rho, \gamma = t$

$= -\rho_0^2(\tanh(3\alpha^2)) \quad \rightarrow \quad \mu = \rho, \gamma = \rho$

$= -\rho_0^2(\tanh(3\alpha^2)) \quad \mu = z, \gamma = z$

$$\mathrm{R}^{\mu}_{\phi\phi\gamma} = \Gamma^{\mu}_{\sigma\phi}\Gamma^{\sigma}_{\phi\gamma} - \Gamma^{\mu}_{\sigma\gamma}\Gamma^{\sigma}_{\phi\phi} - \Gamma^{\mu}_{\phi\phi,\gamma}$$

For $\mu = \gamma = t$:

$$\mathrm{R}^{t}_{\phi\phi t} = \Gamma^{t}_{\phi t}\left(\Gamma^{t}_{\phi t} + \Gamma^{\phi}_{\phi\phi}\right) - \Gamma^{t}_{\phi\phi,t} = 0$$

For indexing like

$$\begin{pmatrix}\mu = \phi, & \gamma = \rho \\ & \gamma = t\end{pmatrix}; \begin{pmatrix}\mu = \rho, \gamma = t \\ \gamma = \rho\end{pmatrix}; \ or \ (\mu = z, \gamma = z)$$

have

$$\mathrm{R}^{\mu}_{\phi\phi\gamma} = -\Gamma^{\mu}_{\phi\phi,\gamma}.$$

So,

$$\frac{d^2\eta^\mu}{d\phi^2} + \mathrm{R}^\mu_{\phi\phi\gamma}\Big|_G \eta^\gamma = 0$$

becomes

$$\frac{d^2\eta^z}{d\phi^2} + \rho_0^2(\tanh(3\alpha^2))\eta^z = 0$$

(Eqn. 7-137)

with the simple oscillatory solution:

$$\eta^z = a_1 e^{i\rho_0\sqrt{\tanh(3\alpha^2)}\phi} + a_2 e^{-i\rho_0\sqrt{\tanh(3\alpha^2)}\phi}.$$

(Eqn. 7-138)

The other components are now easily obtained as well, but let's summarize first in Sec. 7.4.4 to follow.

7.4.4 Blue shift and divergent energy density

Have geodesic deviation equation on closed null geodesic 'G':

$$\frac{d^2\eta^\mu}{d\phi^2} + R^\mu_{\phi\phi\gamma}\Big|_G \eta^\gamma = 0,$$

(Eqn. 7-139)

where:

$R^z_{\phi\phi z} = \rho_0^2(\tanh(3\alpha^2))$

$R^\rho_{\phi\phi\rho} = R^z_{\phi\phi z}$

$R^\rho_{\phi\phi t} = -\rho_0(4\alpha\tanh(3\alpha^2))$

$R^\phi_{\phi\phi t} = -\rho_0[(4\alpha)^2\{\mathrm{sech}^2(3\alpha^2) - \tanh^2(3\alpha^2)\} + 2\tanh(3\alpha^2)]$

$R^\phi_{\phi\phi\rho} = -3[4\alpha\tanh(3\alpha^2)]$

(Eqn. 7-140)

Recall

$$ds^2 = f(-dt^2 + (\rho d\phi)^2) - (1 - f^2)dt\rho d\phi + d\rho^2 + dz^2$$

with

$$f = 1 - \frac{\mathrm{sech}([\rho - \rho_0] + z^2 + t^2 - \alpha^2)}{\mathrm{sech}(3\alpha^2)}$$

with cng:

$$e(\lambda) = \delta(\rho = \rho_0)\delta(z)\delta(t + 2\alpha)\delta(\phi - f(\lambda))$$

alternatively:

$$\nabla_{\vec{u}}\vec{u} = 0,$$

where $\vec{u}$ is the tangent vector in a coordinate system where only u^ϕ is nonzero:

$$u^\phi = \exp\left\{\left(-\Gamma^\phi_{\phi\phi}\right)_G\right\} = \exp[(4\alpha\rho_0\tanh(3\alpha^2))\phi].$$

(Eqn. 7-141)

For the full solution we then have:

$$\frac{d^2\eta^z}{d\phi^2} + \underbrace{\rho_0^2(\tanh(3\alpha^2))}_{\omega_z^2}\eta^z = 0 \quad \rightarrow \quad \eta^z = a_z e^{i\omega_z\phi} + b_z e^{-i\omega_z\phi} \quad as\ before.$$

and

$$\frac{d^2\eta^t}{d\phi^2} = 0 \quad \rightarrow \quad \eta^t = a_t + b_t\phi$$

and

$$\frac{d^2\eta^\rho}{d\phi^2} + \underbrace{\rho_0^2(\tanh(3\alpha^2))}_{\omega_z^2}\eta^\rho + \underbrace{\rho_0(4\alpha\tanh(3\alpha^2))}_{-c}\eta^t = 0$$

Regrouping:

$$\frac{d^2\eta^\rho}{d\phi^2} + \omega_z^2\eta^\rho = c\eta^t = ca_t + cb_t\phi$$

or,

$$\eta^\rho = \underbrace{a_\rho e^{i\omega_z\phi} + b_\rho e^{-i\omega_z\phi}}_{\text{homogeneous sol'n}} + \underbrace{\frac{ca_t}{\omega_z^2} + \frac{cb_t}{\omega_z^2}\phi}_{\text{inhom. sol'n}}$$

So,

η_z oscillates with frequency $\omega_z = \rho_0\sqrt{\tanh(3\alpha^2)}$;

η^t grows as $b_t\phi$; and

η^ρ grows (and oscillates) as $\frac{cb_t}{\omega_z^2}\phi$.

Since the blue-shifting is exponential in ϕ, while the above divergence field grows linearly at best, we get a divergent overall energy density near the cng.

7.4.5 Verification on application to CTC space-time

We have:

$$ds^2 = -fdt^2 + f(\rho d\phi)^2 - (1 - f^2)dt\rho d\phi + d\rho^2 + dz^2$$

$$g_{\mu\nu} = \begin{pmatrix} \begin{bmatrix} -f & -\frac{1}{2}\rho(1-f^2) \\ -\frac{1}{2}\rho(1-f^2) & f\rho^2 \end{bmatrix} & \begin{matrix} 0 & 0 \\ 0 & 0 \end{matrix} \\ \begin{matrix} 0 & 0 \\ 0 & 0 \end{matrix} & \begin{matrix} 1 & 0 \\ 0 & 1 \end{matrix} \end{pmatrix}$$

Using only $f(\rho, z, t) = 0 \qquad (\rho|_G = \rho_0\,, z|_G = 0\,, t|_G = -2\alpha)$:

$$g_{\mu\nu}\Big|_G = \begin{pmatrix} 0 & -\frac{1}{2}\rho_0 & 0 & 0 \\ -\frac{1}{2}\rho_0 & 0 & 0 & 0 \\ 0 & 0 & 1 & 0 \\ 0 & 0 & 0 & 1 \end{pmatrix}, \quad g^{\mu\nu}|_G = \begin{pmatrix} 0 & -\frac{2}{\rho_0} & 0 & 0 \\ -\frac{2}{\rho_0} & 0 & 0 & 0 \\ 0 & 0 & 1 & 0 \\ 0 & 0 & 0 & 1 \end{pmatrix}$$

$g_{\mu\nu},_\phi = 0$

$$g_{\mu\nu,i}\Big|_G : \begin{bmatrix} -\frac{\partial f}{\partial x^i} & -\frac{1}{2}\frac{\partial \rho}{\partial x^i} \\ -\frac{1}{2}\frac{\partial \rho}{\partial x^i} & \rho^2 \frac{\partial f}{\partial x^i} \end{bmatrix}_G \quad \text{is from the bracket in } g_{\mu\nu} \text{ above}$$

$i = \rho, z, t$

$$g_{\mu\nu,ij}\Big|_G : \begin{bmatrix} -\frac{\partial^2 f}{\partial x^i \partial x^j} & \rho \frac{\partial f}{\partial x^j}\frac{\partial f}{\partial x^i} \\ \rho_0 \frac{\partial f}{\partial x^j}\frac{\partial f}{\partial x^i} & \left(2\rho_0 \frac{\partial \rho}{\partial x^j}\frac{\partial f}{\partial x^i} + \rho_0^2 \frac{\partial^2 f}{\partial x^i \partial x^j} + 2\frac{\partial f}{\partial x^j}\rho_0\frac{\partial \rho}{\partial x^i}\right) \end{bmatrix}_G$$

The cng curve is $\tau(\lambda) = \delta(\rho - \rho_0)\delta(z)\delta(t + 2\alpha)\delta(\phi - f(\lambda))$, so:
$\nabla_{\vec{u}}\vec{u} = 0 \Rightarrow$ only have u^ϕ coord.' s so, u^ϕ

$$= \exp\left[-\Gamma^\phi_{\phi\phi}\Big|_G \phi\right] \begin{pmatrix} \text{choose} \\ u^\phi = 1 \text{ at} \\ \phi = 0 \end{pmatrix}$$

So,

$$u^\phi = \exp\left[-\rho_0 \left(\frac{\partial f}{\partial t}\right)\Big|_G \phi\right]$$

as before. For the geodesic deviation equation:

$$\frac{d^2\eta^2}{d\tau^2} + R^\mu_{\alpha\beta\gamma}K^\alpha K^\beta \eta^\gamma = 0,$$

since there is only a K^α comp. (mentioned above as u^ϕ) we have $\left(K^\nu = \frac{\partial x^\nu}{\partial \tau}\right)$

$$\frac{d^2\eta^\mu}{d\phi^2} + R^\mu_{\phi\phi\gamma}\Big|_G \eta^\gamma = 0$$

Now, $R^{\mu}_{\phi\phi\gamma} = \Gamma^{\mu}_{\phi\phi\gamma} + \Gamma^{\mu}_{\sigma\phi}\Gamma^{\sigma}_{\phi\gamma} - \Gamma^{\mu}_{\sigma\gamma}\Gamma^{\sigma}_{\phi\phi}$, let's consider $\Gamma^{\mu}_{\phi\gamma}$ to begin:

$$\underline{\Gamma^{\mu}_{\phi\gamma}} = g^{\mu\delta}\frac{1}{2}\left(g_{\delta\phi,\gamma} + g_{\delta\gamma,\phi} - g_{\phi\gamma,\delta}\right)$$

$$\Gamma^{\mu}_{\phi\gamma}\Big|_G = \frac{1}{2}g^{\mu\delta}\Big|_G\left(g_{\delta\phi,\gamma}\Big|_G - g_{\phi\gamma,\delta}\Big|_G\right)$$

$\underline{\mu = t \text{ case:}}$

$$\Gamma^{t}_{\phi\gamma}\Big|_G = \frac{1}{2}g^{t\phi}\Big|_G\left(g_{\phi\phi,\gamma}\Big|_G\right)$$

$$\boxed{\Gamma^{t}_{\phi i}\Big|_G = \rho_0\left(\frac{\partial f}{\partial x^i}\right)_G}$$

$\underline{\mu = \phi \text{ case:}}$

$$\Gamma^{\phi}_{\phi\gamma}\Big|_G = \frac{1}{2}g^{\phi t}\Big|_G\left(g_{t\phi,\gamma}\Big|_G - g_{\phi\gamma,t}\Big|_G\right)$$

$$\Gamma^{\phi}_{\phi\phi}\Big|_G = \frac{1}{2}\left(-\frac{2}{\rho_0}\right)\left(-\rho_0^2\frac{\partial f}{\partial t}\Big|_G\right) = \underline{\rho_0\frac{\partial f}{\partial t}\Big|_G}$$

$$\Gamma^{\phi}_{\phi i}\Big|_G = \left(-\frac{2}{\rho_0}\right)\left(g_{t\phi,i}\Big|_G - g_{\phi i,t}\Big|_G\right) \qquad \begin{array}{c} i = t \Rightarrow 0 \\ i = z \Rightarrow \dfrac{\partial \rho}{\partial z} = 0 \end{array}$$

$$\boxed{\Gamma^{\phi}_{\phi\rho}\Big|_G = \frac{1}{2\rho_0}}$$

So, $$\boxed{\Gamma^{\mu}_{\phi\gamma}: \begin{cases} \Gamma^{t}_{\phi i}\Big|_G = -\rho_0\left(\frac{\partial f}{\partial x^i}\right)_G \\ \Gamma^{\phi}_{\phi\phi}\Big|_G = \rho_0\left(\frac{\partial f}{\partial t}\right)_G \\ \Gamma^{\phi}_{\phi\rho}\Big|_G = \frac{1}{2\rho_0} \end{cases}}$$

$$\Gamma^{\mu}_{\phi\sigma}\Gamma^{\sigma}_{\phi\gamma} = \Gamma^{\mu}_{\phi\phi}\Gamma^{\phi}_{\phi\gamma} + \Gamma^{\mu}_{\phi t}\Gamma^{t}_{\phi\gamma} = \Gamma^{\phi}_{\phi\phi}\Gamma^{\phi}_{\phi\phi}\Big|_{\substack{\mu=\phi\\ \gamma=\phi\\ \text{zero otherwise}}} + \overbrace{\Gamma^{t}_{\phi t}\Gamma^{t}_{\phi i}}^{\substack{\text{zero others}\\ \mu=t\ \gamma=i\\ \uparrow\ \ \uparrow}}$$

$$\boxed{\Gamma^{\mu}_{\phi\sigma}\Gamma^{\sigma}_{\phi\gamma}: \begin{cases} \left.\left(\Gamma^{\phi}_{\phi\sigma}\Gamma^{\sigma}_{\phi\phi}\right)\right|_G = \rho_0^2\left(\frac{\partial f}{\partial t}\right)_G\left(\frac{\partial f}{\partial t}\right)_G \\ \left(\Gamma^{t}_{\phi\sigma}\Gamma^{\sigma}_{\phi i}\right)_G = \rho_0^2\left(\frac{\partial f}{\partial t}\right)_G\left(\frac{\partial f}{\partial x^i}\right)_G \end{cases}}$$

$$\Gamma^{\mu}_{\sigma\gamma}\Gamma^{\sigma}_{\phi\phi} = \Gamma^{\mu}_{\phi\gamma}\Gamma^{\phi}_{\phi\phi}$$

$$\boxed{\Gamma^{\mu}_{\sigma\gamma}\Gamma^{\sigma}_{\phi\phi}: \begin{cases} \Gamma^{t}_{\sigma i}\Gamma^{\sigma}_{\phi\phi}\big|_G = -\rho_0^2\left(\frac{\partial f}{\partial t}\right)_G\left(\frac{\partial f}{\partial x^i}\right)_G \\ \Gamma^{\phi}_{\sigma\phi}\Gamma^{\sigma}_{\phi\phi}\big|_G = \rho_0^2\left(\frac{\partial f}{\partial t}\right)_G\left(\frac{\partial f}{\partial t}\right)_G \\ \Gamma^{\phi}_{\sigma\rho}\Gamma^{\sigma}_{\phi\phi}\big|_G = \frac{1}{2}\left(\frac{\partial f}{\partial t}\right)_G \end{cases}}$$

$$\Gamma^{\mu}_{\phi\phi} = g^{\mu\nu}\frac{1}{2}\left(g_{\nu\phi,\phi} + g_{\nu\phi,\phi} - g_{\phi\phi,\nu}\right) = -\frac{1}{2}g^{\mu\nu}g_{\phi\phi,\nu}$$

$$\underline{\Gamma^{\mu}_{\phi\phi,\gamma}} = \frac{1}{2}g^{\mu\nu}_{,\gamma}g_{\phi\phi,\nu} - \frac{1}{2}g^{\mu\nu}g_{\phi\phi,\nu\gamma}$$

Need $g^{\mu\nu}$ and then $g^{\mu\nu}_{,\gamma}$

$$g^{\mu\nu} = \begin{pmatrix} \frac{1}{\left[-\frac{1}{4}\rho^2(1-f^2)^2\right]} & \begin{bmatrix} f\rho^2 & \frac{1}{2}(1-f^2)\rho \\ \frac{1}{2}(1-f^2)\rho & -f \end{bmatrix} & \begin{matrix} 0 & 0 \\ 0 & 0 \end{matrix} \\ & \begin{matrix} 0 & 0 \\ 0 & 0 \end{matrix} & \begin{matrix} 1 & 0 \\ 0 & 1 \end{matrix} \end{pmatrix}$$

$$g^{\mu\nu}_{,i}: \ \frac{\partial}{\partial x^i}\left\{\frac{1}{\left[-\frac{1}{4}\rho^2(1-f^2)^2\right]}[f\rho^2 \ldots]\right\}$$

$$= +4\left(\frac{2\rho\frac{\partial\rho}{\partial x^i}(1-f^2)^2+2\rho^2(1-f^2)2f\frac{\partial f}{\partial x^i}}{[\rho^2(1-f^2)^2]^2}\right)[\ldots.] + \frac{1}{[\ \ ]}\frac{\partial}{\partial x^i}[\ldots]$$

$$= \frac{8}{\rho^3}\frac{\partial\rho}{\partial x^i}\begin{bmatrix} 0 & \frac{1}{2}\rho \\ \frac{1}{2}\rho & 0 \end{bmatrix} +$$

$$\frac{1}{\left(-\rho^2/_4\right)}\begin{bmatrix} \left(\frac{\partial f}{\partial x^i}\rho^2 + 2\rho\frac{\partial\rho}{\partial x^i}f\right) & \left(\frac{1}{2}\frac{\partial\rho}{\partial x^i}(1-f^2) + \frac{1}{2}\rho\left(-2f\frac{\partial f}{\partial x^i}\right)\right) \\ (\ldots) & -\frac{\partial f}{\partial x^i} \end{bmatrix}$$

$$= \frac{1}{\left(\rho^2/_4\right)}\begin{bmatrix} 0 & \frac{\partial\rho}{\partial x^i} \\ \frac{\partial\rho}{\partial x^i} & 0 \end{bmatrix} - \frac{1}{\left(\rho^2/_4\right)}\begin{bmatrix} \rho^2\left(\frac{\partial f}{\partial x^i}\right) & \frac{1}{2}\frac{\partial\rho}{\partial x^i} \\ \frac{1}{2}\frac{\partial\rho}{\partial x^i} & -\frac{\partial f}{\partial x^i} \end{bmatrix}$$

$$= \frac{1}{\left(\rho^2/_4\right)}\begin{bmatrix} -\rho^2\left(\frac{\partial f}{\partial x^i}\right) & \frac{1}{2}\frac{\partial\rho}{\partial x^i} \\ \frac{1}{2}\frac{\partial\rho}{\partial x^i} & \frac{\partial f}{\partial x^i} \end{bmatrix}$$

$$g^{\mu\nu}_{,\gamma}\, g_{\phi\phi,\nu} = g^{\mu\phi}_{,\gamma}\, g_{\phi\phi,\phi} + g^{\mu t}_{,\gamma}\, g_{\phi\phi,t} + g^{\mu\rho}\ldots + g^{\mu z}\ldots$$

$$g^{tt}_{,i}\, g_{\phi\phi,t} = \frac{1}{\left(\rho^2/_4\right)}\left(-\rho^2\left(\frac{\partial f}{\partial x^i}\right)\right)\left(\rho^2\frac{\partial f}{\partial t}\right) = -4\rho^2\left(\frac{\partial f}{\partial x^i}\right)_G\left(\frac{\partial f}{\partial t}\right)_G$$

$$g^{\phi t}_{,\rho}\, g_{\phi\phi,t} = \frac{1}{\left(\rho^2/_4\right)}\left(\frac{1}{2}\right)\left(\rho^2\left(\frac{\partial f}{\partial t}\right)\right) = 2\left(\frac{\partial f}{\partial t}\right)_G$$

$g_{\phi\phi,\nu\gamma} \neq 0$ only for $v, \gamma \neq \phi$

$$g_{\phi\phi,ij} = \frac{\partial}{\partial x^j}\left(2\rho\frac{\partial\rho}{\partial x^i}f + \rho^2\frac{\partial f}{\partial x^i}\right)$$

$$g_{\phi\phi,ij}\big|_G = 2\rho_0\frac{\partial\rho}{\partial x^i}\frac{\partial f}{\partial x^j} + 2\rho_0\frac{\partial\rho}{\partial x^j}\frac{\partial f}{\partial x^i} + \rho_0^2\frac{\partial^2 f}{\partial x^i\partial x^j}$$

$$g^{\mu\nu}g_{\phi\phi,\nu\gamma} = g^{\mu i}g_{\phi\phi,i\gamma} = g^{\mu t}g_{\phi\phi,t\gamma} + g^{\mu\rho}g_{\phi\phi,\rho\gamma} + g^{\mu z}g_{\phi\phi,z\gamma}$$

$$
g^{\mu\nu} g_{\phi\phi,\nu\gamma}: \quad \begin{cases}
g^{\phi t} g_{\phi\phi,t\rho} = \left(-\dfrac{2}{\rho_0}\right)\left[2\rho_0\left(\dfrac{\partial f}{\partial t}\right) + \rho_0^2\left(\dfrac{\partial^2 f}{\partial t \partial \rho}\right)\right]_G \\
g^{\phi t} g_{\phi\phi,tz} = \left(-\dfrac{2}{\rho_0}\right)\left[\rho_0^2\left(\dfrac{\partial^2 f}{\partial t \partial z}\right)\right]_G \\
g^{\phi t} g_{\phi\phi,tt} = \left(-\dfrac{2}{\rho_0}\right)\left[\rho_0^2\left(\dfrac{\partial^2 f}{\partial t^2}\right)\right]_G \\
g^{\rho\rho} g_{\phi\phi,\rho t} = \left[2\rho_0\dfrac{\partial f}{\partial t} + \rho_0^2\left(\dfrac{\partial^2 f}{\partial \rho \partial t}\right)\right]_G \\
g^{\rho\rho} g_{\phi\phi,\rho\rho} = \left[4\rho_0\dfrac{\partial f}{\partial t} + \rho_0^2\left(\dfrac{\partial^2 f}{\partial \rho^2}\right)\right]_G = \left[3\rho\left(\dfrac{\partial f}{\partial \rho}\right) + \rho^2\left(\dfrac{\partial^2 f}{\partial \rho^2}\right.\right. \\
g^{\rho\rho} g_{\phi\phi,\rho z} = \left[2\rho_0\left(\dfrac{\partial f}{\partial z}\right) + \rho_0^2\left(\dfrac{\partial^2 f}{\partial \rho \partial z}\right)\right]_G \\
g^{zz} g_{\phi\phi,zz} = \left(\rho_0^2\dfrac{\partial^2 f}{\partial z^2}\right)_G \\
g^{zz} g_{\phi\phi,z\rho} = \left[2\rho_0\left(\dfrac{\partial f}{\partial z}\right) + \rho_0^2\left(\dfrac{\partial^2 f}{\partial \rho \partial z}\right)\right]_G \\
g^{zz} g_{\phi\phi,zt} = \left(\rho_0^2\dfrac{\partial^2 f}{\partial \rho \partial t}\right)_G
\end{cases}
$$

$$R^{\mu}_{\phi\phi\gamma} = -\Gamma^{\mu}_{\phi\phi,\gamma} + \Gamma^{\mu}_{\sigma\phi}\,\Gamma^{\sigma}_{\phi\gamma} - \Gamma^{\mu}_{\sigma\gamma}\,\Gamma^{\sigma}_{\phi\phi}:$$

$$R^{\phi}_{\phi\phi\rho} = -\left(\frac{\partial f}{\partial t}\right)_G - \rho_0\left(\frac{\partial^2 f}{\partial t \partial \rho}\right)_G$$

$$R^{t}_{\phi\phi t} = -2\rho_0^2\left(\frac{\partial f}{\partial x^i}\right)_G\left(\frac{\partial f}{\partial t}\right)_G + \rho_0^2\left(\frac{\partial f}{\partial x^i}\right)\left(\frac{\partial f}{\partial t}\right) + \rho_0^2\left(\frac{\partial f}{\partial t}\right)\left(\frac{\partial f}{\partial x^i}\right) = 0$$

$$R^{\phi}_{\phi\phi\phi} = \rho_0^2\left(\frac{\partial f}{\partial t}\right)_G^2 - \left(\frac{\partial f}{\partial t}\right)_G^2 = 0$$

$$R^{\phi}_{\phi\phi z} = -\rho_0\left(\frac{\partial^2 f}{\partial t \partial z}\right)_G$$

$$R^{\phi}_{\phi\phi t} = -\rho_0\left(\frac{\partial^2 f}{\partial t^2}\right)_G$$

$$R^{\rho}_{\phi\phi t} = -\rho_0\left(\frac{\partial f}{\partial t}\right)_G + \frac{1}{2}\rho_0^2\left(\frac{\partial^2 f}{\partial \rho \partial t}\right)_G$$

$$R^{\rho}_{\phi\phi\rho} = \frac{3}{2}\rho_0\left(\frac{\partial f}{\partial \rho}\right)_G + \frac{1}{2}\rho_0^2\left(\frac{\partial^2 f}{\partial \rho^2}\right)_G$$
$$R^{\rho}_{\phi\phi z} = \rho_0\left(\frac{\partial f}{\partial z}\right)_G + \frac{1}{2}\rho_0^2\left(\frac{\partial^2 f}{\partial \rho \partial z}\right)_G$$
$$R^{z}_{\phi\phi z} = \frac{1}{2}\rho_0^2\left(\frac{\partial^2 f}{\partial z^2}\right)_G$$
$$R^{z}_{\phi\phi t} = \frac{1}{2}\rho_0^2\left(\frac{\partial^2 f}{\partial z \partial t}\right)_G$$

Do the results check for $f = 1 - \frac{\operatorname{sech}([\rho-\rho_0]^2+z^2+t^2-\alpha^2)}{\operatorname{sech}(3\alpha^2)}$?

$$\frac{\partial f}{\partial x^i} = \frac{\operatorname{sech}(\dots)}{\operatorname{sech}(3\alpha^2)}\tanh(\dots)\frac{\partial}{\partial x^i}\{[\rho-\rho_0]^2+z^2+t^2\}\Big|_{\substack{\rho=\rho_0 \\ z=0 \\ t=-2\alpha}}$$
$$= 2(-2\alpha)\tanh(3\alpha^2) \quad i = t\text{, zero otherwise}$$

$$\frac{\partial^2 f}{\partial x^i \partial x^j} = -\tanh^2(3\alpha^2)\underbrace{\frac{\partial}{\partial x^j}\{\ \}\frac{\partial}{\partial x^i}\{\ \}}_{\uparrow}$$
$$+\operatorname{sech}^2(3\alpha^2)[\dots.] + \tanh(3\alpha^2)\frac{\partial^2 f}{\partial x^i \partial x^j}\{\ \}$$
$$= [\operatorname{sech}^2(3\alpha^2) - \tanh^2(3\alpha^2)](4\alpha)^2 + 2\tanh(3\alpha^2) \quad i = j = t$$
$$= 2\tanh(3\alpha^2) \quad i = j = \rho \text{ or } z \text{ rest zero}$$

So,

$$R^{\phi}_{\phi\phi\rho} = -(-4\alpha\tanh(3\alpha^2)) = 4\alpha\tanh(3\alpha^2)$$
$$R^{\phi}_{\phi\phi t} = -\rho_0\{(4\alpha)^2\{\operatorname{sech}^2(3\alpha^2) - \tanh^2(3\alpha^2)\} + 2\tanh(3\alpha^2)\}$$
$$R^{\rho}_{\phi\phi t} = \rho_0(4\alpha)^2\tanh(3\alpha^2)$$
$$R^{\rho}_{\phi\phi\rho} = \frac{1}{2}\rho_0^2(2\tanh(3\alpha^2)) = \rho_0^2\tanh(3\alpha^2) = R^{z}_{\phi\phi z}$$
$$R^{t}_{\phi\phi t} = 0$$

This checks with the previous result. And on cng:

$$R^{\phi}_{\phi\phi\rho}\Big|_G = -\left(\frac{\partial f}{\partial t}\right)_G - \rho_0\left(\frac{\partial^2 f}{\partial t\partial\rho}\right)_G$$
$$R^{\rho}_{\phi\phi t}\Big|_G = \rho_0\left(\frac{\partial f}{\partial t}\right)_G + \frac{1}{2}\rho_0^2\left(\frac{\partial^2 f}{\partial t\partial\rho}\right)_G$$
$$R^{\phi}_{\phi\phi t}\Big|_G = -\rho_0\left(\frac{\partial^2 f}{\partial t^2}\right)_G$$
$$R^{\rho}_{\phi\phi\rho}\Big|_G = 3\rho_0\left(\frac{\partial f}{\partial\rho}\right)_G + \frac{1}{2}\rho_0^2\left(\frac{\partial^2 f}{\partial\rho^2}\right)_G$$
$$R^{z}_{\phi\phi z}\Big|_G = \frac{1}{2}\rho_0^2\left(\frac{\partial^2 f}{\partial z^2}\right)_G$$
$$R^{\phi}_{\phi\phi z}\Big|_G = -\rho_0\left(\frac{\partial^2 f}{\partial t\partial z}\right)_G$$
$$R^{\rho}_{\phi\phi z}\Big|_G = \rho_0\left(\frac{\partial f}{\partial z}\right)_G + \frac{1}{2}\rho_0^2\left(\frac{\partial^2 f}{\partial\rho\partial z}\right)_G$$
$$R^{z}_{\phi\phi t}\Big|_G = \frac{1}{2}\rho_0^2\left(\frac{\partial^2 f}{\partial z\partial t}\right)_G$$

(Eqn. 7-142)

A general analysis for the various possible signs of the above and the subsequent behavior will be done next, but before moving on, let's first summarize the general solution to.

$$\frac{d^2\eta^\mu}{d\phi^2} + R^{\mu}_{\phi\phi\gamma}\Big|_G\eta^\gamma = 0$$

and denote $\dot{\eta}^\mu = \frac{d\eta^\mu}{d\phi}$:

$$\ddot{\eta}^\phi + R^{\phi}_{\phi\phi\rho}\Big|_G\eta^\rho + R^{\phi}_{\phi\phi t}\Big|_G\eta^t + R^{\phi}_{\phi\phi z}\Big|_G\eta^z = 0$$
$$\ddot{\eta}^\rho + R^{\rho}_{\phi\phi\rho}\Big|_G\eta^\rho + R^{\rho}_{\phi\phi t}\Big|_G\eta^t + R^{\rho}_{\phi\phi z}\Big|_G\eta^z = 0$$
$$\ddot{\eta}^z + 0 - \frac{2}{\rho_0}R^{\phi}_{\phi\phi z}\Big|_G\eta^t + R^{z}_{\phi\phi z}\Big|_G\eta^z = 0$$
$$\ddot{\eta}^t + 0 + 0 + 0 = 0$$

(Eqn. 7-143)

Since $\ddot{\eta}^t = 0$ we get $\eta^t = a + b\phi$ thus no oscillatory behavior in η^t.

For η^z we get $\eta^z = c + d\phi + fe^{-i\omega_z\phi} + ge^{+iw_z\phi}$, where ω_z could be complex, depending on the sign of $R^{z}_{\phi\phi z}\Big|_G$.

For η^ρ we get $\eta^\rho = \alpha + \beta\phi + \gamma e^{-i\omega_z\phi} + \delta e^{i\omega_z\phi} + \mu e^{-i\omega_\rho\phi} + \nu e^{i\omega_\rho\phi}$, with ω_ρ also possibly complex

So, η^t goes as $a + b\phi$ and for large ϕ, η^z and η^ρ are either dominated by a linear term in $\phi, d\phi$ or $\beta\phi$, or an increasing exponential term in ϕ. But regardless of the possible exponential behavior in η^z or η^ρ since η^t is only linear the energy density for an exponentially blue-shifting field with ϕ will diverge as $\phi \to \infty$.

7.5 Generalization of Sech space-time

7.5.1 The first generalization

Let's now consider the null vector comprising our critical cng with more care, by first generalizing the form of the null vector under consideration. So returning to the comparison of Minkowski and the toroidal CTC space-time:

$$ds^2 = -dt^2 + (\rho d\phi)^2 + d\rho^2 + dz^2$$
$$ds^2 = -dt\rho d\phi + d\rho^2 + dz^2$$

For $ds^2 = -dt^2 + \cdots$ the future oriented vector is (unit norm.)

$$(X,X) = -1 \Rightarrow X = \frac{\partial}{\partial t}$$

For $s^2 = -dt\rho d\phi + \cdots$ the future oriented vector is:

$$(X,X) = -1 \Rightarrow X = \frac{1}{\sqrt{2}}\left(\frac{\partial}{\partial t} + \frac{1}{\rho}\frac{\partial}{\partial \phi}\right)$$

which checks:

$$(X,X) = \frac{1}{2}\left(\frac{\partial}{\partial t}\cdot\frac{\partial}{\partial t} + \frac{2}{\rho}\frac{\partial}{\partial t}\cdot\frac{\partial}{\partial \phi} + \frac{1}{\rho^2}\frac{\partial}{\partial \phi}\cdot\frac{\partial}{\partial \phi}\right) = \frac{2}{2\rho}(-\rho) = -1$$

Vector orthogonal to X, Y, is $Y = \frac{1}{\sqrt{2}}\left(\frac{\partial}{\partial t} - \frac{1}{\rho}\frac{\partial}{\partial \phi}\right)$ with $(Y,Y) = 1$ spacelike.

So, consider the general null vector $K^a = t^a + z^a \cos\alpha + \sin\alpha\,(\cos\beta\, X^a + \sin\beta\, Y^a)$, where

$$t^a = \frac{1}{\sqrt{2}}\left(\frac{\partial}{\partial t} + \frac{1}{\rho}\frac{\partial}{\partial \phi}\right)$$
$$z^a = \frac{1}{\sqrt{2}}\left(\frac{\partial}{\partial t} - \frac{1}{\rho}\frac{\partial}{\partial \phi}\right)$$

$$X^a = \frac{\partial}{\partial\rho}$$
$$Y^a = \frac{\partial}{\partial z}$$

The critical $R_{\mu v}K^\mu K^v$ term is now:

$$R_{\mu v}K^\mu K^v = R_{\mu v}\left(\frac{1}{\sqrt{2}}\left(\frac{\partial}{\partial t}+\frac{1}{\rho}\frac{\partial}{\partial\phi}\right)+\cos\alpha\frac{1}{\sqrt{2}}\left(\frac{\partial}{\partial t}-\frac{1}{\rho}\frac{\partial}{\partial\phi}\right)\right.$$
$$\left.+\sin\alpha\cos\beta\frac{\partial}{\partial\rho}+\sin\alpha\sin\beta\frac{\partial}{\partial z}\right)^2$$

$$R_{\mu v}K^\mu K^v = R_{\mu v}\left\{\frac{(1+\cos\alpha)}{\sqrt{2}}\frac{\partial}{\partial t}+\frac{(1-\cos\alpha)}{\sqrt{2}}\frac{1}{\rho}\frac{\partial}{\partial\phi}+\sin\alpha\cos\beta\frac{\partial}{\partial\rho}\right.$$
$$\left.+\sin\alpha\sin\beta\frac{\partial}{\partial z}\right\}^2$$

$$= R_{\mu v}\left\{\frac{(1+\cos\alpha)^2}{2}\left(\frac{\partial}{\partial t}\right)^2+\frac{(1-\cos^2\alpha)}{\rho}\frac{\partial}{\partial t}\frac{\partial}{\partial\phi}\right.$$
$$+\frac{2(1+\cos\alpha)}{\sqrt{2}}\sin\alpha\cos\beta\frac{\partial}{\partial t}\frac{\partial}{\partial\rho}$$
$$+\frac{2(1+\cos\alpha)}{\sqrt{2}}\sin\alpha\sin\beta\frac{\partial}{\partial t}\frac{\partial}{\partial z}$$
$$+\frac{(1-\cos\alpha)^2}{2}\frac{1}{\rho_0}\left(\frac{\partial}{\partial\phi}\right)^2$$
$$+\frac{2(1-\cos\alpha)}{\sqrt{2}}\frac{\sin\alpha\cos\beta}{\rho}\frac{\partial}{\partial\phi}\frac{\partial}{\partial\rho}$$
$$\left.+\frac{2(1-\cos\alpha)}{\sqrt{2}}\frac{\sin\alpha\sin\beta}{\rho}\frac{\partial}{\partial\phi}\frac{\partial}{\partial z}+\sin^2\alpha\cos^2\beta\left(\frac{\partial}{\partial z}\right)^2\right\}$$

$$= R_{tt}\frac{(1+\cos\alpha)^2}{2}+R_{t\phi}\frac{(1-\cos^2\alpha)}{\rho_0}+R_{t\rho}\sqrt{2}(1+\cos\alpha)\sin\alpha\cos\beta$$
$$+\sqrt{2}(1+\cos\alpha)\sin\alpha\sin\beta\, R_{tz}+R_{\phi\phi}\frac{(1-\cos\alpha)^2}{2\rho_0^2}$$
$$+\sqrt{2}(1-\cos\alpha)\frac{\sin\alpha\cos\beta}{\rho_0}R_{\phi\rho}$$
$$+\sqrt{2}(1-\cos\alpha)\frac{\sin\alpha\sin\beta}{\rho_0}R_{\phi z}+\sin^2\alpha\cos^2\beta\, R_{\rho\rho}$$
$$+2\sin^2\alpha\sin^2\beta\cos\beta\, R_{\rho z}+\sin^2\alpha\sin^2\beta\, R_{zz}$$

Let $x_1 = \cos\alpha, x_2 = \sin\alpha\cos\beta, x_3 = \sin\alpha\sin\beta$, $x_1^2 + x_2^2 + x_3^2 = 1$

$$= \frac{R_{tt}}{2}(1 + 2x_1 + x_1^2) + \frac{R_{t\phi}}{\rho_0}(1 - x_1^2) + R_{t\rho}\sqrt{2}(1 + x_1)x_2 + R_{tz}\sqrt{2}(1 + x_1)x_3 + \frac{R_{\phi\phi}}{2\rho_0^2}(1 + x_1)^2 + \frac{R_{\phi\rho}\sqrt{2}}{\rho_0}(1 + x_1)x_2 + \frac{R_{\phi z}\sqrt{2}}{\rho_0}(1 + x_1)x_3 + R_{\rho\rho}x_2^2 + 2x_2x_3R_{\rho z} + R_{zz}x_3^2$$

$$= \left(\frac{R_{tt}}{2} + \frac{R_{t\phi}}{\rho_0} + \frac{R_{\phi\phi}}{2\rho_0^2}\right) + \left(R_{tt} - \frac{R_{\phi\phi}}{\rho_0^2}\right)x_1 + \left(\frac{R_{tt}}{2} + \frac{R_{t\phi}}{\rho_0} + \frac{R_{\phi\phi}}{2\rho_0^2}\right)x_1^2 + \left(R_{\phi\rho}\sqrt{2} + \frac{R_{\phi\rho}\sqrt{2}}{\rho_0}\right)x_2 + \left(R_{tz}\sqrt{2} + \frac{R_{\phi z}\sqrt{2}}{\rho_0}\right)x_3 + \left(R_{t\rho}\sqrt{2} - \frac{R_{\phi\rho}\sqrt{2}}{\rho_0}\right)x_1x_2 + \left(R_{tz}\sqrt{2} + \frac{R_{\phi z}\sqrt{2}}{\rho_0}\right)x_1x_3 + (2R_{\rho z})x_2x_3 + R_{\rho\rho}x_2^2 + R_{zz}x_3^2$$

So, $R_{\mu v}K^{\mu}K^{v} = A + Bx_1 + Cx_2 + Dx_3 + Ex_1x_2 + Ex_1x_3 + Gx_2x_3 + Hx_1^2 + Ix_2^2 + Jx_3^2$

(Eqn. 7-144)

With $A = \left(\frac{R_{tt}}{2} + \frac{R_{t\phi}}{\rho_0} + \frac{R_{\phi\phi}}{2\rho_0^2}\right), B = \left(R_{tt} - \frac{R_{\phi\phi}}{\rho_0^2}\right), C = \left(R_{t\rho}\sqrt{2} - \frac{R_{\phi\rho}\sqrt{2}}{\rho_0}\right)$

$$D = \left(R_{tz}\sqrt{2} + \frac{R_{\phi z}\sqrt{2}}{\rho_0}\right), E = \left(R_{t\rho}\sqrt{2} - \frac{R_{\phi\rho}\sqrt{2}}{\rho_0}\right), F = \left(R_{tz}\sqrt{2} + \frac{R_{\phi z}\sqrt{2}}{\rho_0}\right)$$

$$G = 2R_{\rho z}, H = \left(\frac{R_{tt}}{2} + \frac{R_{t\phi}}{\rho_0} + \frac{R_{\phi\phi}}{2\rho_0^2}\right), I = R_{\rho\rho}, J = R_{zz}$$

(Eqn. 7-145)

Given the above general form:
(1) Does the cng satisfy WEC for some *f*?
(2) Does the entire space-time satisfy WEC. for some *f*?

7.5.2 Evaluation of the space-time curvature using the Cartan Method

Let's evaluate the $R_{\mu v}$ using Cartan's method:

$$\begin{aligned} ds^2 &= f(-dt^2 + (\rho d\phi)^2) - (1 - f^2)dt\rho d\phi + d\rho^2 + dz^2 \\ &= (fdt + \rho d\phi)(-dt + f\rho d\phi) + d\rho^2 + dz^2 \\ &= \omega^1\omega^2 + (\omega^3)^2 + (\omega^4)^2 \end{aligned}$$

(Eqn. 7-146)

where

$$\begin{aligned} \omega^1 &= fdt + \rho d\phi \\ \omega^2 &= -dt + f\rho d\phi \\ \omega^3 &= d\rho \\ \omega^4 &= dz \end{aligned}$$

(Eqn. 7-147)

$$d\omega^\mu = -\omega^\mu_v \wedge \omega^v$$

$$g_{\mu v} = \begin{pmatrix} 0 & \frac{1}{2} & 0 & 0 \\ \frac{1}{2} & 0 & 0 & 0 \\ 0 & 0 & 1 & 0 \\ 0 & 0 & 0 & 1 \end{pmatrix} = \text{const}$$

So, $dg_{\mu v} = 0$ thus, $\omega_{\mu v} = -\omega_{v\mu}$ i.e., there are only 6 1-forms.

$\omega^3 = d\rho \rightarrow d\omega^3 = 0$ ... compare with $d\omega^3 = -\omega^3_v \wedge \omega^v$.

$d\omega^3 = -[\omega^3_1 \wedge \omega^1 + \omega^3_2 \wedge \omega^2 + \omega^3_3 \wedge \omega^3 + \omega^3_4 \wedge \omega^4]$.

So, $\omega^3_k \propto \omega^k$, or $\omega^3_k = 0$, or in more complicated ways. Similarly for $d\omega^4$ So assume:

$$\omega^3_k = 0, \omega^4_j = 0$$

Since

$$\omega^1 = fdt + \rho d\phi$$

$$d\omega^1 = \frac{\partial f}{\partial \rho} d\rho \wedge dt + \frac{\partial f}{\partial z} dz \wedge dt + d\rho \wedge d\phi = ?$$

Note that

$$\omega^1 + f\omega^2 = \rho(1 + f^2)d\phi \rightarrow d\phi = \frac{1}{\rho(1+f^2)}[\omega^1 + f\omega^2]$$

and

$$f\omega^1 - \omega^2 = (f^2+1)dt \rightarrow \quad dt = \frac{1}{(1+f^2)}[f\omega^1 - \omega^2],$$

So,

$$d\omega^1 = \left(\frac{\partial f}{\partial \rho}\right)\omega^3 \wedge \left\{\frac{1}{(1+f^2)}[f\omega^1 - \omega^2]\right\} + \left(\frac{\partial f}{\partial z}\right)\omega^4 \wedge \left\{\frac{1}{(1+f^2)}[f\omega^1 - \omega^2]\right\} + \omega^3 \wedge \left\{\frac{1}{\rho(1+f^2)}[\omega^1 + f\omega^2]\right\}$$

$$d\omega^1 = \left(\frac{\partial f}{\partial \rho}\right)\frac{1}{(1+f^2)}[f\omega^3 \wedge \omega^1 - \omega^3 \wedge \omega^2] + \left(\frac{\partial f}{\partial x}\right)\frac{1}{(1+f^2)}[f\omega^4 \wedge \omega^1 - \omega^4 \wedge \omega^2] + \frac{1}{\rho(1+f^2)}[\omega^3 \wedge \omega^1 + f\omega^3 \wedge \omega^2]$$

$$d\omega^1 = \frac{1}{(1+f^2)}\left[f\left(\frac{\partial f}{\partial \rho}\right) + \frac{1}{\rho}\right]\omega^3 \wedge \omega^1 + \frac{1}{(1+f^2)}\left(\frac{f}{\rho} - \frac{\partial f}{\partial \rho}\right)\omega^3 \wedge \omega^2 + \left(\frac{\partial f}{\partial z}\right)\frac{f}{(1+f^2)}\omega^4 \wedge \omega^1 - \left(\frac{\partial f}{\partial z}\right)\frac{1}{(1+f^2)}\omega^4 \wedge \omega^2$$

Likewise for $d\omega^2$:

$$d\omega^2 = \frac{\partial f}{\partial \rho}d\rho \wedge (\rho d\phi) + \frac{\partial f}{\partial z}dz \wedge (\rho d\phi) + \frac{\partial f}{\partial t}dt \wedge (\rho d\phi) + f d\rho \wedge d\phi$$

$$= \left(f + \rho\frac{\partial f}{\partial \rho}\right)d\rho \wedge d\phi + \rho\left(\frac{\partial f}{\partial z}\right)dz \wedge d\phi + \rho\left(\frac{\partial f}{\partial t}\right)dt \wedge d\phi$$

$$= \left(f + \rho\frac{\partial f}{\partial \rho}\right)\omega^3 \wedge \left[\frac{1}{\rho(1+f^2)}(\omega^1 + f\omega^2)\right] + \rho\left(\frac{\partial f}{\partial z}\right)\omega^4 \wedge \left[\frac{1}{\rho(1+f^2)}(\omega^1 + f\omega^2)\right] + \rho\left(\frac{\partial f}{\partial t}\right)\frac{1}{(1+f^2)}[f\omega^1 + \omega^2] \wedge \left[\frac{1}{\rho(1+f^2)}[\omega^1 + f\omega^2]\right]$$

$$= \frac{-1}{\rho(1+f^2)}\left(f + \rho\frac{\partial f}{\partial \rho}\right)\omega^1 \wedge \omega^3 - \frac{1}{\rho(1+f^2)}\left(f + \rho\frac{\partial f}{\partial \rho}\right)\omega^2 \wedge \omega^3$$
$$- \frac{1}{(1+f^2)}\left(\frac{\partial f}{\partial z}\right)[\omega^1 \wedge \omega^4 + f\omega^2 \wedge \omega^4]$$
$$+ \left(\frac{\partial f}{\partial t}\right)\frac{1}{(1+f^2)}[f^2\omega^1 \wedge \omega^2 - \omega^2 \wedge 1]$$

$$= \frac{-1}{\rho(1+f^2)}\left(f + \rho\frac{\partial f}{\partial \rho}\right)\omega^1 \wedge \omega^3 - \frac{f}{\rho(1+f^2)}\left(f + \rho\frac{\partial f}{\partial \rho}\right)\omega^2 \wedge \omega^3$$
$$- \frac{1}{(1+f^2)}\left(\frac{\partial f}{\partial z}\right)\omega^1 \wedge \omega^4 \frac{-f}{(1+f^2)}\left(\frac{\partial f}{\partial z}\right)\omega^2 \wedge \omega^4$$
$$- \left(\frac{\partial f}{\partial t}\right)\frac{1}{(1+f^2)}\omega^1 \wedge \omega^2$$

Thus,

$$C^2_{13} = \frac{1}{\rho(1+f^2)}\left(f + \rho\left(\frac{\partial f}{\partial \rho}\right)\right) \qquad C^2_{23}$$
$$= \frac{f}{\rho(1+f^2)}\left(f + \rho\frac{\partial f}{\partial \rho}\right) \qquad C^2_{14} = \frac{1}{(1+f^2)}\left(\frac{\partial f}{\partial z}\right)$$
$$C^2_{24} = \frac{f}{(1+f^2)}\left(\frac{\partial f}{\partial z}\right) \qquad C^2_{12} = \frac{1}{(1+f^2)}\left(\frac{\partial f}{\partial t}\right)$$

(Eqn. 7-148)

Let's redo the analysis using compact notation and verify the same result:

$$d\omega^\mu = \omega^\mu_v \wedge \omega^v = -\Gamma^\mu_{v\lambda}\omega^\lambda \wedge \omega^v$$
$$= -\Gamma^\mu_{v\lambda}\left(\frac{1}{2}\left[\omega^\lambda \otimes \omega^v - \omega^v \otimes \omega^\lambda\right]\right)$$
$$= -\frac{1}{2}\left[\Gamma^\mu_{v\lambda}\omega^\lambda \otimes \omega^v - \Gamma^\mu_{v\lambda}\omega^v \otimes \omega^\lambda\right]$$
$$= -\frac{1}{2}\left(\Gamma^\mu_{v\lambda} - \Gamma^\mu_{\lambda v}\right)\omega^\lambda \otimes \omega^v = -\Gamma^\mu_{[v\lambda]}\omega^\lambda \otimes \omega^v$$
$$= \frac{1}{2}\mathrm{C}^\mu_{v\lambda}\left(\omega^\lambda \otimes \omega^v\right) = \frac{1}{2}\left(\mathrm{C}^\mu_{v\lambda} - \mathrm{C}^\mu_{\lambda v}\right)\frac{1}{2}\left(\omega^\lambda \otimes \omega^v\right)$$
$$= \frac{1}{2}\mathrm{C}^\mu_{\lambda v}\omega^\lambda \wedge \omega^v$$
$$= -\mathrm{C}^\mu_{|\lambda v|}\omega^\lambda \wedge \omega^v$$

↑ summation only over $\lambda < v$

$$\underline{d\omega^\mu = -\mathrm{C}^\mu_{|\lambda v|}\omega^\lambda \wedge \omega^v}$$

Now to verify:

$$d\omega^4 = 0 \;\Longrightarrow\; C^4_{|\lambda\nu|} = 0 \;\rightarrow\; C^4_{\nu\lambda} = 0$$

$$d\omega^1 = \frac{-1}{(1+f^2)}\left[f\left(\frac{\partial f}{\partial\rho}\right)+\frac{1}{\rho}\right]\omega^1\wedge\omega^3 - \left(\frac{\partial f}{\partial z}\right)\frac{f}{(1+f^2)}\omega^1\wedge\omega^4$$
$$- \frac{1}{(1+f^2)}\left[\frac{f}{\rho}-\frac{\partial f}{\partial\rho}\right]\omega^2\wedge\omega^3 + \left(\frac{\partial f}{\partial z}\right)\frac{1}{(1+f^2)}\omega^2\wedge\omega^4$$

So,

$$C^1_{13} = \frac{1}{(1+f^2)}\left[f\left(\frac{\partial f}{\partial\rho}\right)+\frac{1}{\rho}\right], C^1_{14} = \left(\frac{\partial f}{\partial z}\right)\frac{f}{(1+f^2)},$$
$$C^1_{23} = \frac{1}{(1+f^2)}\left[\frac{f}{\rho}-\frac{\partial f}{\partial\rho}\right], C^1_{24} = \left(\frac{\partial f}{\partial z}\right)\frac{f}{(1+f^2)}.$$

Since

$$\omega_{\mu\nu} = \frac{1}{2}\left(C_{\mu\nu\alpha}+C_{\mu\alpha\nu}-C_{\nu\alpha\mu}\right)\omega^\alpha$$

We have:

$$C^1_{13} = \frac{1}{(1+f^2)}\left[\frac{1}{\rho}+f\left(\frac{\partial f}{\partial\rho}\right)\right] = 2C_{132}\,, C^2_{13} = \frac{1}{(1+f^2)}\left[\frac{f}{\rho}+\left(\frac{\partial f}{\partial\rho}\right)\right]$$
$$= 2C_{131}\,,$$
$$C^1_{23} = \frac{1}{(1+f^2)}\left[\frac{f}{\rho}-\left(\frac{\partial f}{\partial\rho}\right)\right] = 2C_{232}\,, C^2_{23} = \frac{f}{(1+f^2)}\left[\frac{f}{\rho}+\left(\frac{\partial f}{\partial\rho}\right)\right]$$
$$= 2C_{231}.$$
$$C^1_{24} = \frac{1}{(1+f^2)}\frac{\partial f}{\partial z} = C^2_{14} = 2C_{242} = 2C_{141}$$

$$C^1_{14} = C^2_{24} = fC^1_{24} = fC^2_{14} = 2C_{142} = 2C_{241} = 2fC_{242} = 2fC_{141}$$

$$C^2_{12} = \frac{1}{(1+f^2)}\left(\frac{\partial f}{\partial t}\right)$$
$$= 2C_{121} \quad same\ for \quad C_{121}, C_{131}, C_{132}, C_{141}, C_{142}, C_{231}, C_{232}, C_{241}, C_{242}$$

(Eqn. 7-149)

$$\omega_{12} = -\omega_{21} = \frac{1}{2}\left(C_{12\alpha}+C_{1\alpha2}-C_{2\alpha1}\right)\omega^\alpha$$
$$= \tfrac{1}{2}\left(C_{121}\omega^1 + C_{132}\omega^3 + C_{142}\omega^4 - C_{231}\omega^3 - C_{241}\omega^4 - C_{211}\omega^1\right)$$
$$= \tfrac{1}{4(1+f^2)}\left(2\left(\tfrac{\partial f}{\partial t}\right)\omega^1 + \tfrac{(1+f^2)}{\rho}\omega^3\right)$$

$$\omega_{13} = -\omega_{31} = \frac{1}{2}(C_{13\alpha} + C_{1\alpha 3} - C_{3\alpha 1})\omega^\alpha$$
$$= \frac{1}{2}(C_{131}\omega^1 + C_{132}\omega^2 - C_{311}\omega^1 - C_{321}\omega^2)$$
$$= \frac{1}{2(1+f^2)}\begin{pmatrix} \left(\frac{f}{\rho} + \left(\frac{\partial f}{\partial \rho}\right)\right)\omega^1 + \frac{1}{2}\left(\frac{1}{\rho} + f\left(\frac{\partial f}{\partial \rho}\right)\right)\omega^2 \\ + \frac{1}{2}\left(\frac{f^2}{\rho} + f\left(\frac{\partial f}{\partial \rho}\right)\right)\omega^2 \end{pmatrix}$$
$$= \frac{1}{2(1+f^2)}\left[\left(\frac{f}{\rho} + \left(\frac{\partial f}{\partial \rho}\right)\right)\omega^1 + \frac{1}{2}\left(\frac{(1+f^2)}{\rho} + 2f\left(\frac{\partial f}{\partial \rho}\right)\right)\omega^2\right]$$

$$\omega_{14} = -\omega_{41} = \frac{1}{2}(C_{14\alpha} + C_{1\alpha 4} - C_{4\alpha 1})\omega^\alpha$$
$$= \frac{1}{2}(C_{141}\omega^1 + C_{142}\omega^2 - C_{411}\omega^1 - C_{421}\omega^2)$$
$$= \frac{1}{2}(2C_{141}\omega^1 + (C_{142} + C_{241})\omega^2)$$
$$= \frac{1}{2(1+f^2)}\left(\frac{\partial f}{\partial z}\right)\omega^1 + \frac{f}{2(1+f^2)}\left(\frac{\partial f}{\partial z}\right)\omega^2$$

$$\omega_{23} = -\omega_{32} = \frac{1}{2}(C_{23\alpha} + C_{2\alpha 3} - C_{3\alpha 2})\omega^\alpha$$
$$= \frac{1}{2}(C_{231}\omega^1 + C_{232}\omega^2 - C_{312}\omega^1 - C_{322}\omega^2)$$
$$= \frac{1}{2}\left((C_{231} + C_{132})\omega^1 + 2C_{232}\omega^2\right)$$
$$= \frac{1}{2(1+f^2)}\left[\frac{1}{2}\left(\frac{(1+f^2)}{\rho} + 2f\left(\frac{\partial f}{\partial \rho}\right)\right)\omega^1 + \left(\frac{f}{\rho} - \left(\frac{\partial f}{\partial \rho}\right)\right)\omega^2\right]$$

$$\omega_{24} = -\omega_{42} = \frac{1}{2}(C_{241}\omega^1 + C_{242}\omega^2 - C_{412}\omega^1 - C_{422}\omega^2)$$

$$= \frac{1}{2}\left((C_{241} + C_{142})\omega^1 + 2C_{242}\omega^2\right)$$
$$= \frac{1}{2(1+f^2)}\left[f\left(\frac{\partial f}{\partial z}\right)\omega^1 + \left(\frac{\partial f}{\partial z}\right)\omega^2\right]$$

$$\omega_{34} = -\omega_{43} = \frac{1}{2}(C_{34\alpha} + C_{3\alpha 4} - C_{4\alpha 3}) = 0$$

We thus have:

$R^\mu_\nu = d\omega^\mu_\nu + \omega^\mu_\alpha \wedge \omega^\alpha_\nu \qquad (R^{\mu\nu} = -R^{\nu\mu})$

$$R^{\mu\nu} = d\omega^{\mu}_{\nu} + \omega^{\mu}_{\alpha} \wedge \omega^{\alpha\nu} = R^{\mu\nu}_{|\alpha\beta|}\omega^{\alpha} \wedge \omega^{\beta}$$

$$\omega^{12} = \omega_{21}g^{11}g^{12} = 4(-\omega_{12}) = \frac{-1}{(1+f^2)}\left[2\left(\frac{\partial f}{\partial t}\right)\omega^1 + \frac{(1+f^2)}{\rho}\omega^3\right]$$

$$\omega^{13} = \omega_{23}g^{21} = \frac{1}{(1+f^2)}\left[\frac{1}{2}\left(\frac{(1+f^2)}{\rho} + 2f\left(\frac{\partial f}{\partial \rho}\right)\right)\omega^1 + \left(\frac{f}{\rho} - \left(\frac{\partial f}{\partial \rho}\right)\right)\omega^2\right]$$

$$\omega^{14} = \omega_{24}g^{21} = \frac{1}{(1+f^2)}\left(\frac{\partial f}{\partial z}\right)[f\omega^1 + \omega^2]$$

$$\omega^{23} = \omega_{13}g^{12} = \frac{1}{(1+f^2)}\left[\left(\frac{f}{\rho} + \left(\frac{\partial f}{\partial \rho}\right)\right)\omega^1 + \frac{1}{2}\left(\frac{(1+f^2)}{\rho} + 2f\left(\frac{\partial f}{\partial \rho}\right)\right)\omega^2\right]$$

$$\omega^{24} = \omega_{14}g^{12} = \frac{1}{(1+f^2)}\left(\frac{\partial f}{\partial z}\right)[\omega^1 + f\omega^2]$$

$$\omega^{34} = 0$$

Let's proceed by considering $R_{\mu\nu}$ *on* G:

$$R_{tt} = \left(\rho^2\left(\frac{\partial^2 f}{\partial \rho^2}\right) + \rho^2\left(\frac{\partial^2 f}{\partial z^2}\right) - \rho\left(\frac{\partial f}{\partial \rho}\right)\right)\Big/2\rho^2 = \frac{1}{2\rho}\left(\rho\left[\left(\frac{\partial^2 f}{\partial \rho^2}\right) + \rho^2\left(\frac{\partial^2 f}{\partial z^2}\right)\right] - \left(\frac{\partial f}{\partial \rho}\right)\right)\Bigg|_G$$

all evaluations on G, so dropping notation of such from here.

$$R_{t\phi} = \left(\rho^2\left[\left(\frac{\partial f}{\partial \rho}\right)^2 + \left(\frac{\partial f}{\partial z}\right)^2\right] - 2\rho^2\left(\frac{\partial^2 f}{\partial t^2}\right)\right)\Big/(2\rho)$$

$$= \frac{1}{2}\left(\rho\left[\left(\frac{\partial f}{\partial \rho}\right)^2 + \left(\frac{\partial f}{\partial z}\right)^2\right] - \rho\left(\frac{\partial^2 f}{\partial t^2}\right)\right)$$

$$R_{tz} = \left(\frac{\partial f}{\partial z}\right)\left(\frac{\partial f}{\partial t}\right)$$

$$R_{t\rho} = \left(\frac{\partial f}{\partial \rho}\right)\left(\frac{\partial f}{\partial t}\right)$$

$$R_{\phi\phi} = -\left(\rho^2\left(\frac{\partial^2 f}{\partial \rho^2}\right) + \rho^2\left(\frac{\partial^2 f}{\partial z^2}\right) + 3\rho\left(\frac{\partial f}{\partial \rho}\right)\right)/2$$

$$R_{\phi z} = -\rho\frac{\partial^2 f}{\partial z \partial t}$$

$$R_{\phi\rho} = -\rho\frac{\partial^2 f}{\partial \rho \partial t} - \frac{\partial f}{\partial t}$$

$$R_{zz} = \frac{2f\left(\frac{\partial^2 f}{\partial z^2}\right)}{(f^2+1)} = 0$$

$$R_{z\rho} = 0$$

$$R_{\rho\rho} = \frac{1}{2\rho^2}$$

Returning to our prior (pre-Cartan) discussion, on geodesic $R_{\mu\nu}K^{\mu}K^{\nu}$ for the genral null vector: (only $f = 0$ has been used ($\rho \to \rho_0$).)

$$A = \frac{R_{tt}}{2} + \frac{R_{t\phi}}{\rho_0} + \frac{R_{\phi\phi}}{2\rho_0^2}$$

$$= \frac{1}{4\rho}\left(\rho\left(\frac{\partial^2 f}{\partial \rho^2}\right) + \rho\left(\frac{\partial^2 f}{\partial z^2}\right) + 2\rho\left(\frac{\partial f}{\partial \rho}\right)^2 + 2f\left(\frac{\partial f}{\partial z}\right)^2 - 2\rho\left(\frac{\partial^2 f}{\partial t^2}\right)\right) -$$
$$\rho\left(\frac{\partial^2 f}{\partial \rho^2}\right) - \rho\left(\frac{\partial^2 f}{\partial z^2}\right) - 3\left(\frac{\partial f}{\partial \rho}\right).$$

$$\boxed{A = \frac{1}{2\rho_0}\left(\rho_0\left[\left(\frac{\partial f}{\partial \rho}\right)^2 + \left(\frac{\partial f}{\partial z}\right)^2\right] - \rho_0\left(\frac{\partial^2 f}{\partial t^2}\right) - 2\left(\frac{\partial f}{\partial \rho}\right)\right)}$$

(Eqn. 7-150)

$$B = R_{tt} - \frac{R_{\phi\phi}}{\rho_0^2}$$

$$= \frac{1}{2\rho_0}\left(\rho_0\left[\left[\left(\frac{\partial^2 f}{\partial \rho^2}\right) + \left(\frac{\partial^2 f}{\partial z^2}\right)\right] - \left(\frac{\partial f}{\partial \rho}\right)\right] + \rho_0\left[\left(\frac{\partial^2 f}{\partial \rho^2}\right) + \left(\frac{\partial^2 f}{\partial z^2}\right)\right] + 3\left(\frac{\partial f}{\partial \rho}\right)\right)$$

(Eqn. 7-151)

$$\boxed{B = \left[\left(\frac{\partial^2 f}{\partial \rho^2}\right) + \left(\frac{\partial^2 f}{\partial z^2}\right)\right] + \frac{1}{\rho_0}\left(\frac{\partial f}{\partial \rho}\right)}$$

(Eqn. 7-152)

$$\boxed{C = \sqrt{2}\left(R_{t\rho} - \frac{R_{\phi\rho}}{\rho_0}\right) = \sqrt{2}\left[\left(\frac{\partial f}{\partial \rho}\right)\left(\frac{\partial f}{\partial t}\right) - \left(\frac{\partial^2 f}{\partial \rho \partial t}\right) - \frac{1}{\rho_0}\left(\frac{\partial f}{\partial t}\right)\right]}$$

(Eqn. 7-153)

$$\boxed{D = \sqrt{2}\left(R_{tz} - \frac{R_{\phi z}}{\rho_0}\right) = \sqrt{2}\left[\left(\frac{\partial f}{\partial z}\right)\left(\frac{\partial f}{\partial t}\right) - \left(\frac{\partial^2 f}{\partial z \partial t}\right)\right]}$$

(Eqn. 7-154)

$$\boxed{E = \sqrt{2}\left(R_{t\rho} - \frac{R_{\phi\rho}}{\rho_0}\right) = \sqrt{2}\left[\left(\frac{\partial f}{\partial \rho}\right)\left(\frac{\partial f}{\partial t}\right) + \left(\frac{\partial^2 f}{\partial \rho \partial t}\right) + \frac{1}{\rho_0}\left(\frac{\partial f}{\partial t}\right)\right]}$$

(Eqn. 7-155)

$$\boxed{F = \sqrt{2}\left(R_{tz} - \frac{R_{\phi z}}{\rho_0}\right) = \sqrt{2}\left[\left(\frac{\partial f}{\partial z}\right)\left(\frac{\partial f}{\partial t}\right) + \left(\frac{\partial^2 f}{\partial z \partial t}\right)\right]}$$

(Eqn. 7-156)

$$\boxed{G = 0}$$

(Eqn. 7-157)

$$\boxed{H = \frac{1}{2\rho_0}\left(\rho_0\left(\frac{\partial^2 f}{\partial t^2}\right) - \rho_0\left[\left(\frac{\partial f}{\partial \rho}\right)^2 + \left(\frac{\partial f}{\partial z}\right)^2\right] - 2\left(\frac{\partial f}{\partial \rho}\right)\right)}$$

(Eqn. 7-158)

$$\boxed{I = \frac{1}{2\rho_0}}$$

(Eqn. 7-159)

$$\boxed{J = 0}$$

(Eqn. 7-160)

A, B, C, D, E, F, H, I are 8 indep. variables depending on the various derivatives of *f*, and already have the simplification $G = J = 0, and\ I > 0$.

Let's return to the specific case of

$$f(\rho, z, t) = 1 - \frac{\operatorname{sech}(x^2 - \alpha^2)}{\operatorname{sech} 3\alpha^2} \qquad x^2 = (\rho - \rho_0)^2 + z^2 + t^2$$

thus

$$\left(\frac{\partial f}{\partial x^i}\right)_{x=-2\alpha} = \frac{\operatorname{sech}(x^2-\alpha^2)\tanh(x^2-\alpha^2)}{\operatorname{sech} 3\alpha^2} \frac{dx^2}{dx^i}\bigg|_{x=-2\alpha} =$$

$$\tanh(3\alpha^2)\left(\frac{dx^2}{dx^i}\right)_{x=-2\alpha\,:\ t=-2\alpha,\rho=\rho_0,z=0}.$$

So:

$$\left(\frac{\partial f}{\partial z}\right) = 0 \ and \ \left(\frac{\partial f}{\partial \rho}\right) = 0 \ \ and \ \left(\frac{\partial f}{\partial t}\right) = (-4\alpha)\tanh(3\alpha^2)$$

Similarly,

$$\left(\frac{\partial^2 f}{\partial x^i \partial x^j}\right)_{x=-2\alpha} = \frac{\operatorname{sech}(x^2 - \alpha^2)\tanh^2(x^2 - \alpha^2)}{\operatorname{sech} 3\alpha^2} \frac{dx^2}{dx^j}\frac{dx^2}{dx^i}\bigg|_{x=-2\alpha}$$

$$+ \frac{\operatorname{sech}(x^2-\alpha^2)}{\operatorname{sech} 3\alpha^2} \frac{dx^2}{dx^j}\frac{dx^2}{dx^i}\bigg|_{x=-2\alpha}$$

$$+ \frac{\operatorname{sech}(x^2-\alpha^2)\tanh(x^2-\alpha^2)}{\operatorname{sech} 3\alpha^2} \frac{dx^2}{dx^j}\frac{dx^2}{dx^i}\bigg|_{x=-2\alpha}$$

$$\frac{\partial^2 f}{\partial t^2}\bigg|_x = -\tanh^2(3\alpha^2)\,(-4\alpha)^2 + \operatorname{sech}^2(3\alpha^2)\,(-4\alpha)^2 + \tanh(3\alpha^2)\,2.$$

$$\frac{\partial^2 f}{\partial \rho^2}\bigg|_x = 2\tanh(3\alpha^2).$$

$$\frac{\partial^2 f}{\partial z^2}\bigg|_x = 2\tanh(3\alpha^2).$$

$$\left(\frac{\partial^2 f}{\partial \rho \partial t}\right)\bigg|_x = 0 = \frac{\partial^2 f}{\partial z \partial t}.$$

Since $A + Bx_1 + Cx_2 + Dx_3 + Ex_1x_2 + Fx_1x_3 + Hx_1^2 + Ix_2^2 \geq 0$ for general x's, we must have $A, H, I > 0$. From the above we have that $A = -H$, thus we can't satisfy the WEC relation throughout the space-time. The relation $A = -H$ blocks the possibility of satisfying the weak energy principle for the family of space-times considered.

7.6 The Second Sech space-time generalization

Can we modify the family of spacetime solutions to have $A = H > 0$? Recall:

$$A + Bx_1 + Cx_2 + Dx_3 + Ex_1x_2 + Fx_1x_3 + Hx_1^2 + Ix_2^2 \geq 0.$$

(Eqn. 7-161)

Want A, H, I to each be > 0. Since $I = \frac{1}{2\rho_0^2}$ so no problem there. If $A > 0$ and $H > 0$ then $A + H > 0$ so

$$-\left(\frac{\partial f}{\partial \rho}\right) > 0 \Rightarrow \left(\frac{\partial f}{\partial \rho}\right) < 0 \quad \underline{\text{not zero}}$$

(Eqn. 7-162)

If we choose $\left(\frac{\partial f}{\partial z}\right) = 0$ then $D = F = 0$. Recall $G = J = 0$ already from before, So

$$A = \frac{1}{2}\left(\frac{\partial f}{\partial \rho}\right)^2 - \frac{1}{\rho_0}\left(\frac{\partial f}{\partial \rho}\right) - \frac{1}{2}\left(\frac{\partial^2 f}{\partial t^2}\right)$$

$$H = -\frac{1}{2}\left(\frac{\partial f}{\partial \rho}\right)^2 - \frac{1}{\rho_0}\left(\frac{\partial f}{\partial \rho}\right) + \frac{1}{2}\left(\frac{\partial^2 f}{\partial t^2}\right)$$

Sufficient condition for $\left(\frac{\partial f}{\partial \rho}\right) < 0$:

$$-\frac{1}{\rho_0}\left(\frac{\partial f}{\partial \rho}\right) > \frac{1}{2}\left[\left(\frac{\partial f}{\partial \rho}\right)^2 + \left|\left(\frac{\partial^2 f}{\partial t^2}\right)\right|\right]$$

(Eqn. 7-163)

So can now choose $\left(\frac{\partial f}{\partial \rho}\right)$ to make both A and H positive as long as $\left(\frac{\partial f}{\partial \rho}\right)$ doesn't get too large.

So……

Consider

$$f(\rho, z, t) = 1 - \frac{\operatorname{sech}(x^2 - \alpha^2)\tanh(\rho)}{\operatorname{sech}(3\alpha^2)\tanh(\rho_0)}$$

(Eqn. 7-164)

$f|_{cng} = 1 - 1 = 0$ $\qquad$ $x^2 = (\rho - \rho_0)^2 + z^2 + t^2$

$f|_{x\, large} \cong 1$ $\qquad$ $f_{\substack{x>0 \\ \rho<\rho_0 \\ t=-2\alpha}} < 0$ generally for large

$\rho_0 (i.e., \ \rho_0 \gg \alpha)$

$$\left(\frac{\partial f}{\partial x^i}\right)_G = \tanh(3\alpha^2)\left(\frac{\partial x^2}{\partial x^i}\right)_G - \frac{\operatorname{sech}^2(\rho_0)}{\tanh(\rho_0)}\left(\frac{\partial \rho}{\partial x^i}\right)_G$$

$$\left(\frac{\partial f}{\partial \rho}\right) = \frac{-\operatorname{sech}^2(\rho_0)}{\tanh(\rho_0)} \quad \left(\frac{\partial f}{\partial z}\right) = 0 \quad \left(\frac{\partial f}{\partial t}\right) = (-4\alpha)\tanh(3\alpha^2)$$

$$\begin{aligned}\left(\frac{\partial^2 f}{\partial x^i \partial x^j}\right)_G = {} & (\operatorname{sech}^2(3\alpha^2) - \tanh^2(3\alpha^2))\left(\frac{\partial x^2}{\partial x^i}\right)_G \left(\frac{\partial x^2}{\partial x^j}\right)_G \\ & + \tanh(3\alpha^2)\left(\frac{\partial^2 x^2}{\partial x^i \partial x^j}\right)_G \\ & + \tanh(3\alpha^2)\left(\frac{\partial x^2}{\partial x^i}\right)_G \frac{\operatorname{sech}^2(\rho_0)}{\tanh(\rho_0)}\left(\frac{\partial \rho}{\partial x^j}\right)_G \\ & + 2\operatorname{sech}^2(\rho_0)\left(\frac{\partial \rho}{\partial x^i}\right)_G \left(\frac{\partial \rho}{\partial x^j}\right)_G\end{aligned}$$

$$\left(\frac{\partial^2 f}{\partial t^2}\right)_G = 2\tanh(3\alpha^2) + (4\alpha)^2[\operatorname{sech}^2(3\alpha^2) - \tanh^2(3\alpha^2)]$$

$$\left(\frac{\partial^2 f}{\partial \rho \partial t}\right)_G = (-4\alpha)\tanh(3\alpha^2)\frac{\operatorname{sech}^2(\rho_0)}{\tanh(\rho_0)} \quad \left(\frac{\partial^2 f}{\partial z \partial t}\right) = 0$$

$$\left(\frac{\partial^2 f}{\partial z^2}\right)_G = 2\tanh(3\alpha^2)$$

$$\left(\frac{\partial^2 f}{\partial \rho^2}\right)_G = 2\tanh(3\alpha^2) + 2\operatorname{sech}^2(\rho_0)$$

And our new variables become:

$$\begin{aligned}A = {} & \frac{1}{2}\left(\frac{\operatorname{sech}^2(\rho_0)}{\tanh(\rho_0)}\right)^2 + \frac{1}{\rho_0}\left(\frac{\operatorname{sech}^2(\rho_0)}{\tanh(\rho_0)}\right) \\ & - \frac{1}{2}[2\tanh(3\alpha^2) + (4\alpha)^2(\operatorname{sech}^2(3\alpha^2) - \tanh^2(3\alpha^2))]\end{aligned}$$

$$\begin{aligned}H = {} & -\frac{1}{2}\left(\frac{\operatorname{sech}^2(\rho_0)}{\tanh(\rho_0)}\right)^2 + \frac{1}{\rho_0}\left(\frac{\operatorname{sech}^2(\rho_0)}{\tanh(\rho_0)}\right) \\ & + \frac{1}{2}[2\tanh(3\alpha^2) + (4\alpha)^2(\operatorname{sech}^2(3\alpha^2) - \tanh^2(3\alpha^2))]\end{aligned}$$

$$B = 2\tanh(3\alpha^2) + 2\,\mathrm{sech}^2(\rho_0) - \frac{1}{\rho_0}\frac{\mathrm{sech}^2(\rho_0)}{\tanh(\rho_0)}$$

$C =$ etc.

Let $\rho_0 \ll 1$, $\underline{\rho_0 \ll \alpha}$, $\alpha \ll 1$

$$\left(\frac{\partial f}{\partial \rho}\right)_G \cong -\frac{1}{\rho_0} \qquad \left(\frac{\partial f}{\partial t}\right)_G \cong -12\alpha^3$$

$$\left(\frac{\partial^2 f}{\partial t^2}\right)_G \cong 6\alpha^2 + 16\alpha^2 \approx 22\alpha^2$$

$$\left(\frac{\partial^2 f}{\partial \rho \partial t}\right)_G \simeq \frac{-12\alpha^3}{\rho_0}$$

$$\left(\frac{\partial^2 f}{\partial z^2}\right)_G = 6\alpha^2$$

$$\left(\frac{\partial^2 f}{\partial \rho^2}\right)_G = 6\alpha^2 + 2 \simeq 2$$

$$A = \frac{1}{2}\left(\frac{1}{\rho_0}\right)^2 - 11\alpha^2 + \left(\frac{1}{\rho_0}\right)^2 \cong \frac{3}{2}\frac{1}{\rho_0^2}$$

$$H \simeq -\frac{1}{2}\left(\frac{1}{\rho_0}\right)^2 + \left(\frac{1}{\rho_0}\right)^2 \simeq \frac{1}{2}\frac{1}{\rho_0^2}$$

$$B = 2 - \frac{1}{\rho_0^2} \simeq -\frac{1}{\rho_0^2}$$

$$C = \sqrt{2}\left[\frac{12\alpha^3}{\rho_0} + \frac{12\alpha^3}{\rho_0} + \frac{12\alpha^3}{\rho_0}\right] \simeq 36\sqrt{2}\frac{\alpha^3}{\rho_0}$$

$$D = 0 = F = G = J$$

$$E \cong -12\sqrt{2}\frac{\alpha^3}{\rho_0}$$

$$I = \frac{1}{2\rho_0^2}$$

where $g(x_1, x_2) = A + Bx_1 + Cx_2 + Ex_1x_2 + Hx_1^2 + Ix_2^2 \geq 0$

$$g(x_1, x_2) \cong \frac{3}{2\rho_0^2} + \left(-\frac{1}{\rho_0^2}\right)x_1 + \left(36\sqrt{2}\frac{\alpha^3}{\rho_0}\right)x_2 + \left(-12\sqrt{2}\frac{\alpha^3}{\rho_0}\right)x_1x_2\left(\frac{1}{2\rho_0^2}\right)x_1^2 + \left(\frac{1}{2\rho_0^2}\right)x_2^2 \geq 0$$

(Eqn. 7-165)

Thus,

$$\left(\frac{3}{2\rho_0^2} + 36\sqrt{2}\frac{\alpha^3}{\rho_0}x_2 + \frac{1}{2\rho_0^2}x_2^2\right) + \left(-\frac{1}{\rho_0^2} - 12\sqrt{2}\frac{\alpha^3}{\rho_0}x_2\right)x_1 + \frac{1}{2\rho_0^2}x_1^2 \geq 0$$

$$M = \frac{3}{2\rho_0^2} + 36\sqrt{2}\frac{\alpha^3}{\rho_0}x_2 + \frac{1}{2\rho_0^2}x_2^2 > 0 \; for \;\; \rho_0 \ll \alpha, etc.$$

So, need to satisfy

$$M + \left(-\frac{1}{\rho_0^2} - 12\sqrt{2}\frac{\alpha^3}{\rho_0}x_2\right)x_1 + \frac{1}{2\rho_0^2}x_1^2 \geq 0,$$

Which, again, is positive if

$$4\left(\frac{1}{2\rho_0^2}\right)\left(\frac{3}{2\rho_0^2} + \frac{36\sqrt{2}\alpha^3}{\rho_0}x_{2,min} + \frac{1}{2\rho_0^2}x_{2,min}^2\right) - \left(\frac{1}{\rho_0^2} + \frac{12\sqrt{2}\alpha^3}{\rho_0}x_{2,min}\right)^2 \geq 0$$

or

$$\frac{3}{\rho_0^4} + \frac{72\sqrt{2}\alpha^3}{\rho_0^3}x_{2,min} + \frac{x_{2,min}^2}{\rho_0^4} - \frac{1}{\rho_0^4} - \frac{24\sqrt{2}\alpha^3}{\rho_0^3}x_{2,min} + \frac{288\alpha^3}{\rho_0^2}x_{2,min}^2 \geq 0$$

If $x_{2,min}$ is $< \frac{1}{\rho_0}$ then dominant term is $\frac{2}{\rho_0^4} \gg 0$

If $x_{2,min}$ is $> \frac{1}{\rho_0}$ then dominant term is $\frac{x_{2,min}^2}{\rho_0^4} \gg 0$

Thus, we can generally, in this enlarged family of space-times, find α, ρ_0 to satisify the weak energy condition ***on the geodesic***. Does this choice of $f(\rho, z, t)$ satisfy the weak energy condition on all of the space-time?

We have:

$$A = \frac{R_{tt}}{2} + \frac{R_{t\phi}}{\rho_0} + \frac{R_{\phi\phi}}{2\rho_0^2} \quad and \ H = \frac{R_{tt}}{2} - \frac{R_{t\phi}}{\rho_0} + \frac{R_{\phi\phi}}{2\rho_0^2}$$

where:

$$\begin{aligned} R_{tt} = \Bigg\{ & f^7 + f^6 \left[\rho^2 \left(\frac{\partial^2 f}{\partial \rho^2} + \frac{\partial^2 f}{\partial z^2}\right) + 3\rho \left(\frac{\partial f}{\partial \rho}\right)\right] \\ & + f^5 \left[2\rho^2 \left[\left(\frac{\partial f}{\partial \rho}\right)^2 + \left(\frac{\partial f}{\partial z}\right)^2\right] - 1\right] \\ & + f^4 \left[3\rho^2 \left(\left(\frac{\partial^2 f}{\partial \rho^2}\right) + \left(\frac{\partial^2 f}{\partial z^2}\right)\right) + 5_\rho \left(\frac{\partial f}{\partial \rho}\right)\right] \\ & + f^3 \left[4\rho^2 \left[\left(\frac{\partial f}{\partial \rho}\right)^2 + \left(\frac{\partial f}{\partial z}\right)^2 - \left(\frac{\partial^2 f}{\partial t^2}\right)\right] - 1\right] \\ & + f^2 \left[3\rho^2 \left(\frac{\partial^2 f}{\partial \rho^2} + \frac{\partial^2 f}{\partial z^2}\right) + 8\rho^2 \left(\frac{\partial f}{\partial t}\right)^2 + \rho \left(\frac{\partial f}{\partial \rho}\right)\right] \\ & + f \left[2\rho^2 \left[\left(\frac{\partial f}{\partial \rho}\right)^2 + \left(\frac{\partial f}{\partial z}\right)^2\right] - 4\rho^2 \left(\frac{\partial^2 f}{\partial t^2}\right) + 1\right] \\ & + \rho^2 \left[\frac{\partial^2 f}{\partial \rho^2} + \frac{\partial^2 f}{\partial z^2}\right] - \rho \left(\frac{\partial f}{\partial \rho}\right) \Bigg\} / [2\rho^2 (f^2 + 1)^3] \end{aligned}$$

$$
\begin{aligned}
R_{\phi\phi} = -\Bigg\{ & f^7 + f^6\left[\rho^2\left(\frac{\partial^2 f}{\partial\rho^2}+\frac{\partial^2 f}{\partial z^2}\right)+3\rho\left(\frac{\partial f}{\partial\rho}\right)\right] \\
& + f^5\left[2\rho^2\left[\left(\frac{\partial f}{\partial\rho}\right)^2+\left(\frac{\partial f}{\partial z}\right)^2\right]-1\right] \\
& + f^4\left[3\rho^2\left(\frac{\partial^2 f}{\partial\rho^2}+\frac{\partial^2 f}{\partial z^2}\right)+9\rho\left(\frac{\partial f}{\partial\rho}\right)\right] \\
& + f^3\left[4\rho^2\left(\left(\frac{\partial f}{\partial\rho}\right)^2+\left(\frac{\partial f}{\partial z}\right)^2-\frac{\partial^2 f}{\partial t^2}\right)-1\right] \\
& + f^2\left[3\rho^2\left(\frac{\partial^2 f}{\partial\rho^2}+\frac{\partial^2 f}{\partial z^2}\right)+8\rho^2\left(\frac{\partial f}{\partial t}\right)^2+9\rho\left(\frac{\partial f}{\partial\rho}\right)\right] \\
& + f\left[2\rho^2\left(\left(\frac{\partial f}{\partial\rho}\right)^2+\left(\frac{\partial f}{\partial z}\right)^2\right)-4\rho^2\left(\frac{\partial^2 f}{\partial t^2}\right)+1\right] \\
& + \rho^2\left[\frac{\partial^2 f}{\partial\rho^2}+\frac{\partial^2 f}{\partial z^2}\right]-3\rho\left(\frac{\partial f}{\partial\rho}\right)\Bigg\}/[2(f^2+1)^3]
\end{aligned}
$$

$$
\begin{aligned}
R_{\phi t} = -\Bigg\{ & f^7\left[\rho^2\left(\frac{\partial^2 f}{\partial\rho^2}+\frac{\partial^2 f}{\partial z^2}\right)+2\rho\left(\frac{\partial f}{\partial\rho}\right)\right] \\
& + f^6\left[\rho^2\left(\left(\frac{\partial f}{\partial\rho}\right)^2+\left(\frac{\partial f}{\partial z}\right)^2\right)-2\right] \\
& + f^5\left[3\rho^2\left(\left(\frac{\partial^2 f}{\partial\rho^2}\right)+\left(\frac{\partial^2 f}{\partial z^2}\right)\right)+4\rho\left(\frac{\partial f}{\partial\rho}\right)\right] \\
& + f^4\left[\rho^2\left(\left(\frac{\partial f}{\partial\rho}\right)^2+\left(\frac{\partial f}{\partial z}\right)^2\right)-2\left(\frac{\partial^2 f}{\partial t^2}\right)\right] \\
& + f^3\left[3\rho^2\left(\left(\frac{\partial^2 f}{\partial\rho^2}\right)+\left(\frac{\partial^2 f}{\partial z^2}\right)\right)+4\rho^2\left(\frac{\partial f}{\partial\rho}\right)^2+2\rho\left(\frac{\partial f}{\partial\rho}\right)\right] \\
& + f^2\left[\rho^2\left(\left(\frac{\partial f}{\partial\rho}\right)^2+\left(\frac{\partial f}{\partial z}\right)^2\right)+2\right] \\
& + f\left[\rho^2\left(\frac{\partial^2 f}{\partial\rho^2}+\frac{\partial^2 f}{\partial z^2}+4\left(\frac{\partial f}{\partial t}\right)^2\right)\right] \\
& + \rho^2\left[2\left(\frac{\partial^2 f}{\partial t^2}\right)-\left(\frac{\partial f}{\partial\rho}\right)^2-\left(\frac{\partial f}{\partial z}\right)^2\right]\Bigg\}/[2\rho(f^2+1)^3]
\end{aligned}
$$

Again, in order that A, H, I > 0 we find that A > 0 and $H > 0 \Rightarrow A + H > 0$, so,

$$A + H = R_{tt} + \frac{R_{\phi\phi}}{\rho^2} \geq 0$$

(Eqn. 7-166)

Thus,

$$\left\{-4f^4\rho\left(\frac{\partial f}{\partial \rho}\right) - 8f^2\rho\left(\frac{\partial f}{\partial \rho}\right) - 4\rho\left(\frac{\partial f}{\partial \rho}\right)\right\}/[2\rho^2(f^2+1)^3] \geq 0$$

As before, we satisfy this for

$$-\left(\frac{\partial f}{\partial \rho}\right) \geq 0$$

Thus $\left(\frac{\partial f}{\partial \rho}\right) < 0$ everywhere in the spacetime in order to satisfy the weak energy condition.

If instead we use the forms:

$$A = -\frac{1}{\rho}\left(\frac{\partial f}{\partial \rho}\right)\frac{1}{(f^2+1)} + \frac{R_{\phi t}}{\rho} \quad and\ H = \frac{1}{\rho}\left(\frac{\partial f}{\partial \rho}\right)\frac{1}{(f^2+1)} - \frac{R_{\phi t}}{\rho}$$

We also find the condition

$$-\left(\frac{\partial f}{\partial \rho}\right) > (f^2+1)|R_{\phi t}|$$

(Eqn. 7-167)

Can this be satisfied in any reasonable situation?

We have:

$$R_{\phi t} = -\left\{(f^2+1)^2\rho^2(f^3+f)\left[\frac{\partial^2 f}{\partial \rho^2} + \frac{\partial^2 f}{\partial z^2}\right] + 2\rho f^3(f^2+1)^2\left(\frac{\partial f}{\partial \rho}\right)\right.$$
$$+ \rho^2[f^4(f^2+1) + (f^2-1)]\left[\left(\frac{\partial f}{\partial \rho}\right)^2 + \left(\frac{\partial f}{\partial z}\right)^2\right]$$
$$- 2f^2(f^4-1) - 2(f^4-1)\rho^2\left(\frac{\partial^2 f}{\partial t^2}\right)$$
$$\left. + 4\rho^2 f(f^2-1)\left(\frac{\partial f}{\partial t}\right)^2\right\}/2\rho(f^2+1)^3$$

$$= -\left\{ \begin{array}{c} \frac{\rho f}{2}\left[\left(\frac{\partial^2 f}{\partial \rho^2}\right)+\left(\frac{\partial^2 f}{\partial z^2}\right)\right]+\frac{f^3}{(f^2+1)}\left(\frac{\partial f}{\partial \rho}\right)+\frac{\rho}{2}\frac{[f^4(f^2+1)+(f^2-1)]}{(f^2+1)^3}\left[\left(\frac{\partial f}{\partial \rho}\right)^2+\left(\frac{\partial f}{\partial z}\right)^2\right] \\ -\frac{f^2(f^2-1)}{\rho(f^2+1)^2}-\frac{(f^2-1)}{(f^2+1)^2}\rho\left(\frac{\partial^2 f}{\partial t^2}\right)+2\rho\frac{f(f^2-1)}{(f^2+1)^3}\left(\frac{\partial f}{\partial t}\right)^2 \end{array} \right\}$$

As $\rho \to \infty$ or 0 denote the dominant ρ factor in f by $f \propto \rho^n$:

So, $\frac{\partial f}{\partial t} \propto \rho^{n-1}$, $\frac{\partial^2 f}{\partial \rho^2} \propto \rho^{n-2}$, $\frac{\partial^2 f}{\partial z^2} \propto \frac{\partial f}{\partial z} \propto \frac{\partial^2 f}{\partial t^2} \propto \frac{\partial f}{\partial t} \propto f \propto \rho^n$

Using the notation $O(\rho^n) = O(n)$:

$$O_\rho(R_{\phi t}) = O(1+n+n-2)+O(1+n+n)+O(n+n-1) \\ +O(6n+1-6n+2n-2) \\ +O(6n+1-6n+2n)+O(4n-4n-1)+O(2n-4n+1+n) \\ +O(1+n+2n-6n+2n)$$

$$O_\rho(R_{\phi t}) = O(2n-1)+O(2n+1)+O(2n-1)+O(2n-1) \\ +O(2n+1) \\ +O(-1)+O(-n+1)+O(-n+1)$$

Want $O(n-1) > O(R_{\phi t})O(2n)$ to satisfy

$$-\left(\frac{\partial f}{\partial \rho}\right) > (f^2+1)|R_{\phi t}|$$

and this is not generally possible.

So, although $A + H > 0$ can be chosen and then both A and $H > 0$ in same region, on the cng say, we can't have A and $H > 0$ throughout the spacetime since we can't satisfy

$$-\left(\frac{\partial f}{\partial \rho}\right) > (f^2+1)|R_{\phi t}|$$

throughout, as outlined above.

Thus, since the weak energy cond. must be violated, it subsequently must be violated somewhere on the future Cauchy horizon. So, Hawking's CPC statement [28] is true in this spacetime.

Alternatively, the key question is, can the condition $-\left(\frac{\partial f}{\partial \rho}\right) > (f^2+1)\left|R_{\phi t}\right|$ be satisfied everywhere?

$$R_{\phi t} = -\left\{\frac{\rho f}{2}\left[\left(\frac{\partial^2 f}{\partial \rho^2}\right)+\left(\frac{\partial^2 f}{\partial z^2}\right)\right]+\frac{f^3}{(f^2+1)}\left(\frac{\partial f}{\partial \rho}\right)\right.$$
$$+\frac{\rho}{2}\frac{[f^4(f^2+1)+(f^2-1)]}{(f^2+1)^3}\left[\left(\frac{\partial f}{\partial \rho}\right)^2+\left(\frac{\partial f}{\partial z}\right)^2\right]$$
$$-\frac{f^2(f^2-1)}{\rho(f^2+1)^2}-\frac{(f^2-1)}{(f^2+1)^2}\rho\left(\frac{\partial^2 f}{\partial t^2}\right)$$
$$\left.+2\rho\frac{f(f^2-1)}{(f^2+1)^3}\left(\frac{\partial f}{\partial t}\right)^2\right\}$$

$f_{cng} = 0$

$f_{x\ large} \simeq 1$

Consider $\left(\frac{\partial f}{\partial \rho}\right) \alpha \rho^n$ where $\rho \to \infty$, of course $f \approx 1$ as $\rho \to \infty$, then

$$O(\rho^n) > \underbrace{O(\rho^n)+O(\rho)+O(\rho^n)+O(\rho^{2n+1})+O(\rho)+O(\rho^{-1})+O(\rho)+O(\rho)}_{\text{Highest order is}}$$

$2n+1$ for $n>0$

$or\ 1$ for $n<0$

$O(\rho^n) > O(\rho^{2n+1})$ for $n>0$

$or\ O(\rho^1)\ for\ n<0$

To satisfy $n > 2n+1$ for $n>0$

Thus $n<-1$ for $n>0$ a contradiction,

Or $n>1$ for $n<0$ a contradiction.

So, the inequality cannot be solved for the full range of ρ.

Consider $\left(\frac{\partial f}{\partial \rho}\right) \propto \rho^n$ as $\rho \to \infty$

$O(\rho^n) > O(\rho^n)+O(\rho)+O(\cdots)$ etc.

$> O(\rho^{2n+1})$ for $n<-1$ $\quad n>2n+1 \quad n<-1$

$> O(\rho^{-1})$ for $n>-1$ $\quad n>-1$

So, the $\rho \to 0$ limit cannot be satisfied.

Summary, the condition $-\left(\frac{\partial f}{\partial \rho}\right) > (f^2 + 1)|R_{\phi t}|$ can only be satisfied in the entire spacetime, notably for the full range of ρ, if the ρ-order of $\left(\frac{\partial f}{\partial \rho}\right)$ satisfies certain properties. Regarding the restrictions as $\rho \to 0$ we see that the condition is not immediately disallowed. However, considering $\rho \to \infty$, regardless of the ρ-order properties of the *z, t* differentials of *f*, we must satisfy

$$O(\rho^n) \geq O(\rho^n) + O(\rho^{2n+1}) + O(\rho^{-1})$$

This is only possible for $n = -1$, as well as some restrictions on the forms of $\partial f \ / \ \partial z, \partial f \ / \ \partial t$, etc. Therefore, to have $\left(\frac{\partial f}{\partial \rho}\right) \propto \rho^{-1}$ as $\rho \to \infty$ is to have *f* unbounded via a logarithmic divergence. This is counter to the original assumption of asymptotically flat space-time. Thus we can't satisfy the weak energy condition in the space times studied when CTCs are present.

7.7 Exercises

(Ex. 7.1) Rederive Eqn. 7-4 and confirm consistency between equation (1) and (2).
(Ex. 7.2) Rederive Eqn. 7-6.
(Ex. 7.3) Rederive Eqn. 7-11.
(Ex. 7.4) Rederive Eqn. 7-16.
(Ex. 7.5) Rederive Eqn. 7-19.
(Ex. 7.6) Rederive Eqn. 7-24.
(Ex. 7.7) Rederive Eqn. 7-25.
(Ex. 7.8) Rederive Eqn. 7-26.
(Ex. 7.9) Rederive Eqn. 7-27.
(Ex. 7.10) Rederive Eqn. 7-28.
(Ex. 7.11) Rederive Eqn. 7-31.
(Ex. 7.12) Rederive Eqn. 7-44.
(Ex. 7.13) Rederive Eqn. 7-46.
(Ex. 7.14) Rederive Eqn. 7-47.
(Ex. 7.15) Rederive Eqn. 7-49.
(Ex. 7.16) Rederive Eqn. 7-52.
(Ex. 7.17) Rederive Eqn. 7-53, 7-54, 7-55, and 7-56.
(Ex. 7.18) Rederive Eqn. 7-63.
(Ex. 7.19) Rederive Eqn. 7-68.
(Ex. 7.20) Finish the calculation at end of Sec. 7.4.1.
(Ex. 7.21) Rederive Eqn. 7-90.
(Ex. 7.22) Rederive Eqn. 7-92.
(Ex. 7.23) Rederive Eqn. 7-97.

(Ex. 7.24) Rederive Eqn. 7-98, 7-99, and 7-103.
(Ex. 7.25) Rederive Eqn. 7-109.
(Ex. 7.26) Rederive Eqn. 7-123.
(Ex. 7.27) Rederive Eqn. 7-126.
(Ex. 7.28) Rederive Eqn. 7-127.
(Ex. 7.29) Rederive Eqn. 7-134.
(Ex. 7.30) Rederive Eqn. 7-135.
(Ex. 7.31) Rederive Eqn. 7-136.
(Ex. 7.32) Rederive Eqn. 7-137.
(Ex. 7.33) Rederive Eqn. 7-138.
(Ex. 7.34) Rederive Eqn. 7-142.
(Ex. 7.35) Rederive Eqn. 7-143.
(Ex. 7.36) Rederive Eqn. 7-144 and 7-145.
(Ex. 7.37) Rederive Eqn. 7-146 and 7-147.
(Ex. 7.38) Rederive Eqn. 7-148 and 7-149.
(Ex. 7.39) Rederive Equations 7-150 – 7-160.
(Ex. 7.40) Rederive Eqn. 7-162.
(Ex. 7.41) Rederive Eqn. 7-163.
(Ex. 7.42) Rederive Eqn. 7-166.
(Ex. 7.43) Rederive Eqn. 7-167.

Chapter 8. Black Holes

8.1 Black Holes Overview

In this chapter we seek solutions to Einstein's Field equations that are spherically symmetric and asymptotically flat. The first such solution was discovered in 1917 by Schwarzschild [40], for a Black Hole of mass M, and has metric:

$$ds^2 = -\left(1 - \frac{2M}{r}\right) dt^2 + \left(1 - \frac{2M}{r}\right)^{-1} dr^2 + r^2(d\theta^2 + \sin^2\theta\, d\varphi^2)$$

(Eqn. 8-1)

This form of the metric has singularities (they all do), so we will begin by discussing the singularities following the discussion of [24], so further details can be found there.

There is a singularity at $r = 2M$ (the 'horizon'), but this is a 'coordinate' singularity, i.e., there is not an actual curvature singularity at the horizon, and a simple change of coordinates to patch over this region would suggest there is nothing special at the $r = 2M$, 'locally', although 'globally' it is crossing the point of no return, even for light. (Note that $r = 2M$ is 'special', locally, when a quantum field is placed on the manifold, the effect seen externally being Hawking Radiation [41], as is discussed in Book 5 [9]).

In a tensor theory we want to speak of scalar invariants, so when probing the nature of the true singularities of the theory, we are interested in scalar curvature invariants constructed from the key metric and curvature tensors: $g_{\alpha\beta}$, $\epsilon_{\alpha\beta\mu\nu}$, $R_{\alpha\beta\mu\nu}$. For the Kretschmann invariant we have:

$$\lim_{r\to 0} R_{\alpha\beta\mu\nu}\, R^{\alpha\beta\mu\nu} \sim \frac{M^2}{r^6} \to \infty$$

(Eqn. 8-2)

thus a true curvature singularity as we approach $r = 0$. This describes scalar curvature singularities that are polynomial.

In a differential geometry we also want to speak of tensor constructs parallel transported along a path. To probe the singularity at $r = 0$ we, thus, also consider the Riemann tensor $R_{\alpha\beta\mu\nu}$ evaluated (component-wise) in some basis that is parallel transported on a path intersecting the origin. In [24] these are referred to as parallel propagated curvature singularities. We saw in Ch. 7 on CTCs that such evaluations can lead to singularities that are exponential (not polynomial like the scalar divergences), for parallel transport along a path approaching a CTC geodesic.

So far we've spoken of the highly symmetric Schwarzschild solution, and consequences of such. Are we certain that such symmetric, and singular, end-states, will occur in a generic asymmetric collapse? The proof that such singular end-states are ubiquitous was given by Penrose [42]. The proof introduces the Manifold property of geodesic completeness. A manifold that is not geodesically complete is one in which a geodesic exists that can't be extended to every value of its affine parameter (because it intersects a singularity, for example). In this terminology, Penrose then found that incomplete null geodesics will exist on a manifold if:

(1) The null convergence condition is satisfied ($R_{\alpha\beta}k^\alpha k^\beta \geq 0$), for every null vector k^α. Since we are working in Einstein General Relativity, then some form of Einstein's Equations (with or without cosmological constant) will hold, which (for null vectors) then allows:

$$R_{\alpha\beta}k^\alpha k^\beta \geq 0 \quad \rightarrow \quad T_{\alpha\beta}k^\alpha k^\beta \geq 0.$$

(Eqn. 8-3)

Recall that the Weak Energy Condition (WEC) has $T_{\alpha\beta}u^\alpha u^\beta \geq 0$ for every non-spacelike u^α, thus including every null k^α.

(2) The manifold has a non-compact Cauchy surface.

(3) The manifold has a closed trapped surface: a closed (compact without boundary) spacelike 2-surface, where ingoing and outgoing null rays converge towards each other.

A more general description of singularity formation, using a time-like convergence condition, was devised by Hawking and Penrose in 1970 [43]. A singularity associated with incomplete time-like and null geodesics will occur in a manifold if that manifold satisfies:

(1) The time-like convergence condition: $R_{\alpha\beta}u^\alpha u^\beta \geq 0$ for every non-spacelike u^α. Similar use of the Einstein relation and this can be rewritten as:

$$T_{\alpha\beta}u^\alpha u^\beta \geq u_\alpha u^\alpha\left(\frac{1}{2}T - \frac{1}{8\pi}\Lambda\right)$$

(Eqn. 8-4)

If no cosmological constant, then:

$$T_{\alpha\beta}u^\alpha u^\beta \geq \frac{1}{2}T(u_\alpha u^\alpha)$$

(Eqn. 8-5)

which is known as the Strong Energy Condition.
(2) Every nonspacelike geodesic satisfies the 'generic condition' that somewhere on every geodesic we have:

$$u_{[\sigma}R_{\alpha]\beta\gamma[\delta}u_{\tau]}u^\beta u^\gamma \neq 0.$$

(Eqn. 8-6)

Although global in application, this is a local evaluation. Compare with the Penrose proof, where a global property must be known (the existence of a noncompact Cauchy surface). Thus, this formulation is more generally applicable in a practical sense.
(3) There are no closed time-like curves (CTCs). We saw in Ch. 7 on CTCs that there is a strong association of existence of CTCs with the violation of the Weak Energy Condition (WEC). Thus, (3) might be a weaker form of (1), where the Strong energy Condition is already required.
(4) There exists a closed trapped surface. Again, the existence of such in the Penrose theory with the above WEC requirement (minimally), also needed a noncompact Cauchy surface whereas now the proof construct requires the generic condition.

Central to both of the (Black Hole) singularity theorems, as with CTC exclusion (chronology protection conjecture [28]) is the enforcement of the WEC.

Suppose we work with a spacetime for a star collapse that will form a Black Hole, and suppose it is a spacetime where WEC is satisfied such that the singularity theorems are applicable. Let's consider a 2-sphere surface contracting onto the Black Hole Horizon that is about to form. Condition (3) of Penrose's Theorem, to have a closed trapped surface, will be met when the 2-sphere contracts past any Horizon. If we require WEC to hold, but also that there be no singularity problem, then this would indicate that there can be no selection of 2-spheres inside a Horizon formation – i.e., that the Horizon is a boundary of the Manifold. Theories where there is no 'inside' to the Black Hole, only complex boundary conditions, are suggested by both field theory (with prediction of Hawking Radiation) and string theory (the Fuzzball solution). Thus, Penrose's theorem may relate to a condition (a closed trapped surface) that is never allowed to exist in a manifold theory with a quantum spinor field placement -- where consistency of a manifold with spinor field allows the positive energy condition to be proven assuming WEC [12].

Even if we determine that field theory prevents a causally disconnected part of the manifold to come into existence, the Penrose and Hawking Theorems are still applicable in the limit of taking a closed surface to being almost trapped (at the horizon), and as such, still predict the formation of a Black hole as seen by an external observer, but now without probing past the horizon boundary other than to require WEC (at the Horizon boundary, with AWEC outside the horizon). So again, the central role of WEC in the formulation.

Let's now turn the Penrose singularity theorem on its head, since we have ample observational proof of Black Holes perhaps the utility of the proof is not in proving the existence of something we have now seen observationally (unlike in 1965) but as an indication of the validity of the core construct in the proof that such singularities would form – e.g., the WEC hypothesis. BHs occur frequently and are long-lived (part of galactic evolution), thus the required enforcement of WEC consistent with their ubiquitous formation can be taken as a partial validation of the WEC hypothesis itself. As seen in Ch. 7, however, there's really no choice, because allowing WEC violation allows CTCs which is to allow an inconsistency in more ways that time-paradox, it allows infinite energy singularity formation of exponential-type.

8.2 The Schwarzschild Black Hole solution

The Schwarzschild Black Hole solution to Einstein's Relativity equations [44] has metric:

$$ds^2 = -\left(1 - \frac{2M}{r}\right) dt^2 + \left(1 - \frac{2M}{r}\right)^{-1} dr^2 + r^2(d\theta^2 + \sin^2\theta\, d\varphi^2)$$

(Eqn. 8-7)

with coordinate singularity at $r = 2M$, and M is the mass of the black hole. In a Penrose Diagram we see that the Schwarzschild coordinates only cover part of the fully extended spacetime (see Fig. 8.1) in two patches ($r < 2M$) and ($r > 2M$). Both patches can be covered with a single patch if using ingoing Eddington-Finkelstein coordinates. If we do the following coordinate transformation, $u(r,t)$ and $v(r,t)$, due to Kruskal and Szekeres [45] we have another coordinatization of the complete space time:

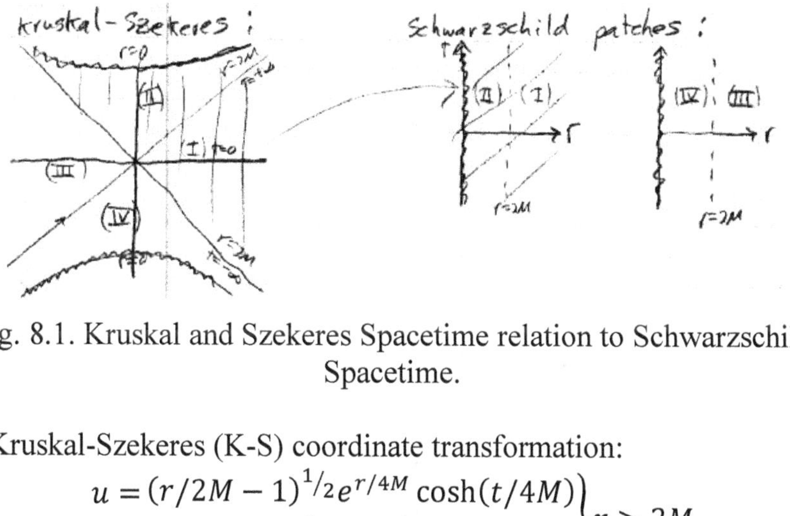

Fig. 8.1. Kruskal and Szekeres Spacetime relation to Schwarzschild Spacetime.

The Kruskal-Szekeres (K-S) coordinate transformation:

$$\left.\begin{aligned} u &= (r/2M - 1)^{1/2} e^{r/4M} \cosh(t/4M) \\ v &= (r/2M - 1)^{1/2} e^{r/4M} \sinh(t/4M) \end{aligned}\right\} r > 2M$$

$$\left.\begin{aligned} u &= (1/2M - 1)^{1/2} e^{r/4M} \sinh(t/4M) \\ v &= (1/2M - 1)^{1/2} e^{r/4M} \cosh(t/4M) \end{aligned}\right\} r > 2M$$

(Eqn. 8-8)

For which we have $(r/2M - 1)e^{r/2M} = u^2 - v^2$ and the new metric is:

$$ds^2 = \frac{32M^3}{r} e^{-r/2M}(-dv^2 + du^2) + r^2(d\theta^2 + \sin^2\theta\, d\varphi^2).$$

(Eqn. 8-9)

A section that follows will analyze K-S in detail, but first we analyze the Eddington-Finkelstein (E-F) solution since it is part of what will be the K-S solution.

8.3 The Eddington-Finkelstein solutions

Consider freely falling photons (first devised by Eddington in 1924 [46], rediscovered by Finkelstein in 1958 [47]). Consider ingoing or outgoing photons (no angular degrees of freedom) travelling along null geodesics $ds^2 = 0$:

$$0 = -\left(1 - \frac{2M}{r}\right)dt^2 + \left(1 - \frac{2M}{r}\right)^{-1} dr^2.$$

So, we need to solve: $\left(\frac{dr}{dt}\right)^2 = (1 - 2M/r)^2$, start with:

$$\frac{dr}{dt} = \pm(1 - 2M/r) \quad \rightarrow \quad dt = \pm\frac{dr}{1 - \frac{2M}{r}}.$$

(Eqn. 8-10)

This is solved to give:

$$t = \pm(r + 2Mln|r/2M - 1|) + \mathrm{C} \qquad \rightarrow \qquad t = \pm r^* + \mathrm{C},$$

(Eqn. 8-11)

where $r^* = r + 2Mln|r/2M - 1|$.

For ingoing Eddington-Finkelstein (E-F) we switch to the new coordinates $(r, t) \Rightarrow (r, \tilde{V})$, i.e., we replace 't' with $t = \tilde{V} - r^*$ (for outgoing we have $t = \tilde{U} + r^*$). Let's see how the coordinate change affects the metric:

$$t = \tilde{V} - r^* \quad \rightarrow \quad dt = d\tilde{V} - \frac{dr}{1 - \frac{2M}{r}}.$$

(Eqn. 8-12)

Thus,

$$ds^2 = -\left(1 - \frac{2M}{r}\right)\left(d\tilde{V} - \frac{dr}{1 - \frac{2M}{r}}\right)^2 + \left(1 - {}^{2M}/{}_{r}\right)^{-1} dr^2 + r^2 d\Omega^2.$$

So,

$$ds^2 = -\left(1 - {}^{2M}/{}_{r}\right)d\tilde{V}^2 + 2d\tilde{V}dr + r^2 d\Omega^2,$$

(Eqn. 8-13)

and notice that $r = 2M$ is no longer a coordinate singularity. So, with infalling E-F coordinates (a Black Hole) we cover Schwarzschild patches I and II in Fig. 8.1, while with outfalling E-F (a White Hole) we cover Schwarzschild patches I and IV.

8.4 The Kruskal-Szekeres Solution

Is there a coordinatization that covers I, II, and IV, and the soon to be discovered wormhole solution due to III (see Fig. 8.1)? The answer is yes, and it is known as the Kruskal-Szekeres solution, which consists of doing both coordinate transformations indicated by E-F and combine them to entirely shift from the (r, t)coordinates to the new $(\tilde{U}, \tilde{V})$ coordinates where:

$$\tilde{V} - \tilde{U} = 2r^* \quad and \quad \tilde{V} + \tilde{U} = 2t \ .$$

(Eqn. 8-14)

The metric is now:

$$ds^2 = -\left(1 - {}^{2M}/_{r}\right) d\tilde{U} d\tilde{V} + r^2 (d\theta^2 + \sin^2\theta \, d\varphi^2)$$

(Eqn. 8-15)

This is no longer singular at $r = 2M$, but it is now pathological in that the metric loses all dependency on $d\tilde{U}$ and $d\tilde{V}$. To identify a further shift in the coordinatization to resolve this pathology note that

$$\exp\left[\left(\tilde{V} - \tilde{U}\right)/4M\right] = \exp(r^*/2M) = (r/2M - 1)\exp(r/2M).$$

(Eqn. 8-16)

Suggesting that we let

$$\tilde{u} = -e^{-\frac{\tilde{U}}{4M}} \quad and \quad \tilde{v} = +e^{+\frac{\tilde{V}}{4M}},$$

(Eqn. 8-17)

for which we get:

$$ds^2 = -\left({}^{32M^3}/_{r}\right) e^{-r/2M} d\tilde{v} d\tilde{u} + r^2 d\Omega^2$$

(Eqn. 8-18)

$$(r/2M - 1)e^{r/2M} = -\tilde{u}\tilde{v}.$$

(Eqn. 8-19)

The last step in the coordinate transformation, to give the K-S transformation, is to shift off null coordinates to one space like and one time like:

$$u = \frac{1}{2}(\tilde{v} - \tilde{u}) = (r/2M - 1)^{1/2} \ e^{r/4M} \ \cosh(t/4M),$$

(Eqn. 8-20)

$$v = \frac{1}{2}(\tilde{v} + \tilde{u}) = (r/2M - 1)^{1/2} \ e^{r/4M} \ \sinh(t/4M).$$

(Eqn. 8-21)

The specific form given is appropriate to region (I), otherwise a slightly different transform will be needed. Note the following relation is valid in all quadrants:

$$(r/2M - 1)e^{r/2M} = u^2 - v^2.$$

(Eqn. 8-22)

8.5 The Novikov Solution

Instead of freely falling photons one may derive the complete Schwarzschild coordinate extension by using freely falling particles, as was done by Novikov in 1963 [48]. Consider radially moving (infalling) test particles in the commoving coordinate. As with the photons we are attempting to physically connect the $r < 2M$ region to the $r > 2M$ region. Consider the trajectory of a test particle that gets ejected from the singularity at $r = 0$, flies radially through $r = 2M$, reaches maximum radius $r_{\max}$ at proper time $\tau = 0$ (set clock appropriately and coord. time $t = 0$, and then falls back.

$$ds^2 = -\left(1 - \frac{2M}{r}\right)dt^2 + \left(1 - \frac{2M}{r}\right)^{-1} dr^2 + r^2 d\Omega^2$$

(Eqn. 8-23)

$$g_{oo} = -\left(1 - \frac{2M}{r}\right) \quad g_{rr} = \left(1 - \frac{2M}{r}\right)^{-1}$$

(Eqn. 8-24)

In Einstein's Geometric Theory of Gravitation we have $g^{\alpha\beta} P_\alpha P_\beta + m^2 = 0$, and there is a tensor connection between energy and momentum. The metric is unaffected by translations in t and φ, thus we have the conserved quantities $P_o = -E,\ P_\varphi = \pm L,\ and\ \ P^\theta = \frac{d\theta}{d\lambda} = 0$:

$$g^{\alpha\beta} P_\alpha P_\beta + m^2 = \frac{-E^2}{(1 - 2M/r)} + \left(1 - \frac{2M}{r}\right) P_r^2 + \frac{L^2}{r^2} + m^2 = 0$$

(Eqn. 8-25)

$$P^r = \frac{dr}{d\tau} \Longrightarrow P_r = g_{rr} P^r \qquad P^o = \frac{dt}{d\tau}$$

(Eqn. 8-26)

$$\frac{-E^2}{(1 - 2M/r)} + \frac{1}{(1 - 2M/r)}\left(\frac{dr}{d\lambda}\right)^2 + \frac{L^2}{r^2} + m^2 = 0.$$

(Eqn. 8-27)

Let's shift the constants for convenience with $\tilde{E} = {}^{E}/_{m}, \tilde{L} = {}^{L}/_{m}$:

$$\left(\frac{dr}{d\tau}\right)^2 = \tilde{E}^2 - (1 - 2M/r)\left(1 + \tilde{L}^{\,2}/_{r^2}\right)$$

(Eqn. 8-28)

For radial motion $\tilde{L} = 0$, and for ingoing this becomes:

$$\left(\frac{dr}{d\tau}\right) = -\sqrt{\tilde{E}^2 - \left(1 - \frac{2M}{r}\right)}.$$

(Eqn. 8-29)

If we let $R \equiv 2M/\left(l - \tilde{E}^2\right)$ this becomes:

$$\tau = \int \frac{-dr}{\sqrt{2M/r - 2M/R}}$$

(Eqn. 8-30)

A solution to this can be written in parametric form:

$$r = \frac{R}{2}(1 + \cos\eta) \quad and \quad \tau = \frac{R}{2}\left(\frac{R}{2M}\right)^{1/2}(\eta + \sin\eta).$$

(Eqn. 8-31)

Note that in this form we see that $R = r_{\max}$. To show this consider

$$\tau = \int_{r=R}^{r=r_0} \frac{-dr}{\sqrt{2M/r - 2M/R}} = -\int_{\eta=0}^{\eta=\eta_0} \frac{-\frac{R}{2}\sin\eta\, d\eta}{\sqrt{2M}\sqrt{\frac{1}{\frac{R}{2}(1+\cos\eta)} - \frac{1}{R}}}$$

$$= +\frac{R}{2}\sqrt{\frac{R}{2M}}\int \frac{\sin\eta\, d\eta}{\sqrt{\frac{2}{(1+\cos\eta)} - 1}},$$

(Eqn. 8-32)

which simplifies to:

$$\tau = +\frac{R}{2}\sqrt{\frac{R}{2M}}\int_{\eta=\pi}^{\eta=\eta_0}(1+\cos\eta)d\eta = +\frac{R}{2}\sqrt{\frac{R}{2M}}\big((\eta + \sin\eta)\big) + \text{constant}$$

(Eqn. 8-33)

Since $\eta = 0 \Rightarrow r = R$ we have $\tau = 0$ at $\eta = 0$ as well, the constant is equal to 0:

$$\tau = \frac{R}{2}\left(\frac{R}{2M}\right)^{1/2}(\eta + \sin\eta)$$

(Eqn. 8-34)

Next, let's examine the $t(\tau(\eta))$ relation:

$$\frac{dt}{d\tau} = P^o = g^{oo}\left(-\tilde{E}\right) = \frac{\tilde{E}}{1 - 2M/r}.$$

(Eqn. 8-35)

Using the chain rule this becomes:

$$\frac{dt}{dr} = \frac{\tilde{E}}{(1 - 2M/r)} \frac{1}{\sqrt{\tilde{E}^2 - \left(1 - \frac{2M}{r}\right)}}.$$

(Eqn. 8-36)

Upon substitution with $r = \frac{R}{2}(1 + \cos\eta)$ integration yields:

$$t(\eta) = \left[\left(\frac{R}{2} + 2M\right)\left(\frac{R}{2M} - 1\right)^{1/2}\right]\eta + \frac{R}{2}\left(\frac{R}{2M} - 1\right)^{1/2} \sin\eta$$
$$+ 2M\, ln \left|\frac{(R/2M - 1)^{1/2} + \tan(\eta/2)}{(R/2M - 1)^{1/2} - \tan(\eta/2)}\right|$$

(Eqn. 8-37)

(solution above first found by Khuri in 1957 [49]).

So, to recap:

$$r = \frac{1}{2} r_{\max}(l - \cos\eta)$$
$$\tau = (r_{\max}^3/8M)^{1/2}(\eta + \sin\eta)$$
$$t = 2M\, ln \left|\frac{r_{\max}(2M - 1)^{1/2} + \tan(\eta/2)}{r_{\max}(2M - 1)^{1/2} - \tan(\eta/2)}\right|$$
$$+ 2M(r_{\max}/2M - 1)^{1/2}\left(\eta + \frac{r_{\max}}{4M}(\eta + \sin\eta)\right)$$

Each trajectory in Novikov coord.'s is labeled by Novikov's R* variable which depends on $r_{\max}$:

$$R^* = (r_{\max}/2M - 1)^{1/2} \quad \Rightarrow \quad r_{\max} = 2M\left(R^{*^2} + 1\right).$$

(Eqn. 8-38)

So, with $\tau = M\left(R^{*^2} + 1\right)^{3/2}(\eta + \sin\eta)$ we get:

$$r(\tau, R^*) = M\left(R^{*^2} + 1\right)(1 + \cos\eta(\tau, R^*))$$

(Eqn. 8-39)

$$t(\tau, R^*) = 2M\, ln \left|\frac{R^* + \tan(\eta/2)}{R^* - \tan(\eta/2)}\right|$$
$$+ 2M\, R^*\left(\eta + \left[R^{*^2} + 1\right]\frac{1}{2}(\eta + \sin\eta)\right)$$

(Eqn. 8-40)

$$ds^2 = -d\tau^2 + \left(\frac{R^{*^2} + 1}{R^{*2}}\right)\left(\frac{\partial r}{\partial R^*}\right)^2 dR^{*^2} + r^2 d\Omega^2$$

(Eqn. 8-41)

Novikov:

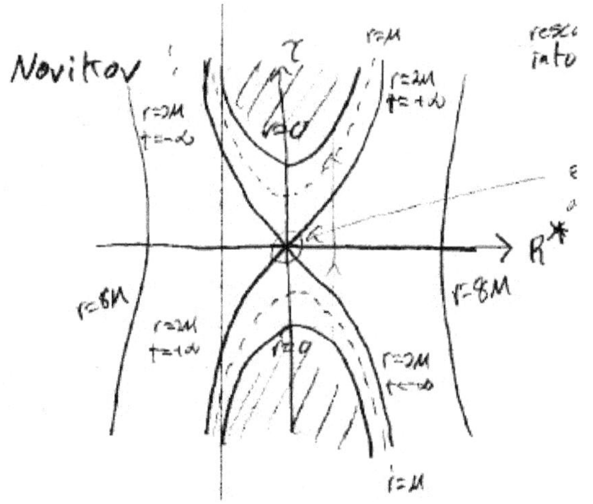

Fig. 8.2. Novikov Spacetime. Constant radius foliation shown.

Note that rescaling time parameter draws geodesies into cusp. The evolutionary cusp is for $r = 2M, t = finite$ and (not shown) one at $t = \pm\infty, r = 0$. The foliation is by surfaces $0 \leq r_L \leq 2M$; $-\infty < t_R < t_\alpha$; $2M \leq r_u \leq 0 + \infty < t_L < -\infty$. Further analysis of the Novikov coordinate system and equations of motion will follow in Sec. 8.9.

Kruskal has a similar foliation:

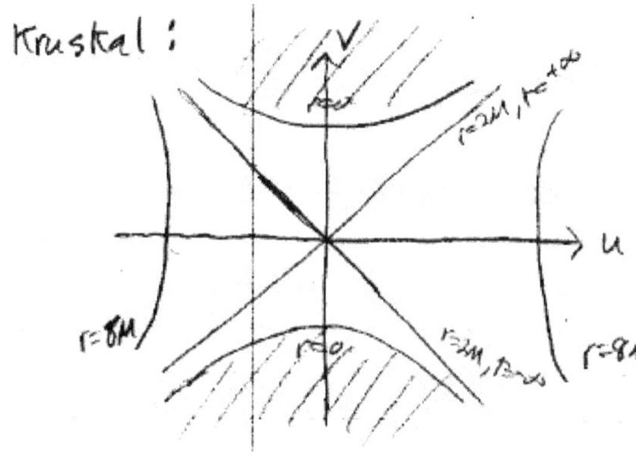

Fig. 8.3. Kruskal-Szekeres Spacetime. Similar foliation to Novikov.

A description of the Einstein field equation for the wormhole solution is now considered from a variational perspective starting with a simple choice for the variational action. This description allows a direct transition to a quantum model and will hopefully clarify the simplest representation to use in attempts at quantization in Book 4 [3].

8.6 The Kantowski-Sacks foliation and related Minisuperspace model

We want the constant "r" foliation. How to make sense of the strip in the foliation for the $r > 2M$ surfaces (having non-constant r however)? Recall

$$S_{Ein} = \frac{4\pi}{M}\int\left\{r[(r')^2 - (\acute{r})^2] - \frac{64M^4}{r}e^{-r/2M}\right\}dudv$$

which was derived in the Kruskal coordinates The resulting Kruskal equation of motion is:

$$2r(\ddot{r} - r'') + [(\acute{r})^2 - (r')^2] + \frac{32M^3}{r}e^{-r/2M}\left(l + \frac{2M}{r}\right) = 0$$

(Eqn. 8-42)

If we consider the labeling of constant r surfaces by the throat variable and here consider the dynamics of the throat variable within the $(r/2M - 1)e^{r/2M} = u^2 - v^2$ Minisuperspace reduction:

$$r|_{u=0} = \theta; \quad \left(\frac{\theta}{12M} - 1\right)e^{\theta/2M} = -v^2 \quad \Rightarrow \quad \frac{\theta}{2M}e^{\theta/2M}\left(\frac{\partial\theta}{\partial v}\right) = -2v$$

(Eqn. 8-43)

$$\left(\frac{r}{(2M)^2}\right)e^{r/2M}\left(\frac{\partial r}{\partial v}\right) = -2v \quad \Rightarrow \quad \left(\frac{\theta}{(2M)^2}\right)e^{\theta/2M}\left(\frac{\partial r}{\partial v}\right)_{u=0} = -2v$$

$$\leftarrow\nwarrow \left(\frac{\partial r}{\partial v}\right)_{u=0} = \left(\frac{\partial\theta}{\partial v}\right)$$

(Eqn. 8-44)

Similarly $\left(\frac{\partial r}{\partial u}\right)_{u=0} = 0$

$$\left(\frac{r}{(2M)^2}\right)e^{r/2M}\left(\frac{\partial^2 r}{\partial v^2}\right) + \left(\frac{r}{(2M)^2}\right)\left(1 + \frac{r}{2M}\right)e^{r/2M}\left(\frac{\partial r}{\partial v}\right)^2 = -2$$

(Eqn. 8-45)

$$\left(\frac{\partial^2 r}{\partial v^2}\right)_{u=0} = -\frac{1}{\theta}\left(1 + \frac{\theta}{2M}\right)\left(\frac{\partial r}{\partial v}\right)^2 - \frac{8M^2}{\theta}e^{-\theta/2M} \text{ and } \left(\frac{\partial^2 r}{\partial u^2}\right)_{u=0} = \frac{8M^2}{\theta}e^{-\theta/2M}$$

(Eqn. 8-46)

$$S_{Ein} = \frac{4\pi}{M}\int\left\{r[(r')^2 - (\acute{r})^2] - \frac{64M^4}{r}e^{-r/2M}\right\}dudv$$

Since $\left(\frac{\partial\theta}{\partial\tau}\right)^2 = \left(\frac{2M}{\theta} - 1\right)$ and $\left(\frac{\partial\tau}{\partial v}\right)^2 = \frac{32M^3}{r}e^{-r/2M}$

$$\left(\frac{r}{2M} - 1\right)e^{r/2M} = u^2 + \left(\frac{\theta}{2M} - 1\right)e^{\theta/2M}$$

$$\frac{\partial r}{\partial v} = \frac{\partial\theta}{\partial v}$$

$$\frac{r}{(2M)^2}e^{r/2M}\left(\frac{\partial r}{\partial u}\right) = 2u \quad\Rightarrow\quad \left(\frac{\partial r}{\partial u}\right)^2 = \frac{4(2M)^4}{r^2}e^{-r/2M}u^2$$

(Eqn. 8-47)

The definition of θ depends on foliation, to arrive at the $R^* = 0$ observer correspondence one must be speaking of the following foliation described by Kantowski-Sachs: each folium is entirely described by the θ variable, i.e. each folium has constant "radius" variable. This is a foliation for which the observer hypersuperspace labeling is preserved under the symmetry (Lorentz) operations, thus a coordinate independent geodesic observer/foliation correspondence is achieved.

We know the $R^* = 0$ observers (or the $r_{\text{max}} = 2M$) are described in a simple Coulombic way. If each of the local regions of all the $r_{\text{max}} = 2M$ observers are regarded and patched together, the resulting congruence provides a labeling of hypersuperspace that evolves locally according to the geodesic derivative and patching the locales together the hypersurfaces must evolve according to that exact same Coulombic dynamics.

So, 1-D Minisuperspace Models require certain types of isometries if they are to be conveniently correlated with the point particle dynamics of an imbedded geodesic. The Kantowski-Sachs calculation won't say anything about wormholes, *per se*, but will describe how the "throat variable" evolves. Different hypersurfaces will evolve in generally more complicated fashion than the K-S one, however at late times the agreement should coincide as far as the singularity is concerned. Nambu and Sasaki examined this problem in 1988 [50]. They consider the BH inner geometry and mention the Kantowski and Sachs analysis in a Minisuperspace analysis.

We have the derivation

$$\left.\left(\frac{\partial r}{\partial \tau}\right)^2\right|_{u=0} = -\left.\left(1-\frac{2M}{r}\right)\right|_{u=0}$$

(Eqn. 8-48)

$$\left(\frac{\partial r}{\partial \tau}\right)^2 = -\left(1-\frac{2M}{r_{\min}}\right)$$

(Eqn. 8-49)

$r_{\min} = \theta$ = threat for the foliation chosen (r = const).

The equation regards the "inner" black hole region only, and as such may be discussed with one of the Schwarzschild "patches":

$$ds^2 = -d\tau^2 = -\left(1-\frac{2m}{r}\right)dt^2 + \left(1-\frac{2m}{r}\right)^{-1} dr^2 + r^2(d\Omega^2)$$

$$(-1) = \left(1-\frac{2m}{r}\right)^{-1}\left(\frac{\partial r}{\partial \tau}\right)^2$$

$$\left(\frac{\partial r}{\partial \tau}\right)^2 = -\left(1-\frac{2m}{r}\right)$$

(Eqn. 8-50)

Here $r = r_{\min}$ due to the constant r hypersurfaces assumed in the foliation that respects the (r, t) coordinates. Now, aside from infinite domain factors it is desired to arrive at the dynamical equation $\left(\frac{\partial r}{\partial \tau}\right)^2 = -\left(1-\frac{2m}{r}\right)$ but respective to a Minisuperspace reduction from the S_{Ein} action where the dynamical equation is expressed in L_{Ein} and described the evolution of the hypersurfaces. The non-compact property of the hypersurfaces leads to an infinite domain factor, however, compacting the spacetime eliminates this complication.

Inner Schwarzschild (full analysis in next section)

$$ds^2 = -\left(1-\frac{2m}{r}\right)dt^2 + \left(1-\frac{2m}{r}\right)^{-1} dr^2 + r^2(d\theta^2 + \sin^2\theta\, d\varphi^2)$$

$$(r < 2M) = -\left(\frac{2M}{r}-1\right)^{-1} dr^2 + \left(\frac{2M}{r}-1\right)dt^2 + r^2 d\Omega^2$$

(Eqn. 8-51)

(inner Schwarzschild or Kantowski-Sachs spacetime geometry)

$$S_{Ein}\text{ (inner Schw)} = 4\pi V \int L_{Ein}\text{ (inner schw)}\, dt$$

The Minisuperspace model will be based on the metric:

$$ds^2 = -\left(\frac{2M}{R(r)} - 1\right)^{-1} dr^2 + \left(\frac{2M}{R(r)} - 1\right) dt^2 + r^2 d\Omega^2$$
$$= -e^{-2a(r)} dr^2 + e^{2a(r)} dt^2 + r^2 d\Omega^2$$

Outer Kantowski-Sachs

Using the geodesic Lagrangian method once again:

$$\dot{a} = \frac{\partial a}{\partial t}, \quad a' = \frac{\partial a}{\partial r}, \quad \breve{a} = \frac{\partial a}{\partial \lambda}$$
$$\breve{a} = \frac{\partial a}{\partial \lambda} = \frac{\partial a}{\partial t}\frac{\partial t}{\partial \lambda} + \frac{\partial a}{\partial r}\frac{\partial r}{\partial \lambda} = a'\breve{r}$$

Starting with

$$I = \int g_{\mu\nu}\, \dot{x}^\mu \dot{x}^\nu d\lambda$$

Vary $t(\lambda)$ first: (let $\dot{a} = \breve{a}$ notationally in what follows

$\delta I = \int (e^{2a})\, \dot{t}\delta\dot{t} d\lambda = -\int (e^{2a}\dot{t})^{\cdot}\, \delta t d\lambda$

$$= -\int \{e^{2a}(2\dot{a})\dot{t} + e^{2a}\ddot{t}\}\, \delta t d\lambda$$

$$= -\int e^{2a}\{\ddot{t} + (2a')\dot{r}\dot{t}\}\, st d\lambda$$

$$\Rightarrow \ddot{t} + 2a'\dot{r}\dot{t} = 0 \quad \Rightarrow \quad \Gamma^t_{rt} = a'$$

Vary $\theta(\lambda)$: $\quad \delta I = \int r^2 \left(\dot{\theta}\delta\dot{\theta} + 2\sin\theta\cos\theta\, \delta\theta\dot{\varphi}^2\right) d\lambda =$
$\int \left(-\left(r^2\dot{\theta}\right)^0 + 2r^2 \sin\theta\cos\theta\, \delta\theta\dot{\varphi}^2\right) \delta\theta d\lambda$

$$\Rightarrow -2rr\theta - r^2\ddot{\theta} + r^2 \sin\theta\cos\theta\, \dot{\varphi}^2 = 0$$

$$\Rightarrow \underline{\Gamma^\theta_{r\theta} = \frac{1}{r}}, \quad \underline{\Gamma^\theta_{r\psi} = -\sin\theta\cos\theta}$$

For $\psi(\lambda)$: $\quad \delta I = \int r^2 \sin^2\theta\, \dot{\psi}\delta\dot{\psi} d\lambda = \int -\left[\left(2r\dot{r}\sin^2\theta + 2r^2\sin\theta\cos\theta\, \dot{\theta}\right)\psi + r^2\sin^2\theta\, \ddot{\psi}\right] d\lambda$

$$\Rightarrow \underline{\Gamma^\theta_{\theta\psi} = \cot\theta}, \quad \underline{\Gamma^\psi_{r\psi} = \frac{1}{r}}$$

Vary $r(\lambda)$:

$$\delta I = 0 = \int \{(2a'\delta r e^{-2a})\dot{r}^2 + 2(-e^{-2a})\dot{r}\delta\dot{r} + 2(2a'\delta r e^{2a})\dot{t}^2 + 2r\delta r\left(\dot{\theta}^2 + \sin^2\theta\, \dot{\psi}^2\right)\} d\lambda$$

$$= \int \{(2a'e^{-2a})\dot{r}^2 - 4e^{-2a}a'\dot{r}^2 + 2e^{-2a}\ddot{r} + 2a'e^{2a}\dot{t}^2 + 2r(\dot{\theta}^2 + \sin^2\theta\,\dot{\varphi}^2)\}\,d\lambda$$

$$= \int \{\ddot{r} - a'\dot{r}^2 + a'e^{4a}\dot{t}^2 + re^{2a}(\dot{\theta}^2 + \sin^2\theta\,\dot{\varphi}^2)\}\,d\lambda\delta r$$

$$\Rightarrow \ddot{r} - a'\dot{r}^2 + a'e^{4a}\dot{t}^2 + re^{2a}\dot{\theta}^2 + re^{2a}\sin^2\theta\,\dot{\varphi}^2 = 0$$

$$\Rightarrow \Gamma^r_{rr} = -a', \qquad \Gamma^t_{tt} = a'e^{4a}, \qquad \Gamma^r_{\theta\theta} = re^{2a}, \qquad \Gamma^r_{\varphi\varphi} = r\sin^2\theta\, e^{2a}$$

Thus,

$$\Gamma^t_{rt} = -\Gamma^r_{rr} = a'$$
$$\Gamma^r_{tt} = +a'e^{4a}$$
$$\Gamma^r_{\varphi\varphi} = \sin^2\theta\,\Gamma^r_{\theta\theta} = +r\sin^2\theta\, e^{2a}$$
$$\Gamma^\theta_{r\theta} = \Gamma^\varphi_{r\varphi} = \frac{1}{r}$$
$$\Gamma^\theta_{\varphi\varphi} = \sin\theta\cos\theta$$
$$\Gamma^\varphi_{\theta\varphi} = \cot\theta$$

8.7 Inner Schwarzschild

$$ds^2 = -e^{-2a(r)}dr^2 + e^{2a}dt^2 + r^2 d\Omega^2$$
$$= -\left(\frac{2M}{R(r)} - 1\right)^{-1} dr^2 + \left(\frac{2M}{R(r)} - 1\right) dt^2 + r^2 d\Omega^2$$

$R(r) = r$ for Minkowski, left undetermined for variational analysis leading to Minkowski model. Thus,

$$g_{rr} = -e^{-2a} \qquad g^{rr} = -e^{2a}$$
$$g_{tt} = e^{2a} \qquad g^{tt} = e^{-2a}$$
$$g_{\theta\theta} = r^2 \qquad g^{\theta\theta} = r^{-2}$$
$$g_{\varphi\varphi} = r^2\sin^2\theta \qquad g^{\varphi\varphi} = r^{-2}\sin^{-2}\theta$$

$$G' = G + \frac{1}{\sqrt{-g}}\frac{\partial}{\partial\theta}\left[\sqrt{-g}\left(g^{iK}\Gamma^\theta_{iK} - g^{i\theta}\Gamma^\ell_{i\ell}\right)\right]$$

$$G = g^{iK}\left(\Gamma^m_{i\ell}\Gamma^\ell_{Km} - \Gamma^\ell_{iK}\Gamma^m_{\ell m}\right)$$

$$g^{iK}\Gamma^m_{i\ell}\Gamma^\ell_{Km} = g^{rr}\Gamma^m_{r\ell}\Gamma^\ell_{rm} + g^{tt}\Gamma^m_{t\ell}\Gamma^\ell_{tm} + g^{\theta\theta}\Gamma^m_{\theta\ell}\Gamma^\ell_{\theta m} + g^{\varphi\varphi}\Gamma^m_{\varphi\ell}\Gamma^\ell_{\varphi m}$$

$$= (-e^{+2a})\left[2(a')^2 + \frac{2}{r^2}\right] + (+e^{+2a})[+2(a')^2e^{4a}] + (r^{-2})[+2e^{2a} + \cot\theta] + (r^{-2}\cos^{-2}\theta)[+2\sin^2\theta\, e^{2a} - 2\sin\theta\cos\theta\cot\theta]$$

$$= +2r^{-2}e^{2a} - r^{-2}\cot^2\theta$$

$$g^{iK}\Gamma^\ell_{iK}\Gamma^m_{\ell m} = g^{rr}\Gamma^\ell_{rr}\Gamma^m_{\ell m} + g^{tt}\Gamma^\ell_{tt}\Gamma^m_{\ell m} + g^{\theta\theta}\Gamma^\ell_{\theta\theta}\Gamma^m_{\ell m} + g^{\varphi\varphi}\Gamma^\ell_{\varphi\varphi}\Gamma^m_{\ell m}$$

$$= (-e^{+2a})(-a')\left[\frac{2}{r}\right] + (+e^{-2a})(+a'e^{4a})\left[\frac{2}{r}\right] + (r^{-2})(+re^{2a})\left[\frac{2}{r}\right]$$
$$+ (r^{-2}\sin^{-2}\theta)(+r\sin^2\theta\, e^{2a})\left[\frac{2}{r}\right]$$
$$+ (r^{-2}\sin^{-2}\theta)(-\sin\theta\cos\theta)\cot\theta$$
$$= +4r^{-2}e^{2a} - r^{-2}\cot^2\theta + 4r^{-1}a'e^{2a}$$

$$G = -4(a')r^{-1}e^{2a} - 2r^{-2}e^{2a}$$

Compared with outer schw it's the same except (a') enters only to 1st power in the G expression.

$$g^{iK}\Gamma^{\theta}_{iK} = g^{\varphi\varphi}\Gamma^{\theta}_{\varphi\varphi} = (r^{-2}\sin^{-2}\theta)(-\sin\theta\cos\theta) = -r^{-2}\cot\theta$$
$$g^{i\theta}\Gamma^{\ell}_{i\ell} = g^{\theta\theta}\Gamma^{\ell}_{\theta\ell} = (r^{-2})\cot\theta$$
$$\sqrt{-g}\frac{\partial}{\partial\theta}\left(\sqrt{-g}\left(g^{iK}\Gamma^{\theta}_{iK} - g^{i\theta}\Gamma^{\ell}_{i\ell}\right)\right)$$
$$= \frac{1}{r^2\sin\theta}\frac{\partial}{\partial\theta}\left(r^2\sin\theta\left[-2r^{-2}\cot\theta\right]\right) = \left(\frac{2}{r^2}\right)$$
$$S_{Ein} = \int \sqrt{-g}\, G'\, d\theta d\varphi dr dt$$
$$= 4\pi\int r^2\left(-4(a')^2e^{2a} - 2r^{-2}e^{2a} - 2r^{-2}\right)drdt$$
$$= -8\pi\int \left(2(a')^2e^{2a} + e^{2a} + 1\right) drdt$$

$$\sqrt{-g}R = \sqrt{-g}g^{iK}R_{iK} = \sqrt{-g}g^{iK}\left[\Gamma^{\ell}_{iK,\ell} - \Gamma^{\ell}_{i\ell,K} + \Gamma^{\ell}_{iK}\Gamma^{m}_{\ell m} - \Gamma^{m}_{i\ell}\Gamma^{\ell}_{Km}\right]$$
$$= \sqrt{-g}G + \frac{\partial}{\partial x^m}\left[\sqrt{-g}\left(g^{iK}\Gamma^{m}_{iK} - g^{im}\Gamma^{\ell}_{i\ell}\right)\right]$$

Have already calculated $\frac{\partial}{\partial\theta}$ term, $\frac{\partial}{\partial\varphi}$ term = 0 as does $\frac{\partial}{\partial t}$ term. Thus

$$g^{iK}\Gamma^{r}_{iK} = g^{rr}\Gamma^{r}_{rr} + g^{tt}\Gamma^{r}_{tt} + g^{\theta\theta}\Gamma^{r}_{\theta\theta} + g^{\varphi\varphi}\Gamma^{r}_{\varphi\varphi}$$
$$= (-e^{2a})(-a') + (e^{-2a})(+a'e^{4a}) + (r^{-2})(+re^{2a})$$
$$+ (r^{-2}\sin^{-2}\theta)(+r\sin^2\theta\, e^{2a})$$
$$= +\frac{2}{r}e^{2a} + 2a'e^{2a}$$

$$g^{ir}\Gamma^{\ell}_{i\ell} = g^{rr}\Gamma^{\ell}_{r\ell} = (-e^{+2a})\left(\frac{2}{r}\right) = -\frac{2}{r}e^{+2a}$$

$$R = G' + \frac{1}{\sqrt{-g}}\frac{\partial}{\partial r}[\ldots]$$
$$= (-4r^{-1}a'e^{2a} - 2r^{-2}e^{2a} + 2r^{-2}) + (4r^{-2}e^{2a} + 2a'e^{2a}r^{-1}) + (8a'r^{-1}e^{2a} + 4(a')^2e^{2a} + 2a''e^{2a})$$
$$= 2a''e^{2a} + 4(a')^2e^{2a} + 2r^{-2}e^{2a} + 8a'r^{-1}e^{2a} + 2r^{-2}$$

$$R = 2\left(\frac{2M}{r^3}e^{-2a} - \frac{1}{2}\left(\frac{2M}{r^2}\right)^2 e^{-4a}\right)e^{2a} + 4\left(\frac{1}{2}\left(\frac{2M}{r^2}\right)^2 e^{-4a}\right)e^{2a} + 2r^{-2}e^{2a} - 4\left(\frac{2M}{r^2}\right)r^{-1} + 2r^{-2}$$

$$R = \frac{4M}{r^3} - \frac{8M}{r^3} + 2r^{-2}\left(\frac{2M}{r} - l\right) + 2r^{-2} = 0$$

which is in agreement with the derivation for Outer Schwarzschild. So,

$$G' = -4r^{-1}a'e^{2a} - 2r^{-2}e^{2a} + 2r^{-2} \neq 0$$
$$S_{Ein} = \int \sqrt{-g}\, G' d\theta d\varphi dr dt = -8\pi \int (2ra'e^{2a} + e^{2a} - 1)\, drd$$
$$= -8\pi \int dt\big(2ra'e^{2a} + (e^{2a} - 1)\big)\, dr$$

$$L_{Ein} = 2ra'e^{2a} + (e^{2a} - 1)$$

Euler Lagrange: $\delta \int \left[\left(2r\frac{\partial a}{\partial r}e^{2a}\right) + (e^{2a} - 1)\right] dr = 0$

$$\int \{2\delta r e^{2a} + (a') + 2a'\delta r e^{2a} + 4r(a')^2\delta r e^{2a} + 2ra''e^{2a}\} = 0$$

Equation of motion: $a'' + 2(a')^2 + 2a'r^{-1} = 0$.

So, $e^{2a} = \left(\frac{2M}{R} - 1\right)$ is a solution of $L_{Ein} = 2ra'e^{2a} + (e^{2a} - 1)$

$$L_{Ein} = -2M\left(\frac{r}{R}\right)\left(\frac{R'}{R}\right) + \frac{2M}{R} - 2$$

Let's compare inner and outer Schwarzschild to get a handle on this problem.

Outer:

$$ds^2 = -\left(l - \frac{2M}{r}\right) dr^2 + \left(l - \frac{2M}{r}\right)^{-1} dr^2 + r^2 d\Omega^2$$
$$= -e^{2a} dt^2 + e^{-2a} dr^2 + r^2 d\Omega^2$$

Inner:

$$ds^2 = -\left(\frac{2M}{r} - l\right) dt^2 + \left(\frac{2M}{r} - l\right)^{-1} dr^2 + r^2 d\Omega^2$$
$$= e^{2a} dt^2 + e^{-2a} dr^2 + r^2 d\Omega^2$$

Outer Variation	Inner Variation
$t(\lambda)$: $\Gamma^t_{rt} = a'$	Same
$\theta(\lambda)$: $\Gamma^\theta_{r\theta} = \frac{1}{r}, \Gamma^\theta_{\varphi\varphi} = \sin\theta\cos\theta$	Same
$\varphi(\lambda)$: $\Gamma^\varphi_{r\varphi} = \frac{1}{r}, \Gamma^\varphi_{\theta\varphi} = \cot\theta$	Same

$r(\lambda)$: $(-2a'e^{-2a})\dot{r}^2 - (2e^{-2a}\dot{r})^0 + (-2a'e^{-2a})\dot{t}^2 + 2r\left(\dot{\theta}^2 + \sin^2\theta\,\dot{\varphi}^2\right) = 0$

vs.

$(2a'e^{2a})\dot{r}^2 + (2e^{-2a}\dot{r})^0 + (2a'e^{2a})\dot{t}^2 + \sin\theta = 0$

and

$\ddot{r}(-2e^{-2a}) + 2a'e^{-2a}\dot{r}^2 - 2a'e^{2a}\dot{t}^2 + 2r\left(\dot{\theta}^2 + \sin^2\theta\,\dot{\varphi}^2\right) = 0$ vs.

$\ddot{r}(2e^{-2a}) + 2a'e^{-2a}\dot{r}^2 + 2a'e^{2a}\dot{t}^2 + \sin\theta = 0$

$\Gamma^t_{rr} = a', \qquad \Gamma^t_{tt} = a'e^{4a}$ *Same*

Inner vs outer:

$$\Gamma^r_{\theta\theta} = -re^{2a},\ \Gamma^r_{\theta\varphi} = -r\sin^2\theta\, e^{2a}$$
$$vs$$
$$\Gamma^r_{\theta\theta} = -re^{2a},\ \Gamma^r_{\theta\varphi} = +r\sin^2\theta\, e^{2a}$$

Outside

$$G' = G + \frac{1}{\sqrt{-g}}\frac{\partial}{\partial\theta}\left[\sqrt{-g}\left(g^{iK}\Gamma^\theta_{iK} - g^{i\theta}\Gamma^\ell_{i\ell}\right)\right]$$

$$G = g^{iK}\left(\Gamma^m_{i\ell}\Gamma^\ell_{Km} - \Gamma^\ell_{iK}\Gamma^m_{\ell m}\right) - 2\Gamma^t_{rr}\Gamma^r_{tt}$$

$$g^{iK}\Gamma^m_{i\ell}\Gamma^\ell_{Km} = g^{rr}\Gamma^m_{r\ell}\Gamma^\ell_{rm} + g^{tt}\Gamma^m_{t\ell}\Gamma^\ell_{tm} + g^{\theta\theta}\Gamma^m_{\theta\ell}\Gamma^\ell_{\theta m} + g^{\varphi\varphi}\Gamma^m_{\varphi\ell}\Gamma^\ell_{\varphi m}$$

$$= (+e^{+2a})\left(\frac{2}{r^2} + 2(a')^2\right) + (-e^{+2a})(2(a')^2e^{4a})$$
$$+ (r^{-2})(-2e^{2a} + \cot^2\theta)$$
$$+ (2r^{-2}\sin^{-2}\theta)(-\sin^2\theta\, e^{2a} - \cos^2\theta)$$
$$= -\frac{2}{r^2}e^{2a} - \frac{1}{r^2}\cot^2\theta$$

$$g^{iK}\Gamma^{\ell}_{iK}\Gamma^{m}_{\ell m} = \left(g^{rr}\Gamma^{\ell}_{rr} + g^{tt}\Gamma^{\ell}_{tt} + g^{\theta\theta}\Gamma^{\ell}_{\theta\theta} + g^{\varphi\varphi}\Gamma^{\ell}_{\varphi\varphi}\right)\left(\Gamma^{r}_{\ell r} + \Gamma^{t}_{\ell t} + \Gamma^{\theta}_{\ell\theta} + \Gamma^{\varphi}_{\ell\varphi}\right)$$

$$= (e^{2a})(-a')\left(\Gamma^{r}_{rr} + \Gamma^{t}_{rt} + \Gamma^{\theta}_{r\theta} + \Gamma^{\theta}_{r\varphi}\right)$$
$$+(-e^{-2a})(+a'e^{4a})\left(\Gamma^{r}_{rr} + \Gamma^{t}_{rt} + \Gamma^{\theta}_{r\theta} + \Gamma^{\theta}_{r\varphi}\right)$$
$$+(r^{-2})(-re^{2a})\left(\Gamma^{r}_{rr} + \Gamma^{t}_{rt} + \Gamma^{\theta}_{r\theta} + \Gamma^{\theta}_{r\varphi}\right)$$
$$+(r^{-2}\sin^{-2}\theta)(+r\sin^2\theta\, e^{2a})\left(\Gamma^{r}_{rr} + \Gamma^{t}_{rt} + \Gamma^{\theta}_{r\theta} + \Gamma^{\theta}_{r\varphi}\right)$$
$$+ (-\sin\theta\cos\theta)\left(\Gamma^{r}_{\theta r} + \Gamma^{t}_{\theta t} + \Gamma^{\theta}_{\theta\theta} + \Gamma^{\theta}_{\theta\varphi}\right)$$
$$= \left[(-2a'e^{2a} - 2r^{-1}e^{2a})\left(\frac{2}{r}\right) - \frac{1}{r^2\sin\theta}(\sin\theta\cos\theta)(\cot\theta)\right]$$

$$= \left(-\frac{4}{r}a'e^{2a} - \frac{4}{r^2}e^{2a} - \frac{1}{r^2}\cot^2\theta\right)$$

$$G = +\frac{2}{r^2}e^{2a} + \frac{4}{r}(a')e^{2a}$$

$$G = \frac{2}{r^2}e^{2a} + \frac{4}{r}(a')e^{2a}$$

$$g^{iK}\Gamma^{\theta}_{iK} = g^{\varphi\varphi}\Gamma^{\theta}_{\varphi\varphi} = (r^{-2}\sin^{-2}\theta)(-\sin\theta\cos\theta) = -r^{-2}\cot\theta$$

$$g^{iK}\Gamma^{\ell}_{i\ell} = g^{\theta\theta}\Gamma^{\ell}_{\theta\ell} = (r^{-2})(\cot\theta)$$

$$\sqrt{-g}\frac{\partial}{\partial\theta}\left(\sqrt{-g}\left(g^{iK}\Gamma^{\theta}_{iK} - g^{i\theta}\Gamma^{\ell}_{i\ell}\right)\right)$$
$$= \frac{1}{r^2\sin\theta}\frac{\partial}{\partial\theta}\left[r^2\sin\theta\,(-2r^{-2}\cot\theta)\right] = +2r^{-2}$$

$$G' = 4\frac{1}{r}a'e^{2a} + \frac{2}{r^2}e^{2a} + 2r^{-2}$$

Thus

$$R = G' + \frac{1}{\sqrt{-g}}\frac{\partial}{\partial r}\left(\sqrt{-g}\left(g^{iK}\Gamma^{r}_{iK} - g^{ir}\Gamma^{\ell}_{i\ell}\right)\right)$$
$$= -2a''e^{2a} - 4(a')^2e^{2a} - 2r^{-2}e^{2a} - 8a'r^{-1}e^{2a} + 2r^{-2}$$

in agreement, aside from appropriate sign flips, with Inner Schw. If we examine ^{3}R:

$$^3R = -2\left(-\frac{2M}{r^3}e^{-2a} - \frac{1}{2}\left(\frac{2M}{r^2}\right)^2 e^{-4a}\right)e^{2a} - 4\left(\frac{1}{4}\left(\frac{2M}{r^2}\right)^2 e^{-4a}\right)e^{2a}$$

$$- 2r^{-2}e^{2a} - 8\left(+\frac{1}{2}\left(\frac{2M}{r^2}\right)e^{-2a}\right)r^{-1}e^{2a} + 2r^{-2}$$

$$= \frac{+4M}{r^3} + \left(\frac{2M}{r^2}\right)^2 e^{-2a} - \left(\frac{2M}{r^2}\right)^2 e^{-2a} - 2r^{-2}e^{2a} + 4\left(\frac{2M}{r^2}\right)\frac{1}{r} + \frac{2}{r^2}$$

$$= \frac{-4M}{r^3} - \frac{2}{r^2}\left(1 = \frac{2M}{r}\right) + \frac{2}{r^2} = 0$$

8.8 Action for the K-S solution for Inner Schwarzschild

Recall:

$$L_{Ein}^{New} = -\frac{2M}{R}\left(r\left(\frac{R'}{R}\right) - 1\right)$$

If $L_{Ein}^{New} = 0$: $r\frac{R'}{R} - 1 = 0$:

$$\frac{R'}{R} = \frac{1}{r}$$

$$\frac{dR}{R} = \frac{dr}{r}$$

$$\ell_n R = \ell_n r + const$$

Thus,

$\boxed{R = Kr}$ → different solution $\left(\frac{2M}{Kr} - 1\right)$ metric terms, thus the correspond to $\boxed{M_{New} = M_{Old}/K}$

Suppose $r\left(\frac{R'}{R}\right) - 1 = 0$, then:

$$R' = r^{-1}R$$

$$(R')^2 = (r^{-1}R)^2$$

$$\left(\frac{\partial R}{\partial r}\right)^2 \left(\frac{\partial r}{\partial \tau}\right)^2 = (r^{-1}R)^2 e^{2a}$$

$$\left(\frac{\partial R}{\partial r}\right)^2 = \left(\frac{R}{r}\right)^2 \left(\frac{2M}{R} - 1\right)$$

So far have chosen Minisup:

$$ds^2 = \left(\frac{2M}{R} - 1\right)dt^2 - \left(\frac{2M}{R} - 1\right)^{-1} dr^2 + r^2 d\Omega^2$$

Switch to Minisup.:

$$ds^2 = \left(\frac{2M}{R} - 1\right) dt^2 - \left(\frac{2M}{R} - 1\right)^{-1} dr^2 + \underline{R^2 d\Omega^2}$$

Now $L_{Ein}^{New} = -\frac{2M}{R}(R' - 1) \overset{L=0}{\Rightarrow} \underline{R = r + K}$

$$R' - 1 \Rightarrow \underline{\left(\frac{\partial R}{\partial r}\right) = \left(\frac{2M}{R} - 1\right)^{1/2}}$$

If

$$L_{Ein}^{New} = -8\pi V_t \int -\frac{2M}{R(r)}\left(\frac{\partial R}{\partial r} - 1\right)\left(\frac{\partial r}{\partial \tau}\right) d\tau$$

$$= -8\pi V_t \int -\frac{2M}{R(r)}\left(\frac{\partial R}{\partial \tau} - \left[\frac{2M}{R} - 1\right]^{1/2}\right) d\tau$$

$$\left(\frac{\partial R}{\partial \tau}\right) = \left[\frac{2M}{R} - 1\right]^{1/2} \quad and \quad \boxed{R = r_{\min}}$$

Precisely as expected.

So, the expected Lagrangian factor exists for:

$$\left(\frac{\partial r_{\min}}{\partial \tau}\right)^2 = \left(\frac{2M}{r_{\min}} - 1\right)$$

Regarding $R(\tau)$ as a wavefunction and viewing $\frac{\partial R}{\partial \tau} = \left[\frac{2M}{R} - 1\right]^{1/2} = 0$ as equivalent in pos. R solutions to:

$$\left(\frac{\partial R}{\partial \tau}\right)^2 = \left[\frac{2M}{R} - 1\right]$$

(Eqn. 8-52)

Call $P_R = m\frac{\partial R}{\partial \tau}$:

$$\frac{1}{m^2} P_R^2 = \left(\frac{2M}{R} - 1\right) \qquad P_R \Rightarrow \frac{\delta}{\delta R}$$

and we arrive at a minisuperspace wavefunction equation:

$$\frac{1}{m^2}\frac{\delta^2}{\delta R^2}\Psi = \frac{2M}{R}\Psi - \Psi \quad \text{let } m = 1$$

Thus,

$$\boxed{\frac{\partial^2}{\partial R^2}\Psi - \frac{2M}{R}\Psi = \Psi} \qquad 0 \le \mathrm{R} \le 2\mathrm{M}$$

(Eqn. 8-53)

which is the truncated hydrogen solution. Notes:

(1) The Action for the Kantowski-Sachs solution for the inner Schwarzschild region does describe a dynamical equation like that for the Novikov $R^* = 0$ observer if there are no surface terms. The non-existence of surface terms means that the infinite number of (geodesic) comoving observers describing congruences through $r = 2M$ all contribute identically the result being an infinite integral factor preceding a Lagrangian valid for the classical motion of an R^* observer.

(2) The comparison of the Kant-Sachs geometries dynamics to a geodesic observer is possible when there is a simply transitive symmetry group acting on the 3-surfaces that foliate the spacetime. So, this could easily be extended to a discussion of FRW, etc, in terms of Minisuperspace.

(3) Now, in order to discuss wormhole dynamics a different foliation must definitely be chosen (otherwise no wormhole in the first place). Furthermore, the classical equation of motion, that's fitted to the foliation chosen, strongly depends on that foliation. Upon Quantization the multiple representations leads to incompatible quantum evolutionary histories. So a choice must be made. If the chosen foliation is constant t (asymptotic time parameter) hypersurfaces tan $r_{\text{min}} = 2M$ and only much later hypersurfaces have noticeable contribution to the $r \cong 2M$ surface. If the potential energy grows great enough, i.e. for small enough Black Holes, there will be substantial "reflection" of this in the hypersurface dynamics.

(4) Before a constant t foliation is studied let's start with a simpler case, one that probes the wormhole dynamics anyway (even though it may not be "physical" in the same sense that t = const surfaces are). Since the hypersurfaces outside the BH aren't transitive there will not be a simple choice of foliation as with inside. Now it is clear that Novikov coord's are no better than Kruskal in the exterior, since Kruskal is simpler, the analysis continues in that system. (A wormhole dynamic will be described by a major subclass of the foliations.)

8.9 Novikov coordinate analysis

$$ds^2 = -d\tau^2 + \left(\frac{R^{*^2}+1}{R^{*2}}\right)\left(\frac{\partial r}{\partial R^x}\right)^2 dR^{*^2} + r^2(d\theta^2 + \sin^2\theta\, d\varphi^2)$$

(Eqn. 8-54)

$r(\tau, R^*)$:

$$\frac{\tau}{2M} = \pm(R^{*^2}+1)\left[\frac{r}{2M} - \frac{(r/2M)}{R^{*^2}+1}\right]^{1/2} + (R^{*^2}+1)^{3/2}\cos^{-1}\left[\left(\frac{(r/2M)}{R^{*^2}+1}\right)^{1/2}\right]$$

(Eqn. 8-55)

$$ds^2 = -d\tau^2 + e^{2D(r)}dR^{*^2} + r^2 d\Omega^2$$

$$\dot{\boldsymbol{D}} = \frac{\partial D}{\partial \tau}, \qquad D' = \frac{\partial D}{\partial R^*}, \quad \dot{D} = \frac{\partial D}{\partial \lambda}$$

(Eqn. 8-56)

Note that there are two types of dot, the time dot will be made bold:

$$I = \int g_{\mu\nu}\, \dot{x}^\mu \dot{x}^\nu d\lambda$$

$$= \int\left\{\left(\frac{\partial\tau}{\partial\lambda}\right)^2 + e^{2D}\left(\frac{\partial R^*}{\partial\lambda}\right)^2 + r^2\left[\left(\frac{\partial\theta}{\partial\lambda}\right)^2 + \sin^2\theta\left(\frac{\partial\varphi}{\partial\lambda}\right)^2\right]\right\} d\lambda$$

(Eqn. 8-57)

$$\tau(\lambda): \delta I = 0 = \int\{-2\dot{\tau}\delta\dot{\tau} + 2\dot{\boldsymbol{D}}\delta\tau\dot{R}^{*^2}e^{2D} + e^{2D}2\dot{R}^*\delta\dot{R}^* + 2r\dot{r}\delta\tau[\dot{\theta}^2\sin^2\theta\,\dot{\varphi}^2]\}\, d\lambda$$

$$2\ddot{\tau} + 2\dot{\boldsymbol{D}}e^{2D}\dot{R}^{*^2} + 2r\dot{\boldsymbol{r}}[\dot{\theta}^2\sin^2\theta\,\dot{\varphi}^2] = 0$$

$$\Gamma^\tau_{R^*R^*} = \dot{\boldsymbol{D}}e^{2D}, \ \ \Gamma^\tau_{\theta\theta} = r\dot{\boldsymbol{r}} = \sin^{-2}\theta\,\Gamma^\tau_{\varphi\varphi}$$

$$R^*(\lambda): \delta I = 0 = \int\left\{2D'\delta R^*(\dot{R}^*)^2 e^{2D} + (e^{2D}2\dot{R}^*)\delta\dot{R}^* + 2rr'\delta R^*[\dot{\theta}^2\sin^2\theta\,\dot{\varphi}^2]\right\} d\lambda$$

$$-2\ddot{R}^*e^{2D} - 2\dot{R}^*\left(2e^{2D}(\acute{D}R^* + \dot{\boldsymbol{D}}\dot{\tau})\right) + 2\acute{D}(\dot{R}^*)^2 e^{2D} + 2rr'[\dot{\theta}^2\sin^2\theta\,\dot{\varphi}^2] = 0$$

$$\Gamma^{R^*}_{R^*R^*} = D', \quad \Gamma^{R^*}_{\tau R^*} = \dot{D}, \quad \Gamma^{R^*}_{\theta\theta} = rr'e^{-2D} = \sin^{-2}\theta\,\Gamma^{R^*}_{\varphi\varphi}$$

$$\theta(\lambda): \delta I = 0 = \int r^2\left(2\dot{\theta}\delta\dot{\theta} + 2\sin\theta\cos\theta\,\delta\theta\dot{\varphi}^2\right)d\lambda$$

$$2r\left(\dot{r}\dot{\tau} + r'\dot{R}^*\right)\dot{\theta} + r^2\ddot{\theta} - r^2\sin\theta\cos\theta\,\dot{\varphi}^2 = 0 \Rightarrow \Gamma^{\theta}_{\varphi\varphi} = -\sin\theta\cos\theta$$

$$\Gamma^{\theta}_{\tau\theta} = \frac{\dot{\boldsymbol{r}}}{r}, \qquad \Gamma^{\theta}_{R^*\theta} = \frac{r'}{r}$$

$$\varphi(\lambda): \delta I = 0 = \int r^2\sin^2\theta\,2\dot{\varphi}\delta\dot{\varphi}d\lambda$$

$$2r\left(\dot{\boldsymbol{r}}\dot{\tau} + r'\dot{R}^*\right)\dot{\theta} + r^2\sin^2\theta\,\ddot{\varphi} + 2r^2\sin\theta\cos\theta\,\dot{\theta}\dot{\varphi} = 0 \Rightarrow \Gamma^{\varphi}_{\tau\varphi} = \frac{\dot{\boldsymbol{r}}}{r}, \quad \Gamma^{\varphi}_{R^*\varphi} = \frac{r'}{r}$$

$$\Gamma^{\varphi}_{\theta\varphi} = \cot\theta$$

$$\Gamma^{\tau}_{R^*R^*} = \dot{\boldsymbol{D}}e^{2D} \qquad \Gamma^{\tau}_{\theta\theta} = \sin^{-2}\theta\,\Gamma^{\tau}_{\varphi\varphi} = r\dot{\boldsymbol{r}}$$

$$\Gamma^{R^*}_{R^*R^*} = D' \qquad \Gamma^{R^*}_{\theta\theta} = \sin^{-2}\theta\,\Gamma^{R^*}_{\varphi\varphi} = rr'e^{-2D}$$

$$\Gamma^{R^*}_{\tau R^*} = \dot{\boldsymbol{D}} \qquad \Gamma^{\theta}_{\tau\theta} = \Gamma^{\varphi}_{\tau\varphi} = \frac{\dot{\boldsymbol{r}}}{r}, \quad \Gamma^{\theta}_{R^*\theta} = \Gamma^{\varphi}_{R^*\varphi} = \frac{r'}{r}$$

$$\Gamma^{\theta}_{\varphi\varphi} = \sin\theta\cot\theta$$

$$\Gamma^{\varphi}_{\theta\varphi} = \cot\theta$$

$$g_{\mu v} = diag(-1, e^{2D}, r^2, r^2\sin^2\theta)$$

$$g^{\mu v} = diag(-1, e^{-2D}, r^{-2}, r^{-2}\sin^{-2}\theta)$$

$$R = G + \frac{1}{\sqrt{-g}}\frac{\partial}{\partial x^v}\left[\sqrt{-g}\left(g^{iK}\Gamma^{v}_{iK} - g^{iv}\Gamma^{\ell}_{i\ell}\right)\right]$$

$$G = g^{iK}\left(\Gamma^{m}_{i\ell}\Gamma^{\ell}_{Km} - \Gamma^{\ell}_{iK}\Gamma^{m}_{\ell m}\right)$$

$$G' = G + \frac{1}{\sqrt{-g}}\frac{\partial}{\partial\theta}\left[\sqrt{-g}\left(g^{iK}\Gamma^{\theta}_{iK} - g^{i\theta}\Gamma^{\ell}_{i\ell}\right)\right]$$

Shifting back to dot meaning time derivative (not bold)

$$g^{iK}\Gamma^{m}_{i\ell}\Gamma^{\ell}_{Km} = g^{\tau\tau}\Gamma^{m}_{\tau\ell}\Gamma^{\ell}_{\tau m} + g^{R^*R^*}\Gamma^{m}_{R^*\ell}\Gamma^{\ell}_{R^*m} + g^{\theta\theta}\Gamma^{m}_{\theta\ell}\Gamma^{\ell}_{\theta m} + g^{\varphi\varphi}\Gamma^{m}_{\varphi\ell}\Gamma^{\ell}_{\varphi m}$$
$$= \dot{D}^2e^{-2D}(D')^2 + 2r^{-2}(\dot{r}^2 - (r')^2e^{-2D}) - r^{-2}\cot^2\theta$$

$$g^{iK}\Gamma^{\ell}_{iK}\Gamma^{m}_{\ell m} = \left(g^{\tau\tau}\Gamma^{\ell}_{\tau\tau} + g^{R^*R^*}\Gamma^{\ell}_{R^*R^*} + g^{\theta\theta}\Gamma^{\ell}_{\theta\theta} + g^{\varphi\varphi}\Gamma^{\ell}_{\varphi\varphi}\right)\Big(\Gamma^{\tau}_{\ell\tau} + \Gamma^{R^*}_{\ell R^*} + \Gamma^{\theta}_{\ell\theta} + \Gamma^{\varphi}_{\ell\varphi}\Big)$$

$$= \left(\dot{D} + 2\frac{\dot{r}}{r}\right)^2 + e^{-2D}\left(D' - 2\frac{r'}{r}\right)\left(D' + 2\frac{r'}{r}\right) - r^{-2}\cot^2\theta.$$

$$G = 4\dot{D}\left(\frac{\dot{r}}{r}\right) - 2\left(\frac{\dot{r}}{r}\right)^2 + 2\left(\frac{r'}{r}\right)^2 e^{-2D}$$

$$\frac{1}{\sqrt{-g}}\frac{\partial}{\partial\theta}\left[\sqrt{-g}\left(g^{iK}\Gamma^{\theta}_{iK} - g^{\theta\theta}\Gamma^{\ell}_{\theta\ell}\right)\right] =?$$

$$g^{iK}\Gamma^{\theta}_{iK} = g^{\varphi\varphi}\Gamma^{\theta}_{\varphi\varphi} = (r^{-2}\sin^{-2}\theta)(-\sin\theta\cos\theta) = -r^{-2}\cot\theta$$

$$g^{\theta\theta}\Gamma^{\varphi}_{\theta\varphi} = (r^{-2})(\cot\theta)$$

$$\frac{1}{r^2\sin\theta}\frac{\partial}{\partial\theta}[r^2\sin\theta\{-2r^{-2}\cot\theta\}] = \underline{2r^{-2}}$$

$$\boxed{G' = 2\left\{\left(\frac{r'}{r}\right)^2 e^{-2D} - \left(\frac{\dot{r}}{r}\right)^2 - 2\dot{D}\left(\frac{\dot{r}}{r}\right) + r^{-2}\right\}}$$

$$\frac{1}{r^2}\frac{\partial}{\partial R^*}\left[r^2\left(g^{iK}\Gamma^{R^*}_{iK} - g^{R^*R^*}\Gamma^{\ell}_{R^*\ell}\right)\right] =?$$

$$g^{iK}\Gamma^{R^*}_{iK} = \big(e^{-2D}D' + (r^{-2})(-rr'e^{-2D}) + (r^{-2}\sin^{-2}\theta)(-\sin^2\theta\, rr'e^{-2D})\big)$$

$$= e^{-2D}\left(D' - 2\left(\frac{r'}{r}\right)\right)$$

$$g^{R^*R^*}\Gamma^{\ell}_{R^*\ell} = e^{-2D}\left(D' - 2\left(\frac{r'}{r}\right)\right)$$

$$\frac{1}{r^2}\frac{\partial}{\partial R^*}\left\{r^2\left(-4\left[\frac{r'}{r}\right]e^{-2D}\right)\right\} = \frac{1}{r^2}\frac{\partial}{\partial R^*}(-4rr'e^{-2D})$$

$$= \frac{1}{r^2}(-4(r')^2e^{-2D} - 4rr''e^{-2D} + 8rr'D'e^{-2D})$$

$$\frac{1}{r^2}\frac{\partial}{\partial\tau}\{r^2(g^{iK}\Gamma^{\tau}_{iK} - g^{\tau\tau}\Gamma^{\ell}_{\tau\ell})\}$$
$$= \frac{1}{r^2}\frac{\partial}{\partial\tau}\left\{r^2\left[\left(e^{-2D}\dot{D}e^{2D} + 2\left(\frac{\dot{r}}{r}\right)\right) + \left(\dot{D} + 2\left(\frac{\dot{r}}{r}\right)\right)\right]\right.$$
$$= 2r^{-2}\frac{\partial}{\partial\tau}(r^2\dot{D} + 2r\dot{r}) = 2r^{-2}(2r\dot{r}\dot{D} + r^2\ddot{D} + 2(\dot{r})^2 2r\ddot{r})$$

Thus,

$$R = -2\left(\frac{r'}{r}\right)^2 e^{-2D} - \left(\frac{\dot{r}}{r}\right)^2 + 2r^{-2} - 4\frac{r''}{r}e^{-2D} + 8\left(\frac{r'}{r}\right)D'e^{-2D} + 2\ddot{D} + 4\frac{\ddot{r}}{r}$$

(Eqn. 8-58)

Verification that $R = 0$ by calculation, r', $\acute{r}$, etc. will be left as an exercise.

The Novikov EOM

The expression for G' is simpler than its Kruskal coordinate counterpart if $\underline{\dot{D} = 0}$, i.e. if $(\dot{r}') = 0$.

$r(\tau, R^*)$:

$$\frac{\tau}{2M} = \pm(R^{*^2} + 1)\left[\frac{r}{2M} - \frac{(r/2M)^2}{R^{*^2}+1}\right]^{1/2} + (R^{*^2} + 1)^{3/2}\cos^{-1}\left[\left(\frac{(r/2M)^2}{R^{*^2}+1}\right)^{1/2}\right]$$

(Eqn. 8-59)

Does $(\dot{r}') = 0$?

$$\frac{1}{2M} = \pm\frac{1}{2}(R^{*^2} + 1)\left[\frac{r}{2M} - \frac{(r/2M)^2}{R^{*^2}+1}\right]^{1/2}\left(\frac{\dot{r}}{2M} - \frac{2\left(\frac{1}{2M}\right)^2 r\acute{r}}{R^{*^2}+1}\right) \mp (R^{*^2} + 1)^{3/2}\left(1 - \frac{(r/2M)^2}{R^{*^2}+1}\right)^{1/2}\left(\frac{1}{R^{*^2}+1}\right)^{1/2}\left(\frac{1}{2M}\right)^{1/2} r^{-1/2}\dot{r}\,\frac{1}{2}$$

$$= \pm(R^{*^2} + 1)\left[\frac{r}{2M} - \frac{(r/2M)^2}{R^{*^2}+1}\right]^{-1/2}\left(\frac{1}{2}\frac{\dot{r}}{2M} - \left(\frac{1}{2M}\right)^2 r\acute{r}(R^{*^2} + 1)^{-1} \mp \frac{1}{2}\frac{\acute{r}}{2M}\right)$$

$$\frac{1}{2M} = \mp\left[\frac{r}{2M} - \frac{(r/2M)^2}{R^{*^2}+1}\right]^{-1/2}\left(\frac{1}{2M}\right)^2 r\dot{r}.$$

$$\dot{r} = \mp\left(\frac{2M}{r}\right)\left[\frac{r}{2M} - \frac{(r/2M)^2}{R^{*^2}+1}\right]^{1/2} = \left[\frac{2M}{r} - \left(\frac{1}{R^{*^2}+1}\right)\right]^{1/2}.$$

$$\Rightarrow \left.\frac{\partial r}{\partial \tau}\right|_{R^*=0} = \left(\frac{2M}{r} - 1\right)^{1/2}$$

$$\dot{r}' = \frac{1}{2}\left[\frac{2M}{r} - \left(\frac{1}{R^{*2}+1}\right)\right]^{-1/2}\left(\frac{2M}{r}\left(\frac{r'}{r}\right) + \left(R^{*^2}+1\right)^{-2}(2R^*)\right)$$

$\neq 0$ in general.

8.10 Analysis of Collapsing Star

Model a collapsing star by a uniform spherical ball of dust. Assume hat the matter is initially static and determine its subsequent evolution. Suggested method: for the interior of the star use a portion of the closed Friedmann geometry with metric:

$$ds^2 = -d\tau^2 + a^2(\tau)(dx^2 + \sin^2 x\, d\Omega^2),$$

(Eqn. 8-60)

where $a(\tau)$ is given parametrically by

$$a = \frac{A}{2}(1 + \cos\eta) \quad and \quad \tau = \frac{A}{2}(\eta + \sin\eta).$$

(Eqn. 8-61)

The star's surface is always at the comoving coordinate $x=x_S$, but its area is shrinking after $\tau = 0$. The exterior metric is a portion of the Schwartzschild geometry, with surface at $r(\tau) = R_S(t)$ in Schwarzschild coordinates. Write the junction conditions, find M in terms of A and x_S, and find $R_S(t)$.

Solution

Model for a collapsing star using a uniform spherical ball of dust. We have:

$$\begin{Bmatrix} \text{spherical symmetry} \\ + \\ \text{uniform density} \end{Bmatrix} \text{equivalent to assuming} \begin{Bmatrix} \text{isotropy} \\ + \\ \text{homogeneity} \end{Bmatrix}$$

So, the interior metric is a FRW solution:

$$ds^2 = -dt^2 + a^2(t)[dx^2 + \Sigma^2(d\theta^2 + \sin^2\theta\, d\phi^2)]$$

where:

$$\Sigma = \sin x \Rightarrow k = 1 \text{ positive spatial curvature}$$
$$\Sigma = x \Rightarrow k = 0 \text{ zero spatial curvature}$$
$$\Sigma = \sinh x \Rightarrow k = -1 \text{ negative spatial curvature}$$

Applying the Einstein field equation to the above metric for dust we get a constraint eqn. and an evolution eqn. The constraint eqn. reads:

$$\left(\frac{a_{,t}}{a}\right)^2 = -\frac{k}{a^2} + \frac{8\pi}{3}\rho\,.$$

(Eqn. 8-62)

Since we are assuming that the matter is initially static, $\left(a_{,t}\right)_0 = 0$, and of course $\rho > 0$, then only the $k = 1$ solution is viable. Thus, only a closed FRW is possible for the prescribed initial conditions. Closed FRW has $ds^2 = -d\tau^2 + a^2(\tau)(dx^2 + \sin^2 x\, d\Omega^2)$. The constraint eqn. actually yields the dynamical eqn. when use is made of conservation of energy: $(\rho a^3) = \text{const}$. So,

$$\left(a_{,t}\right)^2 = -1 + \frac{8\pi}{3}\frac{(\text{const.})}{a} = -1 + \frac{a_{max}}{a}$$

(Eqn. 8-63)

It is convenient to introduce the arc parameter η: $dt = a(t)d\eta$

$$\frac{da}{dt} = \frac{da}{d\eta}\frac{d\eta}{dt} = \frac{1}{a}\frac{da}{d\eta} \quad and\ \dot{a} = \frac{da}{d\eta}$$

Now

$$(\dot{a})^2 = -a^2 + \tfrac{8\pi}{3}(const)_a \qquad a_{,t} =$$

0 when $a = a_{max}$

$$= a(a_{max} - 1) \qquad (const) =$$

$\frac{3}{8\pi}a_{max}$

Thus

$a = a_{max}Cf(\eta)$ and $\dot{f}^2 = f(\eta)\left(\frac{1}{C} - f(\eta)\right)$.

$$f = 1 + \cos\eta \Rightarrow \sin^2\eta = (1 + \cos\eta)\left(\frac{1}{C} - 1 - \cos\eta\right)$$

$$= -\cos^2\eta + \left(-1 + \frac{1}{C}\right)\cos\eta + \left(\frac{1}{C} - 1\right)$$

$$a(\eta) = a_{max}\frac{1}{2}(1 + \cos\eta) \quad \Rightarrow \quad \tau = \int_0^\eta a(\tau)d\eta$$

$$= \frac{a_{max}}{2}(\eta + \sin\eta) \quad (\tau = 0 \text{ at } \eta = 0)$$

From Birkhoff's theorem we know the metric exterior to the dust (star) is Schwarzschild:

$$ds^2 = -\left(1 - \frac{2M}{r}\right)dt^2 + \left(1 - \frac{2M}{r}\right)^{-1} dr^2 + r^2 + d\Omega^2$$

Since there are no pressure gradients acting on the dust at the surface of the star ($p = 0$ throughout) then any ball of dust there must move along radial geodesics according to the exterior Schwarzschild geometry. From [21] we have for a ball beginning at rest at $R = R_i$, at time $t = 0$:

$$R = (R_i/2)(1 + \cos\eta)$$

with proper time

$$\tau = (R_i^3/8M)^{1/2}(\eta + \sin\eta)$$

Does this agree with the evolution of the surface prescribed by matching at the boundary? To begin lets compare the interior 3-geometries:

Intrinsic 3-geometry of Interior metric at surface: $(x = x_0)$

$$^{(3)}ds^2 = a^2(\eta)(-d\eta^2 + \sin^2 x_0\, d\Omega^2)$$
$$= \left(\frac{a_m}{2}[1 + \cos\eta]\right)^2 (-d\eta^2 + \sin^2 x_0\, d\Omega^2)$$

Intrinsic 3-geometry of Exterior metric at surface: $(r = R_i)$

$$^{(3)}ds^2 = d\tau^2 + R^2(\tau)d\Omega^2$$
$$= -\left(\frac{R_i^3}{8M}\right)(1 + \cos\eta)^2 d\eta^2 + \left(\frac{R_i}{2}\right)^2 (1 + \cos\eta)^2 d\Omega^2$$

The matching of Intrinsic 3-geometries across the junction is straightforward:

$\left(\frac{a_m}{2}\right)^2 = \left(\frac{R_i^3}{8M}\right)$ and $\left(\frac{a_m}{2}\right)^2 \sin^2 x_0 = \left(\frac{R_i}{2}\right)^2$ give $R_i = \frac{2M}{\sin^2 x_0}$ thus

$$2M = a_m \sin^3 x_0$$

and

$$R_i = a_m \sin x_0$$

Problem expresses $a_m = A$, so,

$$M = \frac{1}{2} A \sin^3 x_5 \qquad \Rightarrow R$$
$$R(t = 0) = A \sin x_5$$

Since there is no delta function singularity in the mass of dust, we must also equate the extrinsic curvature between the interior and exterior metrics at the boundary. This then will yield $R(t)$:

Extrinsic curvature for a surface: $K_{ij} \equiv -\vec{e}_i \cdot \nabla_j \vec{n}$ where $\vec{n}$ is the unit normal to the surface.

Interior: $\vec{n} = a^{-1}\frac{\partial}{\partial x}$ (since surface $x = const$)

$K_{ij} \equiv -\frac{1}{2}£_{\vec{n}}g_{ij}$ for $i,j = \{\tau,\theta,\phi\}$

Since $g_{in} = \frac{1}{a}g_{ix} = 0$ this simplifies to just:

$$K_{ij} \equiv -\frac{1}{2}g_{ij,n} = -\frac{1}{2a}g_{ij,x}$$

We have

$g_{\theta\theta} = a^2\sin^2 x$ and $g_{\phi\phi} = a^2\sin^2 x\sin^2\theta$

thus,

$k_{\theta\theta} = \frac{1}{2}a\sin x\cos x$ and $k_{\phi\phi} = -a\sin x\cos x\sin^2\theta$. Thus we have:

$$\left.\begin{aligned} k_{\phi\phi} &= \sin^2\theta\, k_{\theta\theta} \\ k_{\theta\theta} &= \frac{1}{2}A(1+\cos\eta)\sin x_0\cos x_0 \end{aligned}\right\} \text{at the surface.}$$

Exterior:

From the 4-velocity: $\vec{u} = u^t\vec{e}_t + u^r\vec{e}_r$ (for surface with no angular motion) and want normal vector, $\vec{n}\cdot\vec{u} = 0$ and $\vec{n}\cdot(angular) = 0$, also $\vec{n}\cdot\vec{n} = 1$, so with

$$\vec{n} = n^t\vec{e}_t + n^r\vec{e}_r$$

we have

$$\vec{u}\cdot\vec{n} = u^t n^t g_{tt} + u^r n^r g_{rr} = u^t n_t + u^r n_r$$

and $n^t = u_r$, $n^r = u_t$. If we again choose coordinate τ,θ,ϕ for i, j, then since $g_{in} = \vec{u}\cdot\vec{e}_i = 0$ from before we directly have $K_{ij} \equiv -\frac{1}{2}g_{ij,n}\ and\ \left(\frac{\partial}{\partial\tau}\right) = \vec{u}$. As before $K_{\phi\phi} = \sin^2\theta\, K_{\theta\theta}$, the rest zero, and we have:

$$K_{\theta\theta} = -\frac{1}{2}(r^2)_{,n} = n_s^t\frac{\partial}{\partial t}\left(-\frac{1}{2}r^2\right) + n_s^r\frac{\partial}{\partial r}\left(-\frac{1}{2}r^2\right) = -n^r r$$

Thus

$$(n^r)_{surface} = (-u_t)_{surface} = -g_{tt}u^t$$

So

$$R_s(t)u_t = -\frac{1}{2}A(1+\cos\eta)\sin x_5\cos x_5$$

For radial infall: (using $u\cdot u = -1\ and\ u\cdot\frac{\partial}{\partial t} = u_t$)

$$-1 = \left(1-\frac{2M}{r}\right)(u^t)^2 + \left(1-\frac{2M}{r}\right)^{-1}(u^r)^2$$

$$= \left[-\left(1-\frac{2M}{r}\right) + \left(1-\frac{2M}{r}\right)^{-1}\left(\frac{dr}{dt}\right)^2\right](u_t)^2\left(1-\frac{2M}{r}\right)^2$$

For motion from rest at $r = R$:

$$-1 = -\left(1 - \frac{2M}{R_i}\right)^{-1} (u_t)^2 \quad \rightarrow \quad u_t = -\left(1 - \frac{2M}{R_i}\right)^{1/2}$$

Since $2M = A\sin^3 x_0$, $R_i = A\sin x_0$ $(x_5 = x_0)$ we get:
$u_t = -\cos x_0$, thus

$$R_s(\tau) = +\frac{1}{2}A(1+\cos\eta)\sin x_0 = \frac{1}{2}R_i(1+\cos\eta)$$

Then, using

$$-1 = -\left(1 - \frac{2M}{r}\right)\left(\frac{\left(1-\frac{2M}{R_i}\right)^{1/2}}{\left(1-\frac{2M}{r}\right)}\right)^2 + \left(1 - \frac{2M}{r}\right)^{-1}\left(\frac{dr}{d\tau}\right)^2$$

or

$$-\left(1-\frac{2M}{r}\right) + \left(1-\frac{2M}{R_i}\right) = \left(\frac{2M}{r} - \frac{2M}{R_i}\right) = \left(\frac{dr}{d\tau}\right)^2$$

thus

$$d\tau = \frac{dr}{\left(\frac{2M}{r}-\frac{2M}{R_i}\right)^{1/2}} \text{ and using } \begin{aligned} r &= \tfrac{1}{2}R_i(1+\cos\eta) \\ dr &= -\tfrac{1}{2}R_i\sin\eta\, d\eta \end{aligned}$$

$$= \frac{dr}{\left[\frac{(2M)}{R_i}\left(\frac{1}{1+\cos\eta}-1\right)\right]^{1/2}} = \left(\frac{R_i}{8M}\right)^{1/2} \frac{\sin\eta\, d\eta}{\left(\frac{1-\cos\eta}{1+\cos\eta}\right)^{1/2}}$$

Thus

$$d\tau = \left(\frac{R_i}{8M}\right)^{1/2} d\eta \left(\frac{2\sin{}^{\eta}/_{2}\cos{}^{\eta}/_{2}}{\left(\frac{\sin^2{}^{\eta}/_{2}}{\cos^2{}^{\eta}/_{2}}\right)^{1/2}}\right) = \left(\frac{R_i}{8M}\right)^{1/2} d\eta 2\left(\cos{}^{n}/_{2}\right)^2$$

$$= \left(\frac{R_i}{8M}\right)^{1/2} d\eta(1+\cos\eta)$$

and we get

$$\tau = \frac{1}{2}A(\eta + \sin\eta).$$

(Eqn. 8-64)

8.11 Area Increase Theorem for BH Horizons

(a) Use the area increase theorem to show that the energy E radiated in the collision of two Schwarzschild black holes of mass M_1 and M_2 is bounded by

$$E \leq M_1 + M_2 - \sqrt{M_1^2 + M_2^2}\,.$$

(Eqn. 8-65)

(b) Conclude that no more than a fraction $\left(1 - \frac{1}{\sqrt{2}}\right) \approx 29\%$ of the total mass of the black hole binary system can be radiated away in collapse.

Solution

(a)

$$A_1 = 4\pi r_H^2 = 4\pi(2M_1)^2 = 16\pi M_1^2 \text{ ,} \textit{and similarly } A_2 = 16\pi M_2^2$$

At coalescence $A_T \geq A_1 + A_2 = 16\pi(M_1^2 + M_2^2)$. Thus $A_T \geq 16\pi M_T^2$, where $M_T^2 = M_1^2 + M_2^2$. The net mass before coalescence is $M_{Net} = M_1 + M_2$. Thus, the energy radiated must be bounded by the difference $E \leq M_{Net} - M_T$, thus:

$$E \leq M_1 + M_2 - \sqrt{M_1^2 + M_2^2}$$

(where maximum energy radiated is equal to the net system mass minus that required by the 2nd law).

(b) $E = M_1 + M_2 - \sqrt{M_1^2 + M_2^2}$ is the maximum possible energy radiated. For a binary we have $E = 2M - \sqrt{2}M = 2M\left(1 - \frac{1}{\sqrt{2}}\right) \cong 2M(.29) =$ 29% of the total mass of the binary system.

8.12 Kepler's Third Law holds in the BH Exterior

Let's show that Kepler's Law, $\Omega^2 = \frac{M}{r^3}$, holds exactly for particles in circular orbit about a Schwarzschild black hole, where r is the radial coordinate.

Solution

$$ds^2 = -\left(1 - \frac{2M}{r}\right)dt^2 + \left(1 - \frac{2M}{r}\right)^{-1} dr^2 + r^2(d\theta^2 + \sin^2\theta\, d\phi^2)$$

Consider the geodesic equation:

$$\frac{d^2x^\mu}{d\lambda^2} + \Gamma^\mu{}_{\gamma\delta} u^\gamma u^\delta = 0.$$

For a circular orbit (choose $\theta = \frac{\pi}{2}$) u^t and u^ϕ are constant and u^r, u^θ are zero. The r component of the equation above reads:

$$\Gamma_\mu \gamma \delta u^\gamma u^\delta = \Gamma_{rH}\left(\frac{dt}{d\lambda}\right)^2 + 2\Gamma^a_{rt\phi}\left(\frac{dt}{d\lambda}\right)\left(\frac{d\phi}{d\lambda}\right) + \Gamma_{r\phi\phi}\left(\frac{d\phi}{d\lambda}\right)^2 = 0$$

Thus

$$\left(\frac{M}{r^2}\right)\left(\frac{dt}{d\lambda}\right)^2 - r\left(\frac{d\phi}{d\lambda}\right)^2 = 0 \rightarrow \left(\frac{d\phi}{dt}\right)^2 = \frac{M}{r^3}.$$

Using $\frac{d\phi}{dt} = \Omega$ we have:

$$\Omega^2 = \frac{M}{r^3}.$$

Example. Find Ω_H and K for Kerr black hole.

Solution

Kerr is expressed in Boyer-Lindquist coordinates (with rotation in ϕ direction):

$$ds^2 = -\left(\Delta/_{P^2}\right)[dt^2 - a\sin^2\theta\, d\phi]^2 + \left(\sin^2\theta/_{P^2}\right)[(r^2+a^2)d\phi - adt]^2 + \left(P^2/_{\Delta}\right)dr^2 + P^2 d\theta^2$$

$$\Delta = r^2 2\mu_r + a^2 \quad , \quad P^2 = r^2 + a^2\cos^2\theta$$

We have

$$\Omega_H = -\frac{t\cdot\phi}{\phi\cdot\phi}\bigg|_H = -\frac{g_{t\phi}}{g_{\phi\phi}}\bigg|_H$$

Since

$$g_{t\phi} = \frac{a\sin^2\theta\left(\Delta - (r^2+a^2)\right)}{P^2} = \frac{-2Mra\sin^2\theta}{P^2}$$

and

$$g_{\phi\phi} = \frac{\sin^2\theta}{P^2}\left((r^2+a^2)^2 - a^2\Delta\sin^2\theta\right)$$

then

$$\Omega_H = \frac{2Mra}{(r^2+a^2)^2 - \Delta a^2\sin^2\theta}\bigg|_H = \frac{2Mr_t a}{(2Mr_t)^2} = \frac{a}{2Mr_t} = \frac{J}{2M^2 r_t}$$

Using $M = \Omega_H J_H + \frac{KA}{4\pi}$:

$$K = (M - \Omega_H J_H)4\pi.$$

8.13 Wormholes

Consider an embedding of the equatorial 2-surfaces t=constant, theta=pi/2 of the vacuum Schwarzschild space-time, where the 2-surface is written in isotropic coordinates:

$$ds^2 = \left(1 + \frac{M}{2\bar{r}}\right)^4 \left(d\bar{r}^2 + \bar{r}^2 d\varphi^2\right).$$

Show that for $0 < \bar{r} \ll M/2$ is an asymptotically flat space (as is $\bar{r} \gg M/2$). A connection between these two flat regions (discovered by Ludwig Flamm 1916 [21]) would then describe a 'wormhole', examine the properties of the wormhole.

(1) $ds^2 = \left(1 + \frac{M}{2\bar{r}}\right)^4 \left(d\bar{r}^2 + \bar{r}^2 d\varphi^2\right) = \underbrace{dz^2 + dr^2 + r^2 d\varphi^2}_{metric\ in\ embedding\ space}$

(2) Circumference $= 2\pi r = 2\pi\,\bar{r}\left(1 + \frac{M}{2\bar{r}}\right)^2 \rightarrow r = \bar{r} + M + \frac{M^2}{4\bar{r}}$

(3) As r goes from 0 to ∞, r goes from $+\infty$ down to $r(\bar{r} = M/2) = 2M$ and then back to $+\infty$.

(4) If we set $\tilde{r} \equiv \frac{M^2}{4\bar{r}}$, so $\bar{r} = \frac{M2}{4\tilde{r}}$, then

$$\frac{M}{2\bar{r}} = \frac{\tilde{r}}{2M}.$$

The line element becomes

$$ds^2 = \left(1 + \frac{2\tilde{r}}{M}\right)^4 \left(\frac{M^4}{16\tilde{r}4} d\tilde{r}^2 + \frac{M^4}{16\tilde{r}4}\tilde{r}^2 d\varphi^2\right)$$
$$= \left(1 + \frac{M}{2\tilde{r}}\right)^4 \left(d\tilde{r}^2 + \tilde{r}^2 d\varphi^2\right)$$

(Eqn. 8-66)

This is identical in form to (1), but with $r \rightarrow \tilde{r}$. Thus, the region $\vec{r} < M/2$ is identical in intrinsic geometry.

(5) The embedding diagram at $\bar{r} > M/2$ will be the same as at $r > 2M$ in Schwarzschild coordinates:

$$z = \int_0^r \frac{dr}{(r/2M - 1)^{1/2}} = \sqrt{8M(r - 2M)}$$

(Eqn. 8-67)

(6) At $\bar{r} > M/2$ the embedding will be the same, except z$\rightarrow -z$: Thus

$$z = \sqrt{8M(r-2M)}, r = \bar{r} + M + \frac{M^2}{4\bar{r}} \quad \text{for } \bar{r} > \frac{M}{2}$$
$$z = -\sqrt{8M(r-2M)}, r = \bar{r} + M + \frac{M^2}{4\bar{r}} \quad \text{for } \bar{r} < \frac{M}{2}$$

(Eqn. 8-68)

8.14 Exercises

(Ex. 8.1) Derive Eqn. 8-2.
(Ex. 8.2) Derive Eqn. 8-3.
(Ex. 8.3) Derive Eqn. 8-4.
(Ex. 8.4) Derive Eqn. 8-9.
(Ex. 8.5) Derive Eqn. 8-11.
(Ex. 8.6) Derive Eqn. 8-13.
(Ex. 8.7) Derive Eqn. 8-37.
(Ex. 8.8) Derive Eqn. 8-42.
(Ex. 8.9) Derive Eqn. 8-47.
(Ex. 8.10) Derive Eqn. 8-52 and 8-53.
(Ex. 8.11) Derive Eqn. 8-58, and show R=0.
(Ex. 8.12) Derive Eqn. 8-68.

Chapter 9. Cosmology

In this Chapter we consider the cosmological results and implication of Einstein's General Relativity. In Sec. 9.1, Einstein's equations of motion for GR are derived for various matter and geometric tensors. We then 'go local' with consideration of individual observer geodesics in Sec. 9.2 followed by making this 'global' by considering a collection of geodesics acting as cosmological world lines (and using the equivalent observer postulate) in Sec. 9.3. The notation ready we then proceed with the implications of large-scale cosmological homogeneity and isotropy. The Universes having this property are known as the 'FRW' Universes, derived in Sec. 9.4 along with the scalar curvature R. The scalar curvature is then rederived the fast way using the Cartan Method and the EOM's are then examined (Sec. 9.5).

More results pertaining to FRW Universes are then examined in Sec. 9.6 (Cosmological Red Shift); Sec. 9.7 (FRW with a perfect fluid); Sec. 9.8 (Thermodynamic considerations); and Sec,. 9.9 (the spherical dust universe). In Sec. 9.10 we examine the very small limit on the cosmological constant Λ indicated by observational data. In Sec. 9.11 we, therefore, focus on the Friedman Models ($\Lambda = 0$) with examination of the behavior when radiation dominated and matter dominated. In the Big Bang, for example, we begin with radiation dominated and eventually transition to matter dominated in the current state.

In Sec. 9.12 we examine the nucleosynthesis implications for the FRW Universes indicated. In Sec. 9.13 we broaden to consider FRWL Universes with non-zero Λ (the LeMaitre Models). Global boundary issues are considered in Sec. 9.14 on particle horizons and event horizons (in FRW models).

The exact numerical particle composition of the Standard Models impacts the number of species in thermodynamics evaluations within the FRW Model chosen. The result of this is that the composition of the Standard Model is partly revealed (along with mass orderings according to phase transitions observed). Standard Model related issues are discussed in Sec.'s 9.15-9.18.

Sec. 9.19 describes Inflation. Section 9.20 a description of perturbations in early Galaxy formation is discussed. There is clear evidence for Dark Matter in this setting. If we turn to the Standard Model this Dark matter could be explained by a minimal extension to the standard model [51], in the form of a right-handed sterile neutrino.

9.1 Einstein Equations from Variation on the Hilbert-Einstein Action

Dynamics will be found by variational principle with action:

$$S = S_G + S_M \ , \ S_G = \frac{1}{16\pi G}\int d^4x\sqrt{-g}\,(R - 2\Lambda) \ , \ S_M \to \delta S_M \, ,$$

(Eqn. 9-1)

where we have the general action that includes cosmological constant, and the matter action is whatever gives us the matter action variation specified next (i.e., choice of energy momentum tensor $T^{\mu\nu}$). With standard variation (not Palatini, for example), so have variation $g_{\mu\nu} \to g_{\mu\nu} + \ \delta g_{\mu\nu}$ with $\delta g_{\mu\nu} = 0$ on boundary. Thus,

$$\delta S_G = -\frac{1}{16\pi G}\int d^4x\sqrt{-g}\left(R^{\mu\nu} - \frac{1}{2}g^{\mu\nu}R + \Lambda g^{\mu\nu}\right)\delta g_{\mu\nu} = 0$$

(Eqn. 9-2)

and

$$\delta S_M = \frac{1}{2}\int d^4x\sqrt{-g}\ T^{\mu\nu}\ \delta g^{\mu\nu}.$$

(Eqn. 9-3)

So, the Einstein equation (c=1) follows from $\delta S = 0$:

$$R_{\mu\nu} - \frac{1}{2}g_{\mu\nu}R + \Lambda\, g_{\mu\nu} = 8\pi G\, T_{\mu\nu} \, ,$$

(Eqn. 9-4)

where we will find that the Newtonian limit requires that Λ be very small.

9.1.1 Variation for $\mathcal{L} = 1$

Let's re-derive the δS_G, starting with the Λ term, e.g., let's consider

$$S_G = \int d^4x\,\sqrt{-g}\,\mathcal{L} = \int d^4x\,\sqrt{-g}$$

where

$$\delta S_G = \int d^4x\,\delta\sqrt{-g}$$

So, $\delta\sqrt{-g}$ =?

g^λ_ρ is a 4×4 matrix, thus $\det g = \frac{1}{4!}\delta^{\mu\nu\rho\sigma}_{\alpha\beta\gamma\delta}\ g^\alpha_\mu\ g^\beta_\nu\ g^\gamma_\rho\ g^\delta_\sigma$:

$$(g^{-1})^\mu_\alpha(\det g) = \frac{1}{4!}\ \delta^{\mu\nu\rho\sigma}_{\alpha\beta\gamma\delta}\ \underbrace{(g^{-1})^\mu_\alpha\ (g^{-1})^\alpha_\mu}_{=4}\ g^\beta_\nu\ g^\gamma_\rho\ g^\delta_\sigma$$

$$= \frac{1}{3!}\ \delta^{\mu\nu\rho\sigma}_{\alpha\beta\gamma\delta}\ g^\beta_\nu\ g^\gamma_\rho\ g^\delta_\sigma$$

So, $d\{\ln|\det g|\} = \frac{d(\det g)}{\det g} = \frac{(g^{-1})^{\mu}_{\alpha}}{\frac{1}{3!}\left(\delta^{\mu\nu\rho\sigma}_{\alpha\beta\gamma\delta}\right) g^{\beta}_{\nu}\, g^{\gamma}_{\rho}\, g^{\delta}_{\sigma}}\, d(\det g)$

$$= (g^{-1})^{\mu}_{\alpha}\, \frac{4\frac{1}{4!}\delta^{\mu\nu\rho\sigma}_{\alpha\beta\gamma\delta}\, d(g^{\alpha}_{\mu}) g^{\beta}_{\nu}\, g^{\gamma}_{\rho}\, g^{\delta}_{\sigma}}{\frac{1}{3!}\delta^{\mu\nu\rho\sigma}_{\alpha\beta\gamma\delta}\, g^{\beta}_{\nu}\, g^{\gamma}_{\rho}\, g^{\delta}_{\sigma}}$$

$$= (g^{-1})^{\mu}_{\alpha}\, d\left(g^{\alpha}_{\mu}\right)$$

$$= g^{\mu\alpha} dg_{\alpha\mu}$$

And, since

$$d\left(\sqrt{-g}\right) = \frac{1}{2}\frac{1}{\sqrt{-g}}\, d(-g) = \frac{1}{2}\frac{1}{\sqrt{-g}}\, d\left|\det g_{\mu\nu}\right|$$

$$= \frac{1}{2}\frac{1}{\sqrt{-g}}(-g)d\{\ln(-g)\} = \frac{1}{2}\frac{1}{\sqrt{-g}}\, g^{\mu\alpha} dg_{\alpha\mu}$$

We have,

$$\delta\sqrt{-g} = \frac{1}{2}\frac{1}{\sqrt{-g}}\, g^{\mu\nu}\delta g_{\nu\mu}.$$

Thus

$$\delta S_G = \frac{1}{2}\int d^4x\sqrt{-g}\; g^{\mu\nu}\delta g_{\mu\nu}.$$

(Eqn. 9-5)

9.1.2 Variation for $\boldsymbol{\mathcal{L}} = \left(\frac{1}{16\pi G}R - \frac{\Lambda}{8\pi G}\right)$

Consider $S_G = \int d^4x\sqrt{-g}\;\left(\frac{1}{16\pi G}R - \frac{\Lambda}{8\pi G}\right)$.

We already know $\delta\sqrt{-g}$ from part (2) what is δR? $R = g^{\mu\nu}R_{\mu\nu}$ so what is $\delta R_{\mu\nu}$? We know $\delta R_{\mu\nu}$ from the Palatini identity: $\delta R_{\mu\nu} = \left(\delta\Gamma^{\lambda}_{\mu\lambda}\right)_{;\nu} - \left(\delta\Gamma^{\lambda}_{\mu\nu}\right)_{;\lambda}$. Thus:

$$\delta S_G = \int d^4x\left\{\delta\sqrt{-g}\;\left(\frac{1}{16\pi G}R - \frac{\Lambda}{8\pi G}\right) + \frac{\sqrt{-g}\,\delta g^{\mu\nu}R_{\mu\nu}}{16\pi G} + \underbrace{\frac{\sqrt{g}\,g^{\mu\nu}\delta R_{\mu\nu}}{16\pi G}}\right\}$$

pure divergence so integrates to zero.

$$= \int d^4x\left\{\frac{1}{2}\sqrt{-g}\,g^{\mu\nu}\delta g_{\mu\nu}\left(\frac{1}{16\pi G}R - \frac{\Lambda}{8\pi G}\right) + \frac{\sqrt{-g}\,R_{\mu\nu}\delta g^{\mu\nu}}{16\pi G}\right\}$$

Note $\delta(\delta_\lambda^\sigma) = \delta(g_{\lambda\gamma} g^{\gamma\sigma}) = \delta g_{\lambda\gamma} g^{\gamma\sigma} + g_{\lambda\gamma} \delta g^{\gamma\sigma} = 0 \Rightarrow \delta g^{\mu\nu} = -g^{\mu\rho} g^{\gamma\sigma} \delta g_{\rho\sigma}$

So,

$$\delta S_G = \frac{-1}{16\pi G} \int d^4x \sqrt{-g} \left(R^{\mu\nu} - \frac{1}{2} g^{\mu\nu} R \right) \delta g_{\mu\nu} - \frac{\Lambda}{16\pi G} \int d^4x \sqrt{-g}\, g^{\mu\nu} \delta g_{\mu\nu}$$

(Eqn. 9-6)

9.1.3 Variation with $\mathcal{L}_M$ a scalar density (S_M a scalar) under coordinate transformations

Let's now examine the matter variation, which notably is defined to have scalar action, with the form of stress-energy tensor obtained by definition. So, to re-derive our definition would be tautological, here let's examine what the given variation for matter suggests under coordinate transformation. Since the matter action is scalar, the variation must be zero, which will give rise to conservation laws.

Consider $\delta S_M = \frac{1}{2} \int d^4x\, T^{\mu\nu}(x) \delta g g_{\mu\nu}(x)$ (defines $T^{\mu\nu}$). If S_M is a scalar then it should be invariant under coordinate transformations: $\delta S_M = 0$ for $x'^\mu = x^\mu + \delta x^\mu$. Let $\delta x^\mu = \varepsilon^\mu(x)$. Now have:

$$g'_{\mu\nu}(x) = g'_{\mu\nu}(x') + \frac{\partial g'_{\mu\nu}(x)}{\partial x^\lambda} (-\varepsilon^\lambda)$$

$$= g'_{\mu\nu}(x') - \frac{\partial g_{\mu\nu}(x)}{\partial x^\kappa} (\varepsilon^\lambda) + 0(|\varepsilon|)$$

$$g'_{\mu\nu}(x) = g_{\mu\nu}(x) - \left\{ \frac{\partial g_{\mu\nu}(x)}{\partial x^\lambda} \varepsilon^\lambda + g_{\lambda\nu} \frac{\partial \varepsilon^\lambda}{\partial x^\mu} + g_{\mu\lambda} \frac{\partial \varepsilon^\lambda}{\partial x^\nu} \right\}$$

So,

$$\delta g_{\mu\nu}(x) = -\left\{ \frac{\partial g_{\mu\nu}(x)}{\partial x^\lambda} \varepsilon^\lambda + g_{\lambda\nu} \frac{\partial \varepsilon^\lambda}{\partial x^\mu} + g_{\mu\lambda} \frac{\partial \varepsilon^\lambda}{\partial x^\nu} \right\}$$

Thus,

$$\delta S_M = \frac{1}{2} \int d^4x \sqrt{-g}\, T^{\mu\nu}(x) \{\ldots\} = 0$$

with integration by parts then gives:

$$\delta S_M = \int d^4x \left[-\frac{1}{2}\left(\frac{\partial g_{\mu\nu}}{\partial x^\lambda}\right)\sqrt{-g}\, T^{\mu\nu} + \frac{\partial}{\partial x^\mu}\left[\frac{1}{2}\sqrt{-g}\, T^{\mu\nu} g_{\lambda\nu}\right] + \frac{\partial}{\partial x^\nu}\left[\frac{1}{2}\sqrt{-g}\, T^{\mu\nu} g_{\mu\lambda}\right]\right]\varepsilon^\lambda$$

$$= \int d^4x \left[-\frac{1}{2}\left(\frac{\partial g_{\mu\nu}}{\partial x^\lambda}\right)\sqrt{-g}\, T^{\mu\nu} + \frac{\partial}{\partial x^\mu}\left[\sqrt{-g}\, T^{\mu\nu} g_{\nu\lambda}\right]\right]\varepsilon^\lambda$$

Since, $\varepsilon^\lambda(x)$ is arbitrary:

$$\left(\sqrt{-g}\, T^{\mu\nu}\lambda\right)_{,\mu} - \frac{1}{2}\, g_{\mu\nu,\lambda}\, T^{\mu\nu}\sqrt{-g} = 0$$

Since,

$$T^\mu_{\nu;\mu} = \frac{1}{\sqrt{-g}}\frac{\partial\left(\sqrt{-g}T^\mu_\nu\right)}{\partial x^\mu} - \Gamma^\lambda_{\mu\nu}T^\mu_\lambda$$

We have

$$T^\mu_{\lambda;\mu} + \Gamma^\alpha_{\mu\lambda}\, T^\mu_\alpha - \frac{1}{2}g_{\mu\nu,\lambda}T^{\mu\nu} = 0$$

In detail:

$$T^\mu_{\lambda;\mu} + g^{\alpha\gamma}\Gamma_{\gamma\mu\lambda}\, T^\mu_\alpha - \frac{1}{2}g_{\mu\nu,\lambda}T^{\mu\nu} = 0$$

$$T^\mu_{\lambda;\mu} + \frac{1}{2}\left(g_{\gamma\mu,\lambda} + g_{\gamma\lambda,\mu} - g_{\mu\lambda,\gamma}\right)T^{\mu\gamma} - \frac{1}{2}g_{\mu\nu,\lambda}T^{\mu\nu} = 0$$

$$T^\mu_{\lambda;\mu} + \frac{1}{2}\left(g_{\gamma\lambda,\mu} - g_{\mu\lambda,\gamma}\right)T^{\mu\gamma} = 0$$

↑ ↑ symmetric on μ,γ

Antisymmetric on μ,γ.

So,

$$T^\mu{}_{\lambda;\mu} = 0 \;\rightarrow\; \nabla_\mu T^{\mu\lambda} = 0$$

(Eqn. 9-7)

9.1.4 Variation of S_G (also a scalar) under coordinate transformations

Similarly, since S_G is a scalar:

$$\delta S_G = \frac{-1}{16\pi G}(-1)\int d^4x\sqrt{-g}\left(R^{\mu\nu} - \frac{1}{2}g^{\mu\nu}R + \Lambda g^{\mu\nu}\right)\{g_{\mu\nu,\lambda}\varepsilon^\lambda + g_{\lambda\nu}\varepsilon^\lambda_{,\mu} + g_{\mu\lambda}\varepsilon^\lambda_{,\nu}\}.$$

Thus

$$\sqrt{-g}\, g_{\mu\nu,\lambda}\left(R^{\mu\nu} - \frac{1}{2}g^{\mu\nu}R + \Lambda g^{\mu\nu}\right) + 2\left[\sqrt{-g}\left(R^{\mu\nu} - \frac{1}{2}g^{\mu\nu}R + \Lambda g^{\mu\nu}\right)g_{\mu\lambda}\right]_{,\mu} = 0$$

As before $R^{\mu}_{\lambda;\mu} = \frac{1}{\sqrt{-g}}\left(\sqrt{-g}\,R^{\mu}_{\lambda}\right)_{,\mu} - \Gamma^{\alpha}_{\lambda}R^{\beta}_{\alpha} = \frac{1}{\sqrt{-g}}\left(\sqrt{-g}\,R^{\mu}_{\lambda}\right)_{,\mu} - \frac{1}{2}g_{\alpha\beta,\lambda}R^{\alpha\beta}$. So:

$$\sqrt{-g}\,g_{\mu\nu,\lambda}R^{\mu\nu} - g\,\frac{1}{2}g_{\mu\nu,\lambda}g^{\mu\nu}R + \Lambda\sqrt{-g}\,g^{\mu\nu}g_{\mu\nu,\lambda}$$
$$-2\sqrt{-g}\left(R^{\mu}_{\lambda,\mu} - \frac{1}{2}\delta^{\mu}_{\lambda}R_{,\mu} + \Lambda\delta^{\mu}_{\lambda,\mu}\right)$$
$$-2\left(R^{\mu\nu} - \tfrac{1}{2}g^{\mu\nu}R + \Lambda g^{\mu\nu}\right)g_{\nu\lambda}\left(\sqrt{-g}\right)_{,\mu} = 0$$

Thus

$$g_{\mu\nu,\lambda}R^{\mu\nu} + \left(-2R^{\mu}_{\lambda;\mu} - g_{\alpha\beta,\lambda}R^{\alpha\beta} + R^{\mu}_{\lambda}g^{\alpha\beta}g_{\alpha\beta,\mu}\right) + \delta^{\mu}_{\lambda}R_{,\mu}$$
$$- g^{\alpha\beta}g_{\alpha\beta,\mu}R^{\mu}_{\lambda} = 0$$

And we get:

$$-2R^{\mu}_{\lambda;\mu} + \delta^{\mu}_{\lambda}R_{;\mu} = 0$$

(since $R_{,\mu} = R_{;\mu}$ with R scalar). Thus:

$$\left[R^{\mu}_{\lambda} - \tfrac{1}{2}\delta^{\mu}_{\lambda}R\right]_{;\mu} = 0$$

(Eqn. 9-8)

which is the contacted Bianchi identity.

9.1.5 Variation with $\mathcal{L}_M$ a scalar density, for particulate matter, under coordinate transformations

Let's now examine the Matter variation, that gave rise to the $\nabla_{\mu}T^{\mu\lambda} = 0$ relation in general, now in the context of particulate motion for which a Lagrangian of the form $L(g_{\mu\nu}, \varphi, \varphi_{,\mu})$ exists, where φ is the scalar matter field with delta-distributions to represent the particulate matter.

Since S_M is scalar and δS_M vanishes under $\varphi \to \varphi + \delta\varphi$ and $x^{\mu} \to x^{\mu} + \delta x^{\mu}$ (i.e. variation of matter variables) this implies that

$$\nabla_{\mu}T^{\mu\nu} = 0.$$

Show that if

$$S_M = \int d^4x\,\sqrt{-g}\;\; L(g_{\mu\nu}, \varphi, \varphi_{,\mu}) \text{ then } \nabla_{\mu}T^{\mu\nu} = 0$$

Hint, under infinitesimal coordinate transformation $x^{\mu} \to x^{\mu} + \delta x^{\mu}$ we know that $\delta g_{\mu\nu}$

$$\mathcal{L}_{\varepsilon}g_{\mu\nu} = g_{\mu\nu,\alpha}\varepsilon^{\alpha} + g_{\beta\nu}\varepsilon^{\beta}_{,\mu} + g_{\mu\beta}\varepsilon^{\beta}_{,\nu}$$

(Eqn. 9-9)

Similarity, get an identity from the S_G part also, since S_G is a scalar:

$$\nabla_\mu\left(R^{\mu v}-\frac{1}{2}g^{\mu v}R+\Lambda g^{\mu v}\right)=0$$

(Eqn. 9-10)

Consider a free particle at position $x^\mu(\tau)$:

$$S_M=\frac{1}{2}m\int_{-\infty}^{\infty}\left(g_{\mu v}\frac{dx^\mu}{dp}\frac{dx^v}{dp}\right)^{1/2}dp,$$

where p an arbitrary parameter. Take p to be an affine parameter with proper time τ, then from
$\delta S_M=0$ when $x^\mu\rightarrow x^\mu+\delta\lambda^\mu$ we get:

$$\frac{d^2x^\mu}{d\tau^2}+\Gamma^\mu_{v\lambda}\frac{dx^v}{d\tau}\frac{dx^\lambda}{d\tau}=0.$$

So,

$$\delta S_M=\frac{1}{2}\int d^4x\sqrt{-g}T^{\mu v}\delta g_{\mu v}$$

$$=\frac{1}{2}m\int_{-\infty}^{\infty}d\tau\frac{dx^\mu}{d\tau}\frac{dx^v}{d\tau}$$

$$=\frac{1}{2}m\int d^4y\sqrt{-g(y)}\int_{-\infty}^{\infty}\frac{d\tau}{\sqrt{-g(y)}}\delta^4(y-x(\tau))\frac{dx^\mu}{d\tau}\frac{dx^v}{d\tau}\delta g_{\mu v}(y).$$

So,

$$T^{\mu v}(y)=\frac{1}{\sqrt{-g(y)}}m\int_{-\infty}^{\infty}d\tau\,\delta^4(y-x(\tau))\frac{dx^\mu}{d\tau}\frac{dx^v}{d\tau}$$

(Eqn. 9-11)

Let $\tau\rightarrow x^0$

$$T^{\mu v}(y)=\frac{1}{\sqrt{-g(y)}}m\int_{-\infty}^{\infty}dx^0\frac{d\tau}{dx^0}\delta^3(\vec{y}-\vec{x}(x^0))\frac{dx^\mu}{d\tau}\frac{dx^v}{d\tau}\delta(y^0-x^0)$$

$$=\frac{1}{\sqrt{-g(y)}}m\frac{d\tau}{dy^0}\delta^3(\vec{y}-\vec{x}(y^0))\frac{dx^\mu}{d\tau}\frac{dx^v}{d\tau}=\rho(y)u^\mu(y^0)u^v(y^0)$$

where the RHS is a scalar (and $u^\mu=\frac{dx^\mu}{d\tau}$). Thus:

$$\rho(y)=\frac{1}{\sqrt{-g(y)}}m\frac{d\tau}{dy^0}\delta^3(\vec{y}-\vec{x}(y^0)).$$

(Eqn. 9-12)

The local inertial frame with particle at rest has $\rho(y) = m\delta(\vec{y} - \vec{x}(y^0))$ which is the mass density of the particle. So, in general, ρ is the proper energy density. A dust cloud has $T^{\mu\nu}(x) = \rho(x)u^\mu(x)u^\nu(x)$ and a perfect fluid has $T^{\mu\nu} = pg^{\mu\nu} + (\rho + p)u^\mu\, u^\nu$.

Since matter should satisfy $\nabla_\mu\, T^{\mu\nu} = 0$, we can now get the equations of motion.

Note that $\nabla_\mu\, T^{\mu\nu} = 0$ holds only if the equation of motion for the particles is true, $\frac{d^2x^\mu}{d\tau^2} + \Gamma^\mu_{\nu x} \ldots = 0$; question: Does it follow that we get the geodesic equation if $\nabla_\mu\, T^{\mu\nu} = 0$? (is there an iff statement in other words.) We have $U^\nu\, \nabla_\nu\, U^\mu = 0$ as another way to write geodesic equation Consider the key term in the perfect fluid case, if it satisfies: $\nabla_\mu\, (\rho(x)U^\mu U^\nu) = 0$ then

$$U^\mu U^\nu\, \nabla_\mu\, \rho(x) + p(x)\left(U^\nu\, \nabla_\mu U^\mu + U^\mu \nabla_\mu U^\nu\right) = 0$$

So,

$$0 = \left[U^\nu\, \nabla_\mu(\rho u^\mu) + \rho u^\mu \nabla_\mu U^\nu\right] = \nabla_\mu(\rho u^\mu) + \rho u^\mu\, \underbrace{U_\nu \nabla_\mu U^\nu}$$

where $U_\nu \nabla_\mu U^\nu = 0$ follows only if $\nabla_\mu(\rho u^\mu) = 0$. So, we only get the geodesic equation for some matter.

9.2 An examination of geodesics

9.2.1 Geodesic Defined using parallel transport

A geodesic is a curve whose tangent vectors are parallel transported along itself. We need not demand that the vector maintain the same length, in which case, for tangent vector T^a we have: $T^a \nabla_a T^b = \alpha T^b$. But, since we are free to parameterize the curve as we wish we can choose a parameterization such that the tangent vector is constant in which case:

$$T^a \nabla_a T^b = 0 \rightarrow \text{affinely parameterized.}$$

Denote the affine parameter as λ and the coordinate system as x where $x^\mu(C)$ is the path of the geodesic (freely falling particle). So,

$$T^a = \frac{d}{d\lambda} = \frac{dx^\mu}{d\lambda}\frac{\partial}{\partial x^\mu} \Rightarrow T^\mu = \frac{dx^\mu}{d\lambda}$$

$$T^a \nabla_a T^b = 0 \Rightarrow T^\beta_{\ ;\alpha} T^\alpha = \left(T^\beta_{\ ,\alpha} + \Gamma^\beta_{\gamma\alpha}\, T^\gamma\right) T^\alpha = 0$$

$$0 = \frac{\partial}{\partial x^\alpha}\left(\frac{dx^\beta}{d\lambda}\right)\left(\frac{dx^\alpha}{d\lambda}\right) + \Gamma^\beta_{\gamma\alpha}\frac{dx^\gamma}{d\lambda}\frac{dx^\alpha}{d\lambda}$$

$$\frac{d^2x^\beta}{d\lambda^2} + \Gamma^\beta_{\gamma\alpha}\frac{dx^\gamma}{d\lambda}\frac{dx^\alpha}{d\lambda} = 0.$$

(Eqn. 9-13)

9.2.2 Geodesic Defined using variational principle

Alternatively, the geodesic equation can also be formulated as a variational principle. Consider the proper time elapsed when a particle falls from A to B:

$$T_{AB} = \int_A^B d\tau = \int_A^B \left\{-g_{\mu\nu}\frac{dx^\mu}{d\lambda}\frac{dx^\nu}{d\lambda}\right\}^{1/2} d\lambda$$

with $-d\tau^2 = g_{\mu\nu}dx^\mu dx^\nu$. So,

$$T_{AB} = \int_A^B \{-\vec{U}\cdot\vec{U}\}^{1/2}\, d\lambda$$

where $U^\mu = \frac{dx^\mu}{d\lambda}$ has norm so that $U^2 = -1$. So,

$$\delta T_{AB} = \int_A^B -\vec{U}\cdot\delta\vec{U}d\lambda = \int_A^B (-\vec{U}\cdot\nabla_a\vec{U})\delta x^a\, d\lambda$$

Since $\vec{U}\cdot\nabla_a\vec{U} = 0 \rightarrow \nabla_{\vec{U}}\vec{U} = 0$, which is true for $U^\alpha = \frac{dx^\alpha}{d\lambda}$, λ affine parameter if

$$\frac{d^2x^\beta}{d\lambda^2} + \Gamma^\beta_{\gamma\alpha}\frac{dx^\gamma}{d\lambda}\frac{dx^\alpha}{d\lambda} = 0$$

as before. However, $\nabla_a\vec{U}$ is a new vector intermediate requiring definition of covariant derivative, so perhaps not satisfactory. Here's the conventional variational method using full index gymnastics:

$$\delta T_{AB} = \frac{1}{2}\int_A^B \left\{-g_{\mu\nu}\frac{dx^\mu}{d\lambda}\frac{dx^\nu}{d\lambda}\right\}^{1/2}\left\{-\frac{\partial g_{\mu\nu}}{\partial x^\lambda}\delta x^\lambda\frac{dx^\mu}{d\lambda}\frac{dx^\nu}{d\lambda} - 2g_{\mu\nu}\frac{d\delta x^\mu}{d\lambda}\frac{dx^\nu}{d\lambda}\right\}d\lambda$$

$$= -\int_A^B \left\{\frac{1}{2}\frac{\partial g_{\mu\nu}}{\partial x^\lambda}\frac{dx^\mu}{d\lambda}\frac{dx^\nu}{d\lambda} - \frac{\partial g_{\lambda\nu}}{\partial x^\sigma}\frac{dx^\sigma}{d\lambda}\frac{dx^\nu}{d\lambda} - g_{\lambda\nu}\frac{d^2x^\nu}{d\lambda^2}\right\}\delta x^\lambda d\lambda$$

Since δx^λ is arbitrary must be zero:

$$g_{\lambda\nu}\frac{d^2x^\nu}{d\lambda^2} + \frac{1}{2}\left\{\frac{\partial g_{\lambda\nu}}{\partial x^\sigma} + \frac{\partial g_{\lambda\sigma}}{\partial x^\nu} - \frac{\partial g_{\mu\nu}}{\partial x^\lambda}\right\}\frac{dx^\sigma}{d\lambda}\frac{dx^\nu}{d\lambda} = 0$$

$$g_{\lambda v}\frac{d^2r^v}{d\lambda^2}+ g_{\lambda v}\Gamma^v_{\sigma\alpha}\frac{dx^\sigma}{d\lambda}\frac{dx^\alpha}{d\lambda}=0 \rightarrow \quad \frac{d^2x^v}{d\lambda^2}+ g_{\lambda v}\Gamma^v_{\sigma\alpha}\frac{dx^\sigma}{d\lambda}\frac{dx^\alpha}{d\lambda}=0$$

9.3 Geodesics as Cosmological World Lines -- Equivalent observers postulate

Assume that there are a set of fundamental world lines w.r.t which the universe presents the same history and appearance (with a large enough average or coarse graining, that of background radiation.)
Imagine a standard clock. Thus we have a cosmic time and space coordinate. Take the space coord's as labels x' on each fundamental world line which remain unchanged on a given world line, put clocks on those world lines. The clocks on different fundamental world lines are synchronized such that the universe has the same appearance at any such point at a given time. So far must have homogeneity, not necessarily isotropy. Let's now consider specifically when there is isotropy, i.e., let's consider Isotropic Universes, particularly the following:

- Cosmic time
- Exchange of signals and entropy
- FRW geometry

Assume there exists a set of fundamental lines w.r.t which the universe looks essentially the same.
For two events occurring on the same world line: $dx^i = 0$ and $dt = d\tau$.
Since we have,

$$ds^2 = g_{\mu v}dx^\mu\, dx^v = -d\tau^2 = -dt^2,$$

we already know that $g_{00} = -1$ in such an instance. Next, let's consider the exchange of light signals between world lines, where worldliness are label by position x^i or $x^i + dx^i$, world line x^i emits signal at $t - dt_A$ that is received at world line $x^i + dx^i$ at time t. Similarly, world line $x^i + dx^i$ emits signal at $t - dt_B$ that is received at world line x^i at time t. By isotropy and homogeneity we must have

$$dt_A = dt_B.$$

Along the null rays of both signals we have that

$$ds^2 = -dt^2 + 2g_{0i}dtdx^i + g_{ij}dx^i\, dx^j.$$

Specifically:
Along null ray $\quad$ A: $ds^2 = 0 = -dt_A^2 + 2g_{0i}dtdx^i + g_{ij}dx^i\, dx^j$

B: $0 = -dt_B^2 + 2g_{0i}dt(-dx^i) +$
$g_{ij}dx^i\, dx^j$

So, $0 = 4g_{0i}dtdx^i \Rightarrow g_{0i} = 0$, thus we have: $ds^2 = dt^2 + g_{ij}dx^i\, dx^j$.
In other words, because of isotropy and homogeneity, we must have that

$$g_{ij}(\vec{x},\tau) = a^2(\tau)\tilde{g}_{ij}(\vec{x}).$$

From: $g_{ij}(\vec{x},\tau_2)dx^i\, dx^j$ at τ_2 and $g_{ij}(\vec{x},\tau_2)dx^i\, dx^j$ at τ_1 is also the notion that the ratio can't depend on $\vec{x}$ by homogeneity (form of cross ratio theorem?), else we could have a non-homogeneous function. Thus:

$$\frac{g_{ij}(\vec{x},\tau_2)}{g_{ij}(\vec{x},\tau_1)} = f(\tau_2,\tau_1).$$

(Eqn. 9-14)

Thus, isotropy demands that the relation above not depend on separation dx^i.

So, consider again (from the start) the form indicated if there is homogeneity and isotropy:

$$ds^2 = dt^2 + a^2(\tau)\,\tilde{g}_{ij}(\vec{x})dx^i dx^j.$$

9.4 Derivation of scalar curvature $\tilde{R}$ for homogenous isotropic metrics --> the FRW Universes

Let's now do the following:

(1) Write the most general isotropic line element of the indicated form.
(2) Calculate its scalar curvature $\tilde{R}$.
(3) By homogeneity $\tilde{R}$ is constant, verify.

Parts (2) and (3) comprise a differential equation whose solutions comprise the FRW Universes.

Consider a the purely spatial line element: $d\ell^2 = \tilde{g}_{ij}(\vec{x})dx^i dx^j$. If thee is isotropy we should be able to choose the coordinates x^i in such a way that the isotropy w.r.t. a given fundamental world line $(x^i = 0)$ is manifestly evident. Therefore, assume the x^i can be chosen in such a way that they transfer under rotations just as Cartesian coordinates do. Then $d\ell^2$ must be found from rotational invariant constructs from the x^i and the dx^i. The invariants are: $r^2 = \vec{x}\cdot\vec{x}$, $\vec{x}\cdot d\vec{x} = rdr$, $d\vec{x}\cdot d\vec{x}$.
The most general $d\ell^2$ that can be constructed from these invariants is:

$$d\ell^2 = A(r)(d\vec{x} \cdot d\vec{x}) + B(r)(\vec{x} \cdot d\vec{x})^2.$$

In spherical coordinates:

$$d\ell^2 = A(r)(dr^2 + r^2\, d\theta^2 + r^2\, sin^2\theta d\varphi^2) + B(r) r^2 dr^2$$
$$= C(r)dr^2 + A(r) r^2 (sin^2\, \theta d\varphi^2 + d\theta^2)$$

Redefine: $r' = A^{1/2}(r)r$. Then:

$$d\ell^2 = D(r)dr^2 + r^2(sin^2\, \theta d\varphi^2 + d\theta^2)$$

What results is a 3-D analysis for the Ricci tensor and scalar curvature, one finds that:

$$\tilde{R}_{rr} = \frac{D'(r)}{rD(r)}, \qquad \tilde{R}_{\theta\theta} = 1 + \frac{rD'(r)}{2D^2(r)} - \frac{1}{D(r)}, \quad \tilde{R}_{\varphi\varphi} = sin^2\theta\, \tilde{R}_{\theta\theta}$$

(Eqn. 9-15)

(2) Consider $d\ell^2 = D(r)dr^2 + r^2(\sin^2\theta d\varphi^2 + d\theta^2)$

$3D\ Analysis$: $g_{rr} = D(r);\ g_{\varphi\varphi} = r^2 \sin^2\theta;\ g_{\theta\theta} = r^2$.

Starting with:

$$\Gamma_{\mu\beta\gamma} = \frac{1}{2}\left(g_{\mu\beta,\gamma} + g_{\mu\gamma,\beta} - g_{\beta\gamma,\mu}\right),$$
$$\Gamma^{v}_{\beta\gamma} = g^{v\mu}\Gamma_{\mu\beta\gamma},$$

and

$$R_{\mu v} = \Gamma^{\alpha}_{\mu v,\alpha} - \Gamma^{\alpha}_{\mu\alpha,v} + \Gamma^{\alpha}_{\beta\alpha}\, \Gamma^{\beta}_{\mu v} - \Gamma^{\alpha}_{\beta v}\, \Gamma^{\beta}_{\mu\alpha}.$$

We have:

$$\Gamma_{rrr} = \frac{1}{2} g_{rr,r} = \frac{1}{2} D'(r) \rightarrow \Gamma^{r}_{rr} = \frac{1}{2}\frac{D'}{D}.$$
$$\Gamma_{\varphi\varphi r} = \frac{1}{2} g_{\varphi\varphi,r} = r\sin^2\theta \rightarrow \Gamma^{\varphi}_{\varphi r} = \Gamma^{\varphi}_{r\varphi} = \frac{1}{r}.$$
$$\Gamma_{r\varphi\varphi} = -\frac{1}{2} g_{\varphi\varphi,r} = -r\sin^2\theta \rightarrow \Gamma^{r}_{\varphi\varphi} = -\frac{r\sin^2\theta}{D}.$$
$$\Gamma_{\varphi\varphi\theta} = -\Gamma_{\theta\varphi\varphi} = r^2 \sin\theta \cos\theta \rightarrow \Gamma^{\varphi}_{\varphi\theta} = \frac{\cos\theta}{\sin\theta}\ \ and\ \ \Gamma^{\theta}_{\varphi\varphi} =$$
$$- \sin\theta \cos\theta.$$
$$\Gamma_{\theta\theta r} = -\Gamma_{r\theta\theta} = \frac{1}{2} g_{\theta\theta,r} = r \rightarrow \Gamma^{\theta}_{\theta r} = \frac{1}{r}\ \ and\ \ \Gamma^{r}_{\theta\theta} = \frac{-r}{D}.$$

So,

$$\Gamma^{\alpha}_{\mu v,\alpha} = \Gamma^{r}_{\mu v,r} + \Gamma^{\theta}_{\mu v,\theta} + \Gamma^{\varphi}_{\mu v,\varphi}$$

but we only have

$$\Gamma^{\alpha}_{\mu\mu,\alpha} = \Gamma^{r}_{\mu\mu,r} + \Gamma^{\theta}_{\mu\mu,\theta}$$

As for the other relation needed:

$\Gamma^{\alpha}_{\mu\alpha,v} = \Gamma^{r}_{\mu r,v} + \Gamma^{\theta}_{\mu\theta,v} + \Gamma^{\varphi}_{\mu\varphi,v}$

For $\mu = r\ we\ have\ \Gamma^{\alpha}_{\mu\alpha,v} = \Gamma^{r}_{rr,v} + \Gamma^{\theta}_{r\theta,v} + \Gamma^{\varphi}_{r\varphi,v}$ → only $v = r$ gives nonzero.

For $\mu = \theta\ we\ have\ \Gamma^{\alpha}_{\mu\alpha,v} = \Gamma^{\varphi}_{\mu\varphi,v}$ → only $v = \theta$ gives nonzero.

So,

$\Gamma^{\alpha}_{\mu v,\alpha} \neq 0$ only for $\mu = v$.

$\Gamma^{\alpha}_{\mu\alpha,v} \neq 0$ only for $\mu = v = r$ or θ.

Thus, using

$\Gamma^{\alpha}_{\mu v,\alpha} - \Gamma^{\alpha}_{\mu\alpha,v} = \Gamma^{\alpha}_{\mu\mu,\alpha} - \Gamma^{\alpha}_{\mu\alpha,\mu}$

we now get:

$$\Gamma^{\alpha}_{\mu v,\alpha} - \Gamma^{\alpha}_{\mu\alpha,v} = \begin{cases} \frac{2}{r^2} \quad for \quad \mu = v = r \\ \left[-\frac{1}{D} + \frac{rD'}{D^2} + \frac{\cos^2\theta}{\sin^2\theta} + 1\right] \quad for \quad \mu = v = \theta \\ \sin^2\theta\left[-\frac{1}{D} + \frac{rD'}{D^2}\right] - \cos^2\theta + \sin^2\theta \ for \quad \mu = v = \varphi \end{cases},$$

and the rest are zero.

An evaluation of

$\Gamma^{\alpha}_{\beta\alpha} = \Gamma^{r}_{\beta r} + \Gamma^{\mu}_{\beta\varphi} + \Gamma^{\theta}_{\beta\theta}$

reveals that:

$\beta = r: \quad \Gamma^{\alpha}_{\beta\alpha} = \frac{1}{2}\frac{D'}{D} + \frac{1}{r} + \frac{1}{r},$

$\beta = \theta: \quad \Gamma^{\alpha}_{\beta\alpha} = \frac{\cos\theta}{\sin\theta},$

and the rest zero.

Similarly,

$\Gamma^{r}_{\mu v} = \Gamma^{r}_{\mu\mu}$, the rest zero; and $\Gamma^{\theta}_{\mu v}$ has $\Gamma^{\theta}_{\varphi\varphi}$ and $\Gamma^{\theta}_{\theta r}$ nonzero.

Let's use the above to expand:

$$\Gamma^{\alpha}_{\beta v}\,\Gamma^{\beta}_{\mu\alpha} = \Gamma^{\alpha}_{\beta v}\,\Gamma^{\beta}_{\alpha\mu} = \Gamma^{r}_{\beta v}\,\Gamma^{\beta}_{r\mu} + \Gamma^{\theta}_{\beta v}\,\Gamma^{\beta}_{\theta\mu} + \Gamma^{\varphi}_{\beta v}\,\Gamma^{\beta}_{\varphi\mu}$$

To get:

$$\Gamma^{\alpha}_{\beta v}\,\Gamma^{\beta}_{\mu\alpha} = \underbrace{\Gamma^{r}_{rv}\,\Gamma^{r}_{r\mu}}_{\mu=v=r} + \underbrace{\Gamma^{r}_{\theta v}\,\Gamma^{\theta}_{r\mu}}_{\mu=v=\theta} + \underbrace{\Gamma^{r}_{\varphi v}\,\Gamma^{\varphi}_{r\mu}}_{\mu=v=\varphi} + \underbrace{\Gamma^{\theta}_{rv}\,\Gamma^{r}_{\theta\mu}}_{\mu=v=\theta} + \underbrace{\Gamma^{\theta}_{\theta v}\,\Gamma^{\theta}_{\theta\mu}}_{\mu=v=r} + \underbrace{\Gamma^{\theta}_{\varphi v}\,\Gamma^{\varphi}_{\theta\mu}}_{\mu=v=\varphi} + \underbrace{\Gamma^{\varphi}_{rv}\Gamma^{r}_{\varphi\mu}}_{\mu=v=\varphi} + \underbrace{\Gamma^{\varphi}_{\theta v}\Gamma^{\theta}_{\varphi\mu}}_{\mu=v=\varphi} + \underbrace{\Gamma^{\varphi}_{\varphi v}\Gamma^{\varphi}_{\varphi\mu}}_{\substack{\mu=v=\theta \text{ or} \\ r \text{ or } mix}}$$

So,

$$\Gamma^{\alpha}_{\beta v}\Gamma^{\beta}_{\mu\alpha} = \begin{cases} (\Gamma^{r}_{rr})^2 + \left(\Gamma^{\theta}_{\theta r}\right)^2 + \left(\Gamma^{\varphi}_{\varphi r}\right)^2 & \mu = v = r \\ \left(\Gamma^{\varphi}_{\varphi\theta}\right)\left(\Gamma^{\varphi}_{\varphi r}\right) & \mu = r, v = \theta \text{ or } \mu = \theta, v = r \\ 2(\Gamma^{r}_{\theta\theta})(\Gamma^{\theta}_{r\theta}) + \left(\Gamma^{\varphi}_{\varphi\theta}\right)^2 & \mu = v = \theta \\ 2\left(\Gamma^{r}_{\varphi\varphi}\Gamma^{\varphi}_{r\varphi}\right) + 2\left(\Gamma^{\theta}_{\varphi\varphi}\Gamma^{\varphi}_{\theta\varphi}\right) & \mu = v = \varphi \end{cases}$$

Since

$$R_{\mu v} = \underbrace{\Gamma^{\alpha}_{\mu v,\alpha} - \Gamma^{\alpha}_{\mu\alpha,v}}_{\text{only } \mu=v} + \underbrace{\Gamma^{\alpha}_{\beta\alpha}\Gamma^{\beta}_{\mu v}}_{\substack{\text{only } \mu=v \\ \text{and } \mu=\theta,v=r \\ v=\theta,\mu=r}} - \underbrace{\Gamma^{\alpha}_{\beta v}\Gamma^{\beta}_{\mu\alpha}}_{\substack{\text{only } \mu=v \\ \text{and } \mu=r,v=\theta \\ \mu=\theta,v=r}}$$

We can evaluate:

$$R_{\theta r} = \Gamma^{\varphi}_{\theta\varphi}\Gamma^{\theta}_{\theta r} - \left(\Gamma^{\varphi}_{\varphi\theta}\right)\left(\Gamma^{\varphi}_{\varphi r}\right) = \Gamma^{\varphi}_{\theta\varphi}\left(\Gamma^{\theta}_{\theta r} - \Gamma^{\varphi}_{\varphi r}\right) = 0\,.$$

$$R_{rr} = \left(\frac{2}{r^2}\right) + \left(\frac{1}{2}\frac{D'}{D} + \frac{2}{r}\right)\left(\frac{1}{2}\frac{D'}{D}\right) - \left\{\left(\frac{1}{2}\frac{D'}{D}\right)^2 + \left(\frac{1}{r}\right)^2 + \left(\frac{1}{r}\right)^2\right\} = \frac{D'}{rD}.$$

$$R_{\theta\theta} = \left\{-\frac{1}{D} + \frac{rD'}{D^2} + \frac{\cos^2\theta}{\sin^2\theta}\right\} + \left(\frac{1}{2}\frac{D'}{D} + \frac{2}{r}\right)\left(-\frac{r}{D}\right) - \left\{2\left(-\frac{r}{D}\right)\left(\frac{1}{r}\right) + \left(\frac{\cos\theta}{\sin\theta}\right)^2\right\} = -\frac{1}{D} + \frac{rD'}{2D^2} + 1\,.$$

$$R_{\varphi\varphi} = \sin^2\theta\left[1 + \frac{rD'}{D^2} - \frac{1}{D}\right] = \sin^2\theta\, R_{\theta\theta}.$$

So, the only nonzero $R_{\mu v}{}'s$ are:

$$\left\{\begin{aligned} R_{rr} &= \frac{D'}{rD} \\ R_{\theta\theta} &= 1 - \frac{1}{D} + \frac{rD'}{D^2} \\ R_{\varphi\varphi} &= \sin^2 R_{\theta\theta} \end{aligned}\right\}.$$

(Eqn. 9-16)

Let's calculate $\tilde{\Gamma}^{i}_{jk}$ and $\tilde{R}_{ij}$ for this metric and arrive at $\tilde{R}_{ij}$ by Cartan form analysis.

Solution: By homogeneity we must have $\tilde{R} = const.$ (scalar curvature). So, let $\tilde{R} = K$, then:

$$\tilde{R} = \frac{2D'}{rD^2} + \frac{2}{r^2} - \frac{2}{r^2 D} = K$$

put $E = D^{-1}$, then,

$$-K + \frac{2}{r^2} = \frac{2}{r}E' + \frac{2}{r^2}E,$$

now make $E = C + br^2 \Rightarrow E = 1 - \frac{1}{6}Kr^2$. We have a specific solution, need the general solution (homogenous equation): $0 = \frac{2}{r}E_0^1 + \frac{2}{r^2}E_0 \Rightarrow E_0 = Cr^{-1}$. The most general solution is then

$$E = 1 - \frac{1}{6}Kr^2 + Cr^{-1} \quad \rightarrow \quad D = E^{-1}$$

Putting back into $\tilde{R}_{rr}$, $\tilde{R}_{\theta\theta}$ we have:

$$\tilde{R}_{rr} = \left(\frac{1}{3}K + Cr^{-3}\right)D(r)$$

$$\tilde{R}_{\theta\theta} = \frac{1}{3}Kr^2 - \frac{1}{2}Cr^{-1}$$

This means that if $C \neq 0$ then $\tilde{R}_{rr}$, $\tilde{R}_{\theta\theta}$ are singular at $r = 0$, e.g., $\tilde{R}_{ij}\,\tilde{R}ij \rightarrow \infty$ at $r = 0$. (Measure behaves funny at $r = 0$ at start anyway, so won't address those such methods.). So we want $c = 0$.

$$D(r) = \left(\frac{1}{1 - \frac{1}{6}Kr^2}\right).$$

At this point it is conventional to write $K = 6k$ thus have $\tilde{R} = K = 6k$ with:

$$ds^2 = -dt^2 + a^2(\tau)\,\tilde{g}_{ij}(\vec{x})dx^i dx^j = dt^2 + a^2(\tau)d\ell^2$$

where

$$d\ell^2 = D(r)dr^2 + r^2(\sin^2\theta d\varphi^2 + d\theta^2), \quad and \quad D(r) = \frac{1}{1 - kr^2}.$$

Let's now calculate $R_{\mu v}$ and R for:

$$ds^2 = dt^2 + a^2(\tau)\left\{\frac{dr^2}{1 - kr^2} + r^2 d\theta^2 + r^2\sin^2\theta d\varphi^2\right\}$$

Rescale $r's$: $r \rightarrow ar$, then $r' = a^{-1}r$ and $dr' = a^{-1}dr$ and:

$$ds^2 = -dt^2 + \left\{ \frac{dr^2}{1 - \left(\frac{k}{a^2(t)}\right) r^2} + r^2 d\theta^2 + r^2 \sin^2\theta d\varphi^2 \right\}$$

Notice that the spatial scalar curvature is now $\frac{k}{a^2(t)}$. Since $\tilde{R}_{ij} = 2K\tilde{g}_{ij}$ the $R_{ij}^{\prime s}$ are simply rescaled by ${}^3R_{ij} = \frac{2k}{a^2}\tilde{g}_{ij}$. We can rescale r and $a(t)$ so that $|k| = 1$: let $r' = |k|^{1/2} r$ and $\underline{a}(t) = |k|^{1/2} a(t)$:

$$ds^2 = -dt^2 + \underline{a}^2(t) \left\{ \frac{dr'^2}{1 - \frac{k}{|k|} r'^2} + r'^2 dG^2 + r'^2 \sin^2\theta d\varphi^2 \right\}$$

So with $\frac{k}{|k|} \to \mathcal{K}$ we have

$$ds^2 = -dt^2 + a^2(t) \left\{ \frac{dr^2}{1 - \mathcal{K} r^2} + r^2 \sin^2\theta d\varphi^2 + r^2 d\theta^2 \right\}$$

(Eqn. 9-17)

$$\text{with} = \begin{cases} 1 & {}^{(3)}R = \frac{6}{a^2} \\ 0 & \quad = 0 \\ -1 & \quad = -\frac{6}{a^2} \end{cases} .$$

(Eqn. 9-18)

9.5 Analysis of ^{4}R for Standard FRW form of metric using Forms (Cartan Analysis) and EOMs

$$ds^2 = -dt^2 + a^2(t) \left\{ \frac{dr^2}{1 - kr^2} + r^2 d\theta^2 + r^2 \sin^2\theta \, d\varphi^2 \right\}$$

(Eqn. 9-19)

Using forms:

$$ds^2 = -(\omega^t)^2 + (\omega^r)^2 + \left(\omega^\theta\right)^2 \tau \, (\omega^\varphi)^{2,}$$

(Eqn. 9-20)

thus:

$\omega^t = dt$

$\omega^r = \dfrac{a}{\sqrt{1 - kr^2}} dr$

$\omega^\theta = ard\theta$

$\omega^\varphi = ar \sin\theta \, d\varphi.$

Now, $d\omega^t = d^2 t = 0$ and $de_\mu = e_v \omega_\mu^v$, and consider:

$dP = e_\mu \omega^\mu.$
$d^2P = de_\mu \wedge \omega^\mu + e_\mu d\omega^\mu = e_\mu(\omega_v^\mu \wedge w^v + d\omega^\mu).$
Thus: $d\omega^\mu = -\omega_v^\mu \wedge \omega^v$ and we have: $d\omega^t = -\omega_v^t \wedge \omega^v$.

Also $dg_{\mu v} = \omega_{\mu v} + \omega_{v\mu} = 0 \Rightarrow \omega_{\hat{t}}^{\hat{t}} = -\omega_{\hat{t}\hat{t}} = 0$, so $d\omega^t = -\omega_k^t \wedge w^k = 0 \rightarrow$ is satified if $\omega_k^t \propto w^k$ or in more complicated ways, or just $\omega_k^t = 0$. So, we have:

$$d\omega^r = \frac{\dot{a}}{\sqrt{1-kr^2}} dt \wedge dr = \left(\frac{\dot{a}}{a}\right) \omega^t \wedge \omega^r = -\left(\frac{\dot{a}}{a}\right) \omega^r \wedge \omega^t$$

and when compared with:
$d\omega^r = -\omega_\mu^r \wedge \omega^\mu = -\omega_t^r \wedge \omega^t - \omega_\theta^r \wedge w^\theta - \omega_\varphi^r \wedge w^\varphi$

Guess $\omega_t^r = (\dot{a}/a)w^r$ and the others zero, this is consistent with $\omega_r^t \wedge \omega^r = 0 = \omega_{xt} \wedge \omega^x$ ... Next:

$$d\omega^\theta = \dot{a}rdt \wedge d\theta + adr \wedge d\theta = \dot{a}r\omega^t \wedge \frac{w^\theta}{ar} + a\left(\frac{\sqrt{1-kr^2}}{a}\right)\omega^r \wedge \frac{w^\theta}{ar}$$
$$= -\left(\frac{\dot{a}}{a}\right)\omega^\theta \wedge \omega^t - \frac{\sqrt{1-kr^2}}{ar} - \omega^\theta \wedge w^r = -\omega_t^\theta \wedge \omega^t - \omega_r^\theta \wedge \omega^r - \omega_\varphi^\theta \wedge w^\varphi$$

Guess $\omega_t^\theta = \left(\frac{\dot{a}}{a}\right)\omega^\theta$, $\omega_r^\theta = \frac{\sqrt{1-kr^2}}{ar}\omega^\theta$, and $\omega_\theta^\theta = 0$.
Next:
$$d\omega^\varphi = \dot{a}r \sin\theta \, dt \wedge d\varphi + a \sin\theta \, dr \wedge d\varphi + ar \cos\theta \, d\theta \wedge d\varphi$$
$$= \left(\frac{\dot{a}}{a}\right)\omega^t \wedge \omega^\varphi + a \sin\theta \frac{\sqrt{1-kr^2}}{a}\omega^r \wedge \frac{\omega^\varphi}{ar \sin\theta} + \omega^\theta \wedge \cos\theta \frac{\omega^\varphi}{ar \sin\theta}$$
$$= -\left(\frac{\dot{a}}{a}\right)\omega^\varphi \wedge \omega^t - \frac{\sqrt{1-kr^2}}{a}\omega^\varphi \wedge \omega^r - \frac{\cot\theta}{ar}\omega^\varphi \wedge \omega^\theta = \omega_t^\varphi \wedge \omega^t - \omega_r^\varphi \wedge \omega^r - \omega_\theta^\varphi \wedge \omega^\theta$$

Guess:
$$\omega_t^\varphi = \frac{\dot{a}}{a}\omega^\varphi, \; \omega_r^\varphi = \frac{\sqrt{1-kr^2}}{ar}\omega^\varphi, \text{ and } \omega_\theta^\varphi = \frac{\cot\theta}{ar}\omega^\varphi.$$

which are consistent with earlier results, so we have a solution:

$$\omega_t^k = \omega_k^t = \left(\frac{\dot{a}}{a}\right)\omega^k$$
$$\omega_r^\theta = -\omega_\theta^r = \frac{\sqrt{1-kr^2}}{ar}\omega^\theta = \sqrt{1-kr^2}d\theta$$
$$\omega_r^\varphi = -\omega_\varphi^r = \frac{\sqrt{1-kr^2}}{ar}\omega^\varphi = \sqrt{1-kr^2}\sin\theta \, d\varphi$$

$$\omega^{\varphi}_{\theta} = -\omega^{\theta}_{\varphi} = \frac{\cot\theta}{ar}\omega^{\varphi} = \cos\theta\, d\varphi$$

Specifically, for the k values 0, ± 1:

k = 0: $ds^2 = -dt^2 + a^2[dr^2 + r^2 d\theta^2 + r^2 \sin^2\theta\, d\varphi^2]$

$\omega^{k}_{t} = \left(\frac{\dot{a}}{a}\right)\omega^{k}$

$\omega^{\theta}_{r} = d\theta$

$\omega^{\varphi}_{r} = \sin\theta\, d\varphi$

$\omega^{\varphi}_{\theta} = \cos\theta\, d\varphi$

k = ±1: let $r = \frac{1}{\sqrt{k}}\sin x\,, a = \sqrt{k}a_{prev.}$, then the line element is:

$ds^2 = -dt^2 + a^2(t)[dx^2 + \sin^2 x\,(d\theta^2 + \sin^2\theta\, d\varphi^2)]$

and we get:

$\omega^{k}_{t} = \left(\frac{\dot{a}}{a}\right)\omega^{k}$

$\omega^{\theta}_{r} = \cos x\, d\theta$

$\omega^{\varphi}_{r} = \cos x \sin\theta\, d\varphi$

$\omega^{\varphi}_{\theta} = \cos\theta\, d\varphi$

Let's consider the curvature 2-form (same convention as MTW on pg. 354):

$$R^{\mu}_{\nu} = d\omega^{\mu}_{\nu} + \omega^{\mu}_{\alpha} \wedge \omega^{\alpha}_{\nu}.$$

Then we have:

$R^{t}_{r} = d\omega^{t}_{r} + \omega^{t}_{\theta} \wedge \omega^{\theta}_{r} + \omega^{t}_{\varphi} \wedge \omega^{\varphi}_{r} = d\left(\frac{\dot{a}}{a}\omega^{r}\right) = d\left(\frac{\dot{a}}{\sqrt{1-kr^2}}dr\right).$

$R^{t}_{r} = \frac{\ddot{a}}{\sqrt{1-kr^2}}dt \wedge dr = \left(\frac{\ddot{a}}{a}\right)\omega^{t} \wedge \omega^{r}$

$R^{r}_{\theta} = \underbrace{d\omega^{r}_{\theta}} + \omega^{r}_{t} \wedge \omega^{t}_{\theta} + \omega^{r}_{\varphi} \wedge \omega^{\varphi}_{\theta} = \frac{(k + \dot{a}^2)}{a^2}\omega^{r} \wedge \omega^{\theta}$

Since we are considering an isotopic universe all space directions in the orthonormal frame ω^{μ} are algebraically equivalent, so:

$R^{t}_{k} = (\ddot{a}/a)\omega^{t} \wedge \omega^{k}$

$R^{k}_{\ell} = a^{-2}(K + \dot{a}^2)\omega^{k} \wedge \omega^{\ell}$

We have the relation $R^{\mu\nu} = R^{\mu\nu}_{\ |\alpha\beta|}\omega^{\alpha} \wedge \omega^{\beta}$ (MTW pg. 358), so

$$\left.\begin{aligned} R^t_k &\Rightarrow R^t_{ktk} = \left(\ddot{a}/a\right) \\ R^k_\ell &\Rightarrow R^k_{\ell k \ell} = a^{-2}(K + \dot{a}^2) \end{aligned}\right\} \text{and other R's that can be related by}$$

symmetries of Riemann Tensor.
the rest are zero. So

$R^t_t = 3\left(\frac{\ddot{a}}{a}\right) \longrightarrow R_{tt} = -3\,\ddot{a}/a.$

$R^k_k = g^{kj} R^\mu_{j\mu k} = 2g^{kj} g_{jk} a^{-2}(K + \dot{a}^2) + g^{kj} g_{jk}\left(\frac{\ddot{a}}{a}\right) = g^k_k\left[a^{-2}(K + \dot{a}^2)2 + \left(\frac{\ddot{a}}{a}\right)\right].$

$R_{kk} = a^{-2}[2K + 2\dot{a}^2 + a\ddot{a}]g_{kk} = [2K + 2\dot{a}^2 + a\ddot{a}]\tilde{g}_{kk}$

Thus,

$R = R^\mu_\mu = 3\left(\frac{\ddot{a}}{a}\right) + 3\left[a^{-2}(K + \dot{a}^2)2 + \left(\frac{\ddot{a}}{a}\right)\right] = 6\left[\frac{\ddot{a}}{a} + a^{-2}(K + \dot{a}^2)\right].$

To summarize:

$$\begin{cases} ds^2 = -dt^2 + a^2(t)\left\{\dfrac{dr^2}{1 - kr^2} + r^2 d\theta^2 + r^2 \sin^2\theta\, d\varphi^2\right\} = -dt^2 + a^2(\\ \left.\begin{aligned} R_{ij} &= (a\ddot{a} + 2\dot{a}^2 + 2K)\tilde{g}_{ij} \\ R_{tt} &= -3\left(\frac{\ddot{a}}{a}\right) \end{aligned}\right\} \text{the rest are zero} \\ R = 6\left[\dfrac{\ddot{a}}{a} + \dfrac{\dot{a}^2}{a^2} + \dfrac{k}{a^2}\right] \end{cases}$$

(Eqn. 9-21)

9.6 Geodesics of FRW are x_i = constant world lines

Let's show x_i = constant (the fundamental world line) are geodesics of the FRW space time (as originally posed).

9.6.1 Geometric Solution

Consider a 2-D sphere imbedded in a 3-D space. Let $\vec{x}$ be a vector projected onto the y, x plane. The unit sphere is then given by: $\vec{x}\cdot\vec{x} + z^2 = 1$. The distance between $(\vec{x}, z)$ and $(\vec{x} + d\vec{x}, z + dz)$ is

$$d\ell^2 = d\vec{x}\cdot d\vec{x} + dz^2,$$

and using $\vec{x}\cdot\vec{x} + z^2 = 1$ constraint we get:

$$d\ell^2 = d\vec{x}\cdot d\vec{x} + \frac{(\vec{x}\cdot d\vec{x})^2}{1 - \vec{x}\cdot\vec{x}}$$

Now consider $\vec{x}$ to be (x_1, x_2, x_3) a point on a 3-D sphere in 4-D space. We get the same $d\ell^2$ with the broader meaning. Now write in spherical coordinates using:

$$x_1 = r\sin\theta\cos\varphi,\ \ x_2 = r\sin\theta\sin\varphi,\ \ x_3 = r\cos\theta$$

and

$$\vec{x}\cdot d\vec{x} = rdr, \quad d\vec{x}\cdot d\vec{x} = dr^2 + r^2\sin^2\theta d\varphi^2 + r^2 d\theta^2,$$
$$and \quad \vec{x}\cdot\vec{x} = r^2.$$

Thus,

$$d\ell^2 = dr^2\left(1+\frac{r^2}{1-r^2}\right) + r^2\sin^2\theta d\varphi^2 + r^2 d\theta^2.$$

This is a $k = 1$ FRW space. For $k = -1$ we use the constraint $\vec{x}\cdot\vec{x} - z^2 = 1$.

9.6.2 Tensor Solution

Are the $x_i = const$ lines geodesics?

$$ds^2 = -dt^2 + a^2(t)\left\{\frac{dr^2}{1-kr^2} + r^2 d\theta^2 + r^2\sin^2\theta\, d\varphi^2\right\}$$
$$= -dt^2 + a^2\tilde{g}_{ij}$$

$$\Gamma^t_{ij} = a\dot{a}\tilde{g}_{ij} \quad \left(\Gamma^t_{ij} = \frac{1}{2}g^{tt}(-g_{ij,t})\right)$$

$$\Gamma^i_{tj} = \frac{\dot{a}}{a}\delta^i_j \quad \left(\Gamma^i_{tj} = \frac{1}{2}g^{ik}[g_{kj,t}] = \frac{\dot{a}}{a}\delta^i_j\right)$$

$$\Gamma^i_{jk} = \tilde{\Gamma}^i_{jk}$$

Geodesic equation:

$$\frac{d^2x^\alpha}{d\lambda^2} + \Gamma^\alpha_{\mu\gamma}\frac{dx^\mu}{d\lambda}\frac{dx^\gamma}{d\lambda} = 0$$

If $x_i = const$ lines are geodesics then they satisfy the above equation, thus we have:

$\frac{d^2x^i}{d\lambda^2} + \Gamma^i_{\mu\gamma}\frac{dx^\mu}{d\lambda}\frac{dx^\gamma}{d\lambda} = 0 \rightarrow \frac{dx^\mu}{d\lambda}$ has only the $\frac{dx^t}{d\lambda}$ component nonzero.

Thus $\Gamma^i_{00}V^0V^0 = 0$, which is satisfied if $\Gamma^i_{00} = 0$. So, the $x_i = const$ lines are geodesics.

9.6.3 Cosmological Red Shift

Consider two FRW observers, one at $r = 0$ and one at $r = 1$. By isotropy the rays move only radially towards $r = 0$ from $r = 1$, so $d\theta = 0, d\varphi = 0$. For one of the light ways the interval between two events on the light way we have: $ds^2 = 0 = -dt^2 + a^2(t)\frac{dr^2}{1-kr^2}$. Since $dr < 0$ when $dt > 0$:

$$\frac{dt}{a(t)} = \frac{-dr}{\sqrt{1-kr^2}}.$$

For the first light ray:

$$\int_{t_1}^{t_0} \frac{dt}{a(t)} = -\int_{r_1}^{0} \frac{-dr}{\sqrt{1-kr^2}} \equiv f(r_1) = \begin{cases} \sin^{-1} r_1 & k=1 \\ r_1 & k=0 \\ \sinh^{-1} r_1 & k=-1 \end{cases}$$

Similarly for a second ray:

$$\int_{t_1+\delta t_1}^{t_0+\delta t_0} \frac{dt}{a(t)} = -\int_{r_1}^{0} \cdots = f(r_1)$$

We want to be able to claim that

$$\int_{t_1+\delta t_1}^{t_0+\delta t_0} \frac{dt}{a(t)} - \int_{t_1}^{t_0} \frac{dt}{a(t)} = 0 \Rightarrow \text{assuming } \delta t_0 \text{ small}$$

in order to have:

$$0 = \frac{\delta t_0}{a(t_0)} - \frac{\delta t_1}{a(t_1)}.$$

with $c = 1$ understood (used all along). It can be shown that

$$\int_{t_1+\delta t_1}^{t_0+\delta t_0} \frac{dt}{a(t)} - \int_{t_1}^{t_0} \frac{dt}{a(t)} = 0$$

if for δt small we get:

$$\int_{t_1+\delta t_1}^{t_1} \frac{dt}{a(t)} - \int_{t_0}^{t_0+\delta t_0} \frac{dt}{a(t)} = 0$$

So,

$$\frac{(\delta t_1)}{a(t_1)} + \frac{\delta t_0}{a(t_0)} = 0 \Rightarrow \delta t_0 = \delta t_1 \left(\frac{a(t_0)}{a(t_1)}\right).$$

We now have $\frac{\delta t_0}{a(t_0)} = \frac{\delta t_1}{a(t_1)}$, and since $c = \lambda v$:

$$\delta t_1 = \frac{\delta \lambda}{c} = \lambda_1,$$

thus $\delta t_1 = \lambda_1$ in lab frequency emitted by atom at r_1. Similarly:

$$\delta t_0 = \lambda_0 = \lambda_1 \left(\frac{a(t_0)}{a(t_1)}\right)$$

So if the universe is expanding we get a red shift:

$$\boxed{Z = \frac{\lambda_0 - \lambda_1}{\lambda_1} = \frac{a(t_0)}{a(t_1)} - 1}$$

(Eqn. 9-22)

If r_1 is not too large the rotation is simply that of the usual Doppler shift:

$$v_0 = v_1 \frac{a(t_0)}{a(t_1)}$$

Suppose $t_0 - t_1$ is not large:

$$v_0 = v_1 \left[1 - \frac{\dot{a}(t_0)}{a(t_1)}(t_0 - t_1) + \cdots\right]$$

if we assume small velocities and omit v^2 terms, this is the familiar Doppler formula:

$$v_0 = v_1 (1 - v),$$

where $v \equiv \frac{\dot{a}}{a}[distance]$. This is in agreement with $v_0 \cong v_1 \sqrt{\frac{(1-v)}{(1+v)}}$ relationship formula to first order.

This was the form originally written down by Hubble from observational data with:

$$H = \frac{\dot{a}(t_0)}{a(t_0)}$$

(Eqn. 9-23)

now known as Hubble constant.

To recap:

$$Z = \frac{\lambda_0 - \lambda_1}{\lambda_1} = \frac{a(t_0)}{a(t_1)} - 1,$$

where $t_1 =$ time of origin of light and $t_0 =$ time received, and we have

$$v_0 = v_1 \left[\frac{a(t_0)}{a(t_1)}\right] = v_1 \left[1 - \underbrace{\frac{\dot{a}(t_0)}{a(t_0)}}(t_0 - t_1) + (\text{next term}) + \cdots\right]$$

↑ ↑ deceleration parameter

Hubble parameter

Current experimental observations give:

$$50(km/sec)Mpc^{-1} < H_0 < 100(km/sec^{-1})Mpc^{-1}$$

where $1Mpc = 3.26 \times 10^6$ light year.

So, $H_0 = 75 \pm 25\ km\ sec^{-1} Mpc^{-1}$

$$= 2.4 \times 10^{-18} sec^{-1} \pm 0.81 \times 10^{1\delta} sec^{-1}$$
$$= 8.0 \times 10^{29} cm^{-1} \pm 2.7 \times 10^{-29} cm^{-1} \text{ with } c = 1$$

9.7 Einstein Equations for FRW and for FRWL with a perfect fluid

For the dynamics of the FRW spacetime:

$$ds^2 = -dt^2 + a^2(t)\left\{\frac{dr^2}{1-kr^2} + r^2 d\theta^2 + r^2 \sin^2\theta d\varphi^2\right\}$$

we have obtained:

$$\Gamma^t_{ij} = a\dot{a}\tilde{g}_{ij}$$

$$\Gamma^i_{tj} = \Gamma^i_{jt} = \frac{\dot{a}}{a}\,\delta^i_j \quad and \quad \Gamma^i_{jk} = \tilde{\Gamma}^i_{jk}$$

Thus

$$\left.\begin{matrix} R_{ij} = \tilde{R}_{ij} + (2\dot{a}^2 + a\ddot{a})\tilde{g}_{ij} \\ R_{ij} = 2k\tilde{g}_{ij} \end{matrix}\right\} \quad \rightarrow \quad R_{ij} = (a\ddot{a} + 2\dot{a}^2 + 2k)\tilde{g}_{ij}\,.$$

Since $R_{tt} = -3\ddot{a}/a$ and rest zero:

$$R = 6\left(\frac{\ddot{a}}{a} + \frac{\dot{a}^2}{a^2} + \frac{k}{a^2}\right)$$

Thus,

$$R_{\mu\nu} - \frac{1}{2}g_{\mu\nu}R + \Lambda g_{\mu\nu} = 8\pi G\, T_{\mu\nu}$$

and from the 00 component:

$$\left(\frac{\dot{a}}{a}\right)^2 + \frac{k}{a^2} - \frac{1}{3}\Lambda = \frac{8\pi G}{3}\,T_{00}$$

and from the i,j *components*:

$$[-2a\ddot{a} - \dot{a}^2 - k + \Lambda a^2]\tilde{g}_{ij} = 8\pi G T_{ij}.$$

The last two equations comprise what are known as the Einstein equations for FRW.

If we describe the matter and radiation by a perfect fluid:

$$T^{\mu\nu} = p g^{\mu u} + (p+\rho)U^\mu U^\nu,$$

where U^μ is the 4-velocity of unit fluid element, then consistency with FRW demands that p and ρ be homogeneous, thus they only depend on time. Furthermore, since there is no preferred direction of time (isotropy), we know that U^μ has no spatial components. So, with proper time of fluid element that of the fundamental world lines, the t of the fluid element (proper) and that of FRW analysis $d\tau = dt$, so $v^0 = 1, v^1 = 0$ in these coordinates (FRW coordinates). Thus, with

$$ds^2 = -dt^2 + a^2(t)\left\{\frac{dr^2}{1-kr^2} + r^2 d\theta^2 + r^2 \sin^2\theta d\varphi^2\right\}$$

and $T^{00} = T_{00} = \rho$; $T_{ij} = pg_{ij}$; and $T_{ij} = pa^2\tilde{g}_{ij}$, we have:

$$\left(\frac{\dot{a}}{a}\right)^2 + \frac{k}{a^2} - \frac{1}{3}\Lambda = \frac{8\pi G}{3}\rho \text{ and } -\frac{2\ddot{a}}{a} - \left(\frac{\dot{a}}{a}\right)^2 - \frac{k}{a^2} + \Lambda = 8\pi Gp.$$

(Eqn. 9-24)

The above describes the Einstein equations for all FRWL universes with a perfect fluid.

Recall that local conservation follows from Bianchi as well as local conservation of energy law: $T^{\mu v}{}_{;v} = 0$. Let's examine this specifically in the context of matter consistng of a perfect fluid. Let's show that $T^{0v}{}_{;v} = 0$ gives

a) $\frac{d}{dt}(\rho a^3) + p\frac{d}{dt}(a^3) = 0$
b) and $T^{iv}{}_{;v} = 0$ is an identity

$$\begin{aligned}\nabla_v T^{0v} &= \nabla_v\{pg^{0v} + (\rho+p)U^0U^v\}\\ &= \nabla_v pg^{0v} + \nabla_v(\rho+p)U^0U^v\\ &+ (\rho+p)\{(\nabla_v U^0)U^v + (\nabla_v U^v)U\}\end{aligned}$$

where $(\nabla_v U^0) = 0$ because fluid elements move on fundamental world lines which are geodesics. Since $\nabla_v U^v = \frac{1}{\sqrt{-g}}\partial_v\left(\sqrt{-g}\, U^v\right)$, in a coordinate system we have:

$$\nabla_v U^v = \frac{1}{(a^3)}\frac{d}{dt}\left(a^3(t)\right) = \frac{3\dot{a}}{a},$$

Thus

$0 = \nabla_v T^{0v} = \frac{dp}{dt} + \frac{d}{dt}(\rho+p) + (\rho+p)\left(3\frac{\dot{a}}{a}\right) \Rightarrow \frac{d}{dt}(pa^3) + p\frac{d}{dt}(a^3) = 0$ as in (a).

We can arrive at (a) from thermodynamic considerations if we assume an equation of state of the form
$p = p(\rho)$.

If $T^{0v}_{;v} = 0 = \nabla_v T^{0v} = \nabla_v\{pg^{0v} + (\rho+p)u^0u^v\}$

$$0 = \nabla_v pg^{0v} + \nabla_v(\rho+p)u^0u^v + (\rho+p)\{\nabla_v u^0 u^v + (\nabla_v u^v)u^0\}$$

$$0 = -\frac{dp}{dt} + \frac{d}{dt}(\rho+p) + (\rho+p)\left(3\frac{\dot{a}}{a}\right)$$

$$0 = \frac{d\rho}{dt} + (\rho + p)3\frac{\dot{a}}{a}$$
$$0 = \frac{d}{dt}(\rho a^3) + p\frac{d}{dt}(a^3)$$

And $T^{iv}_{;v} = 0$ is supposed to be an identity:

$$T^{iv}_{;v} = \nabla_v\{pg^{iv} + (\rho + p)u^i u^v\}$$
$$= \nabla_v p g^{iv} + \nabla_v(\rho + p)u^i u^v$$
$$+ (\rho + p)\{\nabla_v u^i u^v + (\nabla_v u^v)u^i\}$$
$$T^{iv}_{;v} = \nabla_v p g^{iv} = \nabla_i p g^{ii} = 0$$

9.8 Thermodynamic considerations

Just as with fluid dynamics in Book 2 [13], we here find thermodynamics considerations to be unavoidable at the most fundamental level of the model. This is because the Einstein equations and the continuity equations are not independent, with state variables related by a thermodynamic equation of state. As previously, we use a minimal amount of thermodynamics to proceed (for now), that being the 1st Law of thermodynamics (Conservation of Energy in the thermodynamic context). Consider a small volume element: $\Delta r \Delta\theta \Delta\varphi$ (coordinate volume). The actual physical volume is:

$$V = \sqrt{g_{rr} g_{\theta\theta} g_{\varphi\varphi}}\ \Delta r \Delta\theta \Delta\varphi.$$

In the FRW spacetime this becomes:

$$V = \frac{a^3(t)}{\sqrt{1 - kr^2}} r^2 \sin\theta \Delta r \Delta\theta \Delta\varphi$$

By isotropy the net heat flow on any surface must be zero. So, we can use the 1st law of thermodynamics:

$$dU = dQ - dW$$

where dQ is the heat influx to box, and dW is the work done by fluid box. With $W = PdV$ and $U = \rho V$ and $dQ = 0$ (no heat flow in the co-moving volume):

$$0 = \frac{d}{dt}(\rho V) + p\frac{dV}{dt}$$

Thus

$$0 = \frac{d}{dt}(\rho a^3) + p\frac{d}{dt}(a^3).$$

Let's now consider the "Newtonian Picture", and recall that for fluid in isotropic FRW universe the equations of motion gave:

$$\left.\begin{array}{ll}(1)\ (^{oo}\text{comp}) & \left(\frac{\dot{a}}{a}\right)^2+\frac{k}{a^2}-\frac{1}{3}\Lambda=\frac{8\pi G}{3}\rho \\ (2)\ (^{oj}\text{comp}) & -2\frac{\ddot{a}}{a}-\left(\frac{\dot{a}}{a}\right)^2-\frac{k}{a^2}+\Lambda=8\pi Gp \\ (3)\ (\text{conservation})\ T^{\mu\nu}_{;\nu} & \rightarrow\frac{d}{dt}(\rho a^3)+p\frac{d}{dt}(a^3)=0\end{array}\right\}\text{not all indep.}$$

(Eqn. 9-25)

9.9 Spherical Dust Universe

From thermodynamics in an isotropic, homogeneous universe, for co-moving volume $\Delta Q=0\Rightarrow\Delta U=-\Delta W$, thus:

$$\Delta(\rho v)=-p\Delta V$$

which also yield 3rd equation as shown previously. So, to get the Einstein relations in a simple fashion we need only get (1) or (2), this can be done by considering a spherical dust medium and at distance d_0 from an $r=0$ geodesic at time t_0:

$$d(t)=\frac{a(t)}{a(t_0)}d_0,$$

$$and \qquad P.E.=V(t)=\frac{-Gm\left(\frac{4}{3}\pi d^3(t)\rho\right)}{d(t)}$$

$$=-\frac{4}{3}\pi Gm\rho(t)\frac{a^2(t)}{a^2(t_0)}d_0^2$$

The kinetic energy is:

$$K.E.=\frac{1}{2}mv^2=\frac{1}{2}m\dot{d}^2=\frac{1}{2}m\frac{\dot{a}^2(t)}{a(t_0)^2}d_0^2.$$

Isotropy eliminates the pressure calculations, so we simply have:

$$E=T+V=\frac{1}{2}m\frac{d_0^2}{a(t_0)^2}\left[\dot{a}(t)^2-\frac{8}{3}\pi G\rho(t)a^2(t)\right].$$

Let $E=-\frac{1}{2}m\frac{d_0^2}{a(t_0)^2}k$, then:

$$-k=\dot{a}(t)^2-\frac{8}{3}\pi G\rho(t)a^2(t),$$

$k>0\rightarrow E<0$ bound state
$k<0\rightarrow E>0$ unbound state
$k=0\rightarrow E=0$ unbound

So, we have $\frac{\dot{a}^2}{a^2}+\frac{k}{a^2}=\frac{8\pi G}{3}\rho\rightarrow$ same with $\Lambda=0$ (if k isn't ±1 or 0 just rescale to match).

9.10 Limit on Λ

Let's now proceed similarly to get a limit on Λ from known validity of universe square law (Newtonian law). Start by postulating a constant "vacuum" energy density: $\rho\Lambda = +\frac{\Lambda}{8\pi G}$. Introduce in previous arguments to have: $\rho = \rho_{matter} + \rho_\Lambda$ and $p = p_{matter} + P_\Lambda$. We already know p_{matter} satisfies $\Delta U = -\Delta W$ all by itself, so this must carry through, thus the relation between P_Λ and ρ_Λ is as with regular matter:

$$\frac{d}{dt}(\rho_\Lambda a^3) + P_\Lambda \frac{d}{dt}(a^3) = 0$$

but ρ_Λ is constant, so $P_\Lambda = -\rho_\Lambda$. The energy momentum tensor due to the cosmological vacuum energy is then:

$$T_\Lambda^{\mu\nu} = \rho_\Lambda g_{\mu\nu} + (\rho_\Lambda + P_\Lambda)U_\mu V_\nu = \rho_\Lambda g_{\mu\nu}\,.$$

Thus, the cosmological energy density shares the symmetries of the metric. We then have:

$$V(t) = \frac{-GmV(\rho_m + \rho_\Lambda)}{d(t)} = -\frac{GmV(\rho_m)\left(1 + \frac{\rho_v}{\rho_m}\right)}{d(t)} = \frac{-GmV(\rho_m)}{d(t)\left(1 - \frac{\rho_v}{\rho_m}\right)}.$$

Using $D(t) = d(t)\left(1 - \frac{\rho_v}{\rho_m}\right) \cong d(t)e^{-\rho_v/\rho_m\, d(t)}$.

The systems of lowest avg. density ρ_{matter} which obey Newton's gravity consist of small dusters of galaxies. For then $\rho_{matter} = 10^{-29}\ gm/cm^3$. In units where $\hbar = c = 1 \Rightarrow \rho_{matter} = 10^8\ cm^{-4}$. Thus, we must have

$|\rho_\Lambda| \ll 10^{-29}\ gm/cm^3$ or $10^8\ cm^{-4}$ or $|\Lambda| < 8\pi G\ |\rho_\Lambda| = 1.7 \times 10^{-35}\ sec^{-2}\ (cgs)$, or:

$$|\Lambda| \le 1.9 \times 10^{-56} cm^{-2}.$$

Friedman: ~ 1922, considers a dust filled universe with $\Lambda = 0$ (subclass of R-W). Einstein and de Sitter considered radiation filled universe with $\Lambda = 0$. Note that nonzero Λ at very early times are possible with inflationary models.

9.11 Friedman Models ($\Lambda = 0$)

So, we have:

$$\left.\begin{aligned} \left(\frac{\dot{a}}{a}\right)^2 + \frac{k}{a^2} = \frac{8\pi G}{3}\rho \\ -2\,\ddot{a}/a - \left(\dot{a}/a\right)^2 k/_{a^2} = 8\pi G p \\ \frac{d}{dt}(\rho a^3) + p\frac{d}{dt}(a^3) = 0 \end{aligned}\right\}$$

$$with\ \rho, p, a\text{: 3 unknowns, but have only 2 indep. equation}$$

(Eqn. 9-26)

The 3rd independent equation is the equation of state: $p = p(\rho)$.

9.11.1 Radiation filled universe

(radiation or hot particles obeying relativist equation of state).

Then $p = \frac{1}{3}\rho$ $(c = 1)$ and:

$$\frac{d}{dt}(\rho a^3) + \frac{1}{3}\rho\frac{d}{dt}(a^3) = 0$$

$$\frac{d\rho}{dt}a^3 + \frac{4}{3}\rho\frac{d}{dt}a^3 = 0$$

So,

$$\frac{d}{dt}\ln\rho + \frac{4}{3}\frac{d}{dt}\ln a^3 = 0 \rightarrow \ln\rho = - - \frac{4}{3}\ln a^3 + C$$

With solution

$$\rho = Aa^{-4}$$

9.11.1.1 Radiation dominated: k=0

Consider spatially flat universe, $k = 0$:

$$\left(\frac{\dot{a}}{a}\right)^2 = \frac{8\pi G}{3}\rho_0\frac{a_0^4}{a^4} \rightarrow \left(\frac{1}{2}d(a^2)\right)^2 = \frac{8\pi G}{3}\rho_0 a_0^4 \rightarrow \frac{1}{2}d(a^2)$$

$$= \sqrt{\frac{8\pi G}{3}\rho_0\, a_0^2}$$

Thus

$$a^2 = 2\sqrt{\frac{8\pi G}{3}\rho_0\, a_0^2} + \text{const}$$

choose $a^2|_{t=0} = 0 \rightarrow \text{const} = 0$ and get:

$$a = \left(\frac{8\pi G}{3}\rho_0\right)^{1/4}\sqrt{2}\, a_0 t^{1/2}$$

When $a = 0$ we have a true singularity in R anyway so we've initialized time to that instant.

Could we be living in a radiation dominated universe?

$$\rho_{\mu\, wave} = \sigma T^4 \qquad T = 2.7^o k \ (CMBR)$$
$$= 4 \times 10^{-34}\, gm/cm^3$$

Visible matter (including haloes, etc.):

$$\rho_{galaxies} \geq 2 \times 10^{-31} \ gm/cm^3$$

So universe is in matter dominated at this time. Also note:

$$\rho \text{ intergaladic gas}, \rho \text{ cosmic rays } \leq 10^{-33} \ cm/cm^3$$
$$\rho \text{ magnitic field } \leq 10^{-35} \ gm/cm^3 \ (B \leq 10^{-6} \text{ Gauss})$$

Let' consider the equation of state $p = \alpha\rho$ and show that

(a) $\rho = \rho_0 \frac{a_0^{3(\alpha+1)}}{a(t)^{3(\alpha+1)}}$

(b) $a(t) = [6\pi(\alpha+1)^2 G\rho_0]^{\frac{1}{3(\alpha+1)}} a_0 +^{\frac{2}{3(\alpha+1)}}$ for $\alpha \neq -1$
(cosmological constant case)

Solution

(a) With the equation of state $p = \alpha\rho$ and using the Einstein equation's solve for ρ and $a(t)$.

Solution: this is solving the Friedman model ($\Lambda = 0$) with the indicated equation of state:

$\frac{d}{dt}(\rho a^3) + p\frac{d}{dt}(a^3) = 0 \rightarrow \frac{d}{dt}(\rho a^3) + \alpha\rho\frac{d}{dt}(a^3) = 0 \rightarrow a^3\frac{d\rho}{dt} +$
$(\alpha+1)\rho\frac{d}{dt}a^3 = 0.$

So,

$$\frac{d(\ln\rho)}{dt} + (\alpha+1)\frac{d(a^3)}{dt} = 0 \rightarrow \ln\rho = -(\alpha+1)\ln a^3 + const$$

from which we get:

$$const. = \rho_0 a_0^{3(1+\alpha)}$$

Thus, we have $\rho = \rho_0\left(\frac{a_0}{a(t)}\right)^{3(1+\alpha)}$.

(b) Using $\left(\frac{\dot{a}}{a}\right)^2 = \frac{8\pi G}{3}\rho \Rightarrow \frac{\dot{a}^2}{a^2} = \frac{8\pi G}{3}\rho_0 a_0^{3(1+\alpha)} a_0^{-3(1+\alpha)}$

Thus

$$a^{\frac{3}{2}(1+\alpha)} = \frac{3}{2}(1+\alpha)\sqrt{\frac{8\pi G}{3}\rho_0 a_0^{3(1+\alpha)}} + const$$

let $a = 0$ at $t = 0$, then $const = 0$ in the above:

$$a(t) = (6\pi(1+\alpha)G\rho_0)^{1/3(1+\alpha)} a_0 t^{2/3(1+\alpha)}$$

For $\alpha = -1$; $\frac{\dot{a}}{a} = \frac{8\pi G}{3}\rho_0 \quad (\rho = \rho_0)$. So, $H_0 = const \Rightarrow H_0 = \frac{8\pi G}{3}\rho_0$ and:

$$a(t) = \exp\left[\left(\frac{8\pi G\rho_0}{3}\right)^{1/2} t\right] \qquad a(0) = 1$$

For cosmological constant we had $\rho_\Lambda = \frac{\Lambda}{8\pi G}$ if we consider $\rho_0 = p_\Lambda$:

$$a(t) = \exp\left[\left(\frac{1}{3}\Lambda\right)^{1/2} t\right] \to \text{DeSitter}$$

universe

For radiation dominated: $p = \frac{1}{3}\rho$, then

$$p = \frac{1}{3}\rho \Rightarrow p(t) = \rho_0 \frac{a(t_0)^4}{a(t)^4}$$

And $a(t) = \left(\frac{32\pi G}{3}\rho_0\right)^{1/4} a_0 t^{1/2} \to t_0 = \left(\frac{32\pi G}{3}\rho_0\right)^{-1/2}$. At the present time the avg. matter density of the universe appears to dominate the density of radiation (photons, maintenance, gravities, and very light particles). In the early universe it is likely that the universe was radiation dominated (photons$\sim a^{-3} \times a^{-1}$ blueshift while matter$\sim a^{-3}$). Axions, scalar particles used for CP violation, with small mass (if exist), could swing the balance either way. Suppose for a moment that there were enough "hidden" relativistic matter to dominate even today. The Hubble constant at present is $H_0 = \dot{a}(t_0)/a(t_0)$, and Einstein's equation. gives

$$H_0^2 = \frac{8\pi G}{3}\rho_0 \Rightarrow \rho_0 = \frac{3H_0^2}{8\pi G} \text{ and } t_0 = \frac{1}{2H_0}$$

with H_0 approx. measured: $50 \leq H_0 \leq 100\frac{tm}{\sec Mpc}$, the largest possible age of the universe with this model is then: $H_0 \approx 55 \to \; t_0 \leq 9 \times 10^9 \, yrs$, which is nearly violated, and also get $\rho_0 \gtrsim 5.6 \times 10^{-30} \, gm/cm^3$, which is violated (μwave disobeys).

9.11.1.2 Radiation dominated: Positive spatial curvature (k=1)

$$\left.\begin{matrix} \rho = \rho_0 \dfrac{a_0^4}{a_0^4} \quad \text{for radiation} \qquad p = \dfrac{1}{3}\rho \\ \rho = \rho_0 \dfrac{a_0^3}{a_0^3} \quad \text{for dust} \qquad p = 0 \end{matrix}\right\} p = \alpha\rho$$

Now take $k = 1, \Lambda = 0$:

$$\left(\frac{\dot{a}}{a}\right)^2 + \frac{1}{a^2} = \frac{8\pi G}{3}\rho$$

Introduce a new coord. time η such that $dt^2 = a^2(\eta)d\eta^2$, then:
$ds^2 = -dt^2 + a^2(t)d\ell^2,\ d\ell^2 = \frac{dr^2}{1-r^2} + r^2 d\theta^2 + r^2\sin^2\theta d\varphi^2 = a^2(\eta)(-d\eta^2 + d\ell^2)$. Then

$$\left(\frac{\dot{a}}{a}\right)^2 + \frac{K}{a^2} = \frac{8\pi G}{3}\rho, \qquad K = 0, \pm 1$$

and

$$\frac{d}{dt}(\rho a^3) + p\frac{d}{dt}(a^3) = 0, \quad p = p(\rho)$$

Define η time by $dt^2 = a^2(\eta)d\eta^2$ (conformal time)

$$\dot{a} = \frac{1}{a}\frac{da}{d\eta} \Rightarrow \frac{a'^2}{a^4} + \frac{1}{a^2} = \frac{8\pi}{3}G\rho$$

$$\frac{a'^2}{a^4} + \frac{1}{a^2} = \frac{8\pi G}{3}\rho_0\frac{a_0^4}{a^4} \quad \rightarrow \quad a'^2 + a^2 = \frac{8\pi G\rho_0 a_0^4}{3}$$

Try $a = A\sin\eta$, works for

$$a(\eta) = \sqrt{\frac{8\pi G\rho_0 a_0^4}{3}}\sin\eta \qquad \left(at \begin{matrix}\eta = 0, a = 0\\ \eta = \pi, a = 0\end{matrix}\right) \quad 0 < \eta < \pi$$

Thus

$$t = \int_0^\eta a(\eta')d\eta' = A\int_0^\eta \sin\eta\, d\eta = A(1 - \cos\eta) \qquad 0 < t < 2A$$

and

$$a(t) = A\sin\left\{\cos^{-1}\left(1 - \frac{t}{A}\right)\right\} =$$
$$A\sqrt{1 - \left(1 - \frac{t}{A}\right)^2} = A\sqrt{-\left(\frac{t}{A}\right)^2 + 2\left(\frac{t}{A}\right)}$$

So, to show

$$a(t) = \left(\frac{32\pi G\rho_0}{3}\right)^{1/4} a_0 t^{1/2} \sqrt{\left(\frac{32\pi G\rho_0}{3}\right)^{-1/2} t}$$

from

$$a(t) = A^{1/2} t^{1/2} \sqrt{2} \sqrt{1 - \frac{1}{2}\frac{t}{A}}$$

this becomes

$$a(t) =$$
$$\left(\frac{32\pi G\rho_0}{3}\right)^{1/4} a_0 t^{1/2} \sqrt{1 - \left(\frac{32\pi G\rho_0 a_0^4}{3}\right)^{-1/2} t}.$$

(Eqn. 9-27)

The maximum is $\eta = \frac{\pi}{2}$, when $a_{max} = A$. When $t \ll A^{1/2}$, $a(t) \to K = 0$ a radiation dominated solution.

9.11.1.3 Radiation dominated ($\mathbf{k = -1}$)

Have

$a = A\sinh\eta$; $a(t) = A\sinh\left\{\cosh^{-1}\left(1 + \frac{t}{A}\right)\right\}$; $t = A(\cosh\eta - 1)$

Show:

$$a(t) = \left(\frac{32\pi G\rho_0}{3}\right)^{1/4} a_0 t^{1/2} \sqrt{1 + \left(\frac{32\pi G\rho_0 a_0^4}{3}\right)^{-1/2} t}$$

(Eqn. 9-28)

9.11.2 Matter dominated

Start with

$$\frac{d}{dt}(\rho a^3) + p\frac{d}{dt}(a^3) = 0$$

Have $\rho(t) = \rho_0 \frac{a_0^3}{a(t)^3}$ since the equation of state has $p = 0$. Now:

$$\left(\frac{\dot{a}}{a}\right)^2 = \frac{8\pi G}{3}\rho_0 \frac{a_0^3}{a(t)^3}$$

So,

$$\dot{a} = \sqrt{\frac{8\pi G}{3}\rho_0 a_0^3}\ \frac{1}{a(t)^{1/2}} \quad \rightarrow \quad a(t) = (6\pi G\rho_0)^{1/3}\, a_0 t^{2/3}$$

where $t_0 = (6\pi G\rho_0)^{-1/2}$. Using the $\left(\frac{\dot{a}}{a}\right) = \frac{8\pi G}{3}\rho(t)$ relation as before:

$$\rho_0 = \frac{3H_0^2}{8\pi G} = 5.6 \times 10^{-30}\ gm/cm^3 \ \text{(if } H_0 \approx 55\text{)}.$$

Note that the closure density is roughly the same as before, but t_0 is different:

$$t_0 = \frac{2}{3H_0} = 12 \times 10^9\ yrs.$$

9.11.2.1 Matter dominated (k=0)

Radiation dominated $p = \frac{1}{3}\rho$ $\qquad \rho(t) = \rho_0 \left(\frac{a_0}{a(t)}\right)^4$

Matter dominated $p = 0$ $\qquad \rho(t) = \rho_0 \left(\frac{a_0}{a(t)}\right)^3$

General $p = \alpha p, where\ \alpha \neq -1{:}\ \rho(t) = \rho_0 \left(\frac{a_0}{a(t)}\right)^{3(\alpha+1)}$ $;\ a(t) = [6\pi(\alpha +$

9.11.2.2 Matter dominated (k=1)

Have

$$\frac{a'^2}{a^4} + \frac{1}{a^2} = \frac{8\pi G}{3}\rho_0 \frac{a_0^3}{a^3}$$

and

$$a'^2 + a^2 = \left(\frac{8\pi G}{3}\frac{\rho_0 a_0^3}{B}\right) a$$

Try $a = \alpha + \beta\cos\eta$, which works to be:

$$a = \alpha(1 - \cos\eta) = \left(\frac{4\pi G}{3}\rho_0 a_0^3\right)(1 - \cos\eta) \quad for \quad 0 < \eta < 2\pi.$$

Since

$$t = \int_0^\eta a(\eta')d\eta' = \alpha\int_0^\eta (1 - \cos\eta')\, d\eta' = \alpha(\eta - \sin\eta)$$

We have:

$$t = \left(\frac{4\pi G}{3}\rho_0 a_0^3\right)\ (\eta - \sin\eta)$$

where

$$0 < t < \left(\frac{4\pi G}{3}\rho_0 a_0^3\right)\ 2\pi$$

9.11.2.3 Matter Dominated (k = −1)

$$a(\eta) = \left(\frac{4\pi G}{3}\rho_0 a_0^3\right)\ (\cosh\eta - 1)$$

(Eqn. 9-29)

and

$$t = \int_0^\eta a(\eta')d\eta' = \left(\frac{4\pi G}{3}\rho_0 a_0^3\right)(\sinh\eta - \eta).$$

(Eqn. 9-30)

9.12 Nucleosynthesis

A key transition in the analysis of nucleosynthesis is concerned with the binding energy of the deuteron (p-n binding) which is at $T \simeq 2 \times 10^{10} K$. For $T \gg 2 \times 10^{10\,o} K$ (up to about 1.1 sec after Big Bang) we still have $p, n, e^+, e^-, \gamma, \nu_e, \bar{\nu}_e$ in thermal equilibrium (where μ^+, μ^- have annihilated near the cutoff temperature, so ignoring muons and muon neutrinos). The interactions keeping n, p in equilibrium:

$$n + \nu_e \leftrightarrow p + e^- \quad , \quad n + e^+ \leftrightarrow p + \bar{\nu}_e \quad , \quad n \leftrightarrow p + e^- + \bar{\nu}_e,$$

where only the last equilibrium is left for $T \ll 10^{7\,o}K$.

Given the above, we must compare reaction rates with expansion rates to see when the reactions slip out of equilibrium. For the early-time analysis the value of K not important, so take $K = 0$ (and $\Lambda = 0$) to have the form (radiation dominated):

$$\left(\frac{\dot{a}}{a}\right)^2 = \frac{8\pi G}{3}\rho \quad \rightarrow \quad \frac{1}{2t} = \frac{\dot{a}}{a} \quad \rightarrow \quad t = \sqrt{\frac{3}{32\pi G\rho}}$$

We want t in terms of T, at $T = 2 \times 10^{10\,o}K$ in particular. For early times dominated by radiation (where p and n are nonrelativistic at $2 \times 10^{10\,o}K$, while the electrons are still relativistic $2 \times 10^{10\,o}K$, so we can neglect the p and n:

$$\rho = \frac{9}{2}\underline{a}T^4$$

where $\rho = \rho_{\nu_e} + \rho_{\bar{\nu}_e} + \rho_{\nu_\mu} + \rho_{\bar{\nu}_\mu} + \rho_{e^+} + \rho_{e^-} + \rho_\gamma$ and $\underline{a}$ is the Stefan-Boltmann constant:

$$\underline{a} = \frac{\pi^2 k^4}{15 c^3 \hbar^3} = 7.5041 \times 10^{-15} erg\ cm^{-3\,o}K^{-4}.$$

Also:

$\rho_\gamma = \underline{a}T^4$

$\rho_e = \frac{7}{8}\underline{a}T^4$ (for equilibrium density using Fermi-Dirac statistics instead of B.E., with 2 spin states)

$\rho_{\nu_e} = \frac{1}{2}\cdot\frac{7}{8}\underline{a}T^4$ (with only 1 spin state)

Thus,

$$t = \left(\frac{c^2}{48\pi G\underline{a}T^4}\right) = 1.09\,\text{sec}\left[\frac{T}{10^{10\,o}K}\right]^{-2}$$

Want to compare $\frac{1}{t}$ with the $n \rightarrow p$ reaction rates:

$$\lambda(n \rightarrow p) \simeq \lambda(p \rightarrow n) = 0.361\,\text{sec}^{-1}\left(\frac{T}{10^{10\,o}K}\right)^5$$

The

$$\frac{\text{reaction rate}}{\text{expansion rate}} \simeq \frac{\lambda}{(1/t)} = \lambda t$$

when small system no longer in equilibrium. $\lambda t \cong 10$ when $T \simeq 3 \times 10^{10\,o}K$ $(t = 1.1 sec)$. After $t = 1.1 sec$ the system begins to fallout of equilibrium. So, for $T > 3 \times 10^{10}$ we can calculate the fraction of neutrons present. For $T > 3 \times 10^{10\,o}K$ we have equilibrium of p and n. For the fraction of neutrons present, X_n, we have:

neutron fraction

↓ ↓ number density of neutrons ↓ F.D.

$$X_n = \frac{n_n}{n_n + n_p} = \frac{\left(1 + e^{m_n/KT}\right)^{-1}}{(1 + e^{m_n/KT})^{-1} + \left(1 + e^{m_p/KT}\right)^{-1}}$$

Thus,

$$X_n \cong \frac{e^{-m_n/KT}}{e^{-m_n/KT} + e^{-m_p/KT}} = \left(1 + e^{Q/KT}\right)^{-1} \text{ for } T > 3 \times 10^{10\,o}K.$$

$Q = m_n - m_p = 1.293 MeV$

$Q = kT$ when $T = 1.5 \times 10^{10} K^o$

At $T \simeq 3 \times 10^{10\,o}K$: $X_n \simeq 0.38$ and $t = 0.1299\ sec.$

Consider

$$\frac{dX_n}{dt} = \lambda(p \to n)\overbrace{(1 - X_n)}^{X_p} - \lambda(n \to p)X_n$$

A numerical solution provides:
$T = 1 \times 10^{10\,o}K \Rightarrow X_n = 0.241 \quad (t = 1.1\ sec)$.
At this point the rate equation stops to work when e^+, e^-annihilation closing weak channels.

Then,
$T = 3 \times 10^{9\,o}K \Rightarrow X_n = 0.170 \ \ (t = 13.8 \text{ sec})$
At this point all that is left to change X_n after $T = 5 \times 10^9 K$ is neutron decay (lifetime is 1013 sec). A more precise numerical simulation provides $N = 0.164$, and we have the neutron percent then evolving as follows:

$$X_n(t) = N \exp\left(\frac{-t}{1013\text{sec}}\right).$$

9.12.1 Neutron Fraction Example

For $T \geq 5 \times 10^{9\,o}K$ take $X_n = \left(1 + e^{Q/KT}\right)^{-1}$ and for $T \leq 5 \times 10^{9\,o}K$ take $X_n = Ne^{-\frac{t}{1013\text{sec}}}$. Get N by requiring X_n to be continuous at $T = 5 \times 10^{9\,o}K$.

Typically temperature for He^4 formation is $T = 1 \times 10^{9\,o}K$, assume this and calculate X_n at that temp. then infer the He^4 fraction. Have 894 sec for neutron He ($8.5 \times 10^9 eV$). Fraction by weight of He^4:

$$d + d \leftrightarrow He^3 + n \leftrightarrow H^3 + p$$

$$H^3 + d \leftrightarrow He^4 + n \qquad Y = \frac{4N_{He}}{4N_{He} + N_p}$$

The abundance of He^4 from Nucleosynthesis

When T is large: ($k^{-1} = 11605\ {}^oK/eV$ say):

$$X_n = \frac{n_n}{n_n + n_p} \simeq \left(1 + e^{Q/KT}\right)^{-1} \simeq .146 \quad using \ \begin{cases} Q = 1.293 MeV \\ T = 8.5 \times 10^{9\,o}K \\ kT = .7324 MeV \end{cases}$$

When T is sufficiently small we have simple neutron decay reducing He^4:

$$X_n = N \exp\left[\frac{-t}{894\ sec}\right]$$

(Neutron Lifetime is 894 sec). Thus,

$$t = 1.09 \sec\left[\frac{T}{10^{10}K}\right]^{-2} + const.$$

More accurate relation from $\frac{\pi^2}{15}T^4\left(1+3\frac{7}{8}\left(\frac{4}{11}\right)^{4/3}\right) = \frac{3}{32\pi G}\left(\frac{C}{T}\right)^2$

provides:

$t = 1.77 \sec\left[\frac{T}{10^{10}K}\right]^{-2}$, where the neutrino's have decoupled (using this relation going forward).

T at start of decay is $T = 8.5 \times 10^{9\,o}K \Rightarrow T = 2.45 sec.$

T at end of decay is $T = 1 \times 10^{9\,o}K \Rightarrow t = 177 sec.$

So, $t = t_1 - t_0 = 177 - 2.45 = 175$, thus $X_n = (.146)(.823) = .120$, and we have:

$$Y = 2X_n = 24\%$$

9.12.2 Evaluating the parameters of the Friedman model for p=0 matter dominated

Consider

$$\Omega = \frac{\rho}{\rho_c} \ and\ H^2 = \left(\frac{\dot{a}}{a}\right) = \frac{8\pi G}{3}\rho_c$$

Galaxy counts seem to indicate $\Omega = 1$ (volume density of galaxies)

$\rho_c = 1.87 \times 10^{-29}\ gm/cm^3$ with $H = 100$

1.1% to 12% of ρ_c is in baryons.

$1 \times 10^{-31}\ gm/cm^3 < \rho_0 < 10^{-30}\ gm/cm^2$

Parameters used to describe Friedman models:

- Hubble constant: $H_0 = \frac{\dot{a}(t_0)}{a(t_0)}\ where\ t_0 = now.$
- $\Omega = \rho/\rho_c$
- Deceleration parameter: $q_0 \equiv -\frac{\ddot{a}(t_0)a(t_0)}{\dot{a}(t_0)^2} = -\frac{\ddot{a}}{a}\frac{1}{H_0^2}$

Can measure in principle since:(3) $R_0 = 6\frac{K}{a_0^2}\ with\ a_0 = a(t)$. Need ρ_0 and much may be dark.

Einstein equations: $\left(\frac{\dot{a}}{a}\right)^2 + \frac{K}{a^2} - \frac{1}{3}\Lambda = \frac{8\pi G}{3}\rho$ and $-2\frac{\ddot{a}}{a} - \left(\frac{\dot{a}}{a}\right)^2 - \frac{K}{q^2} + \Lambda = 8\pi G\rho$. Suppose $\Lambda = 0$.

Let's show

$$\boxed{\begin{aligned} &\frac{1}{6}{}^{(3)}R_0 = \frac{K}{a_0^2} = H_0^2(2q_0 - 1) \\ &\text{and} \\ &2q_0H_0^2 = \frac{8\pi G}{3}\rho_0 \end{aligned}}$$

(Eqn. 9-31)

Using the Einstein equations with $\Lambda = 0$:

$$\left(\frac{\dot{a}}{a}\right)^2 + \frac{k}{a^2} = \frac{8\pi G}{3}\rho \quad and \quad -\frac{2\ddot{a}}{a} - \left(\frac{\dot{a}}{a}\right)^2 - \frac{k}{a^2} = 8\pi G\rho$$

Rename variables:

$$q_0 = \frac{-\ddot{a}(t)}{a(t)}\frac{1}{H_0^2} \qquad H_0 = \frac{\dot{a}(t_0)}{a(t_0)} \qquad {}^{(3)}R_0 = \frac{6K}{a(t_0)^2} \qquad t_0 = now$$

Then

$H_0^2 + \frac{1}{6}R_0 = \frac{8\pi G}{3}\rho_0$ and $2q_0H_0^2 - H_0^2 - \frac{1}{6}R_0 = 8\pi G p_0$.

To proceed further we need a state equation, so using as state equation $p_0 = 0$:

$2q_0H_0^2 = H_0^2 + \frac{1}{6}R_0 = \frac{8\pi G}{3}\rho_0$ and $\frac{1}{6}R_0 = 2q_0H_0^2 - H_0^2 - H_0^2(2q_0 - 1)$.

Thus,

$$\frac{1}{6}{}^{(3)}R_0 = H_0^2(2q_0 - 1)$$

$$2q_0H_0^2 = \frac{8\pi G}{3}\rho_0$$

For spatially flat case we have: $H_0^2 = \frac{8\pi G}{3}\rho_0$ and $q_0 = \frac{1}{2}$. So closure density is: $\rho_c = \frac{3H_0^2}{8\pi G}$. In matter dominated case ($p = 0$) then $\Omega = \frac{\rho_0}{\rho_c} = 2q_0$.

Measured values:

$H_0 = 75 \pm 25\ km\ \text{sec}^{-1}\ Mpc^{-1}$

In units with $\hbar = c = 1$:

$H_0 = 8.0 \times 10^{-29} cm^{-1} \pm 2.7 \times 10^{-29} cm^{-1}$

Thus

$\rho_0 = \frac{3}{4\pi G} q_0 H_0^2$ $with\ G = 2.6 \times 10^{-66} cm^2$. Let $0.66 \leq h_0 \leq 1.34$, then:

$$\rho_0 = q_0 (5.9 \times 10^8 cm^{-4}) h_0^2 = q_0 h_0^2 \left(2.1 \times 10^{-29} \frac{gm}{cm^3}\right)$$

and

$$\rho_c = (h_0^2) \times 10^{-29} \frac{gm}{cm^3}.$$

Suppose $\Lambda = 0$ and rad. dom.: $p = \frac{1}{3}\rho$ as before, show

$$\frac{K}{a_0^2} = H_0^2 (q_0 - 1)$$

$$q_0 H_0^2 = \frac{8\pi G}{3} \rho_0$$

(Eqn. 9-32)

9.12.3 Evaluating the parameters of the Friedman model for $p = \rho/3$ matter dominated

Consider:

$$\left(\frac{\dot{a}}{a}\right)^2 + \frac{K}{a^2} - \frac{1}{3}\Lambda = \frac{8\pi G}{3}\rho \ and \ -2\frac{\ddot{a}}{a} - \left(\frac{\dot{a}}{a}\right)^2 - \frac{K}{a^2} + \Lambda = 8\pi G \rho$$

Conservation equation: $\partial_t(\rho a^3) + p\partial_t(a^3) = 0$:

$p = 0$: $\rho_M = \rho_M a_0^3 a^{-3}$, and

$p = \frac{1}{3}\rho$: $\rho_{rad} = \rho_{ro} a_0^4 a^{-4}$.

One can consider $\frac{1}{3}\Lambda = \frac{8\pi G}{3}\rho_V$ ('V' for vacuum energy):

$$\left[\frac{d}{dt}\left(\frac{a}{a_0}\right)\right]^2 + \frac{K}{a_0^2} - \frac{1}{3}\Lambda\left(\frac{a}{a_0}\right)^2 = \frac{8\pi G}{3}\left[\rho_{\mu_0}\left(\frac{a_0}{a}\right) + \rho_{ro}\left(\frac{a_0}{a}\right)^2\right]$$

Let $X = \frac{a}{a_0}$:

$$\underbrace{\dot{X}^2 + V(X) = \frac{K}{a_0^2}}_{\text{familiar potential problem}} \quad and \ V(X) = \frac{8\pi G}{3}\left[\rho_{Mo} X^{-1} + \rho_{ro} X^{-2}\right] - \frac{1}{3}\Lambda X^2$$

For $\Lambda > 0, V(X)$ has a max, which is smaller for larger Λ. If Λ is very large as in inflation....

The value of Λ at which $V_{max} = -\frac{1}{a_0^2}$ is called $\Lambda_{\text{critical}}$.

9.13 FRW Models $\{\Lambda, K\}$ and $\Lambda \neq 0$ LeMaitre Models

Cases

$\Lambda = 0$ $\quad K = -1$ $\quad \dot{a} = \text{const. at } t \to \infty$

$a \propto t \text{ as } t \to \infty$ unlimited expansion

$K = 0$ $\quad \dot{a} = 0 \text{ at } t \to \infty$ unlimited expansion

$K = 1$ $\quad$ expands from singularily and recontracts in finite time.

$\Lambda < 0$ $\quad$ all cases have expansion then contraction in finite time.

$\Lambda > 0$ $\quad K = -1$ $\quad$ unlimited expansion accelerating as $t \to \infty$.

For large $t - \Lambda X^2$ dominates with solution

$$a(t) = \exp\left[\left(\frac{\Lambda}{3}\right)^{1/2}(t - t_1)\right] \qquad t_1 = const$$

$K = 0$: same as $K = -1$

$K = 1$:

$\Lambda > \Lambda_{out}$ same as $K = -1$ $\left(V_{max} < -\frac{1}{a_0^2}\right)$ (can hesitate, "hesitation Universe")

$\Lambda < \Lambda_{crit}$ $\quad$ (i) can have expand and contract

(ii) or can have contraction from ∞ and then reexpand, no singularity. (only solution without singularity thus far)

$\Lambda = \Lambda_{crit}$ $\quad$ (i) expands, infinite time to reach peak

(ii) contracts, infinite time to reach peak

(iii) at apex: Einstein static universe.

Models with cosmological constant (LeMaitre models):

$$\left(\frac{\dot{a}}{a}\right)^2 + \frac{K}{a^2} - \frac{1}{3}\Lambda = \frac{8\pi G}{3}\rho$$

$$\frac{d}{dt}(\rho a^3) + p\frac{d}{dt}(a^3) = 0$$

$$-2\frac{\ddot{a}}{a}-\left(\frac{\dot{a}}{a}\right)^2-\frac{K}{a^2}+\Lambda= 8\pi G\rho$$

Have $K=1, E=-\frac{1}{a_0^2}$ unstable solution, the Einstein static universe, with $\Lambda{=}\Lambda_{crit}$ where $a=a_0$ is constant:

$a=a_0$: $\frac{1}{a_0^2}-\frac{1}{3}\Lambda=\frac{8\pi G}{3}\rho \Rightarrow a_0\left[\frac{8\pi G}{3}\rho+\frac{1}{3}\Lambda\right]^{-1/2}$

For equation of state of dust: $p=0, so\ a_0=\Lambda^{-1/2}$: $\Lambda=\frac{8\pi G}{3}\rho+\frac{1}{3}\Lambda\rightarrow$ $\Lambda_{crit}=4G\pi\rho$. With the discovery of expansion, however, these models became less interesting.

Models with no radiation or matter:

$$V(X)=\frac{8\pi G}{3}(\rho_{Mo}X^{-1}+\rho_{Ro}X^{-2})-\frac{1}{3}\Lambda X^2$$

and for $\rho_{Mo}=0, \rho_{Ro}=0 \rightarrow V(X)=-\frac{1}{3}\Lambda X^2$:

$$\dot{X}^2+V(X)=-\frac{K}{a_0^2} \quad\rightarrow\quad \dot{X}^2-\frac{1}{3}\Lambda X^2=-\frac{K}{a_0^2}$$

For spatially flat $K=0$: No solution for $\Lambda<0$.

$\Lambda=0 \Rightarrow$ Minkowski spacetime ($a=const$)

$\Lambda>0 \Rightarrow$ de Sitter: $\left(\frac{\dot{a}}{a}\right)^2-\frac{1}{3}\Lambda=0\rightarrow\ a=$ $const.\, e^{[(\Lambda/3)^{1/2}t]}$

Positive spatial curvature: $K=1$, and $\left(\frac{\dot{a}}{a}\right)^2+\frac{1}{a^2}-\frac{1}{3}\Lambda=0$ only has a solution for $\Lambda>0$:

$$a=\left(\frac{1}{3}\Lambda\right)^{-1/2}\cosh\left[\left(\frac{1}{3}\Lambda\right)^{1/2}t\right],$$

which is all of de Sitter. At $t=0$: $a=a_{max}=\left(\Lambda/3\right)^{-1/2}$.

$K=-1$:

$\dot{a}^2-1=\frac{1}{3}\Lambda a^2$:	$\Lambda<0$:	$a=\lvert\Lambda/3\rvert^{-1/2}\sin\left[\left(\lvert\Lambda\rvert/3\right)^{1/2}t\right]$
	$\Lambda>0$	$a=\lvert\Lambda/3\rvert^{-1/2}\sinh\left[\left(\Lambda/3\right)^{1/2}t\right]$
	$\Lambda>0$	$a=t$ (Milne spacetime)

Example 9.1. Consider a model with $K = 0, \Lambda > 0$ and $\rho_{mo} = 0$ but $\rho_{ro} \neq 0$

(a) Write the 1st order Einstein equation

(b) Find the solution

(c) Find the corresponding solution with $\Lambda < 0$.

Solution

We have $K = 0,\ \Lambda > 0,\ \rho_{Mo} = 0,\ \rho_{ro} \neq 0$:

(a) $\left(\frac{\dot{a}}{a}\right)^2 + \frac{k}{a^2} - \frac{1}{3}\Lambda = \frac{8\pi G}{3}\rho$

$$\left.\frac{d}{dt}(\rho a^3) + p\frac{d}{dt}(a^3) = 0\right\} \quad \xrightarrow[\rho_{mo} = 0]{\rho_{ro} \neq 0} \quad \rho(t) = \rho_{ro}\left(\frac{a_0}{a(t)}\right)^4$$

So,

$$\left(\frac{\dot{a}}{a}\right)^2 - \frac{1}{3}\Lambda = \frac{8\pi G}{3}\rho_{ro}\frac{a_0^4}{a(t)} \rightarrow \left(\frac{d}{dt}\left[\frac{a}{a_0}\right]\right)^2 - \frac{1}{3}\Lambda\left(\frac{a}{a_0}\right)^2 = \frac{8\pi G}{3}\rho_{ro}\left(\frac{a_0}{a}\right)^2$$

Let $X = \frac{a}{a_0}$:

$$\dot{X}^2 + \left(-\frac{1}{3}\Lambda X^2 - \frac{8\pi G}{3}\rho_{ro}X^{-2}\right) = 0 \rightarrow X^2\dot{X}^2 = \frac{1}{3}\Lambda X^4 + \frac{8\pi G}{3}\rho_{ro} \rightarrow \left(X\dot{X}\right)^2 = \left[\frac{d}{dt}\left(\frac{1}{2}X^2\right)\right]^2$$

Want positive real X so choose root appropriately (let $Y = X^2$):

$$\frac{d}{dt}(Y^2) = 2\sqrt{\frac{1}{3}\Lambda Y^2 + \frac{8\pi G}{3}\rho_{ro}} \rightarrow \frac{dY}{2\sqrt{\frac{1}{3}\Lambda Y^2 + \frac{8\pi G}{3}\rho_{ro}}} = dt$$

Integrate:

$$\int_{t_0}^{t} dt = \int_{Y_0}^{Y}\frac{dY}{2\sqrt{\frac{1}{3}\Lambda Y^2 + \frac{8\pi G}{3}\rho_{ro}}} = \frac{1}{2\sqrt{\frac{1}{3}\Lambda}}\int_{Y_0}^{Y}\frac{dY'}{2\sqrt{Y'^2 + \frac{8\pi G}{\Lambda}\rho_{ro}}} = \frac{1}{2\sqrt{\frac{1}{3}\Lambda}}\sinh^{-1}\left(\frac{Y'}{\frac{8\pi G}{\Lambda}\rho_{ro}}\right)\Bigg|_{Y_0}^{Y}$$

So,

$$t = t_0 = \frac{1}{2\sqrt{\frac{1}{3}\Lambda}}\left(\sinh^{-1}\left(\frac{Y}{\frac{8\pi G}{\Lambda}\rho_{ro}}\right) - \sinh^{-1}\left(\frac{1}{\frac{8\pi G}{\Lambda}\rho_{ro}}\right)\right).$$

Let $\alpha = (8\pi G/\Lambda)\rho_{ro}$:

$$\sinh^{-1}\left(\frac{Y}{\alpha}\right) = \sqrt{\frac{4}{3}\Lambda}(t - t_0) + \sinh^{-1}\left(\frac{1}{2}\right) \rightarrow Y$$

$$= \alpha \sinh\left[\sqrt{\frac{4}{3}\Lambda}(t - t_0) + \sinh^{-1}\left(\frac{1}{2}\right)\right]$$

(b) $X = Y^{1/2}$ take positive root:

$$\left.\begin{array}{r} K = 0 \\ \Lambda > 0 \\ \rho_{mo} = 0 \\ \rho_{ro} \neq 0 \end{array}\right\}$$

$$\boxed{a(t) = a_0\sqrt{\frac{8\pi G\rho_{ro}}{\Lambda}}\sinh^{1/2}\left\{\sqrt{\frac{4}{3}\Lambda}(t - t_0) + \sinh^{-1}\left(\frac{\Lambda}{8\pi G\rho_{ro}}\right)\right\}}$$

For $\rho_{ro} \rightarrow 0$ does this yield de Sitter?

$$e.g., is\ it\ \left(a(t) \propto \exp\left(\frac{1}{3}\Lambda\right)^{1/2} t\right)$$

$$\lim_{\rho_{ro}\rightarrow 0} a(t) \cong a_0\sqrt{\frac{8\pi G\rho_{ro}}{\Lambda}} \times \sqrt{\sinh\left[\sqrt{\frac{4}{3}\Lambda}(t - t_0)\right]\cosh\left[\sinh^{-1}\left(\frac{\Lambda}{8\pi G\rho_{ro}}\right)\right]}$$

$$+ \cosh\left[\sqrt{\frac{4}{3}\Lambda}(t - t_0)\right]\sinh\left[\sinh^{-1}\left(\frac{\Lambda}{8\pi G\rho_{ro}}\right)\right].$$

$$\simeq a_0\sqrt{\frac{8\pi G}{\Lambda}}\sqrt{\rho_{ro}}\sqrt{\left\{\sinh\left[\sqrt{\frac{4}{3}\Lambda}(t - t_0)\right] + \cosh\left[\sqrt{\frac{4}{3}\Lambda}(t - t_0)\right]\right\}\left(\frac{\Lambda}{8\pi G\rho_{ro}}\right)}$$

$$\simeq a_0 \exp\left(\sqrt{\frac{\Lambda}{3}}(t - t_0)\right)$$, so this agrees with de Sitter.

(c) Now with $\Lambda < 0$ we have:

$$\frac{dY}{dt} = 2\sqrt{\frac{8\pi G}{3}\rho_{ro} - \frac{1}{3}|\Lambda|Y^2}$$

So,

$$\int_{t_0}^{t} dt = \frac{1}{2\sqrt{\frac{1}{3}|\Lambda|}}\int_{Y_0}^{Y_0}\frac{dY'}{\sqrt{\frac{8\pi G}{|\Lambda|}\rho_{ro}-Y'^2}} = \frac{1}{2\sqrt{\frac{1}{3}|\Lambda|}}\sinh^{-1}\left(\frac{Y'}{\frac{8\pi G}{\Lambda}\rho_{ro}}\right)\Bigg|_{Y_0}^{Y}$$

Thus

$$Y = \left(\frac{8\pi G}{|\Lambda|}\rho_{ro}\right)\sin\left[\sqrt{\frac{4}{3}}|\Lambda|(t-t_0)+\sin^{-1}\left(\frac{|\Lambda|}{8\pi G\rho_{ro}}\right)\right]$$

and we have:

$$a(t) = a_0\sqrt{\frac{8\pi G}{|\Lambda|}\rho_{ro}}\ \sin^{1/2}\left\{\sqrt{\frac{4}{3}}|\Lambda|(t-t_0)+\sin^{-1}\left(\frac{|\Lambda|}{8\pi G\rho_{ro}}\right)\right\}$$

Example 9.2. For $K = 1$, Radiation dominated, (a) find $a(t)$ and (b) repeat for $K = -1$:

(a) So have $\left(\frac{\dot{a}}{a}\right)^2 + \frac{1}{a^2} = \frac{8\pi G}{3}\rho$ and $p = \frac{1}{3}\rho$ with energy conservation: $\frac{d}{dt}(\rho a^3) + p\frac{d}{dt}(a^3)$, which yields $\rho(t) = \rho_0\left({a_0}/{a(t)}\right)^4$. So,

$$a'^2 + a^2 = \frac{8\pi G\rho_0 a_0^4}{3}$$

with solution:

$$a(\eta) = \sqrt{\frac{8\pi G\rho_0 a_0^4}{3}}\sin\eta$$

Solving for t:

$$t = \int_0^{\eta} a(\eta')\,d\eta' = \sqrt{\frac{8\pi G\rho_0 a_0^4}{3}}(1-\cos\eta) \qquad A = \sqrt{\frac{8\pi G\rho_0 a_0^4}{3}}$$

So,

$$a(t) = A\sin\left\{\cos^{-1}\left(1-\frac{t}{A}\right)\right\} = A\sqrt{1-\left(1-\frac{t}{A}\right)^2} =$$

$$\left(\frac{32\pi G\rho_0}{3}\right)^{1/4} a_0 t^{1/2}\sqrt{1-\left(\frac{32\pi G\rho_0 a_0^4}{3}\right)^{-1/2} t}$$

For $K = 1$, Matter dominated, using conformal time:

$$a'^2 + a^2 = \left(\frac{8\pi G}{3}\rho_0 a_0^3\right)a \qquad B = \frac{8\pi G}{3}\rho_0 a_0^3$$

Try $a = \alpha + \beta\cos\eta$, works with $\beta = -\alpha$, $\alpha = \frac{1}{2}B$, so have:

$$a = \left(\frac{4\pi G}{3}\rho_0 a_0^3\right)(1-\cos\eta)$$

$$t = \int_0^\eta a(\eta')\,d\eta' = \left(\frac{4\pi G}{3}\rho_0 a_0^3\right)(\eta - \sin\eta)$$

parametric form.

(b) For $K = -1$, radiation dominated have $a'^2 - a^2 = A^2$, where A as before, conformal time used, using substitution $a = A\sinh\eta$ for a solution and, subsequently, $t = A(\cosh\eta - 1)$, thus:

$$a(t) = A\sinh\left\{\cosh^{-1}\left(1+\frac{t}{A}\right)\right\} = A\sqrt{\left(1+\frac{t}{A}\right)^2 - 1} ==$$

$$\left(\frac{32\pi G\rho_0}{3}\right)^{1/4} a_0 t^{1/2}\sqrt{1+\left(\frac{32\pi G\rho_0 a_0^4}{3}\right)^{-1/2} t}.$$

For matter dominated: $a'^2 - a^2 = Ba$. Guess $a = \frac{1}{2}B(\cosh\eta - 1)\quad a' = B\sinh\eta$:

$$a'^2 - a^2 = \frac{B^2}{4}\sinh^2\eta - \frac{B^2}{4}\cosh^2\eta + \frac{2B^2}{4}\cosh\eta - \frac{B^2}{4} = Ba,$$

so works, and we have:

$$t = \int_0^\eta a(\eta')\,d\eta' = \frac{1}{2}B^2[\sinh\eta' - \eta']_0^\eta = \left(\frac{4\pi G}{3}\rho_0 a_0^3\right)(\sinh\eta - \eta).$$

9.14 Particle Horizons and Event Horizons in FRW models

We have:

$$ds^2 = -dt^2 + a^2(t)\left\{\frac{dr^2}{1-Kr^2} + r^2 d\theta^2 + r^2\sin^2\theta\,d\varphi^2\right\}_{K=0,\pm1}$$

Particle Horizons: a fundamental world line is said to have a particle horizon at time *t* (cosmic time *t*), if at time *t* there are other fundamental world lines which are not causally connected to it.

For the general FRW model, for the light ray $ds^2 = 0$ and we have:

$$\frac{dt'}{a(t')} = -\frac{dr}{(1-Kr^2)^{1/2}}$$

Thus,

$$\int_{t_{min}}^{t}\frac{dt'}{a(t')} = \int_0^{r_p(t)}\frac{dr}{(1-Kr^2)^{1/2}} \equiv f\left(r_p(t)\right) = \begin{cases}\sin^{-1} r_p(t) & K=1\\ r & K=0\\ \sinh^{-1} r_p(t) & K=-1\end{cases}$$

For a given $a(t)$, with $K = 0$ or -1, if $\int_{t_{min}}^{t} \frac{dt'}{a(t')}$ does not converge then $r_p(t) \to \infty$ and there is no particle horizon. For $K = 1$ $\int_{t_{min}}^{t} \frac{dt'}{a(t')}$ need only be greater the $2\left(\frac{\pi}{2}\right) = \pi$ for there to be no particle horizon: $r_p(t) = \sin f\left(r(t)\right)$ and consider a spatial section with $a = 1$, where angle is $F\left(r_p(t)\right)$, when $f \to \frac{\pi}{2}$. If $\int_{t_{min}} > \pi$ then we can cover the entire range of $\sin^{-1} r_p(t)$ so no particle horizon.

<u>First example</u>: de Sitter:

$$\int_{t_{min}=-\infty}^{t} \frac{dt'}{a(t')} = C^{-1} \int_{-\infty}^{t} e^{-Ht'}\, dt' = \frac{1}{CH}\left(e^{-H(-\infty)} - e^{-Ht}\right) = \infty.$$

Consider $K = 0$ Friedman dust dominated with $a \propto t^{2/3}$:

$$d\rho(t) = t^{2/3} \int_{0}^{t} \frac{dt'}{(t')^{2/3}} = 3t$$

Radiation dominated: $d\rho(t) = 2t$. With $K = -1$ or $+1$ and $ad\eta = dt$:

$$d\rho(t) = a \int_{\eta_{min}}^{\eta(t)} d\eta' = (a)(\eta - \eta_{min})$$

So, $d\rho(t) = a(t)\eta(t)$ $(\eta_{min} = 0)$. For $K = -1$ radiation dominated:

$$\cosh\eta = A^{-1}t + 1$$

$$\text{where } A = \left(\frac{8\pi G}{3}\rho_0 a_0^4\right)^{1/2}$$

$$\eta = \cosh^{-1}(A^{-1}t + 1)$$

So, $d\rho(t)_{K=-1} = a(t)\cosh^{-1}(\ldots)$

$$= \sqrt{2At}\,[1 + (2A)^{-1}t]^{1/2} \cosh^{-1}(A^{-1}t + 1)$$

Similar for matter dominated.

$d\rho(t)_{K=1} = similar$

Consider

$d\rho(t) = a(t)\eta(t)\ with\ K = 1$ and with radiation. When $\eta(t) \geq \pi$ there is no particle horizon:

$$a = A\sin\eta \to \eta = \pi \Rightarrow a = 0,$$

so there are particle horizons up to the instant of re-collapse.

Example 9.3. Consider the FRW universe with Negative Curvature, and linearly expanding:

$$ds^2 = -dt^2 + t^2\left\{\frac{dr^2}{1 - Kr^2} + r^2(d\theta^2 + \sin^2\theta\, d\varphi^2)\right\}$$

Using $t > 0 \quad r \geq 0 \quad a(t) = t$:

$$R_{tt} = -3\frac{\ddot{a}}{a} = 0 \quad , \quad R_{ti} = R_{it} = 0$$

$$R_{ij} = (a\ddot{a} + 2\dot{a}^2 - 2)\tilde{g}_{ij} = 0$$

So, $R_{\mu\nu} = 0\ ,\ R = 0\ ,\ in\ fact\ R^{\lambda}_{\mu\nu\sigma} = 0.$

Let $r' = rt \quad and \quad t' = t(1 + t^2)^{1/2}$:

$(dr')^2 = r^2dt^2 + t^2dr^2 + 2rtdrdt$

$(dt')^2 = (1 + r^2)dt^2 + t^2dr^2(1 + r^2)^{-1}dr^2 + 2rtdrdt$

So,

$$-dt'^2 = dr'^2 = -dt^2 + \frac{t^2dr^2}{1 + r^2}$$

Thus,

$ds^2 = -dt'^2 + dr'^2 + r'^2(d\theta^2 + \sin^2\theta\, d\varphi^2)$

For $r' > 0$ and $t' > 0$ also $\left|\frac{t'}{r'}\right| = \frac{(1+r^2)^{1/2}}{r} > 1$. So future light cone of $r' = t = 0$ in Minkowski.

Paradox: Spatial hypersurface has negative curvature but it can be imbedded in Minkowski? – No problem we just have a negative curvature. Milne hypersurface with $t - const$:

$$t'^2 - r'^2 = t^2(1 + r^2) \Rightarrow a\ hyperbola.$$

A conformal transformation on any of the FRW can result in a spacetime that can be imbedded in Minkowski. A conformal transformation on the negative curvature case can be imbedded in the future light cone of Minkowski spacetime. For the positive curvature case what section is chosen?.....

There should be an imbedding in “right” side for appropriate $a(t)$

$ds^2 = -dt^2 + a^2(t)$ {neg curve case}

$$= \Omega^2(t')[-(dt')^2 + t'^2\{neg.\}] = \Omega^2(t)[future\ lightcone\ of\ Mink].$$

9.15 Stages in Evolution of the Universe: The Standard Model

At what temperature/time did the matter ρ equal the radiation ρ? When did the density of non-relativistic matter equal that of relativistic matter? (There is dark-matter also.) Assume matter remains non-relativistic back to the transition time above.

Today:

$\rho_\gamma = \rho_{\mu\ wave} = 4 \times 10^{-34}\ gm/cm^3$

$10^{-29}\ gm/cm^3 \geq \rho_{matter} \geq 2 \times 10^{-31}\ gm/cm^3$ (including dark matter)

Note: $\hbar c^{-1} = 3.50 \times 10^{-38} gm\ cm.$

Also, $H_0 = (35Km\ \sec^{-1} Mpc^{-1})h_0\ for\ \frac{2}{3} < h_0 < \frac{4}{3}.$

Have:

$\rho_{matter}(t) = \rho_{M_0} a_0^3 a(t)^{-3}\ where\ \rho_{M_0} = \rho_{matter}$ at $t = 0$ (present)

$\rho_{rad}(t) = \rho_{r_0} a_0^4 a(t)^{-4}$

Equate for some t_e: $a_e = a(t_e)$:

$\frac{a_e}{a_0} = \frac{\rho_{r_0}}{\rho_{m_0}}$ with Red shift from that time: $z = \frac{\lambda_e - \lambda_0}{\lambda_0} = \frac{a_e}{a_0} - 1 = \frac{\rho_{r_0}}{\rho_{m_0}} - 1.$

We expect $\rho_{r_0} = 10^{-33}\ gm/cm^3$, thus

$$z_e = (1000)\frac{\rho_{m_0}/10^{-30}\ gm/cm^3}{\rho_{r_0}/10^{-33}\ gm/cm^2} - 1$$

At what Temp. T_e do ρ_m and ρ_r become equal?

For photons:

$$\rho_\gamma c^2 = bT^4 \qquad b = \frac{8\pi^5 K_B^4}{15h^3c^3} = 7.56 \times 10^{-15} erg\ cm^{-3} \times \deg^{-4}$$

So, $\rho_\gamma = \rho_{r_0} a_0^4 a^{-4} = bT^4$

$$T = b^{-1/4}\rho_{\gamma 0}^{1/4} a_0 a^{-1} = T_0 a_0 a^{-1} \text{ (where } T_0 \text{ is temp today)}$$

$$T_e = T_0 \frac{a_0}{a_e} = T_0 \frac{\rho_{m_0}}{\rho_{r_0}} \Rightarrow \boxed{\frac{T_e}{T_0} = \frac{\rho_{m_0}}{\rho_{r_0}}}$$

$$T_e \approx T_0(1000)\left(\frac{\rho_{m_0}/10^{-30}}{\rho_{r_0}/10^{-33}}\right) \simeq (2700^oK)\left(\frac{\rho_{m_0}/10^{-30}}{\rho_{r_0}/10^{-33}}\right)$$

2700^oK is just a little below the recombination temperature so results are inconclusive. Let T_{equiv} be the temperature at which the matter and radiation densities are equal. If $T_{recomb} \ll T_{equiv}$ then universe transparent at decoupling:

$$\frac{\rho_m(t_e)}{\rho_r(t_e)} = \left(\frac{\rho_{m_0}}{\rho_{r_0}}\right)^4 \qquad \rho_m(t_e) = \rho_r(t_e)$$

$$= \left(\frac{\rho_{m_0}}{\frac{10^{-30}}{gm/cm^3}}\right)^4 \left(\frac{\rho_{r_0}}{\frac{10^{-33}}{gm/cm^3}}\right)^{-3} 10^{-2} gm\ cm^{-3}$$

All of this is independent of curvature (K) and of whether or not there is a nonzero Λ etc.

Using rad dom. form up to t_e:

$$a(t) = A^{1/2} a_e t^{1/2} \sqrt{1 - K\frac{t}{A}} \qquad A = \left(\frac{32\pi G}{3}\rho_{re}\right)^{1/2}$$

$$a(t_e) = a_e$$

Have $1 = A^{1/2} t_e^{1/2} \sqrt{1 - K\frac{t_e}{A}}$, solve for t_e,

$H_0^2(q_0 - 1) = K/a_0^2 \;\; obs.: 0 < q_0 < 2$

For $|K| = 1$ gives $a_0 \geq 1 \times 10^{28} cm$.

$$\rho_{re} a_0^4 = \rho_{r_0} a_0^4 \geq \left(\frac{\rho_{r_0}}{10^{-33}\ gm/cm^{-3}}\right)(2.9 \times 10^{116}) \qquad \hbar = c = 1$$

Example 9.4. Show that

$$\left(\frac{32\pi G}{3}\rho_{r_0} a_0^4\right)^{1/2} \gtrsim (5.3 \times 10^{15}\text{sec})\left(\frac{\rho_{r_0}}{10^{-33} gm/cm^{-3}}\right)$$

Is $t_e \simeq \left(\frac{32\pi G}{3}\rho_{re}\right)^{1/2}$ small enough to neglect the curvature term in $a(t)$?

Solution

$$\left(\frac{32\pi G}{3}\rho_{ro} a_0^4\right)^{1/2} \gtrsim \left(\frac{32\pi G}{3}\rho_{ro}(1 \times 10^{28} cm)^4\right)^{1/2}$$

$$\gtrsim$$

$$\left(\frac{32\pi}{3}\,\frac{6.67\times10^{-8} cm^3}{gm\ sec^2}\,\frac{(10^{-33} gm\ cm^{-3})(1\times10^{28} cm)^4}{\underbrace{[3\times10^{10} cm/sec]^4}_{\substack{C=1 \\ \text{to regain units}}}}\right)^{1/2} \left(\frac{\rho_{ro}}{10^{-33} gm\ cm^{-3}}\right)^{1/2}$$

$$\gtrsim (2.76 \times 10^{31} sec^2)^{1/2} \left(\frac{\rho_{ro}}{10^{-33} gm\ cm^{-3}}\right)^{1/2}$$

$$\gtrsim (5.3 \times 10^{15} sec)^{1/2} \left(\frac{\rho_{ro}}{10^{-33} gm\ cm^{-3}}\right)^{1/2}$$

Example 9.5. Show that

$$t_e = (2.1 \times 10^{13}\text{sec})\left(\frac{\rho_{m_0}}{10^{-30} gm/cm^3}\right)^{-2}\left(\frac{\rho_{r_0}}{10^{-33} gm/cm^3}\right)^{3/2}$$

Solution

t_e

$$\simeq \left(\frac{32\pi G}{3}\rho_{re}\right)^{-1/2}$$

$$a(t) \simeq A^{1/2} a_e t^{1/2}$$
$$a(t_e) \simeq a_e$$

$$A = \left(\frac{32\pi G\rho_{re}}{3}\right)^{1/2}$$

$$t_e^{1/2} = A^{-1/2} = \left(\frac{32\pi G}{3}\rho_{re}\right)^{-1/4}$$

$\rho_{mo} \simeq 10^{-30}$

$\rho_{ro} \simeq 10^{-33}$

$\rho_{matter}(t) = \rho_{mo} a_0^3 a(t)^{-3}$

$\rho_{rad}(t) = \rho_{ro} a_0^4 a(t)^{-4}$

At t_e: $\rho_{matter} = \rho_{rad}$

Let $a_e = a(t_e)$, then $\frac{a_e}{a_0} = \frac{\rho_{ro}}{\rho_{mo}}$ So,

$$\rho_{rad}(t_e) = \rho_{re} = \rho_{ro}\left(\frac{a_o}{a_e}\right)^4 = \rho_{mo}^4\, \rho_{ro}^{-3}$$

$$\rho_{re} = \left(\frac{\rho_{mo}}{10^{-30}_{gm\ cm^{-3}}}\right)^4 \left(\frac{\rho_{ro}}{10^{-33}_{gm\ cm^{-3}}}\right)^{-3} \times \left(10^{-21}_{gm\ cm^{-3}}\right)$$

and

$t_e \simeq$

$$\left(\frac{32\pi}{3}\frac{6.67\times10^{-8}}{gm\ cm^{-3}sec^2}(10^{-21}gm\ cm^{-3})\right)^{-1/2} \left(\frac{\rho_{mo}}{10^{-30}gm\ cm^{-3}}\right)^{-2} \left(\frac{\rho_{ro}}{10^{-33}gm\ cm^{-3}}\right)^{3/}$$

.

$$\cong (2.1\times10^{13}sec)\left(\frac{\rho_{mo}}{10^{-30}gm\ cm^{-3}}\right)^{-2} \left(\frac{\rho_{ro}}{10^{-33}gm\ cm^{-3}}\right)^{3/2}.$$

So we can neglect root in calc. t_e, and t_e is:

$$t_e = (6.6\times10^5\text{yr})\left(\frac{\rho_{m_0}}{10^{-30}\ g/cm^3}\right)^{-2}\left(\frac{\rho_{r_0}}{10^{-33}\ gm/cm^3}\right)^{3/2}$$

So we have to good approximation throughout the radiation dominated era that:

$$a(t) \simeq \left(\frac{32\pi G}{3}\rho_{r_0}a_0^4\right)^{1/4} t^{1/2} \quad \text{for} \quad K = 0, \pm1.$$

Temperature and density evolution during rad. dom. stage:

$T = T_0 a_0 a^{-1}$ for all t

For $t < t_e$ we get

$$T = T_0\left(\frac{32\pi G}{3}\rho_{r_0}\right)^{-1/4} t^{1/2} \qquad T_0 = 2.7^oK \ \text{ for (photons)}$$

This is a good approximation back to $\sim 10^{12\,o}K \Leftrightarrow 10^{-4}sec$ where the nucleons annihilate.

9.16 Decoupling of Neutrinos from matter:
The $e^+, e^-, \nu_e, \nu_\mu, \mu^+, \mu^-$ are kept in equil. by

$$e^- + \mu^+ \rightleftarrows \nu_e + \nu_\mu$$
$$\nu_\mu + e^- \rightleftarrows \nu_e + \mu^-$$
$$\nu_\mu + \mu \rightleftarrows \nu_e + e^+ \text{ etc.}$$

Masses and rates: $m_e c^2 = .511\ MeV$
$m_\mu c^2 = 207\ Me$

Lifetime $\sim 2 \times 10^{-6} sec$ for $\mu^- \to e^- + \bar{\nu}_e + \nu_\mu$

$$\left.\begin{matrix}\pi^0 \\ \pi^+, \pi^-\end{matrix}\right\} muon \quad \begin{matrix} m_{\pi^0} = 264 m_e & lifeline < 4 \times 10^{-16} sec(em\ decay\ \pi^0 \to \gamma \\ m_{\pi^+} = 273 m_e & lifeline < 2.6 \times 10^{-8} sec(weak\ decay)\end{matrix}$$

$k_B = 1.38 \times 10^{-16} erg\ {}^oK^{-1}$
$k_B^{-1} m_e c^2 = 5.9 \times 10^{9\,o}K$
$k_B^{-1} m_\mu c^2 = 1.2 \times 10^{12\,o}K$
$k_B^{-1} m_\pi c^2 = 1.6 \times 10^{12\,o}K$
$m_\tau c^2 = 1{,}784\ MeV = 3{,}492 m_e$
Mean lifetime = $3.4 \times 10^{-13} sec.\ for\ \tau^- \to \mu^- + \bar{\nu}_\mu + \nu_\tau$ 18%
$\tau^- \to e^- + \bar{\nu}_e + \nu_\tau$ 16%
Most $\tau \to hadrons$ 50%

If GWS is correct then the main interaction which keep ν_τ and $\bar{\nu}_\tau$ in equil with the rest are $\tau^- + \mu^+ \rightleftarrows \nu_\tau + \bar{\nu}_\mu\ etc.$ Let's consider the situation at $T < 1.3 \times 10^{12\,o}K$ amongst the $\mu^+\mu^-e^+e^-\nu_e\bar{\nu}_e\nu_\mu\bar{\nu}_\mu\nu_\tau\bar{\nu}_\tau\gamma, \gamma$.
At $T \cong 6 \times 10^{9\,o}K$ we reach the e^+e^- annihilation energy (from EM: $e^+ + e^- \rightleftarrows \gamma + \gamma$). And we have the Weak interactions: $e^+ + e^- \rightleftarrows \nu_e + \bar{\nu}_e$

$$e^+ + e^- \rightleftarrows \nu_\mu + \bar{\nu}_\mu$$
$$e^+ + e^- \rightleftarrows \nu_\tau + \bar{\nu}_\tau$$

To see until what temperature the weak interactions are able to maintain equilibrium (as we lower the temperature) we have to compare the reaction rate per particle $r = n\sigma\langle v\rangle$ with the expansion rate $H = \frac{\dot{a}}{a}$, ($where\ T \leq 1.2 \times 10^{12\,o}K\ \ is\ the\ \mu^+\mu^-$ annihilation temp.) The expansion rate:

$$H = \frac{1}{2t} \simeq \left(\frac{T_\gamma}{2 \times 10^{10\,o}K}\right)^2 \sec^{-1}.$$

Consider $\langle v \rangle \simeq c$; and n: the lepton number distribution used. The lepton number per unit volume with momentum in range p to $p + dp$:

$$n(p)dp = g4\pi h^{-3} p^2 dp \left\{\exp\left[\frac{E-\mu}{k_B T}\right] + 1\right\}^{-1}.$$

The chemical potentials: $\mu_\gamma = 0$ and $\mu_{e^-} = -\mu_{e^+}$, $|\mu_e| \ll K_B T$. Likewise, we will take $|\mu_\nu| \ll KT$ for all types of neutrons. Also, have $E = \sqrt{p^2c^2 + m^2c^4} \simeq pc$ for $T > 6 \times 10^{9\,o}K$. We have $g = 2$ for e^-, e^+, and $g = 1$ for ν (left handed massless neutrinos)

$$n = \int_0^\infty n(p)\, dp = g4\pi h^{-3} \int_0^\infty \frac{dp\, p^2}{\exp\left[\frac{E-\mu}{KT}\right] + 1}$$

$$= g4\pi h^{-3} \left(\frac{KT}{c}\right)^3 \int_0^\infty d\tau \frac{x^2}{e^x + 1}$$

So,

$n = g(0.061)\left(\frac{kT}{\hbar c}\right)^3.$

Thus,

$n_e = .12\left(\frac{KT}{\hbar c}\right)^3.$

Since $\sigma_{weak} \cong g_{weak}^2 (KT)^2 (\hbar c)^{-4}$ ($g_{weak} = 1.4 \times 10^{-49} erg\ cm^3$):

$$r = n\sigma\langle v \rangle = (0.1)\left(\frac{KT}{\hbar c}\right)^3 g_w^2 (KT)^2 (\hbar c)^{-4} c = 9.6 \times 10^{-53} \frac{1}{sec}\left(\frac{T}{1^o K}\right)^5$$

$$r \simeq \frac{1}{\tau_{collision}}$$

Important ratio is:

$$\frac{r}{H} = \frac{2t}{\tau_{coll}} \cong 9.6 \times 10^{-53} \left(\frac{T}{1^o K}\right)^5 2 \times \left(\frac{1.4 \times 10^{10\,o}K}{T}\right)$$

$$\cong 3.8 \times 10^{-32} \left(\frac{T}{1^o K}\right)^3 \cong \left(\frac{T}{3 \times 10^{10\,o}K}\right)^3$$

So, at $T \sim 1.2 \times 10^{12}$ μ^+ and μ^- annihilate but the weak interactions keep the υ's and γ's in equilibrium via the e's. When $T \sim 3 \times 10^{10\,o}K$ the v's decouple from the e's.

Now, what happens to the υ distribution function after decoupling (is Black Body distribution maintained?). At the time of decoupling we have the Fermi-Dirac distribution for υ:

$$n(p_D)dp_D = 4\pi h^{-3} p_D^3 dp_D \left\{\exp\left[\frac{p_D^2}{KT_D}\right] + 1\right\}^{-1}$$

where T_D is the decoupling temp. As the universe expands the number per unit volume in the p to $p + dp$ range which corresponds to p_D to $p_D + dp_D$ at decoupling:

$$\underline{n}(p)dp = \left(\frac{a}{a_D}\right)^{-3} n(p_D)dp_D$$

(Note that low frequency modes violate the adiabticity condition. Get parametric excitation of (cut off) particle production of low frequency modes.)

Also, $\lambda \propto a(t)$ (wavelength of particle)
Since $p \sim \frac{h}{\lambda}$ we have $p = \left(\frac{a}{a_0}\right)^{-1} p_0$

So,

$$\underline{n}(p)dp = 4\pi h^{-3} p^2 dp \left\{\exp\left[\frac{pc}{KT}\right] + 1\right\}^{-1}$$

With T redefined as $T \equiv \left(\frac{a}{a_0}\right)^{-1} T_D$. So, $\underline{n}(p)$ is a black body distribution.

After $T \sim 3 \times 10^{10\,o}K$ we have $T_\nu = \left(\frac{a}{a_0}\right)^{-1} T_D$. At $T \sim 6 \times 10^{9\,o}K$ the e^+e^- annihilate and the temperature of the photons, T_γ, increases, but the temp. of the ν's, T_ν, does not increase. So after that, T_γ is larger than T_ν.

When $T \sim 6 \times 10^{9\,o}K$: $n_\ell(p)dp = \frac{1}{e^{E/K_BT}+1}$ becomes small (when $kT \ll m_0C^2$). The process is reversible so adiabatic, entropy doesn't change. Entropy of e^+, e^-, γ system remains constant.

Consider a volume V with total entropy S and energy E. Regard S and E as functions of T and V. We have the following:

- The 2nd Law: $TdS = dE + PdV \ \ where\ P = pressure.$
- Entropy, S, in a comoving volume is conserved.
- Baryon number is also conserved: $N_B = \#_{baryons} - \#_{antibaryons}.$
- $ratio = \frac{N_B}{S} = \frac{N_B/V}{S/V} = \frac{n_B}{s}$ is conserved.
- Dimensionless: $\frac{K_B N_B}{S}$ is conserved.

Most of entropy is in massless or relativistic particles today:

$$S = S_{\nu_\tau} + S_{\overline{\nu}_\tau} + S_{\nu_\mu} + S_{\overline{\nu}_\mu} + S_{\nu_e} + S_{\overline{\nu}_e} + (S\ gravators) = 6S_\nu + S_\gamma$$

Each ν has a temp. $\simeq 1.9^oK$:

$$s_\nu = \frac{S_\nu}{V} = \frac{7}{12} bT_\nu^3 \qquad b = \frac{8\pi^5 K^4}{15h^3c^3} = 7.56 \times 10^{-15} erg\ cm^{-3\,o}K$$

$$S_\gamma = \frac{4}{3} bT_\gamma^3$$

So,

$S_\nu = 3.0 \times 10^{-14} erg\ {}^oK^{-1} cm^{-3}$
$S_\gamma = 2.0 \times 10^{-13} erg\ {}^oK^{-1} cm^{-3}$
$k_B = 1.38 \times 10^{-16} erg\ {}^oK^{-1}$
$s = 3.8 \times 10^{-13} erg\ {}^oK^{-1} cm^{-3}$
${}^{s}/_{k_B} = 2.75 \times 10^3 cm^{-3}$

\# of photons per unit volume $\approx 140\, {}^{\gamma/s}/_{cm^3}$ (from integrating # density with $T = 2.73^oK$), so, using ${}^{S}/_{K_B} \approx 10n_\gamma$ and that Baryon number $n_B = \frac{\rho_{galaxies}}{m_{nucleon\ (hydrozen)}}$. Thus:

$$10^{-29}\ gm/cm^3 \geq \rho_{matter} \geq 2 \times 10^{-31}\ gm/cm^3$$

$$6 \times 10^{-6} cm^{-3} \geq n_B \geq 1.2 \times 10^{-7} cm^{-3}$$

and

$2.2 \times 10^{-9} \geq \frac{K n_B}{\sigma} \geq 4.4 \times 10^{-11}$ ← conserved baryon # to entropy ratio

9.17 Decoupling of photons

First consider the ratio of the reaction rate per particle, r_{coll} to the expansion rate H. At early times most of the photon scattering is off free charged particles, so starting with the scattering cross-section of te electron:

$$\sigma_e \sim \left(\frac{e^2}{mc^2}\right)^2$$

Note that e^2/mc^2 is the classical radius of electron in the non-relativistic Thomas cross-section for scattering of photon by electron. More precisely:

$$\sigma_\tau = \frac{8\pi}{3}\left(\frac{e^2}{m_e c^2}\right)^2 = 0.665 \times 10^{-24} cm^2.$$

Note that scattering off electrons dominates by 10^6 over scattering off protons. Thus we have $r_{coll} = n_e \sigma_t c$, where n_e is the # density of electrons:

$$n_e(t) = n_{e_0}\left(\frac{a_0}{a(t)}\right)^3 = n_{e_0}\left(\frac{T_\gamma(t)}{T_{\gamma_0}}\right)^3$$

where n_{e_0} is the # density at present time (including bound!).

Suppose that the ratio $\frac{r_{coll}(t)}{H(t)}$ becomes of order 1 when the universe <u>is matter dominated</u>. I.E. assume $t_{Decoupling}$ of photons is greater than t_{equiv} the time of $\rho_\gamma = \rho_M$. Then $a \propto t^{2/3}$ from t_D to t_e: $\left(\frac{a_0}{a(t)}\right) = \left(\frac{t_0}{t}\right)^{2/3}$, for almost spatially flat universe. So, $n_e(t) = n_{e_0}\left(\frac{t_0}{t}\right)^2$ also, for $\propto t^{2/3}$ and $H = \dot{a}/a \Rightarrow H = \frac{2}{3t}$. Hence $\frac{r_{coll}}{H} \simeq \frac{n_e(t)\sigma_T c}{H(t)} = \left(\frac{3}{2} n_{e_0}\sigma_T c t_0^2\right)t^{-1}$.

$$\frac{r_{coll}}{H} \simeq 1 \Rightarrow t_D = \frac{3}{2} n_{e_0}\sigma_T c t_0^2$$

So,

$n_{e_0} = 0.85\frac{\rho_{m0}}{m_N} \nwarrow 10^{-30}\, gm/cm^3$ 80% of n_B are protons and one electron to a proton gives result.

$n_{e_0} \simeq 5 \times 10^{-7} cm^{-3}$

Using $H \equiv 75$:

$$t_D \simeq \frac{3}{2}(5\times10^{-7}cm^{-3})(0.67\times10^{-24}cm^2)(3 \times 10^{10}\,cm/sec)(2.78\times10^{17}sec)^2$$

$$\underline{t_D \simeq 1.2\times10^{15}sec}$$

which is probably greater than t_{equiv} (when $\rho_\gamma = \rho_\mu$):

$$T_\gamma(t_o) = T_{\gamma_0}\frac{a_0}{a(t)} = T_{\gamma_0}\left(\frac{1}{4.2\times10^{-3}}\right)^{2/3} \qquad \frac{a_0}{a(t)} = \left(\frac{t_0}{t}\right)^{2/3} = 104^oK$$

So atom formation happens, causing decoupling, long before the temp. decreases to the 104^oK temp of the time posed for reaction rate being too slow to keep up with the expansion. So, photons decouple from matter when recombination of Hydrogen (called recombination even though it is the 1st combination!). To estimate temp. of recomb. we use law of mass action:

$$\alpha = \frac{n_p}{n} = \frac{n_p}{n_H + n_p} \qquad \begin{array}{l} n_p = \text{density of protons} \\ n_H = \text{density of Hydrogen atom} \end{array}$$

where $n = n_H + n_p$. Using the partition function:

$$\frac{\alpha^2}{(1-\alpha)}n = \left(\frac{m_e kT}{2\pi\hbar^2}\right)^{3/2} \exp\left[-\frac{13.6eV}{kT}\right]$$

where 13.6eV is the binding energy of H atom. When $\alpha = \frac{1}{2}$, half the e^- are bound T at $\alpha = \frac{1}{2}$ is? We have:

$$n \simeq n_{e_0}\left(\frac{a_0}{a}\right)^3 = (5\times10^{-7}cm^{-2})\left(\frac{T}{2.7^oK}\right)^3$$

(Eqn. 9-33)

9.18 Puzzling things in Standard Model

9.18.1 Isotropic CMBR

When the universe becomes transition to photons: Except for certain freq's where they are absorbed by Hydrogen etc. the decoupled photons are red shifted and detected in our microwave detector yielding the 2.73^oK B.B. spectrum. Consider the "surface of last scattering" at the recombination time with photon $T = 4{,}000^oK$, the redshift of this surface will be $z = 1100$, while the z of furthest observed object using non-space-based, a quasar, is $z \simeq 3\ or\ 4$, later with Hubble, and Webb have $z \simeq 20$.

Observation with COBE satellite shows that the cosmic microwave background is a BB spectrum of $2.7°K$ and is isotropic to at least 1 part in 10^5. Can the standard model explain this?

Consider comoving coordinates with $(z = 0)$, singularity at $t = 0$. Our fundamental world line is $x = y = z$. Consider surface of sphere of last scattering (no photons "communicate" after last scattering). Call the coordinate radius of the past light cone when it hits the singularity is r_p (particle horizon). To a good approximation for $t < t_p$:

$$a(t) = At^{1/2} \quad \text{with} \quad A = \left(\frac{32\pi G}{3}\rho_{r_0} a_0^4\right)^{1/4}$$

So for $0 \le t \le t_r$: $ds^2 = -dt^2 + a^2(t)dr^2 + a^2 d\Omega^2$, along light cone $ds^2 = 0, d\Omega^2 = 0$:

$$\int_0^{t_r} \frac{dt}{a(t)} = -\int_{r_p}^{0} dr = r_p \quad (dt = adr)$$

Thus,

$$r_p = A^{-1}\int_0^{t_r} dt\, t^{-1/2} = 2A^{-1}t_r^{1/2}.$$

So, if P and Q are separated by a distance $2r_p$ then they could not have communicated since the big bang. Thus, P and Q aren't coupled if:

$$\left[\left(x_P - x_Q\right)^2 + \cdots\right]^{1/2} > 2r_p = 4A^{-1}t_r^{1/2}$$

Thus, the proper separation (the distance one would measure) between the furthest world lines which could have communicated since the singularity is

$$D = a(t_r)2r_p = At_r^{1/2}4A^{-1}t_r^{1/2} = 4t_r$$

$\underline{D = 4t_r}$

1. Ex. Redo the calc. of D without assuming spatial flatness, i.e.., that t_r was not so early that we could ignore curvature correction in $a(t)$.

$$ds^2 = -dt^2 + a^2(t)\left\{\frac{1}{1 - Kr^2}dr^2 + \cdots\right\}$$

The prop. radius at t_r of the sphere of last scattering.

$$R_{proper} = -a(t_r)\int_{r_r}^{0} \frac{dt}{(1 - r^2)^{1/2}} = a(t_r)\int_{t_r}^{t_0} \frac{dt}{a(t)}$$

Since we are very close to closure density $a(t) \simeq Bt^{2/3}$ (matter dom.)(assume $K = 0$)

$R_{proper} = t_r^{2/3} \int_{t_r}^{t_0} dt\, r^{-2/3} = 3t_r^{2/3}\left[t_0^{1/3} - t_r^{1/3}\right] = 3t_r\left[\left(\frac{t_0}{t_r}\right)^{1/3} - 1\right] = \frac{3}{4}D\left[\left(\frac{t_0}{t_r}\right)^{1/3} - 1\right]$

$T_\gamma(t) = T_{\gamma_0} a_0 a^{-1}(t)$ with $a \simeq Bt^{2/3}$ (matter dom., $K \simeq 0$)

$T_\gamma(t) \propto t^{2/3}$ or $t \propto T_\gamma^{-3/2}$

So,

$$R = \frac{3}{4}D\left[\left(\frac{T_0}{T_r}\right)^{-1/2} - 1\right] = \frac{3}{4}D\left[\left(\frac{2.7}{4{,}000}\right)^{-1/2} - 1\right] = 29D$$

The angle distended is arc D with radius 29D, and scales by the same expansion factor (as universe expands all the proper dist. get mult. by the same factor $\frac{a(t_0)}{a(t_1)}$ thus the angle remains the same), thus remains the same and is

$$\alpha \simeq \frac{1}{29} \text{ radians } \simeq 2^\circ$$

(Eqn. 9-34)

where α is the angle subtended by two regions that aren't thermalized. But we see all angles thermalized. From this problem (the Particle Horizon Problem) we get the hypothesis of the inflationary universe.

9.18.2 The Particle Horizon Problem and the Inflationary Universe

Hesitation, Mixmaster all tried to explain each has a problem. Infl. Model has problem of cosm. constant (spatial flatness).

The spatial flatness question

With $\Lambda = 0, K = 0$, and using $\left(\frac{\dot{a}}{a}\right)^2 = \frac{8\pi G}{3}\rho$ with $\rho_0 = \rho(t_0)$ at present time. Then evaluation of t_0 gives a ρ_0 remarkably close to that observed. Consider Einstein equations with $\Lambda = 0$ but arbitrary *t*.

$$\left(\frac{\dot{a}}{a}\right)^2 + \frac{K}{a^2} = \frac{8\pi G}{3}\rho$$

and

$$-2\frac{\ddot{a}}{a} - \left(\frac{\dot{a}}{a}\right)^2 + \frac{K}{a^2} = \frac{8\pi G}{3}\rho$$

Now, $p = 0, \frac{\dot{a}}{a} = H_0 = 10^{-18}\,\text{sec}^{-1}$, $q_0 = -\left.\frac{\ddot{a}q}{\dot{a}^2}\right|_{t_0}$ between 0 and 2. We have $2q_0H_0^2 = H_0^2 - \frac{K}{a_0^2} = 0$ and $\frac{K}{a_0^2} = (2q-1)H_0^2$. Since $K = \pm 1 \Rightarrow \frac{1}{a_0^2} \lesssim H_0$ provides the best constraint, or $a_0 \gtrsim H_0^{-1} = 10^{28} cm$. This would mean that at very early times that unless $\frac{1}{a(t)} \lll \frac{\dot{a}(t)}{a(t)}$ at very early times we couldnot have $\frac{1}{a_0} \lesssim \frac{\dot{a}_0}{a_0}$ today. This is the spatial flatness problem. Particle production could explain but inflationary model works once again.

To summarize:

1. So, if $K \neq 0$ then why is it such that $|K/a^2| \lll \left(\frac{\dot{a}}{a}\right)^2$ at early times, or is there a mechanism whereby this comes about.
2. Why is the microwave temperature so uniform today?
3. Why is the Λ so small?

Inflation explains (1) and (2) but make (3) more of a problem.

9.19 Inflation

Inflation is a consequence of the GUT theories if the Higgs Mechanism is used. Weak, e-m, and strong forces are mediated by vector bosons:

e-m: photons, $m = 0$, long range force.
Weak and strong: $m \neq 0$, short range force $\Rightarrow$ vector bosons massive.

In unified theories all bosons are originally massless due to symmetry. The symmetry is spontaneously broken by the Higgs mechanism. In the Higgs mechanism there is a scalar field φ with a self interaction such that the field acquires a non-zero vacuum experimental value.

$$D_\mu \varphi D^\mu \varphi \Rightarrow \varphi_0^2 A_\mu A^\mu \Rightarrow \underbrace{\langle \varphi_0^2 \rangle}_{mass.} A_\mu A^\mu$$

Theory remains renormalizable since $m = 0$. The part of the Lagrangian of interest to us is

$$\mathcal{L} = -\frac{1}{2} g^{\mu\nu} \partial_\mu \varphi \partial_\nu \varphi - V(\varphi)$$

For the Higgs mechanism we need a $V(\varphi)$ like:

$$V(\varphi) = \underbrace{-\frac{1}{2}\mu^2\varphi^2}_{\substack{wrong\ sign\\ mass\ term}} + \underbrace{\frac{1}{4}\lambda\varphi^4}_{self\ interaction} + const.$$

If φ is a doublet we consider a hot due to phase freedom. The lowest energy has $\varphi = \varphi_0 \neq 0$. To get the contribution of φ to $T_{\mu\nu}$ consider

$$S_\varphi = \int d^4x\sqrt{-g}\mathcal{L}(\varphi) \to \delta S_\varphi = \int d^4x \frac{\partial}{\partial g^{\mu\nu}}\left(\sqrt{-g}\mathcal{L}\right)\delta g_{\mu\nu}(x)$$

$$\delta S_\varphi = \frac{1}{2}\int d^4x\sqrt{-g}T^{\mu\nu}(x)\delta g_{\mu\nu}(x)$$

So, $T^{\mu\nu} = \frac{2}{\sqrt{-g}}\frac{\partial}{\partial g^{\mu\nu}}\left(\sqrt{-g}\mathcal{L}\right)$. Since $\frac{2}{\sqrt{-g}}\frac{\partial}{\partial g_{\mu\nu}}\sqrt{-g} = \frac{\partial}{\partial g_{\mu\nu}}\ln(-g) = \frac{1}{g}\frac{\partial}{\partial g_{\mu\nu}}g = g^{\mu\nu}$, we have:

$$T^{\mu\nu} = g^{\mu\nu}\mathcal{L} + 2\frac{\partial \mathcal{L}}{\partial g_{\mu\nu}}$$

$$= g^{\mu\nu}V(\varphi) - \frac{1}{2}\, g^{\mu\nu}g^{\lambda\sigma}\partial_\lambda\varphi\partial_\sigma\varphi$$

$$+ 2\frac{\partial}{\partial g_{\mu\nu}}\left(-\frac{1}{2}\, g^{\lambda\sigma}\partial_\lambda\varphi\partial_\sigma\varphi\right)$$

Have $g^{\sigma\nu}g_{\mu\nu} = \delta^\sigma_\mu$

Exercize: Show $\frac{\partial g^{\alpha\sigma}}{\partial g_{\mu\nu}} = -\frac{1}{2}(g^{\alpha\mu}g^{\sigma\nu} + g^{\alpha\nu}g^{\sigma\mu})$

Then:

$$T^{\mu\nu} = -g^{\mu\nu}V(\varphi) + \partial^\mu\varphi\partial^\nu\varphi - \frac{1}{2}g^{\mu\nu}\partial^\lambda\varphi\partial_\lambda\varphi$$

Inflation in grand unified theories

$$\mathcal{L} = -\frac{1}{2}g^{\mu\nu}\partial_\mu\varphi\partial_\nu\varphi - V(\varphi) \quad with \quad V(\varphi)$$

$$= -\frac{1}{2}\mu^2\varphi^2 + \frac{1}{4}\lambda\varphi^4 + const$$

Then have:

$$T^{\mu\nu} = -g^{\mu\nu}V(\varphi) - \frac{1}{2}g^{\mu\nu}g^{\lambda\sigma}\partial_\lambda\varphi\partial_\sigma\varphi + 2\frac{\partial}{\partial g_{\mu\nu}}\left(-\frac{1}{2}g^{\lambda\sigma}\partial_\lambda\varphi\partial_\sigma\varphi\right)$$

$$= -g^{\mu\nu}V(\varphi) + (\partial^\mu\varphi)(\partial^\nu\varphi) - \frac{1}{2}g^{\mu\nu}\partial^\lambda\varphi\partial_\lambda\varphi$$

Suppose φ is nearly constant at its ground state value of φ_0, i.e., suppose $\langle T^{\mu v}\rangle = -g^{\mu v}V(\varphi_0)$ now. Normally in a grand unified theory with vector bosons of Mass 10^{15} GeV, one expects $V(\varphi_0)$ would "normally" be very large. In Einstein's Equations, this would contribute:

$$R^{\mu v} - \frac{1}{2}g^{\mu v}R = 8\pi G\,\langle T^{\mu v}\rangle = -8\pi G V(\varphi_0)g^{\mu v}$$

So, the effect of $V(\varphi_0)$ would be to give a large cosmological constant (generally) since:

$$\Lambda = 8\pi G V(\varphi_0)$$

and would mess up the Newtonian approximation. So we must fine tune such that $V(\varphi_0) \approx 0$!

As we go back in time the universe becomes "hot", so we have thermal expectation values: $\langle T^{\mu v}\rangle_T$, and one finds that the $\lambda\varphi^4$ term causes a contribution proportional to $T^2\varphi^2$ so

$$V(\varphi) = -\frac{1}{2}\mu^2\varphi^2 - \frac{1}{4}\lambda\varphi^4 + (\cdots)T^2\varphi^2 + const.$$

Thus, as we go back in the early universe we see the phase transition occurring at T_c. In early universe the ground state has $\langle\varphi\rangle = 0$ (a "false vacuum") and $\langle T^{\mu v}\rangle = -g^{\mu v}V(0)$ and in the Einstein equations have $\Lambda = 8\pi G V(0)$ (which is large and positive $\Rightarrow$ deSitter inflation). Have $\rho_\Lambda = -p_\Lambda = V(0)$.

The Inflationary Era gives large particle horizons so uniformity understandable, and makes the $|K|/a^2$ term much less than the $(\dot{a}/a)^2$ term.

Reheating problem: red-shifting kills radiation would lead to oscillation of φ about min.

9.20 Galaxy formation [52]

9.20.1 The Jean's Instability

Consider a non-relativistic Newtonian fluid of mass density ρ, pressure p, velocity $\vec{V}$, and gravitational $\vec{g}$ (force per unit mass).

Equation of continuity: $\frac{\partial\rho}{\partial t} + \vec{\nabla}\cdot(\rho\vec{V}) = 0$

Euler equation: $(\vec{a} = \vec{F}/m)$ $\quad \frac{\partial\vec{v}}{\partial t} + (\vec{V}\cdot\vec{\nabla})\vec{V} = -\frac{1}{\rho}\vec{\nabla}\text{p} + \vec{g}$

Grav. Field equations:
$$\vec{\nabla} \times \vec{g} = 0 \Rightarrow g = -\vec{\nabla}\varphi$$
$$\vec{\nabla} \cdot \vec{g} = -4\pi G\rho \text{ or } \nabla^2\varphi = 4\pi G\rho$$

9.20.2 Perturbation theory

Take the unperturbed "solution" to be $\rho = \rho_0 = const,\ p = p_0 = const,$ $\vec{V} = \vec{V}_0 = 0,$ and $\vec{g} = 0$:

To next order: $V = V_0 + V_1,\ \rho = \rho_0 + \rho_1,\ p = p_0 + p_1,\ \varphi = \varphi_0 + \varphi_1, and\ \nabla^2\varphi_0 = 4\pi G\rho$. For the adiabatic sound velocity $V_s^2 \equiv \left(\frac{\partial p}{\partial \rho}\right)_{adiabatic}$. Since $V_s^2 \cong p_1/\rho_1 \Rightarrow p_1 = V_s^2\rho_1$.

From:
$$\frac{\partial \rho_1}{\partial t} + \rho_0\vec{\nabla} \cdot \vec{V}_1 = 0$$
$$\frac{\partial \vec{V}_1}{\partial t} + \frac{V_s^2}{\rho_0}\vec{\nabla}\rho_1 + \vec{\nabla}\varphi_1 = 0$$
and
$$\nabla^2\varphi_1 = 4\pi G\rho_1$$
we get:
$$\frac{\partial^2 \rho_1}{\partial t^2} + \rho_0\vec{\nabla}\left(-\frac{V_s^2}{\rho_0}\vec{\nabla}\rho_1 + \vec{\nabla}\varphi_1\right) = 0$$
and
$$\frac{\partial^2 \rho_1}{\partial t^2} = V_s^2\nabla^2\rho_1 + \rho_0 4\pi G\rho_1$$

Solution:
$$\rho_1(\vec{r}, t) = \rho_0 A \exp[-|\vec{K} \cdot \vec{r} + \vec{t}|]$$

$\omega^2 = K^2V_s^2 - 4\pi G\rho_0$ (V_s^2 assumed const.)

Density contrast: $\delta(\vec{r}, t) = \frac{\delta\rho}{\rho} \cong \frac{\rho_1}{\rho_0}$.

When $\omega^2 > 0 \Rightarrow$ solution oscillatory (sound waves) when $|K|$ sufficiently small, $\omega^2 < 0 \Rightarrow$ exponential modes, growing mode increases until $\delta \simeq 1$ then perturbation theory breaks down. Thus, we have:

$$K_{critical} = \frac{\sqrt{4\pi G\rho_0}}{V_s} = K_J.$$

When $K < K_J$ one has gravitational instability (notation is using Jean's wavenumber K_J). For $K \ll K_J$ ρ_1 has a growing mode, which grows exponentially with a time scale. We have

$$\omega_c^2 \approx -4\pi G\rho_0 \Rightarrow \omega_c \sim -i\sqrt{4\pi G\rho_0}$$

and

$$\rho_1 \propto e^{-|\omega_c|^t} = e^{-t/\tau}$$

Thus:

$$\tau = \frac{1}{|\omega_c|} = (4\pi G\rho_0)^{-1/2}.$$

The Jean's mass is defined as the total mass contained in a sphere of radius $\frac{\lambda_J}{2} = \frac{\pi}{K_J}$:

$$M_J = \frac{4}{3}\pi\left(\frac{\pi}{K_J}\right)^3 \rho_0$$

9.20.3 An irreversible collapse heuristic argument:

Two time scales:

(1) the time-scale for gravitational collapse of an object: $\tau_{grav} \approx (G\rho_0)^{-1/2}$.

(2) the time-scale for pressure to restore equilibrium: $\tau_{pressure} \simeq \frac{\lambda}{V_s}$.

When $\tau_{pressure} > \tau_{grav}$ then collapse cannot be prevented

$$\frac{\lambda}{V_s} > (G\rho_0)^{-1/2} \quad \rightarrow \quad \frac{1}{\lambda} < \left(\frac{G\rho_0}{V_s^2}\right)^{1/2} = K_J$$

which is what we had before for K_J.

We want to compare to the time scale for the expansion of the universe, e.g., compare $\frac{\dot{a}}{a}$ with $\frac{1}{\tau_{dyn}} \simeq (G\rho_0)^{1/2}$. Since

$$\left(\frac{\dot{a}}{a}\right)^2 \simeq \frac{8\pi G}{3}\rho_0 \rightarrow \frac{\dot{a}}{a} \approx (G\rho_0)^{1/2}$$

the effect of the expansion on τ_{dyn} will be large. For Rad. Dom the exp. growth of the instability is deduced to a power law:

$$K_J \sim \sqrt{\frac{\rho_0}{V_s^2}} \propto \frac{1}{a(t)}.$$

In an expanding universe $K_J(t) \simeq \sqrt{\frac{4\pi G\rho_0(t)}{V_S^2}}$. V_s also depends on time through the temp. which alters radically at recomb. etc. Define the Jean's mass in a sphere of radius $\frac{\lambda_J}{2}$:

$$M_J = \frac{4}{3}\pi\left(\frac{\lambda_J}{2}\right)^3 \rho_0 = \frac{4}{3}\pi\rho_0\left(\frac{\pi}{K_J}\right)^3$$

From the time of e^+e^- annibilation ($T \simeq 10^{10\,o}K$) to recomb. ($T \simeq 4000^oK$) we can take the main constituents to be hydrogen ions, photons, electrons (neutrinos decoupled). Since the photon entropy density S is large, we can neglect during this stage the matter entropy and pressure.

For the fluid:

$$\rho = nm_H + bT^4$$

where n is the number density per unit volume, and

$$P = \frac{1}{3}bT^4$$

We have as conserved quantity the entropy to baryon ratio:

$$\sigma = \frac{\frac{4}{3}bT^3}{nK_B}$$

For σ constant this implies $n \propto T^3$, so $\delta n = 3n\frac{\delta T}{T}$ thus $(3nm_H + 4bT^4)\frac{\delta T}{T} = \delta\rho$. So

$$V_S^2 = \frac{\delta P}{\delta\rho} == \frac{1}{3}\frac{\sigma T}{\left(\frac{m_H}{K}\right) + \sigma T}.$$

During most of stage $\sigma KT \gg m_H$ so $V_S^2 \simeq {}^1/_3$.

Since $K_J \simeq \sqrt{12\pi G\rho_0}$ and during most of this stage $\rho_0 \simeq bT^4$, so

$$K_J \simeq \sqrt{12\pi Gb}T^2$$

What is the mass of the baryons in the sphere of radius $\frac{\lambda_J}{2}$ (use $n = \frac{4bT^3}{3\sigma K}$)?

$$M_{B-J} = \frac{4}{3}\pi n m_H\left(\frac{\pi}{K_J}\right)^3 = \frac{16\pi^{5/2}}{9(12)^{3/2}}\frac{m_H}{G^{3/2}b^{1/2}\sigma K}T^{-3}$$

For $T \simeq 10^{10\,o}K; M_{B-J} \simeq 10^2 M_\odot$
$T \simeq 10^{6\,o}K; M_{B-J} \simeq 10^{12} M_\odot$ (galaxy or small cluster of galaxies)
$T \simeq 4{,}000^oK; B_{B-J} \simeq 10^{18} M_\odot$

<u>After recomb.</u>
The H is neutral, and the photons decouple, can ignore photons in the Jean's analysis (so treat as a non-rel. gas):

$$\rho = nm_H + \frac{3}{2}nkT$$

$\rho = nm_H + \frac{3}{2}nkT \leftarrow$ T is temp of baryons

$p = nkT$

During post-recomb stage n depends on the temp. of the baryon gas as:

$$n \propto a^{-3}(t) \propto T^{3/2}$$

<u>Proof</u>
Consider comoving volume

$$0 = TdS = dE + pdV = d\left(\frac{3}{2}N + T + Nm_H\right) + \frac{N}{V}kTdV$$

$$0 = \frac{3}{2}dT + T\frac{dV}{V} \Rightarrow V \propto T^{-3/2}$$

and we get

$$a^{-3} \propto T^{3/2}$$

Now for photons:

$$\delta n = \frac{3}{2}n\frac{\delta T}{T}$$

$$\delta\rho = \left(\frac{3}{2}nm_H + \left(\frac{3}{2}\right)\left(\frac{5}{2}\right)nkT\right)\frac{\delta T}{T}$$

$$\delta p = \frac{5}{2}(nkT)\frac{\delta T}{T}$$

Thus

$$V_s^2 = \frac{\delta p}{\partial \rho} = \frac{\frac{5}{2}nkT}{\frac{3}{2}nm_H} = \frac{5}{3}\frac{kT}{m_H},$$

where the temperature is that of the baryons.

Note $T_B \sim a^{-2}$ *while* $t_\gamma \propto a^{-1}$. At recomb: $T_{rec} \equiv T_B = T_\gamma$. After $T_\gamma = T_{rec}\frac{a_{rec}}{a(t)}$ *and* $T_B = T_{rec}\frac{a_{rec}^2}{r^2(\tau)}$.

So,

$$T_B(t) = \frac{T_\gamma^2(t)}{T_{rec}} \quad \rightarrow \quad n = \frac{4b}{3\sigma K} T_\gamma^3(t).$$

Thus for $T < T_\gamma = 4{,}000^0 K, M_{B-J} = \frac{4}{3}\pi n m_H \left(\frac{\pi}{K_J}\right)^2 \propto T_\gamma^3$. So, K_J goes at $M_{B-J} = \frac{4}{3}\pi n m_H \left(\frac{\pi}{K_J}\right)^2 \propto T_\gamma^3\, T^{-3/2} \propto T^{3/2}$. With $\sigma \simeq 10^9$ the value of M_{B-J} drops during recom from an initial value of $10^{18} M_\odot$ to a $10^5 M_\odot$.

9.21 More Examples

Example 9.7. The Bertotti-Robinson solution to the Einstein Field Equation, for when there is a uniform magnetic field, has metric:

$$ds^2 = Q^2[-dt^2 + (\sin t)^2 dz^2 + d\theta^2 + sin^2\theta\, d\varphi^2]$$

with, $Q = const.$, $0 \le t \le \pi$, $-\infty < z < \infty$, $0 \le \theta \le \pi$, $0 \le \varphi \le 2\pi$.

(a) Describe the coordinates.
(b) Is the universe spherically symmetric?
(c) Is the universe cylindrically symmetric?
(d) Is the universe asymptotically flat?
(e) Describe how the geometry changes as t traverses is domain.
(f) Characterize he coordinates.

Solution

a. t is time-like, and z, θ, φ are space-like coordinates.

b. For fixed t and z, the metric is given by

$$ds^2 = Q^2[d\theta^2 + sin^2\theta\, d\varphi^2]$$

The geometry is the same as a sphere of radius Q.
The universe is spherically symmetric.

c. For fixed t and φ, $ds^2 = Q^2[(\sin t_0)^2 dz^2 + d\theta^2]$

The geometry is the same as a cylinder.
The universe in cylindrically symmetric.

d. We cannot construct a coordinate system which approximately leads to:

$$ds^2 = -dx^{02} + dx^{12} + dx^{22} + dx^{32}$$

The universe is <u>not</u> approximately flat.

e. Since the metric is z, φ independent, u_e and u_p are constant along a geodesic. Then, $(z, 0, \varphi) = unit$ is a geodesic.

To discuss the evolution of this universe, examine the motions of particles in the spherical shell $0 < z < \Delta z$, each particle moving along a geodesic of constant (z, θ, φ). At $t = 0$ these particles are compacted together; the thickness of the shell is $\Delta S = \sin t \cdot \Delta z = 0$. Now, as t increases from $0\ to\ \pi/2$, the thickness of the shell expands to a maximum of $\Delta s = z$. Then as t traverses from $\pi/2$ to π it contracts back to zero.

f. $ds^2 = ds_1^2 + ds_2^2$

$ds_1^2 = Q^2[-dt^2 + (\sin t)^2 dz^1]$ → a metric for a 2D hyperboloid.

$ds_2^2 = Q^2[d\theta^2 + \sin^1\theta d\varphi^1]$ → a metric for a 2-sphere.

Therefore the universe is equivalent to a product of a hyperboloid and a sphere.

The physical meaning is as follows:

θ, φ are spherical polar coordinates on the sphere of constant t, z that make up this universe. Those spheres all have the same surface area $4\pi Q^2$.

t is $\frac{1}{Q}x(\text{proper measured by observers which are "at rest" i.e. } z, \theta, \varphi$
$= \text{const})$.

z is a "radial" or "longitudinal" space coordinate that connects the spheres to each other, is constant on the world line of each static observer, and exhibits the translation symmetry of the universe: $\frac{\partial g_{\alpha\beta}}{dz} = 0$. z differs from the proper distance between sphere (proper "radial" in "longitudinal ' distance) by the "expansion factor" of the universe, $\sin t$: $\Delta S = \sin t\ \Delta z$.

Example 9.8. Calculate the gravitational redshift of a photon that is emitted from an atom at rest on the surface of a static star of radius R and mass M, by the following steps:
(i) The emitted photon carries the frequency of the atom's proper frame. So, to start, we need to analyze properties of the photon in the (local) proper reference frame of the atom. So first obtain orthonormal basis vectors in this frame appropriate to an external Schwarzschild coordinate system, where

$$ds^2 = -(1 - 2\,M/r)dt^2 + (1 - 2\,M/r)^{-1}dr^2$$
$$+\,r^2(d\theta^2 + \sin^2\theta\,d\varphi^2), \qquad r \ge R > 2M.$$

(ii) Explain why $h\nu_{em} = -P_{\hat{0}}$.

(iii) Show, in the Schwarzschild time basis, that: $P_t = P\cdot\vec{e}_{\hat{t}} = -h\nu_{em}\sqrt{1 - 2\,M/r}$

(iv) Show that an escaping photon (towards $r = \infty$) has conserved time component.

(v) Show that when observed by an external observer $\nu_{obs} = -P_t/h$.

(vi) Show the redshift is: $\frac{\lambda_{rec}-\lambda_{em}}{\lambda_{em}} = \frac{1}{\sqrt{1-\frac{2M}{R}}} - 1.$

(vii) Evaluate for: Earth surface; Sun's surface, and the surface of a 1.4 solar mass neutron start with radius 10km.

Solution

(i) Since $ds^2 = -(1 - 2\,M/r)dt^2 + (1 - 2\,M/r)^{-1}dr^2 + r^2(d\theta^2 + \sin^2\theta\,d\varphi^2)$

$$g_{tt} = -(1 - 2\,M/r);\, g_{rr} = \frac{1}{(1 - 2\,M/r)};\, g_{\theta\theta} = r^2;\, g_{\varphi\varphi}r^2\sin^2\theta$$

and

$$\vec{e}_t\cdot\vec{e}_t = g_{tt} = -(1 - 2\,M/r)$$

For orthonormal basis vectors $g_{tt} = -1$: $\vec{e}_{\hat{t}}\cdot\vec{e}_{\hat{t}} = -1\ \ and\ \ A\vec{e}_{\hat{t}} = \vec{e}_t$

$$A^2 = (1 - 2\,M/r) \Rightarrow A = \sqrt{1 - 2\,M/r}$$

So,

$$\vec{e}_{\hat{t}} = \frac{1}{A}\vec{e}_t = \frac{1}{\sqrt{1 - 2\,M/r}}\frac{\partial}{\partial t}.$$

All cross terms in the orthonormal basis are zero since they are in the direct basis and we are only renormalizing by the respective metric components. Similarly

$$\vec{e}_{\hat{r}} = \frac{1}{\sqrt{g_{rr}}}\vec{e}_r = \sqrt{1 - 2\,M/r}\frac{\partial}{\partial r}$$

$$\vec{e}_{\hat{\theta}} = \frac{1}{\sqrt{g_{\theta\theta}}}\vec{e}_\theta = \frac{1}{r}\frac{\partial}{\partial r}$$

$$\vec{e}_{\hat{\varphi}} = \frac{1}{\sqrt{g_{\varphi\varphi}}}\vec{e}_\varphi = \frac{1}{r\sin\theta}\frac{\partial}{\partial\varphi}$$

(ii) $h\nu_{em}$ = (particle's energy) since light is a zero-rest-mass particle so it only has energy from the time component of its four momentum.

$$h v_{em} = -P \cdot \vec{e}_{\hat{t}} = -P_{\hat{0}}$$

(iii) $P_t = P \cdot \vec{e}_{\hat{t}} = -\sqrt{1 - 2\,M/r}\,(\vec{P} \cdot \vec{e}_{\hat{t}}) = -h v_{em}\sqrt{1 - 2\,M/r}$.

(iv) Once again we make use of the fact that since $g_{\alpha\beta,t} = 0$ then $P_t \equiv \vec{P} \cdot \frac{\partial}{\partial t} = const.$ Since P_t is constant and we know its value at one instant we know it at every instant. Thus, P_t is conserved.

(v) At infinity $\left(\frac{\partial}{\partial t}, \frac{\partial}{\partial r}, \frac{1}{r}\frac{\partial}{\partial \theta}, \frac{1}{r\sin\theta}\frac{\partial}{\partial \varphi}\right)$ for the static observer's Lorentz frame. The observed frequency of the photon is $h v_{rec} = (p \cdot e_{\hat{0}})$ at infinity, so

$$h v_{rec} = -\left(P \cdot \frac{\partial}{\partial t}\right) = -P_0 \quad \rightarrow \quad v_{rec} = P_0/h$$

(vi) From above, $\frac{\lambda_{rec} - \lambda_{em}}{\lambda_{em}} = \frac{v_{em} - v_{rec}}{v_{rec}} = \frac{1}{\sqrt{1 - \frac{2M}{R}}} - 1.$

(vii)

	M (cm)	R (cm)
redshift		
Earth	0.4438	6.37×10^8
7×10^{-10}		
Sun	1.5×10^5	6.96×10^{10}
2×10^{-6}		
Neutron star	1.5×10^5	10^6
0.195		

Example 9.9. Star Implosion under zero pressure

(a) Show that the covariant time coordinate U_t of the 4-velocity $\vec{U}$ of a particle on the stars surface is conserved on its world line. Evaluate in terms of M and initial Radius.

(b) From 4-velocity normalization show that, in terms of proper time τ at the star's surface, we have:

$$\frac{dR}{d\tau} = -\sqrt{constant + \frac{2M}{R}},$$

and evaluate the constant.

(c) Show that the star implodes through the horizon $R = 2M$ in finite proper time (but infinite coordinate time).
(d) Using Eddington-Finkelstein coordinates, show that the star implodes through the horizon $R = 2M$ in finite proper time *and* in finite coordinate time.
(e) Sketch the world line of the stars surface.

Solution

(a) $g_{\alpha\beta,t} = 0$, indep. of x^t, thus $P_t \equiv \vec{P} \cdot U^t$ is constant, thus U^t is constant.

$$U_t = \vec{U} \cdot \vec{e}_t = \frac{1}{\sqrt{1 - 2M/R_0}} \vec{e}_t \cdot \vec{e}_t = \frac{g_{tt}}{\sqrt{1 - 2M/R_0}} = -\sqrt{1 - 2M/R_0}$$

↗

↗

(b) $\vec{U} \cdot \vec{U} = g_{\alpha\beta} U^\alpha U^\beta = -g_{tt} U^t U^t + g_{rr} U^r U^r + g_{\theta\theta} U^\theta U^\theta + g_{\varphi\varphi} U^\varphi U^\varphi$

$$\vec{u}^2 = -1 = g^{tt} u_t u_t + g^{rr} u_r u_r \qquad g^{tt} = \frac{1}{-(1 - 2M/R)};\ g^{rr}$$
$$= (1 - 2M/r)$$

$$u^r u^r = \frac{-1 - g^{tt} u_t u_t}{g_{rr}}$$

$u^r = \frac{dR}{d\tau} = \pm\sqrt{-(1 - 2M/R) + (1 - 2M/R_0)_2}$ we want the negative root since R is decreasing (implosion)

$$\frac{dR}{d\tau} = -\sqrt{-1 + 1 - {}^{2M}\!/_{R_0} + {}^{2M}\!/_{R}}$$

$$\frac{dR}{d\tau} = -\sqrt{-\frac{2M}{R_0} + \frac{2M}{R}}$$

Newtonian potential $\Phi = -\frac{M}{r}$

(c)

$$\frac{dR}{d\tau} = -\sqrt{-\frac{2M}{R_0} + \frac{2M}{R}} \rightarrow \quad \frac{dR}{-\sqrt{-\frac{2M}{R_0} + \frac{2M}{R}}} = d\tau \rightarrow \quad \tau$$

$$= \int_{R_0}^{2M} \frac{-\sqrt{R}dR}{\sqrt{-\frac{R}{R_0}2M + 2M}}$$

Let $R = x^2$, $dR = 2xdx, and\ have\ \sqrt{R_0} < x < \sqrt{2M}$:

$$\tau = \int_{\sqrt{R_0}}^{\sqrt{2M}} \frac{-2x^2dx}{\sqrt{\frac{2M}{R_0}}\sqrt{-x^2 + R_0}} = -2\sqrt{\frac{R_0}{2M}} \int_{\sqrt{R_0}}^{\sqrt{2M}} \frac{x^2dx}{\sqrt{R_0 - x^2}}$$

Thus,

$$\tau = -2\sqrt{\frac{R_0}{2M}}\left\{-\frac{\sqrt{2M}\cdot\sqrt{R_0 - 2M}}{2} + \frac{R_0}{2}\sin^{-1}\left(\sqrt{\frac{2M}{R_0}}\right) - \frac{R_0}{2}\frac{\pi}{2}\right\}$$

and assuming $R_0 \gg 2M$:

$$\tau \simeq -2\sqrt{\frac{R_0}{2M}}\left(-\frac{R}{2}\frac{\pi}{2}\right) = \frac{\pi}{2}\left[\frac{R_0^3}{2M}\right]^{1/2}$$

(d) $\vec{u} \cdot \vec{u} = g_{\alpha\beta}u^\alpha u^\beta = g_{tt}u^t u^t + g_{rr}u^r u^r + 2g_{tr}u^t u^r = -1$
and,
$g_{rr}(u^r)^2 + 2g_{tr}u^t(u^r) + [g_{tt}u^t u^t + 1] = 0$
and,

$$u^r = \frac{-2g_{tr}u^t \pm \sqrt{4g_{tr}^2(u^r)^2 - 4g_{rr}[g_{tt}u^t u^t + 1]}}{2g_{rr}}u^t$$

So we can write:

$$g_{tt} = -\left(1 - \frac{2M}{r}\right), \qquad g_{tr} = 4M/r, \qquad u_t = -\sqrt{1 - 2M/R_0},$$

$$g_{rr} = \left(1 + {}^{2M}/_{r}\right)$$

Thus, we have the contractions:

$$g_{tr}u^t = g_{tr}g^{tt}u_t = \frac{4M}{r}\frac{-\sqrt{1 - 2M/R_0}}{-\sqrt{1 - 2M/r}}$$

and

$$g_{rr}g_{tt}u^t u^t = g_{rr}g^{tt}u_t u_t = \frac{-(1+2M/r)}{(1-2M/r)}(1-2M/R_0)$$

So,

$$u^r = \frac{4M}{r}\frac{\sqrt{1-2M/R_0}}{\left(1+4M^2/r^2\right)} \pm \sqrt{\left[\frac{4M}{r}\frac{\sqrt{1-2M/R_0}}{\left(1+4M^2/r^2\right)}\right]^2 - \frac{1-(1-2M/R_0)\left(1-\frac{2M}{r}\right)}{(1+2M/r)}}$$

and we must have the negative root so u^r doesn't diverge as $r \to 2M$.
Thus:

$$u^r = \frac{4M}{r}\frac{\sqrt{1-2M/R_0}}{\left(1-4M^2/r^2\right)} - \sqrt{\left[\frac{4M}{r}\frac{\sqrt{1-2M/R_0}}{\left(1+4M^2/r^2\right)}\right]^2 - \frac{1-(1-2M/R_0)\left(1-\frac{2M}{r}\right)}{(1+2M/r)}}$$

Now u^r does not diverge as $r \to 2M$, and the integral is finite.

Now, to show $\dfrac{dR}{d\tau} = -\sqrt{-\dfrac{2M}{R_0} + \dfrac{2M}{R}}$ remains valid to $r = 0$, where we have already established its validity through $r = 2M$ since the integral won't possibly diverge at $r = 2M$ (the integrand is well behaved until $r \to 0$). The time τ to go from R_0 to $r = 0$ on implosion is exactly $\dfrac{\pi}{2}\left[\dfrac{R_0^3}{2M}\right]^{1/2}$ our approximate result from earlier.

(e) For Eddington-Finkelstein:

For Eddington–Finkelstein we have
$\dfrac{d\tilde{t}}{dr} = -1$ for ingoing; and $\dfrac{d\tilde{t}}{dr} = \left[\dfrac{1+2M/r}{1-2M/r}\right]$ for outgoing

etc.

I'll do this gra...

this agrees

choose this initial ray

we'll start at $r = 3M$

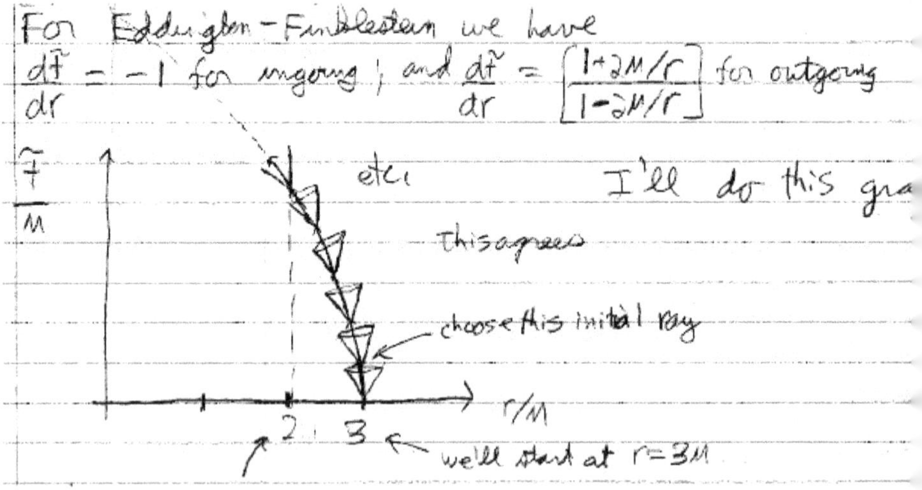

Fig. 9.1. Eddington-Finkelstein Light-Cones.

For Schwarzschild

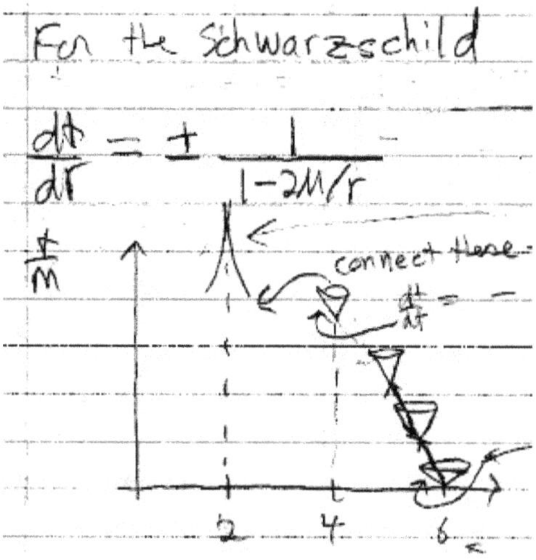

Fig. 9.2. Schwarzschild Light-Cones.

Example 9.10. "Gore at the Singularity"
(a) Show World lines for an imploding star in Eddington-Finkelstein and Schwarzschild coordinates.
(b) Describe the asymptotes of the World lines.
(c) Show that the in-falling observer coordinate basis is related to the Schwarzschild coordinate basis by:

$$\vec{e}_{\hat{0}} = -(2M/r - 1)^{1/2}\frac{\partial}{\partial r}; \quad \vec{e}_{\hat{1}} = (2M/r - 1)^{-1/2}\frac{\partial}{\partial t}; \quad \vec{e}_{\hat{2}} = \frac{1}{r}\frac{\partial}{\partial t}; \quad \vec{e}_{\hat{3}} = \frac{1}{r\sin\theta}\frac{\partial}{\partial\varphi}$$

and evaluate the Riemann components in the in-falling observer's frame.
(d) Describe the tidal forces acting on an in-falling observer as r goes to zero.
(e) Suppose an in-falling astronaut has a 20kg head/helmet and 20kg of feet/magshoes separated by a length/"height" of 2m (as measured in local Lorentz frame). The length is aligned radially. Compute the stretching force as a function of proper time. Let the Black Hole have the mass of 5 billion Suns. Does the in-falling astronaut become gore before getting to the singularity? At what proper time is the singularity reached?

Solution
(a)

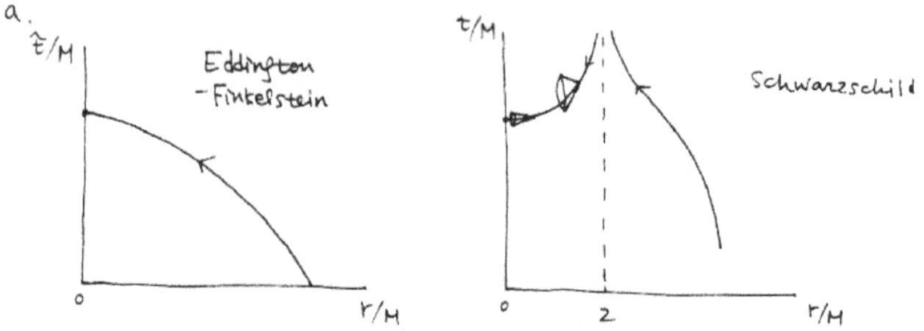

Fig. 9.3. Light-Cone comparison.

This is asymptotic to the curve $\{(t, \theta, \varphi) = \text{const}, r$ varial $p\}$ because $\frac{dt}{dr} = \frac{\pm 1}{1-2M/r}$ in this region, and as $r \to 0$ $\frac{dt}{dr} \to 0$. Also, the light cone closes off in this region both going to $\frac{dt}{dr} = 0$.

(b) The Schwarzschild metric is

$$ds^2 = -(1 - 2M/r)dt^2 + \frac{dr^2}{(1 - 2M/r)} + r^2(d\theta^2 + \sin^2\theta\, d\varphi^2)$$

For $(t, \theta, \varphi) = \text{const}, r < 2M$

$$ds^2 = -\frac{dr^2}{\left(\frac{2M}{r} - 1\right)}$$

thus $ds^2 < 0$ a timelike geodesic.

(c) The in-falling observer's local Lorentz frame is a geodesic. All the components must be orthogonalized:

$$ds^2 = -\frac{dr^2}{(2M/r - 1)} + \left(\frac{2M}{r} - 1\right) dt^2 + r^2(d\theta^2 + \sin^2\theta\, d\varphi^2)$$

Now everything follows directly since $g_{\hat{0}\hat{0}} = (2M/r - 1); g_{\hat{1}\hat{1}} = \left(\frac{2M}{r} - 1\right); g_{\hat{2}\hat{2}} = r^2; etc.$

Thus,

$$\vec{e}_{\hat{0}} = \frac{\vec{e}_1}{\sqrt{g_{\hat{1}\hat{1}}}} = -(2M/r - 1)^{1/2} \frac{\partial}{\partial r}; \qquad \vec{e}_{\hat{1}} = \frac{\vec{e}_0}{\sqrt{g_{\hat{0}\hat{0}}}}$$

$$= (2M/r - 1)^{-1/2} \frac{\partial}{\partial t};$$

and

$$\vec{e}_{\hat{2}} = \frac{1}{r}\frac{\partial}{\partial t}; \qquad \vec{e}_{\hat{3}} = \frac{1}{r \sin\theta} \frac{\partial}{\partial \varphi}$$

(c) We need $R_{\alpha\beta\gamma\delta} = \frac{1}{2}\left(g_{\alpha\delta,\beta\gamma} + g_{\beta\gamma,\alpha\delta} - g_{\alpha\gamma,\beta\delta} - g_{\beta\delta,\alpha\gamma}\right)$ where,

$$g_{\hat{0}\hat{0}} = (2M/r - 1)^{-1};\, g_{\hat{1}\hat{1}} = \left(\frac{2M}{r} - 1\right);\, g_{\hat{2}\hat{2}} = r^2;\, g_{\hat{3}\hat{3}} = r^2 \sin^2\theta.$$

For the external Schwarzschild metric:

$$R_{trtr} = -\frac{2M}{r^3}, R_{t\theta t\theta} = \frac{M}{r^3} r^2\left(1 - {}^{2M}/_{r}\right), R_{t\varphi t\varphi}$$
$$= \frac{M}{r^3} r^2 \sin^2\theta \left(1 - {}^{2M}/_{r}\right)$$

$$R_{\theta\varphi\theta\varphi} = \frac{2M}{r^3} r^4 \sin^2\theta\,, R_{r\theta r\theta} \frac{M}{r^3} \frac{r^2}{(1 - 2M/r)}, R_{r\varphi r\varphi}$$
$$= -\frac{M}{r^3} \frac{r^2 \sin^2\theta}{(1 - 2M/r)}$$

Since the components are still normal internally:-

$$R_{\hat{t}\hat{r}\hat{t}\hat{r}} = \left(\frac{2M}{r} - 1\right)\left(\frac{2M}{r} - 1\right)^{-1} R_{rtrt}$$

$$\boxed{R_{\hat{t}\hat{r}\hat{t}\hat{r}} = R_{rtrt} = -\frac{2M}{r^3}}$$

$$\boxed{R_{\hat{t}\hat{\theta}\hat{t}\hat{\theta}} = \left(\frac{2M}{r} - 1\right) \frac{1}{r^2} R_{r\theta r\theta} = +\frac{M}{r^3}}$$

$$\boxed{R_{\hat{t}\hat{\varphi}\hat{t}\hat{\varphi}} = \left(\frac{2M}{r} - 1\right) \frac{1}{r^2 \sin^2\theta} R_{r\varphi r\varphi} = +\frac{M}{r^3}}$$

$$\boxed{R_{\hat{t}\hat{\theta}\hat{t}\hat{\theta}} = \left(\frac{2M}{r} - 1\right)^{-1} \frac{1}{r^2} R_{t\theta t\theta} = -\frac{M}{r^3}}$$

$$\boxed{R_{\hat{r}\hat{\varphi}\hat{r}\hat{\varphi}} = \left(\frac{2M}{r} - 1\right)^{-1} \frac{1}{r^2 \sin^2\theta} R_{t\varphi t\varphi} = -\frac{M}{r^3}}$$

$$\boxed{R_{\hat{\theta}\hat{\varphi}\hat{\theta}\hat{\varphi}} = \frac{1}{r^2} \cdot \frac{1}{r^2 \sin^2\theta} R_{\theta\varphi\theta\varphi} = \frac{2M}{r^3}}$$

(d) We have $\frac{d^2\varepsilon^{\hat{j}}}{dt^2} = -R^{\hat{j}}_{\hat{0}\hat{k}\hat{0}}\varepsilon^{\hat{k}}$ in a local Lorentz frame. Let ε^{j} be in the $\hat{r}$ direction first:

$$\frac{d^2\varepsilon^{\hat{r}}}{dt^2} = -R^{\hat{r}}_{\hat{0}\hat{r}\hat{0}}\varepsilon^{\hat{r}} = -g^{\hat{r}\hat{r}}R_{\hat{r}\hat{0}\hat{r}\hat{0}}\varepsilon^{\hat{r}} = -g^{\hat{r}\hat{r}}R_{\hat{r}\hat{t}\hat{r}\hat{t}}\varepsilon^{\hat{r}} - \frac{1}{\left(\frac{2M}{r}-1\right)}\left(-\frac{2M}{r^3}\right)\varepsilon^{\hat{r}}$$

$$= \frac{2M}{r^3}\frac{1\cdot\varepsilon^{\hat{r}}}{\left(\frac{2M}{r}-1\right)} = \frac{2M}{r^3}\frac{1}{(2M-r)}\varepsilon^{\hat{r}}$$

As r → 0 $\frac{d^2\varepsilon^{\hat{r}}}{dt^2} \to [\infty]\varepsilon^{\hat{r}}$ you're being stretched infinitely.

$$\frac{d^2\varepsilon^{\hat{\theta}}}{dt^2} = \frac{1}{\left(\frac{2M}{r}-1\right)}R_{\hat{\theta}\hat{t}\hat{\theta}\hat{t}}\varepsilon^{\hat{\theta}} = \frac{M}{r^3}\frac{1\,\varepsilon^{\hat{\theta}}}{\left(\frac{2M}{r}-1\right)} = -\frac{M}{r^2}\frac{1}{(2M-r)}\varepsilon^{\hat{\theta}}$$

$\frac{d^2\varepsilon^{\hat{\theta}}}{dt^2} = [-\infty]\varepsilon^{\hat{\theta}}$ and

$\frac{d^2\varepsilon^{\hat{\varphi}}}{dt^2} = [-\infty]\varepsilon^{\hat{\varphi}}$ you're being squeezed infinitely.

(e) If the two masses were free to move along geodesics, their relative acceleration would be $\frac{D^2}{D\tau^2}h = \frac{2M}{r^3}h$, and their acceleration relations to the center of mass will be half this

$$g = \frac{M}{r^3}h.$$

To counteract this gravitational acceleration the observer's body must exert an inward force $F = \mu g$ on each mass μ:

$$F = \mu g = \mu\frac{M}{r^3}h.$$

Suppose a human being can tolerate $F = 1ton \times w \leq 1 \times 10^4$ Newtons. For $\mu \simeq 20kg$, and
$h \simeq 2m$, and $M = 5 \times 10^9\, M_0$, we have

$$r^3_{critical} = \frac{\mu M h}{F} = \frac{20kg \cdot 5 \times 10^9 M_0 \cdot 2m}{10^4\, newtons} = 2 \times 10^7 M_0\, sec^2.$$

Since $M_o = 1.5 \times 10^{26}\, cm^3/sec^2$, we find

$$r_c \cong 1 \times 10^{11} cm \qquad \left(\frac{r_c}{M} \simeq 10^{-4}\right).$$

So, gored before singularity, then have further in-fall time from $r = r_c$ to $r = 0$ of.

$$d\tau^2 = -\frac{dr^2}{1-2M/r} \simeq \frac{r}{2M}dr^2 \quad \Rightarrow \quad \frac{\Delta\tau}{2M} = \int_0^{r_c/2M} x^{1/2}\,d\gamma$$

$$= \frac{2}{3}\left(\frac{r_c}{2M}\right)^{3/2}$$

Thus

$$\Delta\tau \simeq (2M) \times \left(\frac{2}{3}\right)\left(\frac{r_c}{2M}\right)^{3/2} \simeq 10^{-2} sec$$

Example 9.11. The 3-sphere geometry of a closed universe.

(a) Show that there exists a 3-sphere

$$^{(3)}ds^2 = a^2[(dx)^2 + \sin^2 x\,[(d\theta)^2 + \sin^2\theta\,(d\varphi)^2]$$

is embedded in a 4D Euclidean hypersurface satisfying:

$$w^2 + x^2 + y^2 + z^2 = a^2.$$

(b) Show that the total 3-volumne of the 3-sphere is $2\pi^2 a^3$.

Solution

(a) Consider $w^2 + x^2 + y^2 + z^2 = a^2$ with the substitutions:

$w = a\cos x$

$z = a\cos\theta\sin x$

$y = a\sin\theta\sin\varphi\sin x$

$x = a\sin\theta\cos\varphi\sin x$

Then:

$dw = -a\sin x\,dx$

$dz = -a\sin\theta\sin x\,d\theta + a\cos\theta\cos x\,dx$

$$dy = a\cos\theta\sin\varphi\sin x\,d\theta + a\sin\theta\cos\varphi\sin x\,d\varphi + a\sin\theta\sin\varphi\cos x\,dx$$

$$dx = a\cos\theta\cos\varphi\sin x\,d\theta - a\sin\theta\sin\varphi\sin x\,d\varphi + a\sin\theta\cos\varphi\cos x\,dx$$

and

$dw^2 + dz^2 + dy^2 + dx^2 = {}^{(3)}ds^2$ → our 3 sphere.

Alternatively,

The line element on the surface of the sphere is

$ds^2 = dw^2 + dx^2 + dy^2 + dz^2$, $where\ w = \sqrt{a^2 - R^2},\ R = \sqrt{x^2 + y^2 + z^2}$

Introducing spherical coordinates for x, y, z

$$ds^2 = dw^2 + dR^2(d\theta^2 + \sin^2\theta\, d\varphi^2) \quad , \quad dw = d\left(\sqrt{a^2 - R^2}\right)$$

$$= \frac{-RdR}{\sqrt{a^2 - R^2}}$$

Thus,

$$ds^2 = \frac{dR^2}{1 - {R^2}/{a^2}} + R^2(d\theta^2 + \sin^2\theta\, d\varphi^2).$$

Set $R = a \sin x$. Then, the line element of the 3-sphere is

$$ds^2 = a^2[dx^2 + \sin^2 x\,(d\theta^2 + \sin^2\theta\, d\varphi^2)].$$

(b)

1 – D x→dx

2 – D $x, y \to a d\varphi$

3 – D $x, y, z \to a d\varphi(a \sin\theta\, d\theta)$

4 – D $x, y, z, w \to a dx(a \sin x\, d\theta)(a \sin x \sin\theta\, d\varphi)$

$$\int_0^{\pi}\int_0^{\pi}\int_0^{2\pi} a^3\,(\sin^2 x\, dx)(\sin\theta\, d\theta)d\varphi$$

$$\downarrow \quad \downarrow \quad \downarrow \quad \downarrow$$

$$a^3 \times \left(\frac{\pi}{2}\right) \times 2 \quad 2\pi = 2\pi^2 a^3$$

Or in terms of the differential volume element: $\sqrt{\det\|g_{ij}\|}\, d^3x =$ $a^3 \sin^2 x \sin\theta\, dx d\theta d\varphi$. The total 3-volume of this 3-sphere is:

$$V = \int_0^{\pi} dx \int_0^{\pi} d\theta \int_0^{2\pi} d\varphi\, a^3 \sin^2 x \sin\theta = \frac{\pi}{2} \cdot 2 \cdot 2\pi a^3 = \underline{2\pi^2 a^3}.$$

Example 9.12. Energy Conservation for a Perfect Fluid

Consider a perfect fluid, with $T^{\alpha\beta} = (\rho + p)\mu^\alpha\mu^\beta + p g^{\alpha\beta}$. Let the space-time be arbitrary.

(a) Why does the law of energy conservation for the fluid have the form: $\mu_\alpha T^{\alpha\beta}{}_{;\beta} = 0$?

(b) Show the law of energy conservation is $d\rho/d\tau = -(\rho + p)\vec{\nabla} \cdot \vec{u}$.

(c) Show for fluid at rest in a homogeneous isotropic space-time we get the first law of thermodynamics.

Solution

(a) In the "rest" frame of the fluid, $\vec{u} = \vec{e}_{\hat{0}}$, i.e.: $u^0 = 1, u^j = 0$, and $u_0 = 1, u_j = 0$, thus

$u_\alpha T^{\alpha\beta}{}_{;\beta} = 0$ reduces to $T^{0\beta}{}_{,\beta} = 0$, which represents conservation of time component of 4-momentum (energy) since $T^{00} =$ energy density, and $T^{oj} =$ energy flux along $\bar{e}_j$.

(b) Using the expression for $T^{\alpha\beta}$, we get:

$$u_\alpha T^{\alpha\beta}{}_{;\beta} = u_\alpha\left[(\rho + p)u^\alpha u^\beta + pg^{\alpha\beta}\right]_{;\beta} = 0$$

Thus

$$(\rho + p)_{;\beta} u_\alpha u^\alpha u^\beta + (\rho + p)u^\beta u_\alpha u^\alpha_{;\beta} + (\rho + p)u_\alpha u^\alpha \cdot u^\beta_{;\beta} + P_{;\beta} u_\alpha g^{\alpha\beta} + p \cdot g^{\alpha\beta}_{;\beta} u_\alpha = 0$$

$$\Rightarrow -(\rho + p)_{;\beta} u^\beta - (\rho + p)u^\beta_{;\beta} + P_{;\beta} u^\beta = 0$$

$$\Rightarrow -\rho_{;\beta} u^\beta - (\rho + p)u^\beta_{;\beta} = 0$$

$$\Rightarrow \boxed{\frac{d\rho}{d\tau} = -(\rho + p)\vec{\nabla} \cdot \vec{u}} \;;\; \tau = \text{proper time of fluid.}$$

(c) The equation of mass-energy conservation is: $\frac{d\rho}{d\tau} = -(\rho + p)\vec{\nabla} \cdot \vec{u}$. In the "rest" frame of the universe $\vec{u} = \frac{d}{d\tau}$ reduces to $\vec{u} = \frac{d}{dt} \Rightarrow u^0 = 1, u^2 = 0$. For the Robertson-Walker metric, $\sqrt{-g} = a^3 \sin^2 x \sin\theta$ (where $g = \det(g)$). Thus

$$\vec{\nabla} \cdot \vec{u} = \frac{1}{\sqrt{-g}}\left(\sqrt{-g}u^\mu\right)_{,\mu} = \frac{1}{a^3 \sin^2 x \sin\theta}(a^3 \sin^2 x \sin\theta)_{,0}$$

$$= \frac{1}{a^3} \cdot \frac{da^3}{dt}.$$

for $a = a(t)$. Thus the energy conservation equation becomes,

$$\frac{\partial\rho}{\partial t} = -\frac{(\rho + p)}{a^3} \cdot \frac{\partial a^3}{\partial t} \quad \rightarrow \quad \frac{\partial(\rho a^3)}{\partial t} = -p\frac{\partial a^3}{\partial t}$$

Derivation of some useful formulas:

$$\Gamma^\alpha{}_{\mu\alpha} = \frac{1}{2}g^{\alpha\beta}[g_{\beta\alpha,\mu} + \underbrace{(g_{\beta\mu,\alpha} - g_{\mu\alpha,\beta})}_{Antisym. in\ \alpha\leftrightarrow\beta}] = \frac{1}{2}g^{\alpha\beta}g_{\beta\alpha,\mu}$$

Since $g^{\mu v}$ is the inverse of $g_{\mu v}$, i.e., $g^{\mu v} g_{\mu v} = \delta^{\mu}_{\alpha}$, in terms of metric inverse,

$$g^{\mu v} = \frac{1}{g}[\text{Cofactor of } g_{\mu v} \; in \; \|g_{\mu v}\|]$$

(Eqn. 9-35)

Expanding the determinant by rows which are differentiated.

$$\frac{\partial g}{\partial x^{\mu}} = \frac{\partial}{\partial x^{\mu}}\begin{vmatrix} g_{00} & \cdots & g_{03} \\ \vdots & & \vdots \\ g_{30} & \cdots & g_{33} \end{vmatrix} = \begin{vmatrix} \frac{\partial g_{00}}{\partial x^{\mu}} & \cdots & \frac{\partial g_{03}}{\partial x^{\mu}} \\ \vdots & & \vdots \\ g_{30} & \cdots & g_{33} \end{vmatrix} + \cdots + \begin{vmatrix} g_{00} & \cdots & g_{03} \\ \vdots & & \vdots \\ \frac{\partial g_{30}}{\partial x^{\mu}} & \cdots & \frac{\partial g_{33}}{\partial x^{\mu}} \end{vmatrix}.$$

(Eqn. 9-36)

Thus

$$\frac{\partial g}{\partial x^{\mu}} = \sum_{\alpha,\beta} \frac{\partial g_{\alpha\beta}}{\partial x^{\mu}} [Cofactor \; of \; g_{\alpha\beta} \; in \; \|g_{\alpha\beta}\|] = g_{\alpha\beta,\mu} \cdot (g^{\alpha\beta} g)$$

Thus

$$g_{\alpha\beta,\mu} g^{\alpha\beta} = \frac{1}{g}\frac{\partial g}{\partial x^{\mu}} = 2 \cdot \frac{1}{\sqrt{-g}}\left(\sqrt{-g}\right)_{,\mu}$$

(Eqn. 9-37)

Thus,

$$\Gamma^{\alpha}{}_{\mu\alpha} = \frac{1}{\sqrt{-g}}\left(\sqrt{-g}\right)_{,\mu}$$

(Eqn. 9-38)

Note we also have:

$$\vec{\nabla} \cdot \vec{A} = A^{\mu}_{;\mu} = A^{\mu}_{,\mu} + \Gamma^{\mu}_{\alpha\mu} A^{\alpha} = A^{\mu}_{,\mu} + \frac{1}{\sqrt{-g}}\left(\sqrt{-g}\right)_{,\alpha} A^{\alpha}$$
$$= \frac{1}{\sqrt{-g}}\left(\sqrt{-g} A^{\mu}\right)_{,\mu}.$$

(Eqn. 9-39)

9.22 Exercises

(Ex. 9.1) Derive Eqn.s 9-2, 9-3, and 9-4.
(Ex. 9.2) Derive Eqn. 9-5.
(Ex. 9.3) Derive Eqn. 9-6.
(Ex. 9.4) Derive Eqn. 9-8.
(Ex. 9.5) Derive Eqn. 9-9.
(Ex. 9.6) Derive Eqn.s 9-11 and 9-12.
(Ex. 9.7) Derive Eqn. 9-14.
(Ex. 9.8) Derive Eqn. 9-15.

(Ex. 9.9) Derive Eqn. 9-16.
(Ex. 9.10) Derive Eqn. 9-21.
(Ex. 9.11) Derive Eqn. 9-22.
(Ex. 9.12) Derive Eqn. 9-24.
(Ex. 9.13) Derive Eqn. 9-25.
(Ex. 9.14) Derive Eqn. 9-26.
(Ex. 9.15) Derive Eqn. 9-27.
(Ex. 9.16) Derive Eqn. 9-28.
(Ex. 9.17) Derive Eqn.s 9-29 and 9-30.
(Ex. 9.18) Derive Eqn. 9-31.
(Ex. 9.19) Derive Eqn. 9-32.
(Ex. 9.20) Derive Eqn. 9-33.
(Ex. 9.21) Derive Eqn. 9-34.
(Ex. 9.22) Derive Eqn. 9-37.

Chapter 10. Dust Shell Collapse

10.1 Introduction

In this chapter we examine spherically symmetric dust shell collapse in a full general relativistic formulation. In Sec. 10.1 a brief derivation of the classical non-relativistic solution is given. We then consider the full general relativistic solution for the spherically symmetric asymptotically flat geometry without the shell, e.g., the Schwarzschild solution, in Sections 10.3-10.7. In Sec. 10.8 we describe the topologies consistent with the spherically symmetric asymptotically-flat geometries with dust shell. In Sec 10.9 we calculate the equations of motion (EOM's) indicated in the Sec. 10.8 synopsis beginning with consideration of the most convenient choice of gauge. In Sec. 10.10, we analyze the equations of motion (EOM) solution away from the shell, in the asymptotic Newtonian approximation in particular, verifying the classical form indicated in Sec. 10.2.

A general analysis of the asymptotic properties is given in Sec. 10.11. The reduced phase space formalism is described in Sec. 10.12, where a synopsis of the shell EOM is given. A detailed derivation of the shell EOM in the Kraus and Wilczek ("K&W") [53,54] formalism is given in Sec.s 10.13 and 10.14. The reduced phase space shell EOM is then transformed by a new type of time reparameterization [55], which is described in Sec. 10.15.

In Sec. 10.16 we consider the quantization prospects for the reduced phase space Hamiltonian examined in 10.12-10.15. One of the minisuperspace reductions, from the topology description in Sec. 10.8, has a single canonical coordinate related to the wormhole throat. In Sec. 10.17 we describe the derivation of the throat dynamics indicated by this minisuperspace analysis (a simpler precursor to the shell quantization problem to follow). In Sec. 10.18 a preview is given of the quantization of the shell/geometry viewed as a minisuperspace analysis (this analysis is formally completed using the tools of functional analysis in Book 4 [3]).

10.2 Classical non-relativistic solution

Let's consider dust shell collapse with mass m, when a central mass is present with mass M. As a quick preview, we consider this in the Newtonian approximation, and begin with calculation of the potential energy (P.E.) of a small part, δm, of the dust shell (of non-interacting infinitesimal mass elements):

$$P.E. = \int_r^\infty \frac{G(M+m-\delta m)\delta m}{r^2} dr = -G(M+m-\delta m)\delta m\, r^{-1}.$$

(Eqn. 10-1)

The total potential energy in the shell is thus:

$$P.E. = -\frac{G}{r}\int_0^m (M+m-\mu)d\mu = -\frac{G}{r}\left((M+m)m - \frac{1}{2}m^2\right)$$

$$= -\frac{GMm}{r} - \frac{\frac{1}{2}Gm^2}{r}.$$

(Eqn. 10-2)

If we let $M \to 0$, we find that the potential energy of a gravitating dust shell is:

$$P.E. = -\frac{\frac{1}{2}Gm^2}{r}.$$

(Eqn. 10-3)

A naïve guess might have been $P.E.$ equal to $-Gm^2/r$ without the ½ factor.

Working with the shell P.E., let's consider a simple quantum analysis. We construct the classical Lagrangian/Hamiltonian upon determination of P.E. and K.E. Then, quantum operator substitutions for canonical variables gives the Schrodinger equation for the 1st Quantization of the system (various quantization methods to be described in Book 4 [3]):

$$K.E. = \frac{1}{2}m\dot{r}^2 \qquad and \qquad P.E. = -\left(\frac{1}{2}\right)\frac{Gm^2}{r},$$

where for kinetic energy only radial motion is considered (e.g., no rotation or 'spin'). The energy is thus:

$$E = \frac{1}{2}m\dot{r}^2 - \frac{1}{2}\frac{Gm^2}{r} = \frac{p^2}{2m} + \frac{Gm^2}{2r}.$$

(Eqn. 10-4)

Let's now introduce the radial momentum operator in a shift to the quantum formulation:

$$\hat{P}_r = \frac{\hbar}{i}\frac{1}{r}\frac{\partial}{\partial r}r \rightarrow H\psi = \left\{-\frac{\hbar^2}{2m}\left(\frac{\partial^2}{\partial r^2} + \frac{2}{r}\frac{\partial}{\partial r}\right) + V(r)\right\}\psi = E\psi.$$

(Eqn. 10-5)

If we let $\psi(r) = u(r)/r$ this becomes:

$$\left[-\frac{\hbar^2}{2m}\frac{d^2}{dr^2} + V(r)\right]u(r) = E\,u(r)$$

(Eqn. 10-6)

where the latter form corresponds to a Bohr atom for the case of the spherically symmetric s-wave orbital ($\ell = 0$). The usual Coulomb potential analysis (done in Book 4 [3]) then provides for quantization on energy levels according to:

$$E_N = \frac{-m\left(\frac{1}{2}Gm^2\right)^2}{2\hbar^2(N+1)^2} \qquad N = 0,1,2,\ldots$$

The minimum energy is:

$$E_0 = -\frac{G^2m^5}{8\hbar^2}\frac{c^2}{c^2} = -\frac{mc^2}{8}\left(\frac{m}{m_{Pl}}\right)^4,$$

where m_{Pl} is the Planck mass:

$$m_{Pl} = \left(\frac{\hbar c}{G}\right)^{1/2} = 2.2 \times 10^{-5} g.$$

Notice the indication of quantization, as with the Bohr atom, except the mass of the object itself is not quantized in this (shell) instance. In the Coulomb interaction with charges, the charges were quantized for the Bohr atom electron and nucleus charges, here, with the dust shell, we have no such quantization. Given the freedom to adjust the mass continuously, this naive analysis indicates that a "dust shell" will have no effective quantization. If we consider the parallel with Bohr more closely, in the sense of usage of known elementary particles for the mass (restricting to "spherically symmetric wave-functions), we appear to have mass quantization as before, in the sense of restriction to the elementary particle masses, but there is an inconsistency. This results from the elementary particles having a "spin", e.g., $\ell \neq 0$, so the restriction to an s-wave solution with zero angular momentum is no longer even possible (if the physical approximation being considered was even valid in the first place). Some early work on shell collapse was done by Oppenheimer [56] (which was partly why he was put in charge of the Manhattan Project, since a compressive spherical implosion, a form of spherical collapse, was the key engineering puzzle to solve at that time).

To address the dust-shell collapse problem properly, a full GR analysis for the in-fall is needed, and a full self-adjoint quantization analysis is then needed to determine if a stable spectrum results (shown in Book 4 of the Series [3]).

10.3 Examination of Schwarzschild Spacetime in isotropic form

$$ds^2 = -\left(1+\frac{1}{2}\phi\right)^2\left(1-\frac{1}{2}\phi\right)^{-2} dt^2 + \left(1-\frac{1}{2}\phi\right)^4 [dr^2 + r^2 d\Omega^2],$$

(Eqn. 10-7)

which is known as the Isotropic Schwarzschild equation [21], with $\phi = -\frac{M}{r}$. For $\phi \ll 1$, the metric coefficients expanded to 2nd order in ϕ:

$$ds^2 = -\left(1 + \phi + \frac{1}{4}\phi^2\right)\left(1 - \phi + \frac{1}{4}\phi^2\right)^{-1} dt^2$$
$$+ \left(1 - \phi + \frac{1}{4}\phi^2\right)^2 [dr^2 + r^2 d\Omega^2]$$

(Eqn. 10-8)

and since

$$\frac{1}{\left(1 - \phi + \frac{1}{4}\phi^2\right)} = 1 + \left(\phi - \frac{1}{4}\phi^2\right) + \left(\phi - \frac{1}{4}\phi^2\right)^2 + O(\phi^3)$$
$$= 1 + \phi + \frac{3}{4}\phi^2 + O(\phi^3)$$

We have:

$$\left(1 + \phi + \frac{1}{4}\phi^2\right)\left(1 + \phi + \frac{3}{4}\phi^2 + O(\phi^3)\right) = (1 + \phi)^2 + \phi^2 + O(\phi^3)$$
$$= 1 + 2\phi + 2\phi^2 + O(\phi^3)$$

Also,

$$\left(1 - \phi + \frac{1}{4}\phi^2\right)^2 = (1 + \phi)^2 + \frac{1}{2}\phi^2 + O(\phi^3)$$
$$= 1 - 2\phi + \frac{3}{4}\phi^2 + O(\phi^3)$$

So, in terms of the indicated metric in terms of lapse and shift variables:

$$ds^2 = -{N^t}^2 dt^2 + L^2[dr + N^r dt]^2 + R^2 d\Omega^2$$

Thus, $(N^t)^2 = 1 + 2\phi + 2\phi^2 + O(\phi^3)$

$$L^2 = 1 - 2\phi + \tfrac{3}{2}\phi^2 + O(\phi^3)$$
$$R^2 = r^2 L^2$$
$$N^r = 0$$

$$N^t = \left(1 + \frac{1}{2}\phi\right)\left(1 - \frac{1}{2}\phi\right)^{-1}$$
$$= \left(1 + \frac{1}{2}\phi\right)\left(1 + \frac{1}{2}\phi + \left(\frac{1}{2}\phi\right)^2\right) + O(\phi^3)$$
$$= \left(1 + \tfrac{1}{2}\phi\right)^2 + \tfrac{1}{4}\phi^2 + O(\phi^3)$$
$$= 1 + \phi + \tfrac{1}{2}\phi^2 + O(\phi^3)$$

$L = \left(1 - \frac{1}{2}\phi\right)^2 = 1 - \phi + \frac{1}{4}\phi^2$ exact; $\qquad L^{-1} = 1 + \left(\phi - \frac{1}{4}\phi^2\right) +$
$\left(\phi - \frac{1}{4}\phi^2\right)^2$
$R = rL \qquad\qquad = 1 + \phi - \frac{3}{4}\phi^2 +$
$O(\phi^3)$ not exact
$N^r = 0$

So, for our analysis we choose isotropic coordinates and the gauge where the metric coefficients are expressed in terms a potential function $\phi(r,t)$ such that $\dot{\phi} \ll \phi$ and:

$$\left.\begin{array}{l} N^t = 1 + \phi + \frac{1}{2}\phi^2 \\ L = 1 - \phi + \frac{1}{4}\phi^2 \\ R = rL \\ N^r = 0 \end{array}\right\} \text{(exact in full solution } \phi = -\tfrac{M}{r}\text{)}$$

(Eqn. 10-9)

The "Newtonian gauge" potential function $\phi = \phi(r,t)$ has the properties:

$\dot{\phi} \ll \phi \Rightarrow \dot{\phi} \sim v\phi', v \ll 1$
$\phi' \sim \phi$
$\lim_{r\to\infty}(\phi) = O\left(\frac{1}{r}\right) = 0$
And $\phi \ll 1$

Consider, first, the $\mathcal{H}^{(2)} = 0$ constraint equation to first order in ϕ.

$$\mathcal{H}^{(2)} = \frac{L\pi_L^2}{2R^2} - \frac{\pi_L \pi_R}{r} + \frac{1}{2}\left\{\left(\frac{2RR'}{L}\right)' - \frac{(R')^2}{L} - L\right\} + \mathcal{H}_M$$

where,
$\pi_R = \{(N^r LR)'/N^t - (LR)'/N^t\}$
$\pi_L = -R\left(\dot{R} - N^r R'\right)/N^t$

In the "$\mathcal{N}$" gauge $\qquad \pi_R = (LR)'/N^t = \frac{r\left(1-\frac{1}{2}\phi\right)\dot{\phi}}{\left(1+\phi+\frac{1}{2}\phi^2\right)} \approx r\dot{\phi}$

$$\pi_L = \frac{\frac{1}{2}r^2\left(1-\frac{1}{2}\phi\right)\dot{\phi}}{\left(1+\phi+\frac{1}{2}\phi^2\right)} \approx \frac{1}{2}r^2\dot{\phi}$$

or

$$\left.\begin{aligned}\frac{L\pi_L^2}{2R^2} &\simeq \frac{1}{8}r^2\dot{\phi}^2 \\ \frac{\pi_L\pi_R}{r} &\simeq \frac{1}{2}r^2\dot{\phi}^2\end{aligned}\right\}$$

(Eqn. 10-10)

The above are negligible compared to other terms in $\mathcal{H}^{(2)}$ to second order in ϕ, for higher orders further properties of the "$\mathcal{N}$" gauge would need to be specified (i.e., relation of $|v|$ to $|\phi|$), however that need not be a concern in the analyses to follow here. So,

$$\mathcal{H}^{(2)} \simeq \frac{1}{2}\left\{\left(\frac{2RR'}{L}\right)' - \frac{(R')^2}{L} - L\right\} + \mathcal{H}_M$$

(Eqn. 10-11)

$\mathcal{H}_M = \delta(r-\hat{r})\left(L^{-2}\hat{p}^2 + (4\pi\mu)^2\hat{R}^4\right)^{1/2}$ $\qquad (4\pi\hat{R}^2\mu) = m,$ proper shell mass fixed

$= \delta(r-\hat{r})m$ $\qquad$ Newt $\hat{p}^2 \simeq m^2v^2 \ll m^2$

$= 4\pi r^2\rho \ , \ \rho = \frac{m}{4\pi r^2}\delta(r-\hat{r})$

$= (4\pi\rho)r^2$

(Eqn. 10-12)

$$\left(\frac{RR'}{L}\right)' = [rL + r^2L']' = L + 3rL' + r^2L''$$

$$\frac{(R')^2}{L} = \frac{(L+rL')^2}{L} = L + 2rL' + r^2\frac{L'^2}{L}$$

$$\frac{1}{2}\left\{\left(\frac{2RR'}{L}\right)' - \frac{(R')^2}{L} - L\right\} = r^2L'' + 2rL' - \frac{1}{2}r^2L^{-1}(L')^2$$

Fortunately the expression for L is exact, so fourth order terms need not be calculated to arrive at a second order form for L''.

$$L' = -\phi' + \frac{1}{2}\phi\phi'$$

$$L'' = -\phi'' + \frac{1}{2}\phi\phi'' + \frac{1}{2}(\phi')^2$$

$$L^{-1} = r^2\left[L'' - \frac{1}{2}\frac{(L')^2}{L}\right]$$
$$= r^2\left[-\phi'' + \frac{1}{2}\phi\phi'' + \frac{1}{2}(\phi')^2 - \frac{1}{2}\left(\phi' - \frac{1}{2}\phi\phi'\right)^2\left(1 + \phi - \frac{3}{4}\phi^2\right)\right]$$
$$= -r^2\phi'' \text{ to 1}^{\text{st}} \text{ order in } \phi$$

$$\mathcal{H}^{(2)} - \mathcal{H}_M = -r^2\phi'' - 2r\phi' \text{ to 1}^{\text{st}} \text{ order in } \phi$$
$$= \left(-{}^{(3)}\nabla^2\phi\right)r^2 + O(\phi^2)$$

(Eqn. 10-13)

where $d\ell^2 \cong (1 + 2\phi)(dr^2 + r^2 d\Omega^2)$ and

$${}^{(3)}\nabla^2\phi \simeq \frac{1}{r^2(1+2\phi)^{3/2}}\frac{\partial}{\partial r}\left\{r^2(1+2\phi)^{1/2}\frac{\partial\Psi}{\partial r}\right\}$$
$$\simeq \frac{1}{r^2}\{r^2\phi'' + 2r\phi'\} + O(\phi^2)$$

So,

$$0 = \mathcal{H}^{(2)} = [4\pi\rho - \nabla^2\phi]r^2 + O(\phi^2)$$

(Eqn. 10-14)

This indicates that we obtain the usual constraint equation from Newtonian gravity, Poisson's eqn., and integration with mass M in the interior yields the usual gravitational potential solution:

$$\nabla^2\phi = 4\pi\rho \Rightarrow \phi = -\frac{M}{r},$$

(Eqn. 10-15)

Which agrees with the exact expression of the ϕ in the Iso. Schw. metric.

Re-doing the $\mathcal{H}^{(2)} = 0$ analysis to 2nd order one obtains:

$$\underbrace{\left(1 - \frac{1}{2}\phi\right)\nabla^2\phi}_{\substack{next\ order\ of \\ \phi\ added}} = \underbrace{(1+\phi)4\pi\rho}_{\substack{next\ order\ of \\ \phi\ added}} + \left[\sqrt{p^2 + m^2} - m\right]\delta(r - \hat{r})$$

(Eqn. 10-16)

Continuing in this manner, post-post-Newtonian potentials may be calculated. However, we already have a firm understanding of the kinematics in the full GR. What we seek next is an understanding of the dynamics of the shell. To obtain dynamics we must express the full Hamiltonian, not the Hamiltonian density. This analysis entails the calculation of a surface term at spatial infinity which, for asymptotically flat spacetimes, is the ADM mass of the combined gravitational + matter system.

It seems as if extra information is being extracted from $\mathcal{H}^{(2)} = 0$ in the 1st order kinematical relation $\nabla^2\phi = 4\pi\rho$. However this is precisely because the calculation was first order, as well as suppressing the dynamical properties of the matter directly ($p = 0$) and the dynamical properties of the grav. field indirectly $\left(\dot{\phi} \ll \phi\right)$. If we go to 2nd order we expect a self-grav. term and $p \neq 0$ terms, while still leaving $\dot{\phi} \ll \phi$ i.e., maintaining the Newt. approx., we would expect the Newtonian description of the shell that includes the Newtonian self-gravity term for the shell. What is surprising is that inclusion of all p terms and no higher order ϕ terms should then yield an exact solution for the shell dynamics (Israel [57]). So, starting with

$$H = 0 = \int \mathcal{H}_t^{(2)}\, N^t dr$$

$$= \int \left[-\left\{ \left(\frac{RR'}{L}\right) N^{t\prime} + \frac{N^t}{2}\left(\frac{(R')^2}{L} + L\right) \right\} + \underline{\underline{\left(\frac{RR'N^t}{L}\right)'}} + \mathcal{H}_{tM} N^t \right] dr$$

(Eqn. 10-17)

Second order derivative terms can be eliminated in favor of a surface term which may be uniquely specified when the class of metrics is considered restricted to asymptotically flat ones as in this instance.

For an isotropic coordinate description the gauge potentials, expressed in terms of the potential ϕ, must behave at asymptotic spatial infinity exactly as the metric “potentials” in the isotropic Schwarzschild solution:

$$\lim_{r\to\infty} \phi = \lim_{r\to\infty}\left(-\frac{M}{r}\right)$$

Thus,

$$\lim_{r\to\infty} \left\{ \begin{array}{l} R = r\left(1 + \frac{M}{r}\right) \\ R' = 1 \\ N^t = \left(1 - \frac{M}{r}\right) \\ L = \left(1 + \frac{M}{r}\right) \end{array} \right\}$$

So,

$$\lim_{r\to\infty}\left(\frac{RR'N^t}{L}\right) = (r - M)$$

Stated more rigorously, $\lim_{r\to\infty}\left[\left(\frac{RR'N^t}{L}\right)-r\right]=-M$, and

$$\int_0^\infty \left[\left(\frac{RR'N^t}{L}\right)'-1\right]=-M$$

(since $R(0)=0$). Regrouping the expression for $\int \mathcal{H}^{(2)}\,N^t dr$:

$$0=\int\left\{1-\left[\left(\frac{RR'}{L}\right)N^{t'}+\frac{1}{2}N^t\left(\frac{R'^2}{L}+L\right)\right]\right\}dr-M+\int \mathcal{H}_{tM}N^t\,dr$$

Consider the matter term first:

$$\int \mathcal{H}_{tM}N^t\,dr=\int \underbrace{\left(1+\phi+\frac{1}{2}\phi^2\right)}_{N^t\ to\ O(\phi^2)}\sqrt{p^2L^{-2}+m^2}\,\delta(r-\hat{r})dr$$

$$\begin{aligned}p^2L^{-2}+m^2&=p^2\left(1-2\phi+\frac{3}{2}\phi^2\right)^{-1}+m^2\\&\simeq p^2+m^2\ \text{ since } p^2 \text{ is already } p^2\ll m^2\end{aligned}$$

Since asympt $\int\phi^2\sim\phi$ we need only retain 1st order ϕ term in the matter contribution (due to the δ-function).

$$\int \mathcal{H}_{tM}N^t\,dr\cong(1+\phi)\sqrt{p^2+m^2}$$

Since $\phi p\ll\phi$ we may further simplify $\phi\sqrt{p^2+m^2}$ to $m\phi$:

$$\int \mathcal{H}_{tM}N^t\,dr=\sqrt{p^2+m^2}+m\phi+O(\phi^2)+O(\phi p)$$

Consider now $\left\{1-\left[\left(\frac{RR'}{L}\right)N^{t'}+\frac{1}{2}N^t\left(\frac{R'^2}{L}+L\right)\right]\right\}$, keeping only second order terms:

$$\begin{aligned}\left(\frac{RR'}{L}\right)(N^t)'&=r[L+rL']\underbrace{(\phi'+\phi\phi')}_{+O(\phi^2)}\\&=r\underbrace{\left[1+\phi+r\left(-\phi'+\frac{1}{2}\phi\phi'\right)\right]}_{+O(\phi^2)}((\phi'+\phi\phi'))\\&=r[\phi'+\phi\phi'-\phi\phi'-r(\phi')^2]+O(\phi^2)\end{aligned}$$

From $\nabla^2\phi \simeq 4\pi\rho$ we know that $r\phi' \sim \phi$, thus attention must be generally paid to "r" factors, however here they all pair up with ϕ' somewhere, so ϕ counting still suffices.

$$\frac{1}{2}N^t\left[\frac{(R')^2}{L}+L\right]=\frac{1}{2}\left(1+\phi+\frac{1}{2}\phi^2\right)\left[\frac{(L+rL')^2}{L}+L\right]$$
$$=\frac{1}{2}\left(1+\phi+\frac{1}{2}\phi^2\right)\left[2L+2rL'+r^2\frac{(L')^2}{L}\right]$$
$$=\left(1+\phi+\frac{1}{2}\phi^2\right)\left[\left(1+\phi+\frac{1}{4}\phi^2\right)+r\left(-\phi'+\frac{1}{2}\phi\phi^1\right)\right.$$
$$\left.+\frac{1}{2}r^2\left(-\phi'+\frac{1}{2}\phi\phi'\right)^2\left(1+\phi+\frac{3}{4}\phi^2\right)\right]$$
$$=\left[(1-\phi^2)+\frac{3}{4}\phi^2\right]-r\phi'(1+\phi)+\frac{1}{2}r\phi\phi'+\frac{1}{2}r^2(\phi)^2+O(\phi^3)$$
$$=1-\frac{1}{4}\phi^2-r\phi'-\frac{1}{2}r\phi'\phi+\frac{1}{2}r^2(\phi')^2+O(\phi^3)$$

So,

$$\left\{1-\left[\left(\frac{RR'}{L}\right)N^{t'}+\frac{1}{2}N^t\left(\frac{R'^2}{L}+L\right)\right]\right\}$$
$$=\frac{1}{4}\phi^2+r\phi'+\frac{1}{2}r\phi'\phi-\frac{1}{2}r^2(\phi')^2-r\phi'+r^2(\phi')^2+O(\phi^3)$$
$$=\frac{1}{4}\phi^2+\frac{1}{2}r\phi'\phi+\frac{1}{2}r^2(\phi')^2+O(\phi^3)$$

So,

$$0=H=\int\left[\frac{1}{4}\phi^2+\frac{1}{2}(r\phi'\phi+r^2(\phi')^2)\right]dr-M+\sqrt{p^2+m^2}+m\phi$$

where

$$\int r^2\phi'^2=\int[r^2\phi'\phi]'-r^2(\phi')'\phi=-2r\phi'\phi-r^2\phi\phi''=-r^2\phi\nabla^2\phi$$

So,

$$0=H=\int\left[\frac{1}{4}(r\phi^2)'-\frac{1}{2}\phi r^2\nabla^2\phi\right]dr-M+\sqrt{p^2+m^2}+m\phi$$

(Eqn. 10-18)

Using $\nabla^2\phi=\frac{m}{r^2}\delta(r=\hat{r})$, and use of 1st-order solution allowed since mult. by ϕ. There is also the $(r\phi^2)'$ term which goes to zero asymptotically, yet contributes at the shell for discontinuous ϕ. However, in the Newtonian approx. we can just consider a finite thickness shell and omit the term, while in full G. R. we can just choose a better gauge at the shell such that ϕ is continuous, as that is possible.

So,

$$0 = H = -\frac{1}{2}m\phi(\hat{r}) - M + \sqrt{p^2 + m^2} + m\phi(\hat{r}),$$

(Eqn. 10-19)

where $\frac{\hat{p}}{m} = \frac{\partial \hat{r}}{\partial \tau}$ in Newt. approx.; τ = proper time of shell; and where $\phi(\hat{r})$ is the external shell solution of $\nabla^2\phi = 4\pi\rho$ that is considered continuous through the shell.

The exact solution of the Israel analysis [57] is:

$$M = m\underbrace{\sqrt{1 + \left(\frac{\partial r}{\partial \tau}\right)^2}}_{1+\frac{1}{2}(v)^2 \text{ for } v \ll 1} + \frac{1}{2}m\left(-\frac{m}{r}\right)$$

(Eqn. 10-20)

Thus

$$E = (M - m) = \frac{1}{2}mv^2 - \frac{1}{2}\frac{m^2}{r} \; for\ v \ll 1,$$

(Eqn. 10-21)

in agreement with Sec. 10.2.

10.4 Analysis of Schwarzschild metric using ADM formalism

Analysis of Schwarzschild metric using ADM lapse shift formalism

$$ds^2 = -(1 + 2\phi)dt^2 + (1 + 2\phi)^{-1}dr^2 + r^2 d\Omega^2, \;\; \phi = -M/r$$

Now for an approximate analysis that keeps ϕ to second order:

$ds^2 = -(N^t)^2 dt^2 + L^2 dr^2 + 2L^2 N^r dr dt + R^2 d\Omega^2$

$(N^t)^2 = (1 + 2\phi)$

$N^t = \sqrt{1 + 2\phi} = 1 + \phi - \frac{1}{2}\phi^2$

$L^2 = (1 + 2\phi)^{-1}$

$L = 1 - \phi + \frac{3}{2}\phi^2$

$$N^r = 0 \; and \;\; R = r \;\; \Rightarrow \pi_R, \pi_L = 0$$

and

$$\mathcal{H} = \frac{1}{2}\left\{\left(\frac{2RR'}{L}\right)' - \frac{(R')^2}{L} - L\right\} + \mathcal{H}_M$$

where to lowest order, $\mathcal{H}_M = (4\pi\rho)r^2$, and

$\mathcal{H} - \mathcal{H}_M = \frac{1}{2}\left\{\left(\frac{2r}{L}\right)' - \frac{1}{L} - L\right\} = \frac{1}{2}\left\{2r\left(\frac{1}{L}\right)' + \frac{1}{L} - L\right\}$

$\mathcal{H} - \mathcal{H}_M = r\left(1 + \phi - \frac{1}{2}\phi^2\right)' + \frac{1}{2}\left[\left(1 + \phi - \frac{1}{2}\phi^2\right) - \left(1 - \phi + \frac{3}{2}\phi^2\right)\right]$

$\mathcal{H} - \mathcal{H}_M = r\phi'(1 - \phi) + \phi - \phi^2 = (r\phi)' - \phi(r\phi)'.$

Thus, at first order:

$$\mathcal{H} - \mathcal{H}_M = (r\phi)'$$

So, $0 = \mathcal{H} = (r\phi)' + (4\pi\rho)r^2, 0 = \int \mathcal{H} \Rightarrow where\, r\phi = -m$ at shell, so $\phi = -\frac{m}{r}$.

For the Schwarzschild metric, where R is simply r, the super-Hamiltonian is a first order differential equation, with a straight forward solution. To obtain a Poisson eqn. description a second order diff. eq. is obviously needed, this can be obtained by doing an integration by parts in the full Hamiltonian description.

Evaluation of $\int \mathcal{H}^{(2)}\, N dr = 0$ to lowest order:

$$0 = \int \frac{1}{2}\left\{\left(\frac{2RR'}{L}\right)' - \frac{(R')^2}{L} - L\right\} N^t dr + \int \mathcal{H}_M\, N^t dr$$

Integration by parts to remove as many $\frac{\partial R}{\partial r}$ derivatives as possible.

$$\begin{cases} N^t\left(\frac{2RR'}{L}\right)' = N^t\left(\frac{(R^2)'}{L}\right)' = \left[N^t\left(\frac{(R^2)'}{L}\right)\right]' - \frac{(N^t)'}{L}(R^2)' \\ \qquad\qquad = \left[N^t\left(\frac{(R^2)'}{L}\right)\right]' - \left[\frac{(N^t)'}{L}R^2\right]' + R^2\left(\frac{(N^t)'}{L}\right)' \\ \text{Similarly} \\ \frac{(R')^2}{L}N^t = \left[\frac{(R')N^t}{L}R\right] - R\left(\frac{R'N^t}{L}\right)' \end{cases}$$

Thus,

$$0 = \int \left\{ \frac{1}{2}\left[N^t \left(\frac{(R^2)'}{L} \right) \right]' - \frac{1}{2}\left[\frac{(N^t)'}{L} R^2 \right]' - \frac{1}{4}\left[(R^2)' \frac{N^t}{L} \right]' + \frac{R^2}{2}\left(\frac{(N^t)'}{L} \right)' + \frac{R}{2}\left(\frac{R'N^t}{L} \right)' - \frac{LN^t}{2} \right\} dr + \int \mathcal{H}_M N^t dr$$

$$= - \lim_{r\to\infty} \left[\frac{(N^t)'}{2L} R^2 + \int \left\{ R^2 \frac{R^2}{2}\left(\frac{(N^t)'}{L} \right)' + \frac{R}{2}\left(\frac{R'N^t}{L} \right)' - \frac{LN^t}{2} \right\} dr + \int \mathcal{H}_M N^t dr - \frac{1}{4}(R^2)' \frac{N^t}{L} \right]$$

Evaluate limit: $\lim_{r\to\infty} \left\{ \begin{array}{l} R = r \\ N^t = \sqrt{1 - \frac{2M}{r}} \\ L = (N^t)^{-1} \end{array} \right\}$

So, $r^2 (N^t)' L^{-1} = \frac{1}{2}((N^t)^2)' r^2 = \frac{1}{2}\left(-\frac{2M}{r}\right)' r^2 = M$, while $\frac{1}{4}(R^2)' \frac{N^t}{L} = \frac{1}{2}(r - 2M)$,
and,

$$0 = -\frac{3}{2}M + \int \left\{ \underbrace{\frac{r^2}{r}\phi' + \frac{r}{2}\left(2\frac{1}{2}\right)}_{\frac{1}{2}r = \nabla^2 \phi} - \frac{1}{2} + \frac{1}{2} \right\} dr + \int r^2 \, (4\pi\rho) dr$$

To first order $M = \int r^2 \, (4\pi\rho) dr$, so

$$0 = \int r^2 \, [\nabla^2 \phi - 4\pi\rho] dr \quad \Rightarrow \quad \nabla^2 \phi = 4\pi\rho.$$

Now consider

$$\mathcal{H}_G N^t = \left(\frac{RR'}{L} \right)' N^t - \frac{N^t}{2}\left(\frac{(R')^2}{L} + L \right)$$

Have similar first order analysis:

$N^t = 1 + \phi - \frac{1}{2}\phi^2 \quad N^t = L^{-1} \quad (N^t)^2 = (1 + 2\phi) + O(\phi^3)$

$L = 1 - \phi + \frac{3}{2}\phi^2$

$R = r$

So,

$$
\begin{aligned}
\mathcal{H}_G N^t &= (rN^t)'N^t - \frac{1}{2}N^t(N^t + L) \\
&= r\tfrac{1}{2}\left(N^{t^2}\right)' + (N^t)^2 - \tfrac{1}{2}(N^t)^2 - \tfrac{1}{2}N^t L \\
&= r\phi' + \tfrac{1}{2} + \phi - \tfrac{1}{2} \\
&= (r\phi)' \quad \text{(which is the same for the exact result as well)}
\end{aligned}
$$

$$
0 = \int \mathcal{H}^{(2)}\, N^t dr
$$

$$
= \int \left[(r\phi)' + (1+\phi)\sqrt{p^2 L^{-2} + (4\pi\mu)^2 \hat{R}}\; \delta(r - \hat{r}) \right] dr
$$

For a monotonically decreasing gauge potential ϕ, we have

$$
\int (r\phi)' = \lim_{r\to\infty} (r\phi) = r\left(-\frac{M}{r}\right) = -M
$$

Thus

$$
M = \sqrt{p^2 + m^2} + m\phi.
$$

but not quite right. Consider that there is a boundary term from the shell, so

$$
\int (r\phi)' = -M + \text{shell term}.
$$

The shell term is due to the $\phi \cong -\frac{m}{r}$ solution in the neighborhood exterior to the shell while $\phi = 0$ in the interior (continuity of the metric is maintained by rescaling the coordinate system, we can't just have $\phi = const$ either as we are describing ϕ in terms of the first order ϕ analysis which indicated $\phi = -\frac{m}{r}$ ext, 0 int.) Note, that this complication didn't arise in the isotropic analysis because the 2nd order derivative form of that analysis clearly separated into a surface term, and a term $\frac{1}{2}\phi r^2 \nabla^2 \phi$, where $\nabla^2 \phi$ was, then expressed via the 1st order ϕ approx in terms of the poisson eqn. $\nabla^2 \phi = 4\pi\rho$. In this fashion the impact of the delta function on the analysis was made explicit and the class of solutions of ϕ was larger, allowing $\phi = const$.

Let the shell have finite thickness and, correspondingly, have ϕ then be continuous as well. Now,

$$
\mathcal{H}_M = \frac{1}{\Lambda}\sqrt{p^2 L^{-2} + (4\pi\rho)^2 r^4}\; \theta_-\left(r - \left(R - \frac{1}{2}\Lambda\right)\right) \theta_+\left(r - \left(R + \frac{1}{2}\Lambda\right)\right)
$$

and

$$\int \mathcal{H}_M N^t dr \cong \int (1+\phi)\sqrt{p^2+m^2}\frac{1}{\Lambda}\theta_-\theta_+$$
$$\cong \sqrt{\hat{p}^2+m^2} + \phi_{ext}\left(\frac{1}{2}\frac{\Lambda^2}{\Lambda}\right)\frac{1}{\Lambda}\sqrt{p^2+m^2}$$
$$\simeq \sqrt{\hat{p}^2+m^2} + \frac{1}{2}\phi m$$

Thus,

$$M = \sqrt{p^2+m^2} + \frac{1}{2}m\phi.$$

(Eqn. 10-22)

Thus, we have the generalization to shell motion that is relativistic (taken in the non-relativistic limit and we recover the result of Sec. 10.2 as before).

10.5 $N^r = 0$ gauge, and $L = 1, R = r$ gauge

We now consider some choices of gauge in the ADM formulation. The $N^r = 0$ gauge, with implicit spherical symmetry and asymptotically flat region, has only Minkowski as a solution. The $L = 1, R = r$ gauge, on the other hand, will be seen to offer useful simplifications, so will occur in later analysis.

Consider as usual:

$$ds^2 = -(N^t)^2dt^2 + L^2[dr + N^r dt]^2 + R^2(d\theta^2 + \sin^2\theta\, d\phi^2)$$
$$S = \int dt\, p_i\dot{q}^i + \int dr\, dt(\pi_L\dot{L} + \pi_R\acute{R} - N^tH_t - N^rH_r)$$
$$H_t = \frac{L\pi_L^2}{2R^2} - \frac{\pi_L\pi_R}{R} + \frac{1}{2}\left[\left(\frac{2RR'}{L}\right) - \frac{(R')^2}{L} - L\right] + H_{tM}$$
$$H_r = R'\pi_R - L\pi_L' + H_{rM}$$
$$\pi_R = \frac{1}{N^t}[(N^rLR)' - (LR)^{\cdot}]$$
$$\pi_L = \frac{R}{N^t}(\dot{R} - N^rR')$$

So

$$0 = \frac{R'}{L}H_t + \frac{\pi_L}{RL}H_r = -\mathfrak{M}' + \frac{R'}{L}H_{tM} + \frac{\pi_L}{RL}H_{rM}$$
$$\mathfrak{M} = \frac{\pi_L^2}{2R} + \frac{R}{2}\left[1 - \left(\frac{R'}{L}\right)^2\right]$$

Consider the Gauge $\underline{R' = L}$ (with no matter):

$$H_t = \frac{R'\pi_L^2}{2R^2} - \frac{\pi_L \pi_R}{R} + [0]$$
$$H_r = R'(\pi_R - \pi_L')$$
$$(H_r = 0) \Rightarrow \pi_R - \pi_L' \quad or \;\; R' = 0 \Rightarrow L = 0 \text{ bad}$$
$$H_t = -\tfrac{1}{2}\left(\tfrac{\pi_L^2}{R}\right)' = 0 \Rightarrow \left(\tfrac{\pi_L^2}{R}\right) = constant \text{ as } \underline{\text{above}}.$$
$$\pi_L^2 = CR + Rf(t)$$

The gauge choice $R = r$ yields
$$\mathcal{H}_t = \frac{L\pi_L^2}{2r^2} - \frac{\pi_L \pi_R}{r} + \frac{1}{2}\left[2r\left(\frac{1}{L}\right)' + \frac{1}{L} - L\right]$$

Together with $N^r = 0$ yields $(\pi_L = 0)$
$$\mathcal{H}_t = r\left(\frac{1}{L}\right)' + \frac{1}{2}\left(\frac{1}{L} - L\right) = 0$$
$$\mathcal{H}_r = \pi_R = \frac{R\dot{L}}{N^t} = 0 \Rightarrow \underline{\dot{L} = 0}$$

$L = 1 \Rightarrow Minkowski\,, \mathfrak{M} = 0$
$L \neq 1 \Rightarrow L = f(r)$, and:
$$r\frac{f'}{f^2} + \frac{1}{2}\left(\frac{1}{f} - f\right) = 0$$
$$f' = \frac{1}{2r}(f^3 - f^2)$$
So, Schwarzschild doesn't actually have $R = r, N^r = 0$, rather it has $L = 1, R = r, N^t = 1$, and
$N^r = \sqrt{\frac{2M}{r}}, t_s = t - 2\sqrt{2Mr} = 2M\log|\cdots|$. So the gauge choice $R = r, N^r = 0$ is incompatible with Schw. geometry. Note that for spherically symmetric space-times $R = r$ is natural and if $L = 1$ then the dynamics is completely expressed in lapse N^t and shift N^r. Setting $N^r = 0$ does not allow anything but a comformal factor on $dr \times dt$ and a reparam factor in t, thus restricting to slicings of Minkowski, DeSitter, or other S^2 inv. spaces, but given the asympt. flat restriction we have only Minkowski.

Is the gauge $L = 1, R = r$ still good when matter is present? (e.g., shell)

$$\mathcal{H}_t = \frac{\pi_L^2}{2r^2} - \frac{\pi_L \pi_R}{r} + \sqrt{(p/L)^2 + m^2}\delta(r - \hat{r})$$
$$\pi_L = \frac{rN^r}{N^t}$$
$$\pi_R = \frac{(rN^r)'}{N^t}$$
Thus,
$$\frac{\pi_L^2}{2r^2} - \frac{\pi_L \pi_R}{r} = \left(\frac{1}{N^t}\right)^2 \left[\frac{(rN^r)^2}{2r^2} - \frac{(rN^r)(rN^r)'}{r}\right] = \frac{1}{(N^t)^2}\left[-\frac{(rN^r)^2}{2r}\right]' = 0,$$
Thus
$$(rN^r)^2 = 2r \cdot const. \rightarrow \quad N^r = \sqrt{\frac{M}{2r}}$$

Similarly for $\mathcal{H}_r = \pi_R - \pi_L' - p\delta(r - \hat{r})$
$$\pi_R - \pi_L' = \frac{(rN^r)'}{N^t} - \left(\frac{rN^r}{N^t}\right)' = \frac{rN^r(N^t)'}{(N^t)'} = 0 \Rightarrow N^t = const = \pm 1$$

Some early ideas on the shell discontinuity:
With shell matter present, discontinuities are only avoided if $R \neq r$ so that a 2nd derivative term exists, thus possibly matching the δ by discont. 1st deriv., continuous R.
(1) Want continuous N^r and N^t
(2) What discont. R in 1st deriv. so R'' term matches δ
(3) What L continuous, maybe discont. in 1st deriv.
Such is the case with Iso. Schw. gauge/coord examined in Sec. 10.3, but more useful, perhaps, given spatial hypersurface inextendibility is Kruskal spacetime coordinates for the BH. So let's examine Kruskal next.

10.6 Kruskal needed for maximally extended space-time analysis
Need to use Kruskal spacetime [21,45] so that Israel analysis [57] is possible upon a maximally-extended space-time. Consider an analysis with shell position static, or 'frozen'.

At shell freeze Kruskal is constant through the origin of the spherical system, i.e., it's flat space-time on the interior. For the above choice

$$ds^2 = \frac{(32M^3)}{r} e^{-r/2M}(-dv^2 + du^2) + r^2 d\Omega^2$$

Re-label:
$v \to t$
$u \to r$
$r \to R$
$$ds^2 = \frac{(32M^3)}{R} e^{-R/2M}(-dt^2 + dr^2) + R^2 d\Omega^2$$
$$ds^2 = -(N^t)dt^2 + L^2dr^2 + 2L^2N^r drdt + L^2(N^r)^2dt^2 + R^2d\Omega^2$$
$$R = R(t, r)$$

Note:
Kruskal is relevant for the massless scalar field analysis as well.
Novikov is relevant for the massive scalar field analysis.

$$R = R(t, r)$$
$$N^r = 0$$
$$(N^t)^2 = L^2 \frac{32M^3}{R} e^{-R/2M} = L = N^t$$
where we choose to match with Extended Schwarzschild (Ext. Schw.).
We then get:
$$\pi_R = \frac{1}{N^t}(-LR)' = -\acute{R} - \frac{R\acute{L}}{N^t}$$
$$= -\dot{R} + \dot{R} + \dot{R}\left(\frac{R}{2M}\right)$$
$$= \frac{(R^2)^{\cdot}}{4M}$$

$$\frac{R\acute{L}}{N^t} = \frac{\partial}{\partial t} \ln N^t = \frac{\partial}{\partial t}\left[\ln\left(\frac{32M^3}{R}\right) - \frac{R}{2M}\right]$$
$$= \frac{-\left(\frac{\acute{R}}{R^2}\right)}{\left(\frac{1}{R}\right)} - \frac{\acute{R}}{2M}$$
$$= -\frac{\acute{R}}{R} - \frac{\acute{R}}{2M}$$

$$\pi_L = -R\dot{R}\left(\frac{32M^3}{R}\right)^{-1} e^{R/2M}$$
$$= -\frac{R^2\dot{R}}{32M^3} e^{R/2M} = -\frac{1}{4}\left(\frac{R}{2M}\right)^2 \left(\frac{R}{2M}\right)' e^{R/2M} = -\frac{1}{4}\left(\frac{R}{2M}\right)^2 \left[e^{R/2M}\right]'$$

$$\mathcal{H}_t^G = \frac{L\pi_L^2}{2R^2} - \frac{\pi_L\pi_R}{R} + \frac{1}{2}\left[\left(\frac{2RR'}{L}\right)' - \frac{(R')^2}{L} - L\right]$$
$$= \frac{1}{N^t}\frac{(-RR')^2}{2R^2} + \frac{1}{8}\left(\frac{R}{2M}\right)\left[\left(\frac{R}{2M}\right)^2\right]' \left[e^{R/2M}\right]' + \frac{1}{2}\left[\left(\frac{R^2R'e^{+R/2M}}{16M^2}\right)'\right]$$
$$-\frac{1}{2}\left[\frac{(R')^2R}{32M^3}e^{R/2M} - \frac{32M^3}{R}e^{-R/2M}\right]$$

$$= \frac{1}{2}\left[\frac{(\dot{R})^2 R}{32M^3} - \frac{(R')^2 R}{32M^3}\right] e^{R/2M} + \left[\frac{R^2 \dot{R}}{32M^3}\right] \left[e^{R/2M}\right]' + \left[\left(\frac{R^2 R'}{32M^3}\right) e^{R/2M}\right]'$$
$$- \frac{1}{2}\left[\frac{(R')^2 R}{32M^3} e^{R/2M} - \frac{32M^3}{R} e^{-R/2M}\right]$$

Solution to $\mathcal{H}_t = 0$ is $\left({}^{R}/_{2M} - 1\right)e^{R/2M} = r^2 - t^2$?

$$\frac{\dot{R}}{2M} e^{R/2M} + \left({}^{R}/_{2M} - 1\right)\frac{\dot{R}}{2M} e^{R/2M} - 2t$$
$$\frac{R\dot{R}}{2M} e^{R/2M} = -2t$$
$$\left(\frac{R}{2M}\right)^2 e^{R/2M}\left[(R')^2 - \left(\dot{R}\right)^2\right] = 4(r^2 - t^2) = 4\left({}^{R}/_{2M} - 1\right)e^{R/2M}$$
$$\frac{1}{16M^2}\left[(R')^2 R - \left(\dot{R}\right)^2 R\right] R e^{R/2M} = \left({}^{R}/_{2M} - 1\right)$$

(Eqn. 10-23)

Although interesting, we see that direct application of Kruskal spacetime coordinates, and determination of EOM's therein, leads to complex coupled equations where quantization will be that much more complicated. The search for a better coordinate description continues...

10.7 The Hamiltonian formulation

In order to understand the Schwarzschild solution and Shell dynamics (with "Exterior Schwarzschild" to the exterior of the shell) we now switch to the Hamiltonian formulation and adopt the notation and ADM formulation given by Fischler et al. [58].

The general spherically symmetric metric:

$$ds^2 = \sum_{\alpha,\beta=t,r} g_{\alpha\beta}\,(t,r) dx^\alpha dx^\beta + R(t,r)^2 (d\theta^2 + \sin^2\theta\, d\phi^2)$$

(Eqn. 10-24)

For purposes of canonical quantization Fischler et al [58] use:

$$ds^2 = -N^t(t,r)^2 dt^2 + L(t,r)^2[dr + N^r(t,r)dt]^2$$
$$+ R(t,r)^2(d\theta^2 + \sin^2\theta\, d\phi^2)$$

(Eqn. 10-25)

N^t = lapse
N^r = shift
$L = ds/dr$
R = transverse radius

The action for gravity plus matter (including cosmological const.):

$$S = \frac{1}{16\pi G}\int d^4 x\sqrt{-g}(\overline{R} - 2\Lambda) + \frac{1}{8\pi G}\int K\,(\pm h)^{1/2} d^3x + \int L_m\,(-g)^{1/2} d^4x + C$$

(Eqn. 10-26)

(from Gibbons and Hawking 1977a) PRD 15, 2758, 2752 [23], and C depends only on the boundary metric h and not on the values of g at interior points.

In the Hamiltonian formulation the usual Hamiltonian for Gravity, $H_0 = \int d^3 x\{N(x)\mathcal{H}(x) + N^i(x)\mathcal{N}_1(x)\}$, is supplemented by a surface integral at infinity in the case of aympt. flat spacetime: (Regge + Teitelboim Ann. of phy. 88, 286 [59]):

$$H = H_0 + \oint d^2 S(g_{ik,j} - g_{ij,k})$$

(Eqn. 10-27)

Let's now examine

$$I = \frac{1}{16\pi G}\int d^4 x\sqrt{-g}\,\overline{R}$$

(Eqn. 10-28)

given the metric $ds^2 = -(N^t)^2 dt^2 + L^2[dr + N^r dt]^2 + R^2 d\Omega^2$. (Let $G = 1$ in following).

Note $\sqrt{-g} = N^t L R^2 \sin\theta$.

Choose the following basis 1-forms:
$\theta^0 = N^t dt, \theta^1 = L[dr + N^r dt]\,, \theta^2 = Rd\theta\,, \theta^3 = R\sin\theta\, d\phi$

The metric is then
$g = g_{\mu\nu}\theta^\mu \otimes \theta^\nu\,, \qquad (g_{\mu\nu}) = diag\,(-1,1,1,1)$

Thus the θ basis is orthonormal and $w_{\mu\nu} + w_{\nu\mu} = 0$:
$d\theta^0 = N^{t'} dr \wedge dt$

$$d\theta^1 = \dot{L}dt \wedge dr + (LN^r)'dr \wedge dt$$
$$d\theta^2 = \dot{R}dt \wedge d\theta + R'dr \wedge d\theta$$
$$d\theta^3 = \dot{R}dt \wedge \sin\theta\, d\phi + d\theta^2 = \dot{R}dt \wedge \sin\theta\, d\phi \oplus R\cos\theta\, d\theta \wedge d\phi$$
$$dt = (N^t)^{-1}\theta^0$$
$$dr = (L)^{-1}\theta^1 - N^r dt - (L)^{-1}\theta^1 - \left(\frac{N^r}{N^r}\right)\theta^0$$
$$d\theta^0 = N^{t'}(LN^t)^{-1}\theta^1 \wedge \theta^0$$
$$d\theta^1 = \left[\dot{L}(LN^r)'\right](LN^t)^{-1}\theta^0 \wedge \theta^1$$
$$d\theta^2 = \left[\dot{R}(RN^t)^{-1} - R'N^r(RN^t)^{-1}\right]\theta^0 \wedge \theta^2 + R'(RL)^{-1}\theta^1 \wedge \theta^2$$
$$d\theta^3 = \left[\dot{R}(RN^t)^{-1} - R'N^r(RN^t)^{-1}\right]$$
$$= \theta^0 \wedge \theta^3 + R'(RL)^{-1}\theta^1 \wedge \theta^3 + \frac{\cot\theta}{R}\theta^2 \wedge \theta^3$$

Now to compare with the first structure equation:

$$d\theta^\propto = -w_\beta^\propto \wedge \theta^\beta$$
$$w_1^0 = +w_0^1 = (N^t)'(LN^t)'\theta^0 \text{ from } d\theta^0$$
$$+\left[\dot{L} - (LN^r)'\right](LN^t)'\theta^1 \ \ from\ \ d\theta^1$$

$$w_0^2 = +w_2^0 = \left[\dot{R}(RN^t)^{-1} - R'N^r(RN^t)^{-1}\right]\theta^2$$
$$w_0^3 = +w_3^0 = \left[\dot{R}(RN^t)^{-1} - R'N^r(RN^t)^{-1}\right]\theta^3$$
$$w_1^2 = -w_2^1 = R'(RL)^{-1}\theta^2$$
$$w_1^3 = -w_3^1 = R'(RL)^{-1}\theta^3$$
$$w_2^3 = -w_3^2 = +R^{-1}\cot\theta\,\theta^3$$

These connection forms satisfy the structure equations and as the solution is unique the curvature forms can now be calculated:

$$\mathcal{R}_v^\mu = dw_v^\mu + w_\propto^\mu \wedge w_v^\propto$$
$$\mathcal{R}_1^0 = dw_1^0 + w_\propto^0 \wedge w_1^\propto = dw_1^0$$

$$w_1^0 = \left(\frac{N^{t'}}{LN^t}\right)\theta^0 + \frac{\left[\dot{L} - (LN^r)'\right]}{LN^t}\theta^1$$
$$= \frac{N^{t'}}{LN^t}\theta^0 + \frac{\left[\dot{L} - (LN^r)'\right]}{LN^t}\theta^1$$
$$= \frac{N^{t'}}{L}dt + \frac{\left[\dot{L} - (LN^r)'\right]}{LN^t}(dr + N^r dt)$$

$$dw_1^0 = \left[\frac{N^{t'}}{L} + \frac{N^r}{N^t}\left(\dot{L} - (LN^r)'\right)\right]' dr \wedge dt + \left[\frac{\left(\dot{L} - (LN^r)'\right)}{N^t}\right] dt \wedge dr$$

$$\mathcal{R}_1^0 = \frac{1}{LN^t}\left\{\left[\frac{N^{t'}}{L} + \frac{N^r}{N^t}\left(\dot{L} - (LN^r)'\right)\right]' - \left[\frac{\left(\dot{L} - (LN^r)'\right)}{N^t}\right]'\right\}\theta^1 \wedge \theta^0$$

$$\mathcal{R}_1^0 = dw_2^0 + w_\propto^0 \wedge w_2^\propto = dw_2^0 + w_1^0 \wedge w_2^1$$

$$\begin{aligned} w_2^0 &= +\left[\dot{R}(RN^t)^{-1} - R'N^r(RN^t)^{-1}\right]Rd\theta \\ &= +Ad\theta\,;\ \underline{-A = R'\left(\tfrac{N^r}{N^t}\right) - \dot{R}\left(\tfrac{1}{N^t}\right)} = -\tfrac{DR}{N^t} \\ &= \tfrac{DR}{N^t}d\theta \end{aligned}$$

$$\begin{aligned} dw_2^0 &= A'dr \wedge d\theta + \dot{A}dt \wedge d\theta \\ &= A'\left\{\tfrac{1}{L}\theta^1 - \left(\tfrac{N^r}{N^t}\right)\theta^0\right\} \wedge \tfrac{1}{R}\theta^2 + \dot{A}\tfrac{1}{N^t}\theta^0 \wedge \tfrac{1}{R}\theta^2 \\ &= \tfrac{A'}{LR}\theta^1 \wedge \theta^2 + \left(\tfrac{\dot{A}}{RN^t} - \tfrac{A'N^r}{RN^t}\right)\theta^0 \wedge \theta^2 \end{aligned}$$

$$w_1^0 \wedge w_2^1 = \left(\frac{(N^t)'}{LN^t}\theta^0 + \frac{\left(\dot{L} - (LN^r)'\right)}{LN^t}\theta^1\right) \wedge \left(\frac{-R'}{RL}\right)\theta^2$$

$$\begin{aligned} \mathcal{R}_2^0 = &-\left(\frac{A'}{LR} - \left(\frac{-R'}{LR}\right)\frac{\left(\dot{L} - (LN^r)'\right)}{LN^t}\right)\theta^1 \wedge \theta^2 \\ &+ \left(\frac{\dot{A}}{RN^t} - \frac{A'N^r}{RN^t} - \left(\frac{R'}{RL}\right)\frac{(N^t)'}{LN^t}\right)\theta^0 \wedge \theta^2 \end{aligned}$$

$$\mathcal{R}_3^0 = dw_3^0 + w_\propto^0 \wedge w_3^\propto = dw_3^0 + w_1^0 \wedge w_3^1 + w_2^0 \wedge w_3^2$$

$$\begin{aligned} w_3^0 &= A\sin\theta\, d\phi \\ dw_3^0 &= (dw_2^0 \text{ sub } 3 \to 2) + A\cos\theta\, d\theta \wedge d\phi \\ &= (\cdots) + AR^{-2}\cot\theta\,\theta^2 \wedge \theta^3 \end{aligned}$$

$$\begin{aligned} \mathcal{R}_3^0 = &-\left(\frac{A'}{LR} - \frac{R'}{LR}\frac{\left(\dot{L} - (LN^r)'\right)}{LN^t}\right)\theta^1 \wedge \theta^3 \\ &+ \left(\frac{\dot{A}}{RN^t} - \frac{A'N^r}{RN^t} - \left(\frac{R'}{RL}\right)\frac{N^{t'}}{LN^t}\right)\theta^0 \wedge \theta^3 + AR^2\cot\theta\,\theta^2 \\ &\wedge \theta^3 \\ &- \tfrac{A}{R}\theta^2 \wedge \tfrac{\cot\theta}{R}\theta^3 \end{aligned}$$

$$\mathcal{R}^1_2 = dw^1_2 + w^1_\propto \wedge w^\propto_2 = dw^1_2 + w^1_0 \wedge w^0_2$$

$$w^1_2 = -\frac{R'}{L}d\theta$$

$$\begin{aligned}-dw^1_2 &= \left(\frac{R'}{L}\right)' dr \wedge d\theta + \left(\frac{R'}{L}\right)' dt \wedge d\theta \\ &= \left(\frac{R'}{L}\right)' \left[\frac{1}{L}\theta^1 - \frac{N^r}{N^t}\theta^0\right] \wedge \frac{1}{R}\theta^2 + \left(\frac{R'}{L}\right)' \frac{1}{N^t}\theta^0 \wedge \frac{1}{R}\theta^2 \\ &= \frac{1}{RL}\left(\frac{R'}{L}\right)' \theta^1 \wedge \theta^2 + \frac{1}{N^t R}\left(\left(\frac{R'}{L}\right)' - N^r\left(\frac{R'}{L}\right)'\right)\theta^0 \wedge \theta^2\end{aligned}$$

$$w^1_0 \wedge w^0_2 = \left(\frac{N^{t'}}{LN^t}\theta^0 - \frac{(\dot{L} - (LN^r)')}{LN^t}\theta^1\right) \wedge \frac{A}{R}\theta^2$$

$$\mathcal{R}^1_2 = \left\{\frac{-1}{RL}\left(\frac{R'}{L}\right)' - \frac{A(\dot{L} - (LN^r)')}{R\,LN^t}\right\}\theta^1 \wedge \theta^2 + \left\{\frac{(N^t)'A}{LN^t R} - \frac{\left(\frac{R'}{L}\right)' - N^r\left(\frac{R'}{L}\right)'}{N^t R}\right\} \times \theta^0 \wedge \theta^2$$

$$\mathcal{R}^1_3 = dw^1_3 + w^1_\propto \wedge w^\propto_3 = dw^1_3 + w^1_0 \wedge w^0_3 + w^1_2 \wedge w^2_3$$

$$w^1_3 = \frac{R'}{L}\sin\theta\, d\phi$$

$$dw^1_3 = (dw^1_2 \ \ 2 \to 3 \text{ sub}) + \underbrace{\frac{R'}{LR^2}\cot\theta\, \theta^2 \wedge \theta^3}_{= -w^1_2 \wedge w^2_3}$$

$$\mathcal{R}^1_3 = (\mathcal{R}^1_2\ 2 \to 3 \text{ sub})$$

$$\mathcal{R}^1_3 = \left\{\frac{-1}{RL}\left(\frac{R'}{L}\right)' - \frac{(\dot{L} - (LN^r)')}{LN^t}\right\}\theta^1 \wedge \theta^3 + \left\{\frac{(N^t)'A}{LN^t R} - \frac{\left(\frac{R'}{L}\right)' - N^r\left(\frac{R'}{L}\right)'}{N^t R}\right\} \times \theta^0 \wedge \theta^2$$

$$w_3^2 = -R^{-1}\cot\theta\,\theta^3 = -\cos\theta\,d\phi$$

$$dw_3^2 = \sin\theta\,d\theta\wedge d\phi = R^{-2}\theta^2\wedge\theta^3$$

$$\begin{aligned} w_\propto^2\wedge w_3^\propto &= w_0^2\wedge w_3^0 + w_1^2\wedge w_3^1 \\ &= -A\theta^2\wedge\theta^3 - \left(\frac{R'}{RL}\right)^2\theta^2\wedge\theta^3 \end{aligned}$$

$$\mathcal{R}_3^2 = dw_3^2 + w_\propto^2\wedge w_3^\propto$$

$$\mathcal{R}_3^2 = \frac{1}{R^2}\left\{1-\left(\frac{R'}{L}\right)^2 + \left[\left(\frac{\dot{R}}{N^t}\right)-\left(\frac{R'N^r}{N^t}\right)\right]^2\right\}\theta^2\wedge\theta^3$$

$$\mathcal{R}_\nu^\mu = R^\mu_{\nu|\propto\beta|}w^\propto\wedge w^\beta$$

$$\begin{aligned} R = g^{ik}R_{ik} &= g^{ik}g^{j\ell}R_{ijk\ell} = \\ &= -R_{ioio} - R_{jojo} + R_{jiji} \\ &= +2R^o_{ioi} + R^j_{iji} \\ &= +2(R^0_{101}+R^0_{202}+R^0_{303}) + 2(R^1_{212}+R^1_{313}+R^2_{323}) \end{aligned}$$

$$R^0_{101} = -\frac{1}{LN^t}\left\{\left[\frac{N^{t'}}{L}+\frac{N^r}{N^t}\left(\dot{L}-(LN^r)'\right)\right]' + \left[\frac{\left(\dot{L}-(LN^r)'\right)}{N^t}\right]^{\cdot}\right\}$$

$$R^0_{202} = R^0_{303} = -\left(\frac{\dot{A}}{RN^t} - \frac{A'N^r}{RN^t} - \left(\frac{R'}{RL}\right)\frac{(N^t)'}{LN^t}\right)$$

$$R^1_{212} = R^1_{313} = \frac{-1}{RL}\left(\frac{R'}{L}\right)' - \frac{\left(\dot{L}-(LN^r)'\right)}{LN^t}\frac{A}{R}$$

$$R^2_{323} = R^2\left\{1-\left(\frac{R'}{L}\right)^2 + \left[\left(\frac{\dot{R}}{N^t}\right)-\left(\frac{R'N^r}{N^t}\right)\right]^2\right\}$$

$$I = \frac{1}{16\pi}\int d^4x\sqrt{-g}\,\overline{R}$$

Since there are no angular dependences in $\overline{R} = R^\mu_\mu$, as expected for spherical symmetry, the angular degrees of freedom are directly integrated.

$$I=\frac{1}{2}\int dtdr\underbrace{\frac{1}{2}N^tLR^2\,\overline{R}}_{2\mathcal{L}}$$

$$2\mathcal{L}=-R^2\left\{\left[\frac{N^{t\prime}}{L}+\frac{N^r}{N^t}\left(\dot{L}-(LN^r)'\right)\right]'+\left[\frac{\left(\dot{L}-(LN^r)'\right)}{N^t}\right]'\right\}$$
$$+2LR\left\{\dot{A}-N^rA'-\frac{R'(N^t)'}{L^2}\right\}$$
$$-2N^tR^2\left\{\frac{1}{R}+\left(\frac{R'}{L}\right)'-\frac{A\left(\dot{L}-(LN^r)'\right)}{RN^t}\right\}$$
$$+LN^t\left\{1-\left(\frac{R'}{L}\right)^2+\left[\left(\frac{\dot{R}}{N^t}\right)-\left(\frac{R'N^r}{N^t}\right)\right]^2\right\}$$

$$2\mathcal{L}=+2RR'\left[\frac{N^{t\prime}}{L}-\frac{N^r}{N^t}\left(\dot{L}-(LN^r)'\right)\right]'+2RR\left[\frac{\left(\dot{L}-\dot{(LN^r)'}\right)}{N^t}\right]$$

Total deriv. line $\rightarrow -\left\{R^2\left[\frac{N^{t\prime}}{L}+\frac{N^r}{N^t}\left(\dot{L}-(LN^r)'\right)\right]\right\}'-\left\{R^2\left[\frac{\left(\dot{L}-(LN^r)'\right)}{N^t}\right]\right\}'$
$-2[(LR)'-(LRN^r)']A-2LRR'\frac{(N^t)'}{L^2}+2(N^tR)'\left(\frac{R'}{L}\right)+$
$2RA\left(\dot{L}-(LN^r)'\right)$

Total deriv. line $\rightarrow +(2LRA)-(2LRN^rA)'-\left(2N^tR\frac{R'}{L}\right)'+LN^t-$
$N^t(R')^2/L+\frac{L}{N^t}\left(\dot{R}-R'N^r\right)^2$

Dropping total derivatives:

$$2\mathcal{L}=+2RR'(N^t)'(L^{-1})+N^t(R')^2(L^{-1})+LN^t+\frac{L}{N^t}\left(\dot{R}-R'N^r\right)^2$$
$$+\left(\dot{L}-(LN^r)'\right)2R[A-A]$$
$$+\frac{2}{N^t}\overbrace{\left(\dot{R}-N^rR'\right)}^{A}\;\underbrace{[(LR)'-(LRN^r)']}_{[R(\dot{L}-(LN^r)')+L(\dot{R}-R'N^r)]}$$
$$=LN^t-L\left(\dot{R}-N^rR'\right)^2/N^t+N^t(R')^2/L$$
$$+2RR'(N^t)'/L-\frac{2R}{N^t}\left(\dot{L}-(LN^r)'\right)\left(\dot{R}-N^rR'\right)$$

(Eqn. 10-29)

This agrees with the result found by Kuchař [60].

$$2\mathcal{L} = LN^t - L\left(\dot{R} - N^r R'\right)^2/N^t + N^t\,(R')^2/L + 2RR'\,(N^t)'/L$$
$$- \frac{2R}{N^t}\left(\dot{L} - (LN^r)'\right)\left(\dot{R} - N^r R'\right)$$
$$2RR'\,(N^t)'/L = 2R'\,(RN^t)'/L - 2N^t\,(R')^2/L$$
$$R\left(\dot{L} - (LN^r)'\right) = (LR)' - \dot{R}L - R(LN^r)' = (LR)' - L\left(\dot{R} - N^r R'\right) - (LRN^r)'$$
$$2\mathcal{L} = 2(N^r LR)'\left(\dot{R} - N^r R'\right)/N^t - 2(LR)'\left(\dot{R} - N^r R'\right)/N^t$$
$$+ 2(N^t R)'\,R'/L + LN^t + L\left(\dot{R} - N^r R'\right)^2/N^t$$
$$- N^t\,(R')^2/L$$

(Eqn. 10-30)

This is the result given by Fischler et al. [58].

The above $\mathcal{L}$ is quadratic in velocities, so it can be put in first order form.

$$\pi_R = \frac{\delta\mathcal{L}}{\delta\dot{R}} = \{(N^t LR)'/N^t - (LR)'/N^t\}$$

(Eqn. 10-31)

$$\pi_L = \frac{\delta\mathcal{L}}{\delta\dot{L}} = -R\left(\dot{R} - N^r R'\right)/N^t \rightarrow \text{gives } \dot{R}$$

(Eqn. 10-32)

$$\dot{R} = -N^t \pi_L R^{-1} + N^r R'$$

(Eqn. 10-33)

$$\dot{L} = -N^t \pi_R R^{-1} + (N^r LR)' R^{-1} - L\dot{R}R^{-1}$$
$$= -N^t \pi_R R^{-1} + LN^t \pi_L R^{-2} + (N^r L)'$$

(Eqn. 10-34)

$$S = \frac{1}{2}\int dr\, dt(2\mathcal{L})$$

Switching to a first order form:

$$\dot{R} - N^r R' = -N^t \pi_L R^{-1}$$

$$(LR)' = R\dot{L} + L\dot{R}$$
$$= -N^t \pi_R + LN^t \pi_L R^{-1} - LN^t \pi_L R^{-1} + LN^r R'$$
$$+ R(N^r L)'$$

$$= -N^t\pi_R + (LN^rR)'$$

$$S = \int dr\, dt \Big[(N^rLR)'(-N^t\pi_L R^{-1})/N^t - [-N^t\pi_R + (LN^rR)'](-N^t\pi_L R^{-1})/N^t + (N^tR)'\,R'/L + \frac{1}{2}LN^t + \frac{1}{2}L\,(-N^t\pi_L R^{-1})^2/N^t - \frac{1}{2}N^t\,(R')^2/L\Big]$$

$$= \int dr\, dt \Big[-N^t\pi_R\pi_L R^{-1} + \frac{1}{2}LN^t + \frac{1}{2}LN^t(\pi_L R^{-1})^2 - \frac{1}{2}N^t\,(R')^2/L + (N^tR)'\,R'/L\Big]$$

↓ drop surface term.

Surface term → $(N^t)'\left(\frac{RR'}{L}\right) + N^t\frac{(R')^2}{L} = \left[N^t\left(\frac{RR'}{L}\right)\right]' - N^t\left(\frac{RR'}{L}\right)' + N^t\frac{(R')^2}{L}$

$$= \int dr\, dt \left\{N^t\left[\frac{L\pi_L^2}{2R^2} - \frac{\pi_L\pi_R}{R} - \frac{1}{2}\left(\left(\frac{2RR'}{L}\right)' - \frac{(R')^2}{L} - L\right)\right]\right\}$$

Thus

$$\pi_L\dot{L} = \pi_L[-N^t\pi_R R^{-1} + LN^t\pi_L R^{-2} + (N^rL)']$$
$$\pi_R\dot{R} = \pi_R(-N^t\pi_L R^{-1}) + \pi_R N^rR'$$

Want the form

$$S = \int dr\, dt\big(\pi_L\dot{L} + \pi_R\dot{R} - N^t\mathcal{H}_t - N^r\mathcal{H}_r\big)$$

So,

$$-N^t\mathcal{H}_t - N^r\mathcal{H}_r = N^t\left\{\frac{L\pi_L^2}{2R^2} - \frac{\pi_L\pi_R}{R} - \frac{1}{2}\left[\left(\frac{2RR'}{L}\right)' - \frac{(R')^2}{L} - L\right]\right\} - \pi_L\dot{L} - \pi_R\dot{R}$$

$$= -N^t\left\{\frac{L\pi_L^2}{2R^2} - \frac{\pi_L\pi_R}{R} - \frac{1}{2}\left[\left(\frac{2RR'}{L}\right)' - \frac{(R')^2}{L} - L\right]\right\} - \underbrace{\pi_L(N^rL)'}_{\updownarrow} - \pi_R N^rR'$$

$[\pi_L N^r L]' + \underline{L\pi_L' N^r}$

So,

$$\mathcal{H}_t = \frac{L\pi_L^2}{2R^2} - \frac{\pi_L \pi_R}{R} + \frac{1}{2}\left\{\left(\frac{2RR'}{L}\right)' - \frac{(R')^2}{L} - L\right\}$$

(Eqn. 10-35)

and

$$\mathcal{H}_r = R'\pi_R - L\pi_L'.$$

(Eqn. 10-36)

10.8 Spherically symmetric asymptotically flat with shell – the possible topologies

In classical general relativity, every three-manifold occurs as the spatial topology of a globally hyperbolic vacuum spacetime. In a canonical approach to quantum gravity, the spatial topology is frozen, and one can ask for ground states corresponding to each topology. Even in theories that permit topology change, topologies threaded by electric or magnetic charge in source-free Einstein-Maxwell theory for example (or in higher-dimensional gravity with Kaluza-Klein asymptotic behavior) cannot evolve to Euclidean space. If there is a nonsingular quantum theory of such a system, it must allow a ground state with nonzero asymptotic charge and non-Euclidean topology. Topological geons with half-integral angular momentum in a quantum theory of gravity would similarly be unable to settle down to Euclidean topology.) For these reasons, among others, topology change will not be considered, and we will retain well-behaved asymptotics in the theory, consistent with Birckoff's Theorem [21].

Spherically symmetric minisuperspaces provide simple models for the quantization of geometries with non-Euclidean topology. The spatial topologies consistent with spherical symmetry and asymptotic flatness are R^3, the wormhole $S^2 \times R$ of the extended Schwarzschild geometry with two asymptopias, and the RP^3 geon, a manifold with a single asymptopia obtained by removing a point from the compact manifold RP^3. This last manifold is the space acquired from an extended Schwarzschild geometry by identifying diametrically opposite points on an U + V = constant slice, with U and V the usual Kruskal null coordinates.

For pure Einstein gravity in four spacetime dimensions, spherically symmetric minisuperspace has been considered by several authors [60-67]. For extensions to related theories, including spherically symmetric Einstein-Maxwell theory and lower-dimensional dilatonic theories, see

Refs. [67-72]. For discussions within the Euclidean context, see for example Refs. [73-82] and the references therein. In the present description we add to spherically symmetric Einstein gravity a single degree of freedom by introducing a thin dust shell of (strictly) positive rest mass. A shell of vanishing rest mass has been considered by Kraus and Wilczek [53,54], following an earlier minisuperspace treatment of a bubble wall by Fischler et al . [58]. The work in Ref. [53] also contained an essentially complete derivation of a Hamiltonian for a shell with positive rest mass, using a gauge with at spatial sections outside the shell. However, the resulting Hamiltonian is a complicated function of the shell radius and the conjugate momentum, determined via the solution to a transcendental equation that cannot be explicitly inverted.

In this shell analysis we first (re)derive the Hamiltonian action of Kraus and Wilczek [53]. We then introduce a formalism for reparametrizing time in a Hamiltonian theory with a two-dimensional phase space. Applying this formalism to the shell Hamiltonian, we redefine the coordinate time to coincide with the proper time of the shell, recovering a Hamiltonian that can be given in terms of elementary functions [58]. An alternative choice of the coordinate time yields a Hamiltonian that generalizes to our self-gravitating shell the familiar Hamiltonian of a spherical test shell in Minkowski space.

The minisuperspace derivation of the these Hamiltonians for a self-gravitating shell, via a Hamiltonian reduction and a new type of time reparameterization, was first done in work generally described in [thesis]. Our proper time Hamiltonian for a shell on a wormhole spacetime has been considered previously by Berezin et al [83]. For a flat geometry interior to the shell, our proper time Hamiltonian has been considered classically in [84] and quantum mechanically in [85]. For a flat geometry interior to the shell, a super-Hamiltonian quantization obtained via a standard time reparameterization has been considered in Ref. [86].

The formalism so far describes a shell in a three-geometry with Euclidean topology or wormhole topology. However, this formalism is easily adapted to a shell in a space with topology RP^3. The covering space has wormhole topology, and by lifting shell and geometry to the covering space, one obtains a left-right symmetric geometry with two symmetrically placed shells.

The metric of a spherically symmetric spacetime is written, at least locally, in the Arnowitt-Deser-Misner (ADM) form

$$ds^2 = -N^2dt^2 + \Lambda^2(dr + N^r dt)^2 + R^2 d\Omega^2,$$

where the metric on the unit two-sphere is shown, and the metric coefficients are all functions of t and r. The smoothness properties of the metric and the global properties of the spacetime will be addressed below. The implementation of the global properties will determine the range of r.

The matter consists of a thin shell of dust, with a fixed positive rest mass m. (Note: Carets will be placed to the left in text sections that follow.) We write the trajectory of the shell as r = ^r(t). Denoting by ^N (t), ^N r(t), ^Λ(t), and ^R(t) the values of N, Nr, Λ, and R at r = ^r,

$$\hat{R}(t) := R\ (t, \hat{r}(t)\ )\ \ \text{etc},$$

the Hamiltonian action for the shell is:

$$S_{shell} = \int dt\left(\hat{p}\dot{\hat{r}} - \hat{N}\sqrt{\hat{p}^2\hat{\Lambda}^{-2} + m^2} + \hat{N}^r\hat{p}\right),$$

(Eqn. 10-37)

with ^p being the momentum conjugate to ^r.

The Lagrangian gravitational action for the geometry can be obtained by integrating the Lagrangian density:

$$(16\pi)^{-1}(\,{}^3R - K^{ab}K_{ab} + K^2)\sqrt{-g}$$

(Eqn. 10-38)

over the 2-sphere. After $\dot{\Lambda}$ and $\dot{R}$ are replaced by their conjugate momenta,

$$\begin{aligned}\pi_\Lambda &= -\frac{R}{N}(\dot{R} - N^r R'),\\ \pi_R &= -\frac{\Lambda}{N}(\dot{R} - N^r R') - \frac{R}{N}\left[\dot{\Lambda} - (N^r\Lambda)'\right],\end{aligned}$$

(Eqn. 10-39)

one obtains for the coupled system of shell and geometry the Hamiltonian bulk action

$$S_\Sigma = \int dt\left[\hat{p}\dot{\hat{r}} + \int dr(\pi_\Lambda\dot{\Lambda} + \pi_R\dot{R} - N\mathcal{H} - N^r\mathcal{H}_r)\right],$$

(Eqn. 10-40)

where the super-Hamiltonian H and the supermomentum H_r are given by

$$\mathcal{H} = \frac{\Lambda\pi_\Lambda^2}{2R^2} - \frac{\pi_\Lambda\pi_R}{R} + \frac{RR''}{\Lambda} - \frac{RR'\Lambda'}{\Lambda^2} + \frac{R'^2}{2\Lambda} - \frac{\Lambda}{2}$$
$$+\sqrt{\hat{p}^2\hat{\Lambda}^{-2} + m^2}\delta(r - \hat{r}),$$
$$\mathcal{H}_r = \pi_R R' - \pi_\Lambda'\Lambda - \hat{p}\delta(r - \hat{r}).$$

(Eqn. 10-41)

The bulk action will need to be augmented with boundary terms that depend on the boundary conditions. First, however, we shall address choosing a convenient gauge for the shell description.

10.9 Choice of gauge for shell

In order to study the shell problem in a gauge where we can readily examine the Newtonian limit we now consider various metrics.

We consider Black Hole (BH) metrics of the form $ds^2 = -N^{t^2}dt^2 + L^2[dr + N^r dt]^2 + R^2 d\Omega^2$

Where

$$\mathcal{H}_t^G = \frac{L\pi_L^2}{2R^2} - \frac{\pi_L\pi_R}{R} + \left(\frac{RR'}{L}\right)' - \frac{R'^2}{2L} - \frac{L}{2}$$

And

$$\mathcal{H}_r^G = R'\pi_R - L\pi_L'.$$

And where we now have the junction conditions from the shell:

$$\left\{\begin{matrix} R'(\hat{r}+\epsilon) - R'(\hat{r}-\epsilon) = -\frac{1}{R}\sqrt{p^2 + m^2\hat{L}^2} \\ \pi_L(\hat{r}+\epsilon) - \pi_L(\hat{r}-\epsilon) = -p/\hat{L} \end{matrix}\right\}$$

(Eqn. 10-42)

Schwarzschild:

$$ds^2 = -(1+2\phi)dt^2 + (1+2\phi)^{-1}dr^2 + R^2 d\Omega^2\,, \phi = -\frac{M}{r}$$

Discontinuity in R' means that $R \neq r$ and a second parameter is required in the theory, i.e., ϕ and R.

Similarly for Ingoing Eddington-Finkelstein:

$$ds^2 = -(1 - 2M/r)d\hat{v}^2 + 2d\hat{v}dr + R^2 d\Omega^2$$
$$\hat{v} = t + r^* , r^* = r + 2M \ln(M/r - 1)$$
and Novikov:
$$ds^2 = -d\tau^2 + \left(\frac{{R^*}^2 + 1}{{R^*}^2}\right)\left(\frac{\partial t}{\partial R^*}\right)^2 d{R^*}^2 + R^2 d\Omega^2$$
However, Isotropic coord's do work with Isotropic Schwarzschild:
$$ds^2 = -\left(1 + \frac{1}{2}\phi\right)^2 \left(1 - \frac{1}{2}\phi\right)^{-2} dt^2$$
$$+ \left(1 - \frac{1}{2}\phi\right)^4 [dr^2 + r^2 d\Omega^2] \qquad \left(\phi = -\frac{M}{r}\right)$$

The discontinuity in R' can now be realized via the single parameter ϕ via "R" $= \left(1 - \frac{1}{2}\phi\right)^2 r$.

This will suffice for now, but in more extensive work it will be of value to determine if a BH metric exists for which $R \neq r$ and which extends through the horizons (and not simply through the wormhole throat.)
$$ds^2 = -\left(1 + \frac{1}{2}\phi\right)^2 \left(1 - \frac{1}{2}\phi\right)^{-2} dt^2 + \left(1 - \frac{1}{2}\phi\right)^4 [dr^2 + r^2 d\Omega^2]$$

Isotropic Schwar. solution if $\phi = -\frac{M}{r}$.

Comparison with the Fischler et al. metric: $ds^2 = -{N^t}^2 dt^2 + L^2[dr + N^r dt]^2 + R^2 d\Omega^2$
$$(N^t)^2 = \left(1 + \frac{1}{2}\phi\right)^2 \left(1 - \frac{1}{2}\phi\right)^{-2}$$
$$L^2 = \left(1 - \frac{1}{2}\phi\right)^4$$
$$N^r = 0$$
$$R^2 = r^2 L^2$$

Choose pos. roots: $N^t = \left(1 + \frac{1}{2}\phi\right) / \left(1 - \frac{1}{2}\phi\right)$
$$L = \left(1 - \frac{1}{2}\phi\right)^2$$
$$N^r = 0$$
$$R = rL$$

Desire gauge where $\pi_R = \pi_L = 0$ in exterior $(r > \hat{r})$ and
$\pi_R = \pi_L = const$ in interior $(r < \hat{r})$.

Also require R, L to be continuous throughout.

Can the Newtonian Limit be obtained from $\mathcal{H}$?

Newt. limit applies to exterior so we have $\pi_R = \pi_L = 0$ in above gauage.

$$\mathcal{H} = \frac{1}{2}\left\{\left(\frac{2RR'}{L}\right)' - \frac{(R')^2}{L} - L\right\} + \delta(r - \hat{r})\left(L^{-2}\hat{p}^2 + \underbrace{(4\pi\mu)^2\hat{R}^4}_{m^2}\right)^{1/2}$$

(Eqn. 10-43)

$\left(4\pi\hat{R}^2\mu\right) = m$ since proper shell mass fixed.

Consider mass term. Newt. limit applies in the asymptotic regime where $L \simeq 1 + O(r^{-1})$ and motion is non-relativistic: $\hat{p}^2 \simeq m^2\hat{v}^2 \ll m^2$. Thus, the shell contributes:

$$\mathcal{H}_{shell} = m\delta(r - \hat{r}) = (4\pi r^2)\rho = (4\pi\rho)r^2,$$

for

$$\rho = \frac{m}{4\pi r^2}\delta(r - \hat{r})$$

Consider pure grav. term

$$\mathcal{H}_{Grav}(exterior\ gauge) = \frac{1}{2}\left\{\left(\frac{2RR'}{L}\right)' - \frac{(R')^2}{L} - L\right\}$$

Substituting $R = rL\,, L = \left(1 - \frac{1}{2}\phi\right)^2$:

$$\left(\frac{RR'}{L}\right)' = (r(rL)')' = (rL)' + r(rL)''$$
$$= L + rL' + r(2L' + rL'')$$

$$\frac{(R')^2}{L} = \frac{1}{L}(L + rL')^2 = \frac{1}{L}(L^2 + 2rLL' + r^2(L')^2)$$
$$= L + 2rL' + r^2\frac{1}{L}(L')^2$$

$$\mathcal{H}_{Grav} = \frac{1}{2}\left\{2[L + 3rL' + r^2L''] - \left[L + 2rL' + r^2\frac{1}{L}(L')^2\right] - x\right\}$$
$$= \frac{1}{2}\left\{4rL' + 2r^2L'' - r^2\frac{1}{L}(L')^2\right\}$$

$L = \left(1 - \frac{1}{2}\phi\right)^2$,
$L' = 2\left(1 - \frac{1}{2}\phi\right)\left(-\frac{1}{2}\phi'\right) = -\phi'\left(1 - \frac{1}{2}\phi\right)$, and
$L'' = -\phi''\left(1 - \frac{1}{2}\phi\right) + \frac{1}{2}(\phi')^2$

$$\mathcal{H}_{Grav} = \frac{1}{2}\left\{4r\left[-\phi''\left(1 - \frac{1}{2}\phi\right)\right] + 2r^2\left[\frac{1}{2}(\phi')^2 - \phi''\left(1 - \frac{1}{2}\phi\right)\right] - r^2(\phi')^2\right\}$$
$$= (-2r\phi' - r^2\phi'')\left(1 - \frac{1}{2}\phi\right)$$

In order to express the above in terms of the 3-dim Laplacian I now calculate

$${}^{(3)}\nabla^2\phi \text{ for } d\ell^2 = \left(1 - \frac{1}{2}\phi\right)^4 [dr^2 + r^2 d\Omega^2]$$

$${}^{(3)}\nabla^2\phi = \frac{1}{r^2\left(1 - \frac{1}{2}\phi\right)^6}\frac{\partial}{\partial r}\left\{r^2\left(1 - \frac{1}{2}\phi\right)^6\left(1 - \frac{1}{2}\phi\right)^{-4}\frac{\partial\phi}{\partial r}\right\}$$
$$= \frac{1}{r^2\left(1 - \frac{1}{2}\phi\right)^6}\frac{\partial}{\partial r}\left\{r^2\left(1 - \frac{1}{2}\phi\right)^2\frac{\partial\phi}{\partial r}\right\}$$
$$= \frac{1}{\left(1 - \frac{1}{2}\phi\right)^4}\left(\frac{\partial^2\phi}{\partial r^2} + \frac{2}{r}\frac{\partial\phi}{\partial r}\right) - \frac{1}{\left(1 - \frac{1}{2}\phi\right)^5}\left(\frac{\partial\phi}{\partial r}\right)^2$$

From the condition of asymptotic flatness we know that in the asymptotic regime ϕ has the form

$$\phi \cong \frac{\propto}{r} + O(r^{-2})$$

Thus, $\left(1 - \frac{1}{2}\phi\right) \simeq 1$

$$\left(\frac{\partial\phi}{\partial r}\right)^2 \ll \left(\frac{\partial^2\phi}{\partial r^2}\right) \simeq \left(\frac{1}{r}\frac{\partial\phi}{\partial r}\right)$$

So,

$$\mathcal{H}_{Grav} \simeq -r^2(\phi'' + 2r^{-1}\phi')$$
$${}^{(3)}\nabla^2\phi \simeq \phi'' + 2r^{-1}\phi'$$

Thus,

$$\mathcal{H}_{Grav} \simeq -r^2\left[{}^{(3)}\nabla^2\phi\right]$$

And we obtain:

$0 = \mathcal{H} = [4\pi\rho - \nabla^2\phi]r^2 + O(r^{-2})$

$\Downarrow$

$$\nabla^2\phi_{asympt} = 4\pi\rho \Longrightarrow \phi_{asympt} = -\frac{M}{r} \qquad M = const = M_{ADM}$$

Now to calculate $H = 0 = \int \mathcal{H}_t^{(2)} N^t dr$. From this first integral it is hoped that a dynamical equation for the shell will be obtained.

Again write $\mathcal{H}_t^{(2)}$:

$$\mathcal{H}_t^{(2)} = \frac{L\pi_L^2}{2R^2} - \frac{\pi_L\pi_R}{R} + \frac{1}{2}\left\{\left(\frac{2RR'}{L}\right)' - \frac{R'^2}{L} - L\right\} + \delta(r - \hat{r})(L^{-2}\hat{p}^2 + m^2)^{1/2}$$

(Eqn. 10-44)

2nd order deriv. terms can be eliminated in favor of a surface term when the geometries considered in the variation are restricted to be asymptotically flat. From $\phi_{asympt} = -\frac{M}{r} + O(r^{-2})$ we have

$$N_{asy}^t \simeq \left(1 - \frac{M}{r}\right)$$
$$L_{asy} \simeq \left(1 + \frac{M}{r}\right)$$
$$R_{asy} \simeq r + M$$
$$R'_{asy} \simeq 1$$

So,

$$\left(\frac{RR'N^t}{L}\right)_{asy} = r - M$$

Thus,

$$\lim_{r\to\infty}\left[\left(\frac{RR'N^t}{L}\right) - r\right] = -M$$

$\int_0^\infty \left[\left(\frac{RR'N^t}{L}\right)' - 1\right] = -M$ (since $R(0) = 0$) for regularity at origin

So,

$$0 = -M + \int_0^\infty \left[\left(\frac{RR'}{L}\right)(N^t)' + \frac{1}{2}N^t\left(\frac{(R')^2}{L} + L\right)\right] dr$$

$$+ \int_0^{\hat{r}_-} \left\{\frac{L\pi_L^2}{2R^2} - \frac{\pi_L \pi_R}{R}\right\}_{N^t} dr$$

$$+ \int_0^\infty \delta(r - \hat{r})(L^{-2}\hat{p}^2 + m^2)^{1/2} \, dr$$

Consider matter first:

$$\int \mathcal{H}_{matter} N^t dr = \int_0^\infty N^t \, \delta(r - \hat{r})(L^{-2}\hat{p}^2 + m^2)^{1/2} dr$$

$$= \int_0^\infty \frac{\left(1+\frac{1}{2}\phi\right)}{\left(1-\frac{1}{2}\phi\right)} \delta(r - \hat{r}) \left(\hat{p}^2 \left(1 - \frac{1}{2}\phi\right)^{-4} + m^2\right)^{1/2} dr$$

Recall $\hat{p} = \frac{(4\pi\mu\hat{R}^2)\hat{L}^2\dot{\hat{r}}}{\sqrt{\hat{N}^{t2} - \hat{L}^2\dot{\hat{r}}^2}}$, which follows from:

$$\downarrow -d\tau^2 + \hat{R}^2 d\Omega^2$$

$$S_{matter} = -M \int_{wall} d^3 A = -M \int_{wall} \sqrt{-{}^3 g} \, d\tau d\theta d\phi = 4\pi\mu \int_{wall} \hat{R}^2 \, d\tau$$

$$d\tau^2|_{wall} = \left(\hat{N}^t\right)^2 dt^2 \hat{L}^2 \left[d\hat{r}^2 + \hat{N}^r dt\right]^2$$

$$\left(\frac{d\tau}{dt}\right) dt = \sqrt{\hat{N}^{t^2} - \hat{L}^2 \left(\dot{\hat{r}} + \hat{N}^r\right)^2} \, dt$$

So,

$$S_{matter} = -4\pi\mu \int dt \, \hat{R}^2 \left(\hat{N}^{t^2} - \hat{L}^2 \left(\dot{\hat{r}} + \hat{N}^r\right)^2\right)^{1/2}$$

Drop δ-function for convenience

$$\mathcal{L}_{matter} = -4\pi \hat{R}_\mu^2 \left(\hat{N}^{t^2} - \hat{L}^2 \left(\dot{\hat{r}} + \hat{N}^r\right)^2\right)^{1/2}$$

Using $m = 4\pi\mu\hat{R}^2$, $\hat{N}^r = 0$

$$\mathcal{L}_{matter} = -m\left(\hat{N}^{t^2} - \hat{L}^2 \, \dot{\hat{r}}^2\right)^{1/2}$$

$$\hat{p} = \frac{\partial R_\mu}{\partial \dot{\hat{r}}} = m\hat{L}^2\dot{\hat{r}}\left(\hat{N}^{t^2} - \hat{L}^2\,\dot{\hat{r}}^2\right)^{1/2}$$

$$N^t\mathcal{H}_r = \hat{p}\dot{\hat{r}}\mathcal{L}_M = \frac{m\left(\hat{L}\dot{\hat{r}}\right)^2}{\left(\hat{N}^{t^2} - \left(L\dot{\hat{r}}\right)^2\right)^{1/2}} + m\left(\hat{N}^{t^2} - \hat{L}^2\,\dot{\hat{r}}^2\right)^{1/2}$$

$$= \frac{m\left(\hat{N}^t\right)^2}{\left(\hat{N}^{t^2} - \left(L\dot{\hat{r}}\right)^2\right)^{1/2}}$$

$$1 = \left(\frac{d\tau}{d\tau}\right)^2\Bigg|_{wall} = \left(\hat{N}^t\right)^2\left(\frac{dt}{d\tau}\right)^2 - \hat{L}^2\left(\frac{d\hat{r}}{d\tau}\right)^2$$

$$\left(\frac{d\hat{r}}{dt}\right)^2 = \left(\frac{d\hat{r}}{d\tau}\right)^2\left(\frac{d\tau}{dt}\right)^2 = \left(\frac{d\hat{r}}{d\tau}\right)^2\left(\left(\hat{N}^t\right)^{-2} + \frac{\hat{L}^2}{(N^t)^2}\left(\frac{d\hat{r}}{d\tau}\right)^2\right)^{-1}$$

So,

$$\mathcal{H}_t = \frac{m\left(\hat{N}^t\right)}{\left(\hat{N}^{t^2} - L^2\left(\frac{d\hat{r}}{d\tau}\right)^2\left[\left(\hat{N}^t\right)^{-2} + \frac{\hat{L}^2}{(N^t)^2}\left(\frac{d\hat{r}}{d\tau}\right)^2\right]^{-1}\right)^{1/2}}$$

$$= \frac{m\left(\hat{N}^t\right)\left(\hat{N}^t\right)^{-1}\left[1 + \hat{L}^2\left(\frac{d\hat{r}}{d\tau}\right)^2\right]^{1/2}}{\left(\hat{N}^t\right)^2\left(\hat{N}^t\right)^{-2}\left[1 + \hat{L}^2\left(\frac{d\hat{r}}{d\tau}\right)^2\right] - \left[\left(\frac{d\hat{r}}{d\tau}\right)^2\right.}$$

$$\mathcal{H}_t = m\sqrt{1 + \hat{L}^2\left(\frac{d\hat{r}}{d\tau}\right)^2} \times \delta(r - \hat{r})$$

(Eqn. 10-45)

The "matter" term describes the Kinetic energy term, it encompasses the dynamic aspect.

It is important to note that the local field analysis used in calculating the shell contribution is a static analysis used to evaluate the potential energy of the shell. In contrast, when the asympt. boundary analysis is done for the Grav. field the full solution is considered and while "static" asympt. it encompasses both potential and kinetic energy terms.

10.10 Solutions away from shell, Newtonian

Derivation of solution(s) away from shell: (Static analysis)

The constraint equations are:

$$\mathcal{H}_t = \frac{L\pi_L^2}{2R^2} - \frac{\pi_L\pi_R}{R} + \frac{1}{2}\left\{\left(\frac{2RR'}{L}\right)' - \frac{R'^2}{L} - L\right\} + \delta(r-\hat{r})(L^{-2}\hat{p}^2 + m^2)^{1/2}$$

(Eqn. 10-46)

$$\mathcal{H}_r = R'\pi_R - L\pi_L' - \delta(\hat{r}-r)\hat{p}$$

(Eqn. 10-47)

$$0 = \frac{R'}{L}\mathcal{H}_t + \frac{\pi_L}{RL}\mathcal{H}_r = -\mathcal{M}'$$

(Eqn. 10-48)

$$constant = \mathcal{M} = \frac{\pi_L^2}{2R} + \frac{R}{2}\left[1 - \left(\frac{R'}{L}\right)^2\right]$$

(Eqn. 10-49)

For the exterior, the gauge choice $\pi_L = 0$ has been made, and expressing R and L in terms of the gauge potential ϕ:

$$R = rL\,, L = \left(1 - \frac{1}{2}\phi\right)^2$$

$$\mathcal{M} = M = \frac{rL}{2}\left[1 - \left\{\frac{(L + rL')}{L}\right\}^2\right] = -\frac{rL}{2}\left[\frac{1}{L^2}\left(2rLL' + r^2L'^2\right)\right]$$

$$= -r\left(rL' + \frac{1}{2}r^2\frac{L'^2}{L}\right) \qquad L' = -\left(1 - \frac{1}{2}\phi\right)\phi'$$

$$= -r\left(-r\phi'\left(1 - \frac{1}{2}\phi\right) + \frac{1}{2}r^2(\phi')^2\right)$$

$\phi = -\frac{M}{r}$ is a solution

Check: $M = -r\left(-r\left(\frac{M}{r^2}\right)\left(1 - \frac{1}{2}\left(-\frac{M}{r}\right)\right)\right) + \frac{1}{2}r^2\left(\frac{M}{r^2}\right)^2$

So, for the exterior $\phi = -\frac{M}{r}$

For the interior:

$$\mathcal{M} = \frac{\pi_L^2}{2r}\frac{1}{\left(1 - \frac{1}{2}\phi\right)^2} - r\left(-r\phi'\left(1 - \frac{1}{2}\phi\right) + \frac{1}{2}r^2(\phi')^2\right)$$

Since we would like to consider the interior as flat we need the metric potential to be constant there. In such a case:

$$\phi' = 0 \Rightarrow \mathcal{M} = \frac{\pi_L^2}{2r}\frac{1}{\left(1 - \frac{1}{2}\phi\right)^2}$$

A solution only exists for $\pi_L \propto \pm\sqrt{r}\,, \mathcal{M} = M_-$ or $\pi_L = 0 = \mathcal{M}$. The latter case does not satisfy the junction condition unless a limiting process is defined which essentially encompasses the former solution anyway.

Consider the solution

$$\mathcal{M}_{Inside} = \frac{\pi_L^2}{2r}\left(1 + \frac{M}{2\hat{r}}\right)^{-2}$$

$$\pi_L^2 = \mathcal{M}_I 2r\left(1 + \frac{M}{2\hat{r}}\right)^2$$

$$\underline{\pi_L = \pm\sqrt{r} \cdot \sqrt{2\mathcal{M}_I} \cdot \left(1 + \frac{M}{2\hat{r}}\right)}$$

From the junction condition:

$\pi_L(\hat{r} + \epsilon) - \pi_L(\hat{r} - \epsilon) = \hat{p}/\hat{L}$

So, the correct sign depends on that of $\hat{p}$, i.e., an ingoing or outgiong shell.

$$\sqrt{\hat{r}} \cdot \sqrt{2\mathcal{M}_I}\left(1 + \frac{M}{2\hat{r}}\right) = |\hat{p}|/\hat{L} \qquad \hat{L} = \left(1 + \frac{M}{2\hat{r}}\right)^2$$

So,

$$\underset{(interior)}{\pi_L} = \sqrt{\frac{r}{\hat{r}}}\hat{p}\left(1 + \frac{M}{2\hat{r}}\right)^{-2} \quad and \quad \underset{(exterior)}{\pi_L} = 0$$

From which follows $\pi_R = \frac{L}{R'}\pi_L'$ to give π_R (comes from $\mathcal{H}_r$ constraint away from shell).

$$\underset{(ext)}{\pi_R} = 0 \quad and \quad \underset{(int)}{\pi_R} = \pi_L' = \frac{1}{2r}\pi_L$$

So,

$$H = 0 = \int N^t \mathcal{H}_t^{Grav} dr + \int N^t \mathcal{H}_t^{shell} dr \qquad \underline{N^r = 0}$$

$$\int N^t \mathcal{H}_t^{shell} = \hat{N}_m^t \sqrt{1 + \hat{L}^2 \left(\frac{\partial \hat{r}}{\partial \tau}\right)^2}$$

Consider $\mathcal{H}_t^{Grav}$:
Away from shell and asymptotic boundaries the matter free solution leads to $\mathcal{H}_t^{Grav} = 0$, thus

↓ only contribution is here.

$$\int N^t \mathcal{H}_t^{Grav} dr = -\int_{\hat{r}-\epsilon}^{\hat{r}+\epsilon} N^t \mathcal{H}^{Grav} dr + \int_0^{\infty} N^t \mathcal{H}^{Grav} dr$$

With the gauge chosen $\pi_L = 0 = \pi_R$ for $r > \hat{r}$, and they are const. for $r < \hat{r}$. While L and R are continuous.

(Redone more clearly on next part.)

$$\int_{\hat{r}-\epsilon}^{\hat{r}+\epsilon} N^t \mathcal{H}^{Grav} dr = N^t \left(\frac{RR'}{L}\right)\Bigg|_{\hat{r}-\epsilon}^{\hat{r}+\epsilon}$$

Expressing R and L in terms of the gauge potential $\tilde{\phi}$:

$$R = rL\ , \qquad L = \left(1 - \frac{1}{2}\tilde{\phi}\right)^2$$

$$\tilde{\phi} =$$

$$\begin{cases} \phi & r > \hat{r} \\ K & r < \hat{r}, K = constant \end{cases}$$

$$\left(\frac{RR'}{L}\right) = r\left[\left(1 - \frac{1}{2}\tilde{\phi}\right)^2 + r\left(1 - \frac{1}{2}\tilde{\phi}\right)(-\tilde{\phi}')\right]$$

$$\phi = -M/r$$

is the exterior solution

$$= r\left[1 - \tilde{\phi} + \tfrac{1}{4}\tilde{\phi}^2 - r\tilde{\phi}' + \tfrac{1}{2}r\tilde{\phi}'\tilde{\phi}\right] \qquad K = -M/\hat{r}$$

$$\left(\frac{RR'}{L}\right)\Bigg|_{\hat{r}-\epsilon}^{\hat{r}+\epsilon} = \hat{r}^2\left[\frac{1}{2}\tilde{\phi}'\tilde{\phi} - \tilde{\phi}'\right] = \hat{r}^2\left[-\frac{1}{2}\frac{M^2}{\hat{r}^3} - \frac{M}{\hat{r}^2}\right] = -\frac{M^2}{2\hat{r}} - M$$

So, to be clear:

$$\int_{\hat{r}-\epsilon}^{\hat{r}+\epsilon} N^t \mathcal{H}^{Grav} dr = \int_{\hat{r}-\epsilon}^{\hat{r}+\epsilon} N^t \frac{1}{2}\left\{\left(\frac{2RR'}{L}\right)' - \frac{(R')^2}{L} - L\right\} dr$$

$$= \int_{\hat{r}-\epsilon}^{\hat{r}+\epsilon} \left[\left(\frac{RR'N^t}{L}\right)' - \left\{\left(\frac{RR'}{L}\right)N^{t'} + \frac{N^t}{2}\left(\frac{(R')^2}{L} + L\right)\right\}\right] dr$$

From the continuity of R, N^t, L we then get:

$$\int_{\hat{r}-\epsilon}^{\hat{r}+\epsilon} N^t \mathcal{H}^{Grav} dr = \left(\frac{RR'N^t}{L}\right)\Bigg|_{\hat{r}-\epsilon}^{\hat{r}+\epsilon}$$

Static aspect clear at this substitution:

$$\tilde{\phi} = \begin{cases} \phi = {}^{-m}/_{r} & r > \hat{r} \\ \dot{\phi} = {}^{-m}/_{\hat{r}} & r < \hat{r} \end{cases}$$

and get:

$$R = rL \ , \ L = \left(1 - \tfrac{1}{2}\tilde{\phi}\right)^2 , N^t = \left(1 + \tfrac{1}{2}\tilde{\phi}\right)/\left(1 - \tfrac{1}{2}\tilde{\phi}\right)$$

$$N^t\left(\frac{RR'}{L}\right) = N^t r\left[1 - \tilde{\phi} + \frac{1}{4}\tilde{\phi}^2 - r\tilde{\phi}' + \frac{1}{2}r\tilde{\phi}'\tilde{\phi}\right]$$

$$\left(N^t\frac{RR'}{L}\right)\Bigg|_{\hat{r}-\epsilon}^{\hat{r}+\epsilon} = \widehat{N}^t\left\{\hat{r}^2\left[\frac{1}{2}\phi'\phi - \phi'\right]_{r=\hat{r}}\right\}$$

$$\int_{\hat{r}-\epsilon}^{\hat{r}+\epsilon} N^t \mathcal{H}^{Grav} dr = \widehat{N}^t\left\{-\frac{m^2}{2\hat{r}} - m\right\}$$

Now for the asymptotic term:

$\int_0^\infty \mathcal{H}_t N^t dr = ?$ $\quad \mathcal{H}_t = 0$ away from asympt-∞ boundary

$$\int_0^\infty \frac{1}{2}\left\{\left(\frac{2RR'}{L}\right) - \frac{R'^2}{L} - L\right\} N^t dr$$

$$= \int \left\{\left(\frac{RR'N^t}{L} - r\right)' + \left[1 - \left(\frac{RR'}{L}\right)N^{t'} - \frac{N^t}{2}\left(\frac{(R')^2}{L} + L\right)\right]\right\} dr$$

We already have $\mathcal{H}_t = 0$ away from asympt. boundary, as well as $R(0) = 0$ as a regularity condition giving $R \sim r$ near origin, so

$$\int_0^\infty \mathcal{H}_t \, N^t dr = \lim_{r\to\infty}\left(\frac{RR'N^t}{L} - r\right)$$

For an isotropic coordinate description the gauge potentials, expressed in terms of the potential ϕ, must behave at asymptotic spatial infinity exactly as the matric "potentials" solved for earlier, with new const. $\mathcal{M} = M$:

$$\lim_{r\to\infty} \phi = \lim_{r\to\infty}\left(-\frac{M}{r}\right)$$

(So, the actual $\phi \neq -\frac{m}{r}$ or $-\frac{M}{r}$ once the dynamics is considered, however, near the shell $\phi = -\frac{m}{r}$, and near asympt. ∞ $\phi = -\frac{M}{r}$.) Thus,

$$\lim_{r\to\infty}\begin{cases} R \cong r\left(1 + {}^{M}/_{r}\right) \\ R' \cong 1 \\ N^t \cong \left(1 - {}^{M}/_{r}\right) \\ L = \left(1 + {}^{M}/_{r}\right) \end{cases}$$

So,

$$\underline{\lim_{r\to\infty}\left(\frac{RR'N^t}{L} - r\right) = -M}$$

So, $H = 0 = \hat{N}^t m\sqrt{1 + \hat{L}^2\left(\frac{\partial \hat{r}}{\partial \tau}\right)^2} + \hat{N}^t\left\{-\frac{m^2}{2\hat{r}} - m\right\} - M$

If we take the above form and consider a proper time frame for the shell with line element

$d\ell = dr^2 + r^2 d\Omega^2$

We get:

$$0 = m\sqrt{1 + \left(\frac{\partial \hat{r}}{\partial \tau}\right)^2} - \frac{m^2}{2\hat{r}} - \underbrace{(M + m)}_{\underline{E = \text{total energy}}}$$

In $\left|\frac{\partial \hat{r}}{\partial \tau}\right| < 1$ limit:

$$0 = m\left(1 + \frac{1}{2}\dot{\hat{r}}^2\right) - \frac{m^2}{2\hat{r}} - (M + m)$$

$$M = \frac{1}{2}m\dot{\hat{r}}^2 - \frac{1}{2}\frac{m^2}{\hat{r}} \qquad \underline{M = \text{Energy} - \text{rest mass energy}}$$

Now to consider the Newtonian limit of this problem. In this limit the shell motion in a local Lorentz frame is such that:

$$\left|\frac{d\dot{\hat{r}}}{d\tau}\right| \ll 1.$$

Furthermore, it is also the case that the Gravitational field in the neighborhood of the matter is sufficiently weak that the energy of the matter is very nearly its rest mass energy:

$$0 \simeq M \cong \frac{1}{2}m\dot{\hat{r}}^2 - \frac{1}{2}\frac{m^2}{\hat{r}} \qquad \left|\frac{1}{2}m\dot{\hat{r}}^2\right| \ll m \Rightarrow \frac{1}{2}\frac{m^2}{\hat{r}} \ll m$$

These conditions are only satisfied for non-rel. shell motion in the asymptotic region. In this instance we take $M \ll m$, i.e., ignoring the asympt. term; take $\hat{N}^t \simeq 1$, i.e., choose proper time; take $\hat{L} = 1$ for the proper line element $\left(\hat{R} = r\right)$, and proceed as before:

$$H = \int_{\hat{r}-\epsilon}^{\hat{r}+\epsilon} \mathcal{H}^{Grav}\, dr + \int \mathcal{H}^{shell}\, dr = E = \left(\frac{RR'}{L}\right)_{\hat{r}-\epsilon}^{\hat{r}+\epsilon} + m\sqrt{1 + \left(\frac{\partial\hat{r}}{\partial\tau}\right)^2}$$

$$= -\frac{m^2}{2\hat{r}} - m + m\left(1 + \frac{1}{2}\dot{\hat{r}}^2\right)$$

(Eqn. 10-50)

For the Newtonian limit:

$$\mathcal{H}_t = 0 \Rightarrow \nabla^2\phi = 4\pi\rho \Rightarrow \phi = \frac{\int_0^r 4\pi\rho\, d\hat{r}}{r} = -\frac{M}{r}$$

Where the approx.: $|\phi| \ll 1$ and $\hat{p}^2 \ll m^2$ have been used.

Now to consider the dynamics, as before we expect this information to be in

$$H = 0 = \int \mathcal{H}_t\, N^t dr.$$

However, in this instance (Newt. limit) we are considering a non-rel. shell in the weak field (asymptotic) region. Thus $|\phi| \ll 1$ at the shell and interior to the shell (being const = value at shell) and exterior to the shell it remains $|\phi| \ll 1$, even more strongly so for the most part, as the shell moves non-rel. and thus $\hat{p}^2$ contributes little to the asymptatic mass.

So, working with $\int \mathcal{H}_t\, N^t dr$ where $|\phi| \ll 1, \hat{p}^2 \ll m^2$ the matter term will be examined first:

$$\int \mathcal{H}_t^{shell}\, N^t dr = \int_0^\infty N^t\,(L^{-2}p^2 + m^2)\delta(r - \hat{r})$$

$$\cong \int_0^\infty m(1+\phi)\sqrt{1 + L^{-2}\left(\frac{p}{m}\right)^2}\,\delta(r-\hat{r})$$

So

$$\left\{\begin{matrix} L = \left(1 - \frac{1}{2}\phi\right)^2 \\ N^t = \left(1 + \frac{1}{2}\phi\right)\Big/\left(1 - \frac{1}{2}\phi\right)\end{matrix}\right\}$$

and

$N^t \simeq (1+\phi) + O(\phi^2)$. As $\left(\frac{p}{m}\right)^2 \ll 1$ already, I consider $L = 1 + O(\phi)$, (also $\frac{p}{m} \simeq \frac{\partial r}{\partial \tau}, \tau =$ property of shell):

$\int \mathcal{H}_t^{shell}\, N^t dr \simeq m(1+\phi)\sqrt{1 + \left(\frac{\partial r}{\partial \tau}\right)^2} \simeq m\sqrt{1 + \left(\frac{\partial r}{\partial \tau}\right)^2} + m\phi$ to lowest order

$$\simeq m\left(1 + \frac{1}{2}\left(\frac{\partial r}{\partial \tau}\right)^2\right) + m\phi$$

Now to consider the Grav. field contribution:

$$\mathcal{H}_t^{Grav} = \underbrace{\frac{L\pi_L^2}{2R^2} - \frac{\pi_L \pi_R}{R}} + \frac{1}{2}\left\{\left(\frac{2RR'}{L}\right)' - \frac{R'^2}{L} - L\right\}$$

If the gauge conditions are imposed later as is a concern into quantum analysis, there will be a contribution at the step.

$$\left\{\begin{matrix}\text{In the exterior } \pi_L = 0 = \pi_R, \text{no contribution} \\ \text{In the interior } \pi_R = \frac{L}{R'}\pi_L', \underline{\underline{\pi_L \propto \sqrt{r}}} \Rightarrow \underline{\pi_L' = \frac{1}{2r}\pi_L}\end{matrix}\right.$$

↑ a consistent gauge for the interior

$$\pi_R = \frac{L}{R'}\frac{1}{2r}\pi_L$$

So $\frac{L\pi_L^2}{2R^2} - \frac{\pi_L \pi_R}{R} = \frac{L\pi_L^2}{2R}\left[\frac{1}{R} - \frac{1}{rR'}\right]$, $\underline{R = rL}$

$$= \frac{L\pi_L^2}{2R^2}\left(1 - \frac{L}{(rL)'}\right)$$

$$\frac{L}{(rL)'} = \frac{\left(1 - \frac{1}{2}\phi\right)^2}{\left[\left(1 - \frac{1}{2}\phi\right)^2 + r\left(1 - \frac{1}{2}\phi\right)(-\phi')\right]} = 1$$

Thus, the term $\underline{\frac{L\pi_L^2}{2R^2} - \frac{\pi_L \pi_R}{R} = 0 \text{ off shell.}}$

So, the analysis need only proceed with:

$$\int_0^\infty \mathcal{H}_t N^t dr = \int_0^\infty \frac{1}{2}\left\{\left(\frac{2RR'}{L}\right)' - \frac{(R')^2}{L} - L\right\} N^t dr$$

$$= \int_0^\infty \left\{\left(\frac{RR'N^t}{L} - r\right)' + \left[1 + \left(\frac{RR'}{L}\right)N^{t'} - \frac{N^t}{2}\left(\frac{(R')^2}{L} + L\right)\right]\right\} dr$$

As we are considering the constraint at first order (i.e., the Poisson eqn. $\nabla^2\phi = 4\pi\rho$) we don't obtain $\mathcal{H}_t = 0$ off shell and a full Newt. grav. field analysis results, where the integrand is considered to lowest appropriate order.

For the $\int_0^\infty \left(\frac{RR'N^t}{L} - r\right)' dr$ the analysis proceeds with $= -m$ except now the correct term to "first order" is undetermined. This is because evaluation of the surface term yields a result valid to all orders. The result correct to first order should be the result of the integration: –m, plus a correction term C. It follows from the interpretation of the results that $C = -M$, where M = total energy – rest mass energy as before. Thus the surface term introduces the energy parameter appropriate to the non-rel. problem.

$$\underline{\text{So } \int_0^\infty \left(\frac{RR'N^t}{L} - r\right)' dr\Big|_{1^{st}\ order} = -(M + m)}$$

$$N^t = \left(1+\frac{1}{2}\phi\right)/\left(1-\frac{1}{2}\phi\right)^{-1}$$

$$= \left(1+\frac{1}{2}\phi\right)\left(1+\frac{1}{2}\phi+\left(\frac{1}{2}\phi\right)^2\right)+O(\phi^3)$$

$$= 1+\phi+\frac{1}{2}\phi^2+O(\phi^3)$$

Consider next the term = $\int_0^\infty \left[1+\left(\frac{RR'}{L}\right)N^{t\prime}-\frac{N^t}{2}\left(\frac{(R')^2}{L}+L\right)\right]dr.$

$$\left(\begin{array}{l} N^t = \dfrac{\left(1+\frac{1}{2}\phi\right)}{\left(1-\frac{1}{2}\phi\right)} \\ L = \left(1-\frac{1}{2}\phi\right)^2 \\ R = rL \end{array}\right)$$

(must take to $O(\phi^2)$ since $\int \phi^2 \sim \phi$)

$\left(\frac{RR'}{L}\right)(N^t)' = r(L+rL')(\phi'+\phi\phi')+O(\phi^3)$ (recall $O(r\phi') = O(\phi)$

$$= r\left(1-\phi+r2\left(1-\tfrac{1}{2}\phi\right)\left(-\tfrac{1}{2}\phi'\right)\right)(\phi'+\phi\phi')+O(\phi^3)$$

$$= (r\phi')(1-\phi-r\phi')(1+\phi)+O(\phi^3)$$

$$= (r\phi')(1-\phi^2)(r\phi')^2+O(\phi^3)$$

$$= \underline{(r\phi')(1-r\phi')+O(\phi^3)}$$

$$L^{-2} = \left(1+\frac{1}{2}\phi+\frac{1}{4}\phi^2\right)^2+O(\phi^3)$$

$$= 1+\phi+\tfrac{3}{4}\phi^2+O(\phi^3)$$

$$\frac{1}{2}N^t\left[\frac{(R')^2}{L}+L\right] = \frac{1}{2}\left(1+\phi+\frac{1}{2}\phi^2\right)\left[2L+2rL'+r^2\frac{(L')^2}{L}\right]+O(\phi^3)$$

$$= \left(1+\phi+\frac{1}{2}\phi^2-\phi-\phi^2+\frac{1}{4}\phi^2\right)+\frac{1}{2}r\phi\phi'+\frac{1}{2}(r\phi')^2$$

$$+(1+\phi)(-r\phi')+O(\phi^3)$$

$$= 1 - \frac{1}{4}\phi^2 - r\phi' - \frac{1}{2}r\phi'\phi + \frac{1}{2}(r\phi')^2 + O(\phi^3)$$

Grouping, $\left[1 - \left(\frac{RR'}{L}\right)N^{t'} + \frac{1}{2}N^t\left(\frac{(R')^2}{L} + L\right)\right] = \frac{1}{4}\phi^2 + r\phi' + \frac{1}{2}r\phi'\phi - \frac{1}{2}r^2(\phi)^2 - r\phi' + r^2(\phi')^2 + O(\phi^3)$

$$= \frac{1}{4}\phi^2 + \frac{1}{2}r\phi'\phi + \frac{1}{2}r^2\phi'^2 + O(\phi^3)$$

Thus,

$\int_0^\infty \left[\frac{1}{4}\phi^2 + \frac{1}{2}r\phi'\phi + \frac{1}{2}r^2\phi'^2\right]dr = ?$ using $\phi_{asympt}^{\propto} \frac{1}{r}$ $\phi_{r=0} \propto$ $const$

$$\int \frac{1}{4}(r\phi^2)' = 0, \frac{1}{2}\int r^2\phi'^2 = \frac{1}{2}\int\left\{[r^2\phi'\phi]' - \underline{(r\phi^2)'\phi}\right\}$$

$$-\frac{1}{2}(r^2\phi')'\phi = \frac{1}{2}\{-r\phi'\phi - r^2\phi\phi'\} =$$

$$\underline{-\frac{1}{2}r^2\phi\nabla^2\phi}$$

So,

$$0 = H = \int\left\{-\frac{1}{2}\phi r^2\nabla^2\phi\right\}dr - (M+m) + \left\{m\left(1 + \frac{1}{2}\left(\frac{\partial r}{\partial \tau}\right)^2\right) + m\phi\right\}$$

Now, using $\nabla^2\phi = \frac{m}{r^2}\delta(r - \hat{r})$,

$$0 = -\frac{1}{2}m\phi - (M+m) + \frac{1}{2}m\left(\frac{\partial r}{\partial \tau}\right)^2 + m\phi$$

$$M = \frac{1}{2}m\dot{r}^2 - \frac{1}{2}\frac{m^2}{r} \qquad \underline{M = E = energy}$$

(Eqn. 10-51)

It is interesting to note that replacing $\frac{1}{2}m\dot{r}^2 \rightarrow \left(m\sqrt{1+\dot{r}^2} - m\right)$ yields the full GR solution shown in Sec. 10.9. Thus, in the (post)^n – Newtonian approx. additional terms involving the "grav. potential" eventually cancel, if not at every order, such that the potential remains unchanged. This is certainly understandable as the "full solution" in the asymptotically-free region, and as a regional solution for the GR. It is then <u>the</u> solution since all parameters are local to the shell (*r* being defined in terms of surface area), and thus geometrical.

10.11 Asymptotics

We now turn to the global properties of the geometry. In this section we take the spatial topology to be that of the extended Schwarzschild geometry, $S^2 \times R = S^3\{two\ points\}$, the omitted points being associated with asymptotically flat asymptopias. Other topologies will be discussed as well.

At a general level, restricting the asymptotic behavior of an asymptotically at system allows one to fix the momentum, angular momentum, and mass at spatial infinity. In a quantum theoretic context, to restrict in this way the asymptotic behavior of the operator ${}^3\hat{g}_{ab}$ and its conjugate momentum $\hat{\pi}^{ab}$ is equivalent to restricting the state space to an eigensubspace of fixed total momentum, angular momentum, or mass. In our particular case of spherical symmetry, the angular momentum is necessarily zero. It would be consistent with spherical symmetry to allow a nonzero momentum at infinity (in the classical framework, this would mean allowing boosted Schwarzschild solutions), but for our purposes this freedom does not appear significant, and we shall just set the momentum to zero. However, we shall consider the differences that arise from retaining or discarding the freedom associated with the system's total mass.

We take the coordinate r to have the range $-\infty < r < \infty$. At the asymptopias $r \to \pm\infty$, we introduce the falloff

$$
\begin{aligned}
\Lambda(t,r) &= 1 + O^{\infty}\left(|r|^{-\frac{3}{2}-\alpha}\right) \ , \\
R(t,r) &= |r| + O^{\infty}\left(|r|^{-\frac{1}{2}-\alpha}\right) \ , \\
\pi_{\Lambda}(t,r) &= \sqrt{2M_{\pm}|r|} + O^{\infty}\left(|r|^{-\alpha}\right) \ , \\
\pi_{R}(t,r) &= \sqrt{\frac{M_{\pm}}{2|r|}} + O^{\infty}\left(|r|^{-1-\alpha}\right) \ , \\
N(t,r) &= 1 + O^{\infty}\left(|r|^{-\alpha}\right) \ , \\
N^{r}(t,r) &= \pm\sqrt{\frac{2M_{\pm}}{|r|}} + O^{\infty}\left(|r|^{-\frac{1}{2}-\alpha}\right) \ ,
\end{aligned}
$$

(Eqn. 10-52)

where $M_\pm(t)$ are positive-valued functions of t, and β is a positive parameter that can be chosen at will. O^∞ indicates a quantity that is bounded at infinity by a constant times its argument, with the corresponding behavior for its derivatives.

It is straightforward to verify that the falloff is consistent with the constraints and preserved in time by the dynamical equations. When the equations of motion hold, $M_\pm$ are independent of t, and their values are just the Schwarzschild masses at the two asymptopias. It is easy to show that the existence of two asymptotically at infinities implies that both asymptotic Schwarzschild masses in the classical solutions are necessarily positive, and the restriction $M_\pm(t) > 0$ does therefore not exclude any solutions.

10.12 Reduced phase-space formalism

In the absence of the shell, the Hamiltonian reduction of our theory with a technically different but qualitatively similar falloff at the infinities was discussed in Ref. [60] (see also Refs. [62,63]). When the asymptotic masses are not fixed, it was found that the reduced phase space is two-dimensional, whereas if one asymptotic mass is fixed, the reduced phase space has dimension zero. As the shell brings in one new canonical pair but no new constraints, one therefore expects that the reduced phase space of our theory is four-dimensional when the asymptotic masses are not fixed, and two-dimensional if one asymptotic mass is fixed. In this section we shall verify this expectation by an explicit reduction.

10.12.1 Gauge transformations and the Hamiltonian reduction formalism

In the Hamiltonian theory formulated thus far, the variables $\{\Lambda, R, \pi_\Lambda, \pi_R, \hat{r}, \hat{p}\}$ constitute a canonical chart on the phase space, while N and N_r act as Lagrange multipliers enforcing the constraints. As the Poisson bracket algebra of the constraints closes, we have a first class constrained system [87]. By gauge transformations, we shall refer to the transformations that the constraints generate on the constraint hypersurface in the phase space. Denoting the smearing functions by N(r) and $N_r(r)$, the smeared Hamiltonian constraint transforms an initial data set $\{\Lambda, R, \pi_\Lambda, \pi_R, \hat{r}, \hat{p}\}$ by the time evolution associated with N, and the smeared momentum constraint transforms the initial data set by the spatial diffeomorphism associated with N_r. The smearing functions must fall off so fast that the transformations become trivial at the infinities and the indicated falloff is preserved.

Consider the constraints. Away from the shell, a solution to the constraints is an initial data set for the vacuum Einstein equations with spherical symmetry. Any vacuum data set has a unique time evolution, and, by Birkhoff's theorem, the resulting subspacetimes left and right of the shell are isometric to regions of two Kruskal spacetimes with respective masses M− and M+. Solutions to the constraint equations on each side of the shell can thus be regarded as partial spacelike slices of two Kruskal spacetimes. For us, M− is a prescribed constant, but M+ is not, and the constraint hypersurface includes configurations with different values of M+. This means that a path in the reduced phase space need not correspond to a foliation of a single Kruskal geometry right of the shell.

At the shell, we have already seen that the full content of the Hamiltonian and momentum constraints is encoded in equations shown. Away from the shell, an explicit solution to the constraints has the form:

$$\pi_\Lambda = \begin{cases} R\sqrt{(R'/\Lambda)^2 - 1 + 2M_-/R}, & \text{for } r < \hat{r}, \\ R\sqrt{(R'/\Lambda)^2 - 1 + 2M_+/R}, & \text{for } r > \hat{r}, \end{cases}$$

$$\pi_R = \frac{\Lambda \pi_\Lambda'}{R'}, \quad r \neq \hat{r}.$$

(Eqn. 10-53)

The sign in has been chosen so as to agree with the falloff conventions chosen.

10.12.2 Gauge choice

We now restrict the attention to initial data sets with the following property: In the classical spacetime that is the time evolution of the initial data set, the shell trajectory intersects the right-hand-side region of outer communication in the Kruskal geometry right of the shell. The junction conditions (4.11) then imply M+ > M−, and the trajectory intersects the right-hand-side region of outer communication also in the Kruskal geometry left of the shell. We view this as the situation of physical interest for of an observer who observes the shell motion from the right-hand-side infinity and regards the "interior" mass as fixed.

Given an initial data set, consider the classical spacetime that is its time evolution. On this spacetime, we introduce the local chart

$$ds^2 = -dt^2 + \left(dr + \sqrt{\frac{2M_-}{r}}\,dt\right)^2 + r^2 d\Omega^2, \quad 0 < r \leq \hat{r} - l \ ,$$

$$ds^2 = -dt^2 + \left(dr + \sqrt{\frac{2M_+}{r}}\,dt\right)^2 + r^2 d\Omega^2, \quad \hat{r} \leq r < \infty \ ,$$

(Eqn. 10-54)

where l is a positive parameter. The two metrics in are the ingoing righthand-side spatially at charts in Kruskal manifolds with the respective masses M− and M+ [57,88,89]. If taken individually for 0 < r < 1, each of these two charts would cover the upper right half (regions I and II) in the respective full Kruskal diagrams. With the domains indicated in, the combined chart is spatially flat with mass M− to the left of the shell, and (asymptotically) spatially flat with mass M+ to the right of the shell.

Matching the junction conditions we finally obtain:

$$p = \sqrt{2M_-\hat{r}} - \sqrt{2M_+\hat{r}} + \hat{r}\log\frac{1-\sqrt{\frac{2M_-}{\hat{r}}}}{1-\sqrt{\frac{2M_+}{\hat{r}}}+\frac{1}{\hat{r}}\left(\sqrt{\hat{p}^2+m^2}-\hat{p}\right)} \ .$$

(Eqn. 10-55)

In the limit m→0, this reduces to the result obtained in Kraus and Wilczek [53].

Thus:

$$M_+ - M_- = \sqrt{\hat{p}^2+m^2} + \frac{m^2}{2\hat{r}} - \hat{p}\sqrt{\frac{2M_+}{\hat{r}}};$$

(Eqn. 10-56)

and the equations implicitly give M+ and ^p in terms of M−, ^r, and p.

Soon we will radically simplify the formalism by reparametrizing our time. In doing so, we will use a simple relation between ^p, ^r and ^ṙ, namely

$$\dot{\hat{r}} = \frac{\partial H}{\partial p} = \frac{\hat{p}}{\sqrt{\hat{p}^2+m^2}} - \sqrt{\frac{2M_+}{\hat{r}}}.$$

(Eqn. 10-57)

The Hamiltonian becomes

$$H = 2M_- + \frac{m\left[1-\left(\dot{\hat{r}}+\sqrt{\frac{2M_+}{\hat{r}}}\right)\right]\sqrt{\frac{2M_+}{\hat{r}}}}{\sqrt{1-\left(\dot{\hat{r}}+\sqrt{\frac{2M_+}{\hat{r}}}\right)^2}} + \frac{m^2}{2\hat{r}}.$$

(Eqn. 10-58)

10.13 Analysis of K+W Action [53,54]

By reconstruction $K + W$ obtain the following action for a shell at $\hat{r}$ (in a spherically symmetric, asymtotically flat, spacetime):

$$S = \int_{-\infty}^{\alpha} dr\, F(R, R', L, \mathfrak{M}) + \int dt\, \dot{\hat{R}} \left(\frac{\partial F}{\partial R'}\right)\Bigg|_{(\hat{r}-\epsilon)}^{(\hat{r}+\epsilon)} - \int dt\, dr \left(\frac{\partial F}{\partial \mathfrak{M}}\right)\dot{\mathfrak{M}} - \int dt\, \mathfrak{M}\Big|_{-\infty}^{+\infty}$$

(Eqn. 10-59)

Where

$$F = RL\sqrt{(R'/L)^2 - 1 + 2\mathfrak{M}/R} + RR' \log\left|\frac{(R'/L) - \sqrt{(R'/L)^2 - 1 + 2\mathfrak{M}/R}}{\sqrt{|1 - 2\mathfrak{M}/R|}}\right|$$

(Eqn. 10-60)

And $\mathfrak{M} = M_{\gtrless}$ depending on whether to the left or right of the shell.

There are also junction conditions relating $\hat{r} + \epsilon$ terms to $\hat{r} - \epsilon$:

$$\pi_L(\hat{r}+\epsilon) - \pi_L(\hat{r}-\epsilon) = -p/\hat{L}$$

$$R'(\hat{r}+\epsilon) - R'(\hat{r}-\epsilon) = -\frac{1}{\hat{R}}\sqrt{p^2 + m^2\hat{L}^2}$$

$$\left(\pi_L = R\sqrt{(R'/L)^2 - 1 + 2\mathfrak{M}/R} \;\; \text{off shell}\right)$$

(Eqn. 10-61)

A limit process is employed in order to obtain the Lagrangian from the action in a manner consistent with the j.c.'s. The limit involves a choice of gauge at the shell, i.e., whether the geometric description of the shell should be given in terms of an embedding to the left or right of the shell. The different embeddings yield the same dynamical equations. Since we would like to consider an observer at the right-hand side (RHS) asymptotic region, the shell will be described in terms of a RHS embedding.

$$L = \frac{dS}{dt} = \dot{\hat{r}}\hat{R}\hat{L}[\sqrt{<} - \sqrt{>}] - \dot{\hat{R}}\hat{R}\log\left|\frac{R'\,(\hat{r}-\epsilon)/\hat{L} - \sqrt{"\epsilon"}}{R'_{<}/\hat{L} - \sqrt{<}}\right| + \oint_{-\infty}^{\infty} dr\,[\pi_L \dot{\hat{L}} + \pi_R \dot{R}] - (M_{>} - M_{<})$$

(Eqn. 10-62)

Where $\sqrt{"x"} = \sqrt{(R'_x/\hat{L})^2 - 1 + 2\,M_x/\hat{R}}$.

The J.C.'s: $\sqrt{" + \epsilon"} - \sqrt{" - \epsilon"} = -p/[\hat{R}$

$(R'_{+\epsilon}) - (R'_{-\epsilon}) = -\frac{1}{\hat{R}}\sqrt{p^2 + m^2\hat{L}^2}$

The RHS $K + W$ metric: $$d\tau^2 = dt^2 - \left(dr + \sqrt{\frac{2M_{>}}{r}}\,dt\right)^2$$

$$\left(\frac{dr}{dt}\right) = \frac{\left(\frac{dr}{d\tau}\right)\left(1 - \frac{2M_{>}}{r}\right)}{\sqrt{\frac{2M_{>}}{r}}\left(\frac{dr}{d\tau}\right) \pm \sqrt{\left(1 - \frac{2M_{>}}{r}\right) + \left(\frac{dr}{d\tau}\right)^2}}$$

(Eqn. 10-63)

The $K + W$ gauge is chosen throughout: $R = r, L = 1$:

$$L = \dot{r}p_{<} = (M_{>} - M_{<})$$

(Eqn. 10-64)

$$p_{<} = [\sqrt{2M_{<}r} - \sqrt{2M_{>}r}] - r\log\left|\frac{(1 - \sqrt{2M_{>}/r}) + \frac{1}{r}(\sqrt{p^2 + m^2} - p)}{(1 - \sqrt{2M_{<}/r})}\right|$$

(Eqn. 10-65)

And the j.c.'s:

$$\sqrt{2M_{>}/r} + \frac{p}{r} = \sqrt{(R'_{-\epsilon})^2 - 1 + 2M_{<}/r}$$

(Eqn. 10-66)

$$(R'_{-\epsilon}) = 1 + \frac{1}{r}\sqrt{p^2 + m^2}$$

(Eqn. 10-67)

Which may be grouped to eliminate $R'_{-\epsilon}$:

$$(M_> - M_<) = \sqrt{p^2 + m^2} + \frac{m^2}{2r} - p\sqrt{\frac{2M_>}{r}}$$

(Eqn. 10-68)

From $\frac{\partial L}{\partial p} = 0 \Rightarrow \dot{r} = \left(\frac{p}{\sqrt{p^2+m^2}} - \sqrt{\frac{2M_>}{r}}\right) \Rightarrow \left(\frac{p}{m}\right) = \frac{\eta\left(\dot{r}+\sqrt{\frac{2M_>}{r}}\right)}{\sqrt{1-\left(\dot{r}+\sqrt{\frac{2M_>}{r}}\right)^2}}$

Together with the constraint eqn from the j.c.'s:

$$M_> - M_< = \eta m \frac{\left[1 - \left(\dot{r} + \sqrt{\frac{2M_>}{r}}\right)\sqrt{\frac{2M_>}{r}}\right]}{\sqrt{1 - \left(\dot{r} + \sqrt{\frac{2M_>}{r}}\right)^2}} + \frac{m^2}{2r}$$

(Eqn. 10-69)

For a time-like trajectory

$$\left(\frac{d\tau}{dt}\right) = \eta_2 \sqrt{1 - \left(\dot{r} + \sqrt{\frac{2M_>}{r}}\right)^2} > 0$$

(Eqn. 10-70)

So $\eta_2 = +1$ and $\left|\dot{r} + \sqrt{\frac{2M_>}{r}}\right| < 1$. Whichever η value is taken in the $P(\dot{r})$ relation there are the same difficulties, so <u>take $\eta = 1$</u> for concreteness. If $p < 0$, then $(M_> - M_<)$ is positive, so consider $p > 0$:

$$\left(\frac{p}{m}\right) = \frac{\left(\dot{r}+\sqrt{\frac{2M_>}{r}}\right)}{\sqrt{1-\left(\dot{r}+\sqrt{\frac{2M_>}{r}}\right)^2}} > 0 \text{ if } -\sqrt{\frac{2M_<}{r}} \le \dot{r} < 1 - \sqrt{\frac{2M_<}{r}}$$

So, the maximum pos. $\dot{r}$ is $\sim 1 - \sqrt{\frac{2M_<}{r}}$. Take $\dot{r} = 1 - \sqrt{\frac{2M_<}{r}} - \epsilon$, what is $M_> - M_< = ?$

We have

$$M_> - M_<\Big|_{\dot{r}=1-\sqrt{\frac{2M_<}{r}}-\epsilon} = \frac{m\left[1-(1-\epsilon)\sqrt{\frac{2M_>}{r}}\right]}{\sqrt{1-(1-\epsilon)^2}} + \frac{m^2}{2r}$$

$$\equiv m\left[1-\sqrt{\frac{2M_>}{r}}\right]\frac{1}{\sqrt{2\epsilon}}$$

and

$$\sqrt{\frac{2M_>}{r}} < 1\left(\text{from}\left|\dot{r}+\sqrt{\frac{2M_>}{r}}\right| < 1 \text{ and } \dot{r} > 0\right) \Rightarrow \underline{M_> > 0}.$$

If $\eta \equiv -1$ in p relation to $\dot{r}$, then K.E. term is negative definite, indicating negative time, so excluded.

Alternatively,

$$M_> - M_< = \frac{m\left[1-\left(\dot{r}+\sqrt{\frac{2M_>}{r}}\right)\sqrt{\frac{2M_>}{r}}\right]}{\sqrt{1-\left(\dot{r}+\sqrt{\frac{2M_>}{r}}\right)^2}} + \frac{m^2}{2r}$$

(Eqn. 10-71)

Thus, $M_> > 0$ (even for positive $M_<$). How is negative mass information lost?

$$\text{Consider } \acute{r} = \frac{\dot{r}\left(1-\frac{2M_>}{r}\right)}{\sqrt{\frac{2M_>}{r}}\dot{r} \pm \sqrt{\left(1-\frac{2M_>}{r}\right)+\dot{r}^2}}$$

$\uparrow \eta_3 = \pm 1$, choose positive

$$M_> - M_< = m\sqrt{\left(1-\frac{2M_>}{r}\right)+\dot{r}^2} + \frac{m^2}{2r}$$

Note that: $d\tau^2 = dt^2 - \left(dr + \frac{2M_>}{r}dt\right)^2$

Still clear that $M_>$ is positive regardless of sign for $M_<$:

$$M_> - M_< = m\sqrt{\left(1 - \sqrt{\frac{2M_>}{r}}\right) + \dot{r}^2 + \frac{m^2}{2r}}$$

Taken as a recursive relation on $M_>$ now solve:

$$\left(M_> - M_< - \frac{m^2}{2r}\right)^2 = m^2\left(1 - \frac{2M_>}{r}\right) + m^2\dot{r}^2$$

$$(M_> - M_<)^2 - 2(M_> - M_<)\left(\frac{m^2}{2r}\right) + \left(\frac{m^2}{2r}\right)^2$$

$$= m^2 - m^2\left(\frac{2M_>}{r}\right) + m^2\dot{r}^2$$

$$(M_> - M_<)^2 - (M_> - M_<)\left[\frac{m^2}{2r} - \frac{2m^2}{2r}\right]$$

$$= m^2 + m^2\dot{r}^2 - m^2\left(\frac{2M_<}{r}\right) - \left(\frac{m^2}{2r}\right)^2$$

$$(M_> - M_<)^2 + \left(\frac{m^2}{2r}\right)(M_> - M_<) + m^2\left(\frac{m^2}{4r^2} + \frac{2M_<}{r} - 1 - \dot{r}^2\right) = 0$$

$$(M_> - M_<) = -\frac{1}{2}\left(\frac{m^2}{r}\right) \pm \frac{1}{2}\sqrt{\left(\frac{m^2}{r}\right)^2 - 4m^2\left(\frac{m^2}{4r^2} + \frac{2M_<}{r} - 1 - \dot{r}^2\right)}$$

$$(M_> - M_<) = \pm m\sqrt{\left(1 - \frac{2M_<}{r}\right) + \dot{r}^2} - \frac{m^2}{2r}$$

(Eqn. 10-72)

Even with the pos. root this expression is not manifestly positive. Also, in

$$(M_> - M_<) = m\sqrt{\left(1 - \frac{2M_<}{r}\right) + \dot{r}^2} - \frac{m^2}{2r}$$

(Eqn. 10-73)

$\dot{r}$ is still restricted in regards to $M_>$. Thus, some expressions for the EOM have manifestly pos. $M_>$ and some don't.

10.14 Time Reparameterization

We present in Sec. 10.14.1 a formalism for reparametrizing time in a one-dimensional Hamiltonian system. The formalism is used in Sec. 10.14.2 to simplify our reduced Hamiltonian for a self-gravitating dust shell.

10.14.1 General formalism for a one-dimensional Hamiltonian

Consider a 1-dimensional Hamiltonian of the form

$$h = h(q, p).$$

(Eqn. 10-74)

The inverse of the Legendre transformation relating the canonical momentum p to a velocity v is given by

$$v(q, p) = \frac{\partial h}{\partial p}.$$

(Eqn. 10-75)

That is, solutions q(t); p(t) to Hamilton's equations correspond to dynamical solutions q(t); v(t) to the Euler-Lagrange equations associated with the Lagrangian

$$\ell(q, v) = p(q, v)v - h(q, p(q, v)),$$

(Eqn. 10-76)

with

$$v(t) = \frac{d}{dt}q(t).$$

(Eqn. 10-77)

The Hamiltonian, expressed as a function on velocity space, will be denoted by ~h:

$$\begin{aligned} \tilde{h}(q, v) &= h(q, p(q, v)), \\ &= p(q, v)v - \ell(q, v). \end{aligned}$$

(Eqn. 10-78)

If one reparametrizes the time t in the manner

$$dT = Ndt, \quad N(t) = N(q(t), v(t)),$$ (Eqn. 10-79)

the corresponding velocity, V(T), is related to v(t) by

$$v = N(q, v)V.$$

(Eqn. 10-80)

In the Hamiltonian framework, the canonical momentum P associated with (q; V) is given by

$$\frac{\partial P}{\partial p} = N(q, p),$$ (Eqn. 10-81)

where

$$N(q, p) := N(q, v(q, p)).$$

(Eqn. 10-82)

That is, with a new symplectic form dP ^ dq replacing dp ^ dq, and with the Hamiltonian written in terms of P and q,

$$H(q, P) = h(q, p(q, P)), \quad \text{(Eqn. 10-83)}$$

each dynamical path q(t); p(t) of the Hamiltonian h coincides with a reparametrized path q(T); P(T) of H.

We can thus express a family of solutions,

$$P = \int^{p} N(q, p')dp', \quad \text{(Eqn. 10-84)}$$

corresponding to a choice of initial point. The solutions differ from one another by canonical transformations of the form,

$$P \to P + f(q). \quad \text{(Eqn. 10-85)}$$

Given H(q; P), one can recover the reparametrized velocity V from the relation

$$V = \frac{\partial H}{\partial P}. \quad \text{(Eqn. 10-86)}$$

Consider, for example, the motion of a free relativistic particle, parametrized by a Minkowski time t, and described by the Hamiltonian

$$h = \left(p^2 + m^2\right)^{\frac{1}{2}}, \quad \text{(Eqn. 10-87)}$$

with

$$v = \frac{\partial h}{\partial p} = \frac{p}{(p^2 + m^2)^{\frac{1}{2}}}. \quad \text{(Eqn. 10-88)}$$

If we adopt, as a new time parameter, the property time T along each path, then T is related to Minkowski time by Eq. (4.37) with

$$N - \left(1 - v^2\right)^{\frac{1}{2}} = \frac{m}{(p^2 + m^2)^{\frac{1}{2}}}. \quad \text{(Eqn. 10-89)}$$

We then have

$$P = \int N dp = m \sinh^{-1} \frac{p}{m}, \quad \text{(Eqn. 10-90)}$$

with inverse

$$p = m \sinh \frac{P}{m}. \quad \text{(Eqn. 10-91)}$$

Writing H(P; q) = h(q; p(P; q)), with h given above, we obtain the familiar form

$$H = m \cosh \frac{P}{m}.$$
(Eqn. 10-92)

If one starts directly from the Hamiltonian ~h on velocity space, the momentum p(q; v) is given by the relation

$$\frac{\partial p}{\partial v} = \frac{1}{v} \frac{\partial h}{\partial v},$$
(Eqn. 10-93)

(see [55] for details). By writing the change in velocity corresponding to a re-parametrization of time in the manner,

$$v = N(q, V)V,$$
(Eqn. 10-94)

one can re-express the energy as a function of q and V :

$$\tilde{H}(q, V) = \tilde{h}(q, N(q, V)V).$$
(Eqn. 10-95)

The corresponding canonical momentum P(q; V) is again given by

$$\frac{\partial P}{\partial V} = \frac{1}{V} \frac{\partial \tilde{H}}{\partial V},$$
(Eqn. 10-96)

or

$$P = \int dV \frac{1}{V} \frac{\partial \tilde{H}}{\partial V}.$$
(Eqn. 10-97)

P is determined only up to a canonical transformation of the form P(q; V) = P(q; V) + f(q). For each choice of P, we recover a Hamiltonian H(q; P) by inverting P(q; V) to obtain

$$H(q, P) = \tilde{H}(q, V(q, P)).$$
(Eqn. 10-98)

10.14.2 Proper-time parametrization of a self-gravitating dust shell

The formalism presented in the preceding section is now used to obtain a simple Hamiltonian that describes a self-gravitating, spherically symmetric shell of dust, using proper time along the shell's trajectory. We begin with the Lagrangian expressed earlier:

$$\ell = \dot{r}p - h, \quad h = M_+ + M_- = \sqrt{\hat{p}^2 + m^2} + \frac{m^2}{2r} - \hat{p}\sqrt{\frac{2M_+}{r}} + 2M_-$$

(Eqn. 10-99)

where p is the canonical momentum associated with r, and ^p is the radial component of the shell's momentum in an orthonormal frame. The time indicated is the static-gauge time in the left-hand-side geometry. Here M− is a free parameter (not a dynamical variable), while the Hamiltonian h (and h=M+ + M−) gives the relations:

$$\hat{p} = m\frac{\left(\dot{\hat{r}} + \sqrt{\frac{2M_+}{r}}\right)}{\left[1-\left(\dot{\hat{r}} + \sqrt{\frac{2M_+}{r}}\right)^2\right]^{1/2}}.$$

(Eqn. 10-100)

$$\tilde{h} = M_+ + M_- = m\frac{1-\left(\dot{\hat{r}} + \sqrt{\frac{2M_+}{r}}\right)\sqrt{\frac{2M_+}{r}}}{\left[1-\left(\dot{\hat{r}} + \sqrt{\frac{2M_+}{r}}\right)^2\right]^{1/2}} + \frac{m^2}{2r} + 2M_-.$$

(Eqn. 10-101)

Let us now reparametrize the time along each trajectory, replacing t by the proper time, with

$$V = \frac{d\hat{r}}{d\tau} = \frac{1}{N}\dot{r}, \quad N = \left(\frac{dt}{d\tau}\right)^{-1}.$$

(Eqn. 10-102)

Using the 3-metric on the shell's timelike worldsheet, we have

$$d\tau^2 = dt^2 - \left(d\hat{r} + \sqrt{\frac{2M_+}{\hat{r}}}dt\right)^2.$$

(Eqn. 10-103)

implying for the function N the form

$$\frac{1}{N} = \frac{\sqrt{\frac{2M_+}{\hat{r}}} \pm \left[1+\left(1-\frac{2M_+}{\hat{r}}\right)V^{-2}\right]^{1/2}}{1-\frac{2M_+}{\hat{r}}}V.$$

(Eqn. 10-104)

Then

$$v \equiv \dot{r} = NV = \frac{1-\frac{2M_+}{r}}{\sqrt{\frac{2M_+}{r}} \pm \left[1+\left(1-\frac{2M_+}{r}\right)V^{-2}\right]^{1/2}}.$$

(Eqn. 10-105)

By substituting this expression for rand simplifying, we obtain h in terms of V:

$$\tilde{H}(r,V) = \tilde{h}(r,v(r,V))$$
$$= \pm m\left[V^2+1-\frac{2M_+}{r}\right]^{\frac{1}{2}} + \frac{m^2}{2r} + 2M_-.$$

(Eqn. 10-106)

This is easily inverted to obtain an expression dependent on M−:

$$\tilde{H}(r,V) = \tilde{h}(r,v(r,V))$$
$$= \pm m\left[V^2+1-\frac{2M_-}{r}\right]^{\frac{1}{2}} - \frac{m^2}{2r} + 2M_-.$$

(Eqn. 10-107)

We then have

$$\frac{\partial P}{\partial V} = \frac{1}{V}\frac{\partial H}{\partial V} = \pm m\left[V^2+1-\frac{2M_-}{r}\right]^{-\frac{1}{2}}.$$

(Eqn. 10-108)

The canonical momentum P is given in terms of the r-component V of the shell's 4-velocity by

$$P = m\int\frac{du}{\left[u^2+1-\frac{2M_-}{r}\right]^{\frac{1}{2}}} = m\log V + \sqrt{\left(1-\frac{2M_-}{r}\right)+V^2},$$

(Eqn. 10-109)

with inverse

$$V = \sinh(P/m) + \left(\frac{M_-}{r}\right)\exp(-P/m).$$

(Eqn. 10-110)

We find after some calculation a simple expression for the Hamiltonian H(r; P), namely

$$H(r,P) = m\cosh P/m - \frac{m^2}{2r} + M_-(2-(m/r)\exp(-P/m)).$$

(Eqn. 10-111)

10.14.3 An alternative parametrization of dust-shell time.

A different time reparametrization leads to a Hamiltonian for the dust shell that is analogous to the Minkowski-time Hamiltonian

$$H = \sqrt{p^2 + m^2}$$

(Eqn. 10-112)

for a free spherical dust shell in at space. Starting from the proper-time Hamiltonian, we set

$$N = \frac{m}{\sqrt{p^2 + m^2}},$$

(Eqn. 10-113)

with p the momentum associated with the new time,

$$\frac{\partial P}{\partial p} = N(q, p) = \frac{m}{\sqrt{P^2 + m^2}},$$

(Eqn. 10-114)

has solution

$$p = m \sinh \frac{P}{m}.$$

(Eqn. 10-115)

Then

$$\cosh \frac{P}{m} = \frac{\sqrt{P^2 + m^2}}{m},$$

(Eqn. 10-116)

and we have

$$H = \sqrt{p^2 + m^2} - \frac{m^2}{2r} + M_{-}[2 - \frac{1}{r}(\sqrt{p^2 + m^2} - p)].$$

(Eqn. 10-117)

When the interior mass M− vanishes, this is the Hamiltonian used by Kuchar et. al. [60].

Hamiltonians for the RP³ geon

As noted earlier, an asymptotically at, spherically symmetric spacetime M; g with a single asymptotic region can have the spatial topology of the manifold = RP^3 minus a point at infinity. We will refer to such a spacetime, or to an asymptotically at initial data set with this topology, as an RP^3 -geon. Its covering space has the wormhole topology of the extended Schwarzschild geometry.

One can obtain a spherically symmetric vacuum spacetime with this topology from extended Schwarzschild by picking a Kruskal char. Surfaces of constant Kruskal time T = U+V have a throat, a symmetric two-sphere of minimum radius, and they are symmetric under the involution

$$I : (R, \theta, \phi) \rightarrow (-R, \pi - \theta, \phi + \pi),$$

(Eqn. 10-118)

that identifies points on opposite sides of the throat.

An initial data set with a dust shell on the geon, satisfying the constraint equations, is characterized by the circumferential radius of the dust r, a conjugate momentum P, and a mass M describing the geometry interior to the shell. The data can be lifted to a locally identical initial data set with two dust shells on opposite sides of the throat, each at circumferential radius r. Outside the shells, the Schwarzschild masses have the same value M− = M+, and between the shells is a Schwarzschild geometry with a smaller mass. From the equation for the Hamiltonian the two-shell system has the proper-time form,

$$\bar{H} = M_- + M_+ = 2m \cosh P/m - 2\frac{m^2}{2r} + 2[M_-(2 - (m/r)\exp(-P/m))].$$

(Eqn. 10-119)

The corresponding Hamiltonian for the single shell on the RP^3-geon manifold is then

$$H = m \cosh P/m - \frac{m^2}{2r} + M_-(2 - (m/r)\exp(-P/m)),$$

(Eqn. 10-120)

identical in form to that for standard (non-geon) shell collapse. In terms of the alternative time-parametrization, the system has Hamiltonian

$$H = \sqrt{p^2 + m^2} - \frac{m^2}{2r} + M - [1 - \frac{1}{r}(\sqrt{p^2 + m^2} - p)].$$

(Eqn. 10-121)

10.15 Minisuperspaces for spherically symmetric

We first consider the vacuum case before turning to the quantization of the shell + geometry. As noted, one can include in the asymptotic conditions for a wormhole geometry, a particular value for the mass at the left asymptopia. If this is done, then, once the constraint equations are solved, the right-side mass is determined, M+ = M− = M, and there is no remaining degree of freedom. The system is trivial, and its mass, a c-number by construction, is not quantized. If, on the other hand, one retains M− as a dynamical variable, then, after the constraints are satisfied, the Hamiltonian is the mass

$$H = M_- + M_+ = 2M,$$

(Eqn. 10-122)

and the symplectic structure is dpM ^dM. If, as Kuchar does [60], one takes M as the configuration space variable, then, in a Schrodinger representation the operator ^M acts in the manner

$$\hat{M}\Psi(M) = M\Psi(M),$$

(Eqn. 10-123)

and ^M has continuous spectrum.

One can, however, replace the canonical variables M and PM by any other canonical pair. With the choice N = 1, the mass is

$$M = \frac{1}{2}(a\dot{a}^2 + a).$$

(Eqn. 10-124)

A canonical transformation from M, P_M to a, p_a gives

$$H = M = \frac{1}{2}\left(\frac{p_a^2}{a} + a\right),$$

(Eqn. 10-125)

where

$$p_a = sgn(p)\sqrt{2Ma - a^2}.$$

(Eqn. 10-126)

For obvious choices of measure on the configuration space R+ associated with a, factor orderings that make ^M self-adjoint yield a spectrum that, like the spectrum of the nonrelativistic Coulomb Hamiltonian has discrete and continuous parts.

The continuous spectrum that Kuchar obtains for the mass of this system can be obtained in an analogous way for any dynamical system in which the Hamiltonian is not explicitly time-dependent. That is, for any function H with nonvanishing gradient on phase space, one can find a local canonical chart of the form

$$H, q_2, \cdots, q_n, p_H, p_2, \cdots, p_n,$$

a chart in which H is one of the canonical variables. If, as Kuchar does, one takes a Schrodinger representation with Hilbert space H=L_2(R+), with H a Hilbert space for the remaining q's, then ^H acts as a multiplication operator,

$$\hat{H}\psi(H, q_2, \cdots q_n) = H\psi(H, q_2, \cdots, q_n),$$

and its spectrum is R+.

The dynamics of the RP3-geon geometry are described by the formally identical Hamiltonian, for the throat radius with a proper-time parametrization, and the quantizations and corresponding mass spectra are identical. Only the interpretation is altered.

10.16 Kuchar throat derivation

Vacuum Minisuperspace, spherically symmetric, can be examined in two ways: with asymptotically flat hypersurfaces and (i) introduction of embedding variables [60]; or (ii) t=const. slicing [53,54]. As in Kuchař analysis, begin with the hypersurface Lagrangian that evolves the geometry.

$$L_\Sigma = (16\pi)^{-1} N |g|^{1/2} \left(K^{ab} K_{ab} - K^2 + {}^{(3)}R\right),$$

(Eqn. 10-127)

where the vacuum dynamics of the metric field follows from

$$S_\Sigma[g, N, N^a] = \int dt \int_\Sigma d^3x \, L_\Sigma,$$

(Eqn. 10-128)

and boundary terms will be discussed after passing to the Hamiltonian formalism.

The asymptotically flat restriction on hypersurfaces leads to fall-off conditions on the canonical variables, and results in the boundary terms to be expressed in the full Hamiltonian theory. The $\tau = constant \; \epsilon -$ neighbourhood restriction on hypersurfaces is implemented on the Lagrangian action:

$$S_\Sigma = \int dt \left[\int_{-\infty}^{-\epsilon} + \int_{-\epsilon}^{+\epsilon} + \int_{\epsilon}^{\infty} dr \, L_\Sigma \right].$$

(Eqn. 10-129)

In the ϵ-neighborhood, the restriction to $\tau = constant$ hypersurfaces leads to:

(1)

$$\int_{-\epsilon}^{\epsilon} dr\, L_\Sigma|_{\tau=const} = \int_{-\epsilon}^{\epsilon} dr\, N \frac{3}{2}\left(\frac{2M_0}{(1+x^2)^{1/2}} - \frac{r}{(1+x^2)^{1/2}} + (2M_0)\left(\frac{2M_0}{r} - \frac{1}{1+x^2}\right)^{1/2} \cos^{-1}\left[\left(\frac{r/2M_0}{1+x^2}\right)^{1/2}\right]\right) \times \left[\left(\frac{2M_0}{a} - 1\right)^{-1} \dot{a}^2 - 1\right].$$

In eqn. (1) x satisfies the usual Novikov coordinate relation with $\{r, \tau\}$, i.e., $x = x(r, \tau)$. The variable $a \equiv r|_{\substack{throat \\ (x=0)}}$ and $a = \frac{\partial a}{\partial \tau}$, and M_0 is the ADM mass parameter that is constant on the $\tau = const$ slices and their evolution.

Thus,

(2)

$$\int_{-\epsilon}^{\epsilon} dr\, L_\Sigma|_{\tau=const} = \left\{N(t) \int_{-\epsilon}^{\epsilon} dr \frac{3}{2}\left(\frac{2M_0}{(1+x^2)^{1/2}} - \frac{r}{(1+x^2)^{1/2}} + (2M_0)\left(\frac{2M_0}{r} - \frac{1}{1+x^2}\right)^{1/2} \cos^{-1}\left[\left(\frac{r/2M_0}{1+x^2}\right)^{1/2}\right]\right)\right\} \times \left\{\left(\frac{2M_0}{a} - 1\right)^{-1} \dot{a}^2 - 1\right\}$$

Thus

$$\int_{-\epsilon}^{\epsilon} dr\, L_\Sigma|_{\tau=const} = \tilde{N}(t;\epsilon)\left[\left(\frac{2M_0}{a} - 1\right)^{-1} \dot{a}^2 - 1\right].$$

In the limit $\epsilon \to 0^+$ take $N(t)$ such that $\tilde{N}(t;\epsilon)$ is finite $\tilde{N}(t;\epsilon) \to \tilde{N}_0(t)$:

(3)

$$\lim_{\epsilon \to 0^+} \int_{-\epsilon}^{\epsilon} dr\, L_\Sigma|_{\tau=const} = \tilde{N}_0(t)\left[\left(\frac{2M_0}{a} - 1\right)^{-1} \dot{a}^2 - 1\right].$$

Return now to the full action

$$S_\Sigma = \int dt \left[\lim_{\epsilon \to 0^+}\left(\int_{-\infty}^{-\epsilon} + \int_{\epsilon}^{\infty} dr\, L_\Sigma\right) + \tilde{N}_0(t)\left[\left(\frac{2M_0}{a} - 1\right)^{-1} \dot{a}^2 - 1\right]\right]$$
$$= \int dt \left[\oint_{-\infty}^{\infty} dr\, L_\Sigma + \tilde{N}_0(t)\left\{\left(\frac{2M_0}{a} - 1\right)^{-1} \dot{a}^2 - 1\right\}\right]$$

Since no matter has been placed at the throat, $\mathcal{L}_\Sigma$ is continuous at the throat. Thus, the result of these manipulations is simply to add the constraint

$$C(M_0; a, \dot{a}) = \left(\frac{2M_0}{a} - 1\right)^{-1} \dot{a}^2 - 1 = 0,$$

Where $\{a, \dot{a}\}$ parametrize the throat trajectory and M_0 is the ADM mass of the Novikov gauge imposed int he neighbourhood of that trajectory. With the action,

(4) $\qquad S_\Sigma = \int dt \left[\int_{-\infty}^{\infty} dr\, \mathcal{L}_\Sigma + \tilde{N}_0 C(M_0; a, \dot{a}) \right],$

The analysis proceeds in the same manner outlined by Kuchař, there now being a constraint $\tilde{N}_0 C$ to carry along. With embedding variables introduced the full theory is then

(5) $\qquad S[m, p; a, \dot{a}; \tilde{N}_0] = \int dt \left(p\dot{m} - \mathfrak{N}(t)m + \tilde{N}_0 C(M_0; a, \dot{a}) \right).$

$\tilde{N}_0$ is to be varied, thus $C(M_0; a, \dot{a}) = 0$ is a constraint, while $\mathfrak{N}(t)$ is a prescribed function of t. Choose $\mathfrak{N}(t) = 1$. Note that the constraint $M'(t, r) = 0, M = M(t)$, and $\ddot{M}(t) = \frac{\partial h}{\partial p} = 0$, leads to $M = const. = M_0$.

The action thus reduces to

$$S[a, \dot{a}] = -\int dt\, m(a, \dot{a}) \;,\quad m(a, \dot{a}) = \frac{1}{2}(a\dot{a}^2 + a),$$

Where $m(a, \dot{a})$ is understood as an energy functional:

(6) $\qquad E(a, \dot{a}) = \frac{1}{2}(a\dot{a}^2 + a),$

From the $E(a, \dot{a})$ energy functional on throat geodesic one can define a Hamiltonian up to canonical transformations:

$$E(a, \dot{a}) = H\big(a, p_a(a, \dot{a})\big) = \dot{a} p_a - L(a, \dot{a}) = \dot{a}^2 \frac{\partial}{\partial \dot{a}}\left(\frac{L(a, \dot{a})}{\dot{a}} \right)$$

where $\left(p_a = \frac{\partial L}{\partial \dot{a}} \right)$ and:

$$L(a, \dot{a}) = \dot{a}\left\{ \int_c^{\dot{a}} dx \left(\frac{\frac{1}{2}(ax^2 + a)}{x^2} \right) + f(a) \right\}$$

$$p_a(a, \dot{a}) = \int_c^{\dot{a}} dx \left(\frac{\frac{1}{2}(ax^2 + a)}{x^2} \right) + \frac{\frac{1}{2}(a\dot{a}^2 + a)}{\dot{a}} + f(a)$$

$$= \frac{1}{2}a\left[x - \frac{1}{x} \right]_c^{\dot{a}} + \frac{1}{2}a\left[\dot{a} + \frac{1}{\dot{a}} \right] + f(a)$$

$$= a\dot{a} + \left[\frac{1}{2}a\left(c - \frac{1}{c} \right) + f(a) \right] \text{ (last term 0 for appropriate } f(a) \text{)}$$

$$= a\dot{a}$$

The classical Hamiltonian for this theory is then:

(7) $$\boxed{H = \frac{1}{2a} P_a^2 + \frac{1}{2} a}$$

As mentioned, this result can then be transformed to agree with the time slicing results of Kraus+Wilczek. We switch to the latter formalism to eventually accommodate a shell (or any spherically symmetric matter) into the vacuum solution.

10.17 Shell minisuperspace quantization prospects

We will restrict consideration to the proper-time Hamiltonian. When M− = 0, the shell encloses a flat interior with trivial topology, and the Hamiltonian takes the form corresponding to a relativistic particles in a Coulomb potential

$$H(\hat{r}, P) = m \cosh \frac{P}{m} - \frac{m^2}{2\hat{r}}, \qquad \text{(Eqn. 10-130)}$$

discussed by Hajcek [85,86]. One can adopt a Schrodinger representation corresponding to configuration-space variable ^r , R+ and Hilbert space

$$\mathcal{H} = L_2(\mathrm{R}_+, r^\alpha d\hat{r}).$$

(Eqn. 10-131)

With the factor ordering

$$\widehat{\cosh \frac{P}{m}} = \lim_{N \to \infty} \hat{r}^{\frac{\alpha}{2}} \sum_{n=0}^{\infty} \frac{(-1)^n}{(2n)!} \Delta^n \hat{r}^{\frac{\alpha}{2}},$$

(Eqn. 10-132)

where ^H is self-adjoint with domain [90]:

$$D(\hat{H}) = \left\{ f^{(2n)}(0) = 0,\ f^{(n)} \in L_2,\ for all\, n \right\}.$$

(Eqn. 10-133)

For m < 1.9 m_{Planck} H is bounded from below, and its spectrum, like that of the nonrelativistic Coulomb problem, has discrete and continuous parts.

With specified M− > 0, the shell Hamiltonian we expect a spectrum that is partly discrete and bounded below for small m. There is no reason to expect that for large m the additional term will allow the spectrum to have a lower bound. In Book 4 [3] we examine a self-adjoint extension of H with factor ordering corresponding to:

$$\hat{H} = m\cosh\widehat{\frac{P}{m}} - \frac{m^2}{2\hat{r}} + 2M_- - mM_-\hat{r}^{-\frac{1}{2}}\exp(\widehat{-P/m})\hat{r}^{-\frac{1}{2}}$$

(Eqn. 10-134)

When M− is dynamical, the parameter M− in ^H is replaced by an operator ^M−. One can effectively separate variables by first considering the eigenvalue equation for ^ M−. For each eigenspace of ^M−, the shell Hamiltonian then has the form with c-number M−. Again by a choice of factor ordering (or of time) one can obtain either a continuous or discrete spectrum for ^M . That is, as in the vacuum case, if one chooses a representation (^r; M−), with ^M− = M; the spectrum of ^M− is continuous on the domain $L_2(R+;dM-)$; and by choosing variables canonically related to M−; P_M-, we can recover the form

$$\hat{M}_- = \frac{1}{2}\left(\frac{\hat{P}^2_\alpha}{\alpha} + \alpha\right).$$

(Eqn. 10-135)

Note, however, that as presented, this canonical transformation does not have the physical interpretation it had in the vacuum case. One cannot immediately relate the momentum to the radius a of the throat and its conjugate momentum for a proper-time parametrization of the throat.

As noted earlier, the Hamiltonian or a spherically symmetric RP^3-geon geometry with a dust shell is formally identical to that of a wormhole geometry with shell. In the case of a geon, however, the interior mass is not the mass of an asymptotic region, and one does not have the option of treating it as a parameter characterizing the asymptotics rather than a dynamical variable. This is essentially a distinction of interpretation, because one is free to restrict consideration to an eigensubspace of the interior mass regardless.

10.18 Exercises

(Ex. 10.1) Rederive Eqn. 10-3 using a different method than that shown.
(Ex. 10.2) Derive Eqn. 10-9.
(Ex. 10.3) Derive Eqn. 10-10.
(Ex. 10.4) Derive Eqn.s 10-11 and 10-12.
(Ex. 10.5) Derive Eqn.s 10-13 and 10-14.
(Ex. 10.6) Derive Eqn. 10-16.

(Ex. 10.7) Derive Eqn. 10-18.
(Ex. 10.8) Derive Eqn. 10-19.
(Ex. 10.9) Derive Eqn. 10-22.
(Ex. 10.10) Derive Eqn.s 10-35 and 10-36.
(Ex. 10.11) Derive Eqn. 10-42.
(Ex. 10.12) Derive Eqn. 10-43.
(Ex. 10.13) Derive Eqn. 10-52.
(Ex. 10.14) Derive Eqn. 10-56.
(Ex. 10.15) Derive Eqn. 10-58.
(Ex. 10.16) Derive Eqn. 10-71.
(Ex. 10.17) Derive Eqn. 10-111.
(Ex. 10.18) Derive Eqn. 10-117.

Chapter 11. Generalized Geometries

11.1 Geometric algebra – graded algebras

Thus far we've seen generalizations of the derivative multiple times, in a variety of ways, and more such generalization will occur in Book 4 [3] on QM where derivatives generalize to become operators in a Hilbert space. In Book 7 [5] we make use of Cayley algebras, that are mostly non-associative, so further generalization beyond that indicated here will occur both in terms of a gauge covariant derivative and multiplicative property of non-associativity. Specifically, we now consider a different kind of generalization to the fundamental notion of product but that describes algebras that are still associative, known as the Clifford algebras [91].

Consider the standard "dot product" on vectors from vector calculus, this is an operation that takes two grade-1 elements (vectors):

$$u \cdot v = \sigma \cong \frac{1}{2}(uv + vu),$$

where σ is a scalar, and the operation is symmetric (and is written in expressly symmetrized form). This suggests an operation, the "wedge product", that is inherently anti-symmetric, and that rather than contract on abstract indices, "keeps both", arriving at a grade-2 element (a bivector), that is fundamentally anti-symmetric:

$$u \wedge v = \tau_{ab} \cong \frac{1}{2}(uv - vu).$$

Together the dot and wedge product give a new type product:

$$uv = \frac{1}{2}(uv + vu) + \frac{1}{2}(uv - vu) = u \cdot v + u \wedge v,$$

where now "uv" is the new type of product – the geometric product.

Clifford algebras

From the simple associative extension to 'geometric product" we have the basis for describing the Clifford algebras. Let's derive the first few algebras that result from $uv = u \cdot v + u \wedge v$. The different Clifford algebras are denoted C_k, where k is the number of generators e_k acting as an algebra over $\mathbb{R}$. For convenience let $e_0 = 1$. We also choose "1" as a representative element of $\mathbb{R}$ in what follows. For products on generators, the Clifford algebras are constrained such that:

$$(e_k)^2 = -1 \;\; and \;\; e_i e_j = -e_j e_i \;\; when\, i \neq j.$$

$\boldsymbol{C_0} = \mathbb{R}$

For C_0 there are no generators and the u and v values that we place in $u \cdot v + u \wedge v$ draw from the representative set $\{1\}$, thus $u \cdot v + u \wedge v \rightarrow \{1\}$ (a 1D algebra). Thus, $C_0 = \mathbb{R}$.

$\boldsymbol{C_1} = \mathbb{C}$

For C_1 there is one generator, where the u and v values that we place in $u \cdot v + u \wedge v$ draw from the representative set $\{1, e_1\}$, thus $u \cdot v + u \wedge v \rightarrow \{1, e_1\}$ (a 2D algebra). If we rename $e_1 = i$ then we formally arrive at the standard complex numbers. Thus, $C_1 = \mathbb{C}$.

$\boldsymbol{C_2} = \mathbb{Q}$ ***(the Quaternions)***

For C_2 there are two generators, where the u and v values that we place in $u \cdot v + u \wedge v$ draw from the representative set $\{1, e_1, e_2\}$, thus $u \cdot v + u \wedge v \rightarrow \{1, e_1, e_2, e_1e_2 \equiv "e_3"\}$ (a 4D algebra). If we rename $e_1 = i$, $e_2 = j$, and $e_3 = k$ we see that we generate an algebra from the basis set $\{1, i, j, k\}$, e.g., the quaternions.

Def. Grade of algebraic Term

Notice how the Clifford algebra involves elements that build from products of the generators, where no generator factors group as the 0-

grade terms, while terms with one generator factor are grade-1, etc. Thus, for $\boldsymbol{C_2}$ we have one term at grade-0 ("1"), two terms at grade-1 (e_1 and e_2), and one term at grade-2 (e_1e_2). Equations involving graded algebras naturally separate into equations respective to each grade of the algebra. In this way geometric algebra representations can package a lot of information (multiple relations) that can be useful at times (examples to be shown later). The highest grade element is known as a pseudoscalar., the next highest grade referred to as the pseudovectors or axial vectors.

$\boldsymbol{C_3} = \mathbb{Q}\oplus\mathbb{Q}$ ***(the Bi-Quaternions)***
For C_3 there are three generators, where the u and v values that we place in $u \cdot v + u \wedge v$ draw from the representative set $\{1, e_1, e_2, e_3\}$, thus $u \cdot v + u \wedge v \rightarrow \{1, e_1, e_2, e_3, e_1e_2, e_2e_3, e_3e_1, e_1e_2e_3\}$ (an 8D algebra). If we rename $e_1 = i$, $e_2 = j$, and $e_3 = k$ we see that we generate an algebra from the basis set $\{1, i, j, k\}$, e.g., the quaternions. For $\boldsymbol{C_3}$ we have one term at grade-0 ("1"), three terms at grade-1 (e_1, e_2, and e_3), three terms at grade-2 (e_1e_2, e_2e_3, e_3e_1), and one term at grade-3 ($e_1e_2e_3$). Thus we can write:

$$\boldsymbol{C_3} = (\boldsymbol{scalar}) + (\boldsymbol{vector}) + (\boldsymbol{bivector}) + (\boldsymbol{trivector}).$$

A few important results are now evident:
(1) We see that the $(\boldsymbol{scalar}) + (\boldsymbol{bivector})$ parts make a quaternion:

$$\{1, e_1e_2, e_2e_3, e_1e_3\}$$

If we rename $e_1e_2 = i$, $e_2e_3 = j$, and $e_1e_3 = k$ we see that we generate an algebra from the basis set $\{1, i, j, k\}$, e.g., the quaternions. Quaternions describe rotations. Note that the part that we are referring to here is the even sub-algebra comprised of the even grades. For $\boldsymbol{C_3}$ the even subalgebra is denoted $\boldsymbol{C_3^+}$, which has dimension 4, and is isomorphic to the quaternions.

(2) We see that the $(\boldsymbol{vector})$ part gives us 3D translations to go with the quaternion 3D rotations. Thus the Clifford algebra combines Gibbs-like vector information (a "translational" algebra) with the quaternion rotational algebra in a unified mathematical formalism (still separable by grade).

If we continue this process to higher order Clifford algebras we get the following:

k	C_k	k	C_k	k	C_k
0	$\mathbb{R}$	8	$\mathbb{R}(\mathbf{16})$	16	$\mathbb{R}(\mathbf{256})$
1	$\mathbb{C}$	9	$\mathbb{C}(\mathbf{16})$		…..
2	$\mathbb{Q}$	10	$\mathbb{Q}(\mathbf{16})$		
3	$\mathbb{Q}\oplus\mathbb{Q}$	11	$\mathbb{Q}(\mathbf{16})\oplus\mathbb{Q}(\mathbf{16})$		
4	$\mathbb{Q}(\mathbf{2})$	12	$\mathbb{Q}(\mathbf{32})$		
5	$\mathbb{C}(\mathbf{4})$	13	$\mathbb{C}(\mathbf{64})$		
6	$\mathbb{R}(\mathbf{8})$	14	$\mathbb{R}(\mathbf{128})$		
7	$\mathbb{R}(\mathbf{8})\oplus\mathbb{R}(\mathbf{8})$	15	$\mathbb{R}(\mathbf{128})\oplus\mathbb{R}(\mathbf{128})$		

where $\mathbb{Q}(\mathbf{2})$ stands for 2x2 matrices with $\mathbb{Q}$ terms. An 8-fold periodicity (Bott-periodicity [92]) is evident when the columns are aligned as shown. This can be explained using Dirac operators, whereby minimal dimensional Clifford algebra representations give rise to Dirac operators which, in turn, generate stable homotopy groups of the orthogonal group.

Clifford algebras in Emanator Theory
Of particular importance for Emanator theory in Book 7 [5] of the Series is the occurrence of the bi-quaternion algebra and what immediately follows. In particular, can a sum on $\boldsymbol{C_4}$ be written as a sum on complex bi-quaternions? We see that complex bi-quaternions in the Emanator formalism allow for manifest Lorentz invariance [93-95], so not a trivial concern. What we are really asking is if, in some general sense, a matrix can be written as a (outer) product on vectors, or a sum of such. This is precisely what is addressed by the classic singular value decomposition theorem:

Def. singular value decomposition theorem
We can write a rank-k matrix as a sum of k terms (for rank k):

$$(matrix\ rank\ k) = \sum_{i=1}^{k} a_i u_i {v_i}^T.$$

If $k = 1$, then the sum has one term, and we have a matrix equal to an outer product on vectors. In general, however, we can say that a sum on matrices and be written as a sum over outer products, and that's the relations we must respect in the Emanator formalism that entails sums over such matrix operations.

From the above, we see that a sum on $\boldsymbol{C_4} = \mathbb{Q}(\mathbf{2})$ can be written as a sum on complex bi-quaternions:

$$\sum \mathbb{Q}(2) = \sum \mathbb{Q}(\mathbb{C}) \oplus \mathbb{Q}(\mathbb{C}).$$

Likewise, a sum on $\boldsymbol{C}_5 = \mathbb{C}(\boldsymbol{4})$ can be written as a sum on doubly complex bi-quaternions:

$$\sum \mathbb{C}(\boldsymbol{4}) = \sum \mathbb{Q}(\mathbb{C} \times \mathbb{C}) \oplus \mathbb{Q}(\mathbb{C} \times \mathbb{C}).$$

Emanator theory involves a foundation build from maximal information flow for unit-norm propagation in the Cayley algebras [96] (which are, generally, non-associative), consisting of $\mathbb{Q}(\mathbb{C}) \oplus \mathbb{Q}(\mathbb{C})$ and $\mathbb{Q}(\mathbb{C} \times \mathbb{C}) \oplus \mathbb{Q}(\mathbb{C} \times \mathbb{C})$ terms. The process involves sums on those algebras and in doing so, have a selection process to choose emanations that are overall associative, e.g., we find that the base trigintaduonion is an element of $\boldsymbol{C}_5$ as is the effective emanation (associative multiplicative) step.

Utility of Geometric Algebras

Geometric algebras, or Clifford algebras, permit a richer encapsulation on physical theory, where different field relations can be captured as different relations at different grades of the algebra but within the same algebra. This unification is elegant in many uses, and cases of the power of the method in solving problems have been shown, such as with the very difficult Kerr Black Hole solution [97]. It is not surprising the geometric algebra approach would be on a strong footing since it involves exterior products (and exterior derivatives) and as such encapsulates Cartan calculus, whose utility is well-established in Ch. 4. From the brief description relating to Emanator theory, we see it appears there as well.

11.2 Information Geometry, Neuromanifolds, and Entropy

11.2.1 Overview

This Chapter goes into theoretical detail to show modern arguments for the choice of relative entropy as difference measure on distributions. Also shown is a fundamentally derived variant of Expectation Maximization (EM), referred to as "em". Following the derivation of Amari [98-101], geometric representations are considered for two types of statistical information: families of probability distributions and families of Neural Nets. Emphasis is placed on the dually flat formulation of "information geometry" by Amari, but the development is such that other formulations might be considered as well.

Why establish a geometric formulation? Here are some motivations:
(1) to obtain a concise representation of the elements of families of probability distributions,
(2) if a dynamical description is eventually sought, such as in adaptive algorithms that "learn," or optimization, then an established kinematical representation can offer much needed clarity,
(3) if a variational formalism is provided, the spatial and "temporal" (inertial) kinematical structures can be clearly expressed, separate from the unknown dynamical structures.

Once the geometric formulation is established, we stand to gain substantially by systematic exploration of various algorithms, permitting the more efficient algorithms to be isolated. In part, the search for a more efficient algorithm is then simpler in that it may focus effort on the dynamical element of the algorithm, leaving the kinematics (geometry) to the underlying geometric formulation. Nonetheless, there are limitations to what can be done with a geometric representation. In particular, care must be taken to carry the "new orthodoxy" only as far as "natural" geometric interpretation allows—otherwise one begins to encode the information/algorithm in a (geometric) representation that is no more informative than any other.

11.2.2 Amari's Dually Flat Formulation [98-101]

The application of differential geometry methods to the study of statistical models traces back to C.R. Rao in 1945 [102]. Rao noted that families of probability distributions could be described by a manifold and that the Fisher information matrix might be taken as a metric on that manifold.

Probability Distributions:

$$p(x) \geq 0 \;\; \forall x \in \chi \text{ and } \Sigma_{\{x \in \chi\}}\, p(x) = 1.$$

Family of Probability Distributions:

$$S = \{p_\theta = p(x;\; \theta) \mid \theta = [\theta^1,\; \ldots,\; \theta^n] \in \Phi\}$$

θ parametrizes a n-dimensional statistical model.

Fisher Information Matrix:

$$G(\theta) = [g_{ij}(\theta)]; g_{ij}(\theta) = E_\theta[\partial_i l_\theta \partial_j l_\theta],$$
$$where\ \partial_i = \frac{\partial}{\partial\theta^i} and\ l_\theta(x) = \log p(x;\theta)$$

Sufficient Statistic:

Consider

$$p(x;\theta) = p(x|y;\theta)q(y;\theta).$$

Suppose $p(x;\theta) = p(x|y;\theta)\ q(y;\theta)$, with associated spaces $S \equiv \{p(\cdot,\theta)\}$ and $S_F \equiv \{p(\cdot;\theta)\}$, and y is related to x by the constraint $y = F(x)$. If $\forall x \in \chi$ $p(x|y;\theta)$ does not depend on θ, then F is a "sufficient statistic" w.r.t. S and there results $p(x;\theta) = p(x|y)\ q(y;\theta)$. Now, in order to estimate θ, it suffices to know y, and

$$\frac{\partial}{\partial\theta^i}\log p(x;\theta) = \frac{\partial}{\partial\theta^i}\log p(x;\theta) \Rightarrow g_{ij}\ same\ for\ both\ S\ and\ S_F$$

$G(\theta)$ is symmetric and positive semi-definite (positive definite with linear independence in $\partial_1 p,\ldots \partial_n p$). From this we can define inner product on the natural basis $[\theta^i]$ and obtain the "Fisher Metric." First revealed in an obscure paper by N.N. Chentsov (1972) [103] it was shown that the Fisher metric could be uniquely singled out based on invariance with respect to "sufficient statistics." *Also invariant w.r.t. sufficient statistics was a class of connections that Amari refers to as α-connections.*

"Invariance" w.r.t. sufficient statistics:

Here the invariance result of Chentsov is restated following the presentation of Amari. In essence, the Fisher metric and α-connections are uniquely characterized by invariance w.r.t. sufficient statistics on arbitrarily reduced probability spaces (reminiscent of Khinchin's reducibility axiom that entropy $H(p_1,p_2,\ldots p_n,0) = H(p_1,p_2,\ldots p_{n,})$ in establishing the uniqueness of Shannon's entropy).

The formulation is done for a discrete set and then extended to (continuum) probability densities:

Consider events: $X_n \equiv \{0, 1,\ldots,n\}$; probabilities: $P_n \equiv P(X_n)$; and pairs (g_n,∇_n) of metric and connection on $S \equiv \{P_n\}$. Consider $S_F \equiv \{P_m\}$ where $F: \chi_n \to \chi_m$ ($n \geq m$, and F is surjective). If F is sufficient w.r.t. S, then g_{ij} and Γ_{ijk} on S and S_F are the same. If sufficiency holds for all n, m, s and F, then g_n is the Fisher metric on P_n and ∇_n is an α-connection on P_n.

Recall Dual Connections:

$$X\langle Y,Z\rangle = \langle\nabla_X Y,Z\rangle + \langle Y, \nabla^*_X Z\rangle.$$

$$\partial_k g_{ij} = \Gamma_{kij} + \Gamma^*_{kji}$$

$$\Gamma^{(\alpha)}_{ijk} = E_\theta\left[(\partial_i\partial_j l_\theta + \frac{1-\alpha}{2}\partial_i l_\theta \partial_j l_\theta)(\partial_k l_\theta)\right], \qquad and\ l_\theta = \log p(x;\theta)$$

Note : The α-connections and (-α) connection are dual w.r.t. the Fisher metric.

The α-connections are defined as follows:
Consider the Exponential Family of distributions in this context:

$$p(x;\theta) = exp\left[c(x) + \sum\theta^i F_i(x) - \psi(\theta)\right]$$
$$\partial_i l = F_i - \partial_j\psi$$
$$\partial_i\partial_j l = -\partial_i\partial_j\psi \longrightarrow no\ x\ dependence$$

for α=1 we then have $\Gamma^{(1)}_{ijk} = -\partial_i\partial_j\psi E_\theta[\partial_k l_\theta] = 0$. So, $[\theta^i]$ is a $\nabla^{(1)}$–affine coordinate system, S is $\nabla^{(1)}$–flat. Amari refers to $\nabla^{(1)}$ as the exponential connection, or "e-connection".

Consider the Mixture Family of probability distributions in this context:

$$p(x;\theta) = \sum\theta^i p_i + \left(1 - \Sigma\theta^i\right)p_0(x)$$
$$\partial_i l = \frac{p_i - p_0}{p};\ \partial_i\partial_j l = -\frac{(p_i - p_0)(p_j - p_0)}{p^2}$$
$$\Rightarrow \partial_i\partial_j l + \partial_i l\partial_j l = 0 \Rightarrow \Gamma^{(-1)}_{ijk} = 0$$

So, $[\theta^i]$ is a $\nabla^{(-1)}$–affine coordinate system, S is $\nabla^{(-1)}$–flat. Amari refers to $\nabla^{(-1)}$ as the "mixture connection" or "m connection".

Divergence: A triplet (g,∇,∇^*) can be defined locally from a "divergence", where a divergence D is characterized by:

$$D(\bullet \| \bullet) : S \times S \rightarrow \Re, \text{ where } \forall p, \forall q \in S \times S\text{: } D(p\|q) \geq 0, \text{ and } D(p\|q)=0 \text{ iff } p=q.$$

Dually flat spaces: If ∇ and ∇^* are both symmetric then ∇ flat $\leftrightarrow$ ∇^*-flat. Since the α connections are symmetric, S is α-flat $\leftrightarrow$ S is ($-\alpha$)-flat, and, in particular, if $\nabla^{(1)}$ is flat then so is $\nabla^{(-1)}$.

If (S, g, ∇, ∇^*) is dually flat, then there exists ∇-affine coordinates $[\theta_i]$ and ∇^*-affine coordinates $[\eta_j]$. The inner product between the elements of the tangent spaces is a constant on S, from which it is possible to get the relations:

$$\frac{\partial \eta_i}{\partial \theta^k} = g_{ik} \quad and \quad \frac{\partial \theta^i}{\partial \eta_j} = g^{ij}$$

Introduce Potentials: Suppose $\partial_i \psi = \eta_i$, then $\partial_i \partial_j \psi = g_{ij}$, and since g_{ij}, is a metric tensor the partial derivative must describe a positive definite matrix, which in turn implies that ψ is strictly convex. Likewise for $\partial^i \varphi = \theta^i$. In terms of the potentials ψ and φ thus introduced, it is easy to show that the two coordinate systems $\{\theta^i, \eta_j\}$ can be related by Legendre transformation:

$$\varphi = \theta^i \eta_i - \psi$$

Convexity in potentials that define a Legendre transformation leads to some interesting constructions. If we take our dually flat space (S, g, ∇,∇^*) with coordinate systems $\{[\theta^i], [\eta_i]\}$ and their potentials $\{\psi, \varphi\}$, then it can be shown that

$$\varphi(q) = \lim_{p \varepsilon S} \{\theta^i(p)\eta_i(q) - \Psi(p)\}$$

and

$$\Psi(q) = \lim_{q \varepsilon S} \{\theta^i(p)\eta_i(q) - \varphi(p)\}$$

Since we have:

$$\Psi(q) - \lim_{q \varepsilon S} \{\theta^i(p)\eta_i(q) - \varphi(p)\} = 0$$

we have:

$$\Psi(q) - \{\theta^i(p)\eta_i(q) - \varphi(p)\} > 0$$

call this D(p||q):

$$D(p//q) = \psi(p) + \varphi(q) - \theta^i(p)\eta_i(q),$$

and it is easily shown that:

$$D(p//q) \geq 0, and\ D(p//p) = 0 \Leftrightarrow p = q$$

Thus, the Exponential and Mixture Families induce a dually-flat space which leads to the fundamental difference measure on distributions being a divergence. The "divergence" form of "distance" function is also indicated in the Link formalism [101]. Although the divergence family is singled out as fundamental at this point, the selection of a specific divergence (like Euclidean distance in the case of metrics) is yet to be determined. The "simplest" divergence, the Kullback-Leibler (relative entropy) divergence, is selected when maximizing log likelihood during learning, and that this is, fundamentally, because of the shortest path, or projection theorem. ***Implicit in this result is that the proper way to measure the difference between distributions is not the Euclidean distance between them but the Kullback-Leibler Divergence between them.***

11.2.2.1 Generalization of Pythagorean Theorem

Let p, q and r be three points in S. Let γ_1 be the ∇- geodesic connecting p and q, and let γ_2 be the ∇^*-geodesic connecting q and r. If at the intersection q, the curves γ_1, and γ_2 are orthogonal (w.r.t. g), then

$$D(p||r) = D(p||q) + D(q||r)$$

To show this, first consider the γ_1 geodesic:

$$\theta_t^i = t\theta^i(p) + (1-t)\theta^i(q)$$
$$\frac{d}{dt}\theta_t^i\partial_i = [\theta^i(p) - \theta^i(q)] * \partial_i$$

$$\eta_{ti} = t\eta_i(q) + (1-t)\eta_i(r)$$
$$\frac{d}{dt}\eta_{ti}\partial^i = [\eta_i(q) - \eta_i(r)]\partial^i$$

So, prove relation with:

$$D(p||q) + D(q||r) - D(p||r)$$

$$= \Psi(\rho) + \varphi(q) + \theta^i(p)[\eta_i(r) - \eta_i(q)] - \theta^i(q)\eta_i(r)$$
$$= [\theta^i(p) - \theta^i(q)][\eta_i(r) - \eta_i(q)] = < \left(\frac{d\gamma_1(t)}{dt}\right)_i \left(\frac{d\gamma_2(t)}{dt}\right)^j > = 0$$

11.2.2.2 Projection Theorem and relation between Divergence and Link Formalism

Suppose M a, submanifold of S, is ∇^*-autoparallel, then

> D(p||q) = $\min_{\{r \in M\}}$ D(p||r) when the ∇-geodesic connecting p and q is orthogonal to M at q.

Relation to Link formalism, start with:

$$D(\mathrm{p}||\mathrm{q}) = \psi(p) + \varphi(q) - \theta^i(p)\eta_i(q)$$

Use Legendre transformation:

$$\varphi(p) = \theta^i(q)\eta_i(q) - \psi(q)$$

to get:

$$D(\mathrm{p}||\mathrm{q}) = \psi(\rho) - \psi(q) + \left(\theta^i(q) - \theta^i(p)\right)\eta_i(q), \text{ where } \eta_i = \frac{d\psi}{d\theta^i}$$

Thus, with shift in notation $\{f = \frac{\partial F}{\partial \omega}, \omega\} \rightarrow \{g, \theta\}$ we get the Link formalism used in Ch. 9:

$$D(\omega||\omega') = F(\omega) - F(\omega') - (\omega - \omega')\left.\frac{\partial F}{\partial \omega}\right|_{\omega = \omega'},$$
$$where\ \frac{\partial D}{\partial \omega} = f(\omega) - f(\omega').$$

The $\{f, \omega\} \leftrightarrow \{g, \theta\}$ duality corresponds precisely to the dually flat connection construction with potentials φ and ψ that are related via Legendre Transformation to the "coordinates" θ and ω. The link function $f(\omega) = \omega$, is associated with square loss and the gradient descent (GD) learning rule. The link function $f(\omega) = \ln(\omega)$, is associated with divergence loss and the exponentiated gradient descent (EG) learning rule. These will be explored in [101] in the link formalism context, along with an interpolating learning rule between GD and EG given by the link function $f(\omega) = \sinh^{-1}(\omega)$.

11.2.3 Neuromanifolds [98-101]

The information geometry methods described for families of probability distributions (parametrized by $\{\theta^i\}$) can just as easily be applied to neural networks; where now the parameters are the connection weights. The statistical arguments also carry over and are applicable to stochastic neural nets, i.e., neural networks with noisy input or non-deterministic behavior. As Amari states, "even when a network is deterministic, it is sometimes effective to train it as if it were a stochastic network." [99]. With a stochastic network we then have probability distribution $p(x;\theta)$ and/or conditional probability distribution $p(y \mid x;\theta)$.

Complications associated with repeated observation

For a single observation we have the distribution $p(x;\theta)$, an element of the family (space) S. For repeated independent observations there is the joint conditional distribution where θ generally changes from one observation to the next. The joint distribution, thus, is an element of a larger family (space)

$$S_T^* = S_1 \times S_2 \times \ldots \times S_T.$$

Multiple observations thus lead to a joint distribution whose manifold dimension (under direct product) increases with the number of observation, while the underlying parameterization is fixed—being the manifold dimension of the family of distributions considered for the individual distribution. This generally leads to a Curved Exponential family description on the joint distribution manifold (where the individual distribution was in an exponential family) ,i.e., a submanifold of the joint distribution manifold is specified. It is possible to describe repeated observations within the framework of the manifold S without referring to the product space S_T^*, but this holds only in the i.i.d. case. In general

$$\theta_T^* = (\theta_1, \theta_2, \ldots, \theta_T),$$

and

$$\theta_T^* = \theta(x_t, u_t),\ t=1,\ldots,T.,$$

where x_t's are given and the u_t's are the only free parameters (the parameters of the underlying neural network). Suppose all x_t are subject to $p(x;\theta)$, $\theta = \theta_1 = \ldots = \theta_T$, where $p(x;\theta)$ is the exponential family $p(x;\theta) = \exp(\theta \cdot x - \psi)$. The joint distribution is

$$p(x_1,\ldots x_T;\theta) = \exp\{\Sigma x_t \cdot \theta - T\psi\}$$
$$p(\bar{x};\theta) = \exp\{\bar{x} \cdot \theta - \psi\}$$

The maximum likelihood estimator (m.l.e.) $\hat{\theta}$ from the observed data $x_1,\ldots,x_T$ is given by maximizing $p(\bar{x};\theta)$:

$$\bar{x} = \frac{\partial}{\partial \theta}\psi(\theta)|_{\theta=\hat{\theta}} = \hat{\eta}$$

The observed data are then represented by the m.l.e $\hat{\eta}$ in S in the η-coordinate system.

Suppose that all x_t are subject to $p(x;\theta)$ where $p(x;\theta)$ is a curved exponential family. Again the observed data $x_1,\ldots, x_t$ are represented by $\bar{x} = \hat{\eta}$, now, however, we have that $\hat{\eta}$ does not necessarily belong to M. The m.l.e. $\hat{u}$, or corresponding distribution $\theta(\hat{u}) \in M$, is given by maximizing the log likelihood $x \cdot \theta(u) - \psi(\theta(u))$ w.r.t. u. Maximizing the log likelihood is equivalent to simply m-projecting $\hat{\theta}$ to M in the Amari differential geometry formalism, and this is equivalent to minimizing the Kullback-Leibler (KL) divergence $K(\hat{\theta} || \theta(\hat{u}))$ from $\hat{\theta}$ to $\theta(\hat{u}) \in M$. Thus, the KL Divergence is used in the Amari neuromanifold learning process.

EM Algorithm

The following EM/em discussion closely follows Amari [98-101].

Consider $M=\{p(r;\theta(u))\}$ a curved exponential family from which data is regenerated. Data r is observed. Consider $r=r(S_v,S_h)$, a sufficient statistic that includes hidden part S_h. Need to estimate unknown part of r information. Can do this based on observed S_v, and some candidate distribution u′, via the conditional expectation:

E: $$\hat{r}(u') = E[r|s_v;\theta(u')]$$

Now estimate $\log p(\hat{r};\theta(u))$ by conditional expectation also:

E: $$LLH(s_v;\theta(u')) = E[\log p(\hat{r};\theta(u)|\,s_v;\theta(u')] = \theta(u) * \hat{r}(u') - \psi(\theta(u))$$

Now search for better candidate u by maximizing LLH, or equivalently, by minimizing the KL divergence $D(\hat{r}(u')||\theta(u))$ from the guessed data point $\hat{\eta} = \hat{r}(u')$ to μ w.r.t. u.

The algorithm:

Step 0: Initialization step. Guess u_0, the initial guessed distribution $P_0 \in M$ is given by $\theta(u_0)$. Then repeat the following:
Step 1: E-step. Based on candidate probability distribution $P_i \in M$, calculate the conditional expectation of r. This gives the ith candidate for the observed point $Q_i \in D$, whose η-coordinate are $\eta_{(i)}$.
Step2: M-step. Calculate the mle $u_{(i+1)}$ from $Q_i \in D$.

em Algorithm
Search for the pair of points $P \in M$, $Q \in D$ that minimizes the divergence between D and M, that is:

$$D(\widehat{Q}||\widehat{P}) = \lim_{P \in M, Q \in D} D(Q||P)$$

(i) point $\widehat{P} \in M$ that minimizes $D(Q||P)$ is given by the m-projection of Q to M (i.e., by the m-geodesic connecting $\widehat{P}$ and Q that is orthogonal to M at $\widehat{P}$).
(ii) point $\widehat{Q} \in D$ that minimizes $D(Q||P)$ is given by the e-projection of P to D (i.e., the e-geodesic connecting P and $\widehat{Q}$ that is orthogonal to D at $\widehat{Q}$).

The algorithm:

Step 0: Initialization Step. Guess $\hat{u}_0 \Rightarrow \widehat{P}_0 \in M$. Then repeat:
Step 1: e-step. e-project $\widehat{P}_i$ to D, gives $\widehat{Q}_i$.
Step 2: m-step. m-project $\widehat{Q}_i$ to M, gives $\widehat{P}_{i+1}$.

Relation between EM and em:

(1) M-step and m-step are the same

(2) E-step and e-step differ depending on the conditional expectation of r_h given r_v:

Note: Divergence need not be symmetric or satisfy the triangle inequality, but given "orthogonal" learning steps, a Divergence satisfies a generalized Pythagorean theorem, leading to a the same critical Chapman-Kolmogorov-like propagation rule whether learning step assumes a Euclidean notion of distance (GD) or a KL divergence notion of distance (EG).

Which is Better?
EM is more natural from statistical point of view, but the representation via a neural network does not correspond exactly to statistical inference, therefore em serves better as the approximator ideally suited to the neural network representation of the input-output relation. The two algorithms are asymptotically equivalent when T is large (in the framework of exponential families). If framework extended from exponential family to function space they are exactly equivalent. It may be that a hybrid of GD and EG, or of EM and em type learning, is best, and this will be explored further in [101] in an explicit loss bounds analysis on the learning process.

Amari's Dually Flat formulation is a more natural structure for representing exponential families and neural nets than it might appear at first sight. In particular, the Kullback-Leibler measure is naturally represented in terms of the dual coordinates and their relations via Legendre transformation (with the introduction of appropriate potentials). The potentials, in turn, provide a natural formulation that is precisely that exhibited when using the "link formalism." Similarly, the duality between the coordinates (via Legendre transformation) is precisely that exhibited in the neural net learning algorithms.

While the e-projection and m-projection are natural geometric notions and certainly a strength of Amari's program, they also represent a weakness. This weakness is perhaps most clearly illustrated in the learning algorithm version where the projection is determined and then the actual update is interpolated (between old data and estimator and new, projected data and estimator). In this instance we must fall back on the gradient descent algorithm, or some such arguments, in order to have stable updates. One approach might be to use the SA link algorithm since it results in a formalism that interpolates between GD and EG learning according to the weight magnitude. Further details on this approach are given in [101].

11.3 Exercises

(Ex. 11.1) Re-derive the α-connections for $\alpha=1$ and $\alpha=-1$ and show that they are "dually flat".

(Ex. 11.2) Introduce potentials on a dually flat space as described in Sec. 11.2.2, then show how their Legendre transformation induces a natural difference measure that is a divergence.

(Ex. 11.3) Prove the generalization of the Pythagorean theorem for dually flat spaces.

(Ex. 11.4) Show that em learning singles out the Kullback-Leibler Divergence as fundamental in dually flat space (this is akin to a Riemannian manifold giving rise to a fundamental locally flat spacetime reference and local Euclidean distance measure).

Chapter 12 Series Outlook

General relativity (GR) stands apart from the other force fields. All the other force fields are part of an adjoint representation of the standard model vis-à-vis the stability subgroup $U(1)xSU(2)_LxSU(3)$. The form of which is derivable from the chiral T one-sided products described in Book 7 [5]. The standard model is uniquely obtained in this process, and with no mention of GR. Keep in mind, however, that the adjoint representation has operation on some space (hyperspinorial in case of simple octonion right-products, for example). The 'force' due to gravity is that due to manifold curvature, where the manifold construct is possibly emergent on the space of operation. Thus, the origin of the GR force is entirely different, and it will not allow quantization like the other forces, nor will its singular solutions be resolvable via quantum physics alone, as with EM in Books 4&5, but will also need thermal physics (as will be described in Book 6).

The existence of singular GR solutions, outside of specially symmetric cases (the classic Black hole solutions), wasn't firmly established until the Penrose singularity theorem [42] (awarded Nobel prize in Physics for this in 2020). Some of this material was covered to show how the mathematical formalism shifts to differential topology methods to describe the singularities, with examples referencing the Hawking and Ellis classic [39] and using Penrose diagrams. This, in turn, came in handy when describing the classic FRW cosmologies with radiation and matter dominated phases (using notes from Peebles [103], Peebles won the Nobel in Physics in 2019).

The GR development would be remiss if it didn't briefly delve into cosmological models, the classic FRW cosmologies in particular. With the GR tools developed, cosmological results are examined, starting with the entry of the cosmological constant into the formalism (a candidate for Dark energy). Various observational data on galaxy rotations and universe simulations of galaxy cluster formation both indicate the existence of Dark matter. This, then, means we have new matter, non-interacting except gravitationally, and this is actually consistent with the latest observational data on the muon g-2 value [104], where the discrepancy between theory and experiment has grown to 4.2 standard deviations, where an extension in the Standard Model appears to be in the works. This is convenient as Emanator theory (Book 7 [5]), predicts such an extension.

We can thus arrive at field equations for EM, GR, and Yang-Mills Gauge Fields (Strong and weak). We can obtain wave and vortex phenomena (as hinted in fluid dynamics). We show the classical instability for atomic matter (classical EM instability) and classical gravitational instability (leading to black hole formation with singularity). From Lagrangian formulations we can then arrive at a QFT formulation (Book 5). The QFT formulation completes the QM (Book 4) cure of "non-relativistic atomic instability" with the cure of the fully relativistic atomic description of the radiative-collapse instability. Introduction of QFT also leads to new instability or infinities, but these can be eliminated by renormalization for the EM and electroweak formulations, and the Yang-Mills strong formulation, but not the GR (gauge) formulation. The current theoretical formulation in modern physics has one glaring gap, therefore: a quantum theory of gravitation. Perhaps this is not a missing element, however, if geometry/GR is a derivative phenomenon, like the field of statistical mechanics and thermodynamics appeared as derivative phenomenon when the complexified quantum propagator gives rise to a real (quantum) partition function. The hint of a deeper emanator theory suggests emergent structures of geometry and thermodynamics are arrived at in the process of emanation, with the information emanated being that of the renormalizable quantum matter fields. In Book 7 [5] a precise mathematical meaning will be found for describing maximal information emanation.

Appendix

A. Differential Topology

A.1 The Manifold [105]

First recall R^n, the n-dimensional space of vector algebra, with points the n-tuples $(x_1, x_2, \ldots x_n)$ of real numbers. The concept of continuity in R^n is made precise by the study of its topology (where here topology is meant in the local or point-set sense, not the global or algebraic sense).

In the rigorous definition of manifold, to be described later, we find that manifolds have the property that they are "locally R^n ". So it is possible that local structures on R^n might carry over to local structures on a manifold. One such structure turns out to be the notion of continuity, and the study of topology provides this and more.

A quick way to demonstrate that a topology is defined on R^n , i.e. that R^n is a topological space, is to introduce the Euclidean distance: d(x,y). One can then define open sets according to unit radius neighborhoods, establish the Hausdorff (continuum) property, etc., and thereby obtain an induced topology from the definition and completeness of the open sets. Topology is more "primitive" then distance, however, non-"distance" measures such as d(p||q) and d(q||p) (the Kullback-Leibler measure) also induce a topology, and it is the same as the Euclidean d(p,q), as can be seen from their local ($p \approx q$) behavior.

Aside from familiarity with topological spaces, R^n being the primary example, an understanding of Manifolds requires an understanding of Mappings (Fig. A.1). A map from a space M to a space N, $f:M \rightarrow N$, is a rule that associates to an element x of M a *unique* element f(x) of N, x→f(x):

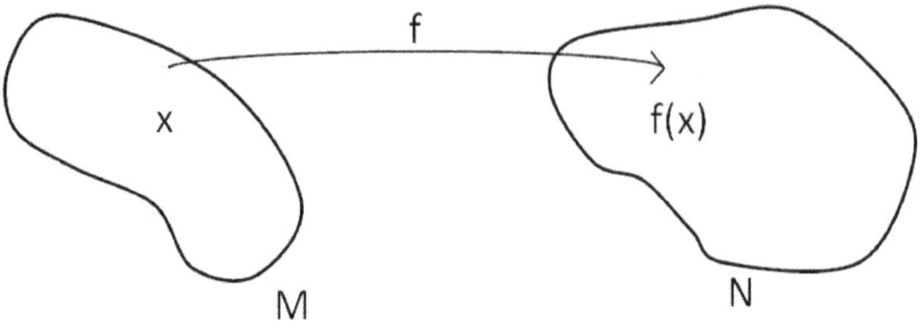

Fig. A.1 Map.

(a) A map is continuous at x in M if any open set of N containing f(x) contains the image of an open set of M containing X. (Presupposes M and N are topological spaces.) Continuity on all of M is obtained iff the inverse image of every open set of N is open in M.

(b) A map f:S→ R^n that is continuous is also described as class C^k if all partial derivatives of f(x) of order less than or equal to k exist.

(c) If f is a 1-1 map of an open set M of R^n to N of R^n , then we can define the Jacobian, $J=\partial(f_1,\ldots,f_n)/\partial(x_1,\ldots,x_n)$. If the Jacobian is nonzero at a point, the inverse function theorem assures that the map f is 1-1 and onto in some neighborhood of x.

Definition for Manifold: A set (of "points") M is defined to be a manifold if each point of M has an open neighborhood which has a continuous 1-1 map onto an open set of R^n for some n. ("M is locally like R^n"). Dimension of M is n, local topology is that of R^n .

Since manifolds are "locally R^n" many of the tools of real analysis, defined on R^n, can be used on manifolds also.

Coordinates: by definition, $P \in M \rightarrow (x_1(P), \ldots , x_n(P))$, with $\{x_1, \ldots , x_n\}$ the coordinates of P under the map.

Chart: The pair consisting of a (bijective) neighborhood and its map (Fig. A.2):

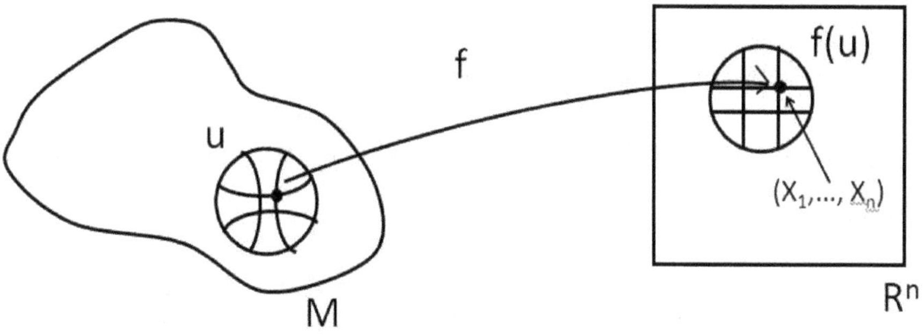

Fig. A.2 Chart.

Overlapping Charts (coordinate transformations), shown in Fig. A.3 below:

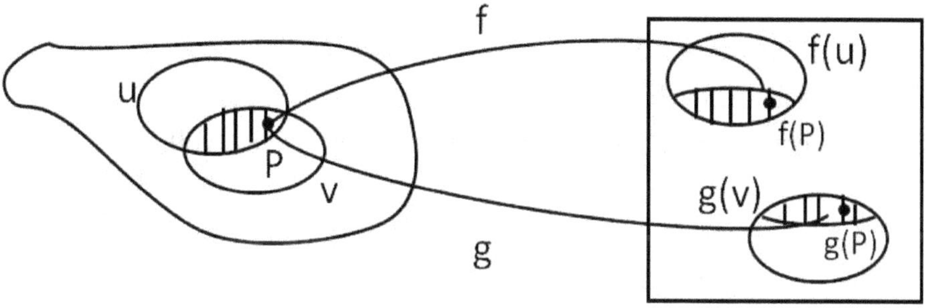

Fig. A.3 Overlapping Charts.

The coordinates are C^k related if partial derivatives of order k or less are continuous.

Atlas: every point in M is in at least one Chart.

C^k manifold: Manifold with Atlas whose every Chart is C^k related on overlaps.

($\Rightarrow$ A Differentiable Manifold if at least C^1)

Natural Structures on a differentiable manifold:

Curve: a differentiable mapping from an open set of R^1 into M (Fig. A.4):

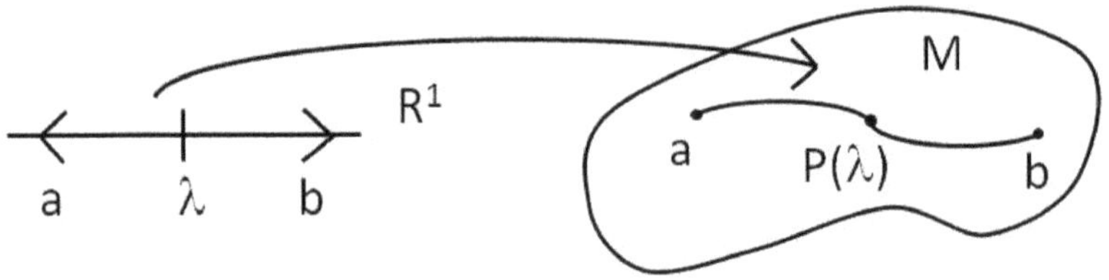

Fig. A.4 Curve. Note: The parameterization λ defines different curves, even if the image is the same.

Function: a function on M is a rule that assigns a real number to each point of M (Fig. A.5). If M maps differentiably to R^n, the function induces a function on R^n, which may be differentiable as well.

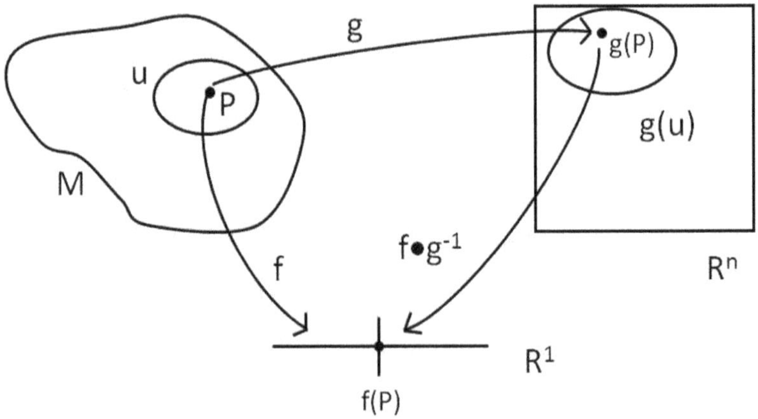

Fig. A.5 Function.

Vectors and Vector Fields: Consider curves and functions together. This allows one to obtain a differentiable function g (λ) which gives f at a point with parameter value λ : $g(\lambda)=f(x^i(\lambda))$. Now have

(i) The space of all tangent vectors at P and the space of all derivatives along curves at P are in 1-1 correspondence.
(ii) Vectors lie, not in M, but in the tangent space to M at P, called "T_P."
(iii) Vector field, a rule for defining a vector at each point of M.

Basis for Tangent Space T_P: Collection of n linearly independent vectors. Coordinate basis $\{x^i\}$ induces T_p basis $\{\partial/\partial x^i\}$.

Tangent bundle (fiber bundle): Consider manifold M combined with its tangent space T_P, call it TM (Fig. A.6).

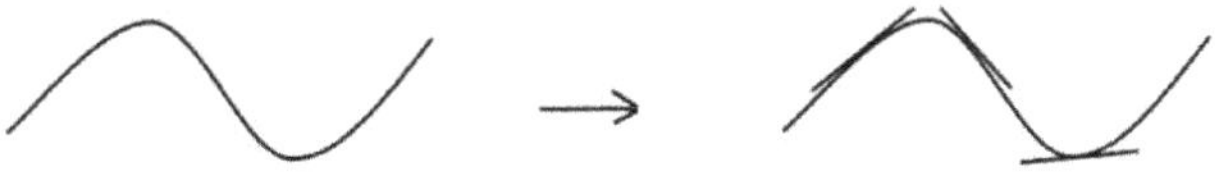

Cross Section

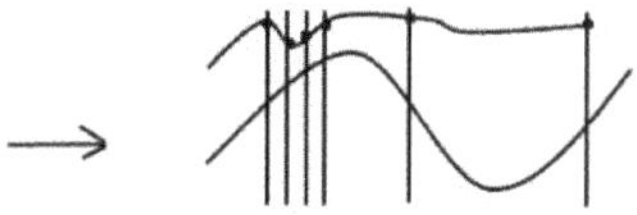

Fig. A.6 Tangent Bundle, where "cross -section" of TM (gives intuitive notion of "continuous" vector field). Note: even a simple manifold like S^2 has a non-trivial vector bundle

Integral Curves exist for vector fields. Due to uniqueness of solutions the curves never cross except where $v^i=0$. Aside from the $v^i=0$ "caustics," the curves can be manifold filling, and if so you have a congruence (Hamiltonian vector fields).

One-forms: linear real-valued functions of vectors. One-form fields can be described in terms of functions on TM (Fig. A.7):

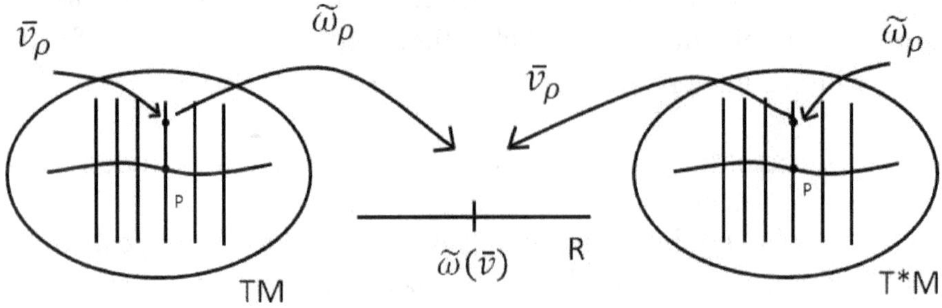

Fig. A.7 One-forms and cotangent bundle. There is a duality between one-forms and vectors, and the fiber bundle for one-forms, T^*M, is referred to as the cotangent bundle.

Gradient can be interpreted as one-form (Fig. A.8):

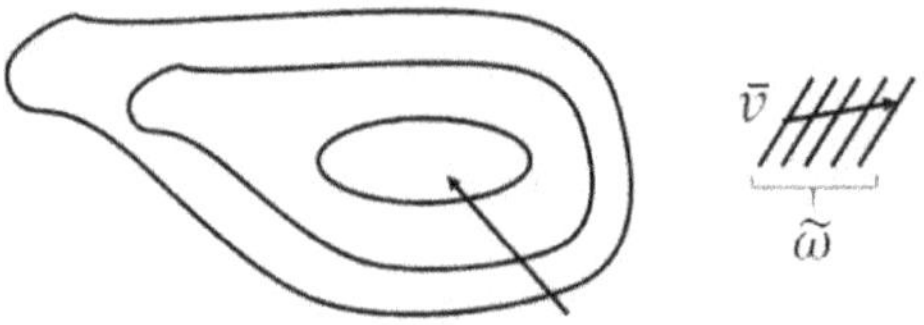

Fig. A.8 Gradient One-form.

The gradient enables a picture of one-form that is complementary to that of a vector:

Tensor and Tensor Field: A tensor at P is defined to be a linear function which takes as arguments N one-forms and N′ vectors and whose value is a real number.

(i) A scalar is a tensor, such as $v^i w_i$. The v^1 component is not a scalar.
(ii) Metric tensor field: A symmetric tensor field with inverse at every point of the manifold.
(iii) Metric tensors reducible to diag $(-1,\ldots,-1,1,\ldots,1) \rightarrow O(n-k,k)$ symmetric group.

A.2 Differential Geometry Overview – Natural Geometric Structures
The following short review is based on much lengthier analysis at [105].

Consider a tensor $T = T^{\beta}{}_{\alpha} e_{\beta} \otimes w^{\alpha}$ (where the "$T^{\beta}{}_{\alpha}$" are the components and $\{e_{\beta}\}$is an orthonormal basis in T_p, $\{w^{\alpha}\}$on orthonormal basis in $T_p{}^*$) If we want to compare T at neighboring points in the manifold we must first move the T at one point to the other point. The intuition from vectors in Euclidean space is that we want to move a vector without rotation, i.e., we want to "parallel-transport" the vector. What that implies here is the need for an additional structure—a rule for parallel-transport (and one that generalizes to tensors).

The definition for (covariant) derivative then follows from the rule for parallel-transport (see Fig. A.9). Consider, for example, the "covariant derivative" $\nabla_u T$ of T along a curve $f(\lambda)$ whose tangent vector is $u = df/d\lambda$:

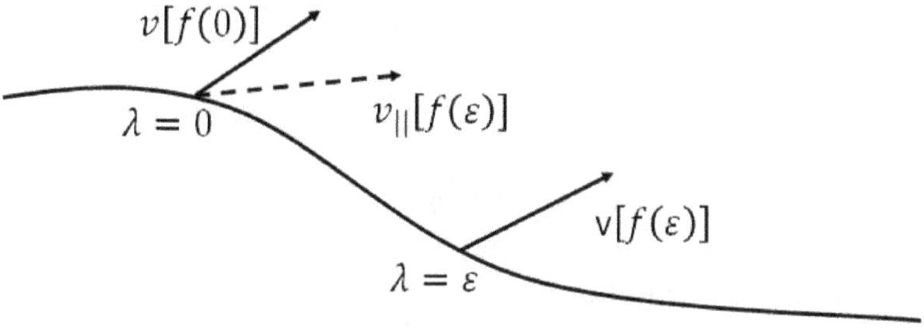

Fig. A.9 Parallel Transport.

Consider the derivation of the parallel transport as indicated in Fig. 8.9:

$$\vec{u} = \left.\frac{df}{d\lambda}\right|_{\lambda = 0} \qquad \vec{\nabla}_u T$$

$$= \lim_{\varepsilon \to 0} \left\{ \frac{T(f(\varepsilon))|\{parallel\ trans.to\ f(0)\} - T(f(0))}{\varepsilon} \right\}$$

$$\begin{aligned} \overline{\nabla} T = \overline{\nabla}(T^{\beta}{}_{\propto} e_{\beta} \otimes \omega^{\propto}) \\ = \overline{\nabla}(T^{\beta}{}_{\propto}) e_{\beta} \otimes \omega^{\propto} + T^{\beta}{}_{\propto}(\overline{\nabla} e_{\beta}) \otimes \omega^{\propto} \\ + T^{\beta}{}_{\propto} e_{\beta} \otimes \overline{\nabla} \omega^{\propto} \end{aligned}$$

Introduce standard connection coefficient notation:

$$\overline{\nabla}_{\lambda} e_{\beta} = \Gamma^{\propto}{}_{\beta\lambda} e_{\propto}$$

$$\bar{\nabla}_\lambda \omega^\propto = -\Gamma^\propto{}_{\beta\lambda}\omega^\beta$$

$$\bar{\nabla}_\lambda T = \left(T^\beta{}_{\propto,\lambda} + T^\mu{}_\propto \Gamma^\beta{}_{\mu\lambda} - T^\beta{}_\mu \Gamma^\mu{}_{\propto\lambda}\right) e_\beta \otimes \omega^\propto$$

$$T^\beta{}_{\propto;\lambda} = T^\beta{}_{\propto,\lambda} + T^\mu{}_\propto \Gamma^\beta{}_{\mu\lambda} - T^\beta{}_\mu \Gamma^\mu{}_{\propto\lambda}$$

Geodesic: A geodesic is a curve that is "straight" and uniformly parameterized as measured in each local R^n Chart. Thus, a geodesic is a curve that parallel-transports its tangent vector u along itself: $\nabla_u u=0$. Simple conceptually, now for concrete implementation. Introduce coordinate system x^α and describe u as $u^\alpha(0)=dx^\alpha(\lambda)/d\lambda|_{\{\lambda=0\}}$:

$$\frac{d^2x^\propto}{d\lambda^2} + \Gamma^\propto{}_{\mu\lambda}\frac{dx^\mu}{d\lambda}\frac{dx^\lambda}{d\lambda} = 0$$

Recap of Global Structures [105]
The review in this section largely draws from [105].

n-frame: a set of vector fields $V_{(1)},V_{(2)},\ldots,V_{(n)}$ defined on some subset of an n-dimensional manifold M such that they are linearly independent at each point of their domain of definition.

A Chart on $U \subset M$ with coordinates $x^1,x^2,\ldots x^n$ determines an n-frame ∂/∂^x on U. In general there is no n-frame defined on all of M.

Parallelizable Manifold: A global n-frame exists. In general, however, it is difficult to find even a single vector field that is defined globally (hedgehog theorem: a smoothly combed hedgehog has at least one point of baldness→i.e., even dim sphere). Cartesian products of parallelizable manifolds are parallelizable. (In GR one usually has $M=R\times\Sigma_3$, and if a 3 dim. manifold is parallelizable it is orientable. Orientability usually assumed, so parallelizability common in GR).

Connections: The information inherent in the connection coefficients, or in parallel translation, can be associated with principle bundles over a manifold. In general, one considers a bundle space P, over base space M, with structural group G and projection map π: P→ M. It is then possible to establish a one-to-one correspondence between 1 forms on P (Lie algebra valued) and connections on P. These are referred to as connection one- forms.

Affine Connection: If M denotes the Manifold and B(M) its bundle of bases, then a connection on B(M) is called an affine connection.

Difference Forms: Let ϕ and ψ be two connection 1-forms on B(M). Define the difference from $\tau = \psi - \phi$.

Connections on parallelizable manifolds: Let α be an R^d valued 1-form on M which gives a parallelization of M. Define three connections associated with ρ as follows:

(i) Direct connection: the one for which the vector fields $\rho^{-1}(x)$, for a fixed $x = R^d$, are parallel along every curve.
(ii) Torsion zero connection: the one with the same geodesic as the direct connection but with torsion zero
(iii) Opposite connection: the one with connection ϕ 2τ where ϕ is the direct connection from and $\phi+\tau$ is the torsion zero connection.

$T_{XY} = [X,Y] - \nabla_X Y + \nabla_Y X$ (torsion)
$R_{XY} = \nabla_{[X,Y]} - \nabla_X \nabla_Y + \nabla_X \nabla_Y$ (failure of ∇ to be a Lie algebra homomorphism)

direct: $T_{XY} = [X,Y]$; $R_{XY} = 0$; $\nabla_X Y = 0$.
torsion zero: $T_{XY} = 0$; $R_{XY} = [[X,Y],Z]/4$; $\nabla_X Y = [X,Y]/2$
opposite: $T_{XY} = -[X,Y]$; $R_{XY} = 0$; $\nabla_X Y = [X,Y]$

Let θ be a 2-form and X,Y,Z, vector fields, then:

$$(\nabla_X\theta)(Y,Z) = X\theta(Y,Z) - \theta(\nabla_X Y,Z) - \theta(Y, \nabla_X Z).$$

For metric connection:

$$X\langle Y,Z\rangle = \langle \nabla_X Y,Z\rangle + \langle Y, \nabla_X Z\rangle.$$

For (g, ∇, ∇^*), with ∇, ∇^* dual connections:

$$X\langle Y,Z\rangle = \langle \nabla_X Y,Z\rangle + \langle Y, \nabla^*_X Z\rangle.$$

With the above concepts reviewed, let's now proceed with the arguments of Amari.

B. Differential Geometry

B.1 Abstract Indices (Penrose [18])

In this section abstract indices are introduced, where a vector object is denoted V^a, which is identical in meaning to $\vec{V}$ (where 'a' is not a concrete index but an abstract index). Contrast this with $\vec{V}$ being a vector in a three-dimensional Cartesian coordinate space, say, with components$(\vec{V})_k$, where $k = 1,2, or\ 3$ corresponding to the components V_k in the x, y, and z directions. Here k is a concrete index (it can and must take on a set of values). The abstract index does not take on a set of values, in many applications, but it will take on the key property of contraction on like-indices (or Einstein summation convention) and tensor product construction when not like-indices (abstract operations inherited from the concrete index formulation). Although a notational convenience, it can be very powerful, as the application of the notation to some problems will demonstrate, and also when describing gauge field theories in Sec. X.

B.2 Vector spaces, Tensors, p-forms

Definition: The real vector space V

A set **V** is a real vector space if it is an abelian group under addition (have inverse) and under scalar multiplication it is associative and distributive with multiplicative identity (but may not have inverse). Thus, we have for V^a and U^a, elements of **V**, with scalars $r, s \in \mathbb{R}$:

$$r(U^a + V^a) = rU^a + rV^a \ ,$$

$$(r+s)U^a = rU^a + sU^a, \qquad 0 + U^a = U^a$$

and

$$r(sU^a) = (rs)U^a \quad , \qquad 1U^a = U^a$$

Note:

If not specified, assume finite-dimensional.

In what follows the dual basis is the roman w instead of the greek ω, to stay consistent with the use of other roman bases u, v, etc.

Example 1

1.a. Let V be a finite-dimensional vector space. Show that a linear map $T: V \to V$ is determined by its action on a basis for V.

1.b. Let V be a Hilbert space with a countable basis, termed a "separable" Hilbert space. (No other Hilbert spaces are used in physics). A linear map $T: V \rightarrow V$ is bounded if $|T(v)| < K|v|$ for some fixed constant K. Show that any bounded linear map T is determined by its action on an orthonormal basis for V.

Solution

(1)(a) V is a finite-dimensional vector space. As such it is spanned by a basis of finitely many linearly independent vectors. In a given basis $\{e_i^a\}, i = 1 \ldots n,$ where $n = \dim(V)$, we can express an arbitrary element of the vector space V as:

$$V^a = e_i^a V^i \quad \text{(summation convention in effect)}$$

A linear map $T: V \rightarrow V$ is defined by

$$T(ru^a + sw^a) = rT(u^a) + sT(w^a)$$

So, for the preceding V^a:

$$T(V^a) = T\left(e_i^a V^i\right) = V^i T(e_i^a)$$

For the case at hand $Domain(T) = V$ and $Range(T) = V$ (from $T: V \rightarrow V$) and the maps $T(e_i^a)$ will determine the mapping of an arbitrary element of V via the expansion above. So, the linear map T is determined by its action on a basis of V.

(1)(b) A Hilbert space is a linear space X together with a mapping $X \times X \rightarrow \mathbb{C}$, denoted $(x, y) \rightarrow \langle x, y\rangle$, for which:

$\langle x, y\rangle = \overline{\langle y, x\rangle}$ (bar denotes complex conjugate),

$\langle \alpha x + \beta y, z\rangle = \alpha\langle x, z\rangle + \beta\langle y, z\rangle$.

Furthermore the mapping $(x, y) \rightarrow \langle x, y\rangle$ must be strictly positive, i.e.,:

$$\langle x, x\rangle \geq 0 \;\; \forall x \in X \text{ and } \langle x, x\rangle = 0 \Rightarrow x = 0.$$

The mapping $(x, y) \rightarrow \langle x, y\rangle$ above induces a norm, thus a Hilbert space is a normed vector space. A norm induces a metric on X, in this sense a Hilbert space is a metric space. This allows us to succinctly state the final property of a Hilbert space – it is complete. Consider a basis for V, $\{e_i^a\}$, where i is the countable index of the basis for the separable Hilbert space being considered. We choose our basis to consist of linearly independent vectors (via Gram-Schmidt process) and thus have $\langle e_i^a, e_j^a\rangle = 0$ for $i \neq j$, where the norm directly induced by the mapping given above has been chosen. Again $T(V^a) = V^i T(e_i^a)$ and $\|T(V^a)\| = \left\|V^i T(e_i^a)\right\| =$

$\sum_{i=1}^{N}|V^i|^2 \langle T(e_i^a), T(e_i^a)\rangle = \sum_{i=1}^{N}|V^i|^2 \|T(e_i^a)\|$. Now T is not simply a linear map as before, here it is bounded, so here we must check the well definedness of $T(V^a) = V^i T(e_i^a)$, i.e., is $V^i T(e_i^a)$ bounded? To recap:

$$\|T(V^a)\| = \sum_{i=1}^{N} |V^i|^2 \|T(e_i^a)\| .$$

Since T is a bounded linear map: $\|T(e_i^a)\| < K\|e_i^a\|$ for some fixed constant K. Thus, T is obviously determined by its action on an orthonormal basis for V as long as such an expansion remains bounded by $K\|V\|$. Since we are considering an orthonormal basis, $\|e_i^a\| = 1$, thus:

$$\sum_{i=1}^{N} |V^i|^2 \|T(e_i^a)\| < K \sum_{i=1}^{N} |V^i|^2 \rightarrow \quad \|T(V^a)\| < K\|V\| ,$$

where $K = \max K_i$.So, $\|V^i T(e_i^a)\| < K\|V\|$ if $\|T(V)\| < K\|V\|$, thus, the statement $T(V^a) = V^i T(e_i^a)$ is well-defined when T is a bounded linear map.

Definition: The dual vector space V* of V

Suppose V is an n-dimensional space. The set of all linear maps of **V** to $\mathbb{R}$ is known as the dual of **V**, and is denoted **V*. V*** is also a vector space, and it is also dimension n.

.

The operative word in the description of the dual vector space in terms of a set of linear maps was that those maps be linear, in other words we have for maps σ and τ that they operate linearly on the vector space with element V^a (they preserve the underlying operations of vector addition and scalar multiplication). Thus, we have:

$$(\sigma + \tau)V = \sigma V + \tau V \quad \textit{and} \quad \sigma(\tau V) = (\sigma\tau)V.$$

The vectors in **V*** are called co-vectors or 1-forms and are written with lower abstract indices, thus $\sigma \rightarrow \sigma_a$ and we have, shifting from map notation to abstract contraction notation:

$$\sigma(V^a) = \sigma_a V^a.$$

Note that when there were repeated indices in the concrete vector and 1-form products described previously (or within any tensor with upper and lower indices), this led to a map to the real numbers $\mathbb{R}$. The abstract indices are capturing this same meaning and operation since the map $\sigma(V^a)$ is to reals, thus contraction (here $\sigma_a V^a$) maps to reals. Similarly in reverse, a vector can be regarded as a linear map from V^* to R: $V^a: \sigma_a \to V^a \sigma_a$.

Example 2. Show that a basis $\{e_i^a\}$ and its dual basis $\{w_b^i\}$ satisfy the completeness relation,

$$e_i^a w_b^i = \delta_b^a.$$

(Regard each side as a map from vector to vectors and check that the maps agree on each basis vector.)

Solution

If the map equality works on a basis and a dual basis it will work for any vector or covector by linearity, so let's check the basis etc.:

$e_j^b(e_i^a w_b^i) = e_i^a(S_b^a) = e_j^a = e_i^a(\delta_j^i)\ iff\ S_b^a = \delta_b^a\ , where\ S_b^a = e_i^a w_b^i$.

Similarly,

$w_a^j(e_i^a w_b^i) = (w_a^j e_i^a) w_b^i = \delta_i^j w_b^i = w_b^j \Rightarrow S_b^a = \delta_b^a$ again.

Thus,

$$e_i^a w_b^i = \delta_b^a.$$

Definition: Tensor

Using abstract indices it is now easy to define what a tensor is. A tensor is a multilinear map from m covectors and n vectors to $\mathbb{R}$ (e.g., the tensor has m upper indices and n lower):

$$T(\sigma_a, \dots \tau_b, u^c, \dots v^d) \to T^{a \dots b}{}_{c \dots d}\, \sigma_a \dots \tau_a\, u^c \dots v^d$$

(analogous to $\sigma(V^a) = \sigma_a V^a$).

Example 3. Using the definitions above, show that the tensor transformation law is equivalent to the multilinearity property when the tensor is regarded as a map. Show that if $\{f_i^a\}$ is a new basis, related to $\{e_i^a\}$, by:

$$f_i^a = a_i^m e_m^a\,,$$

then

$$T'^{ij}{}_k = (a^{-1})_\ell^i (a^{-1})_m^j a_k^n T^{\ell m}{}_n$$

Solution

Denote $T'^{ij}{}_k = T^{ab}{}_c w_a^i w_b^j e_k^c\ ;\ f_i^a = a_i^m e_m^a\ ;\ g_a^j = b_n^j w_a^n$, then

$$g_a^j f_i^a = \delta_i^j = a_i^m g_a^j e_m^a = a_i^m b_n^j (w_a^n e_m^a) = a_i^m b_m^j\,.$$

For this to be the identity, $b_m^i = (a_i^m)^{-1}$, or as a matrix, we have: $b_i^m = (a^{-1})_i^m$. Thus,

$$T'^{ij}{}_k = T^{ab}{}_c[(a^{-1})^i_\ell \sigma^\ell_a][(a^{-1})^j_m \sigma^m_b][a^n_k u^c_n]$$
$$= \{T^{ab}{}_c \sigma^\ell_a \sigma^m_b u^c_n\}(a^{-1})^i_\ell (a^{-1})^j_m a^n_k$$

and we get:

$$T'^{ij}{}_k = (a^{-1})^i_\ell (a^{-1})^j_m a^n_k T^{\ell m}{}_n \, .$$

Example 4.

a. Show that the action of a tensor on its arguments (a set of vectors or convectors), is given in terms of components along a basis by

$$T^{a\ldots b}_{c\ldots d}\sigma_a \cdots \tau_b u^c \cdots v^d = T^{i\ldots j}_{k\ldots\ell}\sigma_i \cdots \tau_j u^k \cdots v^\ell \, .$$

b. Show that any tensor can be written as a sum of its basis tensors,

$$T^{a\ldots b}_{c\ldots d} = T^{i\ldots j}_{k\ldots\ell} e^a_i \cdots e^b_j w^k_c \cdots w^\ell_d \, .$$

and find the dimension of the vector space consisting of all tensors with m up and n down indices (where the space of vectors is d-dimensional).

Solution

(a)

$$T^{a\ldots b}_{c\ldots d}\sigma_a \cdots \tau_b u^c \cdots v^d = T^{a\ldots b}_{c\ldots d}(\sigma_i w^i_a)\cdots(\tau_j w^j_b)(u^k e^a_k)\cdots(v^\ell e^d_\ell)$$
$$= \underbrace{T^{a\ldots b}_{c\ldots d} w^i_a \cdots w^j_b e^a_k e^d_\ell}_{T^{i\ldots j}_{k\ldots\ell}} \sigma_i \cdots \tau_j u^k \cdots v^\ell$$

Thus,

$$T^{a\ldots b}_{c\ldots d}\sigma_a \cdots \tau_b u^c \cdots v^d = T^{i\ldots j}_{k\ldots\ell}\sigma_i \cdots \tau_j u^k \cdots v^\ell \, .$$

(b) Left as an exercise.

Definition: p-form

A p-form $\sigma_{a\ldots b}$ is a totally antisymmetric tensor with p covariant indices:

$$\sigma_{a\ldots b} = \sigma_{[a\ldots b]}$$

thus have $\sigma(u^a, \ldots, v^b) = \sigma([u^a, \ldots, v^b])$.

A contravariant vector at P is equivalent to the tangent to a curve through P

A contravariant vector at P in manifold M can be regarded as the tangent to a curve through P, and is sometimes defined as the equivalence class of all smooth curves that are tangent at P. Let $F = C^\infty(M)$ be the vector space of smooth scalar fields on M.

B.3. Vector and Tensor definition by Map

Definition of a Vector at P given a Map at P

A vector V at P is a map V: f → ℝ that is linear under addition and multiplication by scalars and where that map satisfies Leibnitz rule. We can then define, in terms of a curve $c(\lambda)$ through $P = c(0)$, the linear map:

$$f \to V(f) = \frac{d}{d\lambda} f[c(\lambda)]|_{\lambda=0} \,,$$

with existence of the a limit value at $\lambda = 0$ is due Leibnitz.

Note that a Lie bracket automatically yields a vector that satisfies the Leibnitz condition, whereas simple composition of maps $u[v(g)]$ in and of itself is a mapping for vector fields *u* and *v* that does not satisfy the Leibnitz condition. The Lie bracket, a vector arrived at from two other vectors. but that is guaranteed to satisfy Leibnitz, may prove useful later.

Definition of a Tensor at P given a Multilinear Map at P

Suppose we have manifolds M and N and $\Psi: M \to N$ is a smooth (multilinear) map. Let's consider the dual and its action mapping tensors on M to tensors on N: $\Psi_*: T(M) \to T(N)$. Let's start by considering f, a scalar on N:

$$\Psi_* V(f) = \frac{d}{d\lambda} f[\Psi \circ c(\lambda)]|_{\lambda=0} \,,$$

where we are using $V(f) = \frac{d}{d\lambda} f[c(\lambda)]|_{\lambda=0}$ from the vector-map relation above. Consider the locally isomorphic mapping of M or N to $\mathbb{R}^m$ or $\mathbb{R}^n$. Suppose M has local coordinates x^i and N has local coordinates y^μ. Let $c^i(\lambda) = x^i(c(\lambda)) = x^i \circ c(\lambda)$, and note that

$$\Psi^\mu = (y \circ \Psi \circ x^{-1})^\mu = y^\mu \circ \Psi \circ (x)^{-1}$$

Refer to the figure below for clarification.

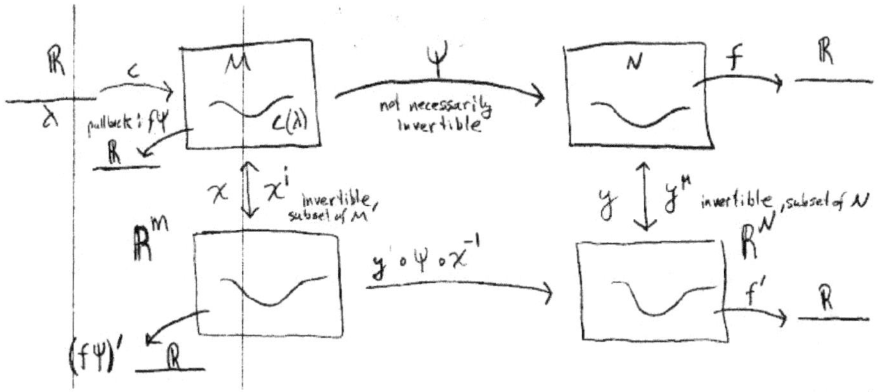

Fig. B.1.

Thus,

$$(f\Psi)'[x(\alpha)] = (f\Psi)(\alpha) = f[\Psi(\alpha)] = f'[y\{\Psi(\alpha)\}]$$

or alternatively:

$$(f\Psi)'[x^i] = f'[y^\mu].$$

So can now write:

$$V(f) = \frac{d}{d\lambda} f[c(\lambda)]|_{\lambda=0}$$

and

$$V^i = \frac{d}{d\lambda} x^i[c(\lambda)]|_{\lambda=0}\,.$$

Note the use of $f = x^i$ with concrete indices (indicated by roman letter). Now let's consider the dual:

$$\Psi_* V(f) = \frac{d}{d\lambda} f[\Psi \circ \mathrm{c}(\lambda)]|_{\lambda=0}$$

and

$$(\Psi_* V)^\mu = \frac{d}{d\lambda} y^\mu[\Psi \circ \mathrm{c}(\lambda)]|_{\lambda=0}$$

Note the use of $f = y^\mu$ with abstract indices (indicated by greek letter).

$$(\Psi_* V)^\mu = \frac{d}{d\lambda}[y^\mu \circ \Psi \circ x^{-1} \circ x \circ \mathrm{c}(\lambda)]|_{\lambda=0} = \frac{d}{d\lambda}\left[\Psi^\mu\big(x \circ \mathrm{c}(\lambda)\big)\right]\Big|_{\lambda=0},$$

thus,

$$(\Psi_* V)^\mu = \frac{\partial \Psi^\mu}{\partial c^i(\lambda)} \frac{\partial c^i(\lambda)}{\partial \lambda}\bigg|_{\lambda=0} = \frac{\partial \Psi^\mu}{\partial x^i} \frac{\partial x^i[c(\lambda)]}{\partial \lambda}\bigg|_{\lambda=0} = \frac{\partial \Psi^\mu}{\partial x^i} V^i\,.$$

Thus,

$$(\Psi_* V)^\mu = \frac{\partial \Psi^\mu}{\partial x^i} V^i .$$

From this we see that Ψ_* is a linear map from vectors at P to vectors at $\Psi(\mathrm{P})$ and as such may be regarded as an object with one covariant index at P and one contravariant index at $\Psi(\mathrm{P})$:

$$\Psi_* \simeq \Psi_a^\alpha \Rightarrow \Psi_* V^a = \Psi_a^\alpha V^a = U^\alpha$$

This is sometimes describes as Ψ "drags along" a vector V^a from M to N (this terminology becomes clearer in later sections when we discuss Lie Derivatives). In general Ψ will not be invertible and one cannot associate a vector on N with a unique vector on M. Co-vectors, however, are naturally "pulled back" by Ψ from M to N. The function $f: N \longrightarrow \mathbb{R}$ is pulled back to $(f\Psi): M \longrightarrow \mathbb{R}$, where $(f\Psi)[M] = f[\Psi(\mathrm{M})]$, and the co-vector $\nabla_\alpha f$ is pulled back to: $\Psi^* \nabla_\alpha f = \nabla_a (f\Psi)$. This means:

$$\nabla_i(f\Psi) = \frac{d}{dx^i}\left[(f\Psi)'(x(\alpha))\right] = \frac{d}{dx^i}\left[(f\Psi)(\alpha)\right] = \frac{d}{dx^i}\left[f(\Psi(\alpha))\right]$$

$$= \frac{d}{dx^i}\left[f'(y \circ \Psi(\alpha))\right] = \frac{d}{dx^i}\left[f'(y \circ \Psi \circ x^{-1} \circ x(\alpha))\right]$$

$$= \frac{d}{dx^i}\left[f'(y \circ \Psi \circ x^{-1}(x))\right] = \frac{\partial f'}{\partial y^\mu} \frac{\partial y^\mu \left(\Psi \circ x^{-1}(x(\alpha))\right)}{\partial x^i}$$

$$= \frac{\partial f'}{\partial y^\mu} \frac{\partial \Psi^\mu(x)}{\partial x^i}$$

Thus

$$\nabla_i(f\Psi) = \partial_\mu f \frac{\partial \Psi^\mu}{\partial x^i} .$$

So, co-vectors W_a at $\Psi(\mathrm{P})$ are pulled back by the same object Ψ_{a}^α to covectors at P.

$$(\Psi_* V)^\mu = \frac{\partial \Psi^\mu}{\partial x^i} V^i \quad \rightarrow \quad \Psi_* V^a = \Psi_a^\alpha V^a = U^\alpha$$

$$\nabla_i(f\Psi) = \frac{\partial \Psi^\mu}{\partial x^i} \partial_\mu f \quad \rightarrow \quad \Psi^* W_a = \Psi_a^\alpha W_\alpha = \sigma_a$$

Generalizing -- contravariant tensors on M are dragged to N:

$$\Psi_* T^{\alpha\ldots\beta} = \Psi^{\alpha}_{a} \ldots \Psi^{\beta}_{b} T^{a\ldots b}$$
and covariant tensors are pulled back to covariant tensors at P:
$$\Psi^* T_{a\ldots b} = \Psi^{\alpha}_{a} \ldots \Psi^{\beta}_{b} T_{\alpha\ldots\beta} \,.$$

A smooth isomorphism Ψ from a manifold M to itself is called a diffeomorphism.

$$(\Psi T)^{ab}{}_{c}(P) = \Psi^{a}_{e} \Psi^{b}_{f} (\Psi^{-1})^{g}_{c} T^{ef}{}_{g}\left(\Psi^{-1}(P)\right)$$

Thus, given a tensor field T on M, Ψ produces a new tensor field (ΨT).

Smooth vector field ⇔ integral curves
Consider V^a, a vector field on M, and $c(\lambda)$, a curve whose tangent vector $U^a(\lambda)$ coincide at each point of c with the value of V^a at that point:
$$U^a(\lambda) = V^a\left(c(\lambda)\right) \;\rightarrow\; \frac{d}{d\lambda} c^i(\lambda) = V^i\left(c(\lambda)\right)$$
Any system of differential equations can be cast in this form, whence, by the existence and uniqueness theorems for ODE's of order N we have the following theorem.

Theorem: Let V^a be a vector field on M, then there is a maximal integral curve $c(\lambda)$ (cannot be extended to a longer integral curve) passing through each point of M and $c(\lambda)$ is unique. Thus, any vector field generates a *1*-parameter group of diffeomorphisms. The orbit of a point M under the action of the group is an integral curve of the vector field.

B.4 Lie Derivatives
Suppose U^a, V^a are two vector fields. A family of diffeomorphisms Ψ_λ are generated by U^a and drag V^a:

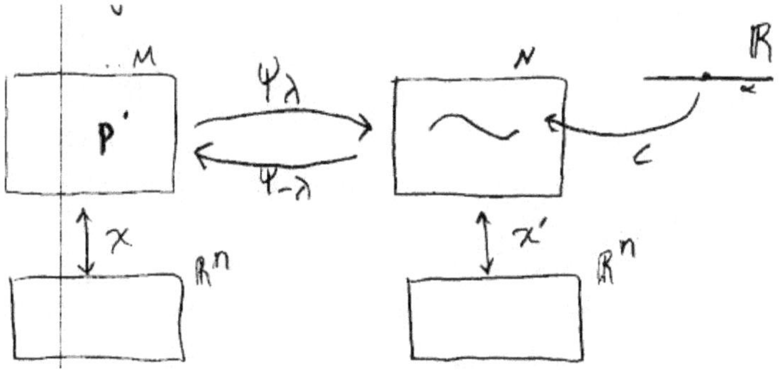

Fig. B.2.

Note that $\Psi_{-\lambda} \bullet c(0) = P$:

$$(\Psi_{-\lambda}V^i)(P) = \frac{d}{d\alpha}x^i[\Psi_{-\lambda} \circ c(\alpha)]|_{\alpha=0}$$

$$= \frac{d}{d\alpha}\left[x^i \circ \Psi_{-\lambda} \circ (x')^{-1} \circ (x') \circ c(\alpha)\right]\Big|_{\alpha=0}$$

$$(\Psi_{-\lambda}V^i)(P) = \frac{d}{d\alpha}\left[\Psi_{-\lambda}^i\left(x^i \circ c(\alpha)\right)\right]\Big|_{\alpha=0} = \frac{\partial \Psi_{-\lambda}^i}{\partial (x^j)'}\frac{\partial (x^j)'}{\partial \alpha}\Bigg|_{\alpha=0}$$

$$= \left(\frac{\partial}{\partial (x^j)'}\Psi_{-\lambda}^i\right)\Big|_{\alpha=0} V^{j'}(\Psi_\lambda(P))$$

$$\frac{d}{d\lambda}(\Psi_{-\lambda}V^i)(P) = \frac{\partial}{\partial (x^j)'}\left(\frac{\partial \Psi_{-\lambda}^i}{\partial \lambda}\right)V^{j'} + \frac{\partial \Psi_{-\lambda}^i}{\partial (x^j)'}\frac{\partial V^{j'}(\Psi_\lambda(P))}{\partial \lambda}$$

Substituting

$$\frac{d}{d\lambda}\Psi_{-\lambda}^i(P) = \frac{d}{d\lambda}x^i(\Psi_\lambda[c(0)]) = -U^i(P)$$

and

$$\frac{\partial x^i}{\partial x^j}\frac{\partial x^j(\Psi_\lambda(P))}{\partial \alpha \partial \lambda} = \frac{\partial V^i(P)}{\partial \alpha} = \frac{\partial V^i}{\partial x^j}\frac{\partial x^j}{\partial \alpha} = U^j\frac{\partial}{\partial x^j}V^i$$

We get:

$$\frac{d}{d\lambda}(\Psi_{-\lambda}V^i)(P) = -\frac{\partial U^i}{\partial x^j}V^j + U^j\frac{\partial V^i}{\partial x^j} = [U,V]^i.$$

Thus, diffeomorphisms are symmetries of the metric iff (if and only if) V^α Lie-derives the metric:

$$\Psi_\lambda g_{\alpha\beta} = 0 \rightarrow \mathcal{L}_V g_{\alpha\beta} = 0 \quad \rightarrow \quad V^\alpha \; is\; a\; Killing\; vector\; field.$$

If V^α is a Killing Vector one can always choose a chart for which a given vector field has components $V^\mu = \delta_0^\mu$. A linear map S from tensor fields $T^{a\ldots b}_{c\ldots d}$ to $\mathbb{R}$ is itself a tensor field iff S is linear under multiplication of T by a scalar field: $S(fT) = fS(T)$. A linear map from tensor fields to tensor fields is itself a tensor field iff it is linear under multiplication by scalar fields. Consider the tensor field T^a_{bc}, it can be regarded as a multilinear map taking one covector field and two vector fields to $\mathbb{R}$; or as a linear map taking one vector field to a tensor field with two covariant indices; or as a linear map taking tensor fields with two up indices to vector fields; or .....

Problem. Show that in a chart, the components of the Lie derivative have the form:

$$\pounds_u T^{i\ldots j}_{k\ldots\ell} = u^m \partial_m T^{i\ldots j}_{k\ldots\ell} + T^{i\ldots j}_{m\ldots\ell}\partial_k u^m + \cdots T^{i\ldots j}_{k\ldots m}\partial_\ell u^m - T^{m\ldots j}_{k\ldots\ell}\partial_m u^i - \cdots$$
$$- T^{i\ldots m}_{k\ldots\ell}\partial_m u^j$$

Solution

Definition of Lie derivative: $\pounds_u T^{a\ldots b}_{c\ldots d} = \frac{d}{d\lambda}\left\{(\Psi_{-\lambda}T)^{a\ldots b}_{c\ldots d}\right\}_{\lambda=0}$

Notice $\pounds_v(f) = \frac{d}{d\lambda}\{\Psi_{-\lambda}f\}|_{\lambda=0}$. If we let the integral curves of V^a be denoted by $c(\lambda)$:

$$\pounds_v(f) = \frac{d}{d\lambda}f(c(\lambda))\big|_{\lambda=0} = V(f)$$

Also, by choosing a chart:

$$\pounds_u V^a = \frac{d}{d\lambda}\{(\Psi_{-\lambda})_* V^a\}_{\lambda=0}$$

Since

$$[(\Psi_{-\lambda})_* V']^i(p) = \frac{d}{d\alpha}x^i[\Psi_{-\lambda}\circ c(\alpha)]_{\alpha=0}$$
$$= \frac{d}{d\alpha}\left[x^i\circ\Psi_{-\lambda}\circ(x')^{-1}\circ(x')\circ c(\alpha)\right]\Big|_{\alpha=0}$$

Thus,

$$[(\Psi_{-\lambda})_* V']^i(p) = \frac{d}{d\alpha}\left[\Psi^i_{-\lambda}x'\circ c(\alpha)\right] =$$
$$\frac{\partial\Psi^i_{-\lambda}(x^j)'}{\partial(x^j)'}\frac{\partial(x^j)'}{\partial\alpha}\Bigg|_{\alpha=0}$$

So, $(\pounds_u V)^i = \frac{d}{d\lambda}\left\{\frac{\partial\Psi^i_{-\lambda}(x')}{\partial(x^j)'}\frac{\partial(x^j)'}{\partial\alpha}\right\}_{\substack{\alpha=0\\ \lambda=0}}$

Now $\left.\frac{d\Psi^i_{-\lambda}(P)}{d\lambda}\right|_{\substack{\lambda=0\\ \alpha=0}} = \frac{d}{d\lambda}\, x^i\,(\Psi_{-\lambda}[c(o)])|_{\substack{\lambda=0\\ \alpha=0}} = -u^i(P)$

And $\frac{d}{d\lambda}\left(\frac{\partial\left[(x^j)'(\Psi_\lambda(P))\right]}{\partial\alpha}\right)_{\substack{\lambda=0\\ \alpha=0}} = \left.\frac{\partial(v^j)}{\partial\alpha}\right|_{\alpha=0} = \left.\frac{\partial(v^j)'}{\partial x^i}\frac{\partial x^i}{\partial\alpha}\right|_{\alpha=0} = u^i\frac{\partial(v^j)'}{\partial x^i}$

Thus

$$(\pounds_u v)^i = \left[\frac{\partial}{\partial (x^j)'} - \left(u^i(P)\right)\right]\left(\frac{\partial (x^j)'}{\partial \alpha}\right) + \left(\frac{\partial x^i}{\partial (x^j)'}\right)\left(u^k \frac{\partial (v^j)'}{\partial x^k}\right)$$
$$= u^j \frac{\partial v^i}{\partial x^j} - v^j \frac{\partial u^i}{\partial x^j} = [u, v]^i$$

And

$$\pounds_u v = [u, v]^a \, .$$

By choosing a convenient coord. system this can be done much more simply. Using $\pounds_v(f) = v(f)$ and $\pounds_u v = [u, v]^a$ and Leibnitz' rule one can generalize to an arbitrary tensor. So, we have:

$$(\pounds_u v)^i = u^j \partial_j v^i - v^j \partial_j u^i \quad and \quad (\pounds_u f) = u^j \partial_j f$$

Need $\pounds_u w$ where w is a l-form. So consider Leibnitz rule:

$$\pounds_u (w_a v^a) = w_a \pounds_u v^a + v^a \pounds_u w_a$$

In a chart:

$$[\pounds_u (w_a v^a)] = w_i u^j \partial_j v^i - w_i v^j \partial_j u^i + v^i (\pounds_u w_a)_i$$

Also

$$\left[\pounds_u \left(w_i v^i\right)\right] = u^j \partial_j \left(w_i v^i\right) = w_i u^j \partial_j v^i + v^i u^j \partial_j w_i$$

So

$$v^i (\pounds_u w_a)_i = w_i v^j \partial_j u^i + v^i u^j \partial_j w_i = v^i \left(u^j \partial_j w_i + w_j \partial_i u^j\right)$$

Thus,

$$(\pounds_u w)_i = u^j \partial_j w_i + w_j \partial_i u^j \, .$$

Now a general tensor can be expressed as a sum of tensor products of vectors and l-forms, so applying the Lie derivative and using the Leibnitz property one gets:

$$\left(\pounds_u T^{a\ldots b}_{c\ldots d}\right)^{i'\ldots j'}_{k'\ldots \ell'} = \pounds_u \left\{T^{i\ldots j}_{k\ldots \ell} e^a_i \cdots e^b_j w^k_c \cdots w^\ell_d\right\}^{i'\ldots j'}_{k'\ldots \ell'} = \left(\pounds_u T^{i'\ldots j'}_{k'\ldots \ell'}\right) +$$
$$T^{i'\ldots j'}_{k'\ldots \ell'} (\pounds_u e^a_i)^{i'} + \cdots T^{i'\ldots j'}_{k'\ldots \ell} \left(\pounds_u w^\ell_d\right)_{\ell'}$$
$$= u^m \partial_m T^{i'\ldots j'}_{k'\ldots \ell'} + T^{i\ldots j}_{k\ldots \ell}\left\{u^m \partial_m e^{i'}_{(i)} - e^m_{(j)} \partial_m u^i\right\} + \cdots +$$
$$T^{i\ldots j}_{k\ldots \ell}\left\{u^m \partial_m w^{(\ell)}_{\ell'} + w^{(\ell)}_m \partial_{\ell'} u^m\right\}$$
$$= u^m \partial_m T^{i'\ldots j'}_{k'\ldots \ell'} - T^{m\ldots j'}_{k'\ldots \ell'} \partial_m u^{i'} \cdots + T^{i'\ldots j'}_{k'\ldots m} \partial_{\ell'} u^m$$

Dropping primes we get the result:

$$\pounds_u T^{i\ldots j}_{k\ldots \ell} = u^m \partial_m T^{i\ldots j}_{k\ldots \ell} + T^{i\ldots j}_{m\ldots \ell} \partial_k u^m + \cdots T^{i\ldots j}_{k\ldots m} \partial_\ell u^m - T^{m\ldots j}_{k\ldots \ell} \partial_m u^i - \cdots$$
$$- T^{i\ldots m}_{k\ldots \ell} \partial_m u^j \, .$$

Problem. Show that if ∇_a is the covariant derivative associated with any metric, the Lie derivative can be written in the form:

$$£_u T^{a...b}_{c...d} = u^e \nabla_e T^{a...b}_{c...d} + T^{a...b}_{e...d} \nabla_c u^c + T^{a...b}_{c...e} \nabla_d u^e - T^{e...b}_{c...d} \nabla_e u^a - T^{a...e}_{c...d} \nabla_e u^b$$

Solution

$£_u T^{a...b}_{c...d} = \left(£_u T^{i...j}_{k...\ell}\right) e^a_i \cdots e^b_j \, w^k_c \cdots w^\ell_d$

$= u^m \partial_m \left(T^{a...b}_{c...d}\right) + T^{a...b}_{m...d} (w^k_c \partial_k) u^m + \cdots - T^{a...m}_{c...d} \left(e^b_j \partial_m u^j\right)$

$= u^e \nabla_e T^{a...b}_{c...d} + T^{a...b}_{e...d} \nabla_c u^c + T^{a...b}_{c...e} \nabla_d u^e - T^{e...b}_{c...d} \nabla_e u^a - T^{a...e}_{c...d} \nabla_e u^b$

This can be done since ∇_a, the covariant derivative associated with any metric reduces to $\partial_i = e^a_i \nabla_a$ when mapped onto a given chart (def. of cov. deriv. on manifold), locally this mapping "e^a_i" is diffeomorphic so $\nabla_a = (e^a_i)^{-1} \partial_i = w^i_a \partial_i$. Since $e^e_m w^m_e = 1$

Problem. Let $T^{\alpha\beta}$ be a symmetric tensor (such as the energy-momentum tensor). Show that $T^{\alpha\beta}$ is divergence-free ($\nabla_\beta T^{\alpha\beta} = 0$) then the current $J^\alpha = T^{\alpha\beta} V_\beta$ associated with the Killing Vector V_β is conserved ($\nabla_\alpha J^\alpha = 0$).

Solution

A vector V^β is a Killing Vector $(\text{K.V.})\ if\ \nabla_\alpha V_\beta = -\nabla_\beta V_\alpha$. We have

$\nabla_\alpha J^\alpha = \left(\nabla_\alpha T^{\alpha\beta}\right) V_\beta + T^{\alpha\beta} \nabla_\alpha V_\beta = \frac{1}{2}\left(T^{\alpha\beta} + T^{\beta\alpha}\right) \nabla_\alpha V_\beta = \frac{1}{2}\left(T^{\alpha\beta} \nabla_\alpha V_\beta + T^{\alpha\beta} \nabla_\beta V_\alpha\right)$

$\nabla_\alpha J^\alpha = \frac{1}{2}\left(T^{\alpha\beta} \nabla_\alpha V_\beta + T^{\alpha\beta}\left(-\nabla_\alpha V_\beta\right)\right) = 0$

Thus, $\nabla_\alpha J^\alpha = 0$, and the current is conserved.

Problem. What conserved quantities correspond to the integral over a space-like plane in Minkowski space,

$$Q = \int n_\alpha J^\alpha dV,$$

where J^α is the current associated with translations, rotations, boosts.

Partial (nonmathematical) Solution

4 translations: timelike K.V. field yields conservation of energy while spacelike K.V. field yields conservation of momentum.

3 rotations: K.V. field yields conservation of angular momentum.
3 boosts → K.V. yields Lorentz transformation along given axis.

Problem 3.2. Show that the expression for the exterior derivative:

$$(d\sigma)_{ab\ldots c} = (p+1)\nabla_{[a}\sigma_{b\ldots c]}$$

is independent of connection. Repeat for Lie derivative:

$$\pounds_u T^{a\ldots b}_{c\ldots d} = u^e \nabla_e T^{a\ldots b}_{c\ldots d} + T^{a\ldots b}_{e\ldots d}\nabla_c u^e + \cdots + T^{a\ldots b}_{c\ldots e}\nabla_d u^e - T^{e\ldots b}_{c\ldots d}\nabla_e u^a - \cdots - T^{a\ldots e}_{c\ldots d}\nabla_e u^a .$$

Solution

We have: $(d\sigma)_{ab\ldots c} = (p+1)\nabla_{[a}\sigma_{b\ldots c]}$ where σ is a p-form. This is independent of connection, i.e., choice of covariant derivative, if

$$\nabla_{[a}\sigma_{b\ldots c]} = \tilde{\nabla}_{[a}\sigma_{b\ldots c]}$$

Also note:

$$\nabla_{[a}\sigma_{b\ldots c]} = 0 \quad if \quad C^f_{[ab}\sigma_{f\ldots c]} = 0 .$$

Since C^f_{ab} is symmetric on its covariant indices antisymmetrization over them will give zero. Thus $\nabla_{[a}\sigma_{b\ldots c]} = \tilde{\nabla}_{[a}\sigma_{b\ldots c]}$, thus the exterior derivative is independent of connection.

Let's now examine

$$\pounds_u T^{a\ldots b}_{c\ldots d} = u^e \nabla_e T^{a\ldots b}_{c\ldots d} + T^{a\ldots b}_{e\ldots d}\nabla_c u^e + \cdots + T^{a\ldots b}_{c\ldots e}\nabla_d u^e - T^{e\ldots b}_{c\ldots d}\nabla_e u^a - \cdots - T^{a\ldots e}_{c\ldots d}\nabla_e u^a$$

Consider the difference of $\pounds_u T^{a\ldots b}_{c\ldots d}$ defined above by covariant derivative ∇: $\pounds^{\nabla}_u$ and that using $\tilde{\nabla}$: $\pounds^{\tilde{\nabla}}_u$, then

$$\left(\pounds^{\nabla}_u - \pounds^{\tilde{\nabla}}_u\right)T^{a\ldots b}_{c\ldots d}$$

$$= u^e\left(\nabla_e - \tilde{\nabla}_e\right)T^{a\ldots b}_{c\ldots d} + T^{a\ldots b}_{e\ldots d}\left(\nabla_c - \tilde{\nabla}_c\right)u^e + \cdots + T^{a\ldots b}_{c\ldots e}\left(\nabla_d - \tilde{\nabla}_d\right)u^e - T^{e\ldots b}_{c\ldots d}\left(\nabla_e - \tilde{\nabla}_e\right)u^a - \cdots - T^{a\ldots e}_{c\ldots d}\left(\nabla_e - \tilde{\nabla}_e\right)u^b$$

$$= u^e\left\{C^a_{fe}T^{f\ldots b}_{c\ldots d} + \cdots + C^b_{fe}T^{a\ldots f}_{c\ldots d} - C^f_{ce}T^{a\ldots b}_{c\ldots d} - \cdots - C^f_{de}T^{a\ldots b}_{c\ldots f}\right\}$$

$$+T^{a\ldots b}_{e\ldots d}C^e_{cg}u^g + \cdots + T^{a\ldots b}_{c\ldots e}C^e_{dg}u^g - T^{e\ldots b}_{c\ldots d}C^a_{eg}u^g - \cdots - T^{a\ldots e}_{c\ldots d}C^b_{eg}u^g$$

$$= u^e\left\{C^a_{fe}T^{f\ldots b}_{c\ldots d} + \cdots + C^b_{fe}T^{a\ldots f}_{c\ldots d} - C^f_{ce}T^{a\ldots b}_{f\ldots d} - \cdots - C^f_{de}T^{a\ldots b}_{c\ldots f}\right\}$$

$$+u^e\{-C^a_{fe}T^{f\ldots b}_{c\ldots d} - \cdots - C^b_{fe}T^{a\ldots f}_{c\ldots d} + C^f_{ce}T^{a\ldots b}_{f\ldots d} + \cdots + C^f_{de}T^{a\ldots b}_{c\ldots f}\}$$

$$= 0$$

So, the Lie derivative is also independent of connection.

Problem 3.3. Prove the following key identities relating Lie derivatives and exterior derivatives. For any p-form $w_{a\ldots c}$ and vector u_a,

$$d\pounds_u w = \pounds_u dw \quad and \quad \pounds_u w = u \cdot dw + d(u \cdot w)\,.$$

Solution

Start with:

$$\pounds_u w_{a\ldots c} = \frac{d}{d\lambda}(\Psi^*_{-\lambda} w_{a\ldots c})_{\lambda=0}$$

We then have

$$d\pounds_u w_{a\ldots c} = \frac{d}{d\lambda}(d\Psi^*_{-\lambda} w_{a\ldots c})_{\lambda=0} = \frac{d}{d\lambda}\left(\Psi^*_{-\lambda}(dw)\right)_{\lambda=0} = \pounds_u dw$$

Thus,

$$d\pounds_u w = \pounds_u dw\,.$$

This step uses the relation $d(\Psi^*_{-\lambda} w) = \Psi^*_{-\lambda}(dw)$ which I will now prove using

$$\Psi^*(df) = d(f\Psi).$$

Going to a coordinate frame where

$$w_{a\ldots c} = w_{i\ldots k} dx^i_a \wedge \cdots \wedge dx^k_c$$

We have

$$dw_{\alpha a\ldots c} = d(w_{i\ldots k})_\alpha \wedge dx^i_a \wedge \cdots \wedge dx^k_c = \left(\frac{\partial w_{i\ldots k}}{\partial x^n}\right) dx^n_\alpha \wedge dx^i_a \cdots \wedge dx^k_c$$

Now,

$$\Psi^*(dw_{\alpha a\ldots c}) = \Psi^*\left\{\frac{\partial w_{i\ldots k}}{\partial x^n} dx^n_\alpha \wedge dx^i_a \cdots \wedge dx^k_c\right\} = \left(\frac{\Psi^*\partial w_{i\ldots k}}{\partial x^n}\right)\{(\Psi^* dx^n_\alpha) \wedge (\Psi^* dx^i_a) \wedge \cdots \wedge (\Psi^* dx^k_c)\}$$

$$= \frac{w_{i\ldots k}(\Psi(P))}{\partial\Psi(x^n)} d\Psi(x^n)_\alpha \wedge d\Psi(x^i)_{\mathrm{a}} \wedge \cdots \wedge d\Psi(x^k)_{\mathrm{c}}$$

Let $y - \Psi(x)$

$$\Psi^*(dw_{\alpha a\ldots c}) = \frac{\partial w_{i\ldots k}(\Psi(P))}{\partial y^n} dy^n_\alpha \wedge dy^i_a \wedge \cdots \wedge dy^k_c$$

$$= d(\Psi^* w_{a\ldots c})$$

The proof is independent of the coordinates used in the definition due to the properties of the exterior derivative on form fields:

$w_{a\dots c} = w_{i\dots k}dx_a^i \wedge \cdots \wedge dx_c^k = w_{i'\dots k'}dx_a^{i'} \wedge \cdots \wedge dx_c^{k'}$

$dw = dw_{i'\dots k'} \wedge dx_a^{i'} \wedge \cdots \wedge dx_c^{k'} \left(\frac{\partial x^i}{\partial x^{i'}} \cdots \frac{\partial x^k}{\partial x^{k'}} A_{i\dots k}\right) \wedge dx_a^{i'} \cdots \wedge dx_c^k =$
$dw_{i\dots k} \wedge dx_a^i \wedge \cdots \wedge dx_c^k$

Consider $£_u w = \frac{d}{d\lambda}(\Psi_{-\lambda}^* w_{a\dots c})_{\lambda=0}$ and choosing a chart:

$$\frac{d}{d\lambda}(\Psi_{-\lambda}^* w_{a\dots c})(P) := \frac{d}{d\lambda}\left(\Psi_{-\lambda}^*\left[w_{i\dots k}(P)dx_a^i \wedge \cdots \wedge dx_c^k\right]\right)$$
$$= \frac{d}{d\lambda}\left(w_{i\dots k}\left(\Psi_{-\lambda}(P)\right)d\left(\Psi_{-\lambda}x^i\right)_a \wedge \cdots d\left(\Psi_{-\lambda}x^k\right)_c\right)$$
$$= \frac{dw_{i\dots k}(\Psi_{-\lambda}(P))}{d\lambda} \wedge d\left(\Psi_{-\lambda}x^i\right)_a \wedge \cdots d\left(\Psi_{-\lambda}x^k\right)_c +$$
$$+w_{i\dots k}\left(\Psi_{-\lambda}(P)\right)d\left(\frac{d}{d\lambda}\Psi_{-\lambda}x^i\right)_a \wedge \cdots \wedge \left(\Psi_{-\lambda}x^k\right)_c + \cdots$$
$$+w_{i\dots k}\left(\Psi_{-\lambda}(P)\right)d\left(\Psi_{-\lambda}x^i\right)_a \wedge \cdots \wedge \left(\frac{d}{d\lambda}\Psi_{-\lambda}x^k\right)_c$$

Also note: $\left.\frac{d(w_{i\dots k})}{d\lambda}\right|_{\lambda=0} = \left.\frac{d(w_{i\dots k})}{d\Psi_{-\lambda}(x)}\frac{d\Psi_{-\lambda}(x)}{d\lambda}\right|_{\lambda=0} = \left.\frac{(dw)m_{i\dots k}}{d\Psi_{-\lambda}(x)}\frac{d\Psi_{-\lambda}(x)}{d\lambda}\right|_{\lambda=0}$
$= (p+1)u^m\nabla_{[m}w_{i\dots k]} = u^m\nabla_m w_{[i\dots k]} +$
$u^m\nabla_i w_{[m\dots k]}$

So, in a chart:

$$\left.\frac{d}{d\lambda}(\Psi_{-\lambda}^* w_{a\dots c})(P)\right|_{\lambda=0}$$
$$= u^m\nabla_m w_{[i\dots k]}dx_a^i \wedge \cdots \wedge dx_c^k + u^m\nabla_i w_{[m\dots k]}dx_a^i \wedge \cdots \wedge dx_c^k$$
$$+w_{i\dots k}\{d\left(u^m\delta_m^i\right) \wedge \cdots \wedge dx_t^k + \cdots + dx^i \wedge d\left(u^m\delta_m^k\right)\}$$
$$= u \cdot dw + \left(u^m\nabla_\ell w_{mj\dots k}dx_a^\ell + w_{mj\dots k}du^m\right)dx_b^i \cdots dx_c^k$$
$$= u \cdot dw + \nabla_\ell\left(u^m w_{mj\dots k}\right)dx_a^\ell \wedge dx_a^i \wedge \cdots dx_c^k$$

Thus

$$£_u w = u \cdot dw + d(u \cdot w)$$

Since the chart was arbitrary the result holds in general.

C. Algebraic Topology

Differential Topology is a branch of geometric topology that studies manifolds.

A topological n-manifold: a Hausdorff topological space with the property that each point has a neighborhood homeomorphic to an open subset of R^n. (Some definitions require it be a separable metric space.) This situation is often, still, too general to obtain specific results, so for this reason topologists often require some additional structure on their manifold. (e.g., piecewise linear structure, or differential structure).

Assuming the additional structure is differential, then the tools of analysis can be used to study manifolds.

<u>Def 1</u>: Let $U \subset \mathbb{R}^n$. A map $f: U \to \mathbb{R}^n$ is smooth if it has continuous partial derivatives of all orders. If $\mathrm{X} \subseteq \mathbb{R}^n$, i.e., not necessarily open, a map $f: \mathrm{X} \to \mathbb{R}^n$ is smooth if we can extend for any
$x \in X\ \exists$ (there exists) an open neighborhood U_x of x in $\mathbb{R}^n$ and a smooth map $F|U_x \to \mathbb{R}^m \to F|_{U_x \cap X} = f$.

In what follows (physics applications of interest) we will be concerned (almost) exclusively with smooth maps. For this reason our spaces will (almost) always be viewed as subsets of $\mathbb{R}^N$ (for some integer N).

<u>Def. 2</u>: A map $f: X \subseteq \mathbb{R}^m \to Y \subseteq \mathbb{R}^n$ is a diffeomorphism
If (a) f is bijective
(b) f is smooth
(c) $f^{-1}$is smooth
i.e., a homeomophism includes differential structure consistent with a mapping that is bijective, e.g., f and f^{-1} are continuous (smooth).

<u>Ex:</u> $f: \mathbb{R}^1 \to Y \subseteq \mathbb{R}^2$ where $f(x) = (x, x^2)$
f bijective
f smooth
Is f^{-1} smooth?
Have: $f^{-1}: Y \to \mathbb{R}^1$ (takes $(x, y) \to x$). So, define $F: \mathbb{R}^2 \to \mathbb{R}^1$ by $F(x, y) = x$. Then, f^{-1} is smooth. So f is a diffeomorphism.

Ex: $f: \mathbb{R}^1 \to z \subseteq \mathbb{R}^2$ where $x \longmapsto (x, |x|)$
f bijective
f not smooth → so not a diffeomorphism.

Ex: Let $f: \mathbb{R}^1 \to \mathbb{R}^1$ by $f(x) = x^3$. Then have $f^{-1}(x) = x^{1/3}$.
f bijective
f smooth
f^{-1} not smooth → so not a diffeomorphism.

Def 3: Let $X \subseteq \mathbb{R}^N$. Then X is a k-dimensional manifold if $\forall x \in X, \exists$ open neighborhood V_x in x in X which is diffeomorphic to some $U \subset \mathbb{R}^k$. A diffeomorphism $\phi: U \to V_x$ is called a parameterization of V_x (a k-dimensional manifold). In this case $\phi^{-1}: V_x \to U$ is called a coordinate system on V_x.

Ex: $S^1 = \{(x, y) \in \mathbb{R}^2 | x^2 + y^2 = 1\}$

Consider 4 separate mappings (charts) to cover S^1 starting with the map:

$$\phi_1: (-1,1) \to S^1 \quad \text{by} \quad \phi_1(x) = \left(x, \sqrt{1 - x^2}\right)$$

This map is a diffeomorphism:
(a) ϕ_1 is surjective onto U,
(b) ϕ_1 is smooth
(c) ϕ_1^{-1} is smooth by projection 'trick'.

Similarly for other patches. So, S^1 is an 1-dim manifold.

Ex: Let $S^n = \{\overline{x} \in \mathbb{R}^{n+1} | \, |\overline{x}| = 1\}$
Show S^n is an n-manifold.

(Exercise)
Suppose $X \subseteq \mathbb{R}^N$ and $Y \subseteq \mathbb{R}^M$. Then we view

$$X \times Y = \left\{ (x, y) \in \underbrace{\mathbb{R}^N \times \mathbb{R}^M}_{\mathbb{R}^{N+M}} \middle| \, x \in X, y \in Y \right\}$$

Theorem 1: Suppose $X^n \subseteq \mathbb{R}^N$ and $Y^m \subseteq \mathbb{R}^M$ are manifolds. Then $X^n \times Y^m$ is an (n + m) – manifold.

Proof: Let $(x, y) \in X^n \times Y^m$. Then there are local parameterizations

$\phi: U \subseteq \mathbb{R}^n \to V_x \subset X^n$ and
$\Psi: S \subseteq \mathbb{R}^m \to W_y \subset X^m$

So, $(\phi, \Psi): U \times S \subset \mathbb{R}^{n+m} \to V_x \times W_y \subset X^n \times Y^m$
This is Smooth, a surjection, what of inverse? If $V_x = \tilde{V}_x \cap X^n$ where $\tilde{V}_x \subset \mathbb{R}^N$ and there is a smooth map $\Phi: \tilde{V}_x \to U$ extending ϕ^{-1} (as promised by def.) and similarly a smooth extension $\Psi': \widetilde{W}_y \to S$ of Ψ^{-1} then $(\Phi', \Psi'): \tilde{V}_x \times \widetilde{W}_y \subseteq \mathbb{R}^{N+M} \to U \times S$ is a smooth extension of $(\Phi, \Psi)^{-1}: V_x \times W_y \to U \times S$. Therefore, $X^n \times Y^m$ is an (n + m) – manifold.

Def. 4: If X and Y are manifolds in $\mathbb{R}^N$ and $X \subseteq Y$ we call X a submanifold of Y.

Ex. If $X \subseteq \mathbb{R}^N$ is a manifold, then X is a submanifold of $\mathbb{R}^N$

Ex. If X and Y are manifolds and $y \in Y$ then $X \times \{y\}$ is a submanifold of $X \times Y$.

Goal – extend the concept of derivative to maps between manifolds, such as 2-sphere living in $\mathbb{R}^3$.

Let $f: \mathbb{R}^n \to \mathbb{R}^m$ (or $f: U \subseteq \mathbb{R}^n \to \mathbb{R}^m$)

Let $x \in \mathbb{R}^n$ and $\vec{h} \in \mathbb{R}^n$, then derivative of f at x in the direction $\vec{h}$ is given by

$$df_x(h) = \lim_{t \to 0} \frac{f(x + th) - f(x)}{t}$$

provided the limit exists. If t is smooth then df_x is defined $\forall h \in \mathbb{R}^n$. In fact $df_x: R^n \to \mathbb{R}^m$ is a linear map which may be represented by the matrix:

$$\begin{bmatrix} \frac{\partial f_1(x)}{\partial x_1} & \frac{\partial f_1(x)}{\partial x_2} & \cdots & \frac{\partial f_1(x)}{\partial x_n} \\ \cdots & \cdots & & \cdots \\ \frac{\partial f_m(x)}{\partial x_1} & \cdots & \cdots & \frac{\partial f_m(x)}{\partial x_n} \end{bmatrix}$$

w.r.t. the standard basis.

The linear map df_x is the best linear approx. to f at x. In fact, we may define f to be differentiable at x if there exist a linear map $\lambda: \mathbb{R}^n \to \mathbb{R}^m \ni$ (such that) ….

f is differentiable at x if $\exists$ a linear map $\lambda: \mathbb{R}^n \to \mathbb{R}^m \ni$

$$\lim_{h\to 0} \left| \frac{f(x+h) - f(x) - \lambda(h)}{h} \right| = 0$$

λ is called the "total derivative of f at x.

<u>Chain Rule</u>: Suppose $f: U \subseteq R^n \to V \subseteq R^m$ and $g: V \to \mathbb{R}^\ell$ are smooth maps then $d(gof)_x = dg_{f(x)} o df_x$

<u>Def. 5</u> (tangent space) Let $X^n \subseteq \mathbb{R}^N$ be a manifold, $x \in X^n$ and $\phi: U \subseteq \mathbb{R}^n \to V_x$ be a local parameterization of X^n at x with $\phi(0) = x$. The image of $(d\phi_0: \mathbb{R}^n \to \mathbb{R}^N)$ is called the tangent space of X at x and is denoted $T_x(X)$. If $V \in T_x(X)$, V is called a tangent vector to X at x. In this situation $\phi_0 + T_x(X)$ is the best "flat approx" to X^n at x.

<u>Lemma</u>: Let $X^n \subseteq \mathbb{R}^N$ be an n-manifold then $T_x(X)$ is well-defined $\forall x \in X$.

Proof: Suppose $\phi: U \to W_x \subset X^n$ and $\Psi: V \to S_x \subset X^n$ are local param. of X at x with $\phi(0) = x = \Psi(0)$. Now have a well-defined map from U to V. Let $W_x' \subset W_x \cap S_x$ and $U' = \phi^{-1}(W_x')$ and $V' = \Psi^{-1}(W_x')$. Then $\phi: U' \to W_x'$ *and* $\Psi: V' \to S_x' = W_x'$ are local parameterizations. Then $h = \Psi^{-1} \circ \phi: U' \to V'$ is a diffeomorphism (local paranm are diffeos $\to$ compositions thereof are diffeos). Clearly $\phi = \Psi \circ \text{h}$, therefore $d\phi_0 = d\Psi_0 \circ dh_0$ (chain rule, classical so holds)

Therefore image $(d\Psi_0) \supseteq$ image $(d\phi_0)$ (because dh_0 may cutback as $d\Psi_0$). Similarly, we get the reverse, so image $(d\Psi_0) =$ image $(d\phi_0)$.

<u>Lemma</u>: If $X^n \subseteq \mathbb{R}^N$ is an n-manifold then $\dim(T_x(X)) = h \, \forall x \in X^n$.

<u>Proof</u>: Let $\phi: U \to W_x$ be a local param of X at x. Then $\exists$ smooth $\Phi': \widetilde{W}_x \to \mathbb{R}^n$ whih extends ϕ^{-1}. Now $U \underset{\to}{\phi} \widetilde{W}_x \underset{\to}{\phi'} \mathbb{R}^n$.

So, $d\Phi_x' \circ d\phi_0 = d(id_U) = id_{\mathbb{R}^n}$

Therefore $\dim(im\ d\phi_0) = n$ (where $d\phi_0 = T_x(X)$)

Now to define derivative ….

Let $f: X^n \to Y^m$ be a smooth map from an n-manifold to one m-manifold.

Q: How to define df_x?

Partial answer: df_x with map $T_x(X) \to T_{f(x)}(Y)$

Idea

$$\begin{array}{ccc} X^n & \overset{f}{\to} & Y^m \\ \phi \uparrow & & \uparrow \Psi \\ U \subseteq \mathbb{R}^n & \to & V \subseteq \mathbb{R}^m \end{array} \quad \to \quad \begin{array}{ccc} T_x(X) & \overset{df_x}{\to} & T_{f(x)}(Y) \\ d\phi_0 \uparrow & & \uparrow d\Psi_0 \\ \mathbb{R}^n & \overset{dh_0}{\to} & \mathbb{R}^m \end{array}$$

<u>Note</u>: If the chain rule is to hold we must define df_x this way.

<u>Def. 6</u>: Let $f: X^n \to Y^m$ be a smooth map between manifolds and $x \in X^n$
Choose a local param. $\Psi: V \to Y^m \ni \Psi(0) = f(x)$ and a local param.
$\phi: U \to X^n \ni \phi(0) = x$ and $f(\phi(u^1)) \subseteq V$.

Let $h = \Psi^{-1} \circ f \circ \phi$

$f_x: T_x(X) \to T_{f(x)}(Y)$ is defined by $df_x = d\Psi_0{}^\circ dh_0 (d\phi_0)^{-1}$

<u>Chain Rule</u> (for functions between manifolds)
Suppose $f: X^n \to Y^m$ and $g: Y^m \to Z^\ell$ are smooth maps, then
$d(g^\circ f)_x = dg_{f(x)}{}^\circ df_x$

<u>Recall</u> If $f: X^n \to Y^m$ is a smooth map between manifolds and $x \in X$, we define $df_x: T_x(X^n) \to T_{f(x)}(Y^m)$ via the diagram:

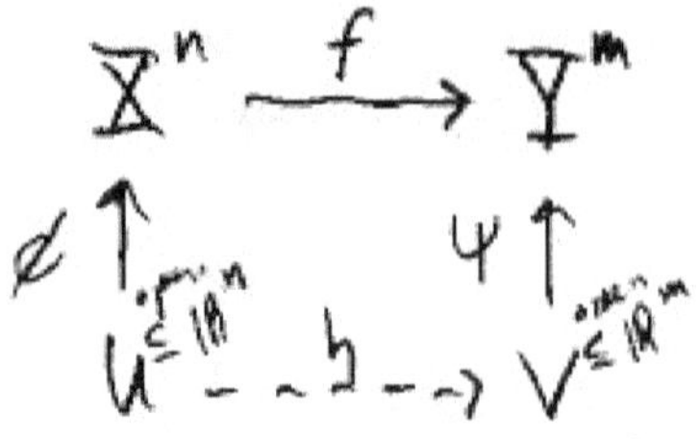

Fig. C.1. $\phi(0) = x, \Psi(0) = f(x),\ f(\phi(u)) \subseteq \Psi(v)$

which leads to:

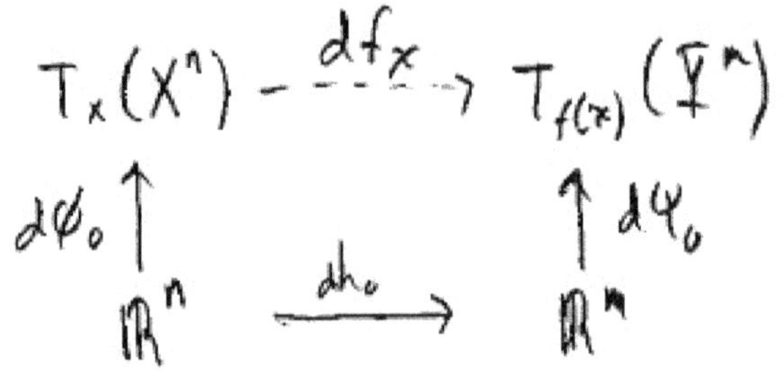

Fig. C.2. $df_x = (d\Psi_0)(dh_0)(d\phi_0)^{-1}$

<u>Lemma</u>: Our definition of df_x is well defined.
<u>Proof</u>: Choose a different param. and show that result is unchanged. Suppose $\Psi': W \subseteq \mathbb{R}^m \to Y$ with $\Psi'(0) = f(x)$ is another local param. Cutback v: choose v sufficiently small so that $\Psi(v) \subseteq \Psi'(w)$. Then we have:

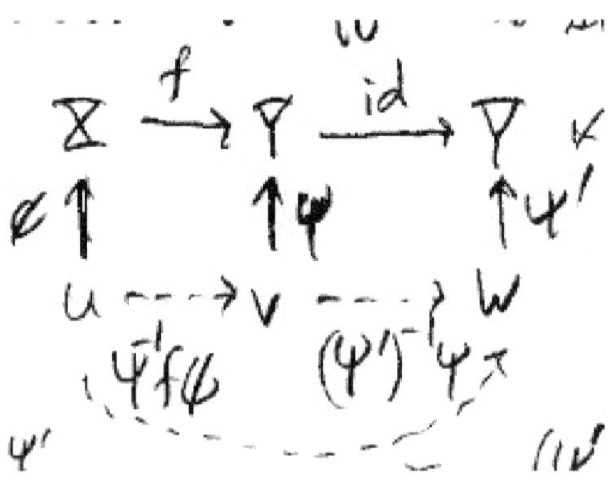

Fig. C.3.

df_x defined with Ψ' → using Ψ' we get $\hat{d}f_x = d(id)_{f(x)} \circ df_x = df_x$,
need the claim $d(id)_{f(x)} = id$, so:
$d(id)_{f(x)} = (d\Psi_0^1)d((\Psi')^{-1} \circ \Psi)_0(d\Psi_0)^{-1} = d(\Psi' \circ \Psi^{-1} \circ \Psi \circ \Psi^{-1}) = d(id) = id$.
Therefore, $\hat{d}f_x = df_x \Rightarrow df_x$ well defined.

Inverse function theorem: (from analysis)
Let $f: U \longrightarrow \mathbb{R}^n$, $U \subset \mathbb{R}^n$, be a smooth map with $x \in U$ where df_x is an isomorphism. (df_x is a linear map from $\mathbb{R}^n \to \mathbb{R}^n$). Then there is an open set $U_x \subseteq U$ with $x \in U_x$ and an open set $V_{f(x)} \subseteq \mathbb{R}^n$ containing $f(x) \ni$
(1) $f|_{U_x}: U_x \to V_{f(x)}$ is a bijection and
(2) $\left(f|_{U_x}\right)^{-1}$ is smooth

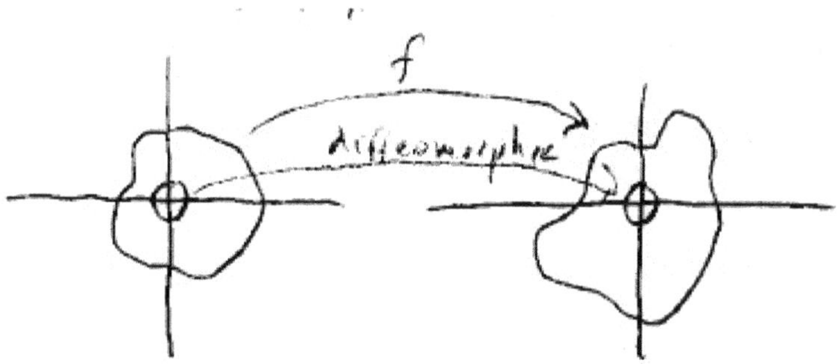

Fig. C.4.

Inverse function theorem for manifolds:
Suppose $f: X^n \to Y^m$ is a smooth map between manifolds, $x \in X$ and df_x is an isomorphism. Then f is a local diffeomorphism.

Proof:
Have by hypothesis:

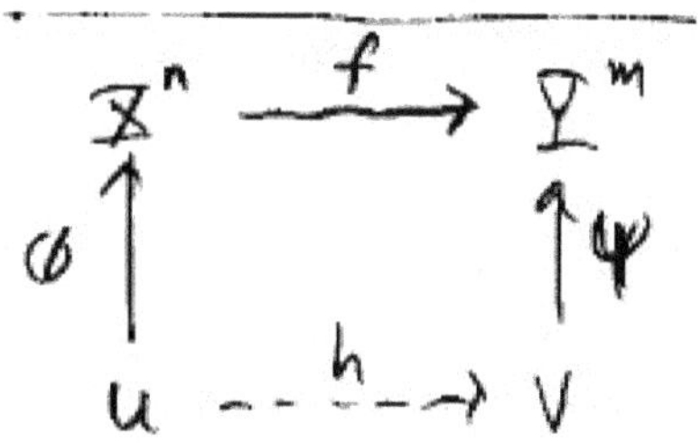

Fig. C.5.

Thus, $df_x = (d\Psi_0) \circ (dh_0) \circ (d\phi_0)^{-1}$ is an isomorphism $\Rightarrow dh_0$ isomorphic $\Rightarrow h$ locally diffeomorphic $\Rightarrow f$ is local diffeo. So, if the derivative is nice at a point, and the function is nice there – then it is diffeomorphic there.

Warning: A function can be a local diffeo at each point and still not be a global diffeo.:

Ex: $\mathbb{R}^1 \to S^1$ where $x \longmapsto (\cos(2\pi x), \sin(2\pi x))$. This is locally diffeo, but globally not.

Remarks: If $f: X^n \to Y^m$ is a local diffeo at x, we can choose local coordinates so that f (or h) apepars to be the identity near x.

Q: What if $f: X^n \to Y^m$ and $n \neq m$? What is "nice local behavior" of this map?

What are the nicest maps from $f: \mathbb{R}^n \longrightarrow \mathbb{R}^m$?

(a) n = m : identity, i.e., $f(x_1, \dots, x_n) = (x_1, \dots, x_n)$
(b) n < m : inclusion, i.e., $f(x_1, \dots, x_n) = (x_1, \dots, \dots, x_n, 0, \dots, 0,)$
(c) n > m : projection, i.e., $f(x_1, \dots, x_n) = (x_1, \dots, x_m)$

(a) derivative : isomorphism
(b) derivative : linear map from $\mathbb{R}^n \longrightarrow \mathbb{R}^m$ (injection) nice if injective
(c) derivative : linear map from $\mathbb{R}^n \longrightarrow \mathbb{R}^m$ (surjection) bad, scrunching coord's

Def. 7 A smooth map $f: X^n \to Y^m$ between manifolds is an immersion at $x \in X$if $df_x: T_x(X^n) \to T_{f(x)}(Y^m)$ is injective. If f is an immersion at each point of x then f is called an immersion.

Local Immersion theorem: Suppose $f: X^n \to Y^m$ is an immersion at $x \in X^n$ and let $f(x) = y$, then there exist local coord's around x y so that $f(x_1, x_2, \dots, x_n) = (x_1, x_2, \dots, x_n, 0,0,0, \dots 0)$

Proof of Local Immersion Theorem:
We have:

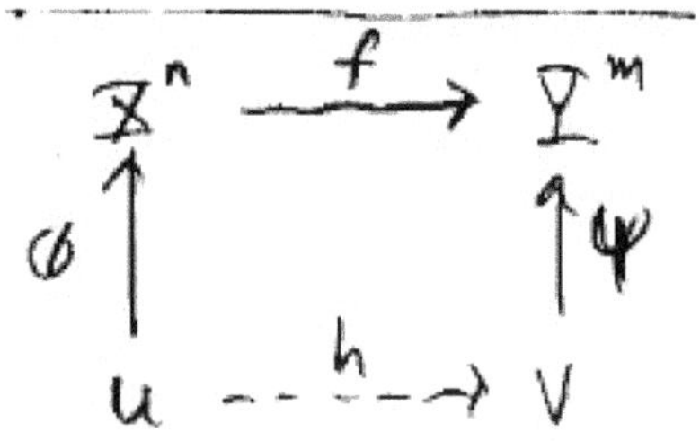

Fig. C.6.

If df_x is injective $\Longrightarrow dh_0$ is injective. $dh_0\colon \mathbb{R}^n \longrightarrow \mathbb{R}^m \rightarrow$ can straighten hyperplane by taking $\mathbb{R}^m \rightarrow \mathbb{R}^m$, i.e., choosing a different Ψ.

There is a linear isometry $\lambda\colon \mathbb{R}^m \rightarrow \mathbb{R}^m$

$$\underbrace{\mathbb{R}^n \overset{dh_0}{\longrightarrow} \mathbb{R}^m \overset{\lambda}{\rightarrow} \mathbb{R}^m}_{\textit{the inclusion.}}$$

Now, change original diagram to:

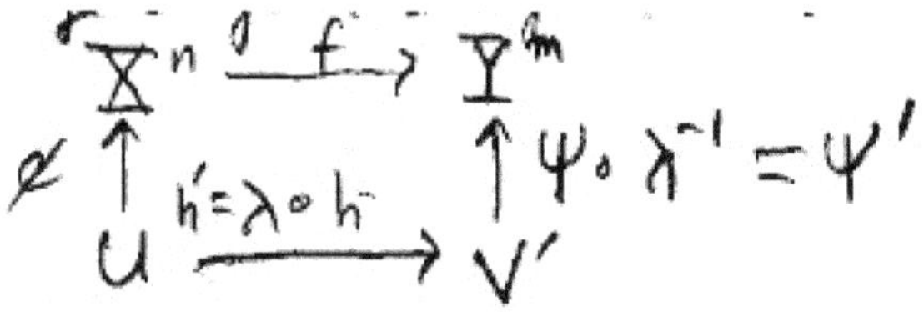

Fig. C.7.

Note: dh_0' is the inclusion map.

Above we chose a local parameterization around x and $f(x)$ and altered it to obtain the first diagram. Now define $H\colon \mathbb{R}^n \times \mathbb{R}^{m-n} \rightarrow \mathbb{R}^m$ by $(x,z) \longmapsto h^1(x) + (0,z)$. Then $dH_0 = I_m$. By Inverse Function Theorem, H is a local diffeo at 0. So,

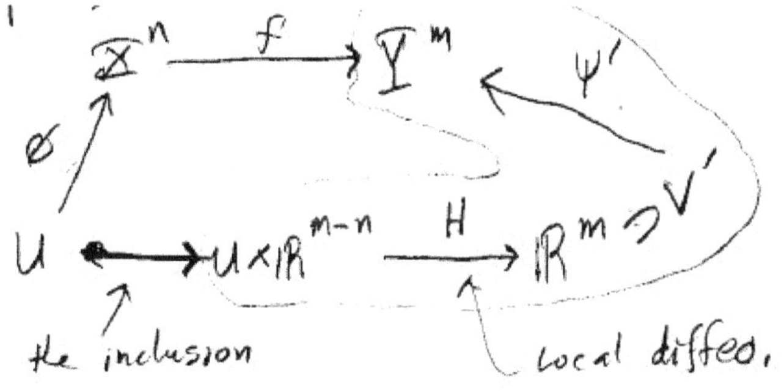

Fig. C.8.

Now choose U' containing 0.

$U' \subset U$ *and* $\in > 0 \ni H|_{U' \times (-\epsilon, \epsilon)^{m-n}}: U' \times (-\epsilon, \epsilon)^{m-n}$ is a diffeo and $H(U \times (-\epsilon, \epsilon)^{m-n}) \subseteq V'$. So the above cutbacks allow us to redraw the diagram as:

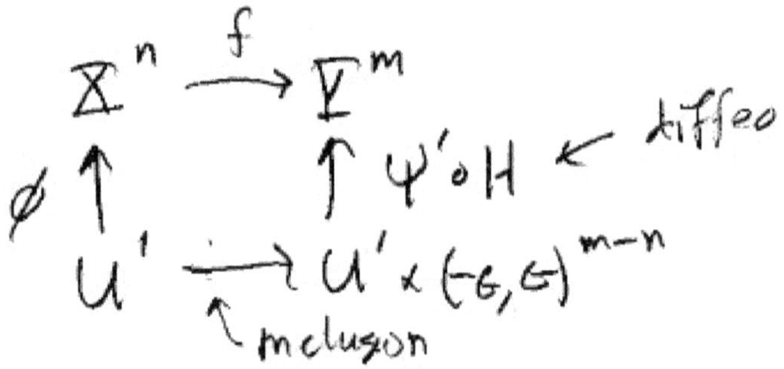

Fig. C.9.

Q: If an immersion is 1 – 1 will its image be a submanifold? NO! "irrational flow on the torus"

Ex. Begin with the map $f: \mathbb{R}^1 \to S^1$ $f(x) = (\cos(2\pi x), \sin(2\pi x))$. Now consider $f \times f: \mathbb{R} \times \mathbb{R} \to S^1 \times S^1$. Consider the image of a line, if we choose the slope to be irrational its image will never hit itself, it will also be dense. Since its dense the image is not a submanifold. It's an immersion but it can't be thought of as a submanifold. So $f \times f|_L: L \to S^1 \times S^1$ is an injective immersion provided the slope of L is irrational (if rational is a submanifold). Also image (L) is dense in $S^1 \times S^1$. Furthermore image (L) is not a submanifold of $S^1 \times S^1$.

Def. A map $f: A \to B$ between top, spaces is proper if $\forall$ compact set $C \subseteq B, f^{-1}(C)$ is compact in A.
Note: assuming A and B are Hausdorff, if A compact, f must be proper.

Def. A smooth map $f: X^n \to Y^m$ is an embedding if
(i) f is an immersion
(ii) f is injective
(iii) f is proper.

Theorem: If $f: X^n \to Y^m$ is an embedding and Z denotes $f(X^n)$, then Z is a submanifold of Y and $f|: X^n \to Z$ is a diffeo.

Proof is topological: First we show that Z is a manifold. Let $y \in Z$. We must find a local param. of Z at y. Let $x = f^{-1}(y)$. By the local immersion theorem, there is a sufficiently small neighborhood U_x of x $\ni$ $f|_{U_x}: U_x \to f(U_x)$ is a diffeo. Problem: How do we know $f(U_x)$ is open in z? Suppose $f(U_x)$ is not open in z, then $\exists$ a sequence $\{y_i\}$ of points in $Z - f(U_x)$ which converge. How do we know $f(U_x)$ contains a neighb. of y in Z. Suppose not then $\exists$ a sequence $\{y_i\}$ of points in $z - f(U_x)$ which converge to y. Note that then $\{y_i\} \cup \{y\}$ is compact. By hypothesis $f^{-1}(\{y_i\} \cup \{y\})$ is compact in X. Let x_i denote $f^{-1}(y_i)$ then for each: $\{x_i\} \cup \{x\}$ is an infinite compact set in X → thus there is a convergent subsequence. Convergent subseqeunce $\{x_{i_k}\} \to z \in Z$. If $x_{i_k} \to z$ then $f(x_{i_k}) \to f(z)$, i.e., $y_{i_k} \to f(z)$. Therefore $f(z) = y = f(x)$ but f is 1 – 1 so $x = z$, but this is a contradiction because the x is on U_x which is open – i.e., consists of only interior points. Therefore, $f(U_x)$ is open. Therefore, there is a trivial param. push on U_x over to $f(U_x)$. So z has a local param at y, i.e., it is a manifold. Lastly, since f is 1 – 1, $f^{-1}: z \to X$ is defined. Furthermore, by the above f^{-1} is a local diffeo. So $f^{-1}_{:X \to Z}$ is smooth. Therefore $f: X \to Z$ is a diffeo.

Def. A smooth map $f: X^n \to Y^m$ between manifolds is a submersion at $x \in X$ if $df_x: T_x(X) \to T_{f(x)}(Y)$ is surjective, if f is a submersion at each point of X we call f a submersion.

Ex: $\mathbb{R}^n \xrightarrow{P} \mathbb{R}^m \quad (n \geq m)$

$(x_1, \ldots x_m, \ldots, x_n) \longmapsto (x_1, \ldots, x_n)$

Note $dp_0 = [\mathrm{I_m} \vdots 0] \leftarrow matrix$

Local submersion theorem: Suppose $f: X^n \to Y^m$ is a local submersion of $x \in X$ and $f(x) = y$. Then there are local coordinates around x and y so that $f(x_1, \ldots x_m, \ldots, x_n) = (x_1, x_2, \ldots, x_n)$.

Proof: begin with the usual set-up:

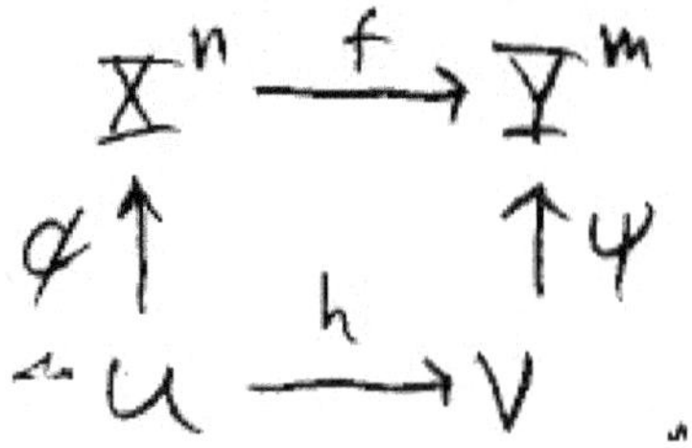

Fig. C.10.

Since dh_0 is surjective $(dh_0: \mathbb{R}^n \to \mathbb{R}^m)$ then precompose $\mathbb{R}^n$ such that

$$\underbrace{\mathbb{R}^n \xrightarrow{\lambda} \mathbb{R}^n \xrightarrow{dh_0} \mathbb{R}^m}_{projection}$$

Thus, we have:
$\lambda: \mathbb{R}^n \to \mathbb{R}^n \ni dh_0 \circ \lambda = dh_0 \circ d\lambda_0 = d(h \circ \lambda_0) = [\mathrm{I_m} \vdots 0]$
So

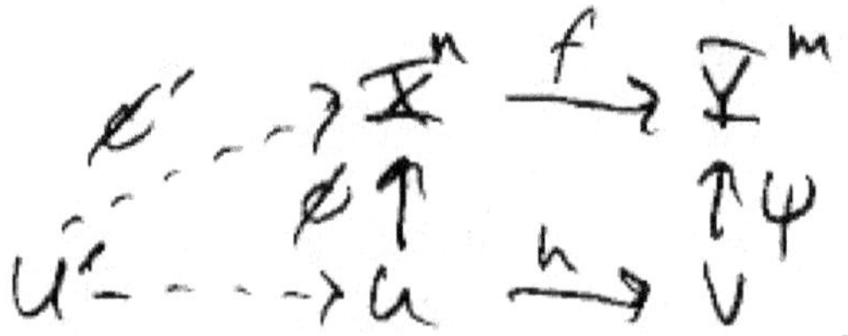

Fig. C.11.

So there is a local parameterization set-up:

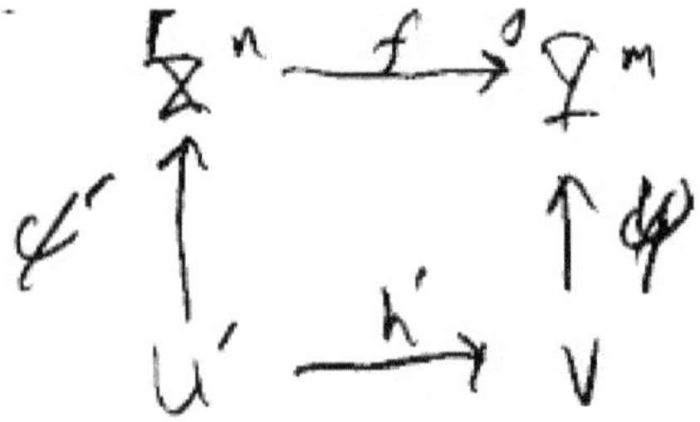

Fig. C.12.

With $dh_0' = [\mathrm{I_m} \vdots 0]$.

Define $H: U' \to V \times \mathbb{R}^{n-m}$ by
$$H(x_1, \dots x_n) = (h(x_1, \dots, x_m), x_{m+1}, \dots, x_n)$$
Note $dH_0 = \begin{bmatrix} I_m & \vdots & 0 \\ 0 & \vdots & I_{n-m} \end{bmatrix} = I_n$

So H is a local diffeomorphism (using inv. func. theorem). By Inv. Fun. Th., H is a local diffeo at 0.

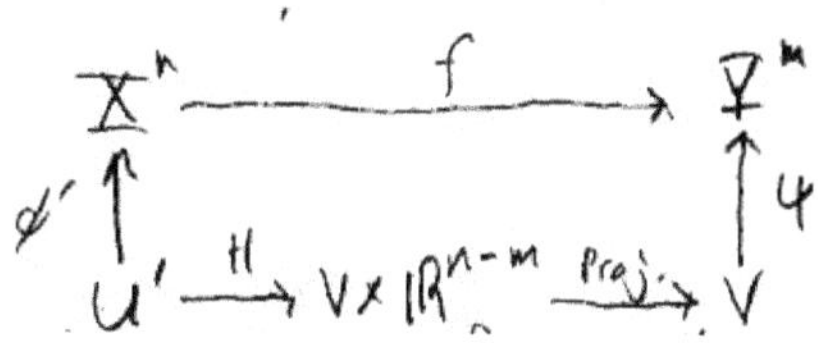
Fig. C.13.

H local but by cutting back to U'', i.e., a sufficiently small subset of $V \times \mathbb{R}^{n-m}$ such that H is diffeo we can write H^{-1} in the relation above, then we can compose to get.:

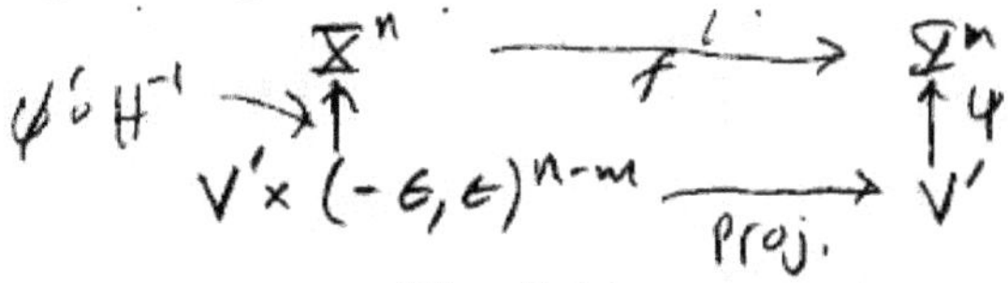
Fig. C.14.

Done.

Def. Let $f: X^n \to Y^m$ (smooth implied)
A point $y \in Y^m$ is a regular value of f if df_x is surjective $\forall x \in f^{-1}(y)$. If y is not a regular value it is called a critical value.

Pre image theorem. Suppose $f: X^n \to Y^m$ and y is a regular value of f. Then $f^{-1}(y)$ is a submanifold of X with dimension of $f^{-1}(y)$ being $n - m$.

Proof: Let $x \in f^{-1}(y)$. By local sub theorem, ∃ a local param diagram set x and y as follows:

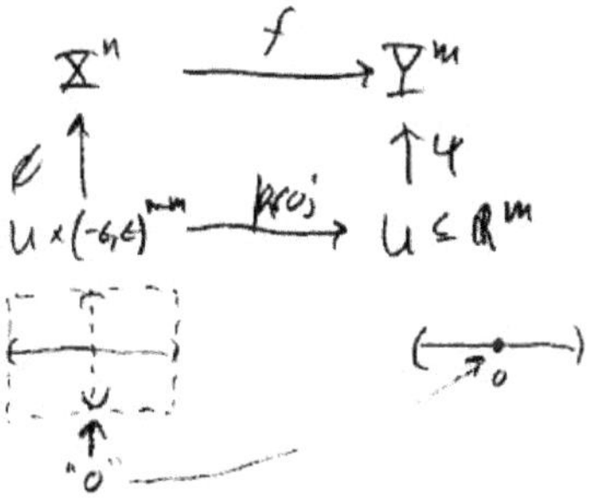
Fig. C.15.

So, $\phi|_{\{0\}\times\{-\epsilon,\epsilon\}^{n-m}}: \{0\} \times (-\epsilon, \epsilon)^{n-m} \to f^{-1}(y)$. This shows that $f^{-1}(y)$ is a manifold of dim. $n - m$.

<u>Ex.</u> Define $\mathbb{R}^n \xrightarrow{f} \mathbb{R}^1$ by $\overline{x} \to ||\overline{x}||^2$, i.e.,

$f(x_1, \dots x_n) = x_1^2 + \cdots + x_n^2$, then $df_{\overline{x}} = [2x_1, 2x_2, \dots, 2x_n]$ is surjective for all $\overline{x} \neq 0$. Then 1 is a regular value. Then $f^{-1}(1) = \{\overline{x} \in \mathbb{R}^n | ||\overline{x}|| = 1\} = S^{n-1}$

<u>Ex.</u> Let $M(n)$ be the set of all $n \times n$ matrices (real entries). (Think of this as $\mathbb{R}^{n^2}$.) Let $O(n) = \{A \in M | AA^t = I_n\}$ be the collection of orthogonal matrices with geometrically – linear isometrics. We can use preimage theorem to show that $O(n)$ is a submanifold of $M(n)$. Let $S(n)$ denote the collection of symmetric $n \times n$ matrices. $S(n)$ is an $\frac{1}{2}n(n+1)$ dimensional sub-manifold of $M(n)$. Define $f: M(n) \longrightarrow S(n)$, have $A \longrightarrow AA^t$ which checks. Then $f^{-1}(I_n) = O(n)$ if I_n is a regular value of f. Therefore, $O(n)$ is a submanifold with dim. $\frac{n(n-1)}{2}$. (still need to show that df is surjective).

We call $O(n)$ a Lie Group, where "Lie Group" is a manifold (smooth implied) L whose group operations are smooth, i.e.,

$$\left.\begin{array}{l} (1)\ L \times L \xrightarrow{product} L \\ (2) L \to L \quad x \to x^{-1} \end{array}\right\} \text{are smooth maps.}$$

<u>Def.</u> Let $f_1, f_2, \dots, f_k$ be smooth functions, i.e., maps from X^n to $\mathbb{R}^1$. We say that these functions are indep. at $x \in X$ if the linear functional $d(f_1)_x, d(f_2)_x, \dots d(f_k)_x$ are linearly independent on $T_x(X)$.

References

[1] Winters-Hilt, S. The Dynamics of Manifolds. (Physics Series: "Physics from Maximal Information Emanation" Book 3.)
[2] Arnowitt, R.; Deser, S.; Misner, C. (1959). "Dynamical Structure and Definition of Energy in General Relativity" (PDF). *Physical Review*. **116** (5): 1322–1330.
[3] Winters-Hilt, S. Quantum Mechanics, Path Integrals, and Algebraic Reality. (Physics Series: "Physics from Maximal Information Emanation" Book 4.)
[4] Winters-Hilt, S. Thermal & Statistical Mechanics, and Black Hole Thermodynamics. (Physics Series: "Physics from Maximal Information Emanation" Book 6.)
[5] Winters-Hilt, S. Emanation, Emergence, and Eucatastrophe. (Physics Series: "Physics from Maximal Information Emanation" Book 7.)
[6] Winters-Hilt, S. Informatics and Machine Learning: from Martingales to Metaheuristics. (2021) Wiley.
[7] Doran, C. and A. Lasenby. Geometric Algebra for Physicsists. CUP 2013.
[8] Hawking, S. W. (1992). "Chronology protection conjecture". *Phys. Rev. D*. **46** (2): 603.
[9] Winters-Hilt, S. Quantum Field Theory and the Standard Model. (Physics Series: "Physics from Maximal Information Emanation" Book 5.)
[10] Francis, M.R. and A. Kosowski. General algebra Techniques for General Relativity. gr-qc/0311007v1 (2003).
[11] Winters-Hilt, S. Classical Mechanics and Chaos. (Physics Series: "Physics from Maximal Information Emanation" Book 1.)
[12] Witten, Edward (1981). "A new proof of the positive energy theorem". *Communications in Mathematical Physics*. **80** (3): 381–402.
[13] Winters-Hilt, S. The Dynamics of Fields, Fluids, and Gauges. (Physics Series: "Physics from Maximal Information Emanation" Book 2.)
[14] Landau, L.D. and E.M. Lifshitz. Fluid Mechanics. 2013.
[15] Feigenbaum, M. J. (1976). "Universality in complex discrete dynamics" (PDF). Los Alamos Theoretical Division Annual Report 1975–1976.

[16] Mandelbrot, Benoît (1982). The Fractal Geometry of Nature. W H Freeman & Co.
[17] Adler, R. Introduction to General Relativity. McGraw-Hill, 1975.
[18] Penrose, R. (1968) in *Battelle Recontres*, eds. C.M. deWitt and J.A. Wheeler, Benjamin, New York.
[19] *On the Hypotheses which lie at the Bases of Geometry*. Bernhard Riemann. Translated by William Kingdon Clifford [Nature, Vol. VIII. Nos. 183, 184, pp. 14–17, 36, 37.]
[61] Barger, Vernon; Marfatia, Danny; Whisnant, Kerry Lewis (2012). *The Physics of Neutrinos*. Princeton University Press. ISBN 978-0-691-12853-5.
[20] Wald, R.M. General Relativity. Univ. of Chicago Press. 1984.
[21] Misner, Charles W., Thorne, K. S., & Wheeler, J. A. Gravitation. Princeton University Press, 2017. ISBN: 9780691177793.
[22] Taub, A.H., Annals of Mathematics, Vol. 53, 1951, pp. 472-490.
[23] Gibbons, G.W. and S. W. Hawking. Action integrals and partition functions in quantum gravity. PRD 15, 2752 (1977).
[24] Cuifolini, I. and J.A. Wheeler. Gravitation and Inertia. Princeton Univ. Press. (1995).
[25] Cartan E. Leçons sur la géométrie des espaces de Riemann. 1928.
[26] Cartan E. Leçons sur la géométrie des espaces de Riemann. 1946.
[27] Manasse, F.K. and C.W. Misner. Fermi Normal Coordinates and Some Basic Concepts in Differential Geometry. J. Math. Phys. 4, 735-745 (1963).
[28] Hawking, S. W. (1992). "Chronology protection conjecture". *Phys. Rev. D*. **46** (2): 603.
[29] Levi-Civita, Sur l'écart géodésique,. Math. Ann 97, 291 (1926)
[30] Synge, J.L. Relativity, The General Theory. Interscience, New York, 1960.
[31] L. P. Eisenhart. Riemannian Geometry. Princeton Univ. Press (1950).
[32] York, J.W. Energy and momentum of the gravitational field, in Essays n General Relativity, ed. F.J. Tipler. Academic Press, New York, 1980.
[33] Schoen, R.; Yau, S.- T. On the proof of the positive mass conjecture in general relativity. Comm. Math. Phys. 65 (1979), no. 1, 45–76.
[34] Schoen, R.; Yau, S.- T. Proof of the Positive-Action Conjecture in Quantum Relativity. Phys. Rev. Lett. 42, 547-48 (1979).
[35] Schoen, R.; Yau, S.- T. Positivity of the Total Mass of a General Space-Time, Phys. Rev. Lett. 43 1457 (1979).

[36] Schoen, R.; Yau, S.- T. The energy and the linear momentum of space-times in general relativity. *Comm. Math. Phys.* **79** (1): 47–51 (1979).
[37] Schoen, R.; Yau, S.- T. Proof of the positive mass theorem. II. Comm. Math. Phys. 79 (1981), no. 2, 231–260.
[38] Deser, S. and C. Teitelboim. Supergravity Has Positive Energy. Phys. Rev. Lett. 39, 249 (1977).
[39] Hawking, Stephen & Ellis, G. F. R. (1973). The Large Scale Structure of Space-Time. Cambridge: Cambridge University Press.
[40] Schwarzschild, K. (1916). "Über das Gravitationsfeld eines Massenpunktes nach der Einsteinschen Theorie". Sitzungsberichte der Königlich Preussischen Akademie der Wissenschaften. 7: 189–196. For a translation, see Antoci, S.; Loinger, A. (1999). "On the gravitational field of a mass point according to Einstein's theory". arXiv:physics/9905030.
[41] Hawking, S. W. (1974-03-01). "Black hole explosions?". Nature. 248 (5443): 30–31.
[42] Penrose, Roger (1965), "Gravitational collapse and space-time singularities", Phys. Rev. Lett., 14 (3): 57.
[43] Hawking, S.W. and R. Penrose. The singularities of gravitational collapse and cosmology. Proc Roy. Acad. A 314, 1519 (1970).
[44] Einstein, A. On the General Theory of Relativity. *Preussische Akademie der Wissenschaften, Sitzungsberichte*, 1915 (part 2), 778–786, 799–801.
[45] Kruskal, M.D. Maximal Extension of Schwarzschild Metric. Phys. Rev. 119, 1743 (1960).
[46] Eddington, A.S. A Comparison of Whitehead's and Einstein's Formulæ" (PDF). Nature. 113 (2832): 192. (1924).
[47] Finkelstein, D. Past-Future Asymmetry of the Gravitational Field of a Point Particle. Phys. Rev. 110 (4): 965–967. (1958).
[48] Novikov, I.D. and Frolov V.P. (1989) Physics of Black Holes. Kluwer Acad. Publ. (Russian version (19 86, Nauka, Moscow).
[49] Khuri, N. N. (1957). Analyticity of the Schrödinger Scattering Amplitude and Nonrelativistic Dispersion Relations. Physical Review 107, 1148.
[50] Nambu, Y. and M. Sasaki. The Wave Function of a Collapsing Dust Sphere inside the Black Hole Horizon. Progress of Theoretical Physics, DOI:10.1143/PTP.79.96 (1988).
[51] Zyla, P.A.; et al. (Particle Data Group) (2020). Heavy neutral leptons. *Prog. Theor. Exp. Phys.* (Report). Particle data listings. Lawrence Berkeley Laboratory. 083C01.

[52] Kolb, E.W and M. S. Turner. The Early Universe. Addison-Wesley (1990).
[53] P. Kraus and F. Wilczek, Nucl. Phys. B433, 403 (1995). (gr-qc/9408003)
[54] P. Kraus and F. Wilczek, Nucl. Phys. B437, 231 (1995). (hep-th/9411219)
[55] Winters-Hilt S. Topics in Quantum Gravity and Quantum field Theory in Curved Spacetime. UWM PhD Dissertation, 1997.
[56] Oppenheimer, J.R. and Snyder H. (1939) Phys. Rev. 56, 455.
[57] W. Israel, in Three Hundred Years of Gravitation, edited by S. W. Hawking and W. Israel (Cambridge University Press, Cambridge, 1987), p. 234.
[58] W. Fischler, D. Morgan, and J. Polchinski, Phys. Rev. D 42, 4042 (1990).
[59] Regge, T., Teitelboim, C. (1974) Role of surface integrals in the Hamiltonian formulation of general relativity. Annals of Physics, 88. 286-318 doi:10.1016/0003-4916(74)90404-7.
[60] K. V. Kuchar, Phys. Rev. D 50, 3961 (1994). (gr-qc/9403003)
[61] W. G. Unruh, Phys. Rev. D 14, 870 (1976).
[62] T. Thiemann and H. A. Kastrup, Nucl. Phys. B399, 211 (1993). (grqc/ 9310012)
[63] H. A. Kastrup and T. Thiemann, Nucl. Phys. B425, 665 (1994). (grqc/ 9401032)
[64] M. Cavaglia, V. de Alfaro, and A. T. Filippov, Int. J. Mod. Phys. D 4, 661 (1995). (gr-qc/9411070)
[65] M. Cavaglia, V. de Alfaro, and A. T. Filippov, "Quantization of the Schwarzschild Black Hole," Report DFTT 50/95, gr-qc/9508062.
[66] S. R. Lau, Class. Quantum Grav. 13, 1541 (1996). (gr-qc/9508028)
[67] T. Thiemann, Int. J. Mod. Phys. D 3, 293 (1994).
[68] T. Thiemann, Nucl. Phys. B436, 681 (1995).
[69] J. Gegenberg and G. Kunstatter, Phys. Rev. D 47, R4192 (1993). (grqc/ 9302006) 121
[70] J. Gegenberg, G. Kunstatter, and D. Louis-Martinez, Phys. Rev. D 51, 1781 (1995). (gr-qc/9408015)
[71] D. Louis-Martinez and G. Kunstatter, Phys. Rev. D 52, 3494 (1995). (grqc/ 9503016)
[72] M. Varadarajan, Phys. Rev. D 52, 7080 (1995). (gr-qc/9508039)
[73] R. Laamme, in: Origin and Early History of the Universe: Proceedings of the 26th Liege International Astrophysical Colloquium (1986), ed. J. Demaret (Universite de Liege, Institut d'Astrophysique, 1987); Ph.D. thesis (University of Cambridge, 1988).

[74] B. F. Whiting and J. W. York, Phys. Rev. Lett. 61, 1336 (1988).
[75] H. W. Braden, J. D. Brown, B. F. Whiting and J. W. York, Phys. Rev. D 42, 3376 (1990).
[76] J. J. Halliwell and J. Louko, Phys. Rev. D 42, 3397 (1990).
[77] G. Hayward and J. Louko, Phys. Rev. D 42, 4032 (1990).
[78] J. Louko and B. F. Whiting, Class. Quantum Grav. 9, 457 (1992).
[79] J. Melmed and B. F. Whiting, Phys. Rev. D 49, 907 (1994).
[80] S. Carlip and C. Teitelboim, Class. Quantum Grav. 12, 1699 (1995). (grqc/ 9312002)
[81] S. Carlip and C. Teitelboim, Phys. Rev. D 51, 622 (1995). (gr-qc/9405070)
[82] G. Oliveira-Neto, Phys. Rev. D 53, 1977 (1996).
[83] V. A. Berezin, N. G. Kozmirov, V. A. Kuzmin, and I. I. Tkachev, Phys. Lett. B 212, 415 (1988).
[84] V. P. Frolov, Zh. Eksp. Teor. Fiz. 66, 813 (1974) [Sov. Phys. JETP 39, 393 (1974)].
[85] P. Hajcek, Commun. Math. Phys. xx, xxxx (1993).
[86] P. Hajcek, B. S. Kay, and K. V. Kuchar, Phys. Rev. D 46, 5439 (1992).
[87] M. Henneaux and C. Teitelboim, Quantization of Gauge Systems (Princeton University Press, Princeton, New Jersey, 1992).
[88] P. Painleve, C. R. Acad. Sci. (Paris) 173, 677 (1921).
[89] A. Gullstrand, Arkiv. Mat. Astron. Fys. 16(8), 1 (1922).
[90] Winters-Hilt, S. and S. Oharu. Preprint.
[91] Hitzer, E. Introduction ot Clifford's Geometric algebra. arXiv:1306,1660.
[92] Bott, Raoul (1957), "The stable homotopy of the classical groups", *Proceedings of the National Academy of Sciences of the United States of America*, **43** (10): 933–5.
[93] Cailler, C. 1917. Archs. Sci. Phys. Nat. ser. 4, 44 p. 237.
[94] P.R. Girard. The Quaternion group and modern physics. Eur. J. Phys. 5 (1984): 25-32.
[95] Synge, J.L. Quaternions, Lorentz Transformations and the Conway-Dirac-Eddington Matrices.
[96] Winters-Hilt, S. Feynman-Cayley Path Integrals select Chiral Bi-Sedenions with 10-dimensional space-time propagation. Advanced Studies in Theoretical Physics, Vol. 9, 2015, no. 14, 667 – 683. dx.doi.org/10.12988/astp.2015.5881.
[97] Francis, M.R. and A. Kosowski. General algebra Techniques for General Relativity. gr-qc/0311007v1 (2003).

[98] Amari, S. and H. Nagaoka. Methods of Information Geometry. Oxford University Press. 2007.
[99] Amari, S. Theory of information spaces: A differential geometrical foundation of statistics. Post RAAG Reports, 1980.
[100] Amari, S. Differential-geometrical methods in statistics. Lecture Notes on Statistics, 28, 1985. s
[101] Winters-Hilt, S. Data Analytics, Bioinformatics, and Machine Learning. 2019.
[102] Rao, C.R. Information and the Accuracy Attainable in the Estimation of Statistical Parameters. Bulletin of Calcutta Mathematical Society, 37, 81-91. (1945).
[102] Chentsov, N.N., Statistical Decision Rules and Optimal Inference [in Russian], Nauka, Moscow (1972).
[103] Peebles, P. J. E. (1980). Large-Scale Structure of the Universe. Princeton University Press.
[104] B. Abi et al. Measurement of the Positive Muon Anomalous Magnetic Moment to 0.46 ppm
Phys. Rev. Lett. 126, 141801 (2021).
[105] Gockeler, M. Differential geometry, guage theories and gravity. CUP (1989).
[106] Jackson, J.D. Classical Electrodynamics, 2nd Edition. Wiley 1975.
[107] Lorentz, Hendrik Antoon (1899), "Simplified Theory of Electrical and Optical Phenomena in Moving Systems" , *Proceedings of the Royal Netherlands Academy of Arts and Sciences*, **1**: 427–442.
[108] D'Alembert, Jean Le Rond (1743). Traité de dynamique.
[109] Laplace, P S (1774), "Mémoires de Mathématique et de Physique, Tome Sixième" [Memoir on the probability of causes of events.], Statistical Science, 1 (3): 366–367.
[110] Winters-Hilt S, I. H. Redmount, and L. Parker, "Physical distinction among alternative vacuum states in flat spacetime geometries," Phys. Rev. D 60, 124017 (1999).
[111] Friedman J. L., J. Louko, and S. Winters-Hilt, "Reduced Phase space formalism for spherically symmetric geometry with a massive dust shell," Phys. Rev. D 56, 7674-7691 (1997).
[112] Louko J and S. Winters-Hilt, "Hamiltonian thermodynamics of the Reissner-Nordstrom-anti de Sitter black hole," Phys. Rev. D 54, 2647-2663 (1996).
[113] Louko J, J. Z. Simon, and S. Winters-Hilt, "Hamiltonian thermodynamics of a Lovelock black hole," Phys. Rev. D 55, 3525-3535 (1997).

[114] Winters-Hilt, S. Unified Propagator Theory and a non-experimental derivation for the fine-structure constant. Advanced Studies in Theoretical Physics, Vol. 12, 2018, no. 5, 243-255.
[115] Winters-Hilt, S. Theory of Trigintaduonion Emanation and Origins of α and π. Researchgate 05/24/20.
[116] Winters-Hilt, S. Fiat Numero: Trigintaduonion Emanation Theory and its Relation to the Fine-Structure Constant α, the Feigenbaum Constant $C\infty$, and π. Advanced Studies in Theoretical Physics, Vol. 15, 2021, no. 2, 71-98.
[117] Winters-Hilt, S. Meromorphic precipitation of quantum matter with dimensionful action. May 2021. DOI:10.13140/RG.2.2.32294.24640.
[118] Winters-Hilt, S. Chiral Trigintaduonion Emanation Leads to the Standard Model of Particle Physics and to Quantum Matter. Advanced Studies in Theoretical Physics, Vol. 16, 2022, no. 3, 83-113.
[119] Winters-Hilt, S. Emanator Theory using split octonions is Manifestly Lorentz Invariant and reveals why the fundamental constant $\hbar$ should be so small. Advanced Studies in Theoretical Physics, 2023.
[120] Landau, Lev D.; Lifshitz, Evgeny M. (1969). Mechanics. Vol. 1 (2nd ed.). Pergamon Press.
[121] Goldstein, Herbert (1980). Classical Mechanics (2nd ed.). Addison-Wesley.
[122] Fetter, A.L and J.D Walecka, Theoretical Mechanics of Particles and Continua, Dover (2003).
[123] Percival, I.C. and D. Richards. Introduction to Dynamics. (1983) Cambridge University Press.
[124] Arnold, V.I. Ordinary Differential Equations. MIT Press. (1978).
[125] Arnold, Vladimir I. (1989). Mathematical Methods of Classical Mechanics (2nd ed.). New York: Springer.
[126] Woodhouse, N.M.J. Introduction to Analytical Dynamics. Springer, 2nd Edition. 2009.
[127] Bender, C.M. and S.A. Orszag. Advanced Mathematical Methods for Scientists and Engineers: Asymptotic Methods and Perturbation Theory. Springer. 1999.
[128] Robert L. Devaney. An Introduction to Chaotic Dynamical Systems. Addison -Wesley.
[129] Landau, Lev D.; Lifshitz, Evgeny M. (1971). The Classical Theory of Fields. Vol. 2 (3rd ed.). Pergamon Press.
[130] Einstein, A. "On a heuristic point of view concerning the production and transformation of light" (Ann. Phys., Lpz 17 132-148).

[131] Balmer, J. J. (1885). "Notiz über die Spectrallinien des Wasserstoffs" [Note on the spectral lines of hydrogen]. Annalen der Physik und Chemie. 3rd series (in German). 25: 80–87.
[132] Werner Heisenberg (1925). "Über quantentheoretische Umdeutung kinematischer und mechanischer Beziehungen". Zeitschrift für Physik (in German). 33 (1): 879–893. ("Quantum theoretical re-interpretation of kinematic and mechanical relations")
[133] Schrödinger, E. (1926). "An Undulatory Theory of the Mechanics of Atoms and Molecules" (PDF). Physical Review. 28 (6): 1049–1070.
[134] Max Born; J. Robert Oppenheimer (1927). "Zur Quantentheorie der Molekeln" [On the Quantum Theory of Molecules]. Annalen der Physik (in German). 389 (20): 457–484.
[135] Dirac, P. A. M. (1928). "The Quantum Theory of the Electron" (PDF). Proceedings of the Royal Society A: Mathematical, Physical and Engineering Sciences. 117 (778): 610–624.
[136] Dirac, Paul Adrien Maurice (1930). The Principles of Quantum Mechanics. Oxford: Clarendon Press.
[137] Dirac, Paul A. M. (1933). "The Lagrangian in Quantum Mechanics" (PDF). Physikalische Zeitschrift der Sowjetunion. 3: 64–72.
[138] Feynman, Richard P. (1942). The Principle of Least Action in Quantum Mechanics (PhD). Princeton University.
[139] Feynman, Richard P. (1948). "Space-time approach to non-relativistic quantum mechanics". Reviews of Modern Physics. 20 (2): 367–387.
[140] Laplace, P S (1774), "Mémoires de Mathématique et de Physique, Tome Sixième" [Memoir on the probability of causes of events.], Statistical Science, 1 (3): 366–367.
[141] Erdeyli, A. Asymptotic Expansions. 1956 Dover.
[142] Erdeyli, A. Asymptotic Expansions of differential equations with turning points. Review of the Literature. Technical Report 1, Contract Nonr-220(11). Reference no. NR 043-121. Department of Mathematics, California Institute of Technology, 1953.
[143] Hawking, S. W. (1974-03-01). "Black hole explosions?". Nature. 248 (5443): 30–31.
[144] Birrell, N.D. and Davies, P.C.W. (1982) Quantum Fields in Curved Space. Cambridge Monographs on Mathematical Physics. Cambridge University Press, Cambridge.
[145] Maldacena, Juan (1998). "The Large N limit of superconformal field theories and supergravity". Advances in Theoretical and Mathematical Physics. 2 (4): 231–252.

[146] Witten, Edward (1998). "Anti-de Sitter space and holography". Advances in Theoretical and Mathematical Physics. 2 (2): 253–291.
[147] Caves, Carlton M.; Fuchs, Christopher A.; Schack, Ruediger (2002-08-20). "Unknown quantum states: The quantum de Finetti representation". Journal of Mathematical Physics. 43 (9): 4537–4559.
[148] Sommerfeld, Arnold (1916). "Zur Quantentheorie der Spektrallinien". Annalen der Physik. 4 (51): 51–52.
[149] Tolkien, J.R.R. (1990). The Monsters and the Critics and Other Essays. London: HarperCollinsPublishers.

Index

B

C

D

E

F

G

J

M

O

P

Q

R

S

U

V

W

X

Y

Z

α

ℏ

www.ingramcontent.com/pod-product-compliance
Lightning Source LLC
Chambersburg PA
CBHW050452190326
41458CB00005B/1253
* 9 7 9 8 9 8 8 8 1 6 0 3 4 *